BASIC FOOD MICROBIOLOGY

George J. Banwart

Professor of Microbiology
The Ohio State University

AVI PUBLISHING COMPANY, INC.
Westport, Connecticut

Library of Congress Cataloging in Publication Data

Banwart, George J
Basic food microbiology.

Includes bibliographical references and index.
1. Food—Microbiology. I. Title.
QR115.B34 664′.001′576 79-10571
ISBN 0-87055-322-4

Printed in the United States of America by Eastern Graphics, Inc.

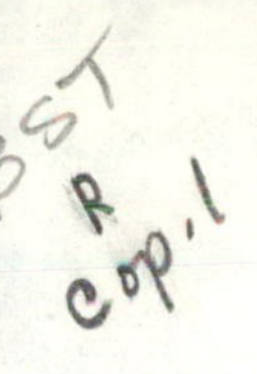

Preface

In the March 1969 issue of Food Technology, five food microbiologists outlined a course for food microbiology. That outline has served as the basis for this text. However, some changes in their outline were found to be desirable from the author's viewpoint.

We are fortunate that one of the problems in the United States is surplus food. In the past, most of the people have been busy producing food. Now, many people are busy criticizing food.

As the population of the world increases, it is essential that food production keep pace. It will be necessary not only to produce more food, but also to maintain this food so that it is safe and wholesome for human consumption. In this endeavor, the food microbiologist will play an important part.

With the variations and constant changes occurring in biological systems, facts are rare. Much of the information that was believed to be factual 10 or 20 years ago has been outmoded due to newer evidence and information. This text was written to bring food microbiology up to date.

It is assumed that the reader either has taken courses or has read books on the subjects of biology, general microbiology and chemistry. Hence, the reader should have some background concerning cellular structure, microbial multiplication, morphology and knowledge of biochemical reactions by means of enzymes.

Many texts attempt to describe the microbiology of a few simple foods. However, processed foods now account for most of the retail sales. It would be impossible to describe the microbiology of each of the nearly 10,000 food items that are available in the supermarket. Textured vegetable protein is added to hamburger; yeast protein is included in some foods. As new foods with various ingredients are developed, the microbiology of processed foods will become more dominant. Thus it is believed that it is more important to concentrate on the basics of food microbiology. This basic information may be helpful in the development of new products and processes, as well as comprehending the microbiological aspects of foods that are developed by someone else.

In order to have a text that is within a reasonable size, it was necessary to limit the discussion of certain subjects. The present text is less than one-half of the size of the original manuscript. In order to aid the reader, many references are included so that further details can be found in the literature.

In 1954, Herrington made the following statement regarding a review article about lipase that he published in the Journal of Dairy Science: "Some may feel that too much has been omitted; an equal number may feel that too much has been included. So be it."

The author is grateful to his family for allowing him to spend the time required for composing this text. He is especially indebted to his partner, Sally, whose assistance in typing, helpful suggestions and proofreading made it all possible.

I would also like to express my appreciation to Dr. Norman W. Desrosier and Barbara J. Flouton and the AVI Publishing Company for their encouragement and assistance in bringing this book into being.

GEORGE J. BANWART

April 1979

Contents

1

General Aspects of Food

IMPORTANCE OF FOOD

Man's basic needs include air that contains an adequate amount of oxygen, water that is potable, edible food and shelter. Food provides man with a source of energy needed for work and for chemical reactions that occur in his body. Food also supplies chemicals needed for growth, for repair of injured or worn out cells, and for reproduction. For man, food consumption can be considered a pleasurable experience, and a time for meeting with family or friends. Food is so necessary for our existence that the search for food has been the main occupation of humans throughout history.

FOOD DEMANDS

In a country with well-stocked supermarkets, it might be assumed that the search for an adequate food supply has been successful. However, there are some people on earth who do not eat an adequate diet or are actually hungry. Many people do not eat enough quality foods, particularly the proteins, needed for a healthy and productive life.

Some reasons given for this hunger problem are unequal distribution of food and money as well as cultural, religious or superstitious beliefs.

It is generally agreed that the main problem facing mankind is the increasing population (Fig. 1.1). In recent years the birth rate has declined, but so has the death rate. The present world populations is over 4 billion and the annual increase is estimated to be 60 to 80 million people.

How will the future part of the population curve appear? How many people can the facilities of our planet support? Will the present standard of living increase or will it be forced down? Can food sources be found to adequately feed the present and future populations? If we cannot adequately feed everyone today, how can the increased number of people be adequately fed in the future?

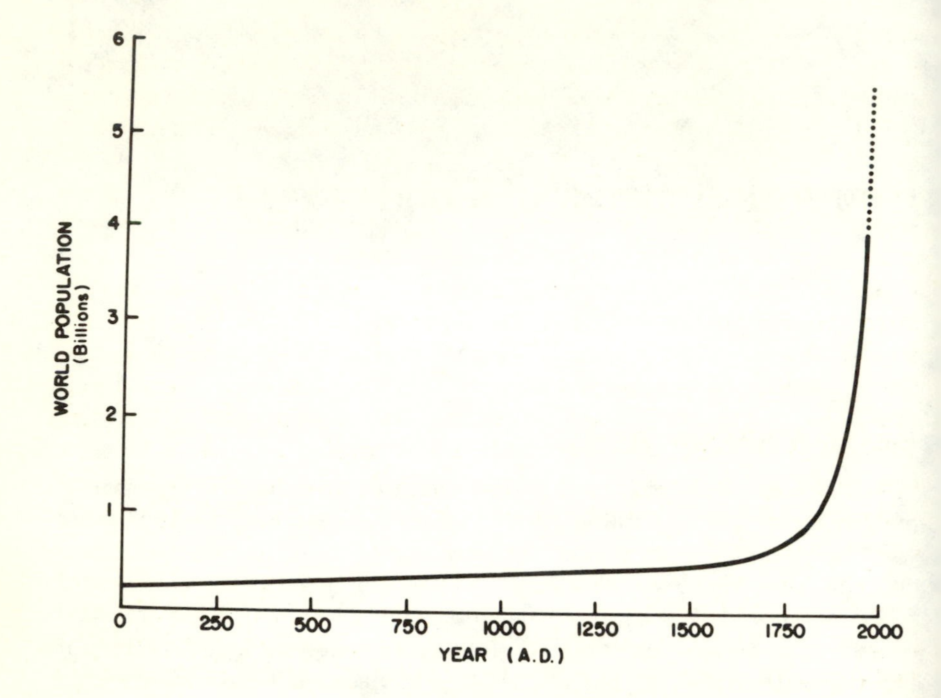

FIG. 1.1. ESTIMATED WORLD POPULATION

How many people can our planet support?

The predictions for the future food supply range from very pessimistic with famines to start in the near future, to the very optimistic viewpoint that there will be plenty of food for a population of almost limitless numbers. The most prominent opinion seems to be that, unless the rate of population increase can be substantially reduced, regardless of the efficiency of production, the demand for food will eventually outrun the supply.

Although world agricultural production has been increasing, the vagaries of nature (floods, droughts, freezing or other adverse climatic conditions) could cause a severe setback. With the expected increase in population, the search for food, especially proteins, will continue to be the endeavor of many food scientists, including microbiologists.

NUTRIENTS IN FOOD

Food contains proteins, carbohydrates, fats, vitamins and minerals as well as water. The primary concern is the supply of proteins. The population is composed of individuals of different weights, ages and sexes doing different types and amounts of work. Therefore, the dietary needs vary. In the U.S., the recommended daily allowance (RDA) for protein ranges from 28 g for infants and children under four years up to 65 g for adults and children over four years of age. If the protein efficiency ratio (PER) is equal to or better than that of casein, the RDA is 20 g and 45 g, respectively.

Protein is used in building muscle and body tissues. It forms an essential part of the hemoglobin molecule of red blood cells and is a major constituent of enzymes, hormones and antibodies.

The digestibility, utilization and presence of essential amino acids determines the quality of a protein. Eight amino acids are absolutely required by the adult human. These are: threonine, tryptophan, valine, lysine, leucine, isoleucine, methionine and phenylalanine. There is evidence that histidine may be required, and infants need arginine (Labuza 1974). The essential amino acids should be present in a balanced mixture. Although other amino acids are considered as nonessential, there must be sufficient nitrogen sources or amino acids in the diet for their synthesis.

Proteins differ in their amino acid content and hence, their quality and nutritional value. A high quality protein is one that, when digested, and its amino acids are absorbed from the intestinal tract, provides the essential amino acids in the proportion required for use in the body.

SOURCES OF FOODS

Our food supply depends upon the photosynthetic reaction between solar energy and plants that contain chlorophyll. Through photosynthesis, carbon dioxide and water are converted to glucose. Further cellular reactions produce the various organic compounds (carbohydrates, fats, proteins, vitamins).

Both the amount and dietary quality of food can be increased. To do this, we can improve the utilization of our land and fishery resources, upgrade the quality of plant proteins, convert waste materials into edible foods and prevent losses or deterioration of our food supplies.

Land Resources

To increase the productivity of land resources will require utilization of more land and higher yields of foods. There is a finite amount of land. Not all of this can be used for agriculture and even less can be used for crop production. It has been estimated that the amount of land used for agriculture can be doubled. This will not necessarily double production. These added lands are marginal and will require increased water, fertilizer, technology and energy to make them productive.

One of our problems is the loss of good agricultural land for the construction of highways, airports, drive-in movies, shopping centers, housing and factories. With an increasing population, there will be an even greater demand for land for these purposes.

Plants.—In the U.S., the harvest has increased 50% from 1950 to 1970. On a worldwide basis there have been increased yields per land unit. Estimates suggest that yields of some crops could be increased by five to ten times the present amounts.

Several protein concentrates have been prepared from plant products. Leaves are one source of concentrates. Although we do eat certain leafy crops such as spinach and kale, most leaves are either wasted or fed to animals.

Over 1.5 mg of protein per hectare has been obtained from leaf crops. Besides special crops, field wastes and processing wastes can be used as a source of leaf protein.

Much research has been accomplished with leaf protein in recent years (Anelli *et al.* 1977; Edwards *et al.* 1975; Kohler and Knuckles 1977; Pirie 1975), but problems still exist (Anon. 1975). It is important that toxins or potentially toxic fractions be removed during the processing of leaves that contain these substances. However, the nutritive value, high yields, and relative simplicity of extraction and preparation indicate that leaves could be a source of protein for human food.

Animals.—It is generally recommended that some animal protein should be included in the human diet. Domestic animals have been improved so that cows give more milk, chickens grow faster, hens lay more eggs and there is an increased efficiency in the conversion of feed to meat, milk or eggs.

Plants vs. Animals.—Since plants are the primary source of food, it has been suggested that humans should consume only plant products and eliminate the inefficiency of raising animals. Although animals compete with man for food, they should not be completely eliminated as food producers. Ruminants such as cattle and sheep can graze the grass on hilly, mountainous, arid or semi-arid land that could not be used for other crops. These animals can produce high grade protein for human food from various wastes and cellulosic materials.

The proteins of man more closely resemble the proteins of animals than those of plants. Hence, animal proteins are of higher biological value than are plant proteins. For example, most grains and legumes are marginal in one or more amino acids (Table 1.1). The biological value of plant proteins can be improved. The limiting amino acids can be synthesized and added, various plant proteins can be mixed, or small amounts of animal protein can be added to improve the protein value. Compared to the cost of food, synthetic amino acids are quite expensive, but usually only a small amount is needed to significantly improve the quality of a plant protein. This fortification is the least expensive method of improving the diet of a population, and it is often more acceptable to the people since it requires little or no change in their normal eating habits. The net effect of fortification is not only a better diet, but also an apparent increase in the amount of protein that is available for consumption.

TABLE 1.1

LIMITING AMINO ACIDS IN PLANT PROTEINS

	Limiting Amino Acids	
Plant Protein	First	Second
Barley	Lysine	Tryptophan
Rice	Lysine	Threonine
Wheat	Lysine	Threonine
Sorghum	Lysine	Threonine
Millet	Lysine	Threonine
Oats	Threonine	
Rye	Lysine	
Corn	Lysine	Tryptophan
Cottonseed	Lysine	
	Threonine	
Sunflower	Lysine	Threonine
Soybeans	Methionine	
Sesame seed	Lysine	Threonine
Peanuts	Methionine	
	Lysine	
Triticale	Lysine	

Fishery Resources

The oceans or seas cover over 70% of the earth's surface, yet less than 10% of our protein comes from fishery resources. It would seem that this resource could be utilized more effectively as a source of food for the future. As on land, the principal food producing organisms in the sea are plants (phytoplankton). These plants use the photosynthetic process to produce food for herbivorous animals in the sea. These, in turn, furnish food for the small carnivorous animals, the larger fish, and ultimately, man.

Not all of the ocean is productive. There is an uneven distribution of phosphates and sources of nitrogen (nitrates, nitrites and ammonium salts). This affects the distribution of organisms. The deep, blue-water area of the ocean contains low amounts of nitrogen compounds and phosphates. Although this region encompasses 90% of the global water, it supplies only 0.1% of the annual harvest from the sea (Poppensiek 1972). The two regions of the sea that are productive are the coastal waters and certain limited areas in the ocean where there are upwelling currents of nutrient-laden deeper water. Although these upwelling areas comprise about 0.1% of the ocean's surface, they produce almost 50% of the harvest. The upwelling of the Humboldt Current off the coast of Peru supplies 10% of the world's total fish harvest (Othmer and Roels 1973). The yield from the fisheries resource will depend on factors such as pollution, and regulations needed to prevent over-fishing of areas. The harvest may be increased by using trash fish presently thrown overboard, by using krill (a species of planktonic shrimp), or using algae or kelp. Vishniac (1971) determined that the maximum harvest could be about 3×10^8 mg of food per year. This yield is limited by the amount of sunshine falling on the oceans as well as other factors.

The seas are, and will continue to be, a source of protein, but this source is not unlimited. The recent reduction of fish harvested from the ocean off Peru reflects the problem of depending too much on the ocean as a source of food. Fish farming is developing and may overcome the hazards of nature (Brown 1973).

Energy Resources

Besides solar energy, fossil fuel is used in the production and harvesting of foods. The energy yield from crops is about two or three times the input of fossil fuel. However, the overall food system (production, processing, transportation, storage and preparation for consumption) requires more energy than is present in the food. Hence, the food system is influenced by the impending shortages of petroleum and our food supply will depend on the development of acceptable alternative sources of energy. None of our basic resources (land, water, fertilizer or energy) needed for food production can be considered to be abundant.

Microorganisms as Food

The use of microorganisms in food products is not a new idea. The action of yeast in the fermentations producing wine and beer and the leavening of doughs has been known for at least 4,000 or 5,000 years. The nature of the active factor in these fermentations did not become established until the latter part of the 19th century when the relation of living yeast cells to fermentation was discovered. Microorganisms are used in various fermentations and are consumed as part of the food. This is especially evident in cheese. *Penicillium roqueforti*, the blue mold of Roquefort cheese, and *Penicillium camemberti*, the white mold of Camembert cheese, are consumed with the cheese. Thus, the concept of using microorganisms as part of the food supply should not be completely objectionable.

Although man can synthesize amino acids and polypeptides commercially, microorganisms produce not only these substances, but proteins, antibiotics, vitamins, steroids and many other products. With their diverse biochemical capabilities, microorganisms are a potential source of food. The development of microorganisms as a food source will involve many disciplines, but food microbiologists will play an important role.

Bacteria, yeasts or molds cannot create foods but they can grow on cellulosic compounds which would be wasted. Algae can utilize solar energy to produce food. The production of microbial protein is discussed in Chapter 9.

Wastes as Food

For every kilogram of food, between five and ten kilograms of waste materials are left in the field or at the processing plant. These are wastes because their economic value is such that it is not profitable to utilize them. Surprisingly, not too long ago, fat was removed from milk, fish and oilseeds for human use, while the more valuable protein was wasted or fed to animals. Hence, the waste products of today may have food value in the future.

Some wastes are not readily usable because they are seasonal, diluted with water or require transportation to amass a large quantity for processing. With waste disposal becoming an ever-increasing problem, and food shortages becoming more crititical, it seems difficult to believe that the problems involved in utilizing acceptable wastes cannot be resolved.

Processes such as reverse osmosis and ultrafiltration can be used to concentrate dilute wastes. With these systems, water is removed from solutions by means of pressure exerted on a membrane. By selection of the porosity of the membrane, molecules of various sizes can be separated.

Progress is being made in the recovery of wastes and the utilization of these materials in foods and feeds (Callihan and Dunlap 1971; Cherry *et al.* 1975; Cooper 1976; Kamm *et al.* 1977; Toyama 1976; USDA 1972).

Legal Aspects.—Food laws and regulations must be considered before any new or novel food can be used. The U.S. Department of Agriculture (USDA) regulates red meat and poultry processing operations. The U.S. Food and Drug Administration (FDA) decides what is an acceptable food or food ingredient in all other cases. Besides these federal agencies, the food laws of states and local jurisdictions must be followed.

If an ingredient is to be used in a food, it cannot create a health hazard.

The Food, Drug and Cosmetic Act provides that a food shall be deemed to be adulterated if it consists in whole or in part of any filthy substances. What will be the reaction to single cell protein obtained from microorganisms grown on these unconventional substrates? If the projected food shortage develops, we may need to reevaluate the aesthetic aspects of our foods. The health criteria for processed wastes were discussed by Taylor *et al.* (1974).

Preventing Losses

The amount of food is usually appraised only in terms of production and consumption. Not enough attention is devoted to the control of deterioration, waste and losses. There is not adequate or reliable data to fully evaluate losses of plant and animal products so that we can determine a complete analysis of their causes. We do know that deterioration, waste and losses occur in almost every step from production to consumption. If we already have a shortage of food, and more food will be needed in the future, an increased effort must be made to protect our food supply from potential losses. Obviously, if losses could be reduced or prevented, our food supply would increase with no additional utilization of land or sea resources.

The main losses of foods are due to the action of microorganisms, insects, rodents, birds, nematodes and the enzymes inherent in the food. Ennis *et al.* (1975) estimated 30% of the worldwide crops are lost to pests. Albrecht (1975) suggested that pests destroy 25% of our crops. He believed that government regulations prevented adequate pest control. Schweigert (1975) estimated food losses from production to consumption to range from 20 to 50%. He found losses were due not only to pests but also to man-made defects. One loss was due to government regulations based solely on aesthetics and not on scientific information. Other defects were due to improper handling, processing or storage as well as human errors.

Production.—Poor farm management can result in low yields, or the production of poor quality foods. The improper use of pesticides may result in products that are hazardous to animals and man, reduce the harvest of adjacent fields, or may affect desirable insects, such as bees, or soil microorganisms, such as those that fix nitrogen.

There are some 1,500 plant diseases, over 200 animal diseases and about 2,000 species of weeds that cause economic losses. Between 6% and 12% of our crops are lost to nematodes (parasitic worms). The annual loss of crops to diseases, weeds and nematodes is over $4 billion.

Rodents, insects and birds cause losses of crops through consumption, contamination and damage. Damaged fruits and vegetables are more subject to microbial deterioration than are sound products.

Harvesting.—The mechanical harvesting of crops can result in damage, loss or waste of product. Bruising or damage during harvesting causes rapid deterioration in subsequent storage. Growth of fungi is accelerated on fruit that is damaged prior to storage. Mechanical harvesting leaves from 3 to 8% of the crop in the field.

The gathering and handling of shell eggs can result in damage to up to 7% of the shells. Although some damaged eggs may be salvaged, they usually are more highly contaminated with microorganisms than are eggs with intact shells. Due to improper handling as much as 10% of the shrimp catch is lost.

Processing.—Most foods are perishable and begin to deteriorate shortly after gathering, harvesting or slaughter. The processing of foods includes washing, removal of inedible parts, removal or destruction of potentially harmful substances and preservation. If processing is conducted improperly, deterioration of the product can occur.

Protein losses have been found due to scalding or chilling of poultry. Extensive contact with water and brines in fluming, washing and blanching can cause nutrient and weight losses of vegetables.

Inspectors condemn and destroy about one-half million kilograms of red meat and meat products each working day because of disease, contamination or spoilage. Approximately a million poultry carcasses are condemned each week. About 10% of these condemned carcasses are a result of contamination, overscalding or other defects occurring during processing. Since red meat and poultry are our main sources of protein, these losses are significant. Even with inspection, there is no guarantee that these foods do not contain organisms potentially pathogenic for man, inasmuch as condemnation is done by gross examination, with essentially no microbial testing of the carcasses. These condemned carcasses may be salvaged and processed for animal food. However. this practice may be a source of potential pathogens for man and animals if the reprocessing is done improperly. An example of contamination is that of organisms of the genus *Salmonella.* It was found that bone meal and meat meal contained *Salmonella* which could infect animals when used as food. These infected animals, when brought to the processing plant, could contaminate the plant, other carcasses, and, when consumed by man, they could cause salmonellosis. Thus, the food microbiologist should be concerned with animal food as well as human food.

Any type of processing, including simple cooking, tends to reduce, to some extent, the nutrient content or quality of foods. Modern processing methods are designed to keep such losses as low as possible. In many instances, nutrients are restored by enrichment after processing. Processing can help prevent losses by inactivating enzymes or microorganisms present in the food.

Storage.—During storage, foods may be affected in many ways. There can be a loss of functional properties, color, flavor, aroma, texture, appearance or nutrients due to oxidation, hydrolysis, or other chemical reactions. Inherent enzymes, microorganisms and pests such as rodents and insects, degrade and cause loss of foods. Rodent and insect contamination is one of the main reasons for recall of foods by the U.S. FDA.

It is difficult to determine the amount of food that is destroyed by insects, rodents or microorganisms. In some tropical countries, crop losses of 40% are thought to be common. Wycoff and Anderson (1970) estimated that 50% of the wheat stored in India is ruined by these entities.

Microorganisms.—Microorganisms reduce man's food supply by creating an unacceptable product. Metabolism of certain microorganisms can produce spoilage or an undesirable appearance. Foods are discarded or, in some cases, reprocessed if unusually high numbers of microorganisms or if certain potential pathogens are found. If microbial toxins are present in the food, it is usually advisable to destroy the food rather than attempt to salvage it.

Preparation and Consumption.—Some food is lost in washing, peeling or trimming during preparation. Cooking reduces the nutrient value. Leaving food on the plate is wasteful. More food is served in some restaurants than can be consumed.

Foods that are left over from a meal can usually be served again if they are handled in an acceptable manner. If not promptly refrigerated, cooked foods can become deteriorated or a health hazard due to microbial growth.

Reasons for Food Preservation

Preservation can be defined as a process by which foods are treated to retard decay or spoilage. There are many reasons for preserving foods. Several plant foods are harvested only once each year. In order to have a supply of these foods throughout the year, rather than only during harvesting, preservation is necessary. In case of a crop failure due to natural disasters such as drought, wind, hail, floods, fire, freezing, or insect and disease infestations or man-made disasters, such as war, the preservation of previously produced excess food becomes paramount.

With preservation, one can obtain a more varied diet, both from the aspect that a crop can be used throughout the year and that crops native to only a small area can be transported and used anywhere in the world.

TABLE 1.2

FOOD PRESERVATION METHODS

Asepsis	Gas or Vacuum Packing
Centrifugation	Acidification
Filtration	Fermentation
Refrigeration	Fumigation
Freezing	Pasteurization
Drying	Cooking
Freeze Drying	Canning
Chemicals	Radiation
Smoking	

One of the reasons developing countries have food shortages is that they do not have facilities for preservation or transportation of foods. Thus, certain areas have a temporary surplus of food while other areas have a shortage.

Flesh foods deteriorate rapidly if held at ambient temperatures. In some countries, although fish is plentiful along the coastal areas, there is a protein shortage in inland areas, because refrigeration and rapid transit are lacking to transport the fish protein without spoilage.

Preservation allows the holding of foods so that they can be used as ingredients for mixed foods. Many of our convenience foods are combinations of various foods.

Some of our systems used to preserve food also destroy many of the organisms and toxic factors that are hazards in food products.

Methods for Food Preservation.—The chief methods of preservation can be listed in four basic categories: asepsis (preventing entry of microorganisms into foods); removal of microorganisms; inhibiting growth by controlling the environment; and the destruction of microorganisms. Various systems have evolved from these basic procedures (Table 1.2).

Most of our methods of preserving food are merely modifications of systems used in ancient times. The addition of salt as a chemical preservative, fermentations, smoking and cold storage have been practiced for over 2,000 years. Comparatively, canning might be considered a modern method, although it was patented by Appert in 1810. With all of our modern technology, the main new idea in food preservation is radiation and, at the present time, it has not been officially approved for this use in the U.S. by the FDA.

FOOD HAZARDS

From the beginning of life until death, a person is subjected to potentially hazardous environments. During one's lifetime, some disappear and others take their place so that the problems of safety are not static. The ingestion of food is no exception. Food may serve as a carrier of chemical and biological substances, either added or acquired as contaminants from soil, water, air, food handlers, equipment and other sources. The possible subtle relationships between these substances in foods and physical vigor, mental alertness, longevity, resistance to infection, and the onset of degenerative diseases are not fully understood. Since we do not have all of the answers regarding the safety of all possible substances, there are many controversies concerning the overall safety of food. Although absolute safety of food is ideal, for all practical purposes, this goal is unattainable. Since man is a biological system, and since biological systems vary, a food that causes no ill effects for one person may cause problems for another person.

Wodicka (1977) listed six principal categories of food hazards. These were: microbiological hazards, malnutrition, environmental contaminants, naturally occurring toxins, pesticides and conscious food additives.

Food Additives

Spicer (1975) stated that food additives have made a substantial contribution to the incidence of foodborne illnesses. Over 2,000 different chemicals are used as direct additives to foods. The use of any chemical additive should be on the basis of risk and benefit. The benefit should be to the consumer, not the producer.

Generally, food additives are ingested in such small quantities that acute poisoning rarely occurs. Any carcinogenic, mutagenic or teratogenic effects or effects on behavior or reproduction are difficult to determine by epidemiological investigations, since these occur too long after the consumption of the agents.

The FDA is constantly reevaluating food additives. It is the responsibility of industry to supply data to the FDA regarding the safety of these chemicals.

Pesticides

Although pesticides in foods have been involved with poisoning and death, usually these incidents have been due to the ignorance of people using the pesticide or the contaminated food. Perhaps the largest incident occurred in Iraq. Seed wheat, treated with a mercury fungicide and dyed pink to distinguish it from edible wheat, was distributed to farmers with a warning that it was poisonous and should not be eaten. In spite of these precautions, many people consumed the wheat. Although official figures show over 400 died, estimates as high as 6,000 deaths were made, with perhaps 100,000 others poisoned and injured.

There have been many reported outbreaks of poisonings in which pesticides or other chemicals were misused or accidentally added to foods, but fortunately none of these have involved as many people as the Iraq incident. How do you legislate against ignorance? The present solution is to legislate against everyone by banning the use of chemicals that are valuable in increasing our food supply but could cause a problem if mishandled. In some cases, pesticides considered to be safe can enter into reactions with components of foods or are altered during food processing, so that an unsafe chemical is produced (Morrison 1976).

Naturally Occurring Toxins

There have been several reviews concerning toxic agents naturally present in foods (Coon 1975; Crook 1975; Hatfield and Brady 1975; Munro 1976). These naturally occurring toxins include estrogens, goitrogens, hemagglutinins, saponins, lathyrogens, tumorigens, carcinogens, cyanogens, as well as pressor amines, antivitamins, antienzymes, seafood toxins, fungal toxins, toxic mushrooms, nutritional inhibitors and antigens that produce allergies. The quantities of these substances are usually low and, during processing, some are altered to reduce their potency. Problems resulting from the natural toxicants are often due to people eating raw foods, too much of only one type of food or mistaking toxic plants for similar edible plants. Potatoes contain the alkaloid solanine, a potent cholinesterase inhibitor that interferes with the transmission of nerve impulses. The amount of potatoes that an average person eats each year contains enough solanine to be fatal if consumed in one dose.

At high levels, even polyunsaturated fats reportedly increase the incidence of tumors and gallstones, increase the requirement of vitamin E and cause premature aging in laboratory animals. When excessively heated, these fats are reported to contain toxic substances. One product, malonaldehyde, is carcinogenic. Potentially carcinogenic lipid peroxides are easily formed from polyunsaturated fats by autoxidation.

Apparently, the people who advocate eating natural rather than processed foods do not realize that processing, such as heating, can reduce, or eliminate many of these toxic factors.

The possible presence of toxic substances will need to be determined for any protein or nutrient derived from a new or novel food source.

Environmental Contaminants

The incidence of toxicity from environmental contaminants is relatively low. There have been a few cases of illness and even death due to ingestion of lead and mercury.

There are other residues besides pesticides that gain entrance to our food. Feed additives, such as antibiotics, have been detected in animal tissues. Plant products are not immune from problems. When grown on soil high in certain elements, these elements have been found in the plant tissue. Although molybdenum is essential as a trace mineral for some plants, the grass grown on soil with a high concentration of this element can be toxic for cattle.

Malnutrition

Malnutrition in the U.S. is due to ignorance or poverty. In other countries, food shortages enter into the problem. With increasing populations, food shortages and malnutrition will become more prevalent.

Microorganisms

The data compiled by the Center for Disease Control (1976) show that microbiological hazards of foods are by far the most important (Table 1.3). Since not all foodborne illnesses are reported, these data are not exact, but they are the most complete data available at the present time.

In every year, over 60% of the foodborne outbreaks are due to bacterial

TABLE 1.3

CONFIRMED FOODBORNE OUTBREAKS, 1973–1975

	Years					
	1973		1974		1975	
Cause	No.	%	No.	%	No.	%
Bacterial	84	66.1	122	60.7	123	64.4
Viral	5	3.9	6	3.0	3	1.6
Subtotal	89	70.0	128	63.7	126	66.0
Parasitic	10	7.9	16	8.0	22	11.5
Chemical	28	22.0	57	28.3	43	22.5
Total	127		201		191	

Data from Center for Disease Control (1976).

etiologies, while less than 30% are due to chemicals from various sources. The Center for Disease Control discontinued listing foodborne outbreaks caused by unknown etiologies. These accounted for 25-30% of the total outbreaks in the past.

Although not listed as causing any foodborne outbreaks, the toxins produced by molds, the mycotoxins, have received much attention in the past 10-15 years.

Other biological entities besides microorganisms present a potential hazard. *Trichinella spiralis*, the agent of trichinosis; *Taenia solium*, the pork tapeworm; *Taenia saginata*, the beef tapeworm; and *Entamoeba histolytica*, the cause of amoebic dysentery, are a few of the agents that have been found in foods.

ROLE OF THE MICROBIOLOGISTS

The food microbiologist is concerned with the biochemical reactions of microorganisms in and on foods. These reactions result in spoilage, public health hazards and fermentation products. Determining the numbers and types of microorganisms associated with food, and knowing the sources of microorganisms, factors affecting their multiplication and systems that can be used for their control are important to food microbiologists. However, we cannot look only at these aspects, but must also examine other facets of foods such as the chemical and physical characteristics and the various attributes which are referred to as quality. We can sterilize a food to destroy all microorganisms, but if the food is no longer edible, or the nutritional value depleted, then the sterilization process is not satisfactory.

With an understanding of food science, a food microbiologist can relate his role to the very important endeavor of providing mankind with an adequate supply of safe, wholesome foods.

BIBLIOGRAPHY

ALBRECHT, J.J. 1975. The cost of government regulations to the food industry. Food Technol. *29*, No. 10, 61-65.

ANELLI, G., FIORENTINI, R., MASSIGNAN, L., and GALOPPINI, C. 1977. The poly-protein process: a new method for obtaining leaf protein concentrates. J. Food Sci. *42*, 1401-1403.

ANON. 1975. LPC...untapped protein reserve. Food Process. *36*, No. 7, 38-41.

BROWN, E.E. 1973. Mariculture and aquaculture. Food Technol. *27*, No. 12, 60-66.

CALLIHAN, C.D., and DUNLAP, C.E. 1971. Construction of a Chemical-microbial Pilot Plant for Production of Single-cell Protein from Cellulosic Wastes. U.S. Environmental Protection Agency, Washington, D.C.

CDC. 1976. Foodborne and Waterborne Disease Outbreaks. Annual Summary 1975. Center for Disease Control, Atlanta, GA.

CHERRY, J.P., YOUNG, C.T., and SHEWFELT, A.L. 1975. Characterization of protein isolates from keratinous material of poultry feathers. J. Food Sci. *40*, 331-335.

COON, J.M. 1975. Natural toxicants in foods. J. Amer. Diet. Assoc. *67*, 213-218.

COOPER, J.L. 1976. The potential of food processing solid wastes as a source of cellulose for enzymatic conversion. Biotechnol. Bioeng. Symp. No. 6, 251-271.

CROOK, W.G. 1975. Food allergy—the great masquerader. Pediat. Clin. No. Amer. *22*, 227-238.

EDWARDS, R.H. *et al.* 1975. Pilot plant production of an edible white fraction leaf protein concentrate from alfalfa. J. Agr. Food Chem. *23*, 620-626.

ENNIS, W.B., JR., DOWLER, W.M., and KLASSEN, W. 1975. Crop protection to increase food supplies. Science *188*, 593-598.

HATFIELD, G.M., and BRADY, L.R. 1975. Toxins of higher fungi. Lloydia *38*, 36-55.

KAMM, R., MEACHAM, K., HARROW, L.S., and MONROE, F. 1977. Evaluating new business opportunities from food wastes. Food Technol. *31*, No. 6, 36-40.

KOHLER, G.O., and KNUCKLES, B.E. 1977. Edible protein from leaves. Food Technol. *31*, No. 5, 191-195.

LABUZA, T.P. 1974. Food for Thought. AVI Publishing Co., Westport, CT.

MORRISON, A.B. 1976. Food safety in the seventies. J. Milk Food Technol. *39*, 218-224.

MUNRO, I.C. 1976. Naturally occurring toxicants in foods and their significance. Clin. Toxicol. *9*, 647-663.

OTHMER, D.F., and ROELS, O.A. 1973. Power, fresh water, and food from cold, deep sea water. Science *182*, 121-125.

PIRIE, N.W. 1975. Leaf protein: a beneficiary of tribulation. Nature *253*, No. 4589, 239-241.

POPPENSIEK, G.C. 1972. Aquaculture and mariculture: protein potentials. J. Amer. Vet. Med. Assoc. *161*, 1467-1475.

SCHWEIGERT, B.S. 1975. Food processing and nutrition—priorities and needed outputs. Food Technol. *29*, No. 9, 36-38.

SPICER, A. 1975. Toxicological assessment of new foods. Brit. Med. Bull. *31*, 220-221.

TAYLOR, J.C., GABLE, D.A., GRABER, G., and LUCAS, E.W. 1974. Health criteria for processed wastes. Fed. Proc. *33*, 1945-1946.

TOYAMA, N. 1976. Feasibility of sugar production from agricultural and urban cellulosic wastes with *Trichoderma viride* cellulase. Biotechnol. Bioeng. Symp. No. 6, 207-219.

USDA. 1972. Proceedings Whey Products Conference. ERRL Publication No. 3779. U.S. Department of Agriculture, Philadelphia, Pa.

VISHNIAC, W. 1971. Limits of microbial productivity in the ocean. *In* Microbes and Biological Productivity. D.E. Hughes and A.H. Rose (Editors). Cambridge University Press, England.

WODICKA, V.O. 1977. Food safety—rationalizing the ground rules for safety evaluation. Food Technol. *31*, No. 9, 75-79.

WYCKOFF, G.H., and ANDERSON, R.D. 1970. Fumigation or famine. J. Amer. Vet. Med. Assoc. *157*, 1828-1834.

2

Estimating the Number of Microorganisms

An important aspect of food microbiology is the examination of food or other materials for microorganisms.

NUMBERS OF MICROORGANISMS IN FOOD

The number of microorganisms as determined by the aerobic plate count (APC) is variable due to the original contamination, increase or decrease of microorganisms during processing, recontamination of processed product and growth or death during storage, retailing and handling. The microbial flora is changing constantly. In foods such as refrigerated fresh meat, the microbial numbers increase during storage, while in dried or frozen foods, the viable organisms tend to decrease in number. The APCs for a food may vary from less than ten to over 100,000,000 microorganisms per gram, depending upon the product, how long it was stored and the temperature of storage. The logarithms of the range of APCs reported for various foods are listed in Table 2.1.

The usual range of most animal products is 1,000 to 10,000 organisms per gram. Ground meat is more contaminated than whole cuts of meat due to the type of meat that is used in the product, the extra handling during grinding and the release of meat juices that allow bacteria to multiply. Foods that received a heat treatment (cooking, pasteurization) have lower microbial numbers than foods not heated. Even then, due to poor quality ingredients, poor sanitation, unsatisfactory heating, recontamination or poor handling and storage, some heated products have high numbers of microorganisms.

An estimate of the number of microorganisms in or on foods is needed in order to determine if a product meets the microbial levels expressed in specifications, guidelines or standards. Spoilage of some foods is imminent when the APC reaches very high numbers (10^7-10^8/g). Hence, the microbial count can be used to help predict the shelf life of certain foods. To a limited extent, the microbial numbers might be used to evaluate the potential safety of foods. The count also might indicate if the product was produced under sanitary conditions, or if the product was mishandled during harvesting, processing or storage. In general, as the microbial count increases, the quality of the food is reduced. This generalization does not apply to fermented foods, since microorganisms are used in their production. There are cases in which the number of microorganisms in a food has little or no relationship to potential shelf life, spoilage or a health hazard. Other factors to be considered include the type of food, the type of microorganisms present and the storage conditions.

TABLE 2.1

AEROBIC PLATE COUNTS OF VARIOUS FOODS AS LOGARITHMS OF BACTERIA PER GRAM UNLESS OTHERWISE NOTED

Food	Overall Range	Usual Range
Animal Products		
Beef (steaks, roasts)	2-6	4
Beef (ground)	3-8	5-7
Pork sausage	4-6	5
Ham	1-8	4
Bacon	3-7	4
Dry sausage	3-7	4-5
Chicken carcasses (cm^2)	2-7	3-4
Fish (fresh)	2-8	4-5
Fish (smoked)	1-7	2-4
Fish sticks or crab cakes	2-6	3-4
Shrimp (raw)	2-7	4-5
Shrimp (raw, breaded)	2-8	4-6
Milk (raw, grade A)	2-5	3
Milk (pasteurized)	2-4	2
Milk (dry)	1-6	2-3
Butter	3-5	4
Plant Products		
Raw		
Almonds	0-4	3
Beans or peas	3-7	4-5
Broccoli or kale	6-7	
Carrots, potatoes or spinach	4-7	
Corn or cucumbers	5-7	
Tomatoes	3-7	
Frozen		
Asparagus, beans or peas	2-5	
Corn	2-7	
Squash	2-4	

TABLE 2.1. *(Continued)*

Food	Overall Range	Usual Range
Dried		
Carrots	2-4	
Garlic	4-6	
Parsley	2-5	
Spices		
Cinnamon	1-5	2-3
Cloves	2-3	3
Ginger	2-7	
Nutmeg	2-4	
Oregano	2-6	3-4
Pepper	6-7	7
Sage	3	
Mixed dried		
Soup (meat-type)	3-5	4
Soup (vegetable-type)	2-5	3-4
Salads		
Chicken or ham	1-7	3-5
Green	3-8	5-6
Macaroni	3-6	4-5
Shrimp	3-7	6
Tuna	2-6	3-4

To produce food with a low number of microorganisms, it is necessary that not only the final food product be assayed, but also such things as ingredients, processing equipment, packaging and environmental samples. These determinations will aid in the evaluation of general sanitary practices prevailing during processing and handling of food, and the potential sources of contamination. The determination of microbial numbers is needed to evaluate the effectiveness of methods of preservation.

The presence of particular types of microorganisms, especially potential pathogens or toxin producers, is more important than the estimate of the total number of microorganisms. In general, the main difference in these analyses is that specific types of microorganisms are determined with selective and/or differential media rather than non-inhibitory media. Thus, for purposes of simplicity, this discussion will be limited to total number estimations. Some of the special procedures are discussed with specific organisms in later chapters of this text.

Although the term "total count" has been used, it should be remembered that no single method or medium is capable of detecting all of the microorganisms in a food. Thus, the counts that are obtained are merely estimates of the actual microbial population. Errors of ± 90% in counts are not unusual when the level is 10,000 to 100,000 per gram (Collins and Lyne 1970). Besides the errors, many assumptions are involved in microbial estimations. Also, there are factors that affect the growth of microorganisms and influence the results when the viability of the cells is involved in the enumeration technique. With all of these considerations, it is essential that the technician doing the testing does not further influence the results due to poor technique.

For microbial analysis, one needs a sample and a system for estimating the number of microorganisms in the sample. After the data from the evaluation are obtained, the information must be reported and, when necessary, follow-up checks should be made; otherwise, the tests are being run for the exercise. If the report is for management, an interpretation of the results might be included. What do they mean? Are the levels of microorganisms acceptable or too high?

THE SAMPLE

If the samples are not delivered to the laboratory, it might be necessary to establish a sampling procedure. The samples of food might be obtained from the processing line, from warehouse storage or from retail shelves. Food is processed as liquid, solid, mixed solid and liquid, or semi-solid, and in many shapes and sizes. Since there are many variables in the food, and places of sampling, several sampling plans will be needed.

The sampling plan should reflect the ultimate use of the analysis, the potential health hazard of the food, or potential for spoilage. If the results are needed to satisfy the requirements of a microbiological standard, the sampling plan as outlined in the standard should be followed. If the results are for the producer's information, a less restrictive sampling plan can be used. Sampling plans have been suggested for microbiological standards (ICMSF 1974) and for salmonellae (health hazard) testing (Olson 1975). Further discussion of these sampling plans is presented in appropriate chapters of this text.

A sample will yield significant and meaningful information only if it represents the mass of material being examined, is collected in such a manner to protect it against microbial contamination, and is protected from changes in the population that might occur between collection and analysis.

Representative Samples

The need for a representative sample cannot be overemphasized. The results of the analysis can be no more reliable than the sample on which they were based. Usually microorganisms are not distributed homogeneously, so that thorough mixing of the product prior to sampling is important. If the food is a liquid, mixing before sampling is possible. Thorough mixing is not as easy for nonliquid foods.

The size of the particles being sampled may influence the sampling procedure, since it is evident that many particles of a product such as powdered milk can be obtained, but if the product were sides of beef, one would have to set up alternate procedures.

Sampling material in motion, such as on a production line, tends to minimize variables and gives a more representative sample than sampling material at rest, such as in stacks in a warehouse or on retail shelves. With on-line sampling, automatic sampling devices might be considered. These devices usually give a more random and reliable sample and at less cost than manual sampling of the product.

The laboratory analysis is usually more expensive than obtaining the sample, so that cutting corners in sampling is not the way to save money.

Number of Samples

The number of samples needed, or the frequency of sampling, depends upon many factors. The uniformity or homogeneity of the product, the size of the many particles, previous knowledge of the material and experience will help dictate the amount of sampling needed. Either too few samples or too many samples waste product, laboratory material and labor.

For lot sampling, the number of packages to be sampled must be determined. In general, the square root of the number of containers is sampled, except that, as the lot size is increased, a percentage of the number of containers is sampled. The amount may vary from 10% for small lots to less than 1% for large lots of product.

If cases or containers are stacked as a lot, the person must randomly select containers throughout the entire pile. If only containers around the edges or in front of the stack are selected, he is introducing a bias into the results of the analysis.

Sampling suggestions for various products can be found in AOAC (1975), APHA (1976) and FDA (1976). For microbiological standards, the number of samples to be obtained and analyzed is included in the standard. One of the prime considerations that influence the number of samples to be analyzed is the potential health hazard of the foods. Statistical sampling schemes will help ensure that the samples will give an acceptable assessment of the microbial condition of the food, ingredient or other substance being analyzed.

Aseptic Collection of Samples

Aseptic technique is needed when samples are collected. To prevent possible contamination, if the samples are in individual containers, such as cans, bottles or boxes of food, they should be taken directly to the laboratory for analysis. On the other hand, if the product is in bulk, or in containers of impractical size to submit directly to the laboratory, representative portions must be transferred to sterile containers using aseptic technique.

Since there is little interest in bacteria associated with sampling devices or sample containers, the instruments must be sterile. If possible, the instruments should be sterilized in the laboratory, rather than at the place of sampling. After cleaning the sampling equipment, the preferred methods of sterilization are: a) steam at 121.5°C in an autoclave for 15-30 min (the time for exposure depends on how bulky the material is and how closely the material is packed in the chamber), or b) the hot air oven. The suggested conditions for hot air sterilization vary from 1 to 3 hr at 160° to 180°C. If protected from recontamination, the sterilized instruments may be stored. Alternative systems for sterilizing are needed where neither an autoclave nor a hot air oven is available. These include: c) exposure to steam (100°C) for 1 hr and use the same day, d) immerse in water at 100°C for 5 min and use immediately, e) immerse in 70% alcohol and flame to burn off alcohol immediately before use, or f) flame with hydrocarbon (propane or butane) torch so that all working surfaces contact the flame before use. Using the alternative systems has been questioned. According to the FDA (FDA (1976), alcohol flaming is unsatisfactory because the instrument does not get hot enough to be effectively sterilized, and the flaming alcohol creates a fire hazard. The FDA recommended using a propane torch. Tansey (1973) suggested using a heavy duty butane lighter rather than an unwieldy torch.

When obtained, the sample should be placed in a sterile container. A wide mouth screw-capped jar is recommended (FDA 1976; APHA 1976); however, plastic bags or other acceptable containers can be used.

The methods of sampling and the types of instruments needed are determined by the substance to be sampled.

Liquids and Small Particles.—These foods can be mixed and sampled with sampling tubes, dippers, teaspoons, tablespoons, spatulas or similar instruments.

Large Materials.—If these substances can be cut, they may be sampled with a knife or cheese trier. For many materials, such as animal carcasses or processing equipment, the surface is sampled.

Surfaces.—To obtain a suspension of organisms from a surface, besides excising tissue with a knife or scalpel, one may wash, rinse or swab a prescribed area or entire piece of food, remove the organisms with sticky-backed cellophane tape or a contact agar system. Pressing a sterile and wetted membrane filter against the surface of meat allows the transfer of organisms from the meat surface to the filter surface. The use of ultrasonics as well as a vacuum probe have been suggested for removal of microorganisms from surfaces for sampling purposes.

The system for sampling surfaces depends upon the type of surface, amount of contamination and the ultimate use of the information. Each system has certain advantages and disadvantages. No one method is best for all of the diversified surfaces of foods and equipment used in the food industry.

Reeves (1973) found mechanical blending of vegetables yielded higher results than merely washing the vegetables in a diluent. He believed that mechanical blending broke up clumps of bacteria on the surface or liberated subsurface contamination. The blending of excised skin of poultry carcasses yields higher bacterial numbers than merely swabbing the area.

Systems using the "drip" from poultry carcasses or exuded tissue juice from seafoods for analysis compare favorably with swab sampling or mechanical blending. Van Schothorst *et al.* (1976) advocated both the drip and rinse water methods for sampling poultry carcasses for salmonellae.

Air.—The two general methods for air sampling are solid or liquid impingement. The systems for solid impingement include the settling plate, slit sampler, the sieve or Anderson sampler and the membrane sampler. Except for the settling plate, specific volumes of air are sampled. With the slit or sieve samplers, the microorganisms are impinged upon an agar surface. The membrane filter containing the organism is removed to a petri dish that contains nutrients so the organism can grow.

Holding of Sample

For best results, the sample should be analyzed immediately. When this is not possible, the sample should be refrigerated to prevent growth of any microorganisms. Alternatively, the sample can be packed in ice or, if shipment to another city is necessary, or if the sample is a frozen product, dry ice should be placed in the package. Refrigeration is preferred to freezing, since freezing may cause death or damage to some cells which may then give erroneous results when the sample is analyzed.

Preparation of Sample

Many of the methods of analysis require some preparation of the sample. The main consideration is to get the bacteria into a homogeneous suspension so they can be pipetted. If a food is a liquid such as milk, an aliquot can be mixed and pipetted, but if the food is a solid, such as hamburger, it is necessary to blend the food with a diluent to obtain a suspension. The rinse or wash samples from surfaces are treated as liquid samples, while swabs are placed in sterile diluent and shaken to suspend the bacteria.

Solid Food.—Solid food is generally mixed with a sterile diluent in a mechanical blender to obtain a homogeneous suspension. To 25 g of food in a sterile blender jar, 225 g of sterile diluent is added, and when mixed, a 1:10 dilution of the food and associated organisms is obtained (Fig. 2.1).

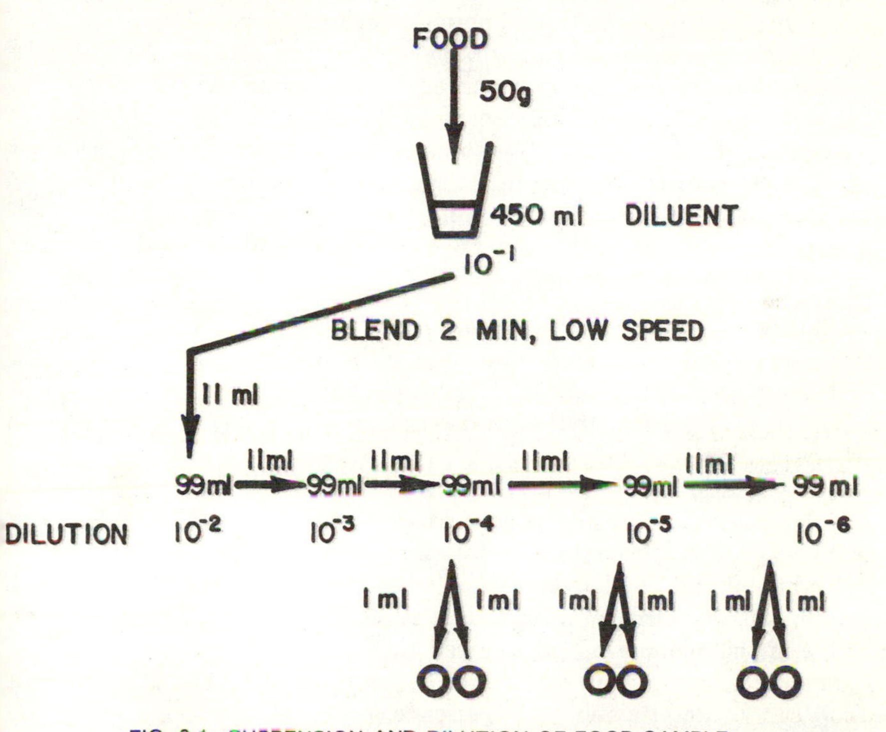

FIG. 2.1. SUSPENSION AND DILUTION OF FOOD SAMPLE FOR MICROBIAL ANALYSIS

This 1:10 dilution also is referred to as a 1/10 or 10^{-1} dilution. A 1:10 dilution means that in 10 g of the mixture, there is 1 g of food, or in 1 g of the mixture, there is 0.1 g of food, with associated organisms. Thus, if 1 g of the 1:10 dilution is analyzed, the microbial count is that of 0.1 g of food. To report the count as the number per gram, it is multiplied by 10.

As an alternative to blending, a sterile plastic bag containing the sample and diluent is placed in a device called a stomacher (Sharpe and Jackson 1972; Tuttlebee 1975). In the stomacher, the compression and shearing forces of the pounding result in a homogeneous suspension of sample and microorganisms.

Whatever method is used for preparation and dilution of the sample, the damage to the microbial cells should be minimized.

Diluents.—Several diluents have been suggested and used. Although AOAC (1975) recommends the use of Butterfield's buffered phosphate, there seems to be an acceptance of 0.1% peptone water. Peptone water is easy to prepare and it protects the organisms during dilution and plating. One disadvantage is that, if the prepared dilution is allowed to remain at room temperature for extended periods, the organisms will multiply. Not more than 20 min should elapse between the first dilution in phosphate buffer until the last plate is poured in the series (APHA 1967). An increase in count up to 10% can be expected in this 20 min period.

According to Harrewijn (1975) some pertinent aspects to be considered are the composition, temperature and pH of the diluent; anaerobic or aerobic condition; carry-over of inhibitors with the food; and any treatments needed to allow the recovery of cells injured during food processing or preparation of the sample.

Dilutions Needed.—For the plate count, only plates between 30 and 300 colonies are considered countable, although this has been disputed. For the tube dilution and MPN systems, the organisms are diluted to extinction. This means that dilutions are needed beyond the 1:10 dilution of the original suspension.

A 1:10 dilution of a 1:10 dilution is a 1:100 dilution. This 1:100 dilution is prepared by aseptically transferring 10 ml of the 1:10 dilution to a screw-capped bottle containing 90 ml of sterile diluent (or 11 ml transferred to 99 ml). The bottle is shaken (25 times through a 30 cm arc in 7 sec) to distribute the organisms homogeneously. Further dilutions can be made in this manner as far as needed.

The dilutions needed to estimate the number of microorganisms in a food can be determined by experience, previous knowledge or by the requirements of standards, guidelines or specifications. If 50,000 organisms per gram are allowed in a specification, a 1:1000 dilution can be used for the plate count. If less than 50 organisms are observed on the incubated plate, the food is within the limit, but if over 50 colonies are observed, it does not meet the requirement. Usually two or three dilutions are analyzed to increase the chances of obtaining an acceptable plate to count and, for the MPN, at least three dilutions are needed.

ANALYSIS

There are several procedures that can be used to estimate a microbial population. These include: microscopic counts, electronic particle counts, plate counts (pour or spread plates), tube dilution, most probable numbers (MPN), membrane filter, roll tube, little plate, Burri strip method, contact plates, reductase tests, chemical indicators, adenosine triphosphate (ATP), respiration with or without the use of radioactive tracers, measuring the mass or volume of cells or spectrophotometric (optical density) tests.

Not all of these procedures are readily adapted to foods, while some adapt to many or a majority of foods. The ideal test should be accurate, rapid, inexpensive and useful for most types of samples.

Total Cell Counts

Some systems make no differentiation of living or dead cells. All microbial cells are counted. Two of these are the direct microscopic count and the electronic particle count.

Direct Microscopic Count (DMC).—With this method, the results are obtained sooner than with most other procedures since no incubation period is needed for the cells to metabolize and multiply.

Liquid foods may be determined directly, but solid foods must be put into a suspension (1:10 dilution) before analysis. A counting chamber can be used, but for food examination, a portion (0.01 ml) of the material (measured with a standard loop or microsyringe) is uniformly spread over a prescribed area on a glass slide (usually 1 cm^2).

For products such as eggs or cream, xylene or other suitable solvent is added prior to staining to remove the fat from the material. After drying, the slide is then fixed by dipping in ethyl alcohol for one to two minutes before staining.

Several stains have been suggested and used in the DMC. The stained films are examined with a microscope, using the oil immersion objective. The number of fields to be examined and counted is inverse to the number of cells and clumps observed in each field.

To calculate the organisms per gram of food, the diameter of the field that is examined must be known. The diameter (d) is measured with a stage micrometer to the nearest 0.001 mm. Since the field is a circle, the area can be calculated ($A = \pi r^2$).

The average number of cells or clumps per field is calculated and divided by the area of the field diameter to obtain the number per mm^2. To determine the number of cells or clumps per cm^2, the number per mm^2 must be multiplied by 100 (since there are 100 mm^2 per cm^2). The resultant number is then multiplied by the dilution factor, which, in the case of liquid food (milk) is 100 (0.01 ml was used), or 1000 for solid food (0.01 ml of a 1:10 dilution).

Values of the DMC.—The DMC is a rapid method since an estimate of the bacterial load is obtained in a short time. This is of value in being able to make on-the-spot alterations or adjustments in the processing operation to remedy any problems.

Other values of the DMC that have been suggested include: a) little work is required; b) the test is not too difficult; c) except for a microscope, very little apparatus or equipment is needed; d) the prepared slide can be stored and maintained as a permanent record; e) one can get some idea as to type of organism present (cocci or rods); f) counts represent the organisms in the original product (if it has been treated, such as by heat); g) preservatives can be added to the sample for holding prior to analysis, for shipment, or to hold for further study, so that organisms do not multiply; h) only a small amount of sample is needed which is of value if the product is expensive.

Since both living and dead cells are counted, there is some question as to the value of the microscopic count. However, high numbers of cells, whether living or dead, in pasteurized products indicate poor quality of product before processing, survival or multiplication of bacteria during processing, or recontamination and/or growth after processing.

The value of the DMC is limited to samples with high counts, and gives little or no information for samples with low numbers of organisms. As the number of organisms increases, fewer fields need to be counted which decreases the work and, with high numbers, the DMC more nearly reflects the microbial condition of the food. With increased technology and efforts in sanitation, the bacterial load in many foods has been reduced. The DMC has little or no value for foods with low microbial loads.

Although the DMC has been used for many foods, including eggs, it was of little value for determining organisms in pasteurized egg products (Hall *et al.* 1971). They observed a reduction in the DMC of as much as 88% during pasteurization of whole egg. It was shown that the lytic enzyme lysozyme, present in egg white, was responsible. The activity of lysozyme increases with the temperature, up to 60° C, which is near the pasteurization temperature of whole egg. The enzyme lysed the cells so that they were no longer counted in the DMC.

Besides using the DMC to evaluate the microbial content of a food, it can be used to evaluate the number of body cells (leucocytes or lymphocytes) in milk. This is especially valuable in indicating mastitis infection in cows.

The value of the DMC depends upon the type of food and the type of organisms associated with the food. For products that have received a treatment such as heat to control the microorganisms, it would be doubtful if the DMC could predict shelf-life of the product. It is doubtful if the DMC would have any value in determining the public health hazard of the product.

Assumptions and Errors.—There are many assumptions and errors in any microbiological method of analysis. There are errors inherent in the analytical procedure, as well as those introduced by the technician doing the test.

There are assumptions regarding the sample. It is assumed: a) the sample is representative of the entire lot of product; b) the subsample used for analysis is representative of the sample; c) that cells are distributed homogeneously in the sample as well as the subsample and, if not originally homogeneous, that operations such as mixing, blending or shaking have produced a homogeneous mixture; d) the weighing or measuring of the subsample, diluents and aliquots is accurate; and e) the sample or subsample has been handled so that there is no contamination or multiplication of cells during sampling or analysis.

The sample may not be either representative or homogeneous, and the subsample that is analyzed will only be as representative as the original sample. Mixing, blending or shaking will produce a more homogeneous subsample, but the extent of homogeneity will vary due to type of product and type of organisms present. The measurement of the amount of subsample analyzed is not likely to be precise. Standard methods (APHA 1967) allows a tolerance of ± 0.025 ml for a 1 ml pipet. Thus, a difference of 5% can exist between the amount delivered by various pipets without even considering the technician's errors during measurement.

These errors in sampling persist not only for the DMC, but also for other microbial procedures.

Errors in the DMC can occur during the preparation and staining of the cells, counting the cells or clumps, or in the calculations involved in converting the raw data into the count per gram of product.

For measuring the 0.01 ml of sample onto the slide, the microsyringe is preferred to the calibrated loop. It is easier to uniformly spread the sample into circular areas than square ones. If the material is not spread uniformly, the cells will not be homogeneously distributed.

It is assumed that the smear on the slide will dry into flat layers of uniform density. However, the smears have been found to vary in thickness from one area to another. For some foods, such as liquid egg, the film on the slide may be of such thickness that it obscures many bacterial cells for counting. Organisms that are lightly stained are difficult to discern, and with an unevenly stained background, it is difficult to distinguish dirt or other particles from bacterial cells. Besides causing errors, this becomes time consuming and causes eye strain. Improper illumination with resulting eye strain and fatigue is a major concern that can cause errors in counting the cells. The staining process washes some cells from the slide or can result in the counting of precipitated stain as cells.

For other errors, see APHA (1967).

Electronic Particle Count.—The electronic counter is based on the principle that cells are poor electrical conductors as compared to an electrolyte solution. A dilute suspension of cells in saline or other suitable electrolyte is drawn through a minute aperture conducting an electric current between two electrodes (usually platinum). Each cell passing through the aperture displaces an equal volume of the electrolyte solution, and causes a momentary increased impedance to the flow of electric current. The resulting voltage pulse is proportional to the size or volume of the particle passing through the aperture. These pulses are amplified and counted. They simultaneously appear on the screen of an oscilloscope.

The electronic counter has threshold dials that can be set for counting only those pulses of a certain magnitude. By varying the threshold settings, particles within a given volume range can be selectively counted. With concentrated solutions, it is possible for two cells to pass through the aperture simultaneously. If this occurs, only one pulse is registered. These coincidences can be corrected for by using a calibration chart.

Since background particles, such as occur in foods, also would produce pulses as they pass the aperture, or could clog the aperture, it is necessary to remove them.

The laser beam flow microfluorometer counts particles coated with immunofluorescent markers. The use of laser light increases the sensitivity of the procedure.

Although electronic counters have been used successfully to count blood cells and mammalian cells such as in tissue culture work, much work needs to be done before they are accepted as a means to determine the microorganisms in food products.

Viable Counts

There are several methods designed to estimate the number of viable microorganisms. Most of the systems are based on the plate count or tube dilution methods.

Plate Count (Pour).—The standard plate count (SPC) has been the usual technique for estimating the living microorganisms in foods. The procedure is relatively simple. When the appropriate dilutions are prepared, they should be planted immediately by transferring a measured aliquot to a sterile petri plate and adding sterile, melted and cooled (42°-45°C) agar. The type of agar used for estimating the viable count should be nutritious and non-inhibitory, unless specific microbial types are being determined. Standard methods agar (plate count agar or tryptone glucose yeast agar) or milk protein hydrolysate glucose agar are the agars of choice for total counts of egg products (AOAC 1975), but other agars, such as tryptone glucose extract agar, nutrient agar, nutritive caseinate agar and brain heart infusion agar have been used for making viable plate counts of various products. The agar should be mixed thoroughly with the inoculum to distribute the cells uniformly. After solidification, the plates are inverted (turned upside down) to prevent condensation of moisture on the agar surface, and then incubated. The temperature and time of incubation will vary depending upon the viable count that the investigator desires (psychrotrophs, mesophiles or thermophiles). A temperature of 32°C for 3 days is used for eggs and egg products, while a temperature of 35°C for 48± 2 hr is listed for frozen, chilled, precooked or prepared foods (AOAC 1975). Huhtanen (1968) found the highest counts in raw milk when the plates were incubated at a temperature of 27°C. However, these counts were not significantly different from those obtained at a range from 10° to 30°C.

During the incubation period, growth and multiplication of cells will occur until a visible colony is formed. These colonies are counted on the plates that contain 30-300 colonies (AOAC 1975; FDA 1976). Cowell and Morisetti (1969) furnished evidence that greater precision is obtained by counting 80-320 colonies on a plate. The number of colonies is multiplied by the dilution factor, and reported as the number of organisms per gram of food. Since not all cells can grow with any one particular set of conditions, the count should be referred to as colony forming units (CFU) per gram of food.

Desirable Characteristics.—Gilchrist *et al.* (1973) stated that the pour plate procedure is simple, it can cover a large concentration range, and, at present, is probably the most precise method for determining those bacteria that will grow in an agar medium. Besides, these virtues, the organisms can be recovered for further study. The results should reflect the level of viable microorganisms in the food at time of sampling.

The data obtained from the pour plate should reveal information such as the source of microorganisms, potential shelf life or possible public health hazards of the product. With the present aerobic plate system, the source of microorganisms generally is not determined (Blankenagel 1976).

In most foods, microbial growth causes undesirable changes. Hence, the plate count might be used as an indicator of potential shelf life or of incipient spoilage. No relationship was found to exist between the bacterial count and potential shelf life of iced shrimp (Cobb *et al.* 1973), or pasteurized milk (Watrous *et al.* 1971). The total CFU system does not differentiate types of bacteria that cause spoilage.

It is generally agreed that any potential health hazard is not determined by the total plate count. Some people believe that a high microbial count indicates improper handling with possible pathogens being present. Quite often the reverse is true and low-count products contain potential pathogens. Microbial toxins can be present after the bacteria are destroyed by processing.

Undesirable Characteristics.—There are many facets of the pour plate system which are undesirable. The things of most concern are time, expense, technical requirements, information obtained and accuracy.

The prepared plates must be incubated so that the organisms can produce a visible colony prior to counting. This incubation period may range from two to ten days. For highly perishable products, or for determining production or processing conditions, it is desirable to obtain the results as soon as possible. If a 10 day incubation period is needed, the potential shelf-life of a food can be determined more easily by incubating it directly.

Since the pour plate system is so common in the United States, we might not realize that it is rather expensive, when compared to other methods. There are countries in which other less expensive methods are used in preference to the plate count. To save the technician's time and lower expenses, automation of the procedure has been suggested (Goss *et al.* 1974; Sharpe *et al.* 1972A). Although some equipment is available to make the pour plate technique more automated, for most of the procedure, hand labor is used. With all of the equipment and material required for the pour plate, it has been described as cumbersome.

The pour plate method seems simple to do, but a trained technician is needed to perform the test. It has been suggested that the technician should be more highly trained and more closely supervised. The accuracy of the pour plate depends upon the ability of the technician as well as assumptions and errors inherent in the technique.

Assumptions and Errors.—The same assumptions and errors due to sampling as discussed for the DMC, apply equally to the pour plate. The technical ability and concern of the technician during cleaning of glassware, preparation of dilutions and media, sampling, plating, counting and calculating can influence the reported CFU.

Two major assumptions of the pour plate system are that: a) microorganisms are in suspension as dissociated single-cell units so that each colony on the plate arises from an individual cell, and b) that all cells that are planted in the culture medium will multiply to produce a visible colony. Neither assumption is accurate.

Not all colonies develop from only one bacterial cell. Bacteria often grow in clusters or chains. Mixing, shaking, or other procedures are not assurance that all of the bacteria are separated into individual cells. Hence, when plated, a colony may arise from not only one, but several bacterial cells.

The environment in which the organisms are placed (medium, temperature, oxygen) as well as previous treatments of the cells (sublethal heating, freezing, radiation) and even the presence of other types of microorganisms will influence the ability of the cell to multiply and produce a visible colony. No one environmental condition will support the growth of all of the types of microbial cells that might be present in a food product.

With all of the errors and assumptions, the plate count may represent somewhere between 10% and 90% of the real number of bacteria present in a food. The main value of a plate count is to be able to compare the results of various samples, at different times, from different laboratories. This is possible only when the results are reproducible. It is important that standardized procedures be followed so that results can be compared.

Plate Count (Surface).—In this system, the sterile, melted and cooled agar is poured into sterile petri plates. After solidification, the plates are preincubated overnight. The incubation dries the surface of the agar so that, when planted, the organisms do not coalesce. Before using, one should observe the dried agar surface for any possible contamination.

Aliquots of dilutions are added to the dry surface and uniformly spread over the agar by means of a sterile glass rod, bent in the shape of a hockey stick. Various amounts of aliquots have been suggested. Inasmuch as we usually work with dilutions in the order of 10, it is much easier to calculate the results per gram of product if 0.1 ml aliquots are used.

For simplification, calibrated loops can be used in place of pipets for preparing dilutions as well as for inoculating the pour plate or the surface of the spread plate.

With the drop plate method, 0.02 ml of inoculum is allowed to drop on the surface so that it spreads over an area of 1.5 to 2.0 cm in diameter. From 6 to 8 drops are placed on an agar surface in a petri dish, with no further manual spreading.

After inoculation, the plates are inverted, incubated and the resultant colonies may be counted as with the pour plate method.

Automatic devices for spreading the sample over the agar surface have been described by Gilchrist *et al.* (1973), Jarvis *et al.* (1977) and Trotman and Byrne (1975). The spiral plate method of Gilchrist *et al.* (1973) is listed as official first action by AOAC (1977).

Surface vs Pour Plates.—The desirable aspects listed for the pour plate are equally applicable to the spread plate.

The surface plating can be more automated than the pour plate method, which may reduce the cost of analysis. With all the colonies on the surface, the spread plates are adapted to counting with electronic colony counters.

It is well recognized that higher counts are obtained by surface spread plates than by pour plates. The possibility of heat sensitive organisms being damaged by hot agar during the preparation of pour plates is overcome by using the spread plate technique. Obligate aerobic organisms will grow faster on the surface than in the depth of agar in pour plates. Surface colonies are always detectable sooner, are much larger and easier to count than colonies in a pour plate.

The undesirable characteristics of the spread plate are similar to those discussed for the pour plate. With the spread plate system, some of the organisms might cling to the glass rod used for spreading. Treating of the glass rod with silicone helps overcome the problem. Reportedly there is better precision with the pour plate than with the spread plate.

Roll Tube.—The basic idea of the roll tube is the same as the pour plate method, except that screw-capped test tubes or bottles are used in place of petri plates. Test tubes are sterilized with 2-4 ml of plate count agar (with 2% agar). When the melted agar is cooled to 42°-45°C, 0.1 ml of the appropriate dilution of the sample is added and the tube rolled in cold water in a horizontal position until the agar is solidified in a thin layer on the inner wall of the tube.

The roll tubes are incubated upside down so that any water that condenses collects below the inoculated agar and does not smear the colonies. After incubation, the colonies that develop are counted with the aid of a low-power magnifier. Multiplying the colony count by the dilution factor yields the number of organisms per gram of food.

Although the basic idea of the roll tube is similar to the plate count, there are obvious differences. Since test tubes are used rather than petri plates, the cost of the procedure may be lower or higher, depending upon the relative cost of these items. Less plate count agar is used in the roll tube method.

Hartman (1968) believed that the roll tube should be considered for use in place of the petri plate. He stated that the roll tube required less space, materials and time, with less risk of contamination, less desiccation of the media in the tubes than in plates during long incubation periods and there is no waiting for agar to solidify to invert and incubate such as in the pour plate system. There are machines for rolling the tubes.

It would seem that the colonies would be more difficult to discern and count in the roll tube than in the pour or spread plate techniques. In his review, Hartman (1968) did not find counting of the colonies to be a problem in the roll tube. There are devices available to assist in the counting of colonies in roll tubes. The roll tube technique can be used to determine anaerobic types of microorganisms in foods (Gray and Johnson 1976).

Burri Strip or Slant.—This method involves the spreading of a sample over an agar slant with a calibrated loop. Test tubes can be used, but the oval tube gives a larger surface for the growth of colonies. The agar surface must be dry to prevent colonies from coalescing. After incubation (32° or 37°C for 24 hr) in a horizontal position, the surface is examined for microbial growth. Colonies may be counted or comparisons can be made as to the extent of growth that occurs so that high and low count products can be distinguished.

The Burri slant method is a simple test for the evaluation of plant sanitation.

Oval Tube.—The oval tube method might be called a combination of the pour plate, roll tube and Burri slant systems. Oval tubes containing sterile melted but cool agar are inoculated with standard loops. The agar and inoculum are mixed and the tube is slanted, or laid flat during solidification of the agar, so that a sheet is formed. The prepared tubes are incubated and the colonies are counted.

The oval tube test is a simple, rapid control test that has merit for determining viable counts of Grade A raw milk for pasteurization.

Little Plates.—Since Frost introduced the little plate system in 1916, many modifications have been proposed. The original procedure was to mix 0.1 ml of milk with about 2 ml of nutrient agar and this was spread uniformly over a 4 cm^2 area on a glass slide. After incubation for 3-8 hr in a moist chamber, the slides were air-dried, flame fixed and stained for counting. The colonies were observed and counted with a microscope.

Modifications have been suggested in the procedure, such as the types of slide used, the method of inoculation and incubation, as well as type of stains. A similar procedure was described by Postgate (1969) to distinguish viable cells from dead cells, since to observe colonies on the slide, the cells must be viable.

This system is a more rapid method than the plate count, since only 3-8 hr of incubation are used. Besides being rapid, an estimate of the viable number of cells is obtained which is not the case with DMC. The little plate, slide plate or microplate methods give results comparable to the plate count.

Membrane Filters.—When fluids are filtered through a membrane filter (MF), all particles, bacteria or cells larger than the pores are retained on the filter surface. The retained microbial cells can be examined and counted with a microscope in a manner similar to the DMC. Staining of the cells and adding immersion oil (to make the filter transparent), aids in the detection of the cells.

The retained microorganisms can be cultured by aseptically transferring the filter to a sterile petri plate containing double strength liquid nutrients. The nutrients diffuse through the porous sheet to supply the microorganisms with growth factors. After incubation for six to eight hours, the microcolonies can be counted with a microscope similarly to the little plate or microplate method.

The cultures can be incubated on the MF or transferred from the MF to an agar surface. After incubation, the colonies are counted as in the surface plate method.

The retained organisms can be exposed to specific fluorescent antibody for about one hour. The filter is then washed and observed for the antigen-antibody reaction by using an ultraviolet light microscope. The presence of fluorescent cells indicates the specific organisms. Due to fluorescence of normal filters, black filters have been developed for this purpose.

The procedure is especially useful for the examination of water, beer or other fluids or air when the microbial count is relatively low. The membrane filter can be used to concentrate organisms from solutions used to wash equipment or food surfaces. By using appropriate MF (1.2 μm and 0.22 μm) yeasts and bacteria can be separated. Food particles will clog the filters. By using prefilters to retain food particles and allow bacteria to pass to the MF total numbers of specific organisms that would be in low numbers in foods can be detected and the number estimated.

Tube Dilution.—The tube dilution method is essentially the aseptic inoculation of a series of tubes of sterile nutrient broth with a series of dilutions of the food. After incubating the inoculated tubes, the broth is observed for turbidity which indicates growth of organisms. If no turbidity is evident, it is assumed that no microorganisms were present or were able to multiply. With broth that appears turbid due to the inoculated food, growth can be detected by streaking on an agar surface and observing growth after a few hours of incubation, or by spreading some turbid broth on a slide and looking for microorganisms with the aid of a microscope.

By using several dilutions and inoculating a separate tube from each dilution, after incubation some tubes may have growth and others no growth. If the tube with the 1:100 dilution showed growth and the tube with 1:1000 had no growth, there were between 100 and 1000 organisms in the food. Sometimes this rough estimate is all that is needed. It only gives an estimate of the range of bacteria that are present.

Most Probable Numbers (MPN).—By using several tubes at each dilution and recording the positive (showing growth) tubes and negative (no growth) tubes, you get a more accurate estimate of the number of organisms present. In the tube dilution example, if you inoculated 10 tubes with 1 ml of the 1:1000 dilution, there would be as much total inoculum as in the 1:100 tube which showed growth. Theoretically, one or more of the 10 tubes with the 1:1000 dilution also should be turbid. The relationship of positive and negative tubes has been determined mathematically and MPN tables have been derived (Tables 2.2, 2.3). To use the MPN system, at least three dilutions are needed. Ideally, the least dilute tubes should all be positive and the most dilute tubes (of the three dilutions) should all be negative. This is not always the case, so the rule that has been established is to select the highest dilution in which all portions tested are positive (no lower dilution giving negative results) and the two succeeding dilutions are then chosen. The more tubes that are used in each dilution, the more accurate is the estimate, but for reasons of convenience, three-tube or five-tube series are adopted. After selecting the three series of dilutions, consult the appropriate MPN table, obtain a most probable number that satisfies the number of positive tubes, and multiply this by the dilution factor to obtain the MPN per gram of product.

Assumptions and Errors (MPN).—The assumptions and errors due to sampling and diluting apply to the MPN technique. It is assumed that a single viable cell inoculated into a tube of broth will multiply so that a change such as turbidity, acid or gas production can be observed. Because dilution to extinction is necessary, good aspetic technique is needed since any contamination during inoculation of the tubes of broth could result in growth. The MPN is less precise than the agar plating methods (Pike *et al.* 1972).

Some people become confused when one gram of sample is added to a tube with nine ml of broth for the MPN series. They feel that since this

TABLE 2.2

MOST PROBABLE NUMBER (MPN) PER GRAM OF SAMPLE, STANDARD ERROR, UPPER AND LOWER 95% CONFIDENCE LIMITS, AND ONE-SIDED UPPER 95% CONFIDENCE LIMITS WHEN 3 DILUTIONS ARE USED WITH 3 TUBES IN EACH DILUTION AT LEVELS OF 1.0, 0.1 AND 0.01 GRAM PER TUBE

Number of Positives			Program Values		2-Sided 95% Conf. Limits		1-Sided Upper 95% Limit
1.0	0.1	0.01	MPN	St. Error	Lower	Upper	
0	0	0	< 0.03	–	–	–	–
1	0	0	0.36	0.36	0.05	2.54	1.85
1	1	0	0.74	0.52	0.18	2.94	2.36
1	1	1	1.12	0.64	0.36	3.47	2.89
2	0	0	0.92	0.65	0.23	3.67	2.94
2	1	0	1.47	0.85	0.47	4.55	3.80
2	1	1	2.05	1.02	0.77	5.46	4.66
2	2	0	2.11	1.05	0.79	5.61	4.79
2	2	1	2.76	1.24	1.15	6.64	5.76
2	2	2	3.48	1.42	1.56	7.74	6.80
3	0	0	2.31	1.33	0.74	7.17	5.98
3	1	0	4.27	2.14	1.60	11.38	9.72
3	1	1	7.49	3.35	3.12	17.99	15.63
3	2	0	9.33	4.17	3.88	22.41	19.47
3	2	1	14.94	6.10	6.71	33.25	29.23
3	2	2	21.46	8.11	10.23	45.02	39.97
3	3	0	23.98	17.41	5.78	99.49	79.15
3	3	1	46.22	17.47	22.03	96.96	86.07
3	3	2	109.89	38.87	54.94	219.82	196.65
3	3	3	>110.00	–	–	–	–

Data Courtesy of Robert J. Parnow (personal communication).

TABLE 2.3

FREQUENTLY OCCURRING MOST PROBABLE NUMBER (MPN) PER GRAM OF SAMPLE, STANDARD ERROR, UPPER AND LOWER 95% CONFIDENCE LIMITS AND ONE-SIDED UPPER 95% CONFIDENCE LIMITS WHEN 3 DILUTIONS ARE USED WITH 5 TUBES IN EACH DILUTION AT LEVELS OF 1.0, 0.1 AND 0.01 GRAMS

Number of Positives			Program Values		2-Sided 95% Conf. Limits		1-Sided Upper 95% Limit
1.0	0.1	0.01	MPN	St. Error	Lower	Upper	
0	0	0	<0.18	–	–	–	–
1	0	0	0.19	0.19	0.03	1.34	0.98
1	1	0	0.40	0.28	0.10	1.60	1.28
1	1	1	0.60	0.35	0.19	1.86	1.55

TABLE 2.3. *(Continued)*

Number of Positives 1.0	0.1	0.01	Program Values MPN	St. Error	2-Sided 95% Conf. Limits Lower	Upper	1-Sided Upper 95% Limit
2	0	0	0.44	0.31	0.11	1.76	1.41
2	1	0	0.68	0.39	0.22	2.11	1.76
2	1	1	0.92	0.46	0.34	2.45	2.09
2	2	0	0.93	0.46	0.35	2.48	2.12
2	2	1	1.17	0.52	0.49	2.81	2.44
2	2	2	1.42	0.58	0.64	3.16	2.78
3	0	0	0.77	0.44	0.25	2.39	1.99
3	1	0	1.07	0.54	0.40	2.85	2.44
3	1	1	1.36	0.61	0.57	3.27	2.84
3	2	0	1.38	0.62	0.57	3.32	2.88
3	2	1	1.69	0.69	0.76	3.76	3.31
3	2	2	2.02	0.76	0.96	4.24	3.76
3	3	0	1.72	0.70	0.77	3.83	3.37
3	3	1	2.05	0.77	0.98	4.30	3.82
4	0	0	1.27	0.64	0.48	3.38	2.89
4	1	0	1.68	0.75	0.70	4.04	3.50
4	1	1	2.11	0.86	0.95	4.70	4.13
4	2	0	2.16	0.88	0.97	4.81	4.23
4	2	1	2.64	1.00	1.26	5.54	4.92
4	2	2	3.17	1.12	1.58	6.34	5.67
4	3	0	2.70	1.02	1.29	5.66	5.03
4	3	1	3.25	1.15	1.62	6.50	5.81
4	3	2	3.86	1.29	2.01	7.42	6.68
4	4	0	3.35	1.18	1.68	6.70	5.99
4	4	1	3.98	1.33	2.07	7.65	6.89
5	0	0	2.31	1.03	0.96	5.55	4.82
5	1	0	3.29	1.34	1.48	7.32	6.44
5	1	1	4.56	1.72	2.17	9.56	8.49
5	2	0	4.93	1.86	2.35	10.34	9.18
5	2	1	6.99	2.47	3.50	13.98	12.50
5	2	2	9.43	3.14	4.91	18.12	16.32
5	3	0	7.92	2.80	3.96	15.84	14.17
5	3	1	10.86	3.62	5.65	20.87	18.79
5	3	2	14.05	4.44	7.56	26.11	23.64
5	3	3	17.49	5.27	9.68	31.58	28.72
5	4	0	12.99	4.33	6.76	24.97	22.48
5	4	1	17.23	5.45	9.27	32.02	28.99
5	4	2	22.11	6.67	12.24	39.92	36.31
5	4	3	27.80	8.02	15.79	48.95	44.70
5	4	4	34.54	9.58	20.06	59.49	54.51
5	5	0	23.97	7.58	12.90	44.55	40.33
5	5	1	34.76	10.48	19.25	62.77	57.08
5	5	2	54.22	15.65	30.79	95.48	87.18
5	5	3	91.78	25.46	53.28	158.09	144.86
5	5	4	160.94	43.06	95.26	271.90	249.92
5	5	5	>1600	–	–	–	–

Data Courtesy of Robert J. Parnow (personal communication).

is a 1:10 dilution, somehow it has to be considered when the dilution factor for the MPN is determined. It does not make any difference if there are 8, 9, 10 or 11 ml of nutrient media per tube. The only consideration is the amount of original sample that is added to the tube (0.01, 0.001, 0.0001 g or whatever).

Advantages of the MPN.—The MPN is, in some ways, easier or simpler to do than the plate count. Broth can be dispensed into tubes with an automatic pipetter. Selective or differential media can be used so that certain types of organisms can be determined. The MPN is particularly useful for samples with only a few organisms and can be used to detect organisms in samples larger than one gram.

Per sample, it may be that more incubator space is needed when the MPN is used as compared to the SPC, depending upon the number of dilutions needed for each test.

Other Viable Counting Methods.—There are other procedures for obtaining estimations of viable microorganisms in foods. These include: agar droplets (Sharpe and Kilsby 1971; Sharpe *et al.* 1972B), bactostrips (O'Toole 1974), microtiter-spot plate test (Fung *et al.* 1976), the plate loop method (Murdock and Hatcher 1976) and a capillary tube system that allows automatic counting of bacteria (Schoon *et al.* 1970).

These systems are based on tests previously described, with an added or different feature to make the test more rapid, less expensive, less cumbersome, less space needed or less labor. A simple system to test water, foods or surfaces was described by Millipore (1976).

Estimations Based on Metabolism

The metabolism of microorganisms is used in general microbiology to determine fermentation of sugars, starch hydrolysis, production of hydrogen sulfide, indole, or nitrate reduction. The metabolism of microorganisms and the production of metabolic products in foods can be measured and used to estimate bacterial populations or the microbial quality of foods.

Reductase Tests.—Organisms obtain energy from chemical reactions involving either organic or inorganic compounds. This involves an oxidation-reduction reaction; the energy source becomes oxidized, while another compound is reduced. Oxygen may or may not be involved since oxidation-reduction reactions concern electron transfers. When a compound loses an electron it becomes oxidized and another compound which accepts this electron is reduced.

Compounds vary in their oxidation-reduction potential, which is the tendency for a compound to give up electrons. Since these reactions consist of electron transfers, they can be measured electrically with a potentiometer and are expressed by the electrical unit, the volt. The oxidation-reduction potential also is called the redox potential.

Besides being determined potentiometrically, the redox potential can be determined with indicators or dyes. Many compounds undergo color changes when oxidized or reduced. If such a compound is added to a substrate containing metabolizing bacteria, electrons may be transferred to the indicator and its color will be altered.

Since the color change of the indicator depends on the metabolic rate of a microbial culture, the larger the number of cells, the sooner the indicator will show a color change. The reduction time is inversely proportional to the number of cells present (Fig. 2.2). Although several oxidation-

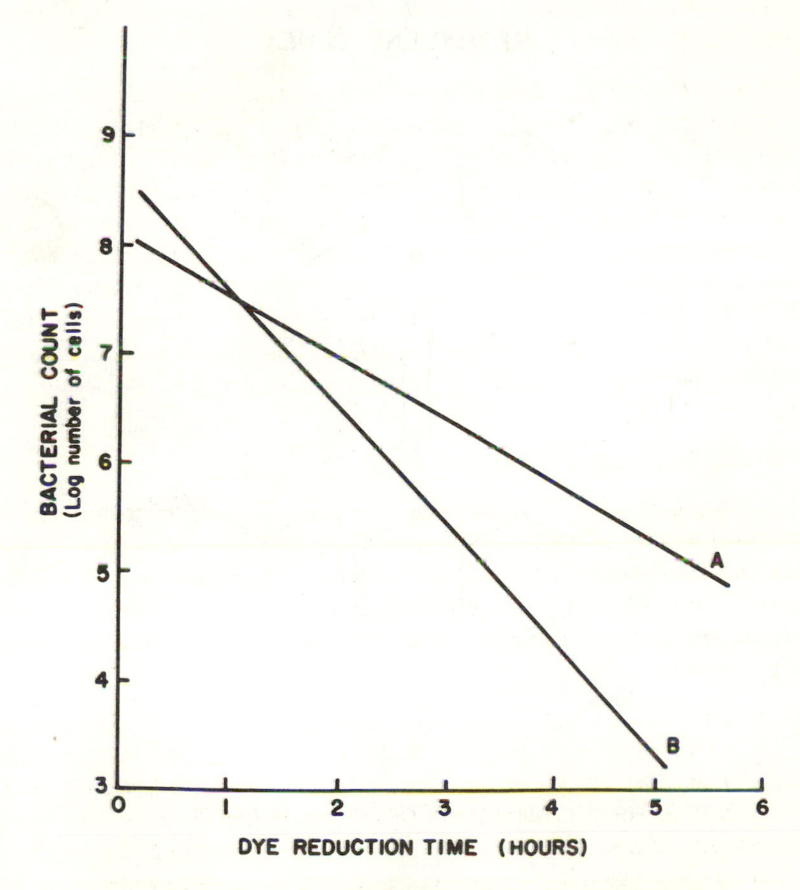

FIG. 2.2. RELATIONSHIP BETWEEN REDUCTION TIME AND MICROBIAL LOAD

The slope of the line depends upon the types of microorganisms that are present.

reduction indicators could be used, methylene blue, resazurin and the tetrazoliums are the ones most often used in food analysis. The reductase tests are usually called dye reduction tests, apparently because the dye, methylene blue, is used. However, resazurin and the tetrazoliums are not dyes, but are indicators (Conn 1961).

Methylene Blue.—The methylene blue solution is prepared and added to the suspended food in the ratio of 1:10. The prepared tube is placed in a water bath (37°C) and observed each 30 or 60 min for dye reduction (Fig. 2.3). The dye is decolorized from blue to colorless. The time required for decolorization is recorded.

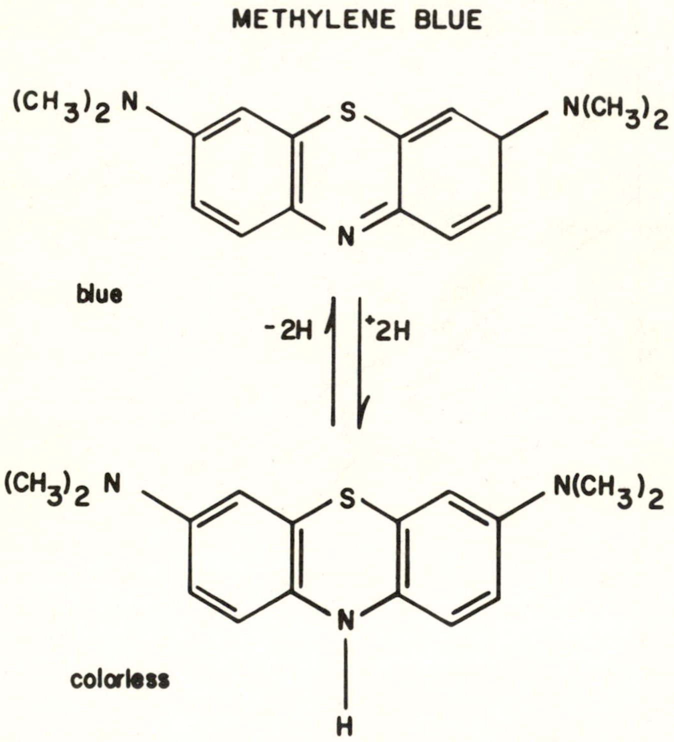

FIG. 2.3. REDUCTION OF METHYLENE BLUE

When raw milk is tested it is necessary that the tubes be inverted when the temperature reaches 36° C and after observation, to distribute the cream into the milk. The bacteria tend to migrate with the cream to the top of the tube. Removing the bacteria from the substrate affects the reduction time. The dye should not be exposed to light, especially direct sunlight, since the dye becomes more toxic to the cells in light than in darkness.

The methylene blue method has been used to determine the bacterial quality of milk and dairy products such as ice cream (Anderson and Whitehead 1974). This reductase test has been suggested as a means to predict the sterility of heated food products (Hall 1971) and to estimate the bacteria in ground beef (Emswiler *et al.* 1976).

Resazurin.—This indicator has been used as a substitute for methylene blue in estimating the microbial quality of raw milk. There are two color changes during reduction (Fig. 2.4). It imparts a blue color to fresh milk, and, as incubation progresses, the indicator goes through various shades of purple and mauve to pink. In the second stage of color change, the indicator becomes colorless.

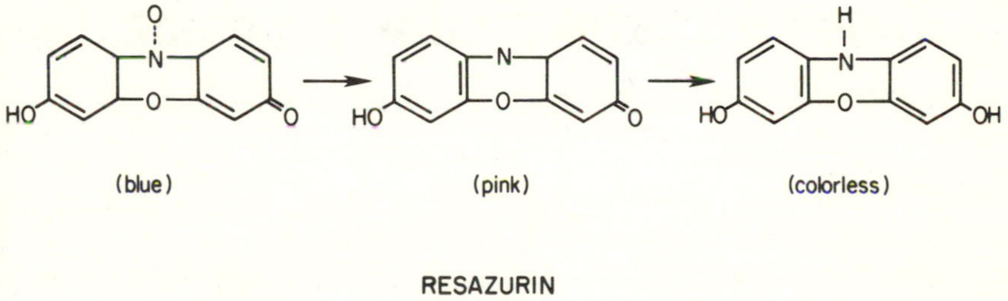

(blue) (pink) (colorless)

RESAZURIN

FIG. 2.4. REDUCTION OF RESAZURIN

The first stage of the reduction of resazurin is not due to an electron transfer, but is due to a loss of an oxygen atom loosely bound to the nitrogen of the phenoxazine nucleus. The change to the pink resorufin is not reversible by atmospheric oxygen and is largely independent of both reduction potential and oxygen content. The second stage of reduction to the colorless state is reversible by atmospheric oxygen.

Besides using resazurin reduction for testing milk and dairy products, this system has been applied to liquid and dried eggs, frozen meat and poultry pies, frozen vegetables and fresh poultry. The resazurin test is described as a simple, relatively rapid, inexpensive and objective test to determine the quality of fresh scallop meat (Webb *et al.* 1972).

Tetrazolium Salts.—Tetrazolium salts have been used as indicators in the reductase test for the bacterial analysis of foods. The tetrazolium salt most often used for food analysis is 2,3,5, triphenyl tetrazolium chloride (TTC), since it is less toxic to bacteria than are the other tetrazolium salts.

TTC is colorless when in the oxidized state, but forms intensely colored pink to red pigments when reduced. The reduced form is a formazan (Fig. 2.5).

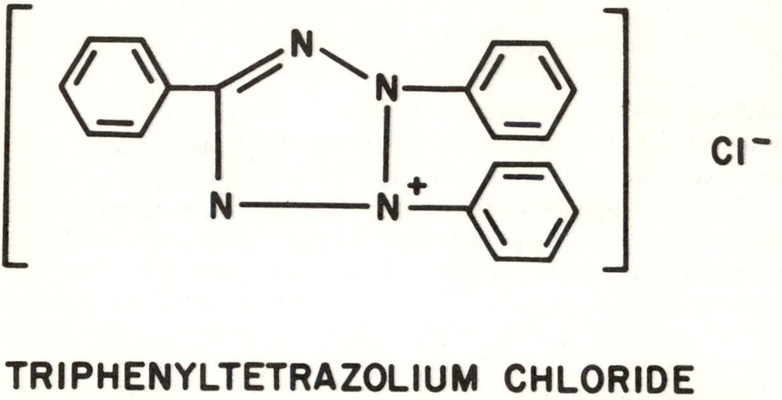

TRIPHENYLTETRAZOLIUM CHLORIDE

FIG. 2.5. FORMULA OF TRIPHENYLTETRAZOLIUM CHLORIDE

The TTC reduction system has been used to predict the potential shelf life of pasteurized milk and cream. Surface contamination can be detected by spraying on a solution of tetrazolium chloride. Development of the red color due to reduction of the tetrazolium indicates sites of bacterial activity.

Comparison of Reductase Tests to Viable Count Tests.—The reductase test generally gives an estimate of the bacterial contamination in a shorter time than the SPC. The information obtained from reductase tests can, at best, be used to obtain a rough estimate of the number of microorganisms present in or on a food.

Not all organisms cause a lowering of the redox potential at the same rate. If a clump or chain of bacteria is plated in agar, a single colony will develop, but the metabolic activity in the reductase test will be the sum of the total number of cells in the clump or chain. This will result in a more rapid color change in the indicator than the plate count would suggest.

For a cell to be counted in the SPC, it must multiply and form a visible colony. Cells may be metabolizing, but not reproducing. These cells could cause a color change in the redox indicator, and not be included in the SPC.

Methylene blue and tetrazolium are inhibitory to certain microorganisms. Sometimes tetrazolium is added after the organisms have grown, such as by flooding an agar surface, due to its potential for inhibiting the cells. It has been suggested that reducing enzymes naturally present in foods can cause color changes of these indicators. In this case, the reductase test would indicate more contamination than is present.

Chemical Indicators of Decomposition.—Food is composed of various chemical compounds which are subject to biochemical changes. These changes may be desirable or undesirable depending upon the food, the microorganisms present and the end products of the reaction. Decomposition of foods with resulting quality deterioration is an undesirable change.

The main reactions occurring in foods are catalyzed by enzymes. These enzymes may be tissue enzymes naturally present in the food, or they may be produced by microorganisms associated with the food. Some oxidative changes occur in foods without specific enzymes to catalyze the reactions. The degree of metabolic activity may or may not be related to the number of organisms present.

The type and amount of metabolic products formed depends upon the kind of food (protein, carbohydrate or fat), the type of microorganism (proteolytic, saccharolytic or lipolytic), the availability of oxygen (aerobic-oxidation, decay or oxidative rancidity; anaerobic-fermentation, putrefaction or hydrolytic rancidity), the temperature (psychrotrophic, mesophilic or thermophilic organisms) and the types of inhibitors that might be present.

It is not feasible to discuss or describe all of the possible chemicals that have been studied as indicators of quality deterioration of food. Fields *et al.* (1968) presented a comprehensive review of chemical indicators.

Organoleptic Evaluations.—Everyone makes organoleptic evaluations of food by sight, smell, taste, or touch.

The food industry also relies on organoleptic tests to determine certain quality attributes of foods.

This type of analysis is very subjective and many arguments can develop between the seller and the buyer. Chemical indicators can be used to evaluate the microbial or other quality of food in a more objective manner.

Criteria for Chemical Indicators.—For a chemical to be a useful indicator it: 1) must be absent or at very low levels in sound food; 2) should be produced by the predominant spoilage flora and not used as a nutrient; 3) should be detected quantitatively with simple and rapid tests and the tests should never yield false positive results; 4) should not have a useful function in the food; and 5) should preferably be able to distinguish poor quality from poor processing operations.

Possible Chemical Indicators.—Some potential chemical indicators for estimating the microbial or other quality of food are listed in Table 2.4. None of these chemical indicators is entirely satisfactory due to variations in a food and its microbial flora. However, the presence of certain indicators in some foods does correlate with the microbial count or organoleptic evaluation. In general, a group of compounds such as volatile reducing substances, total volatile acids or bases gives a better indication of quality than a single indicator such as ammonia, indole or alcohol. One problem with many of these indicators is that by the time there is a significant change in the amount that is present, deterioration of the food is very evident. Perhaps incubation of the food for a few hours prior to analysis could be used to develop the presence of indicators more rapidly. However, this higher temperature could alter the dominant spoilage flora so that the metabolic products would differ from those expected to develop with normal storage conditions.

TABLE 2.4

POTENTIAL CHEMICAL INDICATORS OF FOOD QUALITY

Ammonia	Indole	Histamine
Trimethylamine	Ethanol	Hypoxanthine
Dimethylamine	Furfural	Diacetyl
Total volatile bases	Hydrogen sulfide	Acetylmethylcarbinol
Total volatile acids	Total reducing substances	
Free fatty acids	Volatile reducing substances	
Water insoluble acids (oleic, palmitic)		
Organic acids (acetic, lactic, pyruvic, succinic)		

Physical Tests.—Microbial growth causes alterations in foods such as acid content or pH. As the pH is varied, the water binding capacity of protein changes. This difference can be determined by the extract release volume test.

The fluorescence of liquid egg is related to mustiness and growth of certain bacteria, while pyoverdine, a fluorescent pigment produced by pseudomonads, has been determined in frozen whole egg and on poultry carcasses.

pH.—Since basic compounds, such as ammonia and amines, are formed during deterioration of protein foods, the pH will tend to rise. If carbohydrates are present and fermented, the pH will tend to decline.

During the storage of shrimp at 5° C, Vanderzant and Nickelson (1971) reported an increase in pH from 7.4 to 8.6 in 24 days. The pH of scallops decreased from 7.0 to 5.9 after three days of storage (Groninger and Brandt 1970). It was stated (Shelef and Jay 1970) that from freshness to incipient spoilage, the pH change generally does not exceed 0.3 to 0.5 of a pH unit. They recommended a titrimetric method wherein a 10 g sample of beef, blended with water, is titrated to pH 5.0 with 0.02 N HCl. If more than 2.0 ml of acid is needed, the meat is in some form of incipient spoilage.

Extract Release Volume (ERV).—There are many reports relating water-holding capacity, hydration capacity or extract release volume to microbial quality (Miller and Price 1971; Shelef and Jay 1971; Vanderzant and Nickelson 1971). As meat deteriorates, there is an increase in the amount of water retained and a decrease in ERV. Shelef (1974) reported the relationship between pH and ERV. Regardless of the microbial quality of the meat, the maximum ERV occurred at pH 5.5.

Adenosine Triphosphate (ATP).—During metabolism, cells form high energy phosphate bonds stored in ATP. Not only microbial cells, but also other living cells contain ATP.

When an animal dies and the muscle glycogen is utilized by anaerobic glycolysis, the amount of ATP decreases rapidly. It has been established that ATP in bacterial cells disappears when the cells are killed. In starved bacterial cells, the ATP falls to low levels before loss of viability is evident. Since all living microbial cells contain ATP, it should be possible to estimate the number of cells by quantitating the ATP in a system.

The method for determining ATP is based on the firefly reaction as shown in Fig. 2.6. A purified extract from the firefly, containing luciferin and luciferase, when reacted with ATP in the presence of magnesium ions, causes a light emission. Crude extracts from the firefly, when reacted with adenosine diphosphate (ADP) have caused light emission. When the purified firefly extract is added to bacterial ATP, the light emission can be measured with a photometer. One molecule of ATP yields one photon of light. The amount of ATP per organism is fairly constant during the growth cycle of bacteria. The mean value is 4.7×10^{-10} μg ATP per bacterial cell (Daly 1974).

For the assay, it is necessary to eliminate the nonbacterial ATP and then to release the bacterial ATP to react with the luciferin-luciferase system. One method is to rupture the nonbacterial cells by a selective procedure such as treating with Triton X-100 (Thore *et al.* 1975). The released ATP is hydrolyzed by adding an enzyme. According to Anon. (1974) potato apyrase is preferred. Then the enzyme is denatured by heating for 5 to 12 min at 60° to 100°C. This must be done with care so that the bacterial cells are not affected adversely for the subsequent bioluminescent reaction. After destruction of the nonbacterial ATP and hydrolyzing enzyme, the bacterial ATP is released by rupturing the bacterial cells by physical or chemical means (Alexander *et al.* 1976; Lundin and Thore 1975). This released ATP is assayed with the luciferin-luciferase system and an estimate of the number of bacterial cells is calculated.

$$\underset{\text{Luciferin}}{LH_2} + \underset{\text{Luciferase}}{E} + \underset{\text{Adenosine triphosphate}}{ATP} \xrightleftharpoons{Mg^{++}}$$

$$\underset{\text{Luciferyl adenylate complex}}{E \cdot LH_2 \cdot AMP} + \underset{\text{Pyrophosphate}}{PP}$$

$$E \cdot LH_2 \cdot AMP + O_2 \longrightarrow E + \underset{\text{Oxyluciferin}}{LO} + CO_2 + \underset{\text{Adenosine monophosphate}}{AMP} + \text{Light}$$

FIG. 2.6. REACTIONS INVOLVED IN ADENOSINE TRIPHOSPHATE DETERMINATION

The ATP method is relatively simple, quick and yields an acceptable estimation of bacterial numbers in a sample.

Measurement of Gas Production.—When organisms metabolize compounds, carbon dioxide is produced as a metabolic product and oxygen is consumed. One method for determining CO_2 production uses ^{14}C-labelled glucose and measures the radioactivity of the $^{14}CO_2$.

The detection time for $^{14}CO_2$ is proportional to the logarithm of the original inoculum (Waters 1972). Korsch *et al.* (1971) used the production of radioactive CO_2 to determine coliforms in water, Lampi *et al.* (1974) and Previte (1972) to detect bacteria inoculated into beef loaf, Hatcher *et al.* (1977) for microorganisms in orange juice, Schrot *et al.*

(1973) for analysis of blood samples, and Mafart *et al.* (1978) to analyze liquid foods.

Other Instrumentation or Methods.—Besides those described, other systems have been used by microbiologists to develop rapid methods for microbial detection. A pyrolysis-gas chromatography-mass spectrometry experiment was used to analyze Martian soil for organic compounds that would indicate life on Mars (Klein 1976; Simmonds 1970). Sanders and Parkes (1970) attempted to correlate infrared analysis of swabbings from broiler skin with bacterial numbers.

Gas chromatography or gas-liquid chromatography has been used to detect metabolic products of microorganisms which can be used to differentiate bacterial species or estimate the count or quality (Carlsson 1973; Staruszkiewicz and Bond 1978).

The metabolic activity of cells changes the composition of the growth substrate. This change in the medium influences the electrical impedance, or the resistance to the electrical flow of alternating current. This system has been used to estimate microorganisms in frozen vegetables (Hardy *et al.* 1977).

Fluorescence microscopy has been used to determine microbial levels on surfaces as well as in conjunction with membrane filter systems (Bowden 1977; Hobbie *et al.* 1977; Jones and Simon 1975; Paton and Jones 1973, 1975).

Limulus amebocyte lysate forms a gel in the presence of small amounts of endotoxin from Gram negative bacteria. This assay has been suggested as a system to estimate Gram negative bacteria in various samples (Coates 1977; Evans *et al.* 1978).

The determination of cell weight, cell mass, light scattering, optical density or turbidity are used to estimate microbial numbers in broth cultures. However, these methods are not readily adaptable for the analysis of turbid foods containing mixed cultures.

Comparison of Methods

There are many procedures that can be used to estimate the total microbial population of a food product. The test depends upon the use of the information that is obtained. If the results are to be used to satisfy a microbiological standard, then the specified method, usually the standard plate count, must be used. If the results are for internal quality control, some other method that is simpler, faster and/or less expensive might be used. The procedure selected will depend on the accuracy, reliability or precision that is needed. The laboratory equipment and personnel, both now and potential, will probably dictate the type of analysis that can be conducted. The cost of supplies and materials, instruments and labor must be compared with other factors to determine which method or methods give the desired information with ease, simplicity, speed and low cost.

In selecting the test, not only precision but also accuracy must be considered. Precision is an index of the random error in a group of determinations and is not related to accuracy. The term accuracy considers the relationship of the determined value and the actual value. Since there is no way to know the actual or true number of microorganisms in a sample, it is difficult to assess the accuracy of a method. Quite often, the results of a method are compared to those obtained with the SPC. It is recognized that the results obtained with the SPC may not be accurate, but the method yields acceptable reproducibility or precision.

The DMC gives results in a much shorter time than the APC so, for this purpose, it is valuable. With the plate count, one can isolate the organism for further study, while this is impossible with the DMC.

The major difference is that both dead and live cells are counted with the DMC, while only those cells able to multiply and produce a colony are counted with a plate count. With most products, especially heat treated products, the DMC gives a higher count than the plate count.

Rapid tests such as those based on metabolic products and instrumentation are valuable for control purposes. Time is important in analyzing highly perishable foods. Also, it is desirable to keep the inventory of processed food at a reasonable level. If unacceptable product is shipped from storage it might need to be recalled. A recall of a product with the resultant publicity is not desirable for a processor.

If microbial inhibitors are present in the food, these will be carried over into the growth medium. In these cases, as the sample is diluted, higher counts might be obtained due to dilution of the inhibitors. If a food containing glucose is added to a medium designed to determine lactose fermentation, erroneous results may be obtained due to fermentation of the added glucose.

In processed foods, there may be cells which are sublethally injured. These are of special importance when the types of organisms are determined with selective media.

BIBLIOGRAPHY

ALEXANDER, D.N., EDERER, G.M., and MATSEN, J.M. 1976. Evaluation of an adenosine 5′-triphosphate assay as a screening method to detect significant bacteriuria. J. Clin. Microbiol. *3*, 42-46.

ANDERSON, G.E., and WHITEHEAD, J.A. 1974. The validity of the methylene blue reduction test in the grading of ice cream. J. Appl. Bacteriol. *37*, 487-492.

ANON. 1974. ATP/fast assay for bacteriuria. Lab. Manage. *12*, No. 5, 35-36, 42.

AOAC. 1975. Official Methods of Analysis of the Association of Official Analytical Chemists. W. Horwitz (Editor) 12th Ed. Association of Official Analytical Chemists, Washington, D.C.

AOAC. 1977. Changes in official methods of analysis made at the 90th annual meeting, October 18-21, 1976. J. Assoc. Offic. Anal. Chem. *60*, 460-503, Section 46.C10−46.C16.

APHA. 1967. Standard Methods for the Examination of Dairy Products. 12th Edition. American Public Health Association. New York, N.Y.

APHA. 1976. Compendium of Methods for the Microbiological Examination of Foods. M.L. Speck (Editor). The American Public Health Association, Washington, D.C.

BLANKENAGEL, G. 1976. An examination of methods to assess post-pasteurization contamination. J. Milk Food Technol. *39*, 301-304.

BOWDEN, W.B. 1977. Comparison of two direct-count techniques for enumerating aquatic bacteria. Appl. Environ. Microbiol. *33*, 1229-1232.

CARLSSON, J. 1973. Simplified gas chromatographic procedure for identification of bacterial metabolic products. Appl. Microbiol. *25*, 287-289.

COATES, D.A. 1977. Enhancement of the sensitivity of the *Limulus* assay for the detection of Gram negative bacteria. J. Appl. Bacteriol. *42*, 445-449.

COBB, B.F., VANDERZANT, C., THOMPSON, C.A., JR., and CUSTER, C.S. 1973. Chemical characteristics, bacterial counts, and potential shelf-life of shrimp from various locations on the northwestern Gulf of Mexico. J. Milk Food Technol. *36*, 463-468.

COLLINS, C.H., and LYNE, P.M. 1970. Microbiological Methods, 3rd Edition. Butterworth & Co., London.

CONN, H.J. 1961. Biological Stains, 7th Edition. Williams and Wilkins Co., Baltimore.

COWELL, N.D., and MORISETTI, M.D. 1969. Microbiological techniques—some statistical aspects. J. Sci. Food Agr. *20*, 573-579.

DALY, K.F. 1974. The luminescence biometer in the assessment of water quality and waste-water analysis. Amer. Lab. *6*, No. 12, 38-44.

DRAKE, J.F., and TSUCHIYA, H.M. 1973. Differential counting in mixed cultures with Coulter counters. Appl. Microbiol. *26*, 9-13.

EMSWILER, B.S., KOTULA, A.W., CHESNUT, C.M., and YOUNG, E.P. 1976. Dye reduction method for estimating bacterial counts in ground beef. Appl. Environ. Microbiol. *31*, 618-620.

EVANS, T.M., SCHILLINGER, J.E., and STUART, D.G. 1978. Rapid determination of bacteriological water quality by using *Limulus* lysate. Appl. Environ. Microbiol. *35*, 376-382.

FDA. 1976. Bacteriological Analytical Manual for Foods. Food and Drug Administration, Washington, D.C.

FIELDS, M.L., RICHMOND, B.S., and BALDWIN, R.E. 1968. Food quality as determined by metabolic by-products of microorganisms. Advan. Food Res. *16*, 161-229.

FUNG, D.Y.C. *et al.* 1976. A collaborative study of the microtiter count method and standard plate count method for viable cell count of raw milk. J. Milk Food Technol. *39*, 24-26.

GILCHRIST, J.E. *et al.* 1973. Spiral plate method for bacterial determination. Appl. Microbiol. *25*, 244-252.

GOSS, W.A., MICHAUD, R.N., and MCGRATH, M.B. 1974. Evaluation of an automated colony counter. Appl. Microbiol. *27*, 264-267.

GRAY, W.M., and JOHNSON, M.G. 1976. Characteristics of bacteria isolated by the anaerobic roll-tube method from cheeses and ground beef. Appl. Environ. Microbiol. *31*, 268-273.

GRONINGER, H.S., and BRANDT, K.R. 1970. Some observations on the quality of the weathervane scallop (*Platinopecten caurinus*). J. Milk Food Technol. *33*, 232-236.

HALL, H.E., BROWN, D.F., and READ, R.B., JR. 1971. Effect of pasteurization on the direct microscopic count of eggs. J. Milk Food Technol. *34*, 209-211.

HALL, R.C. 1971. Simple test to predict commercial sterility of heated food products. J. Milk Food Technol. *34*, 196-197.

HARDY, D., KRAEGER, S.J., DUFOUR, S.W., and CADY, P. 1977. Rapid detection of microbial contamination in frozen vegetables by automated impedance measurements. Appl. Environ. Microbiol. *34*, 14-17.

HARREWIJN, G.A. 1975. Preparation of sample material for analysis. Antonie von Leeuwenhoek *41*, 381-382.

HARTMAN, P.A. 1968. Miniature Microbiological Methods. Adv. Appl. Microbiol., Supplement 1. Academic Press, New York.

HATCHER, W.S., DIBENEDETTO, S., TAYLOR, L. E., and MURDOCK, D.I. 1977. Radiometric analysis of frozen concentrated orange juice for total viable microorganisms. J. Food Sci. *42*, 636-639.

HOBBIE, J.E., DALEY, R.J., and JASPER, S. 1977. Use of nucleopore filters for counting bacteria by fluorescence microscopy. Appl. Environ. Microbiol. *33*, 1225-1228.

HUHTANEN, C.N. 1968. Incubation temperatures and raw milk bacterial counts. J. Milk Food Technol. *31*, 154-160.

ICMSF. 1974. Microorganisms in Foods 2. Sampling for Microbiological Analysis: Principles and Specific Applications. International Commission on Microbiological Specifications for Foods. University of Toronto Press, Canada.

JARVIS, B., LACH, V.H., and WOOD, J.M. 1977. Evaluation of the spiral plate maker for the enumeration of micro-organisms in foods. J. Appl. Bacteriol. *43*, 149-157.

JONES, J.G., and SIMON, B.M. 1975. An investigation of errors in direct counts of aquatic bacteria by epifluorescence microscopy, with reference to a new method for dyeing membrane filters. J. Appl. Bacteriol. *39*, 317-329.

KLEIN, H.P. 1976. Microbiology on Mars? ASM News *42*, 207-214.

KORSCH, L.E., YURASOVA, O.I., NIKONOVA, A.G., and MOTOVA, M.A. 1971. Use of C^{14} for rapid *E. coli* counts in water. Hyg. Sanit. *36*, 423-426.

LAMPI, R.A. *et al.* 1974. Radiometry and microcalorimetry—techniques for the rapid detection of foodborne microorganisms. Food Technol. *28*, No. 10, 52-58.

LUNDIN, A. and THORE, A. 1975. Comparison of methods for extraction of bacterial adenine nucleotides determined by firefly assay. Appl. Microbiol. *30*, 713-721.

MAFART, P., BOURGEOIS, C., DUTEURTRE, B., and MOLL, M. 1978. Use of [^{14}C] lysine to detect microbial contamination in liquid foods. Appl. Environ. Microbiol. *35*, 1211-1212.

MILLER, L.S., and PRICE, J.F. 1971. Extract release volume (ERV) responses with aseptic and inoculated pork. J. Food Sci. *36*, 70-73.

MILLIPORE. 1976. Dip-test Microbiology for Food Processors. AB810. Millipore Corp., Bedford, MA.

MURDOCK, D.I., and HATCHER, W.S., JR. 1976. Plate loop method for determining total viable count of orange juice. J. Milk Food Technol. *39*, 470-473.

OLSON, J.C., JR. 1975. Development and present status of FDA *Salmonella* sampling and testing plans. J. Milk Food Technol. *38*, 369-371.

O'TOOLE, D.K. 1974. A modification of the "bacto-strip" technique for counting bacteria. Aust. J. Dairy Technol. *29*, 117.

PATON, A.M., and JONES, S.M. 1973. The observation of micro-organisms on surfaces by incident fluorescence microscopy. J. Appl. Bacteriol. *36*, 441-443.

PATON, A.M., and JONES, S.M. 1975. The observation and enumeration of micro-organisms in fluids using membrane filtration and incident fluorescence microscopy. J. Appl. Bacteriol. *38*, 199-200.

PIKE, E.B., CARRINGTON, E.G., and ASHBURNER, P.A. 1972. An evaluation of procedures for enumerating bacteria in activated sludge. J. Appl. Bacteriol. *35*, 309-321.

POSTGATE, J.R. 1969. Viable counts and viability. *In* Methods in Microbiology, Vol. 1. J.R. Norris and D.W. Ribbons (Editors). Academic Press. New York, N.Y.

PREVITE, J.J. 1972. Radiometric detection of some food-borne bacteria. Appl. Microbiol. *24*, 535-539.

QURESHI, A.A., and DUTKA, B.J. 1976. Comparison of various brands of membrane filters for their ability to recover fungi from water. Appl. Environ. Microbiol. *32*, 445-447.

REEVES, M.P. 1973. Examination of frozen vegetables by two sample preparation procedures. J. Food Sci. *38*, 365-366.

SANDERS, D.H. and PARKES, M.R. 1970. Infrared estimation of microbial population on broiler chicken carcasses during refrigerated storage. Poultry Sci. *49*, 173-178.

SCHOON, D.J., DRAKE, J.F., FREDRICKSON, A.G., and TSUCHIYA, H.M. 1970. Automated counting of microbial colonies. Appl. Microbiol. *20*, 815-820.

SCHROT, J.R., HESS, W.C., and LEVIN, G.V. 1973. Method for radiorespirometric detection of bacteria in pure culture and in blood. Appl. Microbiol. *26*, 867-873.

SHARPE, A.N., BIGGS, D.R., and OLIVER, R.J. 1972A. Machine for automatic bacteriological pour plate preparation. Appl. Microbiol. *24*, 70-76.

SHARPE, A.N., DYETT, E.J., JACKSON, A.K., and KILSBY, D.C. 1972B. Technique and apparatus for rapid and inexpensive enumeration of bacteria. Appl. Microbiol. *24*, 4-7.

SHARPE, A.N., and JACKSON, A.K. 1972. Stomaching: a new concept in bacteriological sample preparation. Appl. Microbiol. *24*, 175-178.

SHARPE, A.N. and KILSBY, D.C. 1971. A rapid, inexpensive bacterial count technique using agar droplets. J. Appl. Bacteriol. *34*, 435-440.

SHELEF, L.A. 1974. Hydration and pH of microbially spoiling beef. J. Appl. Bacteriol. *37*, 531-536.

SHELEF, L.A., and JAY, J.M. 1970. Use of a titrimetric method to assess the bacterial spoilage of fresh beef. Appl. Microbiol. *19*, 902-905.

SHELEF, L.A., and JAY, J.M. 1971. Hydration capacity as an index of shrimp microbial quality. J. Food Sci. *36*, 994-997.

SIMMONDS, P.G. 1970. Whole microorganisms studied by pyrolysis-gas chromatography-mass spectrometry: significance for extraterrestrial life detection experiments. Appl. Microbiol. *20*, 567-572.

STARUSZKIEWICZ, W.F., JR., and BOND, J.F. 1978. Multiple internal standard technique for the gas-liquid chromatographic determination of indole in shrimp. J. Assoc. Offic. Anal. Chem. *61*, 136-138.

TANSEY, M.R. 1973. Use of butane lighter for sterilization of soil sampling instruments. Mycologia *65*, 215-216.

TAYLOR, M.M. 1975. The water agar test: a new test to measure the bacteriological quality of cream. J. Hyg. (Camb.) *74*, 345-357.

THORE, A., ÅNSÉHN, S., LUNDIN, A., and BERGMAN, S. 1975. Detection of bacteriuria by luciferase assay of adenosine triphosphate. J. Clin. Microbiol. *1*, 1-8.

TOBIN, R.S., and DUTKA, B.J. 1977. Comparison of the surface structure, metal binding, and fecal coliform recoveries of nine membrane filters. Appl. Environ. Microbiol. *34*, 69-79.

TROTMAN, R.E., and BYRNE, K.C. 1975. The automatic preparation of bacterial culture plates. J. Appl. Bacteriol. *38*, 61-62.

TUTTLEBEE, J.W. 1975. The stomacher—its use for homogenization in food microbiology. J. Food Technol. *10*, 113-122.

VANDERZANT, C., and NICKELSON, R. 1971. Comparison of extract-release volume, pH, and agar plate count of shrimp. J. Milk and Food Technol. *34*, 115-118.

VAN SCHOTHORST, M., NORTHOLT, M.D., KAMPELMACHER, E.H., and NOTERMANS, S. 1976. Studies on the estimation of the hygienic condition of frozen broiler chickens. J. Hyg. (Camb.) *76*, 57-73.

WATERS, J.R. 1972. Sensitivity of the $^{14}CO_2$ radiometric method for bacterial detection. Appl. Microbiol. *23*, 198-199.

WATROUS, G.H., JR., BARNARD, S.E., and COLEMAN, W.W. 1971. A survey of the actual and potential bacterial keeping quality of pasteurized milk from 50 Pennsylvania dairy plants. J. Milk Food Technol. *34*, 145-149.

WEBB, N.B., THOMAS, F.B., BUSTA, F.F., and KERR, L.S. 1972. Evaluation of scallop meat quality by the resazurin reduction technique. J. Milk Food Technol. *35*, 664-668.

3

Microorganisms Associated with Food

The significance of microorganisms in foods depends upon: 1) the numbers found; 2) the types of microorganisms; 3) the type of food; 4) the treatments to which the food has been exposed; 5) the processing or storage treatments the food will receive; 6) whether the food is to be eaten as is or heated; and 7) the individuals that might consume the food. In this text, it is necessary to limit the discussion of microorganisms in foods to bacteria, fungi and viruses.

Microorganisms may have one or more of four functions in a food. They may have a useful function, cause spoilage, be a health hazard or be inert. The inert microorganisms do not find an environment favorable for growth. In most cases of foodborne illness, spoilage or useful activity, there is growth and multiplication of the microorganisms.

Those organisms that cause foodborne illness are of more concern than are the other types. This does not mean that microorganisms that cause spoilage or are useful are not important.

Spoilage consists of producing undesirable changes in the odor, color, taste, texture or appearance of the food. Some organisms that do not directly cause changes in a food may alter the flora so that spoilage organisms can grow. An example is the bacteriophages that attack useful organisms, so that undesirable organisms can grow and cause spoilage.

Useful organisms are those which produce desirable changes in food, such as converting milk to cheese, sugar to alcohol and cabbage to sauerkraut. These are referred to as fermentations. The microorganism does not always have to be present to have a useful function, since enzymes can be separated from the organism, and the enzymes used to produce the desired reaction. Another useful aspect is the production of single cell protein which can be used as food. There is a similarity between spoilage organisms and useful organisms since they both produce changes in the food. The difference is that one is desirable and the other is undesirable.

Although it is easy to establish categories for microorganisms, it is difficult to place an organism in only one category, since it may have different functions in different foods. An organism may spoil one food but be inert in another food, due to the different characteristics of the foods. However, we do tend to associate certain organisms with particular functions in food so, as much as possible, these will be considered.

DETERMINING THE TYPES OF MICROORGANISMS

The basic systems discussed in Chapter 2 can be used to determine the specific types of microorganisms that are present in or on a food. However, differential or selective media are substituted for the non-inhibitory, non-selective media. Quite often, prior to determining if certain types of organisms are present, an enrichment procedure is used to increase the probability of detection. After these enrichment processes, the organisms are detected on differential or selective agar.

Various tests are used to differentiate organisms isolated from selective or differential agars. The usual staining reactions and morphological characteristics are determined. The required biochemical tests depend upon the organisms. Immunological tests can be done on cellular, flagellar or other antigens. The production of pigments, sensitivity to antibiotics, phage typing and bacteriocin typing can be useful for the differentiation of strains of some organisms. The metabolic products of isolates can be determined by gas-liquid chromatography. This is especially useful for differentiation of *Clostridium* species.

There are two ways to proceed in the testing of isolated microorganisms. In one system, the results of one test will determine the next test that is made. Obviously, this system requires an extended time period before an organism can be identified.

The other system is to simultaneously determine several important characteristics of the isolates, and then decide which organism these results best describe. If only 5 tests with a positive or negative value are conducted, there are 2^5 or 32 possible combinations. Using only 5 tests to characterize an organism is not satisfactory. If 15 different tests are made, there are 32,768 possible combinations of positive and negative values. With this volume of potential information, a computer is a valuable tool.

MICROBIAL TYPES IN FOOD

The microbial analysis of food products yields many diverse types of microorganisms. However, we are concerned with the predominant types and those which may cause spoilage or be a health hazard.

The prevalent bacteria on animal carcasses are species of *Pseudomonas, Micrococcus, Bacillus, Acinetobacter, Lactobacillus* and coryneforms. Coliforms, *Staphylococcus aureus, Clostridium perfringens* and salmonellae are present in many samples. The temperature of holding and the packaging material influence the types of microorganisms that become dominant on fresh meat. Organisms such as pseudomonads tend to become dominant on refrigerated fresh meat. Species of *Micrococcus* and *Lactobacillus* are important on cured meat. Besides these types, *Bacillus, Microbacterium* and yeasts (*Candida, Debaryomyces, Torulopsis* and *Trichosporon*) are prevalent on frankfurters. Gram positive bacteria, Gram negative *Vibrio* species and yeasts have been reported in bacon and bacon curing brines (Dempster 1973).

The same genera found on fresh red meat tend to be present on poultry and fishery products. *Vibrio* species are found on fishery products harvested from the ocean. Other genera found on or in fishery products include *Bacillus, Microbacterium, Micrococcus, Moraxella, Pseudomonas, Arthrobacter, Acinetobacter, Flavobacterium*, as well as organisms designated as coryneforms.

Being animal products, milk and eggs tend to contain bacterial species similar to those of meat and poultry.

Many types of microorganisms are associated with plant products. During processing, the initial microflora is distributed throughout the product and is supplemented with other microorganisms. Most of the bacteria in or on fruit and vegetables neither cause spoilage nor are health hazards.

In green vegetable salads the most common genera are *Lactobacillus* and *Leuconostoc* along with *Klebsiella, Enterobacter* and *Serratia*. The Gram negative *Erwinia* is important in vegetable spoilage. Other genera in vegetables include *Aeromonas, Alcaligenes, Bacillus, Corynebacterium, Flavobacterium, Micrococcus, Pseudomonas* and *Xanthomonas*.

The important organisms on fruits are molds (*Penicillium, Alternaria, Aspergillus*). *Lactobacillus* species cause spoilage of fruit juices.

Molds such as *Aspergillus, Fusarium* and *Penicillium* are found on dry products such as grain and peanuts.

Several genera of yeasts are important in the spoilage of foods, especially fruits and those containing sugar. Also, yeasts are useful in fermentations and the production of single cell protein.

BACTERIA

Compared to the total number of species of bacteria, relatively few have any importance in food and most of these are the common types which are discussed in general microbiology. Individual species of bacteria are discussed with their functions in later chapters, so we will be concerned mainly with the genera that are important.

In *Bergey's Manual of Determinative Bacteriology* (Buchanan and Gibbons 1974), the bacteria are grouped according to the Gram reaction, morphology and relation to growth with or without oxygen. That system is used in this text for the bacteria found in food. Various morphological types of bacteria are shown in Fig. 3.1 and 3.2.

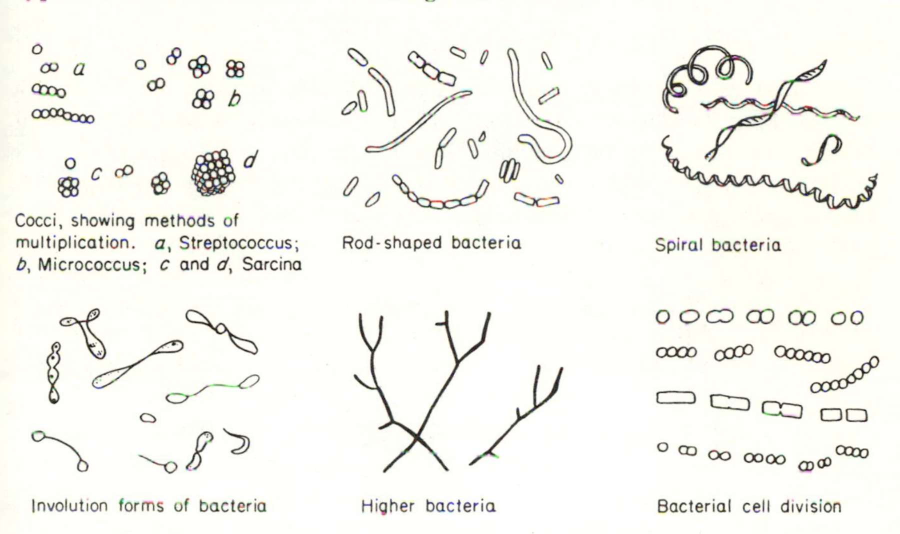

Courtesy of Weiser et al. (1971)

FIG. 3.1. MORPHOLOGICAL FORMS OF BACTERIA

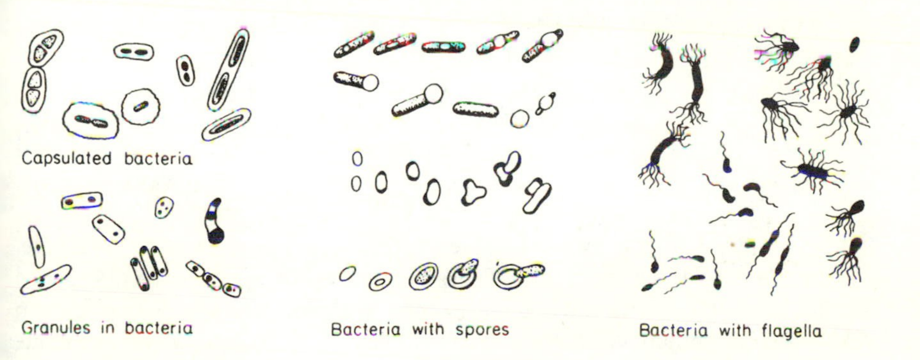

Courtesy of Weiser et al. (1971)

FIG. 3.2. STRUCTURES OF BACTERIA

Besides the general characteristics of bacteria, many strains contain plasmids. These are genetic elements within the cell but outside of the chromosome. In some cases these plasmids can be transferred from one cell to another closely related cell. These plasmids are involved with fertility, drug resistance, radiation and chemical resistance, enzyme coding, as well as unknown activities.

Gram Negative, Aerobic Rods and Cocci

Included in this group are the family Pseudomonadaceae with the genera *Pseudomonas, Xanthomonas* and *Gluconobacter*; the family Halobacteriaceae with the genera *Halobacterium* and *Halococcus*; and three genera (*Alcaligenes, Acetobacter* and *Brucella*) of uncertain affiliation (Buchanan and Gibbons 1974).

Pseudomonas.—This genus is divided into four sections on the basis of growth factors and the accumulation of poly-β-hydroxy-butyrate as an intracellular carbon reserve (Buchanan and Gibbons 1974). The organisms are straight to curved, motile (polar flagella) rods.

The pseudomonads are noted for their biochemical activity, being able to attack a wide variety of organic compounds, including aromatic types, and some man-made chemicals. Some species (*P. aeruginosa* and P. *cepacia*) have been found to grow in distilled water. The pseudomonads have a respiratory, not fermentative metabolism. They produce catalase and most produce oxidase.

Being aerobic, pseudomonads need readily available O_2 for growth. However, King and Nagel (1967) reported that *P. aeruginosa* was not inhibited until 75% of the oxygen in air was replaced with nitrogen.

The pseudomonads produce proteinase enzymes which catalyze proteolytic reactions and contribute to food spoilage. These organisms are very important in the spoilage of refrigerated, fresh animal products. Some species form pigments (green, blue-green, blue, red, yellow, orange or light or dark brown). Their fluorescent pigments, pyocyanin and fluorescein, can be observed in spoiled foods by using an ultraviolet light. Fluorescent pigments may appear as yellow-green, blue or orange, depending upon the species and the environmental factors of the culture. The pigments can cause color defects of food. Some species are plant and animal (including man) pathogens. Some *Pseudomonas* species have been suggested as causative agents of foodborne illness, but there is inconclusive proof (Bryan 1973). *P. aeruginosa* produces an enterotoxin and can cause gastroenteritis (Schook *et al.* 1976). Also, this species is opportunistic, causing infection in patients with lowered resistance.

The pseudomonads are widely distributed in nature. They are found on animal and plant products. It is thought that raw vegetables are an important source and vehicle for colonization of the human intestinal tract with pseudomonads (Pasch 1974). Green *et al.* (1974) suggested that soil is the reservoir of *P. aeruginosa* and from this source, the organism can colonize plants. The continual contamination through animal excrement maintains the organism in soil.

The psychrotropic pseudomonads are found in almost all types of refrigerated and frozen foods. Since they are not very heat resistant, they are not found in heat processed foods unless the food is recontaminated after heating.

The pseudomonads are not very resistant to drying. They are sensitive to heat and gamma irradiation. *P. aeruginosa* is notoriously resistant to quaternary ammonium compounds. Naturally occurring organisms are more resistant to inactivation by quaternary ammonium compounds, chlorine dioxide, acetic acid or activated glutaraldehyde than cells previously subcultured on synthetic media.

Xanthomonas.—The organisms in this genus are plant pathogens. They are straight, motile (polar flagellum) rods. Like the pseudomonads, they have a respiratory, never fermentative metabolism. They are catalase positive and the oxidase reaction is either negative or weak. On agar media, yellow growth is evident.

Being plant pathogens, these organisms have been involved with various types of rots of these products. *X. campestris* produces xanthan gum, which is used in the food industry.

Gluconobacter.—*G. oxydans* represents this genus. This organism is ellipsoidal to rod-shaped, Gram negative to weakly Gram positive in older cultures. The cells occur singly, in pairs, or in chains. They are strictly aerobic and motile (polar flagella) to nonmotile. It is not fermentative, having a respiratory metabolism, and oxidizes ethanol to acetic acid.

G. oxydans is found in various food products such as vegetables, fruits, bakers' yeast, beer, wine, cider and vinegar. It is involved in spoilage, causing the souring of fruits.

Halobacteriaceae.—This family contains two genera, *Halobacterium* and *Halococcus*. These are extreme halophiles requiring a high concentration of salt (about 15%) for growth. The organisms are found in salt produced by the evaporation of sea water (solar salt).

Due to their halophilic character, they cannot grow in most foods. However, in foods preserved by salting, these organisms can produce a red pigment, bacteriorubein.

The high salt concentration is necessary for activity of the enzymes, stability of membranes and ribosomes and for synthesis of protein. A low salt concentration results in *Halobacterium* changing from rods to spherical forms.

Alcaligenes.—The four species in this genus are motile (four to eight peritrichous flagella) rods, coccal rods or cocci. They are strict aerobes, oxidase positive with a respiratory, not fermentative, metabolism. The optimum temperature is 20°-37° C.

They are widespread in nature, being found in soil, water, decaying matter, and the intestinal tract of animals. These organisms are involved with the spoilage of protein foods (eggs and dairy products), but they are not actively proteolytic in casein or gelatin media.

Acetobacter.—These cells are ellipsoidal to straight or slightly curved rods. Young cultures are Gram negative, while older cultures are Gram variable. They are motile (peritrichous flagella) or nonmotile.

The *Acetobacter* are strict aerobes, with respiratory, never fermentative metabolism. They are noted for the oxidation of ethanol to acetic acid. They oxidize acetate and lactate to CO_2 and H_2O.

Species of *Acetobacter* are found on fruits and vegetables and are involved in the souring of fruit juices and alcoholic beverages (beer and wine).

Brucella.—These coccobacilli or short rods are nonmotile, with respiratory metabolism. *Brucella melitensis* (pathogenic for goats and sheep), *B. abortus* (pathogenic for cattle) and *B. suis* (pathogenic for pigs) can affect man, causing brucellosis (undulant fever). The sources of infection are raw milk or dairy products, uncooked meat or sausage products, discharges from infected animals and human carriers.

Since these organisms are susceptible to the heat treatment used for pasteurization of milk and cooking of meat, they should be of limited importance in food.

Annually, there are between 100 and 200 cases of brucellosis in the U.S. The majority of cases involve meat-packing plant workers, and most of these are involved with hog slaughter operations. Most of the other cases are livestock producers, veterinarians and government inspectors.

Gram Negative, Facultative Anaerobic Rods

In this group, the family Enterobacteriaceae contains several genera (*Escherichia, Edwardsiella, Citrobacter, Salmonella, Shigella, Klebsiella, Enterobacter, Hafnia, Serratia, Proteus, Yersinia* and *Erwinia*), the family Vibrionaceae includes the genera *Vibrio* and *Aeromonas* and the genera of uncertain affiliation are *Chromobacterium* and *Flavobacterium.*

Escherichia.—Only one species (*E. coli*) is described for this genus (Buchanan and Gibbons 1974). *E. coli* can be motile (peritrichous flagella) or nonmotile. Most strains ferment lactose, but this may be delayed or absent. It is oxidase and urease negative and does not produce H_2S. When tested with the IMViC (indole, methyl red, Voges-Proskauer and citrate) tests, the organism produces indole from tryptophan, is methyl red positive, does not form acetylmethylcarbinol and citrate is not utilized. Hence, it is ++−−.

E. coli is found in soil, water, on plants, in the intestinal tract of animals and in various foods, especially animal products and those handled by humans. Since the cells are heat sensitive, their presence in heat pasteurized or cooked products indicates recontamination after the treatment.

In a food, the organism may cause spoilage, certain strains are enteropathogenic or enterotoxigenic and it is used as an indicator organism of fecal contamination. Spoilage is due to its ability to grow on a variety of substrates. It produces acid and gas from most carbohydrates and synthesizes its needed vitamins. The cells grow at a temperature from 10° to 45.5°C. Besides gassiness, off-odors and off-flavors described as "barny," are produced in foods. *E. coli* may be associated with several disease syndromes.

Edwardsiella.—The species *E. tarda* is motile with peritrichous flagella. This species has been isolated from human cases of diarrhea and from normal stools. Although causing enteritis, proof that transmission is by means of food is inconclusive (Bryan 1973). The species has been isolated from both warm-blooded and cold-blooded animals.

Citrobacter.—The species in this genus are motile (peritrichous flagella) rods that utilize citrate as sole carbon source and ferment lactose, although this may be delayed or absent. The late lactose fermenters and nonfermenters are often confused with *Salmonella* on differential agars.

They are normal intestinal inhabitants and are found in various foods, especially animal products. They are members of the coliform group of indicator organisms.

They are listed as causing enteritis in man, although proof of transmission by foods is inconclusive (Bryan 1973). They have been associated with various human infections.

Besides being a possible health hazard, *Citrobacter* can cause spoilage of foods.

Cross reactions with antisera of other Enterobacteriaceae indicate there are close relationships between *Citrobacter, Salmonella* and *Escherichia.*

Salmonella.—The genus *Salmonella* is divided into four subgenera (I, II, III and IV). These subgenera are differentiated on the basis of minor biochemical differences. Subgenus I includes the majority of serotypes of *Salmonella*, especially the more important ones, and subgenera III comprises the organisms often referred to as the Arizona group. Most of the different types were segregated on the basis of serology so they are called serotypes. At present, there are more than 1,700 serotypes and variants of *Salmonella* (CDC 1975).

The organisms in this genus are catalase positive and oxidase negative with both respiratory and fermentative metabolism. Most of the serotypes are motile (peritrichous flagella), do not ferment lactose, and grow well in simple media containing glucose, inorganic nitrogen and mineral salts. A number of strains are unable to synthesize certain vitamins and amino acids.

The salmonellae are said to be ubiquitous, being world-wide and found in or on soil, water, sewage, animals, humans, processing equipment, feed and various food products. The natural habitat is the intestinal tract of humans and animals. It is logical that humans, animals and their environments are the primary sources of salmonellae. Some serotypes seem to be localized in a region or a country, but with national and international travel and trade, the organisms are easily disseminated.

The serotypes of *Salmonella* are an important cause of foodborne illness. Further information about these organisms is found in Chapter 6.

Shigella.—The four species in this genus are nonmotile, nonsporeforming rods. The normal habitat is the intestinal tract of humans and other primates. Rarely are these organisms isolated from other animals.

The shigellae are the causative agent of shigellosis, a foodborne gastroenteritis. The number of foodborne illnesses due to these organisms is not very dramatic. However, difficulty in the detection of *Shigella* may play a part in the apparent low incidence of these organisms. Additional information concerning *Shigella* and shigellosis is in Chapter 6.

Klebsiella.—The three species of *Klebsiella* are nonmotile, encapsulated rods, appearing singly, in pairs, or short chains.

These organisms are found in water, sewage, soil and are part of the flora of the mouth, pharynx and intestinal tract. They have been found on grain, fresh produce and frozen foods.

The klebsiellae are thought to be opportunistic pathogens, involved with several syndromes including pneumonia and upper respiratory tract infections.

These organisms are potential health hazards, members of the coliform group of indicator organisms and can cause food spoilage. There is inconclusive proof that transmission is through food to cause gastroenteritis (Bryan 1973).

Enterobacter.—Organisms in the *Enterobacter* are similar to the klebsiellae, except they are motile (peritrichous flagella). Two species are described for this genus, *E. cloacae* and *E. aerogenes.*

The *Enterobacter* can be found in soil, water, sewage, the intestinal tract of man and animals and in various food products. The organisms are important in food as a potential health hazard, an indicator organism and spoilage. Although the organisms cause enteric infection in humans, proof of transmission by food is inconclusive (Bryan 1973). He listed foods involved in enteritis as cream filled pastries, milk and stew.

Hafnia.—The one species *H. alvei* formerly was named *Enterobacter hafnia*. The organism is found in sewage, soil, water, the feces of man and other animals, and in various foods such as dairy products.

Serratia.—The cells of *S. marcescens* are aerobic, mesophilic, motile (peritrichous flagella) rods. Citrate and acetate can be used as the sole source of carbon. Glucose, cellobiose, inositol and glycine are fermented. *S. marcescens* is noted for the production of a red pigment, prodigiosin. The importance of *Serratia* is as a potential spoilage organism in some foods.

Proteus.—Organisms in this genus are highly motile (peritrichous flagella) rods that exhibit pleomorphism. They occur singly, in pairs or short chains. On moist agar surfaces, *P. vulgaris* and *P. mirabilis* tend to swarm. This swarming is more evident at 20°C than at 37°C. At 37°C, motility may be weak or absent. Various chemicals (bile salts, tannic acid, sodium azide, aniline dyes, sodium deoxycholate, *p*-nitrophenol, glycerine and sulfa drugs) suppress or inhibit swarming. Although swarming seems to be involved with motility, swarming can be inhibited without affecting the normal motility of the cells, and many other motile organisms do not swarm. This swarming by *Proteus* makes it difficult to pick isolated colonies from agar for further study.

The organisms are widely distributed. They are common inhabitants of the gastrointestinal tract, and are found in sewage, soil, in decomposing animal protein and in various foods.

The importance of *Proteus* in food includes potential health hazards and spoilage. Although strains of *Proteus* can cause enteric infection in humans, according to Bryan (1973), proof of transmission by food is inconclusive.

Yersinia.—The cells in this genus are ovoid or rod shaped. *Y. pestis* is nonmotile, while *Y. pseudotuberculosis* and *Y. enterocolitica* are motile at 30°C or less.

These organisms are pathogenic for animals and, in some cases, man. *Y. pestis* is the plague organism and is transferred by the rat flea. The other species are found in the mesenteric lymph glands, lymph nodes and intestines of healthy animals as well as humans.

Y. enterocolitica causes gastroenteritis as well as being involved in abscesses, bacteremia, peritonitis and other syndromes (Bissett 1976; Bottone and Robin 1977). The pathogenic mechanism of the organism is not known.

Sakazaki *et al.* (1974) reported that two of three strains of *Y. enterocolitica* cause dilation of rabbit gut loops. The organism apparently affects lymphoid tissue of the digestive tract. This species has been isolated from foods such as red meat, oysters, mussels and ice cream. These organisms may be more prevalent in food than reported. It is difficult to distinguish them from other organisms that do not ferment lactose.

Erwinia.—Organisms in this genus are small, mostly motile (peritrichous flagella), predominantly single, straight rods. They are oxidase negative and catalase positive.

The genus is divided into three groups. The amylovora group contains six species that are primarily plant pathogens. The herbicola group contains three species that have been found associated with plants as well as animals and man. The carotovora group has four species which cause soft rot of plant products during storage.

Although presently in the genus *Erwinia*, an effort has been made to place the carotovora group into *Pectobacterium.*

E. carotovora is the main spoilage organism of stored vegetables. *E. rhapontici* has been associated with the rotting of onions and cucumber slices, crown rot of rhubarb and pink discoloration of wheat. These organisms cause rot of plant products through the inducible enzyme pectate lyase which degrades pectin.

Vibrio.—This genus is in the family Vibrionaceae. There are five species (Buchanan and Gibbons 1974). The cells are short, asporogenous, motile (polar flagellum), curved or straight rods. They have either respiratory or fermentative metabolism. They are oxidase positive and urease negative. Some strains fail to grow without NaCl, and optimum NaCl is about 3.0%.

V. anguillarum is found in a diseased condition of fish. Strains of *V.*

costicola can tolerate NaCl concentrations of 23%. This species is found in cured meats and brines. Vibrios were isolated from bacon by Gardner (1973), but these did not fit the biochemical pattern of *V. costicola.* Gardner (1973) described a medium for enumerating salt requiring *Vibrio* from Wiltshire bacon and brine. Methods for enumeration and phage typing of *V. cholerae* were discussed by Bockemuhl and Meinicke (1976) and Morris *et al.* (1976).

V. cholerae and *V. parahaemolyticus* are important pathogens, causing gastroenteritis in humans. *V. cholerae* is found in the intestinal tract of man and animals, in water and occasionally in food. *V. parahaemolyticus* is found in the ocean, in seafoods and in the intestinal contents of infected humans. This species is discussed further in Chapter 6.

Aeromonas.—The cells are rods with rounded ends to coccoid. They are motile (polar flagella), have both respiratory and fermentative metabolism and are oxidase and catalase positive.

These organisms are frequently mistaken for members of the family Enterobacteriaceae because of the similarity in growth and biochemical characteristics. A positive oxidase test and nitrate reduction help differentiate the *Aeromonas* from Enterobacteriaceae.

The main habitat of *Aeromonas* is water. Some strains cause disease (hemorrhagic septicemia) in fish, eels and frogs. Some strains cause enteritis in humans. One source of this infection may be fish or other sea food. *Aeromonas* may play a role in spoilage of fish and other animal products.

Chromobacterium.—This genus has two species, *C. violaceum* and *C. lividum* (Buchanan and Gibbons 1974). They are violet and dark blue organisms, respectively. The pigment has antibiotic properties.

The organisms are found in water, soil and occasionally cause infections in animals and food spoilage. Media for enumeration of these organisms were described by Ryall and Moss (1975).

Flavobacterium.—This genus contains diverse bacteria. The species include nonmotile rods that produce yellow, orange or greenish-yellow pigment (Hayes 1977). Environmental factors (substrate and temperature) affect the synthesis of pigment and resultant hue. These organisms prefer temperatures below 30°C, although some strains can grow at 37°C.

Flavobacterium species have been isolated from water, soil, animals, humans and various food products. They can produce discoloration on some foods. The organisms have been found on thawing frozen vegetables, fresh vegetables, refrigerated fish and shellfish, cannery environments, meat, meat products and poultry.

Gram Negative, Aerobic Cocci and Coccobacilli

This group includes the family Neisseriaceae, with the genera *Neisseria, Branhamella, Moraxella* and *Acinetobacter.* The first three genera are associated primarily with the mucous membranes of animals and are described by Henriksen (1976). *Neisseria, Moraxella* and *Acinetobacter* have been reported in foods, but no particular roles have been determined for *Neisseria* or *Moraxella.*

Acinetobacter.—These organisms are very short, plump rods or coccobacilli, predominantly in pairs or short chains. They are strictly aerobic, mesophilic saprophytes.

They are found in soil, water and also in animals and man. Organisms identified as *Acinetobacter* have been isolated from various types of raw and prepared foods, including beef and poultry carcasses (Lahellec *et al.* 1975). Besides potential spoilage of foods, *Acinetobacter* has been suggested as a source of single cell protein (Abbott *et al.* 1974).

Gram Positive Cocci

The Gram positive cocci include aerobic or facultative anaerobic bacteria in the family Micrococcaceae with the genera *Micrococcus* and *Staphylococcus* and the family Streptococcaceae which includes the genera *Streptococcus, Leuconostoc, Pediococcus* and *Aerococcus.*

Micrococcus.—These spherical cells are strict aerobes, catalase positive, occur singly or in pairs and characteristically divide in more than one plane to form irregular clusters, tetrads or cubical packets. They can grow in the presence of 5% salt.

The micrococci are found in soil, water, dust and on the skin of man and other animals. They are found in several types of foods, especially milk, dairy products, on animal carcasses and meat products. They are important as potential spoilage organisms.

Staphylococcus.—These nonmotile cells occur singly, in pairs or irregular clusters. They are facultative anaerobes with respiratory and fermentative metabolism. Most strains can grow in 7.5-15% salt. The organisms usually are sensitive to chlorine, chloramine, iodine and iodophors. Although usually sensitive to heat, they are moderately resistant to radiation.

Both *S. aureus* and *S. epidermidis* are commonly found on the skin and mucous membranes of humans and warm blooded animals. They are potential pathogens, being either the primary pathogen or secondary invader.

The staphylococci are found in many types of food products. In general, they are not able to compete very well with other organisms. *S. aureus* produces enterotoxins which are one of the main causes of foodborne illness. *S. aureus* is found in pimples, boils, acne, wound infections and the nose. It is easily transferred to foods by careless food handlers. This species is differentiated from the other two species by its ability to produce coagulase, which clots blood plasma, and a heat-stable nuclease.

Streptococcus.—The bacterial species in this genus have various characteristics and functions. These organisms have been divided into different groups on the basis of growth in various culture conditions (temperature, pH, salt, methylene blue) according to Table 3.1, hemolytic types (alpha, beta and gamma) and serologically on the basis of group antigens (A, B, C, D, E, F, G, H, K, N, and Q).

The streptococci are facultative anaerobes. Hydrogen peroxide may accumulate since the streptococci are catalase negative. The nutrients required for growth vary with species and strains within species. Some strains require certain amino acids, vitamins, purines, pyrimidines, fatty acids and elevated levels of CO_2.

These spherical cells occur in pairs or chains (Fig. 3.3). With the exception of some strains, they are not motile. They have a fermentative metabolism, fermenting glucose primarily by the hexose diphosphate pathway and producing mainly lactic acid. Thus, they are called homofermentative. When other types of carbohydrates are fermented, butyric acid, acetoin, diacetyl and n-butanol are produced.

The fermentation of carbohydrates to lactic acid is desirable in products such as buttermilk, cheese, yogurt, sauerkraut, etc., but causes spoilage of products, such as fresh milk, in which lactic acid is not wanted. They can utilize sugar in an alcohol fermentation, resulting in lower alcohol and higher acid than is wanted. Some strains utilize citric acid and form acetoin and diacetyl. The ability to produce diacetyl in milk is desirable in the manufacture of cultured sour cream, buttermilk and butter.

Lactose metabolism, proteinase activity, hemolysin and bacteriocin production are mediated by plasmids (Anderson and McKay 1977; Oliver *et al.* 1977).

An important problem in the dairy industry is the susceptibility of lactic acid bacteria to phage. The rapid reproduction of phage can reduce or eliminate the streptococci in a fermenting milk so that an unacceptable product is produced.

The streptococci are widely distributed, being found in air, water, sewage, soil, on plants, in the intestinal tract of man and animals, and in various food products. Some streptococci are involved with bovine mastitis and found in raw milk (McDonald and McDonald 1976).

TABLE 3.1

CULTURAL CONDITIONS AFFECTING GROWTH OF THE FOUR GROUPS OF *STREPTOCOCCUS* SPECIES

	Streptococcus Group Number			
	I[1]	II	III	IV
Temperature				
Grows at 10°C	–[2]	–	+	+
Grows at 45°C	–	+	+	–
pH 9.6	–	–	±	–
Salt (6.5%)	–	–	±	–
Methylene blue (0.1%)	–	–	±	+

1) Species are: I. *S. pyogenes, S. equisimilis, S. zooepidemicus, S. equi, S. dysgalactiae, S. sanguis, S. pneumoniae, S. anginosus, S. agalactiae, S. acidominimus*

II. *S. salivarius, S. mitis, S. bovis, S. equinus, S. thermophilus*

III. *S. faecalis, S. faecium, S. avium, S. uberis*

IV. *S. lactis, S. cremoris*

2) (–) = no growth; (+) = can grow in this condition.

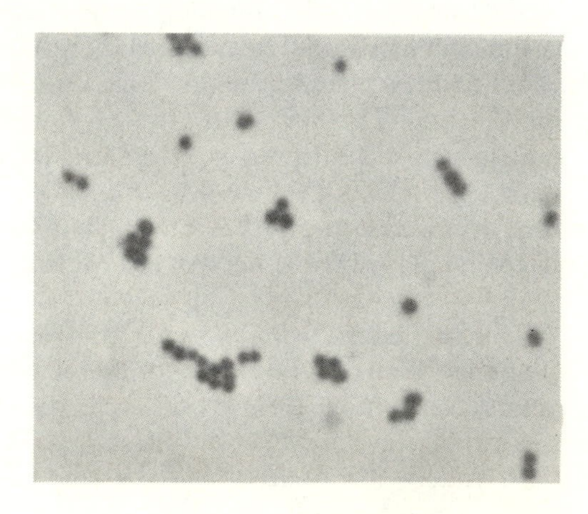

Courtesy of Pederson (1979)

FIG. 3.3. *STREPTOCOCCUS LACTIS*, A GRAM POSITIVE COCCUS USED IN MILK FERMENTATIONS

Some species and strains are rather heat resistant, surviving 60°C for 30 min. The enterococci can survive freezing and frozen storage in food products. This makes these organisms acceptable as potential indicator organisms in frozen foods.

Although some pathogenic types may be transmitted to foods, there is more interest in the streptococci as possible indicators of fecal contamination, as useful fermentative organisms or as potential spoilage bacteria.

Leuconostoc.—These spherical to lenticular cells occur in pairs or chains. They often have complex nutrient requirements such as vitamins, amino acids and a fermentable carbohydrate. They ferment glucose to lactic acid, ethanol and CO_2.

The organisms are important in fermentation and spoilage of foods. They are not pathogenic. *L. mesenteroides* and *L. dextranicum* produce dextrans, resulting in a characteristic slime in sugar solutions. Some strains can produce a flavor defect in orange concentrate. The leuconostocs are important in the manufacture of fermented vegetables and dairy products.

Pediococcus.—This is a genus of microaerophilic cocci that occur as tetrads and sometimes as single cells or in pairs (Fig. 3.4) They show poor surface growth, are homofermentative, and do not reduce nitrate, liquify gelatin or produce catalase. They are nonmotile, chemoorganotrophic, with rather complex nutritional requirements, such as vitamins and amino acids.

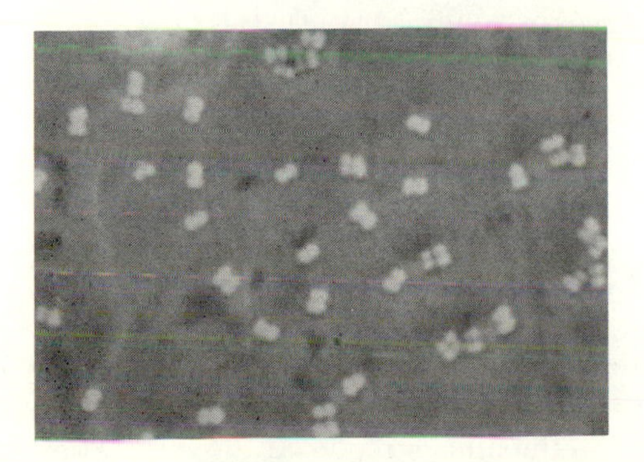

Courtesy of Pederson (1979)

FIG. 3.4. *PEDIOCOCCUS CEREVISIAE*, COMMON IN MANY FERMENTING MATERIALS

These organisms are found in sauerkraut, pickles, wine, beer and other fermenting material. *P. cerevisiae* produces diacetyl, which causes an off-odor and spoilage of beer. Besides their use in fermentations, pediococci can be used for the assay of vitamins and amino acids.

Aerococcus.—This genus has only one recognized species, *A. viridans*, which used to be a *Pediococcus (P. homari)* (Buchanan and Gibbons 1974). The genus is very similar to *Pediococcus*. On blood agar, the colonies are surrounded by a green zone, hence the name viridans (green).

The aerococci have been found in air and dust, human infections, meat curing brines and on raw and processed vegetables.

Endospore-forming Rods and Cocci

This group includes the family Bacillaceae with the genera *Bacillus, Clostridium* and *Desulfotomaculum*.

The spores that these organisms produce are different from the vegetative cells (Fig. 3.5) so that we have another entity to consider. The spores are more refractive, more resistant to staining, heat, radiation, chemicals and other destructive action, and the spores contain dipicolinic acid. The spores are an inactive or dormant state of the organisms.

There are certain stages which the organisms go through to change from a vegetative cell to a spore and back to a vegetative cell. The stages we can consider are sporulation, germination and outgrowth of the cell.

Courtesy of Weiser et al. (1971)

FIG. 3.5. BACILLACEAE SPORES FORM IN THE CENTER OF THE ROD (Magnified)

Bacillus subtilis cells at left, *Clostridium sporogenes* at right.

Sporulation.—The mechanisms that trigger spore formation are not fully known. Most strains of clostridia produce spores when incubated in a good medium under anaerobic conditions, 3° to 8° C below their optimum growth temperature. Saccharolytic species sporulate when a fermentable carbohydrate is present. *C. perfringens* requires special media for sporulation. Methionine is required for sporulation of this organism (Muhammed *et al.* 1975). One strain also needs riboflavin, isoleucine, serine and lysine. Sporulation in a chemically defined medium is not observed in the absence of $CaCO_3$ (Muhammed *et al.* 1975). They stated that it is impossible to demonstrate the complete sporulation requirements of an organism since some nutrients essential for spore formation also may be required for growth.

For all species, not all cells sporulate, regardless of the conditions. There is no information as to why some cells in a culture sporulate and others do not.

There is agreement that spore formation begins after the exponential growth phase, during the stationary phase. As in all activities, the genes of the cells control sporulation. It is thought that during rapid vegetative growth, the spore genome is repressed. When growth slows, as in the stationary phase, the repression is removed somehow, due to an unknown factor.

Spore formation can be arbitrarily divided into seven stages: 1) development of axial chromatin filament; 2) spore septation; 3) engulfment of the spore protoplast; 4) cortex formation; 5) coat formation; 6) maturation; and 7) the free spore stage.

An organism can initiate sporulation without completing the process, and reverting to a vegetative cell. However, there is a stage when the cell becomes irreversibly engaged in sporulation and there is a commitment to complete the process. The composition of the substrate and the temperature of sporulation can affect the resistance characteristics of the resultant spores.

The spore consists of a core surrounded by several layers of mucopeptide and proteinaceous outer coats (Aronson and Fitz-James 1976). Some spores have hairlike fibers on the surface.

The main characteristic that is important to food microbiologists is the resistance of spores to heat, radiation, chemicals, desiccation and freezing (Gould 1977). Quite commonly, the heat resistance of spores is about 10^5 times and radiation resistance is about 10 times more than that of the corresponding vegetative cells.

Spores can be stored for long periods and retain their ability to germinate and produce vegetative cells.

Germination.—Since spores are dormant, they must be converted to vegetative cells to be important in food spoilage or toxin production. The conversion of the heat resistant spore to vegetative cells would allow less severe thermal processes for food preservation.

Spores may need a conditioning treatment prior to germination. This is called activation and may be induced by aging, heating, radiation, altering the pH or with chemicals. Low numbers of viable spores do not always germinate immediately. Thus, canned foods which apparently pass short storage tests to determine potential spoilage, after extensive storage can show spoilage.

Activation is a reversible process and, if conditions do not allow germination, the spore reverts back to its dormant state. Although some spores may germinate without heat shock, low levels of heat treatment, such as 65°-85°C for 10 to 30 min, will activate most spores and induce germination. The heat treatment used depends upon the species and strain. Spores of species such as *C. botulinum* type E strains are heat sensitive and should not be heated above 70°C. Higher temperatures (110°-115°C) for 3-10 min can be used to activate spores of the thermophilic *B. stearothermophilus.* Spores normally needing heat activation can germinate without heat in a medium containing calcium dipicinolate. The addition of calcium dipicinolate also increases the heat resistance of spores. L-alanine induces germination in some species. D-alanine inhibits germination. Sodium nitrite is regarded as an inhibitor of spore germination. Water is needed for activation. Spores are not activated when heated in glycerol.

During germination, the bright, refractile spores become dark; dipicolinic acid is released and the cortex disintegrates. Although respiration in the spore is undetectable, there is an abrupt onset of respiration during germination. There is activity of a variety of enzymes, typical of the vegetative form. The spores lose their resistance to heat, desiccation, chemical agents, electric shock and hydrostatic pressure. The germinating spores show a temporary rise in resistance to ultraviolet light and ionizing radiation followed by a rapid fall in resistance.

Although pasteurized milk supported germination of *B. cereus*, raw milk supports little or none (Wilkinson and Davies 1973). This helps to explain the spoilage defect due to *B. cereus* in pasteurized but not in raw milk. This inability to germinate in raw milk might be due to natural inhibitory agents found in raw milk that are inactivated by pasteurization.

Outgrowth.—The development of a vegetative cell, through the first cell division, is called outgrowth. Outgrowth proceeds by the swelling of the spore, emergence from the spore coat, elongation and cell division. One can follow germination and outgrowth by determining synthesis of RNA, proteins and DNA, in that order.

Outgrowth occurs when germination takes place in a substrate capable of supporting vegetative growth. If the germinating medium is not sufficient to support growth, either development is stopped or the outgrowing cell may form a second spore with no intervening cell division. This cycle of spore, cell, spore, is called a microcycle. Microcycle sporulation can be induced by dilution of an acceptable medium or by suspending germinated spores in a glucose-free medium.

The effects of various factors on the germination and outgrowth is important in food microbiology. It is the ability to determine viable from non-viable spores that allows us to establish thermal processes that will destroy spores in food.

Bacillus.—These organisms are usually Gram positive rods, but older cultures may appear as Gram negative. The majority are motile, produce catalase and produce acid but not gas from glucose.

The cells in this genus vary from strict aerobes to facultative anaerobes. The nutrient requirements vary from simple to complex. There are psychrotrophs, mesophiles and thermophiles. Thus, the minimum temperature varies from −5° to about 45°C and the maximum temperature for some species is 25°C and for others up to 75°C. The minimum pH for growth varies from pH 2.0 for *B. acidocaldarius* to pH 7.5 to 8.0 for *B. alcalophilus*. The salt tolerance is 2% or less for some species while others can grow in 25% salt. Due to the diversity of the species, there have been suggestions that the genus should be divided. Five genera have been proposed, but this differentiation has not been adopted.

Bacillus can be found in soil, water, fecal material, decaying materials and in various foods. Ingredients can serve as a source of *Bacillus*. Spices, flour, starch and sugar have been incriminated as sources of spores for contamination of fermented sausages, bread and canned foods. The resistance of the spores of *Bacillus* to various agents makes these organisms important in food preservation. Some species such as *B. subtilis*, decompose pectin and polysaccharides of plant tissue, causing spoilage of fresh plant products. Low acid canned foods are spoiled by *B. stearothermophilus*, and *B. coagulans* causes spoilage of tomato products. *B. cereus* is involved in foodborne gastroenteritis and *B. anthracis* causes anthrax of both animals and man.

Besides potential spoilage and health hazards, some species of *Bacillus* can be useful in foods. The bacilli are a source of proteolytic enzymes that might be used to clot milk for cheese production. Some bacilli have been suggested for use in the production of single cell protein. Some species are insect pathogens which makes them useful in food production.

Clostridium.—The species in this genus are divided into four groups on the basis of spore position and gelatin liquefaction (Buchanan and Gibbons 1974). The cells are usually Gram positive, at least in the early stages of growth. They are catalase negative and, except for a few aerotolerant species, most are strictly anaerobic.

The nutritional requirements of clostridia vary. *C. butyricum* can grow with ammonium as the source of nitrogen, and the vitamin biotin. On the other hand, *C. perfringens* requires more than 20 amino acids and vitamins. The critical tolerance of NaCl is 2.5-6.5%. Sodium nitrite at 0.5-1.0% inhibits these organisms. The lethal chlorine concentration is 2.5 μg/ml.

The principal metabolic products of a species can be determined by chromatography (column, thin-layer or gas). In conjunction with other tests, gas chromatography is a valuable aid in differentiating the clostridia.

The primary source of clostridia is soil. They are found in the intestinal tract of man and animals, as well as various foods.

The species in this genus include two involved with foodborne illness, several that cause food spoilage and many of no concern to food microbiologists. Some are free-living nitrogen fixing organisms, some cause serious illness (tetanus, gas gangrene) and others are used to produce commercial chemicals such as butyric acid, butanol, acetone and enzymes.

The spores of some species are very heat resistant and may survive the heat treatment of canned foods. If the surviving spores can germinate and the vegetative cells grow, spoilage will result. If *C. botulinum* spores survive the heat treatment, germination and outgrowth of the spores may result in the production of potent toxins.

Spoilage can be manifested as gas production causing swelling of cans, gassiness of foods such as in cheese, or proteolysis by the proteolytic clostridia such as *C. sporogenes*.

Desulfotomaculum.—The species in this genus are similar to the clostridia. However, they are Gram negative and have a higher DNA base composition (GC = 41 to 46 moles % as compared to 23 to 43 moles %). The clostridia do not reduce sulfate whereas these organisms reduce sulfur compounds (sulfates, sulfites and other reducible sulfur compounds) to H_2S.

D. nigrificans is a thermophilic spore former that causes sulfide spoilage of canned foods.

Gram Positive Asporogenous Rod-shaped Bacteria

In this group, the family Lactobacillaceae and the genus *Lactobacillus* are important to food microbiologists.

Lactobacillus.—These organisms are straight to curved rods occurring singly or in chains. The rods vary from long and slender to short coccobacilli. This is shown in Fig. 3.6 and 3.7. Generally they are nonmotile. Although considered to be Gram positive, as the culture ages, the cells may become Gram negative. Growth is enhanced by 5-10% CO_2. These organisms generally have complex nutrient requirements. Both homofermentative and heterofermentative types are in this genus.

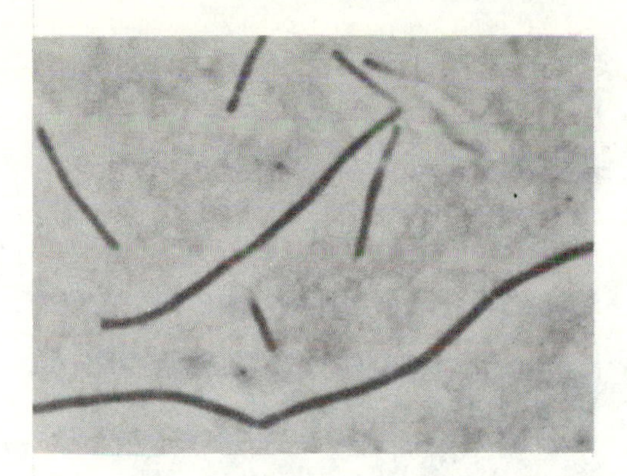

Courtesy of Pederson (1979)

FIG. 3.6. *LACTOBACILLUS BULGARICUS*, THE HIGH ACID-PRODUCING LONG ROD LACTIC ACID BACTERIUM OF SOUR MILK PREPARATIONS SUCH AS YOGURT (Magnified Approx. 2250X)

Lactobacilli are found in plant and animal material, in various places in the body of man and warm-blooded animals (including the intestinal tract) and in various foods (dairy products, grain, meat products, meat curing brines, beer, wine, fruits and fruit juices, sourdough, pickles, sauerkraut and olives).

The presence of lactobacilli in the gut is considered to be desirable and beneficial. Hence, the eating of yogurt, other fermented products, or even cultures, has been suggested. The main beneficial species is *L. acidophilus*. It has been suggested that 10^{11} cells should be ingested every day. This organism has been credited with amazing preventive or curative powers involving an unbelievable number of ills. Part of the therapeutic value has been associated with the interaction and inhibition of other organisms by the lactobacilli.

In food microbiology, the lactobacilli are useful but also cause spoilage. They are useful in fermentations in which lactic acid production is desirable, whether in plant or animal products (sauerkraut, pickles, olives, fermented sausages). Some lactobacilli form slime and spoil sugar cane. Others cause green discolorations of sausage products. Vacuum packaged meat becomes sour due to lactobacilli. Even in fermented vegetables, they may be a problem, such as causing pink sauerkraut or bloater formation in fermented cucumbers. The lactobacilli can cause spoilage of vinegar preserved products, such as catsup and mayonnaise.

Due to their complex nutrient requirements, they are used in assays of foods for vitamins, amino acids and other nutrients. The lysine-excreting

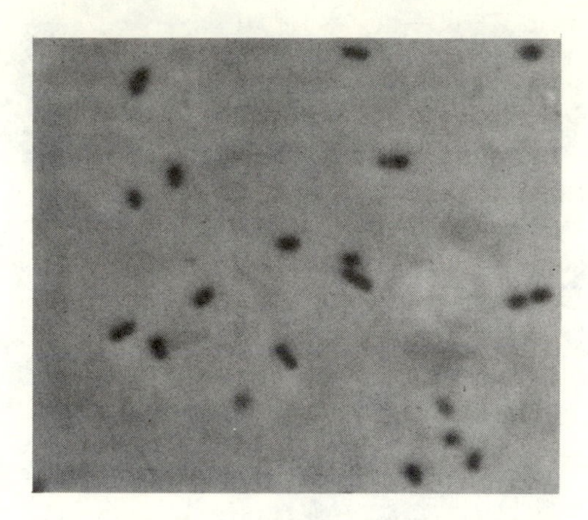

Courtesy of Pederson (1979)

FIG. 3.7. *LACTOBACILLUS BREVIS*, THE COMMON HETEROFERMENTATIVE SPECIES OF THE GENUS (Magnified Approx. 2800X)

mutants of lactobacilli have been suggested for use in food and feed enrichment (Sands and Hankin 1974). Vandercook and Smolensky (1976) suggested using *L. plantarum* to detect adulteration of orange juice with imitation orange beverages.

Actinomyces and Related Organisms

This group of bacteria includes the coryneforms with the genera *Corynebacterium, Arthrobacter* and *Kurthia*; two genera of uncertain affiliation (*Brevibacterium* and *Microbacterium*); the family Propionibacteriaceae with the genus *Propionibacterium*, and the order Actinomycetales are included in this group.

The organisms in the order Actinomycetales include the families and genera as follows: Actinomycetaceae (*Actinomyces*); Mycobacteriaceae (*Mycobacterium*); Nocardiaceae (*Nocardia*); and Streptomycetaceae (*Streptomyces*).

Quite often, the isolated organisms are not characterized beyond a group designation. Thus, in the literature, there is referral to coryneforms without further designation of types.

The organisms in the order Actinomycetales, although different, have some common characteristics. Quite often, these organisms are simply called actinomycetes whether they are *Streptomyces, Nocardia, Myco-*

bacterium or *Actinomyces*. Thus, in this text, they are discussed as one group rather than as individual genera.

Corynebacterium.—This genus is divided into three sections: 1) human and animal parasites and pathogens; 2) plant pathogens; and 3) nonpathogenic.

The coryneform bacteria are characterized by their pleomorphism. The *Corynebacterium* cells are straight to slightly curved rods but have a tendency to form club and pointed shapes (Fig. 3.8). Generally, they are not motile and are Gram positive. The best growth is aerobic but they can grow in anaerobic conditions.

Courtesy of Weiser et al. (1971)

FIG. 3.8. *CORYNEBACTERIUM DIPHTHERIAE*

Characteristic features are the granules in the rods.

These organisms are widely distributed in nature. They are found in water, soil, plants and animals. These organisms can be found in food products derived from both animals and plants. They have been associated with spoiling food, but it is doubtful they are primary spoilage organisms.

Arthrobacter.—Organisms in this genus show considerable pleomorphism. The coccal form may appear as spheres, ovoid or slightly elongated. When large cocci are transferred to a fresh medium, from one to three (seldom four) germination tubes arise from the cells. These develop into rods which vary in size and shape. Upon aging, the rods change almost completely into cocci. The cocci are Gram positive and the rods have Gram positive granules surrounded by Gram negative cellular material. They are catalase positive.

The organisms are found in soil, in and on meat and poultry products, in milk products, dairy waste, activated sludge, brewery and fish slime.

According to Buchanan and Gibbons (1974), the organism referred to as *Brevibacterium linens*, which is important in cheese, is related to and should perhaps be placed in the genus *Arthrobacter*.

Brevibacterium.—Although this genus is listed in Bergey's Manual (Buchanan and Gibbons 1974), no species are recognized. It is believed the formerly recognized species belong in other genera, primarily *Arthrobacter*, and possibly *Corynebacterium.*

Brevibacterium linens is important in flavor production in cheese, especially limburger cheese. Since no other name has been proposed for this organism, *B. linens* will be used in this text.

Microbacterium.—This genus is listed in Bergey's Manual, but no species are recognized (Buchanan and Gibbons 1974). It is thought that organisms in this genus belong to other genera such as *Corynebacterium, Arthrobacter* or *Kurthia.*

Organisms called *Microbacterium* have been isolated from pork, beef, poultry, eggs and dairy products. They cause flavor deterioration of some meat products. Due to their thermoduric nature (survive 72°C for 15 min), they can be found in pasteurized milk. They produce lactic acid which is not desired in fresh pasteurized milk.

Kurthia.—These Gram positive cells are regular, unbranched rods in young cultures, but become coccoid in older cultures by fragmentation of the rods. The cells are strict aerobes.

These organisms are found in intestinal contents, stagnant water, fresh and spoiling meat and meat products, meat processing plants and milk. According to Gardner (1969), *K. zopfii* is not known to cause spoilage of refrigerated meat, but its presence indicates that the meat was exposed to higher than refrigerator temperatures during processing, distribution or retailing. In meat held at 2°C, *Kurthia* is overgrown by *Pseudomonas* species and other organisms. Although present in food, their importance might be only as an indicator of mishandling.

Propionibacterium.—The organisms in this genus are generally pleomorphic, Gram positive, nonsporeforming rods. The rods may be diphtheroid or club-shaped and the cells in some cultures may be coccoid, elongate, bifid or branched.

They are anaerobic to aerotolerant. Certain strains can grow in 6.5% salt.

Propionibacters are found on humans and in the intestinal tract of man and animals. Gray and Johnson (1976) used the anaerobic roll tube method to isolate *P. acnes* from cheddar cheese. During fermentation, proprionibacters produce propionic acid and acetic acid, with lesser amounts of other organic acids. The species of most interest in food microbiology is *P. freudenreichii* subsp. *shermani.* This is used in the manufacture of Swiss cheese. Due to the production of propionic acid and CO_2, the organism is responsible for the characteristic flavor and the eyes in this cheese. These bacteria synthesize large quantities of vitamin B_{12} and produce propionic acid, and can be used in the commercial production of these compounds.

Actinomycetales.—This is an order of bacteria that has organisms with a fungal morphology. They may form a substrate mycelium, an aerial mycelium or both. Some may show complex mycelial structures with conidia and sporangia. Diphtheroid cells or branched rods are common. Although certain growth characteristics might indicate these organisms are fungi, the cells are procaryotic, so are classified with bacteria.

This order of organisms includes aerobic, facultative and anaerobic types. The overall growth range is 10° to 50°C and for pH is 5 to 10. The salt tolerance varies from 1 or 2% up to 10% for some strains.

Many of these organisms, especially the streptomyces, produce antibiotics. Some of the antibiotics, like streptomycin, inhibit bacteria and some, like cycloheximide, inhibit fungi.

The actinomycetes have been reported in flour and in wheat. Lyons *et al.* (1975) isolated Actinomycetales from whole corn and corn products (flour, grits). Of 345 strains isolated, only 3 were not classified as *Streptomyces.*

Mycobacterium tuberculosis causes tuberculosis. To help control this disease, pasteurization of milk was inaugurated. Today, tuberculosis is primarily an airborne disease, rather than foodborne. *Mycobacterium* strains are found in raw milk, oysters, pork and vegetables sprayed with sewage effluent (Hosty and McDurmont 1975; Thoen *et al.* 1975; Van Donsel and Larkin 1977). None of the isolates was identified as *M. tuberculosis*, but other potential human pathogens were detected. The porcine tissue that was contaminated had been condemned by meat inspectors during examination of carcasses.

Besides the possibility of being a health hazard or causing defects in food, some Actinomycetales may be useful. *Streptomyces* produces extracellular α-galactosidase (Lyons *et al.* 1969). According to these workers, this enzyme may be useful during beet-sugar processing (aids in crystallization of sucrose) and be used for determining raffinose and for eliminating the agent in beans that induces flatulence. *Streptomyces* was suggested as a possible source of single cell protein, protease enzymes and glucose isomerase.

Rickettsias

In this grouping, the family Rickettsiaceae includes the genus *Coxiella.*

Coxiella.—These cells are short rods, occasionally appearing as diplobacilli or spheres. There are no flagella or capsules. Although considered to be Gram negative, with certain conditions the cells appear to be Gram positive. They grow in the vacuoles of the host cell, rather than in the cytoplasm or nucleus. They resist drying and elevated temperatures. They do not grow on agar, but will grow on the yolk sac of chick embryos.

Since these organisms do not grow outside of a host cell, they are not important in food spoilage. However, food can serve as a carrier of the organisms so that they may infect humans. *Coxiella burnetii* is the causative agent of Q fever. Infected cows, sheep and goats shed the organism in their milk, which is the food source for human infection if raw milk is consumed. This organism is more heat resistant than the vegetative cells of most pathogens. Therefore, the time and temperature for pasteurization of milk are determined primarily to destroy these organisms. Heating at 62.8°C for 30 min or 71.7°C for 15 sec is adequate to eliminate *C. burnetii* from milk.

Uncertain Organisms

In the 8th Edition of *Bergey's Manual of Determinative Bacteriology* (Buchanan and Gibbons 1974), many organisms that were named in the food literature lost their legitimacy. Some have disappeared, not even being listed in the index. Although some names were said to be incorrect, no new names were proposed.

When possible, the bacteria in this text are named according to the listing in Bergey's Manual. However, due to the lack of certainty, it is necessary to use some names that are not official in the 8th Edition of Bergey's Manual.

MOLDS

Molds are part of a larger group of microorganisms called fungi. The exact number of fungi is not known. Some fungi have not been isolated and identified, while others have been given more than one name. The fungi are ubiquitous, being found everywhere. Soil, air, water and decaying organic matter are prime sources.

Fungi are heterotrophic organisms that lack the definite root, stem or leaves of higher plants. They possess a thallus and are called thallophytes. They are differentiated from the algae and higher plants by their lack of chlorophyll. Hence, they are saprophytic or parasitic. They differ from bacteria by their more complex structure and greater size. The fungi may be multicellular or unicellular.

The fundamental structural units of molds are filaments or tubes called hyphae. By formation of crosswalls or septa, some hyphae form chains of cells that are septate. Others may not form septa and the hyphae are nonseptate or coenocytic. The septa have pores that allow the movement of cytoplasm from one cell to another. As the hyphae elongate, they intertwine. A mass of these intertwined branched hyphae is called a mycelium. Part of the mycelium grows into the substrate and absorbs food. This is known as the vegetative mycelium. The mycelium that remains in the air above the substrate and bears spores is called the aerial or reproductive mycelium. When the spores find a proper substrate, the cycle is repeated.

Molds can reproduce sexually, asexually or by both systems. A fungus that has a sexual phase is known as a perfect fungus, while one that has no sexual phase is an imperfect fungus.

There are several types of spores formed by the fungi. The asexual fungi produce spores directly from or by the mycelium. These are called thallospores, conidiospores or sporangiospores. There are three types of thallospores: 1) blastospores; 2) chlamydospores; and 3) arthrospores. Sexual spores include oospores, zygospores, ascospores and basidiospores.

Enumeration

With bacteria, one colony originates from a single cell, but a fungal colony may result from a single cell, a spore, a piece of mycelium or from a number of cells. Thus, there is a poor one-to-one relationship for fungi. The problems involved with estimating growth by evaluation of the mycelium were discussed by Calam (1969) and Sutton and Starzyk (1972).

In the past, media for enumerating molds and yeasts were acidified to pH 4-5. The low pH inhibited bacteria and allowed the fungi to grow. However, fungi that have experienced a sublethal treatment do not grow on these acidified media. Newer media incorporate antibiotics to inhibit bacteria and the media pH is near 7.0 (Mossel *et al.* 1975; Nelson 1972). Microscopic methods are used to enumerate mold filaments in tomato products and other canned fruits and vegetables (AOAC 1975).

Fungi in Food

Several types of fungi have been isolated from food. The extent of contamination is influenced by the prevalence in the environment. *Penicillium* and *Aspergillus* enjoy favorable conditions throughout the year while some other fungi are limited to warm temperatures. *Cladosporium* and *Alternaria* are prevalent during the summer and early autumn.

Although important in all types of foods, molds are more apt to cause spoilage or become a health hazard in foods such as grain, flour or nuts with low water activity or fruits with a low pH.

Fungi are an important cause of spoilage of stored seeds and grains. Christensen and Kaufman (1974) believe that throughout the world, fungi rank second only to insects in causing loss of stored products. In the developed countries where insects and rodent controls have been implemented, fungi cause more destruction of stored products than any other agent.

Molds isolated from corn in the field belong primarily to the genus *Fusarium* (Hesseltine and Bothast 1977). *Aspergillus* was reported in over 50% of the samples of field corn (Hesseltine *et al.* 1976). Other field molds reportedly include the genera *Alternaria, Helminthosporium, Cladosporium, Rhizopus* and *Absidia.* For growth, these fungi require a moisture content of the grain of 24-25%. The storage fungi are primarily *Aspergillus* and *Penicillium*, which can grow at lower moisture levels (14-18%). The fungal count of corn meal varies from 410 to 200,000 fungi per gram (Vojnovich *et al.* 1972). The most common genera on rice are *Aspergillus* and *Penicillium.* According to Ito *et al.* (1971), the only species of mold that can grow on stored rice with 14-15% moisture is *Aspergillus restrictus.* Rice with 17% moisture allows the growth of several species of *Aspergillus* and *Penicillium.*

The composition of tree nuts and peanuts allows molds to grow. If pecans are gathered as they fall from the tree, they are practically free of molds. If they are left lying on the damp ground, the kernels can become moldy. Schindler *et al.* (1974) identified nine genera of mold from inshell pecans, while Huang and Hanlin (1975) reported 44 genera and 119 species from freshly harvested and in-market pecans. In both surveys, *Penicillium* was the predominant genus.

Importance of Molds

Molds can be considered as spoilage organisms in food products. Although we often associate the appearance of mold as an indication of spoilage, these organisms can degrade products before growth is evident to the naked eye.

Although molds associated with food are not considered to be pathogenic, some produce mycotoxins. These toxic substances pose a potential health hazard to humans.

Molds can be useful in the processing of many foods such as cheese. Their enzyme systems can be isolated and used in many food processes. Some fungi are used to convert wastes into usable food (protein) for animal food or human use. Also, they are a source of vitamins.

The production of antibiotics by molds is well known. Antibiotics have been of great value to humans in medicine and there have been investigations to use antibiotics as food preservatives.

There are obviously many inert molds associated with foods. They simply cannot grow due to an unsatisfactory environment or the overgrowth by bacteria which cause spoilage before the molds get a chance to multiply.

Zygomycetes

The organisms in this group are terrestrial fungi. They produce asexual (sporangiospores and conidiospores) or sexual (zygospores) spores. The hyphae usually contain no septa, although older hyphae may have septa. They are usually saprophytic.

The organisms of interest in food microbiology are in the order Mucorales. The family Mucoraceae includes the genera *Mucor* and *Rhizopus.* The family Thamnidiaceae includes the genus *Thamnidium.*

Mucor.—Like other dimorphic fungi, species of *Mucor* can develop as either individual spherical cells that multiply by budding (as yeasts) or they form typical mycelia of coenocytic hyphae.

In normal aerobic conditions, growth is usually filamentous. A short distance behind the growing hyphal tips, the mycelium differentiates to form sporangiophores. These contain the asexual sporangiospores. Sexual hyphae may form zygophores. Compatible zygophores fuse in mated pairs and, as a result, form zygospores. This sexual spore formation is rarely observed. The zygospores are resting spores and are difficult to germinate.

Stoloniferous growth does not occur in *Mucor* which differentiates this genus from *Rhizopus.* These molds are shown in Fig. 3.9 and 3.10.

Organisms in this genus occur in soil, manure, fruits, vegetables, stored grain and other foods. *Mucor* species are used in the Orient in food fermentations. *M. pusillus* produces an extracellular protease which has milk-clotting activity.

Rhizopus.—These organisms have coenocytic mycelia, and spread by stolons. They usually reproduce asexually. Sporangiophores develop from the nodes at which thick tufts of rhizoids develop. The sporangiophores bear a terminal large globose sporangium containing many spores. When the spores mature, they are released by the rupture of the sporangial membrane.

These are common spoilage organisms of various stored foods. *Rhizopus stolonifer* is the common bread mold. Pectinolytic enzymes are secreted by *Rhizopus* species. The degradation of pectin results in the soft rot of various plant products.

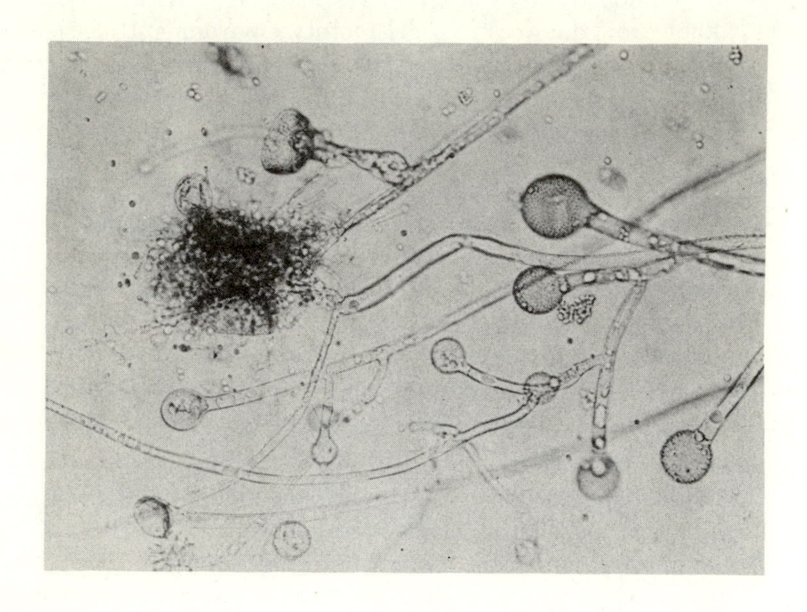

Courtesy of Continental Can Co.

FIG. 3.9. HYPHAE AND SPORANGIA OF *MUCOR*

Due to the heat stability of the pectinolytic enzyme, the infection of fresh apricots with *R. stolonifer* before canning can cause post-processing softening of the canned fruit.

Rhizopus produces high yields of fumaric acid from fermentable sugars. Species are used in the production of fermented foods, such as tempeh. Harris (1970) discussed the use of a *Rhizopus* fermentation to upgrade the nutritional value of cassava flour.

Thamnidium.—Organisms in this genus have coenocytic mycelia. They can form spores on terminal sporangia or on sporangiola. Low temperature and light induce the formation of sporangia as opposed to sporangiola. Zygospores are produced at 6°-7°C, but not at 20°C. *Thamnidium elegans* is used in a patented process to improve the flavor and tenderness of beef.

Thamnidia are found on meat as well as soil and animal excrement. They grow on refrigerated meat causing a defect referred to as whiskers.

Ascomycetes

Both molds and yeasts are included in this class of fungi. The mycelia

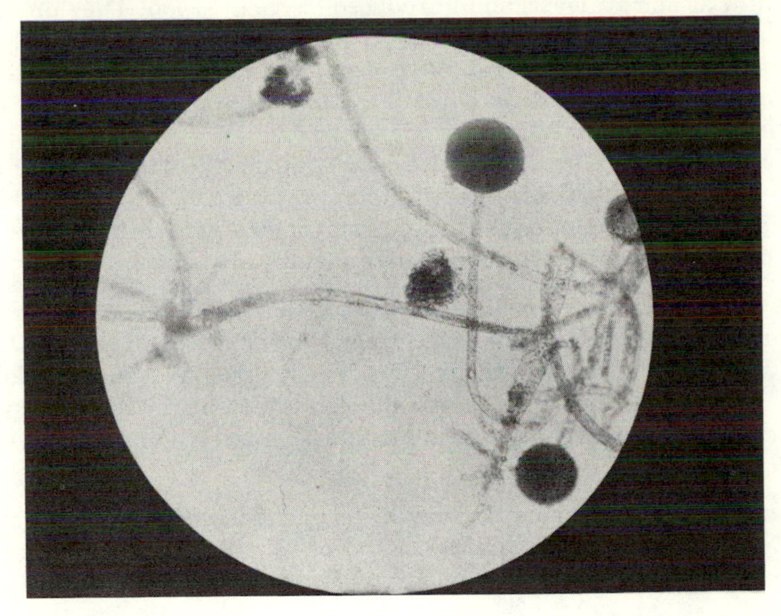

Courtesy of Continental Can Co.

FIG. 3.10. HYPHAE AND SPORANGIA OF *RHIZOPUS*

are septate, but the septa do not prevent the movement of cytoplasm or nuclei from cell to cell.

Ascomycetes develop sexual ascospores in a sac-like structure called an ascus. The spores are the product of the fusion of two nuclei, followed by meiosis.

Besides this sexual phase, they may grow by extension of the hyphal tip, or have an imperfect state in which asexual spores are produced. Those fungi for which no perfect state has been observed are listed in Deuteromycetes (Fungi Imperfecti). When a sexual phase is found for a fungus in Fungi Imperfecti, it is given a name according to the structure and form of the perfect state. This name takes precedence over the imperfect, asexual state. It is thought that from ⅓ to over ½ of the Fungi Imperfecti belong to the Ascomycetes.

There are between 1,900 and 2,000 genera in Ascomycetes. Only a few are of importance in foods.

Byssochlamys.—Besides sexual reproduction, the two species, *B. fulva* and *B. nivea*, reproduce asexually by the formation of chains of conidia derived from phialides on the aerial mycelium.

The ascospores of these organisms are heat resistant. These organisms can grow at low pH levels and in reduced oxygen tension. They produce strong pectolytic enzymes. With this combination of characteristics, the organisms, especially *B. fulva*, have been implicated in the spoilage of canned fruit and fruit juice. Growth is sometimes accompanied by gas which causes a slight bulging of the can.

Besides spoilage, *B. fulva* produces a mycotoxin which is toxic to brine shrimp, chicken embryos and rats (Kramer *et al.* 1976). Chu *et al.* (1973) studied a rennin-like enzyme (byssochlamyopeptidase A) that might be useful in milk clotting for cheese manufacture.

Claviceps.—The species *Claviceps purpurea* is of interest to food microbiologists since it produces toxic alkaloids on cereals. These contain a tetracyclic ring called lysergic acid. When ingested, these alkaloids cause numerous symptoms, especially hallucinations. With improved grain handling, the illness is rare in humans. The last major outbreak was in 1951.

Neurospora.—The name *Neurospora* is derived from the characteristically ribbed ascospores. The ascospores of *N. crassa* remain viable for many years. Heat shock for 20 min at 60°C aids in the germination of the spores.

In asexual reproduction, branched chains of pink conidia develop from upright stalks. Budding of the terminal conidia of the chain produces further conidia. When the terminal conidia form two buds, the chain branches. *Neurospora* is shown in Fig. 3.11.

Neurospora crassa and *N. sitophila* are two well known species of this genus. Wild type strains of *Neurospora* have simple nutrient requirements. They have been used as tools in genetics and biochemical research. *Neurospora* can be found in warm, humid environments. *N. sitophila* causes problems in bakeries and is the red or pink bread mold. *N. sitophila* is used in the fermentation of red or orange ontjom, a fermented peanut press cake used in the Orient.

Courtesy of Weiser et al. (1971)

FIG. 3.11. *NEUROSPORA SITOPHILA*

Deuteromycetes

This heterogeneous group of fungi has branching, septate hyphae, and reproduces asexually by conidia or sclerotia. The organisms can be considered as ascomyces or basidiomyces that do not have sexual (perfect form) reproduction, or it is rare. When sexual reproduction is observed, the organism is assigned a name in Ascomycetes. However, this does not mean that all of the species of a genus in the Fungi Imperfecti are renamed. It is sometimes desirable to retain the imperfect name, as well as the taxonomic name of the perfect form.

There are various systems used to classify the Fungi Imperfecti. Rather than attempt to segregate them, those of importance in food microbiology are discussed in alphabetical order.

Alternaria.—These organisms are characterized by muriform, dark-colored spores. The aerial mycelia are described as wooly, gray, brown and olive-green (Fig. 3.12).

Alternaria is one of the most prevalent molds that causes spoilage of tomatoes in the field, attacking injured or weakened tissue. Due to darkening of the tissues, the defect is called black rot. Late harvested tomatoes are particularly susceptible to *Alternaria.* These organisms also are involved in the development of rancid or off-flavors in dairy products.

Aspergillus.—There are over 100 species in this genus. These organisms, along with penicillia, are known as storage fungi of grain. They discolor infected grain and reduce or destroy germination of the seed. These organisms can cause spoilage of a wide range of food products.

Certain aspergilli are very useful in food microbiology. Species such as *A. oryzae* are used to break down rice starch to glucose in the manufacture of saké and similar alcoholic beverages. Some strains are used in the production of shoyu (soy sauce) and miso (bean paste).

Strains of *Aspergillus* are used in the commercial production of citric, gluconic and gallic acids. Almost all of the commercial citric acid is produced by *A. niger* growing in sucrose solutions. The organisms are a source of amylase and pectinolytic enzymes. A proteolytic enzyme of *Aspergillus* is able to clot milk and might be a substitute for rennet in cheesemaking.

These fungi may be useful as a source of protein. The mycelium of *A. niger* grown on brewery wastes contained 29% crude protein and might be used as a feed supplement (Hang *et al.* 1975). Reade and Gregory (1975) described the use of *A. fumigatus* to convert the starchy root, cassava, to microbial protein for food or feed.

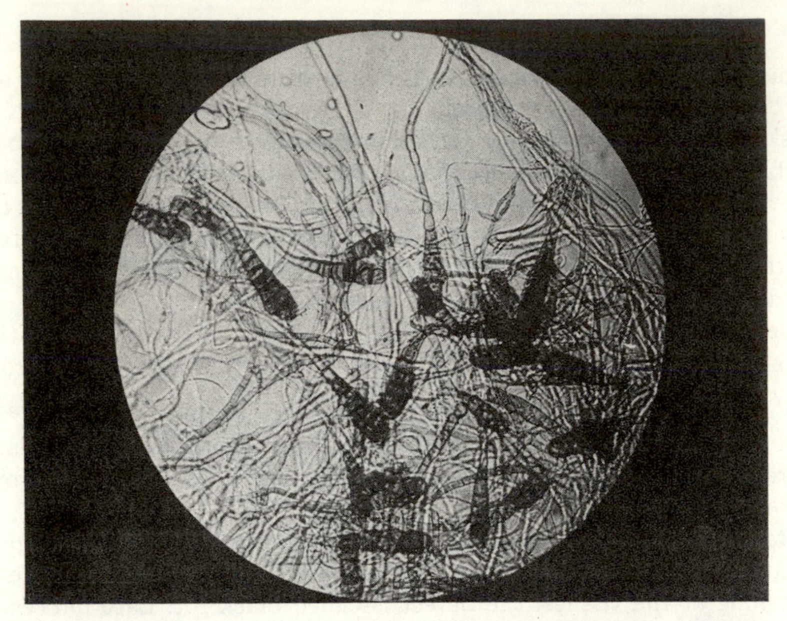

Courtesy of Continental Can Co.

FIG. 3.12. HYPHAE AND SPORES OF *ALTERNARIA*, A CAUSE OF ROT OF TOMATOES

Spores show the typical Indian-club shape.

Several species of *Aspergillus* produce substances that are toxic to other biological systems. Strains of *A. flavus* and *A. parasiticus* produce aflatoxins. There is considerable information that implicates aflatoxins with human illness.

The aspergilli are common contaminants of organic materials and soils. They are found on fruits, vegetables, stored grain, peanuts and other food products.

The appearance of *Aspergillus* is shown in Fig. 3.13.

Botrytis.—The asexual form of *Botrytis* reproduces by conidia. The conidiophores develop from a sclerotium, and are irregularly branched. The conidia occur on short sterigmata. The positions and numbers of sterigmata cause the conidia to appear in grape-like clusters (Fig. 3.14).

Some of the species in this genus are imperfect forms of the sexual form, *Botryotinia*. *B. cinerea* is the common species of *Botrytis*. It is the gray mold of various plants and plant products, especially lettuce, tomato, strawberry, raspberry and grape. It can be classed as a field mold, since it is a common soil contaminant and attacks fruits and vegetables in the field. It enters fruits, such as tomatoes, through cracks and injured areas.

Courtesy of Continental Can Co.

FIG. 3.13. *ASPERGILLUS*, SHOWING MYCELIA AND CONIDIAL HEADS

Cladosporium.—The one or two-celled conidia are formed in branched chains on conidiophores. The conidia can reproduce by budding, which causes the branching. The colonies are fairly thick and velvety. They may be colored green olive-green, black or brown. This organism is shown in Fig. 3.15.

Organisms in this genus are common in soil. They can grow on the connective tissue or fat covering of meat when refrigerated several days. Growth results in black spots on the meat. *C. carpophilum* causes peach scab, which is numerous dark, circular lesions on the fruit. The organisms are associated with stored grains and dairy products.

Colletotrichum.—Molds in this genus are involved with spoilage of foods. *C. circinans* causes onion smudge. Colored onions are resistant. *C. phomoides* causes anthracnose rot of tomatoes. Invasion by this organism is not dependent upon an injury to the fruit.

Fusarium.—The conidia produced by these organisms have various shapes such as cylindrical, oblong, globose, sickle-shaped, pear-shaped, ellipsoidal, spindle-shaped, sausage-shaped, crescent-shaped and oval (Fig. 3.16).

Courtesy of Continental Can Co.

FIG. 3.14. MYCELIA AND SPORES OF *BOTRYTIS*

The taxonomy of *Fusarium* was discussed by Booth (1975) and Joffe and Palti (1975). The colonies may be fluffy and spreading. The colors of the many species of *Fusarium* have been described as white, white-rose, pink, rose, rose-red, carmine, red, red-brown, brown, light-brown, yellow, ochre, peach, orange, blue, violet-blue and purple.

F. moniliforme causes a disease of rice which led to the discovery of gibberellic acid, a plant growth stimulant.

The fusaria are widespread in nature, being found in soil, decaying material and food. Some fusaria are associated with plant diseases. These

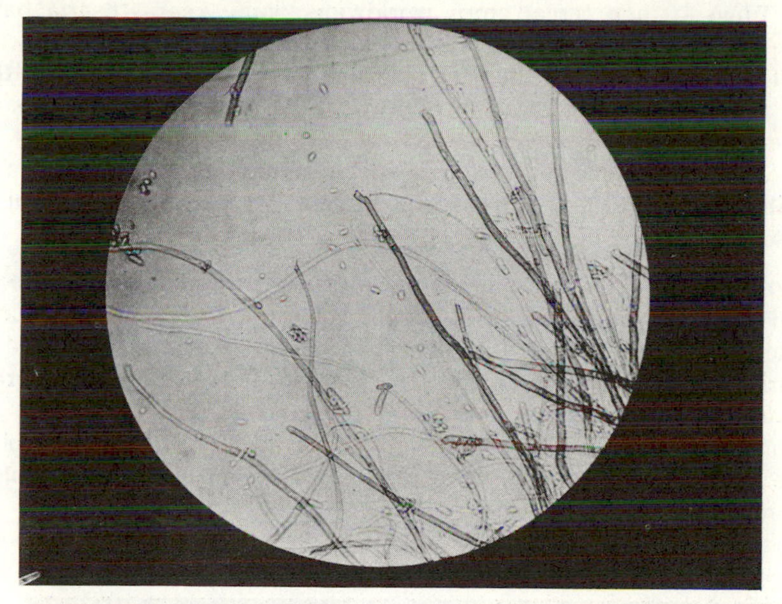

Courtesy of Continental Can Co.

FIG. 3.15. HYPHAE AND SPORES OF *CLADOSPORIUM*

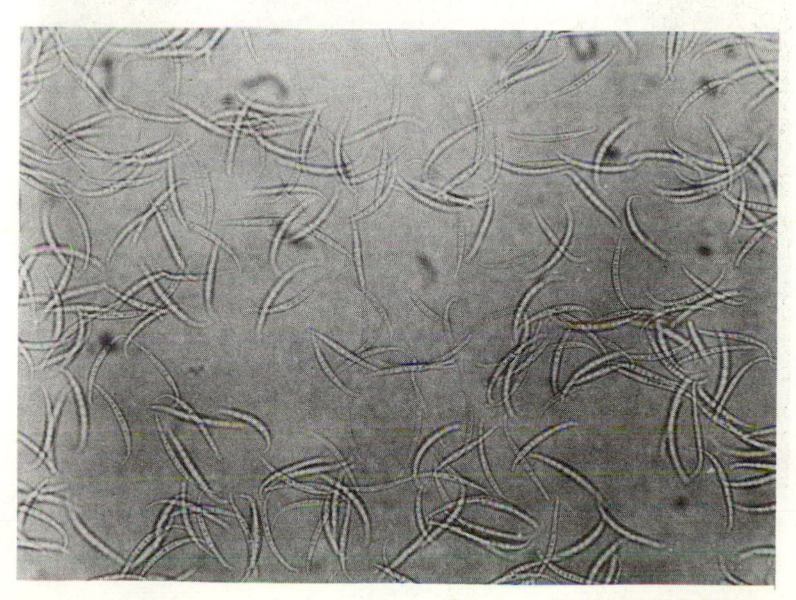

Courtesy of Continental Can Co.

FIG. 3.16. SICKLE-SHAPED CONIDIA OF *FUSARIUM*

field fungi, attack cereal crops worldwide. Some years fusaria have caused a 50% loss of wheat and other crops in Japan. *Fusarium* causes a rot of tomatoes, entering the fruit through insect or other damage on the skin. *F. solani* causes a decay of potatoes called powdery rot, dry rot or white rot.

The fusaria produce mycotoxins which affect various animals and possibly humans. When feed is highly infected with fusaria, the animals refuse to eat.

Geotrichum.—This yeast-like fungus grows rather rapidly at room temperature, forming white to cream colored colonies. The septate, branching hyphae fragment into chains of rectangular, barrel-shaped or spherical arthrospores that readily break apart. These arthrospores are a means of reproduction. *Geotrichum* is depicted in Fig. 3.17.

Geotrichum is a spoilage organism. It has been called "dairy mold" since it is found growing on dairy products. It also causes watery rot, a common spoilage of tomatoes.

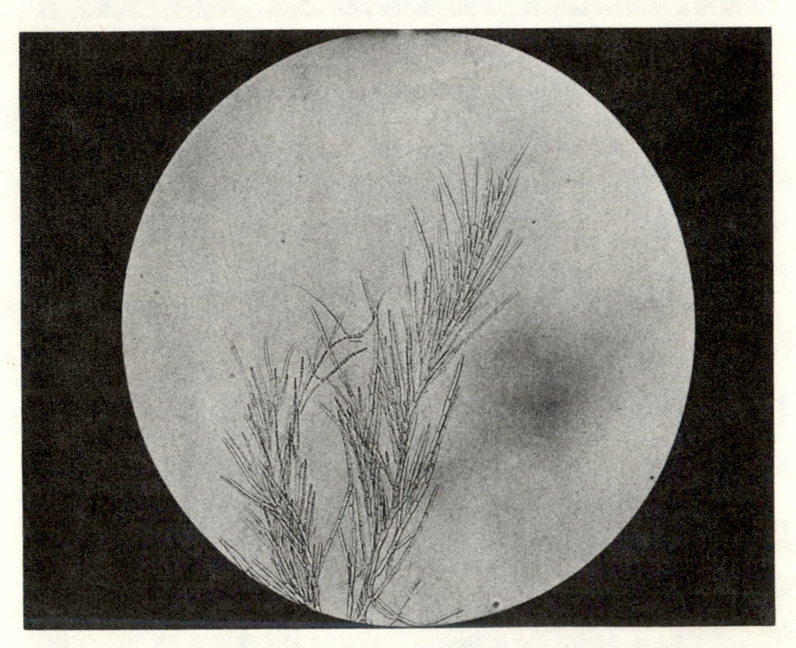

Courtesy of Continental Can Co.

FIG. 3.17. *GEOTRICHUM*—MACHINERY MOLD

G. candidum is called "machinery mold" since it will grow on equipment with attached food particles or juices. As the food product is processed, it becomes contaminated with the mold. Hence, the mold is found in many types of processed foods. The mold is killed by heat used in thermal processing of food, but the hyphae can be determined by microscopic examination. The presence of the mold in canned foods is considered to be an adulterant and indicates inadequate sanitation in the processing plant.

Helminthosporium.—The organisms in this genus have dark mycelia. The conidiophores usually are unbranched and arise in groups. The conidia are dark and ellipsoidal with three or more cells.

Organisms in this genus are common plant parasites, especially of cereal crops. In 1970, *H. maydis* (Race T), the causative agent of southern corn leaf blight, infected much of the corn in the United States. Although apparently not toxic, there was a great economic loss due to lower yield.

Penicillium.—There are many species of *Penicillium.* They are closely related to the aspergilli. The penicillia are characterized by branching of the conidiophore to form a brush-like conidial head (Fig. 3.18). The organisms in this genus can be divided into three groups on the basis of the type of conidiophore branching. The conidia may be white, green, gray-green, blue-green or yellow-green. Some species produce perithecia or sclerotia, which may be pink, yellow or orange.

Courtesy of Continental Can Co.

FIG. 3.18. MYCELIUM AND CONIDIAL HEADS OF *PENICILLIUM*

These organisms are widely distributed in nature and are found on many foods. They are important spoilage organisms. *P. digitatum* causes green mold rot and *P. italicum* causes blue mold rot of citrus fruit. *P. expansum* is a common spoilage (blue mold, soft rot) organism of various fruits (apples, pears, peaches, etc). *P. cyclopium* is a common species isolated from various stored fruits and vegetables.

This is one of the storage fungi of grain. The species most often encountered are *P. cyclopium* and *P. viridicatum*. This is probably because they are able to grow at relatively low levels of moisture and temperature. *P. martensii* and *P. viridicatum* are associated with blue-eyes, a blue-green discoloration of corn germs. Species of *Penicillium* are found growing on the fatty layer or connective tissue of meat that is stored in the refrigerator for several days, and on moldy bread.

Species of *Penicillium* are useful in various ways. Antibiotics produced by penicillia include penicillin. Some species are used in cheese manufacture. *P. camemberti* and *P. caseicolum* are important in Camembert, Brie and similar cheeses, while *P. roqueforti* is used in Roquefort, Gorgonzola and blue-veined cheeses. Schwimmer and Kurtzman (1972) suggested using *P. crustosum* to remove caffein from coffee. The penicillia produce enzymes such as glucose oxidase as well as proteins which can be used by the food industry.

Certain species of penicillia can be a health hazard. Some species have been associated with pulmonary and urinary tract infections. Penicillia produce mycotoxins. *P. islandicum, P. citrinum* and *P. citreoviride* were involved in "yellow rice disease" which caused several deaths of humans.

Although *P. roqueforti* is used in cheese processing, toxin producing strains of these species have been isolated (Scott *et al.* 1977) and roquefortine, a neurotoxin, has been isolated from blue cheese (Scott and Kennedy 1976).

Scopulariopsis.—This genus produces conidia in brush-like clusters similar to *Penicillium*. The colonies appear cottony and may be cream, yellow, brownish, chocolate brown or almost black. They are never green like the *Penicillium*. One characteristic of *S. brevicaulis* is that it produces a poisonous gas (diethylarsine) with a garlic-like odor when grown in the presence of arsenical compounds.

The perfect state (sexual form) is the genus *Microascus*. *S. brevicaulis* is the most common species of *Scopulariopsis*. It is found on all types of decaying matter, and unlike some molds, it grows well on high protein substrates.

Being proteolytic, *Scopulariopsis* is involved with the development of off-flavors in dairy products (especially Camembert cheese), and spoilage of meat. It has been detected on ham (both on the surface and in deeper tissues), on stored eggs and peanuts.

According to Bothast *et al.* (1975), the specific function that *S. brevicaulis* performs in the deterioration or synthesis of foods remains to be determined.

Sporotrichum.—The conidia are borne singly on very short projections that arise near the ends of the septate hyphae and do not occur in chains. The colonies are usually white, but may be yellow, gray, pink, red or green.

Sporotrichum is a common soil inhabitant. *S. carnis* grows at low temperatures (−5° to −8°C) and can cause a defect called "white spot" of refrigerated meat.

Sporotrichosis, caused by *S. schenckii* is a chronic, progressive infection of the skin and subcutaneous tissues.

S. thermophile has an optimum growth temperature near 40°C. It produces cellulase and has been suggested for use in converting cellulose to simple carbohydrates (Coutts and Smith 1976).

The use of *S. pulverulentum* as a potential food source was studied by Eriksson and Larsson (1975).

Trichoderma.—This genus has irregularly branched conidiophores. The conidia are produced in slime balls.

Species are common in soil and on organic matter. The nitrogen content of the cells is about 5% of which 60-70% is protein. Since the species *T. viride* is cellulolytic, and has a high protein content, it has been considered for converting cellulosic wastes into foods.

YEASTS

The yeasts have been defined as fungi in which the usual dominant, or conspicuous form is unicellular. The unicellular form gives yeasts an advantage over the mycelial form of molds. There is a greater surface to volume ratio that allows a higher metabolic activity. Also, the unicellular form is more readily distributed than is the mycelial form. Some yeasts produce a true mycelium by fission of cells that remain attached. A pseudomycelium, formed by budding, has constrictions between the cells.

There are some 350 species recognized and classified into 39 genera (Kreger-van Rij 1969; Lodder 1970). The classification of yeasts is based on morphological, cultural, sexual and physiological characteristics.

On the basis of method of reproduction, the yeasts can be separated into four groups. Only two of these groups contain yeasts involved with foods. One group produces sexual ascospores in asci and is in the class Ascomycetes (Fig. 3.19). These are the so-called true yeasts. The yeasts in the other group form no sexual spores and have no sexual life cycle. These are the false yeasts, Fungi Imperfecti, or Deuteromycetes.

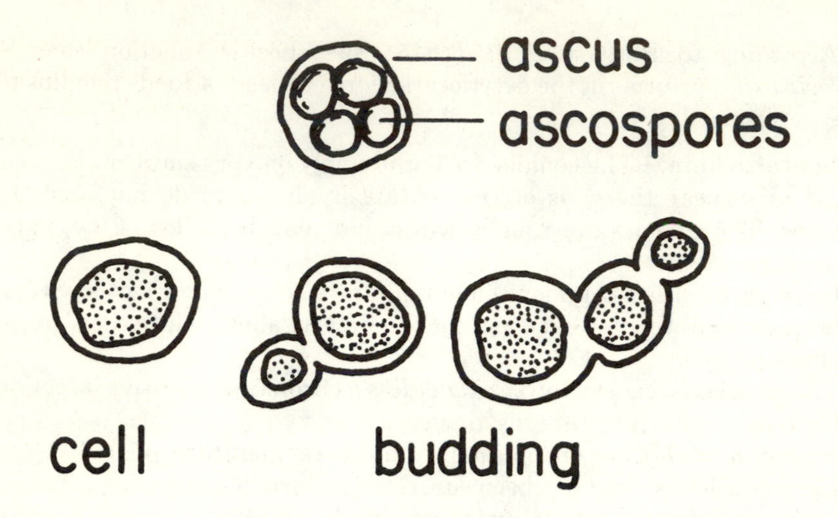

FIG. 3.19. YEAST CELLS SHOWING BUDDING AND ASCOSPORE FORMATION

Vegetative reproduction refers to asexual reproduction. All yeasts can reproduce asexually, and this is the only method for about 50% of the yeasts.

The usual vegetative reproduction is by budding. Some yeasts reproduce by fission or by an intermediate system called bud fission. If the new cell or bud appears at the short end of the mother cell, it is called polar budding. If a bud forms at both ends, it is bipolar budding. When buds appear at any place on the mother cell, it is multilateral budding.

Nearly all of the yeasts that produce sexual ascospores and are associated with food are in the class Ascomycetes, order Endomycetales, and family Saccharomycetaceae. For most yeasts, the maximum number of spores per ascus is four, but a few yeasts can produce eight. Strains of *Kluyveromyces* may produce large numbers (up to 100) of spores per ascus. These ascospores have many shapes, such as spheroidal, ovoid, reniform, elliptical, bean-shaped, cylindrical and needle-shaped. Some spherical or oval spores have a ledge in the middle or to one side which makes them appear as saturn-shaped or hat- or helmet-shaped.

The production of ascospores may play a role in survival or in adaptation to altered environmental conditions. However, the main purpose is the rearrangement of heritable qualities.

The vegetative cells may have shapes similar to those of ascospores.

Young yeast cells may be small and spherical and, as they grow older and larger, they may assume other shapes. The older cells become misshapen and scarred due to budding. Aged cells tend to become shriveled.

Some molds are dimorphic and have a yeast-like phase. Some yeasts produce a pseudomycelium or true mycelium. Hence, these organisms may be confused when a culture is examined, especially for borderline genera. The mold *Geotrichum* resembles the yeast *Trichosporon*. They both form a true mycelium and arthrospores. However, *Trichosporon* can reproduce by budding.

Most yeast cultures are cream, tan or gray. However, some are yellow, pink, red, green or brown.

The physiology of yeasts was discussed by Kreger-van Rij (1969) and MacMillan and Phaff (1973). Information about fermentation, assimilation and nutritive requirements aids in identification of yeast species. This information was compiled in Lodder (1970) and is too extensive to repeat. The heat resistance of various yeasts was reported by Put *et al.* (1976).

Yeasts are associated with nearly all types of food products. Foods such as fresh vegetables, meat, poultry and cheese often contain yeasts, but in these foods, bacteria usually outgrow the yeasts. When bacterial inhibitors are added, yeasts can dominate. Osmophilic yeasts such as *Saccharomyces rouxii* can tolerate high sugar concentrations. These yeasts are found in foods such as honey, molasses, sugar and fruit. Salt tolerant yeasts grow as films on brined food and on salted food and ham.

Methodology

Sublethally stressed yeast is recovered at maximum levels with a medium at pH 8.0 or over (Nelson 1972). Therefore, the use of potato dextrose agar (PDA) acidified to pH 3.5 is not satisfactory for enumeration of yeasts in many food products. To inhibit bacteria and allow yeasts to grow, a combination of antibiotics is used in place of acidification. Fluorescent microscopy can be used to detect yeasts in certain products (Aries 1976; Cranston and Calver 1974). After the yeasts are grown on agar surfaces, identification is sometimes desirable. Simplified, rapid methods are available for identifying yeasts (Huppert *et al.* 1975; Roberts *et al.* 1976; Segal and Ajello 1976).

Functions

Yeasts are involved with spoilage of various food products, particularly those containing sugars, brined foods and fruits (MacMillan and Phaff 1973; Peppler 1977).

Certain strains of yeasts are very important in the fermentation of alcoholic beverages, the baking industry, chemicals, single cell protein, sewage disposal and other uses.

Classification

In this text, the genera and species of yeasts are those listed in Lodder (1970). As more information accumulates regarding the relationships of yeast such as DNA base composition and serological reactions, there may be some revisions in the classification (Price *et al.* 1978).

The yeasts are divided into Ascomycetes (the ascosporogenous yeasts), and Deuteromycetes (the asporogenous yeasts).

Ascomycetes.—These yeasts produce sexual ascospores as well as reproduce asexually. Unless otherwise noted, asexual reproduction is by budding. Certain genera such as *Citeromyces, Dekkera* and *Endomycopsis* have been isolated from food and may be involved with spoilage or fermentation. However, due to space limitations these are not discussed in this text.

Debaryomyces.—The vegetative cells are often spherical to globose and reproduce by multilateral budding. The sexual ascospores, produced by conjugation of the mother cell and bud, are spherical or oval. There are usually one or two spores per ascus. Fermentation is weak, slow or absent. Nitrate is not assimilated.

D. hansenii has a high salt tolerance, being able to grow in media with 18-20% salt. It has been isolated from brined food and salted meat products and can use nitrite as the sole source of nitrogen. *Debaryomyces* species are perhaps the most widely distributed film forming yeasts associated with food brines.

Species of *Debaryomyces* have been associated with spoiled mushrooms, cheeses, cider, white wine, tomato puree, sausage and salted beans.

Hanseniaspora.—These yeast cells are diploid, lemon-shaped, ovoid or long ovoid. These yeasts are the ascospore stage of the genus *Kloeckera.* Budding is bipolar.

They have a vigorous fermentation, but have a low tolerance to alcohol (about 4-6%). All species require inositol and pantothenate for growth. Due to this need, they can be used to assay for these compounds.

These yeasts are found in the soil of orchards and vineyards and in insects. They have been isolated from grape berries, grapes and associated with fermenting spoiled fruit.

Hansenula.—This genus reproduces by multilateral budding and ascospores. A pseudomycelium or a true mycelium may be formed. These yeasts assimilate nitrate. Some species form phosphomannoses which results in mucoid colonies. The colonies may be white, gray-white, cream, slightly pink, pink or red.

Species of *Hansenula* have been isolated from foods such as grain, fruit, shrimp and brines of fermenting cucumbers and olives.

Some species and strains are useful. According to Batra and Millner (1974) certain species are used in India to produce kanji (a beer-like beverage) and murcha (rice wine). A strain of *Hansenula* grows on methanol to produce single cell protein (Levoney 1973).

Kluyveromyces.—These organisms reproduce by multilateral budding. The cells may be spherical ellipsoidal, cylindrical or elongate. Some species produce a red pigment, but the colonies are grayish-white, brownish-cream, yellowish-gray, sometimes tinged with pink. The organisms show a vigorous fermentation. They grow from 5° to 46°C.

This genus has a widespread distribution. They are found in various foods such as fruits, preserves, corn meal, grape must, milk and other dairy products. The species that ferment lactose, such as *K. lactis*, are found in dairy products. Some species are osmophilic. Organisms of this genus cause spoilage of foods, especially figs and dairy products. Certain strains have been suggested as a source of the enzyme polygalacturonase.

Pichia.—The cells have many shapes. Asexual reproduction is by multilateral budding with most species forming a pseudomycelium. There is limited formation of a true mycelium.

The colonies are slimy or pasty and colored yellowish-white, yellowish-brown, tan, white, cream, grayish-white or red.

These are common yeasts found in various sources. Their association with foods includes both fermentations and spoilage. Foods in which these yeasts are found and can cause spoilage include fruit, beer, dairy products, sorghum, and wine. They can occur as films on brined foods.

Saccharomyces.—This genus is considered to be a rather heterogeneous group of organisms. The cells are spheroidal, ellipsoidal, cylindrical or elongate (Fig. 3.20). Vegetative reproduction is by multilateral budding. Pseudomycelium may be formed but a true mycelium is absent.

On agar, the colonies are generally white or cream, with a typical yeasty odor.

The name *Saccharomyces* means sugar fungus. All species have a vigorous fermentation. Some species are among the most useful yeasts due to the production of alcohol (brewing) and carbon dioxide (baking). Strains of *S. uvarum* and *S. cerevisiae* are most often used in these fermentations. Besides these important uses, species and strains have been used to remove glucose from egg white prior to drying, to remove

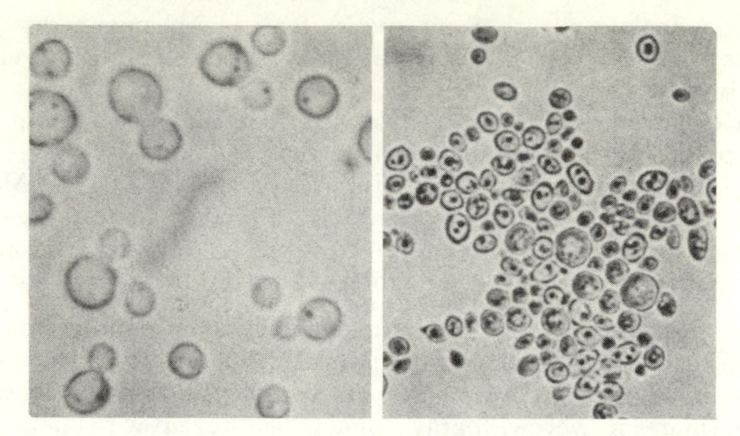

Courtesy of Amerine et al. (1972)

FIG. 3.20. YOUNG AND OLD CELLS OF *SACCHAROMYCES CEREVISIAE*

the mucilage layer from coffee beans during processing, to assay for vitamins (biotin, pyridoxine, pantothenic acid, thiamin), as a source of enzyme (invertase), as autolyzed yeast extract for flavoring in place of meat extracts, as a source of ergosterol and a source of single cell protein.

Sinai *et al.* (1974) fed brewers' yeast (*S. cerevisiae*) to Rhesus monkeys and found that this significantly enhanced the monkeys' resistance to respiratory and enteric infections. They believed the resistance was due to stimulation of phagocytosis.

The organisms in this genus have a widespread distribution. They are associated with a variety of foods and can cause spoilage in fruit and fruit products, sugar, syrup, honey, vinegar, mayonnaise, salad dressing, dairy products (buttermilk, cheese) and fermenting foods such as cucumbers.

S. aceti, found in wine, can convert alcohol to acetic acid. *S. bailii* var *osmophilus* is found in substances with a high concentration of sugar, alcohol, sulfur dioxide or acetic acid. *S. bailii* var *bailii* has been associated with spoiled wine, vinegar, salad dressing and mayonnaise. Two sugar-tolerant (osmophilic) species, *S. rouxii* and *S. bisporus*, are the main spoilage organisms of dates and can cause spoilage of honey, syrup and other foods with 40 to 60% sugar.

At the present time, no pathogenicity has been established for species of *Saccharomyces*.

Schizosaccharomyces.—The cells are spheroidal to cylindrical. Vegetative reproduction is by fission. The absence of budding distinguishes this genus from other yeasts. The cells may form a rudimentary true mycelium which can break up into arthrospores. There are four to eight spores per ascus. Glucose is fermented by all species.

The name *Schizosaccharomyces* indicates the close relationship to *Saccharomyces* (fermentation of glucose to alcohol), but also a difference (fission as compared to budding).

As with other yeasts, these species are widespread. They have been associated with spoilage of prunes, figs, raisins and wine.

Yang (1973) suggested using *S. pombe* in the fermentation to produce wine from grapes with a low pH. The yeast converts malic acid to lactic acid resulting in a less acid wine than when *Saccharomyces cerevisiae* is used. *S. malidevorans* is noted for its ability to convert malic acid to lactic acid.

Deuteromycetes.—These yeasts do not produce sexual spores. Unless otherwise stated, these asporogenous yeasts reproduce by budding. Certain genera, such as *Brettanomyces* and *Kloeckera* are found in foods and may be involved in spoilage or fermentation, but are not discussed in this text.

Candida.—This is a rather large genus of yeasts. This genus contains imperfect forms of ascosporogenous yeasts, such as *Debaryomyces, Hansenula, Kluyveromyces, Pichia* and *Saccharomyces.*

The cells are spheroidal, cylindrical, ovoid or elongate. All species form pseudomycelium and, by fission, some form true mycelium and chlamydospores. They can produce alcohol by fermentation.

Organisms in this genus are widespread. Besides soil, water and air, they are found in plants, insects, higher animals, humans, sewage, on processing equipment and food products.

These organisms have been involved with spoilage of various foods such as frankfurters, fresh fruits, vegetables, dairy products, brines and alcoholic beverages. *C. vini* (formerly *Mycoderma vini*) forms "flowers" on wine and causes spoilage. *C. mycoderma* is a film forming yeast on fermenting olive and cucumber brines. The psychrophilic strains tend to become dominant in refrigerated fruit juices.

This genus contains strains of yeast that are very useful in food and associated industries. They are used as food yeasts, a source of lipid, vitamins, invertase, lactose and lysine. *C. utilis* is used as a fodder yeast for animals and is a potential food yeast for humans.

C. albicans is the etiological agent of thrush, a pathological condition of white patches in the mouth, throat and esophagus. There are other potentially pathogenic species of *Candida* both for humans and animals. However, there is no information that food is the source of these infective agents. *Candida* species are found in the intestinal tract and it was suggested by Gracey *et al.* (1974) that they may cause diarrhea in malnutrition.

Rhodotorula.—The cells (spheroidal, ovoidal, elongate) reproduce by multilateral budding. There may be a rudimentary pseudomycelium. The perfect (sexual) form of this genus is *Rhodosporidium*, a member of Basidiomycetes. The organisms cannot assimilate inositol as the sole source of carbon or ferment carbohydrates. They form red or yellow carotenoid pigments.

The organisms in this genus are widespread. They are involved in the spoilage of a wide range of foods.

Species of this genus may be useful as a source of lipids, cystine and methionine and for degrading wastes in the production of single cell proteins.

Torulopsis.—*Torulopsis* has been defined as a heterogeneous group of imperfect yeasts that cannot be placed in the homogeneous genera. The cells are spherical, oval to elongate and reproduce by multipolar budding. *Torulopsis* can be differentiated from *Candida* by not being able to form a pseudomycelium, from *Cryptococcus* by not producing starch and from *Rhodotorula* by not producing carotenoid pigments.

The colonies are usually white to cream and usually glossy or shiny but may be dull. The tolerance to sodium chloride varies from 2% to 21% ($^w/_v$) for different species. At 30°C, *T. halonitratophilia* is an obligate halophile. There are osmophilic species in this genus.

Species of *Torulopsis* are found in various types of foods. They are credited with causing surface slime on cottage cheese, spoilage of refrigerated beef, cream, butter, sweetened condensed milk and various food brines.

Torulopsis magnoliae has been suggested for use in the production of glycerol. *T. ernobii* is a source of lipase. *T. utilis* is a source of single cell protein. *Torulopsis* strains have been found in sourdough, apparently aiding in the leavening of this product.

Trichosporon.—The cells of *Trichosporon* species are of various shapes. Buds are multilateral. True mycelium and arthrospores are formed. The presence of arthrospores differentiates this genus from *Candida*. Pseudomycelium may be developed by the budding cells.

These organisms are found on various foods such as fresh shrimp, crab, beef, butter, cheese, fruit, fruit juice and rice.

Trichosporon is part of the flora used in the production of idli (Batra and Millner 1974). It has been suggested as a possible source of protein.

VIRUSES

Viruses are obligate intracellular parasites. They generally have a rather limited host range. However, nearly all forms of life are susceptible to some type of virus. To survive and replicate, a virus must contact and invade an acceptable host cell. If a host cell is not available, the virus may become inactivated. This inactivation is influenced by temperature, humidity, pH, substrate composition as well as other factors.

There is some doubt that viruses are alive. Cliver (1971) stated that they are not alive, but borrow life from the host cell and direct it to produce more viruses.

Viruses have a genetic system with either RNA or DNA. This is surrounded by a protein coat, called the capsid. The capsid protects the genetic material from nucleases, serves as the vehicle of transmission from one host cell to another and forms a bond with the host cell during the attachment stage of infection.

For a virus to infect a host cell there must be an attachment site on the host cell. The attached virus injects its nuclear material into the host. This can result in the viral genetic material causing the cell to manufacture new viruses, or the genetic material may attach to the cellular chromosome and the cell becomes lysogenic. If new viral particles are produced, the cell is usually lysed upon release of the new viruses. These viruses must then contact another acceptable host cell.

If the virus material becomes attached to the chromosome of the cell, the genetic material of the virus is reproduced along with that of the cell (Campbell 1976). This lysogenic state may continue for an indefinite period. Eventually some disturbance triggers the viral genetic material to become activated and the cell produces new viruses, which are subsequently released.

Although the viruses are inactive as far as the reaction on food is concerned, they are important. Since they are obligate parasites, and some cause disease in man, the presence of viruses can be considered a potential health hazard. They can be put into the category of an environmental condition that allows spoilage since they, as bacteriophages, can attack useful fermentative organisms, resulting in an unsatisfactory product being developed. Since they are involved with genetic transfer (transduction), the viruses might be considered helpful in the development of better strains of useful microorganisms. However, transduction may be undesirable when it results in unacceptable mutant strains that are less sensitive to inhibitors, are more virulent or lose the characteristics that made them useful. Transduction is known in only a few groups of bacteria.

Intestinal Viruses

The primary viruses that can infect humans and are foodborne are the intestinal or enteric viruses. Cliver (1971) included the enteroviruses, adenoviruses, reoviruses and the agents that cause infectious hepatitis and epidemic viral diarrheas as intestinal viruses.

The enteroviruses are spherical and about 20-30 nm. They have single stranded RNA. These viruses are stable at pH 3.0. This tolerance to low pH allows them to survive passage through the stomach. They replicate in the intestinal tract, are shed in the feces and, if they contaminate food which is ingested, the process is repeated. Wilner (1973) included poliovirus, coxsackievirus A, coxsackievirus B and echovirus as enteroviruses.

The adenoviruses are hexagonal, about 70 to 90 nm diameter. They contain double stranded DNA. They are acid stable. There are 33 serotypes of adenoviruses that infect humans (Wilner 1973).

The reoviruses are hexagonal, about 70 to 80 nm in diameter. They contain double stranded RNA. The reoviruses, like the other intestinal viruses, are stable at pH 3 to 5.

The agents that cause infectious hepatitis were described by Provost *et al.* (1975). They are spherical, about 27 nm in diameter. Viral diarrheal agents have not been isolated or characterized. However, these agents pass through bacterial filters and can be transmitted through several people and still cause diarrhea. If the agent were merely a chemical toxin, this transmission would dilute it beyond the ability to cause illness.

Since these viruses are intestinal, they are found in feces of infected persons and resultant sewage. Larkin *et al.* (1976) pointed out that present sewage treatment does not remove viruses from the effluent or sludge. When sewage wastes are used to irrigate or fertilize crops, the vegetation is contaminated. Poliovirus persisted on radishes and lettuce for up to 36 days (Larkin *et al.* 1976). Such foods are eaten raw, so any contaminating viruses that are not removed by washing are ingested.

The procedure for the detection of viruses in or on foods consists of several parts. First, the viruses are separated from a food suspension by differential filtration or centrifugation. Then, the viruses are concentrated by centrifugation and/or filtration. Bacterial cells are destroyed in the viral concentrate by adding antibiotics. Aliquots of the treated viral suspension are planted on a monolayer of susceptible host cells. After incubation the cells are observed for plaques (holes in the monolayer produced by lysis due to the virus). The number of plaque forming units (PFU) is determined by counting the plaques and multiplying by the calculated dilution factor. Methods for concentration and viral detection are discussed by Gerba *et al.* (1977), Kostenbader and Cliver (1977) and Larkin *et al.* (1975). Fortunately, the viral contamination of food is very low or nonexistent (Kostenbader and Cliver 1977).

Bacterial Viruses (Bacteriophages)

The taxonomy of bacteriophages was discussed by Ackermann and Eisenstark (1974). The phages can have at least four relationships with their host cells. These are: 1) virulent phages cannot combine with the host's chromosomes to establish a lysogenic relationship; 2) pseudolysogenic phages can establish a carrier state in the host and may act as transducing agents; 3) defective temperate phages form particles with morphological attributes of a normal phage, but cannot replicate; and 4) true temperate phages are able to replicate and lysogenize their host cells.

The genetic material of temperate phages can become attached to the chromosome of the host cell. The properties of the phage are modified and it becomes a prophage. The prophage acts as a bacterial gene, dividing with the bacterial chromosome and being transmitted to daughter cells. A bacterium that carries a prophage is called lysogenic. Agents such as UV light, X-rays, organic peroxides, nitrogen mustard and mitomycin C can cause the release of phage material from the chromosome with resultant production of phage and lysis of the cell when the phage is released.

Bacteriophages can be stored by freezing with liquid nitrogen, using or not using glycerol (19%) as a protective additive and holding at −196°C. They can be held for about 2 years in a broth lysate stored at 4°C, with only slight loss of titer. Freeze drying or aerosolization results in a loss of titer of phage (Cox *et al.* 1974). The loss occurs during rehydration of the phage.

Bacteriophages can be enumerated by determining the clearance of susceptible bacterial cultures in broth or by the formation of plaques on lawns of the bacteria (Fig. 3.21).

Perhaps the most important function of phages in foods is their role in destroying cultures of lactic acid bacteria during the fermentation of dairy products.

There is a possibility that phages could be used to control unwanted bacteria in various food products. The main problems appear to be the host specificity of the phages and the development of lysogenic bacteria. The lysogenic bacteria can continue metabolizing and cause spoilage of foods.

Phages can be useful in epidemiological work. Since phages are rather host specific, a set of phages can be devised and used for typing bacterial species or strains. Phage typing can aid in tracing the origin of contamination or infection of foodborne outbreaks of gastroenteritis.

There are reports that prophage may be involved in the production of toxins by some bacteria, such as *Clostridium botulinum* and *Staphylo-*

coccus aureus. If this is true then the phage can be considered as a health hazard by manipulating bacteria.

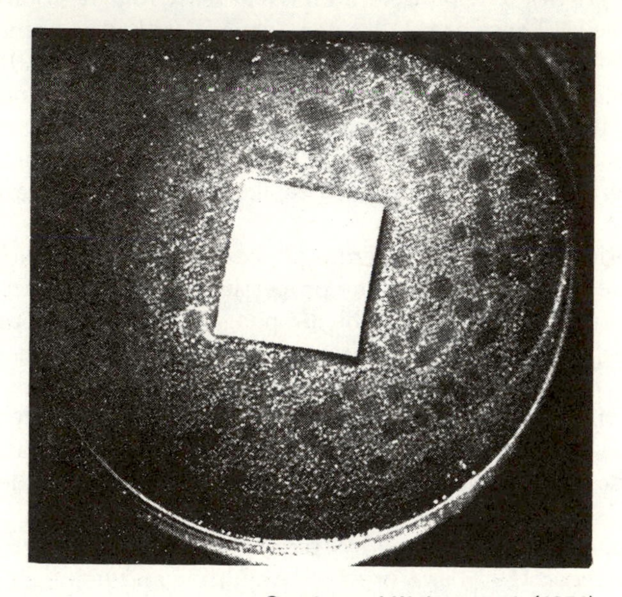

Courtesy of Weiser et al. (1971)

FIG. 3.21. PLAQUES DUE TO BACTERIOPHAGE ON A LAWN OF *STREPTOCOCCUS LACTIS*

One reason that phages can lyse an entire bacterial culture is that they have a high rate of replication. If a bacterium has a generation time of 20 min, in one hour there will be eight cells. A phage may require from 40 to 80 min for replication, but the burst size (number of phages released on lysis of the cell) may be from 2 to over 100. It is obvious that with a large burst size, in only a short time, the phage will outnumber susceptible bacterial cells.

The thermal resistance of streptococci bacteriophages varies with the type of phage, the pH of the substrate and composition of the substrate. The heat sensitive phage may be inactivated in about 10 min at 60°C while the more heat resistant strains require 25 min at 75° or even 30 min at 88°C. As the pH is lowered from 7.0, the phages are more sensitive to heat. Some phages can withstand pH values between 3 and 11 without inactivation.

Fungal Viruses

Lemke and Nash (1974) reviewed the fungal viruses. According to them, viruses have been reported for over 60 species of fungi. They believe the metabolism and genetics of fungi may be influenced by these viruses. The fungal viruses all have double stranded RNA, and most have a polyhedral shape. Koltin and Day (1975) found that the virus-like particles from *Ustilago maydis* infected several other species of *Ustilago* but very few other fungi.

BIBLIOGRAPHY

ABBOTT, B.J., LASKIN, A.I., and MCCOY, C.J. 1974. Effect of growth rate and nutrient limitation on the composition and biomass yield of *Acinetobacter calcoaceticus*. Appl. Microbiol. *28*, 58-63.

ACKERMANN, H.W., and EISENSTARK, A. 1974. The present state of phage taxonomy. Intervirology *3*, 201-219.

AMERINE, M.A., BERG, H. W., and CRUESS, W.V. 1979. The Technology of Wine Making, 4th Edition. AVI Publishing Co., Westport, CT.

ANDERSON, D.G., and MCKAY, L.L. 1977. Plasmids, loss of lactose metabolism, and appearance of partial and full lactose-fermenting revertants in *Streptococcus cremoris* B_1. J. Bacteriol. *129*, 367-377.

AOAC. 1975. Official Methods of Analysis, 12th Edition. Association of Official Analytical Chemists, Washington, D.C.

ARIES, V.C. 1976. An immunofluorescent technique for rapid control of the purity of yeast cultures. Eur. J. Appl. Microbiol. 2, 113-119.

ARONSON, A.I., and FITZ-JAMES, P. 1976. Structure and morphogenesis of the bacterial spore coat. Bacteriol. Rev. *40*, 360-402.

BATRA, L.R., and MILLNER, P.D. 1974. Some Asian fermented foods and beverages, and associated fungi. Mycologia *66*, 942-950.

BISSETT, M.L. 1976. *Yersinia enterocolitica* isolates from humans in California, 1968-1975. J. Clin. Microbiol. *4*, 137-144.

BOCKEMUHL, J., and MEINICKE, D. 1976. Value of phage typing of *Vibrio cholerae* biotype *eltor* in West Africa. Bull. World Health Organ. *54*, 187-192.

BOOTH, C. 1975. The present status of *Fusarium* taxonomy. Amer. Rev. Phytopathol. *13*, 83-93.

BOTHAST, R.J., LANCASTER, E.B., and HESSELTINE, C.W. 1975. *Scopulariopsis brevicaulis:* effect of pH and substrate on growth. Eur. J. Appl. Microbiol. *1*, 55-66.

BOTTONE, E.J., and ROBIN, T. 1977. *Yersinia enterocolitica*: recovery and characterization of two unusual isolates from a case of acute enteritis. J. Clin. Microbiol. *5*, 341-345.

BRYAN, F.L. 1973. Diseases Transmitted by Foods. Department of Health, Education and Welfare Publication No. (HSM) 73-8237. Center for Disease Control, Atlanta, GA.

BUCHANAN, R.E., and GIBBONS, N.E. 1974. Bergey's Manual of Determinative Bacteriology, 8th Edition. Williams and Wilkins Co., Baltimore.

CALAM, C.T. 1969. The evaluation of mycelial growth. *In* Methods in Microbiology. Vol. I. J.R. Norris, and D.W. Ribbons (Editors). Academic Press, London and New York.

CAMPBELL, A.M. 1976. How viruses insert their DNA into the DNA of the host cell. Sci. Amer. *235*, No. 6., 103-113.

CDC. 1975. *Salmonella* surveillance. Report No.125: Center for Disease Control, Atlanta, GA.

CHOPIN, M.C., CHOPIN, A., and ROUX, C. 1976. Definition of bacteriophage groups according to their lytic action on mesophilic lactic streptococci. Appl. Environ. Microbiol. *32*, 741-746.

CHRISTENSEN, C.M., and KAUFMAN, H.H. 1974. Microflora. *In* Storage of Cereal Grains and Their Products, 2nd Edition. C.M. Christensen (Editor). American Association of Cereal Chemists, St. Paul, MN.

CHU, F.S., NEI, P.Y.W., and LEUNG, P.S.C. 1973. Byssochlamyopeptidase A, a rennin-like enzyme produced by *Byssochlamys fulva*. Appl. Microbiol. *25*, 163-168.

CLIVER, D.O. 1971. Transmission of viruses through foods. Critical Rev. Environ. Control *1*, 551-579.

COUTTS, A.D. and SMITH, R.E. 1976. Factors influencing the production of cellulases by *Sporotrichum thermophile*. Appl. Environ. Microbiol. *31*, 819-825.

COX, C.S., HARRIS, W.J., and LEE, J. 1974. Viability and electron microscope studies of phages T3 and T7 subjected to freeze-drying, freeze-thawing, and aerosolization. J. Gen. Microbiol. *81*, 207-215.

CRANSTON, P.M., and CALVER, J.H. 1974. Quantitative fluorescent microscopy of yeast in beverages. Food Technol. Aust. *25*, 15-17.

CROMBACH, W.H.J. 1974. Genetic, Morphological and Physiological Relationships among Coryneform Bacteria. Centre for Agricultural Publishing and Documentation. Wageningen, Netherlands.

ERIKSSON, K., and LARSSON, K. 1975. Fermentation of waste mechanical fibers from a newsprint mill by the rot fungus *Sporotrichum pulverulentum*. Biotechnol. Bioeng. *17*, 327-348.

GARDNER, G.A. 1969. Physiological and morphological characteristics of *Kurthia zopfii* isolated from meat products. J. Appl. Bacteriol. *32*, 371-380.

GARDNER, G.A. 1973. A selective medium for enumerating salt requiring *Vibrio* spp. from Wiltshire bacon and curing brines. J. Appl. Bacteriol. *36*, 329-333.

GERBA, C.P., SMITH, E.M., and MELNICK, J.L. 1977. Development of a quantitative method for detecting enteroviruses in estuarine sediments. Appl. Environ. Microbiol. *34*, 158-163.

GOULD, G.W. 1977. Recent advances in the understanding of resistance and dormancy in bacterial spores. J. Appl. Bacteriol. *42*, 297-309.

GRACEY, M. *et al.* 1974. Isolation of *Candida* species from the gastrointestinal tract in malnourished children. Amer. J. Clin. Nutr. *27*, 345-349.

GRAVES, R.R., ROGERS, R.F., LYONS, A.J., JR., and HESSELTINE, C.W. 1967. Bacterial and actinomycete flora of Kansas-Nebraska and Pacific Northwest wheat and wheat flour. Cereal Chem. *44*, 288-299.

GRAY, W.M., and JOHNSON, M.G. 1976. Characteristics of bacteria isolated by the anaerobic roll-tube method from cheeses and ground beef. Environ. Microbiol. *31*, 268-273.

GREEN, S.K. *et al.* 1974. Agricultural plants and soil as a reservoir for *Pseudomonas aeruginosa.* Appl. Microbiol. *28*, 987-991.

HANG, Y.D., SPLITTSTOESSER, D.F., and WOODAMS, E.E. 1975. Utilization of brewery spent grain liquor by *Aspergillus niger.* Appl. Microbiol. *30*, 879-880.

HARRIS, R.V. 1970. Effect of *Rhizopus* fermentation on the lipid composition of cassava flour. J. Sci. Food Agr. *21*, 626-627.

HAYES, P.R. 1977. A taxonomic study of flavobacteria and related Gram negative yellow pigmented rods. J. Appl. Bacteriol. *43*, 345-367.

HENRIKSEN, S.D. 1976. *Moraxella, Neisseria, Branhamella,* and *Acinetobacter.* Annu. Rev. Microbiol. *30*, 63-83.

HESSELTINE, C.W. and BOTHAST, R.J. 1977. Mold development in ears of corn from tasseling to harvest. Mycologia *69*, 328-340.

HESSELTINE, C.W. *et al.* 1976. Aflatoxin occurrence in 1973 corn at harvest. II. Mycological studies. Mycologia *68*, 341-353.

HOLMBERG, K., and NORD, C.E. 1975. Numerical taxonomy and laboratory identification of *Actinomyces* and *Arachnia* and some related bacteria. J. Gen. Microbiol. *91*, 17-44.

HOSTY, T.S., and MCDURMONT, C.I. 1975. Isolation of acid-fast organisms from milk and oysters. Health Lab. Sci. *12*, 16-19.

HUANG, L.H., and HANLIN, R.T. 1975. Fungi occurring in freshly harvested and in-market pecans. Mycologia *67*, 689-700.

HUPPERT, M., HARPER, G., SUN, S.H., and DELANEROLLE, V. 1975. Rapid methods for identification of yeasts. J. Clin. Microbiol. *2*, 21-34.

ISHII, S., and YOKOTSUKA, T. 1973. Susceptibility of fruit juice to enzymatic clarification by pectin lyase and its relation to pectin in fruit juice. J. Agr. Food Chem. *21*, 269-272.

ITO, H., SHIBABE, S., and IIZUKA, H. 1971. Effect of storage studies of microorganisms on gamma-irradiated rice. Cereal Chem. *48*, 140-149.

JOFFE, A.Z., and PALTI, J. 1975. Taxonomic study of fusaria of the sporotrichiella section used in recent toxicological work. Appl. Microbiol. *29*, 575-579.

JONES, D. 1975. A numerical taxonomic study of coryneform and related bacteria. J. Gen. Microbiol. *87*, 52-96.

KEOGH, B.P. 1972. A re-assessment of the starter rotation system. Aust. J. Dairy Technol. *27*, No. 3, 86-87.

KING, A.D., JR., and NAGEL, C.W. 1967. Growth inhibition of a *Pseudomonas* by carbon dioxide. J. Food Sci. *32*, 575-579.

KOBURGER, J.A., and FARHAT, B.Y. 1975. Fungi in foods. VI. A comparison of media to enumerate yeasts and molds. J. Milk Food Technol. *38*, 466-468.

KOLTIN, Y., and DAY, P.R. 1975. Specificity of *Ustilago maydis* killer proteins. Appl. Microbiol. *30*, 694-696.

KOSTENBADER, K.D., JR., and CLIVER, D.O. 1977. Quest for viruses associated with our food supply. J. Food Sci. *42*, 1253-1268.

KRAMER, R.K., DAVIS, N.D., and DIENER, U.L. 1976. Byssotoxin A, a secondary metabolite of *Byssochlamys fulva.* Appl. Environ. Microbiol. *31*, 249-253.

KREGER-VAN RIJ, N.J.W. 1969. Taxonomy and systematics of yeasts. *In* The Yeasts. Vol. I. Biology of Yeasts. A.H. Rose and J.S. Harrison (Editors). Academic Press, London and New York.

LAHELLEC, C., MEURIER, C., BENNEJEAN, G., and CATSARAS, M. 1975. A study of 5920 strains of psychrotrophic bacteria isolated from chickens. J. Appl. Bacteriol. *38*, 89-97.

LARKIN, E.P., TIERNEY, J.T., and SULLIVAN, R. 1976. Persistence of virus on sewage-irrigated vegetables. J. Environ. Engineering Div. ASCE. *102*, 29-35.

LARKIN, E.P., TIERNEY, J.T., SULLIVAN, R., and PEELER, J.T. 1975. Collaborative study of the glass wool filtration method for the recovery of virus inoculated into ground beef. J. Assoc. Offic. Anal. Chem. *58*, 576-578.

LEMKE, P.A., and NASH, C.H. 1974. Fungal viruses. Bacteriol. Rev. *38*, 29-56.

LEVINE, D.W., and COONEY, C.L. 1973. Isolation and characterization of a thermotolerant methanol-utilizing yeast. Appl. Microbiol. *26*, 982-990.

LODDER, J. 1970. The Yeasts: A Taxonomic Study. North-Holland Publishing Company, Amsterdam.

LYONS, A.J., JR., PRIDHAM, T.G., and HESSELTINE, C.W. 1969. Survey of some *Actinomycetales* for α-galactosidase activity. Appl. Microbiol. *18*, 579-583.

LYONS, A.J., PRIDHAM, T.G., and ROGERS, R.F. 1975. *Actinomycetales* from corn. Appl. Microbiol. *29*, 246-249.

MACMILLAN, J.D., and PHAFF, H.J. 1973. Yeasts. General survey. *In* Handbook of Microbiology. Vol. I. Organismic Microbiology. A.I. Laskin, and H.A. Lechevalier (Editors). CRC Press, Cleveland.

MCDONALD, T.J., and MCDONALD, J.S. 1976. Streptococci isolated from bovine intramammary infections. Amer. J. Vet. Res. *37*, 377-381.

MORRIS, G.K., DEWITT, W.E., GANGAROSA, E.J., and MCCORMACK, W.M. 1976. Enhancement by sodium chloride of the selectivity of thiosulfite citrate bile salts sucrose agar for isolating *Vibrio cholerae* biotype El Tor. J. Clin. Microbiol. *4*, 133-136.

MOSSEL, D.A.A., VEGA, C.L., and PUT, H.M.C. 1975. Further studies on the suitability of various media containing antibacterial antibiotics for the enumeration of moulds in food and food environments. J. Appl. Bacteriol. *39*, 15-22.

MUHAMMED, S.I., MORRISON, S.M., and BOYD, W.L. 1975. Nutritional requirements for growth and sporulation of *Clostridium perfringens.* J. Appl. Bacteriol. *38*, 245-253.

NELSON, F.E. 1972. Plating medium pH as a factor in apparent survival of sublethally stressed yeasts. Appl. Microbiol. *24*, 236-239.

OLIVER, D.R., BROWN, B.L., and CLEWELL, D.B. 1977. Characterization of plasmids determining hemolysin and bacteriocin production in *Streptococcus faecalis* 5952. J. Bacteriol. *130*, 948-950.

PARK, C., and MCKAY, L.L. 1975. Induction of prophage in lactic streptococci isolated from commercial dairy starter cultures. J. Milk Food Technol. *38*, 594-597.

PASCH, J.H. 1974. Food and other sources of pathogenic microorganisms in hospitals. A review. J. Milk Food Technol. *37*, 487-493.

PEDERSON, C.S. 1979. Microbiology of Food Fermentations, 2nd Edition. AVI Publishing Co., Westport, CT.

PEPPLER, H.J. 1977. Yeast properties adversely affecting food fermentations. Food Technol. *31*, No. 2, 62-65.

PRICE, C.W., FUSON, G.B., and PHAFF, H.J. 1978. Genome comparison in yeast systematics: delimitation of species within the genera *Schwanniomyces, Saccharomyces, Debaryomyces,* and *Pichia.* Microbiol. Rev. *42*, 161-193.

PROVOST, P.J. *et al.* 1975. Biophysical and biochemical properties of CR326 human hepatitis A virus. Amer. J. Med. Sci. *270*, 87-91.

PUT, H.M.C., DEJONG, J., SAND, F.E.M.J., and VAN GRINSVEN, A.M. 1976. Heat resistance studies on yeast spp. causing spoilage in soft drinks. J. Appl. Bacteriol. *40*, 135-152.

READE, A.E., and GREGORY, K.F. 1975. High-temperature production of protein-enriched feed from cassava by fungi. Appl. Microbiol. *30*, 897-904.

ROBERTS, G.D., WANG, H.S., and HOLLICK, G.E. 1976. Evaluation of the API 20 C microtube system for the identification of clinically important yeasts. J. Clin. Microbiol. *3*, 302-305.

RYALL, C., and MOSS, M.O. 1975. Selective media for the enumeration of *Chromobacterium* spp. in soil and water. J. Appl. Bacteriol. *38*, 53-59.

SAKAZAKI, R., TAMURA, K., NAKAMURA, A., and KURATA, T. 1974. Enteropathogenic and enterotoxigenic activities on ligated gut loops in rabbits of *Salmonella* and some other enterobacteria isolated from human patients with diarrhea. Jap. J. Med. Sci. Biol. *27*, 45-48.

SANDS, D.C., and HANKIN, L. 1974. Selecting lysine-excreting mutants of lactobacilli for use in food and feed enrichment. Appl. Microbiol. *28*, 523-524.

SCHINDLER, A.F. *et al.* 1974. Mycotoxins produced by fungi isolated from inshell pecans. J. Food Sci. *39*, 213-214.

SCHOOK, L.B., CARRICK, L., JR., and BERK, R.S. 1976. Murine gastrointestinal tract as a portal of entry in experimental *Pseudomonas aeruginosa* infections. Infec. Immunity *14*, 564-570.

SCHWIMMER, S., and KURTZMAN, R.H., JR. 1972. Fungal decaffeination of roast coffee infusions. J. Food Sci. *37*, 921-924.

SCOTT, P.M., and KENNEDY, B.P.C. 1976. Analysis of blue cheese for roquefortine and other alkaloids from *Penicillium roqueforti*. J. Agr. Food Chem. *24*, 865-888.

SCOTT, P.M., KENNEDY, B.P.C., HARWIG, J., and BLANCHFIELD, B.J. 1977. Study of conditions for production of roquefortine and other metabolites of *Penicillium roqueforti*. Appl. Environ. Microbiol. *33*, 249-253.

SEGAL, E., and AJELLO, L. 1976. Evaluation of a new system for the rapid identification of clinically important yeasts. J. Clin. Microbiol. *4*, 157-159.

SINAI, Y., KAPLUN, A., HAI, Y., and HALPERIN, B. 1974. Enhancement of resistance to infectious diseases by oral administration of brewer's yeast. Infec. Immunity *9*, 781-787.

SUTTON, L.M. III, and STARZYK, M.J. 1972. Procedure and analysis of a useful method in determining mycelial dry weights from agar plates. Appl. Microbiol. *24*, 1011-1012.

THOEN, C.O., JARNAGIN, J.L., and RICHARDS, W.D. 1975. Isolation and identification of mycobacteria from porcine tissues: a three-year summary. Amer. J. Vet. Res. *36*, 1383-1386.

VANDERCOOK, C.E., and SMOLENSKY, D.C. 1976. Microbiological assay with *Lactobacillus plantarum* for detection of adulteration in orange juice. J. Assoc. Offic. Anal. Chem. *59*, 1375-1379.

VAN DONSEL, D.J., and LARKIN, E.P. 1977. Persistence of *Mycobacterium bovis* BCG in soil and on vegetables spray-irrigated with sewage effluent and sludge. J. Food Protection *40*, 160-163.

VOJNOVICH, C., ANDERSON, R.A., and ELLIS, J.J. 1972. Microbial reduction in stored and dry-milled corn infected with southern corn leaf blight. Cereal Chem. *49*, 346-353.

WEISER, H.H., MOUNTNEY, G.J., and GOULD, W.A. 1971. Practical Food Microbiology and Technology, 2nd Edition. AVI Publishing Co., Westport, CT.

WILKINSON, G., and DAVIES, F.L. 1973. Germination of spores of *Bacillus cereus* in milk and milk dialysates: effect of heat treatment. J. Appl. Bacteriol. 36, 485-496.

WILNER, B.I. 1973. Viruses of vertebrates. *In* Handbook of Microbiology, Vol. I. Organismic Microbiology. A.I. Laskin, and H.A. Lechevalier (Editors). CRC Press, Cleveland.

YAMADA, K., and KOMAGATA, K. 1972A. Taxonomic studies on coryneform bacteria. IV. Morphological, cultural, biochemical and physiological characteristics. J. Gen. Appl. Microbiol. *18*, 399-416.

YAMADA, K., and KOMAGATA, K. 1972B. Taxonomic studies on coryneform bacteria. V. Classification of coryneform bacteria. J. Gen. Appl. Microbiol. *18*, 417-431.

YANG, H.Y. 1973. Effect of pH on the activity of *Schizosaccharomyces pombe.* J. Food Sci. *38*, 1156-1157.

4

Conditions that Influence Microbial Growth

INTRODUCTION

A knowledge of the factors that influence microbial growth is needed by food microbiologists. Desirable growth conditions are needed for enumeration, fermentations or the production of single cell protein. Undesirable conditions are used for food preservation.

The microenvironments in food products are changing constantly. The reactions catalyzed by enzyme systems naturally inherent in some foods result in heat being generated, atmospheric oxygen used and carbon dioxide and other gases given off. These changes in the food affect the processes of microbial systems. The environment determines the conditions under which the microorganisms exist, and microorganisms influence the conditions prevailing in the environment. In any food environment, certain microbial species survive and become dominant. Organisms that lack the ability to withstand stresses induced by an unfavorable environment will succumb.

Discussions concerning environmental conditions for growth attempt to relate the macroenvironment of the system to the microenvironment in which a microorganism exists. This has been necessary since there is essentially no information about the microenvironments in foods or other products. There are many microenvironments with different conditions in a food product. Some microorganisms can grow in these microenvironments that cannot grow in the measured macroenvironment.

With optimum conditions for growth, some bacteria will reproduce each 15-20 min. Although they may grow in conditions above or below the optimum, the lag phase and/or the generation time may be longer (Fig. 4.1). If the stresses become too severe or are increased in number, the organisms might not grow or fail to survive.

The optimum conditions for microbial growth are influenced by the enzyme systems. The enzymes are organic catalysts that increase the rate of reactions. The rates of metabolic reactions are influenced by the environmental conditions that affect the activity of the enzymes.

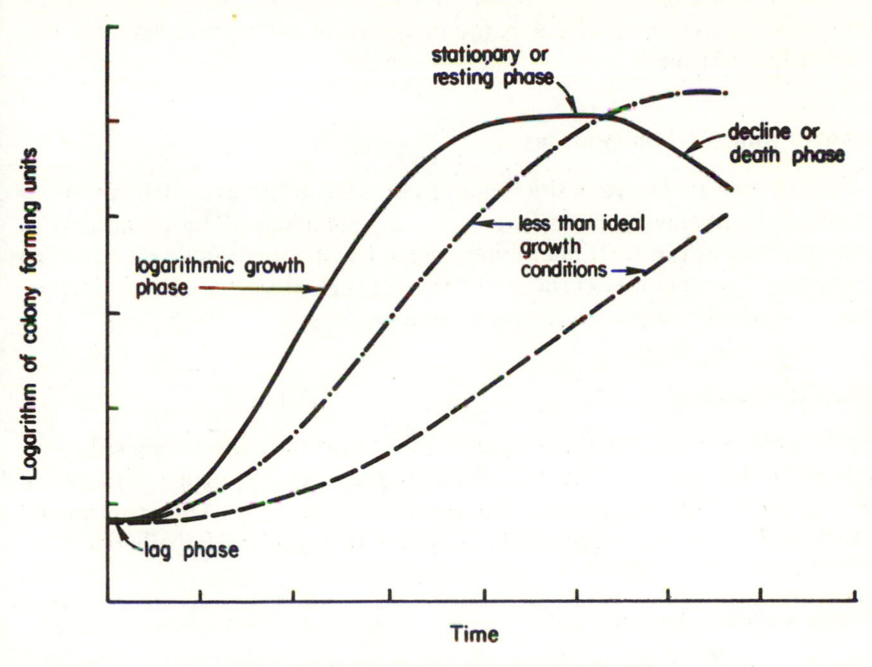

FIG. 4.1. BACTERIAL GROWTH CURVES

With less than optimum conditions, the lag phase and/or generation time may be longer.

The conditions that affect the metabolism and multiplication of microorganisms include nutrients, water, pH, inhibitors, oxygen, light, temperature and time. Since organisms are part of the environment and can alter it, microbial interactions (the effect of one organism on another) are important. The previous stresses (sublethal heating, freezing or radiation) that a microorganism has endured will affect its ability to cope with the environment.

NUTRIENTS

All biosystems require certain chemicals and chemical reactions in order to survive and reproduce. The microbial cell must be able to obtain a source of energy and synthesize cellular protoplasm from its environment. Organisms require a source of carbon and nitrogen, growth factors, such as vitamins, minerals and water. The ability of organisms to utilize compounds and synthesize cellular components depends upon the enzyme systems that the organism can make, according to its genetic code. Since the genetic code can change (mutate) there may be slight or important differences between strains of a species of microorganism. The substrate and other environmental factors affect the amount of enzymes produced by microorganisms. Some enzymes of microorganisms are discussed in Chapter 9.

Energy and Carbon Sources

The distinction between the types of energy sources and carbon sources of microorganisms is made in general microbiology. The organisms of primary concern in food microbiology are the chemoorganotrophs which use organic compounds as their energy and carbon sources. By definition, these microorganisms are heterotrophic.

Nitrogen Source

Amino acids are needed to produce cellular proteins, including enzymes. A number of organisms, such as *Escherichia coli*, can utilize nitrogen in the form of nitrates or ammonia to produce amino acids. Other organisms need one or several amino acids supplied to them in the substrate.

Other Growth Factors

Some organisms need growth factors, such as vitamins, purines or pyrimidines. For some fastidious organisms, metabolites must be supplied as preformed molecules. In general, the Gram positive bacteria require more preformed growth factors than do other microorganisms. Vitamins function in coenzyme systems as listed in Table 4.1.

TABLE 4.1

FUNCTIONS OF VITAMINS

Vitamin	Coenzyme	Function-Reaction
Biotin	Biotin coenzyme	Carboxyl transfer, deamination, CO_2 fixation
Pantothenic acid	Coenzyme A	Acyl transfer or exchange
Folic acid	Tetrahydrofolate	Transfer of one-carbon units, methylation, purine and pyrimidine synthesis

TABLE 4.1. *(Continued)*

Vitamin	Coenzyme	Function-Reaction
Thiamin (B_1)	Thiamin pyrophosphate	Accepts carboxyl groups from decarboxylation of keto acids; TCA cycle, lipid metabolism
Riboflavin (B_2)	Flavin mononucleotide (FMN)	Dehydrogenation, electron transfer-oxidations
	Flavin adenine dinucleotide	Dehydrogenation
Nicotinamide (B_5) (Niacin)	Nicotinamide adenine dinucleotide	Dehydrogenation, energy generation
	Nicotinamide adenine dinucleotide phosphate	Dehydrogenation, hydrogen acceptor, synthesis of fatty acids
Pyridoxine (B_6)	Pyridoxyl phosphate	Deamination, racemization, transamination, decarboxylation of amino acids
Cobamide (B_{12})	Cobamide coenzyme, deoxyadenosyl (B_{12})	Transfer of methyl groups

The obligate parasites, such as viruses, require a living organism to supply the machinery to synthesize their components.

It would be difficult to list the essential metabolites of specific microorganisms. There are shifts in the nutritional requirements of the strains comprising a culture. Not only do species of organisms have different essential metabolites, but also strains within the species differ in their requirements.

Elements and Minerals

Certain elements or minerals found in cellular components are needed in trace amounts by microorganisms. Somewhat larger amounts of sodium, potassium, calcium and magnesium are needed as compared to iron, copper, manganese, zinc, cobalt and molybdenum. The activity of some enzyme systems is enhanced by trace minerals, as listed in Table 4.2. Some elements are needed for the production of toxins or other secondary metabolites.

Besides metals, organisms need elements, such as phosphorus and sulfur. Phosphates are found in the high energy bonds of adenosine triphosphate (ATP).

Foods as Nutrients

Foods contain carbohydrates, fats, proteins, vitamins, minerals, water and other factors which microorganisms need for growth. Since they are derived from biological systems, foods vary in their composition. The components of the soil can affect the chemical composition of plants and the food derived therefrom. Variations in composition are also due to geography, climate, season, maturity, processing methods, variety and market conditions. The type of diet eaten by an animal can affect the composition of the food derived from this source.

TABLE 4.2

SOME MINERALS INVOLVED WITH ENZYME ACTIVITY

Mineral	Enzyme(s)
Calcium	Amylase, proteinase
Cobalt	Peptidases
Copper	Tyrosinase, oxidases
Iron	Cytochromes, electron-transport systems in mitochondria, ferredoxin
Magnesium	Phosphatases, ATP reactions, carboxylases
Manganese	Peptidases, isomerases
Molybdenum	Nitrate reduction, xanthine oxidase
Potassium	Phosphopyruvate transphosphorylase, fructokinase
Zinc	Dehydrogenases, peptidases, carbonic anhydrase

From their general composition, foods are classed as protein, carbohydrate or fat. The protein foods are primarily of animal origin, the carbohydrate foods of plant origin and fats or oils are derived from animals and plants. The composition of selected foods is listed in Tables 4.3 and 4.4. The values in these tables are not exact because of the variations in foods. Murphy *et al.* (1973) described the history of food composition tables and the sources, uses and limitations of the data. Fresh meats contain essentially no carbohydrate (less than 1%), since most is converted to lactic acid by glycolytic reactions during rigor. The vitamin and mineral content of most foods is sufficient for the growth of most organisms. The content of B-complex vitamins of some foods is listed in Table 4.5. The composition of foods influences the types of microorganisms that will grow, as well as the products that are formed during microbial growth and spoilage.

TABLE 4.3

COMPOSITION OF SOME PROTEIN FOODS

Food	Water %	Amount per 100 g		
		Protein g	Fat g	Carbohydrate g
Corned beef	54.2	15.8	25.0	–
Cottage cheese	79	17.0	0.3	2.7
Dried chipped beef	47.7	34.3	6.3	–
Eggs, whole	73.7	12.9	11.5	0.9
Egg white	87.6	10.9	trace	0.8
Egg yolk	51.1	16.0	30.6	0.6
Frankfurter, all meat	56.5	13.1	25.5	2.5
Goose	51.1	16.4	31.5	0
Haddock	80.5	18.3	0.1	0
Ham, cured	56.5	17.5	23.0	0
Hamburger, regular	60.2	17.9	21.2	–
Milk, whole	87.4	3.5	3.5	4.9
Milk, skim	90.5	3.6	0.1	5.1
Milk, dry whole	2.0	26.4	27.5	38.2
Salami, dry	29.8	23.8	38.1	1.2
Swiss cheese	39	27.5	28.0	1.7
Yogurt	89	3.4	1.7	5.2

Adapted from Watt and Merrill (1963).

TABLE 4.4

COMPOSITION OF VEGETABLES, FRUITS, AND MISCELLANEOUS FOOD

Food	Water %	Amount per 100 g		
		Protein g	Fat g	Carbohydrate g
Asparagus	91.7	2.5	0.2	5.0
Beans, lima, immature	67.5	8.4	0.5	22.1
Beans, smap, green	90.1	1.9	0.2	7.1
Cabbage	92.4	1.3	0.2	5.4
Lettuce, head	95.5	0.9	0.1	2.9
Potatoes	79.8	2.1	0.1	17.1
Apples, fresh	84.8	0.2	0.6	14.1
Avocados	74.0	2.1	16.4	6.3
Bananas	75.7	1.1	0.2	22.2
Oranges	86.0	1.0	0.2	12.2
Raisins	18.0	2.5	0.2	77.4
Prunes	2.5	3.3	0.5	91.3
Crackers, soda	4.0	9.2	13.1	70.6
Honey	17.2	0.3	0	82.3
Macaroni	10.4	12.5	1.2	75.2
Molasses, black	24.0	–	–	70.0
Peanut butter	1.8	27.8	49.4	17.2
Potato chips	1.8	5.3	39.8	50.0

TABLE 4.4. (*Continued*)

Food	Water %	Protein g	Fat g	Carbohydrate g
Sugar				
brown	2.1	0	0	96.4
granulated	0.5	0	0	99.5
Wheat flour	12.0	13.3	2.0	71.0

Adapted from Watt and Merrill (1963).

TABLE 4.5

B-VITAMINS IN SELECTED FOODS

Food	Vitamin				
	B_1	B_2	B_5	B_6	B_{12}
	mg per 100 g				
Apples	0.03	0.02	0.10	0.03	0
Asparagus	.18	.20	1.5	.15	0
Bananas	.05	.06	.7	.56	0
Beans, common	.65	.22	2.4	.56	0
Beef, boneless	.07	.15	4.0	.33	0.006
Bread, white	.09	.08	1.2	.04	trace
Bread, whole wheat	.26	.12	2.8	.18	0
Blue cheese	.03	.61	1.2	.17	.001
Swiss cheese	.01	.40	.01	.08	.002
Clam meat	–	–	–	.08	.098
Cod	.06	.07	2.2	.225	.0008
Eggs, whole	.11	.30	.1	.11	.002
Egg, yolk	.22	.44	.1	.30	.006
Grapes	.05	.03	.3	.08	0
Honey	trace	.04	.3	.02	0
Liver, beef	.25	3.26	13.6	.84	.08
Milk, whole	.03	.17	.1	.04	.0004
Onions	.03	.04	.2	.13	0
Orange juice	.09	.03	.4	.06	0
Walnuts	.33	.13	.9	.73	0

Adapted from Watt and Merrill (1963) and Orr (1969).

MOISTURE

Some microorganisms can remain alive in a dried condition, but cannot carry out their normal metabolic activities or multiply without water. Microorganisms can grow only in aqueous solutions. They cannot grow in pure water or in the absence of water. Water dissolves more substances than any other solvent. In this capacity, water is used to bring nutrients into the cells and to dispel waste products. Water is involved in the chemical reactions which break down substrates to usable molecules. These reactions include the hydrolysis of the peptide bonds in protein, the ester bonds in fats and in the conversion of polysaccharides to monosaccharides.

The moisture contents of various foods are listed in Tables 4.3 and 4.4. Although the water content of foods is shown as percentage, it has been adequately documented that this is not a valid determinant of biological activity. The water in a food is both "bound" and "free." Bound water is held by physical forces to macromolecules and is not available to act as a solvent or to participate in chemical reactions. Hence, it is not available to microorganisms for metabolic activity.

When solutes are dissolved in water, the freezing point is lowered, the boiling point is increased, the vapor pressure of the water is reduced and there is potential development of an osmotic pressure.

Water Activity

This is an index of the availability of water for chemical reactions and microbial growth. The biological activity of water, or water activity (a_w), is related to equilibrium relative humidity (ERH) or vapor pressure (VP). The a_w has been defined as the ratio of the VP of water above a material and the VP of pure water at the same temperature, or $a_w = P/P_o$.

Vapor Pressure.—The VP of a pure liquid depends upon the rate of escape of molecules from the surface of the liquid. The escape of water to the air is measured by the ERH. Thus, the VP and ERH are related.

When water is altered by the addition of a solute, the concentration of the water is decreased and the rate of escape from the surface is diminished. In an ideal solution, the partial VP of one component is directly proportional to the fraction of molecules of that component in the mixture.

According to Raoult's law, the partial vapor pressure of the solvent is equal to the vapor pressure of the pure solvent multiplied by the mole fraction of the solvent in the solution. The equation for this relationship is:

$$P = P_o N$$

when P is the partial pressure of water, P_o is the VP of pure water and N is the mole fraction of the solvent. N is usually shown as:

$$N = \frac{n_2}{n_2 + n_1}$$

with n_2 equal to the moles of water and n_1 equal to the moles of solute.

One liter of water weighs 1000 g, and the gram molecular weight of water is 18.016. This means there are 1000/18.016, or 55.51 moles of water in 1000 g. With pure water, the mole fraction is 55.51/55.51 + 0 or 1.00.

If one mole of a solute is added to 1000 g of water, the mole fraction of water becomes 55.51/55.51 + 1, or 0.9823. The ERH of the solution theoretically is 98% and the a_w is 0.98. This type of calculation can be applied only to very dilute, ideal solutions.

The water activity of a solution is defined in terms of VP and ERH by the formulae:

$$a_w = \frac{P}{P_0} = \frac{ERH}{100}$$

Calculated or Measured a_w. Due to non-ideal solutions, the a_w calculated by the mole fraction of water does not correspond to the a_w as measured by VP or humidity determinations. The calculated and measured a_w of electrolyte solutions compare favorably below 0.6 molal concentrations. At a level of 4 molal, the calculated a_w value of a sucrose solution is 0.93, but the measured value is about 0.90. At equivalent molal concentrations, sucrose lowers the a_w more than does glycerol.

Many methods have been suggested for measuring the a_w of various solutions. Seven of the more common methods were compared by Labuza *et al.* (1976). The vapor pressure manometric technique reportedly gave the best results.

Theoretically, in ideal solutions, the a_w is independent of temperature. Foods, as well as most solutions, are not ideal so these factors are not independent (Ross 1975).

Water Activity and Microbial Growth

Microorganisms have a maximum, optimum and minimum a_w for growth. Since the a_w of pure water is 1.00, and microorganisms cannot grow in pure water, the maximum or upper limit for microbial growth is an a_w somewhat less than 1.00. Some foods have a_w values greater than 0.995, which are rounded to 1.00. However, microorganisms can grow in these foods.

The prevention of microbial growth is of importance to food microbiologists. Hence, the minimum a_w at which growth can occur has been of most interest. In general, for growth, bacteria require a higher a_w than yeasts and yeasts require a higher a_w than molds. The minimum a_w values for the growth of various microorganisms are listed in Table 4.6. Even different strains of the same species appear to have different minimum a_w for growth. The limiting a_w for growth is influenced by other environmental factors which should be at optimum values for the organism being tested if the absolute minimum water activity is to be determined. If these factors are not optimum during testing, the resultant minimum a_w for growth of an organism will appear to be higher than the true value.

TABLE 4.6

APPROXIMATE MINIMUM WATER ACTIVITIES FOR GROWTH OF MICROORGANISMS

Organism	Minimum a_w	Organism	Minimum a_w
Most spoilage bacteria	0.90-0.91	*Vibrio parahaemolyticus*	0.93-0.98
Acinetobacter	0.95-0.98	Halophilic bacteria	0.75
Aeromonas	0.95-0.98	Most yeasts	0.87-0.94
Alcaligenes	0.95-0.98	Osmophilic yeasts	0.60-0.78
Arthrobacter	0.95-0.98	Most molds	0.70-0.80
Bacillus	0.90-0.99	Xerophilic molds	0.60-0.70
B. cereus	0.92-0.95	*Aspergillus*	0.68-0.88
Citrobacter	0.95-0.98	*A. glaucus*	0.70-0.75
Clostridium botulinum	0.90-0.98	*A. flavus*	0.80-0.90
Type A	0.95	*A. halophilicus*	0.68
Type B	0.94	*A. niger*	0.80-0.84
Type E	0.97	*Botrytis cinerea*	0.93
Corynebacterium	0.95-0.98	*Debaryomyces*	0.87-0.91
Enterobacter	0.95-0.98	*Fusarium*	0.80-0.92
Escherichia coli	0.94-0.97	*Hansenula*	0.89-0.90
Flavobacterium	0.95-0.98	*Mucor*	0.80-0.93
Klebsiella	0.95-0.98	*Penicillium*	0.80-0.90
Lactobacillus	0.90-0.94	*P. rubrum*	0.67
Leuconostoc	0.96-0.98	*Rhodotorula*	0.89-0.92
Micrococcus	0.90-0.95	*Saccharomyces cerevisiae*	0.90-0.94
M. roseus	0.90-0.93	*S. rouxii*	0.62-0.81
Pseudomonas aeruginosa	0.96-0.98	*Xeromyces bisporus*	0.60-0.61
P. fluorescens	0.95-0.97		
Salmonella	0.93-0.96		
Staphylococcus albus	0.88-0.92		
S. aureus	0.84-0.92		

The solute used to lower the a_w affects the minimum a_w. For example, *Saccharomyces rouxii* is sugar tolerant and can grow near an a_w of 0.62 in a sugar solution, but only above 0.81 if salt is the solute. Organisms tend to grow better in a substrate in which the a_w is altered by desorption rather than by adsorption. In one food system, pseudomonads grew at an a_w of 0.84 (Labuza *et al.* 1972).

The minimum a_w for the growth of bacteria is 0.90-0.91, except for *S. aureus* and the halophilic bacteria. The value for halophiles is listed as 0.75, which is also the a_w of a saturated NaCl solution (Table 4.7). The growth of *S. aureus* at various a_w levels is shown in Figure 4.2. As the a_w is reduced below 0.99, a reduced growth rate, as well as less total growth, becomes evident.

TABLE 4.7

EQUILIBRIUM RELATIVE HUMIDITIES OF SATURATED SALT SOLUTIONS AT 20°C

Chemical	ERH	Chemical	ERH
$NaOH \cdot H_2O$	7.0	$NaNO_3$	74.0
$LiCl \cdot H_2O$	11.0	$NaCl$	75.5
$K(C_2H_3O_2)$	20.0	$Na_2S_2O_3$	78.0
$CaCl_2 \cdot 6H_2O$	32.3	$(NH_4)_2SO_4$	80.6
$MgCl_2 \cdot 6H_2O$	33.6	KCl	85.1
$Zn(NO_3)_2 \cdot 6H_2O$	42.0	$ZnSO_4 \cdot 7H_2O$	90.0
$K_2CO_3 \cdot 2H_2O$	43.5	$BaCl_2$	90.5
KNO_2	45.1	K_2HPO_4	92.0
$LiNO_3 \cdot 3H_2O$	47.0	KNO_3	93.2
$Mg(NO_3)_2 \cdot 6H_2O$	54.9	$NH_4H_2PO_4$	93.2
$NaBr \cdot 2H_2O$	59.0	$Na_2HPO_4 \cdot 12H_2O$	96.2
NH_4NO_3	64.9	K_2SO_4	97.2
$NaNO_2$	65.6	$CuSO_4 \cdot 5H_2O$	98.0
$SrCl_2 \cdot 6H_2O$	71.0	$K_2Cr_2O_7$	98.3

Besides the minimum a_w for bacterial growth, there are other aspects of this parameter that are important. These include: spore formation and germination and the toxin formation by potentially hazardous organisms (such as *S. aureus* and *C. botulinum*).

Pitt and Christian (1968) listed the a_w requirements of 36 molds for germination and for the formation of asexual and sexual sporulation. Often a higher water activity is needed for sporulation than for germination and growth. Sexual sporulation usually requires a higher a_w than asexual sporulation.

At water activities below the minimum, the organism supposedly will not grow. As the a_w is lowered below the optimum, there is an increase in the lag phase and a decrease in the rate of growth. This means that foods at minimum a_w levels could be stored longer with less microbial deterioration than foods with higher water activities.

A very high a_w may limit the growth rate. A typical curve relating the growth rate to a_w is shown in Fig. 4.3. Frazier (1967) listed the optimum a_w for growth of *Streptococcus faecalis* as 0.982 and *Lactobacillus viridescens* as 0.975. *Vibrio costicola* did not grow at a_w values above about 0.98 (Forsyth and Kushner 1970).

Other Effects.—The water activity influences the resistance of microorganisms to physical effects such as heat. In general, lowering the a_w increases the resistance until at extremely low values of a_w, the resistance may be decreased. Spores are most heat resistant at a_w values of 0.2 to 0.4. The a_w also influences the survival time of microorganisms. Gen-

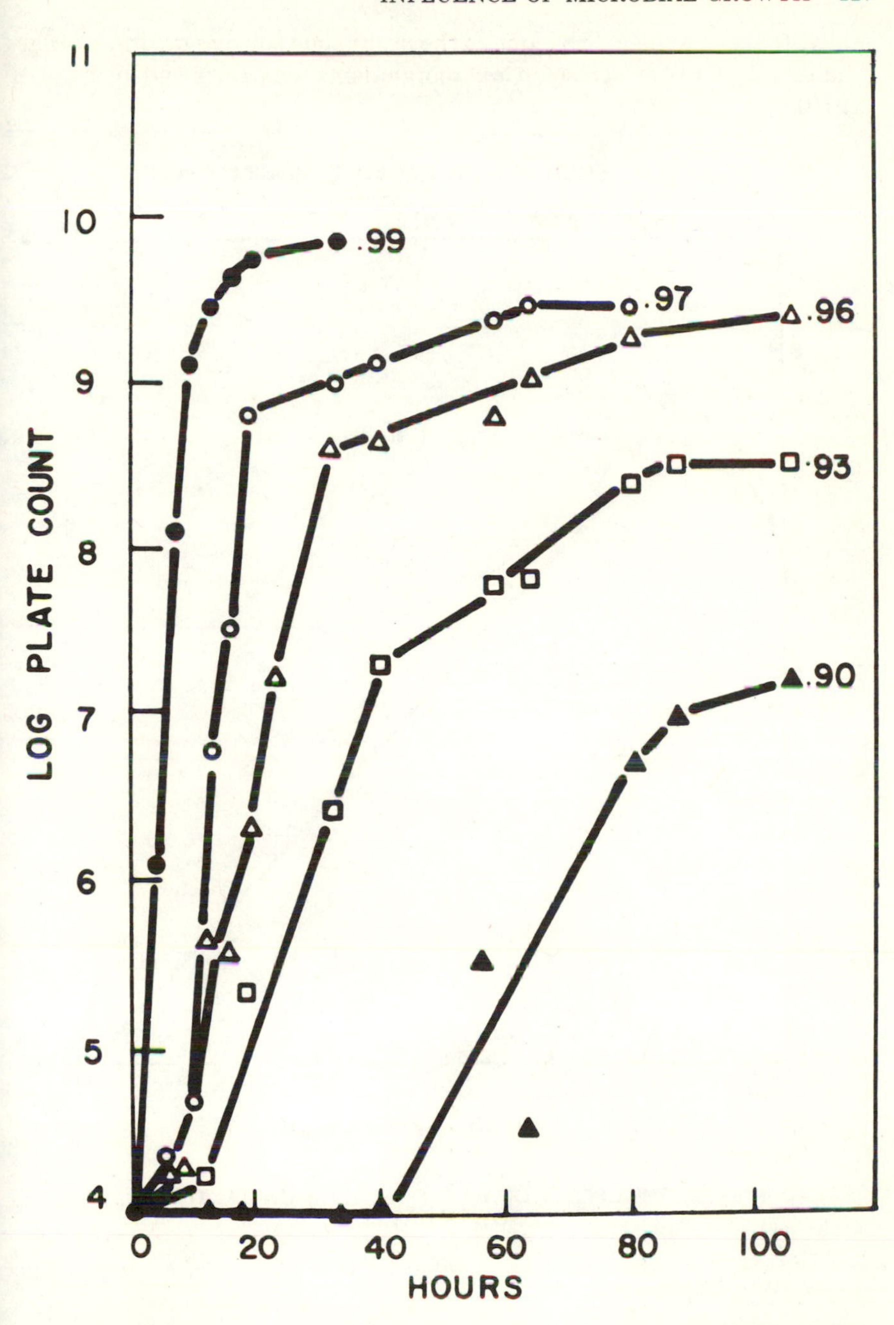

Courtesy of Troller (1971)

FIG. 4.2. GROWTH OF *STAPHYLOCOCCUS AUREUS* AT VARIOUS LEVELS OF WATER ACTIVITY

erally, the lower the a_w, the longer the organisms survive during storage. The effect of water stress on microorganisms was reviewed by Brown (1976).

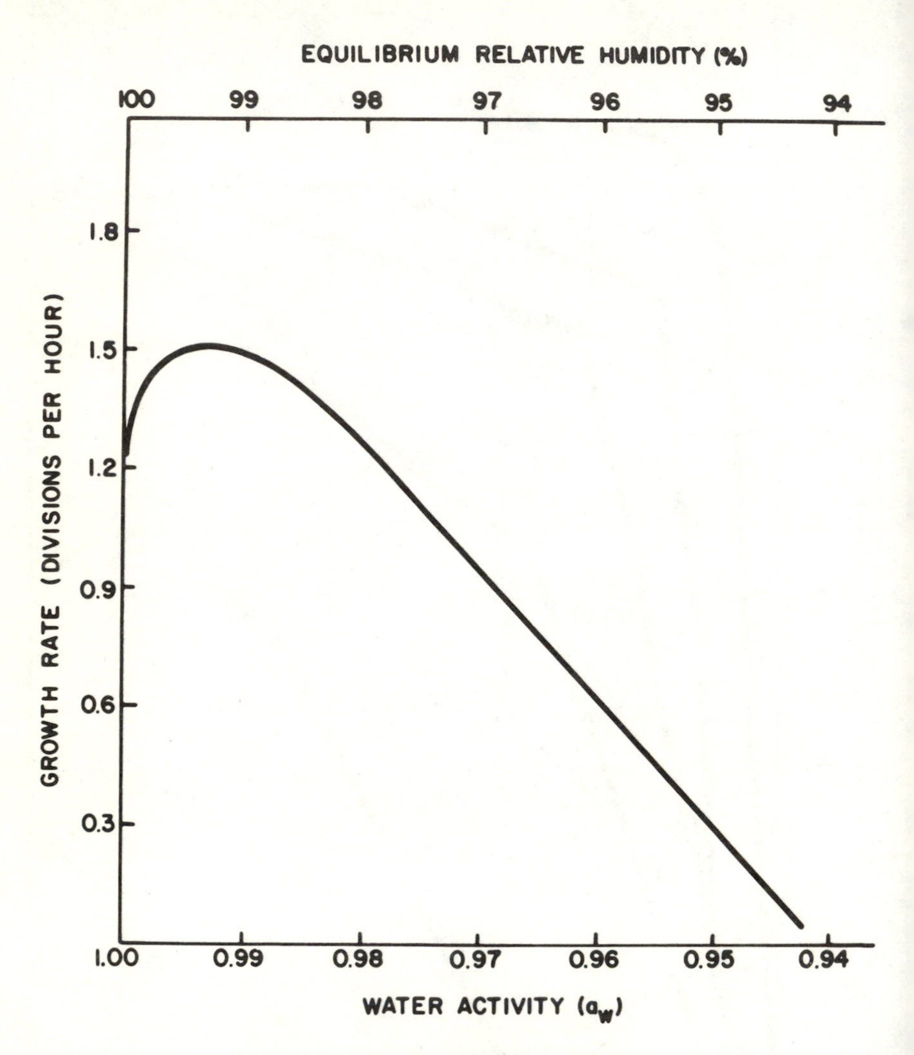

FIG. 4.3. THE EFFECT OF WATER ACTIVITY ON BACTERIAL GROWTH RATES

Water Activity of Food

The a_w of food can be lowered by removing water (dehydration), by adding solutes (sugar, salt) or by freezing.

The usual a_w's of various foods are listed in Table 4.8. Depending upon the system of measurement, Fett (1973) reported variations of a_w of ± 0.003 to ± 0.011, and Labuza *et al.* (1976) reported that different methods of measurement may vary ± 0.02 a_w units.

TABLE 4.8

APPROXIMATE WATER ACTIVITIES OF SELECTED FOODS

Food	a_w	Food	a_w
Fresh fruit or vegetables	0.97-1.00	Jam	0.75-0.80
Fresh poultry or fish	0.98-1.00	Jelly	0.82-0.94
Fresh meats	0.95-1.00	Rice	0.80-0.87
Pudding	0.97-0.99	Fruit cake	0.80-0.87
Eggs	0.97	Flour	0.67-0.87
Juices, fruit and vegetable	0.97	Cake icing	0.76-0.84
Bread	0.96	Molasses	0.76
white	0.94-0.97	Honey	0.54-0.75
crust	0.30	Dried fruit	0.55-0.80
Cheese, most types	0.95-1.00	Chocolate candy	0.69
Parmesan	0.68-0.76	Rolled oats	0.65-0.75
Cured meat	0.87-0.95	Toffee	0.60-0.65
Baked cake	0.90-0.94	Caramels	0.60-0.65
Refrigerated biscuit dough	0.94	Noodles	0.50
Maple sirup	0.90	Dried whole egg	0.40
Salted egg yolk	0.90	Biscuits	0.30
Soft, moist pet food	0.83	Dried whole milk	0.20
		Dried vegetables	0.20
		Cereals	0.10-0.20
		Crackers	0.10
		Sugar	0.19

Sources of information: Fett (1973); Mossel (1971A,B).

Fresh foods, such as fruit, vegetables, meat, poultry and fish have a_w values of 0.98 to over 0.99, that will allow the growth of most microorganisms.

The rate of growth of bacteria is greater than that of yeasts or molds. Hence, with foods of high a_w, bacteria will outgrow the fungi (yeasts and molds) and cause spoilage, while at a_w values that restrict or prevent bacterial growth, the fungi grow and become dominant. Exceptions to these generalizations include fruits which are spoiled by fungi due to the acidity of the product restricting bacterial growth. Products that have low a_w due to sugar content (jams, jellies or honey) will be subject to attack by osmophilic yeasts, while products that contain high salt concentrations (low a_w) will be spoiled by halotolerant or halophilic bacteria. Dried foods generally have a_w values below 0.75. A safe a_w level for storage is usually considered to be 0.70 or less. In foods protected by low a_w, enzymatic changes can occur, although at a slow rate.

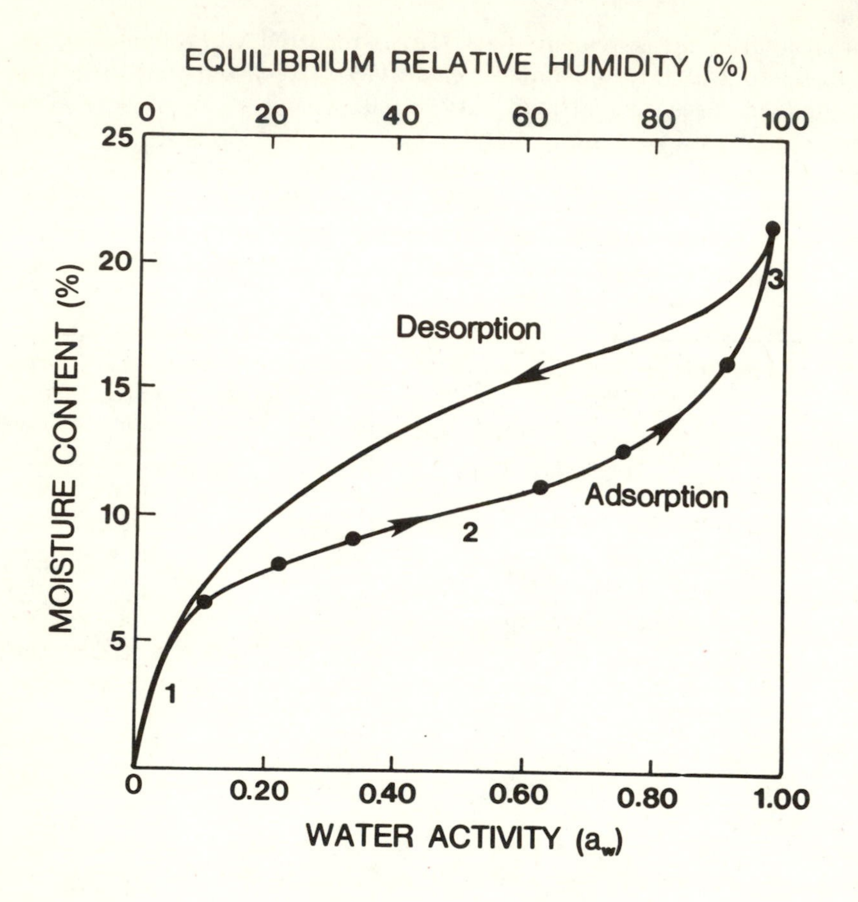

FIG. 4.4. A TYPICAL SORPTION ISOTHERM

1. Monomolecular film of water, no growth of organisms; 2. Some added water to monomolecular film; slight growth in upper range; 3. Water activity is sufficient for microbial growth.

Sorption Isotherms.—Quite often, foods are considered in terms of moisture content. The relationship between a_w and moisture is desirable. This relationship is described by a sorption isotherm which has the shape of a sigmoid curve (Fig. 4.4). Numerous mathematical equations have been derived to describe sorption isotherms (Iglesias and Chirife 1976). The curve may be an isotherm of adsorption (a moistening process), or an isotherm of desorption (a drying process). The isotherms vary with temperature and for different products.

Three areas are apparent in most isotherms. In the lowest part of the curve, the water is in a monomolecular layer on the food product, or so-called "bound" water, and is not available for biological activity. In the middle of the curve, there are additional layers of water added to the monomolecular film, and there is a rapid increase in a_w with a moderate increase in moisture content. In the upper regions of this section of the curve, some water becomes available to support the growth of some molds and yeasts. In the upper part of the isotherm, the moisture content increases rapidly with comparatively small increases in water activity, and the food contains free water that is available for bacterial as well as fungal growth.

If food is not stored in moisture-proof containers, it will equilibrate with the relative humidity of the surrounding air according to its sorption or desorption isotherm. The direction will be determined by the moisture in the food and the relative humidity of the air. Dry foods may pick up moisture to the extent that molds will be able to grow, while foods with high a_w stored in low humidity air may lose moisture, causing shrinkage and other organoleptic changes. The rate of equilibration depends upon the relative humidity of the air, initial a_w of the food and temperature.

In mixtures of foods such as cereals and raisins, or marshmallow topping on a cookie base, the moisture from one food will transfer to the drier product. For raisins in cereals, this results in the cereal becoming less crisp and the raisins becoming hard.

Coating of the raisins with beeswax, oils or sugar has been used to retard the transfer of moisture. Since many of our foods are combination of various ingredients, the a_w values and isotherms of the ingredients should be considered in order to avoid potential problems either with organoleptic quality or possible microbial growth.

Although a product may have an a_w not conducive to growth of microorgamisms, syneresis may occur, increasing the a_w in microenvironments, thus allowing the microbial growth.

Case-hardening, the phenomenon in which the moisture content of the exposed surface is different from the unexposed portion, may act as a barrier for moisture sorption or desorption. This may influence the measurement of the a_w of a food, so that, although the determined a_w of the food would appear in a safe range, the interior of the food will support microbial growth.

Safe Moisture Levels.—Since most microorganisms will not grow below a_w 0.70, if food is reduced to this level, it should be storable at room temperature. Labuza *et al.* (1972) studied a banana mix adjusted to various water activities by desorption and by absorption of water. In those mixes adjusted by desorption, mold spores germinated and grew at 0.68 a_w (at 25°C), but when adjusted by absorption, molds showed a decrease even at 0.90 a_w. With these data, the concept of a_w 0.70 being the safe level should be reevaluated.

pH

The pH unit was defined by Sorenson in 1909. Today, the determination and maintenance of pH are important in quality control and processing in the food industry.

In many texts, pH is defined as the negative logarithm of the hydrogen ion concentration. The pH is actually a function of apparent concentrations or activity of the electrolytes in solution. Thus, pH is the negative logarithm of the activity of the hydrogen ions. Since pH is usually determined by means of a pH meter, it might be defined as a value calculated from the electromotive force (emf) of an electromotive cell in which the substance being tested is one of the electrolytes.

Effect on Microorganisms

Due to the cell membrane being only slightly permeable to hydrogen or hydroxyl ions and a possible buffering effect of the cytoplasm, even in acid or alkaline substrates, the internal cellular pH is near 7.0.

Microorganisms have a minimum, optimum and maximum pH for growth. Most bacteria show optimum growth at a pH near 7.0, while others are favored by an acid environment, probably due to the inhibition of other organisms, thus eliminating microbial competition. The acid formers (*Lactobacillus* and *Streptococcus*) can tolerate moderate acidity, while proteolytic types (*Pseudomonas*) can grow in moderately alkaline substrates. In general, molds can grow at lower pH values than yeasts, and yeasts are more tolerant of low pH values than are bacteria. Bacteria usually grow faster than yeasts in a neutral or slightly acid substrate, but at pH 5.0 or less, yeasts can compete with or outgrow bacteria.

Microorganisms can grow in a wide pH range (Table 4.9). The variations in pH values for growth may be due to different strains of a species or different species in a genus, the type of substrate, the acid or base used to adjust the pH or other factors. There is an interrelationship between pH and other environmental factors.

The lowest pH ever reported for the growth of *C. botulinum* is 4.8 (Townsend *et al.* 1954). Even at pH 4.8, growth was observed in only one (pineapple-rice pudding) of many foods tested. Segner *et al.* (1966) reported that for *C. botulinum* type E spores, at 30°C, outgrowth was observed at pH 5.9, but not at pH 5.7. These workers concluded that growth and toxin production will not occur below pH 4.8, the lowest pH permitting growth of types A and B.

Segner *et al.* (1971) found that terrestrial strains of type C, *C. bot-*

ulinum failed to grow at pH 5.62 or less. Marine strains grew at pH 5.25, but not at pH 4.96.

TABLE 4.9

APPROXIMATE pH RANGES OF MICROBIAL GROWTH

Organism	pH		
	Minimum	Optimum	Maximum
Bacteria (most)	4.5	6.5-7.5	9.0
Acetobacter	4.0	5.4-6.3	—
Aeromonas	5.5	—	9.0
Bacillus subtilis	4.2-4.5	6.8-7.2	9.4-10.
Clostridium	4.6-5.0	—	—
C. botulinum	4.8-5.0	—	—
C. perfringens	—	6.0-7.6	8.5
C. sporogenes	5.0-5.8	6.0-7.6	8.5- 9.0
Erwinia carotovora	4.6	7.1	9.3
Escherichia coli	4.3-4.4	6.0-8.0	9.0-10.
Gluconobacter oxydans	4.0-4.5	5.5-6.0	—
Lactobacillus (most)	3.0-4.4	5.5-6.0	7.2- 8.0
L. acidophilus	4.0-4.6	5.5-6.0	7.0
L. plantarum	3.5	5.5-6.5	8.0
Leuconostoc cremoris	5.0	5.5-6.0	6.5
L. oenos	—	4.2-4.8	—
Pediococcus cerevisiae	2.9	4.5-6.5	7.8
Propionibacterium	—	6.5-7.0	—
Proteus vulgaris	4.4	6.0-7.0	8.4- 9.2
Pseudomonas (most)	5.6	6.6-7.0	8.0
P. aeruginosa	4.4-5.6	6.6-7.0	8.0- 9.0
Salmonella (most)	4.0-5.0	6.0-7.5	9.0
S. typhi	4.0-4.5	6.5-7.2	8.0- 9.6
S. choleraesuis	5.0	7.0-7.6	8.2
Serratia marcescens	4.6	6.0-7.0	8.0
Staphylococcus (most)	4.2	6.8-7.5	9.3
S. aureus	4.0-4.7	—	9.5- 9.8
Streptococcus (most)	—	6.2	—
S. lactis	4.1-4.8	6.4	9.2
Vibrio	6.0	—	9.0
V. cholerae	—	8.6	—
V. parahaemolyticus	4.8	—	7.8
Yeasts	1.5-3.5	4.0-6.5	8.0- 8.5
Candida krusei	1.5-2.0	—	—
C. albicans	2.2	—	9.6
Hansenula	—	4.5-5.5	—
Kluyveromyces	1.5-2.0	—	—
Pichia	1.5	—	—
Saccharomyces cerevisiae	2.0-2.4	4.0-5.0	—
Molds	1.5-3.5	4.5-6.8	8.0-11.
Aspergillus	—	3.0-6.8	—
A. niger	1.2	3.0-6.0	—
A. oryzae	1.6-1.8	—	9.0- 9.3
Botrytis cinerea	2.5	—	7.4
Mucor	—	3.0-6.1	9.2
Penicillium	—	4.5-6.7	—
Rhizopus nigricans	—	4.5-6.0	—

Baird-Parker and Freame (1967) reported that at pH 5.0, type B spores germinated, type E spores germinated very slowly and type A spores did not germinate. They believed the limiting pH for growth of *C. botulinum* type A or B spores is pH 4.8 to 5.0, while type E spores are inhibited at somewhat higher pH.

The outgrowth of *C. botulinum* spores was inhibited at pH 4.8 but not at pH 5.0 (Ito *et al.* 1976). Ten strains of *C. botulinum* types A and B, were tested at various pH levels in tomato juice (Huhtanen *et al.* 1976). Four of the strains failed to grow at pH 5.68 or less, and eight did not grow at pH 5.37 or less. Only one strain grew at pH 5.24. They stated that this was the minimum pH for growth of this strain of *C. botulinum.*

It is the opinion of the U.S.FDA that *C. botulinum* can grow and produce toxin at pH 4.8. Since they desire a safety factor, an upper limit of pH 4.6 has been set for acid foods. Just because a pH of 4.6 is used for control purposes, it should not be construed as the minimum pH value for growth of the organism. Obviously, if the organism will not grow below pH 4.8, it will not grow below pH 4.6 or pH 4.5.

Alteration of the pH of a substrate may relate indirectly to growth. For example, the availability of metallic ions is altered. Although coexisting in the free form at low pH, magnesium and phosphorus form an insoluble complex at higher pH values. In an alkaline medium, ferric, zinc and calcium ions become insoluble.

The pH of the substrate may influence cell permeability. At low pH, the membrane becomes saturated with hydrogen ions, thus limiting passage of essential cations. Conversely, at a high pH, saturation of the membrane with hydroxyl ions will limit the passage of essential anions.

The toxicity of adverse pH values is partly due to the penetration of undissociated molecules of acid or basic substances into the cells. At low pH values, undissociated weak acids can enter the cell, then ionize and alter the internal pH. In alkaline solutions, undissociated weak bases can enter the cell. If the internal pH is changed to basic pH values, amino acid transferase RNA is inhibited and protein synthesis is stopped. The pH of the substrate influences the enzyme system and the products of metabolism of microorganisms. The activity of the enzyme systems of microorganisms is influenced by the pH.

Alteration of pH by Microorganisms.—The environment influences microorganisms and the microorganisms influence the environment. During growth, metabolic products are formed. These can be either acidic or alkaline, depending upon the substrate, the organisms involved and the time allowed for growth. The initial reaction of most organisms is acidic due to the breakdown of carbohydrates and the formation of organic acids. The alteration of pH by production of acids is used in the food fermentation industries. The lactic acid bacteria tend to lower the pH by production of lactic acid, while proteolytic types such as pseudomonads tend to raise the pH by production of ammonia or other basic chemicals.

Survival.—The pH for survival of cells is somewhat different than that for growth. Some organisms show greater survival at slightly acid pH levels 5.6-6.5 and optimum growth at 6.8-7.2. Microorganisms can survive at pH levels too acid or too basic for metabolism and growth. The effect of pH on survival during heating is very evident.

Toxicity.—The early literature indicated that toxin production and virulence were favored by an alkaline pH. However, Idziak and Suvanmongkol (1972) found *Salmonella typhimurium* decreases in virulence in a neutral environment and is more virulent in an acid medium. The production of aflatoxin by *Aspergillus flavus* is favored by an acid environment (Joffe and Lisker 1969). A pH of 4.5 enhances the production of rubratoxin by *Penicillium rubrum* (Emeh and Marth 1976).

pH of Food

The pH of a food, along with other environmental factors, will determine the types of microorganisms that are able to grow and dominate, and eventually cause spoilage, a desired fermentation or a potential health hazard. Besides the microbial aspects of the pH of foods, the acids in foods play other important roles.

The food may be naturally acid, acids may be added or acids may be produced in the food by enzymatic action with or without microbial growth.

The pH of a food is determined by the balance between the buffering capacity and the acids or alkaline substances that it contains. Since protein has a high buffering capacity, protein foods have greater buffering than do fruits or vegetables. This is important in the fermentations producing lactic acid, since the production of small amounts of acid in sauerkraut or pickle processes will lower the pH significantly.

The approximate pH values for selected foods are listed in Table 4.10. The pH values are only approximations, since there is considerable variation in the pH of some foods and the pH of food products can change during ripening, processing or spoilage.

Foods have been categorized according to their pH as follows:

High acid foods	pH below 3.7
Acid foods	pH 3.7-4.6
Medium acid foods	pH 4.6-5.3
Low or non-acid foods	pH over 5.3

TABLE 4.10

APPROXIMATE pH RANGES OF SELECTED FOODS

Food	pH Range	Food	pH Range
Egg white	7.6-9.5	Cabbage	5.2-6.3
Shrimp	6.8-8.2	Turnip	5.2-5.6
Crab	6.8-8.0	Spinach	5.1-6.8
Scallops	6.8-7.1	Asparagus	5.0-6.1
Cod, small	6.7-7.1	Cheeses, most	5.0-6.1
Cod, large	6.5-6.9	Camembert	6.1-7.0
Catfish	6.6-7.0	Cottage	4.1-5.4
Soda crackers	6.5-8.5	Gouda	4.7
Maple syrup	6.5-7.0	Bread	5.0-6.0
Milk	6.3-6.8	Carrots	4.9-6.3
Brussels sprouts	6.3-6.6	Beets	4.9-5.8
Whiting	6.2-7.1	Bananas	4.5-5.2
Haddock	6.2-6.7	Dry sausages	4.4-5.6
Cantaloupe	6.2-6.5	Pimientos	4.3-5.2
Dates	6.2-6.4	Tomato juice	3.9-4.7
Herring	6.1-6.6	Mayonnaise	3.8-4.0
Butter	6.1-6.4	Tomatoes	3.7-4.9
Honey	6.0-6.8	Jams	3.5-4.0
Mushrooms	6.0-6.5	Apricots	3.5-4.0
Cauliflower	6.0-6.7	Apple sauce	3.4-3.5
Lettuce	6.0-6.4	Pears	3.4-4.7
Egg yolk	6.0-6.3	Grapes	3.3-4.5
Corn, sweet	5.9-6.5	Cherries	3.2-4.7
Oysters	5.9-6.6	Pineapple	3.2-4.1
Celery	5.7-6.0	Peaches	3.1-4.2
Peas	5.6-6.8	Rhubarb	3.1-3.2
Turkey	5.6-6.0	Strawberries	3.0-4.2
Chicken	5.5-6.4	Grapefruit	2.9-4.0
Halibut	5.5-5.8	Raspberries	2.9-3.7
Beans, lima	5.4-6.5	Apples	2.9-3.5
Potatoes, Irish	5.4-6.3	Plums	2.8-4.6
Walnuts	5.4-5.5	Oranges	2.8-4.0
Pork	5.3-6.4	Cranberries	2.5-2.8
Beef	5.3-6.2	Lemons	2.2-2.4
Onions	5.3-5.8	Limes	1.8-2.0
Sweet potatoes	5.3-5.6		

Egg White.—This is one of the most alkaline biological solutions. The albumen of a freshly laid chicken egg is approximately pH 7.6. When the egg is stored in air, the carbon dioxide from carbonic acid in the albumen is released through the egg shell. When this occurs, the pH increases and has been reported to reach levels of 9.4-9.7. Although the high pH is desirable from a microbiologist's viewpoint, these high pH levels are associated with thinning of albumen and a decrease in the strength of the vitelline (yolk) membrane, or a general decrease in egg quality. Egg quality is maintained at an albumen pH near 8.2, which is above the maximum for the growth of many microorganisms. Storage in an atmosphere of CO_2 or oiling the egg shell maintains the pH at an intermediate level.

Red Meat.—Living animal tissue is near neutral (pH 7.0-7.2). The circulating blood brings nutrients and oxygen to the cells and removes the waste products of metabolism. When an animal is slaughtered, blood no longer circulates, anaerobic conditions develop and any metabolic products accumulate. The inherent tissue enzymes ferment the muscle glycogen to lactic acid which lowers the pH. When glycogen supplies are abnormally low in the muscle prior to slaughter, there is less acid produced, with a higher than normal ultimate pH. The temperature at which a beef carcass is held determines the rate of glycolysis and pH changes.

It might be expected that a higher glycogen content is present in the muscles of well fed, rested and unexcited animals than in starved, tired or stressed animals.

Stress conditions, such as fighting among pigs immediately prior to slaughter, have been associated with a rapid pH drop post mortem, and formation of pale, soft and exudative (PSE) muscle. In other than normal glycolysis of beef, resulting in a high ultimate pH, there is insufficient oxygen available to convert the muscle hemoglobin (myoglobin) from the reduced state which is purplish colored. This is known as dark-cutting beef. Although it is acceptable beef, the consumer associates the purple color with old animals or possible spoilage.

Immediately after slaughter, the pH of most beef muscles is 6.9-7.4 and 24 hr post mortem most muscles are in the range of pH 5.4-6.4 (Fredeen *et al.* 1974). It is generally believed that pH 5.3 is the limit below which glycolysis is inhibited even though glycogen may be present.

In a survey of pork, most samples had a pH of 5.4-6.6 (Kempster and Cuthbertson 1975).

The pH of commercial dry sausages was reported as ranging from 4.4-5.6 (Acton and Dick 1976). Fermentation is responsible for the lower pH values of sausages as compared to fresh meat. The addition of acid (acetic, lactic, citric) results in a low pH of pickled meats.

Microbiologically, a low pH in fresh meat is desired. The minimum pH for the growth of most pseudomonads which spoil meat is 5.6. If meat has an ultimate pH less than 5.6, it would be expected to have a longer shelf life. On the other hand, higher pH values have been associated with enhanced water-holding capacity, nutrition, juiciness, tenderness and increased emulsifying properties and binding strength of meat, which are desirable attributes.

Chicken.—The pH of chicken muscle varies similarly to that of red meats. Post-slaughter chicken muscle is approximately pH 7.0 which is reduced during glycolysis to 5.5-5.9. A rapid glycolysis at high temperature can result in tough meat. The pH of meat from heat stressed birds was 5.46, from cold stressed, 5.51, and from control birds, the pH was 5.53 (Lee *et al.* 1976). This indicates that this type of stress has little influence on the final pH of chicken meat.

Seafoods.—The pH of fish (7.0-7.3) is lowered during rigor to pH 5.5-6.5, depending on the species of fish and the initial amount of glycogen in the muscle. Fish muscle generally has less glycogen than poultry or red meat.

The pH of cod was found to be lower at the end of the spawning season and large cod had a lower pH than small cod (Love *et al.* 1972). They found that a low pH was related to gaping in cod. This is a condition in which the tissue which holds the muscle segments together becomes weakened and causes slits to appear across the fillets.

Haddock has a lower pH than whiting (Love and Haq 1970). They postulated that the difference may be in their eating habits. The whiting chases food, while the haddock tends to graze. The haddock being a less active fish probably does not struggle as much when caught, so that there would be greater carbohydrate reserve, which results in more lactic acid production post mortem and a lower pH. Of all the fish for which data are available, the halibut has the lowest ultimate pH and the best keeping quality.

Canned crab is usually pH 6.8-7.4, but may be higher. Motohiro and Inoue (1970) observed that the pH of fresh king crab was 6.8 for the hard shell, and 7.4 for the paper shell intermolt stages.

A pH range of 7.1-8.2 was determined on brown shrimp obtained from commercial fishing boats at landing (Cobb *et al.* 1973). The upper limit, pH 8.2, may seem rather high since pH 7.95 has been indicative of shrimp spoilage. At the time of spoilage, Cobb *et al.* (1973) found the pH of brown shrimp to be 7.3-8.5. This indicates that pH 7.95 has little, if any, significance. At a pH of about 8.5, the bacterial numbers were estimated at 10^5 per gram (Flores and Crawford 1973). This level of microorganisms usually does not indicate spoilage. Vanderzant *et al.* (1973) reported the pH of frozen shrimp to range from 6.8 to 7.6.

The pH of fresh oysters ranges from 5.9-6.6. During spoilage, the pH is reduced. Oysters are sour or putrid at pH 5.2 or below.

Since the pH of seafoods tends to be higher than that of red meat or poultry, they are more subject to bacterial deterioration.

Fruits and Vegetables.—Fruits have a lower pH than do most foods. Unripe fruit generally has a lower pH than ripe fruit. As the fruit matures, the moisture and sugars increase and the acids decrease or are diluted. The ripening temperature influences the ultimate pH, generally in a direct relationship. The pH of fruit influences not only the growth of microorganisms, but also quality factors, such as softening and discoloration of the canned product. Since the pH is low, fruits are usually spoiled by mold growth.

The variety, maturity, geographical area and storage conditions influence the pH of the tomato. Peeling and removal of the core result in a lower pH of the product.

Vegetables generally have a higher pH than fruits, and are therefore subject to bacterial spoilage. During heating, the pH is lowered, but not to the level that would prevent bacterial growth. A pH of 6.7-7.0 in green vegetables has some merit since heating these vegetables at lower pH levels results in a change of color due to chlorophyll being converted to pheophytin, which is dull green to brownish-olive.

OXIDATION REDUCTION POTENTIAL

The oxidation-reduction (OR) potential is referred to in Chapter 2 in regard to the use of reductase tests to estimate microbial numbers in foods.

When a substance is oxidized, it loses electrons, and these electrons must be accepted by another substance which then becomes reduced. The OR potential is a measure of the tendency of a reversible system to give or receive electrons.

The electrode potential at a constant pH is described by the Nernst equation:

$$Eh = Eo - \frac{RT}{nF} \ln \frac{(red)}{(ox)}$$

Eo is a constant characteristic of the system; R is the gas constant (8,315 volt-coulombs); T is the absolute temperature; n is the number of electrons involved in the OR process; F is the Faraday (96,500 coulombs); ln is the natural logarithm (base e); and (red) and (ox) are the concentration of the reduced and oxidized states. At 30°C, and constant pH, by substituting the values for R, T, F, and multiplying by 2.302 to convert the natural logs to common logs, the equation becomes:

$$Eh = Eo - \frac{0.06}{n} \log \frac{(red)}{(ox)} \quad \text{or}$$

$$Eh = Eo + \frac{0.06}{n} \log \frac{(ox)}{(red)}$$

When (ox) = (red), the system is 50% oxidized and 50% reduced, the log of 1 equals 0, and Eh = Eo. The Eo values for various indicators are listed in Table 4.11 for pH 7.0 at 30°C. The OR potential is interrelated to temperature and pH. Methylene blue has an Eo of +0.101 volts at pH 5, +0.011 v at pH 7 and −0.050 v at pH 9, when the temperature is 30°C.

TABLE 4.11

OXIDATION–REDUCTION INDICATORS

Indicator	Eo (volts)[1]
Phenol-*m*-sulfonate-indo-2,6-dibromophenol	+0.273
m-Chlorophenol-indo-2,6-dichlorophenol	+0.254
o-Chlorophenol-indophenol	+0.233
Bindschedler's green	+0.224
2,6-Dibromophenol-indophenol	+0.218
2,6-Dichlorophenol-indophenol	+0.217
o-Cresol-indophenol	+0.191
2,6-Dichlorophenol-indo-*o*-cresol	+0.181
Thymol-indophenol	+0.174
m-toluylene-diamine-indophenol	+0.125
1-Napthol-2-sodium sulfonate-indo-2,6-dibromophenol	+0.119
1-Napthol-2-sodium sulfonate-indo-2,6-dichlorophenol	+0.119
Toluylene blue	+0.115
Thionine (Lauth's violet)	+0.062
Cresyl blue	+0.047
Methylene blue	+0.011
Indigo tetrasulfonate	−0.046
Indigo trisulfonate	−0.081
Indigo disulfonate	−0.125
Cresyl violet	−0.167
Phenosafranine	−0.252
Tetramethyl-phenosafranine	−0.273
Induline scarlet	−0.299
Rosindone sulfonate No. 6	−0.385

(1) Eo value at pH 7.0 and 30°C.

As stated in Chapter 2, by the use of these indicators, the OR potential of a system can be estimated. The indicator should give a distinct color to the solution at the pH of the system being studied. Unfortunately, some behave as both OR and pH indicators. They may give an intense color in an alkaline solution, but in neutral and acid solutions yield only a dull tint.

Eh is a measure of intensity, not of capacity of a system. The capacity is the poising ability, which acts in a manner similar to buffers in maintaining pH at a constant value. A system is said to be poised if it resists change in potential.

Effect on Microorganisms

In microbial cultures, the simultaneous oxidations and reductions are the sources of energy for cell processes. Since energy is needed by the cell to function normally, OR reactions and OR potentials are important.

The broad classifications of organisms into aerobes and anaerobes, as discussed in general microbiology, is based on their apparent tolerance to oxygen during growth. The obligate or strict anaerobes can grow only in the absence of oxygen. Facultative anaerobes can grow with or without oxygen. Generally, these organisms grow more rapidly in aerobic conditions. However, the streptococci and lactobacilli do equally well with or without oxygen. Certain streptococci and pediococci are microaerophilic. They grow best in very small amounts of oxygen. The strict or obligate aerobes require oxygen for growth.

In discussing growth limiting effects on anaerobic bacteria, Hentges and Maier (1972) cited references to show that some obligate anaerobic types of organisms failed to grow if exposed to oxygen in the air during the manipulation of diluting and plating. These obligate anaerobes rarely are reported in foods, probably because not enough care is taken to find them. If they are such strict anaerobes, they probably cause few, if any, serious problems in foods.

Loesche (1969) reported that anaerobes differ in their sensitivity to oxygen. Strict or obligate anaerobes included organisms from the genera *Treponema, Clostridium, Selenomonas, Succinivibrio, Butyrivibrio* and *Lachnospira.* This group is called anaerobic with an anoxybionotic (not capable of using atmospheric oxygen in their growth) metabolism. The second group of organisms, designated as moderate anaerobes, included *Bacteroides, Fusobacterium, Clostridium* and *Peptostreptococcus.* A third group of organisms produced less growth in strictly anaerobic conditions than in the presence of small amounts of oxygen. This is typical for microaerophiles. The only genus named was *Vibrio.*

The moderate anaerobes can be plated at room atmosphere and incubated in anaerobic jars. These are the anaerobes which have been of significance in food microbiology.

From preliminary studies, Hentges and Maier (1972) stated that the OR potential is unimportant regarding the multiplication of *Bacteroides* species. When oxygen is excluded from the environment, *Bacteroides* are unaffected by a high OR potential of the medium. Walden and Hentges (1975) studied the growth and survival of *Bacteroides fragilis, Clostridium perfringens* and *Peptococcus magnus* at low and high Eh with and without oxygen. Even at an Eh of −50 mv these organisms did not grow in the presence of oxygen. No inhibition was observed in the absence of oxygen even at an Eh of +325 mv. This indicates that oxygen, rather than a positive potential, was the limiting factor for growth of these anaerobes. Of these three organisms, *P. magnus* was the most sensitive to oxygen and *C. perfringens* the least sensitive.

Onderdonk *et al.* (1976) reported that *B. fragilis* was not affected by an Eh of +300 mv in the absence of oxygen, but cell death occurred with oxygen at +250 mv. They suggested that dissolved oxygen has an inhibitory effect on *B. fragilis*, which may be independent of changes in Eh alone. Similar effects of oxygen at low Eh were noted for sporulation and growth of *Clostridium butyricum* (Douglas *et al.* 1973).

Due to complexities in maintaining and measuring OR potentials, there is not as much information on this parameter as on pH. Also, the results vary considerably from one experiment to another, between strains of species and between different researchers.

Keeney (1973) suggested the ranges for growth of aerobes from +350 to +500 mv, facultative anaerobes from +100 to +350 mv and obligate anaerobes below −150 mv.

Clostridium perfringens is not a strict anaerobe and is known to be aerotolerant. Limiting Eh values for growth range from −125 to +287 mv. Many anaerobic clostridia (*C. botulinum, C. histolyticum, C. sporogenes*) do not need a negative Eh for growth. These organisms will grow at Eh levels from +85 to +160 mv.

Effect of Organisms on Eh

It has been well established that, as organisms metabolize, a lower Eh value is produced in the medium or substrate. The lowering of the potential has been ascribed to the consumption of oxygen or the production of reducing substances.

With aerobic bacteria, there is a slight lowering of the OR potential during the lag phase of growth due to the consumption of oxygen. As the bacteria enter the log phase of growth, this oxygen usage increases and causes a rapid drop in Eh. As the Eh becomes negative, the growth rate of the bacteria decreases. An overall lowering of the potential is between 400 and 500 mv.

At the beginning of growth, facultative anaerobes alter the Eh of the substrate in a manner similar to aerobes, but the rate of reduction may be somewhat slower. After the bacteria enter the log phase of growth, there is a very rapid decrease in the Eh. This overall potential drop may be from 700 to 800 mv or more. As the culture reaches the stationary phase, and as the age of the stationary phase increases, the observed OR potential shifts to more positive values.

Oxygen is removed from media prior to inoculation and growth of anaerobes. As a result, the initial Eh is lower than that for aerobes or facultative anaerobes. During the germination and emergence of spores or the lag phase of growth, the Eh is lowered by 500 to 700 mv (Douglas and Rigby 1974). There is little further reduction during the log phase of growth. Although the total reduction of Eh is generally less for anaerobes than facultative anaerobes, a lower Eh is attained due to the lower initial Eh of the substrate.

Redox Potential of Food

The OR potential of a food depends upon the natural redox potential and poising capacity of the food and on the oxygen tension in the atmosphere and the access that the atmosphere has to the food. Due to the complications of determining the Eh of foods and the variations obtained due to atmospheric influences, the redox potentials of only a few foods are available.

Harrison (1972) stated that in a conglomerate mixture with different systems, not in equilibrium, an overall redox potential cannot be conceived. It is evident that most foods are conglomerate mixtures. Regardless, there have been attempts to measure and compare the redox potential of foods. Living cells tend to have a low OR potential due to -SH groups in animal products and ascorbic acid and reducing sugars in plants. In dairy products, the redox potential is related to fat oxidation. Thus, there is an inverse relation of OR potential and keeping quality. The contamination of milk with cupric or ferric ions tends to increase the OR potential, resulting in less shelf-life of the product.

The OR potential of cheese depends on the type, with Emmenthal (Swiss) cheese varying from −200 to −50 mv and cheddar cheese from +220 to +340 mv (Mossel and Ingram 1955).

The Eh of meat has been reported to range from −150 to +250 mv. There is an oxygen requirement for post-mortem tissue. The myoglobin of muscle can bind oxygen to form oxymyoglobin, or it can be oxidized by oxygen to form metmyoglobin. To oxygenate the pigment requires about 5 μl of oxygen per gram of meat, while oxidation to metmyoglobin requires about 13 μl of oxygen per gram of meat. Other oxygen requirements are tissue respiration, lipid oxidation, tissue fluids with low oxygen tension dissolving O_2 and bacterial demands. DeVore and Solberg (1974) found that tissue respiration accounted for 80% of the oxygen uptake of post-mortem meat, while bacterial demands were insignificant.

The Eh of plant products varies from +383 to +436 mv for fruit juices, +74 mv for spinach, +225 mv for barley and −470 mv for wheat germ (Mossel and Ingram 1955). The Eh of cherries and peaches was determined to be +179 mv and +175 mv, respectively (Ross *et al.* 1953). The addition of ascorbic acid lowered the Eh of the fruit.

Apparently no measurements have been made on the poising capacity of foods. Many fresh foods have thiol-containing amino acids and peptides, reducing sugars and ascorbic acid so that the pO_2 must be changed greatly before the Eh is affected (Mossel 1971A).

The oxygen content of the atmosphere and the access of the atmosphere to the food influences the OR potential. Large pieces of foods, such as carcasses, have less total surface area than do smaller sized foods. Food in a deep vat has less surface exposed than food in shallow vats and liquid at rest in a vat has a lower Eh than liquid being stirred or mixed.

Food that is packaged in a material impervious to oxygen should have a lower OR potential than unpackaged food. Vacuum packaging will change the atmosphere surrounding the food and also prevent the free access of oxygen to the food.

Relationships of Food OR and Microorganisms

The relatively high Eh of fruit juices favors the growth of aerobic microorganisms. Since these products have a low pH, the aerobic yeasts and molds are the predominant organisms that cause spoilage.

In meat, the surfaces exposed to the atmosphere allow the growth of aerobic bacteria, but in deeper tissue, the Eh is lower so that anaerobes can grow. In pre-rigor meat, the Eh is sufficiently high to prevent the growth of anaerobic types, but during rigor, the Eh is reduced to allow the growth of *Clostridium* species. Heating of milk lowers the Eh, so that although clostridia do not grow in fresh raw milk, *Clostridium botulinum* grows in heat-sterilized whole milk (Kaufmann and Marshall 1965).

Although molds are considered to be strictly aerobic, some types can grow in low levels of oxygen. One of the reasons that *Penicillium roqueforti* can grow in the inside of Roquefort-type cheese is its tolerance for low pO_2. *Byssochlamys fulva* grows in grape products at extremely low levels of oxygen.

INHIBITORY SUBSTANCES

When inhibitory substances in foods are mentioned, one might envision some chemical preservative added by the processor. However, microbial inhibitors were present long before man appeared on earth.

Action of Inhibitors

Generally, inhibitors affect microorganisms by acting on the whole cell, cell wall or cell membranes, by interfering with the genetic mechanism of the cell, by interfering with the enzyme systems of the cell or by binding essential nutrients.

Many microbial inhibitors have been reported to be naturally present in foods. The factors involved with the growth of microorganisms, such as pH, that have already been discussed in this chapter, are partially due to chemicals naturally present in foods. Besides these, there are other inhibitors in both animal and plant products.

Inhibitors in Animal Products

Microbial inhibitors have been found in egg white, milk and various tissues from animals.

Egg White.—The growth of bacteria in egg white is limited due to the presence of lysozyme, enzyme inhibitors (such as ovomucoid), avidin, conalbumin and a high pH (Table 4.12).

TABLE 4.12

MICROBIAL INHIBITORS IN EGG WHITE

Inhibitor	Effect on Microorganisms	% Egg White Solids
Lysozyme	Lysis of cell walls (Gram positive bacteria)	3.5
Ovomucoid	Enzyme inhibitor	11.0
Conalbumin	Chelates iron	13.0
Avidin	Binds biotin	0.05
pH	Alkaline conditions restrict growth	–

Lysozyme.—In 1922, Fleming reported a lytic agent in egg white and other biological systems, which was named lysozyme. This enzyme has been referred to by names such as muramidase, N-acetylmuramide glycanohydrolase, glucohydrolase, mucopeptide glycohydrolase and β-glucosaminidase. To help alleviate confusion about the name of this and other enzymes, the Commission on Enzymes has assigned numbers to the various enzymes. Lysozyme is 3.2.1.17.

Lysozyme is found in egg white, milk, many tissues and secretions of animals, some molds and in the latex of various plants. This widespread distribution has made it a very popular substance for study. Lysozymes from different animal species and from different organs of the same animal differ chemically and immunologically, but they possess the same type of biological activity. The lysozyme content varies not only between species, but also between strains of the same species. On a dry weight basis, chicken egg white contains about 3.5% lysozyme. This is the most readily available source, and hence, the most studied of the lysozymes.

Heated lysozyme is unstable in alkaline solutions, while unheated lysozyme is quite stable even at pH values of 8.7 to 9.0. The optimum activity is observed at pH 5.3-6.6.

Bacteria display various sensitivities to the lysis of the cell wall by lysozyme. Certain micrococci are rapidly lysed by 1 μg/ml of lysozyme. *Bacillus megaterium* requires 50 μg/ml and *B. cereus* barely is attacked by 50 μg/ml of lysozyme.

Gram negative bacteria are rather insensitive to lysozyme due to their lipoprotein-lipopolysaccharide layer which acts as a barrier protecting the sensitive mucopolysaccharides of the cell. If the barrier is damaged by certain treatments, the enzyme can penetrate to the mucopeptide layer and cause lysis, or partial lysis, of the cell wall. Starving Gram negative cells at an abnormal pH, treatment with polymyxin B or ethylenediamine tetraacetic acid, as well as reacting with complement, osmotic injury or aerosolization can render these cells susceptible to lysozyme action.

Enzyme Inhibitors.—Enzyme inhibitors found in egg white are the ovomucoids, ovoinhibitors and a ficin-papain inhibitor.

Rhodes *et al.* (1960) divided ovomucoids into four classes according to their inhibitory activities. Some ovomucoids, such as the chicken and goose, inhibit trypsin. Another group inhibits chymotrypsin (golden pheasant). A third group (turkey) inhibits approximately equal amounts of trypsin and chymotrypsin and the fourth group included those from duck and emu which inhibit twice as much trypsin as chymotrypsin. A duck ovomucoid inhibited a commercial bacterial proteolytic enzyme.

Matsushima (1958) discovered an inhibitor in egg white which he termed ovoinhibitor. This substance inhibited fungal proteinase of an *Aspergillus* and a bacterial proteinase prepared from a strain of *Bacillus subtilis.*

There is no evidence that these enzyme inhibitors have any significant effect on microbial growth in egg white.

Avidin.—Each mole of avidin combines with two moles of biotin. Thus, organisms having a strict requirement for biotin as a nutrient are inhibited by this compound.

Conalbumin.—This protein comprises about 12% of the total solids of egg white. It forms a stable complex with two ferric or ferrous ions per molecule and, by this action, it inhibits organisms that require iron. When iron salts are added to egg white, a pink coloration appears due to the formation of the complex. Conalbumin also complexes with copper in a manner similar to iron salts.

Heating conalbumin to 70°-79°C for 3 min results in an 80% loss of activity. When heated at pH 7.6, conalbumin precipitates. The complex of conalbumin and iron is more stable toward proteolysis and thermal denaturation than conalbumin without iron.

Many organisms are inhibited by the presence of conalbumin and lack of iron. These include Gram negative and Gram positive bacteria as well as yeasts.

It is evident that conalbumin does not destroy the microorganisms but merely inhibits the growth as evidenced by an extended lag phase of growth. Feeney and Nagy (1952) postulated that the dissociation of the complex would always allow traces of iron in the albumen which, when used by the organism, would allow the release of further supplies of iron from the complex. This would result in a slower rate of growth of the organisms. To capture iron, microorganisms synthesize their own iron-sequestering compounds.

pH.—The high pH established when carbon dioxide leaves egg albumen inhibits the growth of many microorganisms.

The most important inhibitor in egg white is lysozyme which reacts with Gram positive organisms. Primarily because of this substance, the spoilage of eggs is due to Gram negative bacteria. Conalbumin is the main inhibitor of Gram negative bacteria. It does not prevent growth but does inhibit or delay growth. This is important since it is necessary to have shell eggs in commercial channels for grading, packing and retailing.

Dairy Products.—Several substances may be present in milk that inhibit microorganisms. These include antibiotics, pesticides, bacterial viruses, sanitizing and cleaning compounds and various natural inhibitors. Antibiotics can be defined as microbial metabolic products which have antimicrobial activity.

The naturally occurring microbial inhibitors in raw milk include: lysozyme, cationic proteins, agglutinins or antibodies, leucocytes, lactenins, lactoperoxidase, lactoferrin and fatty acids.

Many of the inhibitors in milk are heat labile and not found in pasteurized milk. However, there are heat stable inhibitors which can interfere with studies determining the effect of heat on organisms in milk since they will inhibit the growth of organisms surviving the heat treatment. Natural inhibitors also pose a problem with tests used to detect the presence of antibiotics in milk. The results must be observed cautiously to determine if inhibition is due to natural inhibitors or due to antibiotics.

The antibiotic, nisin, is produced by certain strains of *Streptococcus lactis*, a common dairy organism, and occurs naturally in small amounts in raw milk and certain milk products.

Lysozyme.—Goat milk has about twice as much lysozyme as cow milk or sheep milk. Human milk has about 400 mg/kg, or 1500 times as much lysozyme as goat milk. Sow milk has essentially no lysozyme.

Due to the relatively low concentration of lysozyme in milk, the effect on microorganisms would not be as great as found for egg white lysozyme.

Cationic Proteins.—These proteins carry a positive charge at physiological pH values and react with the anionic binding sites of cell walls and membranes of bacteria (MacMillan and Hibbitt 1973). This causes an increased permeability of the bacterial membranes resulting in leakage from the cells. Hibbitt *et al.* (1971) isolated cationic proteins from bulk milk samples. The antimicrobial activity was not destroyed by heating to 70°C for 30 min, but was almost completely destroyed at 100°C. These proteins have activity to both Gram positive and Gram negative bacteria.

Antibodies.—Antibodies of various kinds have been found in milk. Vedamutha *et al.* (1971) reported a pseudoglobulin, one of the immune globulins of milk, was inhibitory to several strains and species of propionibacter.

Leucocytes.—Leucocytes or phagocytes are present in freshly drawn milk and continue to ingest microorganisms for an undetermined period. Leucocytes are especially prevalent in milk from cows with mastitis.

Lactenins.—In 1930, the inhibitors in milk were called lactenins. Auclair and Hirsch (1953) found two antistreptococcal principles in milk and described them as lactenin 1 (an agglutinin) and lactenin 2 (lactoperoxidase). Hence, various inhibitory agents have been referred to as lactenins.

Lactoperoxidase.—There are multiple forms of this enzyme in milk. Besides peroxidase, the inhibition requires the presence of thiocyanate and peroxide. Hydrogen peroxide is a metabolic product of microorganisms. Catalase interferes with the inhibitory properties of this system. The addition of hydrogen peroxide to milk increases the antibacterial activity. The action on cells is greatest at pH 5.0 or less (Reiter *et al.* 1976). The exact mechanism of this inhibition is not known at present.

Lactoferrin.—This is an iron binding protein in milk that acts on bacteria similar to conalbumin in egg white. Normal milk contains 0.1 to 0.3 mg/ml. This is increased from 4 to 10 times in mastitic milk and in colostrum (Harmon *et al.* 1975). The addition of citrate reduces the inhibitory effect of lactoferrin (Bishop *et al.* 1976). Ashton *et al.* (1968) suggested that casein, a protein in milk, also is capable of binding iron.

Fatty Acids.—Fatty acids may be inhibitory, stimulatory or have no effect on microorganisms. The action depends upon the type of organism, the type and concentration of the fatty acid, the pH of the growth medium or the presence of certain other compounds. Those fatty acids with six or more carbon atoms are more inhibitory to Gram positive than to Gram negative cells (Table 4.13). Shorter chain fatty acids are about equally effective toward either Gram positive or Gram negative cells. It has been suggested that long chain fatty acids cannot penetrate the lipopolysaccharide layer of Gram negative cells. Reportedly, there are greater biological effects with branched chain fatty acids. *Cladosporium* utilized long chain saturated fatty acids while short chain acids were toxic to the fungus (Teh 1974).

TABLE 4.13

SOME COMMON FATTY ACIDS

Common Name	Systematic Name	Chemical Formula
	Saturated Acids	
Formic acid[1]	Methanoic acid	$HCOOH$
Acetic acid	Ethanoic acid	CH_3COOH
Propionic acid	Propanoic acid	CH_3CH_2COOH
Butyric acid	Butanoic acid	$CH_3(CH_2)_2COOH$
Caproic acid	Hexanoic acid	$CH_3(CH_2)_4COOH$
Caprylic acid	Octanoic acid	$CH_3(CH_2)_6COOH$
Capric acid	Decanoic acid	$CH_3(CH_2)_8COOH$
Lauric acid	Dodecanoic acid	$CH_3(CH_2)_{10}COOH$
Myristic acid	Tetradecanoic acid	$CH_3(CH_2)_{12}COOH$
Palmitic acid	Hexadecanoic acid	$CH_3(CH_2)_{14}COOH$
Stearic acid	Octadecanoic acid	$CH_3(CH_2)_{16}COOH$
Arachidic acid	Eicosanoic acid	$CH_3(CH_2)_{18}COOH$
	Unsaturated Acids	
Crotonic acid	*trans*-2-Butenoic acid	$CH_3CH{=}CHCOOH$
Palmitoleic acid	9-Hexadecenoic acid	$CH_3(CH_2)_5CH{=}CH(CH_2)_7COOH$
Oleic acid	*cis*-9-Octadecenoic acid	$CH_3(CH_2)_7CH{=}CH(CH_2)_7COOH$
Linoleic acid	*cis*-9, *cis*-12-Octadecadienoic acid	$CH_3(CH_2)_3(CH_2CH{=}CH)_2(CH_2)_7COOH$
Linolenic acid	9,12,15-Octadecatrienoic acid	$CH_3(CH_2CH{=}CH)_3(CH_2)_7COOH$
Arachidonic acid	5,8,11,14-Eicosatetraenoic acid	$CH_3(CH_2)_3(CH_2CH{=}CH)_4(CH_2)_3COOH$

[1] Formic, acetic, and propionic acids are included since they are listed as fatty acids in some publications, although others consider butyric acid as the simplest fatty acid.

The mode of action and the type of inhibition depends upon the fatty acid and the concentration. At low concentration, some fatty acids are stimulatory, while at higher concentrations, inhibitory to the same organism. The toxic effect is, in general, bacteriostatic, but for certain organisms at high concentrations, may be bactericidal. The bacteriostatic effect may be due to the blockage of adsorption of essential nutrients. As the incubation period is increased, the cells seem to overcome this difficulty, and have, at times, shown total cell growth greater than when no fatty acids were present. Besides affecting the permeability of the cell membranes and inhibiting transfer mechanisms, fatty acids inhibit the action of enzyme systems.

Certain substances interfere with the inhibitory effect of the fatty acids. These substances include lecithin, cholesterol, alpha-tocopherol, charcoal, serum albumin, starch, bile salts, lumisterol and saponin. Galbraith and Miller (1973) reported the antagonism of alkaline earth metals on the activity of fatty acids.

Although fatty acids are found in various foods containing fats, they have been of special concern to the dairy industry due to their prevalence in butter and cheese. The amount of free fatty acids increases during the aging of cheese. This increase in fatty acids inhibits the growth of Gram positive microorganisms in cheese.

Tissue Foods.—Various natural microbial inhibitors have been found in tissue foods (meat, poultry, fish). Lysozyme, discussed in relation to inhibitors in egg white, is found in various tissues. Tolerances for antibiotic residues in meats have been established by the World Health Organization. Even at low levels, some microorganisms may be inhibited.

A living animal has defense mechanisms to inhibit the invasion of microorganisms. These include various antibacterial substances as well as natural and immune antibodies. Although, after slaughter, the mechanisms for producing these materials is lost, those substances already produced in the tissue will remain. The extent to which they have an antibacterial function depends primarily upon their stability.

Tissues from cattle (brain, spleen, heart, kidney and liver) possess antistaphylococcal substances. Basic polypeptides with antibacterial properties have been extracted from tissues (thymus, spleen and thyroid). Synthetic basic polyamino acids are active against both bacteria and viruses. Spermine and spermidine are polyamines found widely distributed in animal tissues and are inhibitors of many types of microorganisms (Bachrach and Weinstein 1970; Fair and Wehner 1971). The antibacterial action of basic polyamino acids is thought to be disruption of normal cell functions due to their combining with components of the cell wall.

An active substance against Gram positive organisms in lymph nodes is believed to be derived from damaged leucocytes.

Some hormones are bacteriostatic. Progesterone (progesterol) is antibacterial to Gram positive organisms and diethylstilbestrol (DES) has a bactericidal action against *Staphylococcus aureus*, the extent of killing depending upon the concentrations of DES, pH and time of exposure (Yotis and Baman 1969). Various hormones affect the growth of *S. aureus* (Fitzgerald and Yotis 1971). Deoxycorticosterone inhibits Gram positive bacteria, yeasts and molds, but not Gram negative bacteria.

A viral inhibitor was found in ground beef by Konowalchuk and Speirs (1973). The chemical and physical properties of the inhibitor would indicate that it was a viral antibody. Ground beef purchased in supermarkets showed wide variation in the inactivation of coxsackievirus B5. Poliovirus was inactivated by some of the meat extracts. They observed that the virus was released from the inhibitor when treated with hydrochloric acid at pH 3 for one hour. This causes concern about the possible release of bound viruses caused by the pH found in the stomach and also the possibility of finding higher titers of viruses in meats if given an acid treatment prior to analysis.

The antimicrobial free fatty acids can be found in animal products, especially in cured ham and fermented sausages.

Plant Products

Due to the discovery of antibiotics, there have been several surveys in which crude plant extracts or juices were tested for antibacterial activity. The extracts of several thousand plant species have been tested and, depending upon the plants surveyed and the test organisms used, from 30 to 50% of the extracts showed some antimicrobial activity. Juices and/or extracts from cabbage, carrot, green beans, celery, chicory, cucumber, chard, okra, rhubarb, corn and turnip have shown antimicrobial activity to one or more microorganisms.

To be fair, it should be mentioned that some extracts have shown stimulatory effects. Pea medium is used to detect heat stressed anaerobes in canned foods.

Lysozyme is found in some plant products, although in relatively low levels. The trypsin inhibitor of soybeans is well known. Enzyme inhibitors are found in various plant products.

Gossypol, a component of cottonseed, inhibits Gram positive bacteria and some yeasts and molds, but has little or no effect on Gram negative bacteria.

Purothionins, basic polypeptides present in the wheat endosperm and other cereal species, have shown inhibitory properties toward some human pathogens and yeast. DeCaleya *et al.* (1972) tested wheat purothionins against phytopathogenic bacteria and found they inhibited one of seven *Pseudomonas* species tested, two *Xanthomonas* species, *Erwinia amylovora* but not *Erwinia carotovora* or five *Corynebacterium* species. The inhibitor was bactericidal at a concentration approximately twice that showing inhibition.

Benzoic acid and its salts are used as food preservatives and are found naturally in many berries. Cranberries are especially prominent for containing benzoic acid. Other compounds in cranberries with antimicrobial properties were described by Chu *et al.* (1973).

The hop resins, lupulone, humulone and isohumulone, are inhibitory to Gram positive, but not Gram negative bacteria. Isohumulone is less inhibitory than the other two resins (Teuber 1970).

Some vegetable oils contain a sporistatic or sporicidal agent (Dallyn and Everton 1970). They believed a peroxide precursor of a carbonyl is involved in the antimicrobial activity.

Anthocyanin pigments are present in various fruits. These pigments possess antibacterial properties for both Gram positive and Gram negative organisms but, in some cases, show stimulatory properties. Carpenter *et al.* (1967) suggested the inhibitory effects are due to chelation of metal ions or a denaturing effect due to their redox activity.

An antimicrobial effect of cocoa has been observed. The action may be due to anthocyanin pigments. Weissberger *et al.* (1971) reported several organic acids in cocoa beans including free fatty acids in cocoa fat which also might account for some of the antimicrobial activity.

Pectin is a component of plant products. Although pectin shows no inhibitory properties to microorganisms, a product of hydrolysis of pectin, alpha methyl D-galacturonate, inhibits the growth of many Gram negative dysentery bacteria.

A compound, first called lycopersicin and later named tomatin, or α-tomatine, is found in tomatoes. This substance inhibits various fungi and bacteria.

Olives contain oleuropein, a phenolic compound. Juven *et al.* (1972) reported that oleuropein was inhibitory to species of *Lactobacillus, Leuconostoc,* yeasts and molds. An ethyl acetate extract of oleuropein reveals two components (glycoside oleuropein and its phenolic aglycone), which have antibacterial activity. Fleming *et al.* (1973) found oleuropein was not inhibitory to lactic acid bacteria, but two hydrolysis products, the aglycone and elenolic acid were inhibitory for four species of lactic acid bacteria. The acid hydrolysate of oleuropein showed inhibitory properties to species of *Lactobacillus, Pediococcus, Leuconostoc, Staphylococcus, Bacillus, Salmonella, Pseudomonas, Erwinia* and *Xan-*

thomonas, but no inhibition of the yeasts that were used. Whether oleuropein, per se, is a microbial inhibitor seems to be questionable. According to Juven *et al.* (1972), the inhibition is influenced by salt, size of inoculum and the substrate used for growing the cells and testing the inhibitory properties. The inhibition is due to leakage of cell constituents.

Homogenates and extracts of onions and garlic show antibacterial activity to a variety of microorganisms. It is thought that the intact bulbs are not antimicrobial, but, when crushed, enzymes react on compounds containing cysteine sulfoxide and form inhibitory thiosulfinates. The active principle is allicin (thio-2-propene-1-sulfinic acid-5-allyl ester). It is not known if the action is due to enzyme inhibition or to membrane disruption (Bogin and Abrams 1976). The inhibitory activity can result in low microbial plate counts of these foods. Higher counts of bacteria, yeasts and molds are obtained when 0.05% K_2SO_3 is added to the dilution water.

Plants of the mint family contain tannin and non-tannin polyphenol substances with antiviral properties (Herrmann and Kucera 1967). Tannins comprise a group of natural phenolic compounds synthesized by plants. English walnut meats, carob pods, sorghum grain and malt inhibit microorganisms due to the presence of tannins. There is evidence that tannins interfere with enzyme activity as well as absorbing on the cell surface and altering the permeability of the cell wall.

Fruits contain various organic acids that inhibit the growth of microorganisms. Fruit juices have antiviral action (Konowalchuk and Speirs 1978). A natural isoflavone from certain fruits has antifungal activity (Fukui *et al.* 1973). Acetaldehyde is a natural compound in fruit tissue that inhibits yeast (Barkai-Golan and Aharoni 1976).

A phenolic type compound isolated from white potatoes inhibits the growth of *Aspergillus parasiticus* (Swaminathan and Koehler 1976).

Spices were recognized as having food preserving possibilities by the Egyptians some 3,000 years ago. The antibacterial factors are found in the essential oils of the spices. Two of the most effective germicidal spices are cinnamon and cloves. Cinnamon contains cinnamic aldehyde and both contain the ether eugenol.

The essential oils of citrus fruits are found in the peel and can be recovered from this waste product. The disinfectant power of lemon and orange oil against spore-bearing organisms is reportedly greater than that of phenol. Orange oil at 0.1% and/or 0.2% was found to inhibit both Gram positive and Gram negative bacteria (Subba *et al.* 1967). At 0.2 to 0.3% orange and lemon oil suppressed growth of *Aspergillus parasiticus* and aflatoxin production (Alderman and Marth 1976).

Summary

There are many natural microbial inhibitors in foods. These inhibitors give the living animal or plant some protection from invading parasites. The food products, being an extension of the animal or plant, have a residual amount of these agents. Most of these natural compounds inhibit Gram positive more than Gram negative organisms. Perhaps this is one reason that spoilage is more often associated with Gram negative organisms.

The microbial inhibitors naturally present in foods may maintain the food in a satisfactory state for transportation, processing and distribution.

PROTECTIVE BARRIERS

Some foods are protected from direct contact with microorganisms by a natural cover or barrier. Examples of barriers are the shell and shell membrane of eggs, the testa of seeds and the cuticle of intact plant organs.

Egg Shell and Membranes

An invisible, natural, protein-like film on the egg shell, called the cuticle, or bloom, is considered by some to be the first line of defense against microbial penetration of the egg. The second physical barrier is the egg shell. Inside the shell are two membranes, the outer and inner shell membranes, which are the third and fourth barriers to microbial penetration. The structure of the egg and egg shell is shown in Fig. 4.5 and 4.6.

Cuticle.—Washing of eggs, or the use of abrasives to remove dirt, disrupts the cuticle layer and allows easier penetration by microorganisms. There is a diversity of opinion on the effect of washing eggs on the resultant contamination of the interior. The cuticle tends to crack and deteriorate with time. Hence, it might not be much of a microbial barrier for eggs stored for long periods.

Egg Shell.—The egg shell is not a homogeneous structure. It consists of an organic framework of fibers and an interstitial substance of inorganic material. The two main layers of the shell are the outer or spongy layer and the inner or mammillary layer. The outer layer is thicker and contains most of the minerals.

The shell contains from six to eight thousand microscopic pores. These pores allow the exchange of water vapor and gases between the contents of the shell and the outer atmosphere. The pores vary in size with extremes being reported as 1.6 to 74.7 μm. The average pore size is probably between 20 and 45 μm. The size of most pores will allow the passage of many microorganisms. Even yeast cells can be forced through pores and mold mycelia grow through the pores. There is a linear relationship between shell porosity and infection of the eggs.

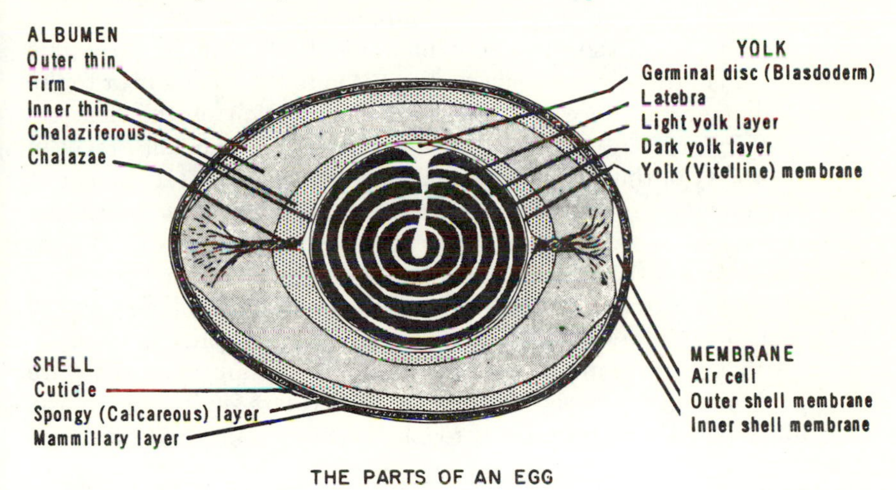

Courtesy of USDA

FIG. 4.5. THE PARTS OF AN EGG

Shell weight and shell thickness do not correlate with penetration of microorganisms. However, these factors can be involved when considering damage to the shell during handling, since thin shells have a tendency to crack more readily than thicker shells. Eggs with damaged shells are more subject to penetration of bacteria and egg spoilage.

If the temperature of the egg is higher than the surroundings, a condition which exists when the egg is laid, there is a greater possibility of penetration and, as the temperature differential increases, the expected infectivity increases. This is due to the cooling of the egg contents with resultant contraction which results in a higher pressure on the outside of the shell than on the inside. This pressure differential will force microorganisms through the pores of the shell.

Moisture on the egg shell, such as occurs during washing of eggs or when a cold egg is brought into a warm and humid room, increases the potential for microbial infection of the egg.

Shell Membranes.—The outer shell membrane is in contact with the egg shell and the inner shell membrane, sometimes called the egg membrane, is in contact with the albumen. They are composed mainly of protein fibers (fibrin and mucin) strengthened with an albuminous cementing material. The membranes have been considered as effective bacterial filters. The inner membrane is more effective than the outer membrane because of its closely knit fibers. Organisms inoculated onto the shell surface are found on the membranes almost instantly, but microorganisms inoculated into the air cell and onto the inner membranes are not recoverable from the albumen for several days. This apparent retention of bacteria on the inner membrane may be partly due to conalbumin in the albumen, since adding iron to the inoculum or to the interior albumen of the egg hastens the penetration of the bacteria through the inner membrane. Wedral *et al.* (1971) found no difference in the permeability of the inner shell membrane before and after passage of either *Pseudomonas aeruginosa* or *Salmonella typhimurium*, indicating that no enzymatic activity is needed for penetration. With the conditions of their experiment, both bacterial species were able to penetrate from the shell surface through the inner membrane within two hours. By 36 hr after shell inoculation, all except one egg showed contaminated albumen. Although there are no pores, per se, in the membranes, there are apparently openings between the intermeshed fibers similar to a maze, which allow passage of bacteria through the membranes.

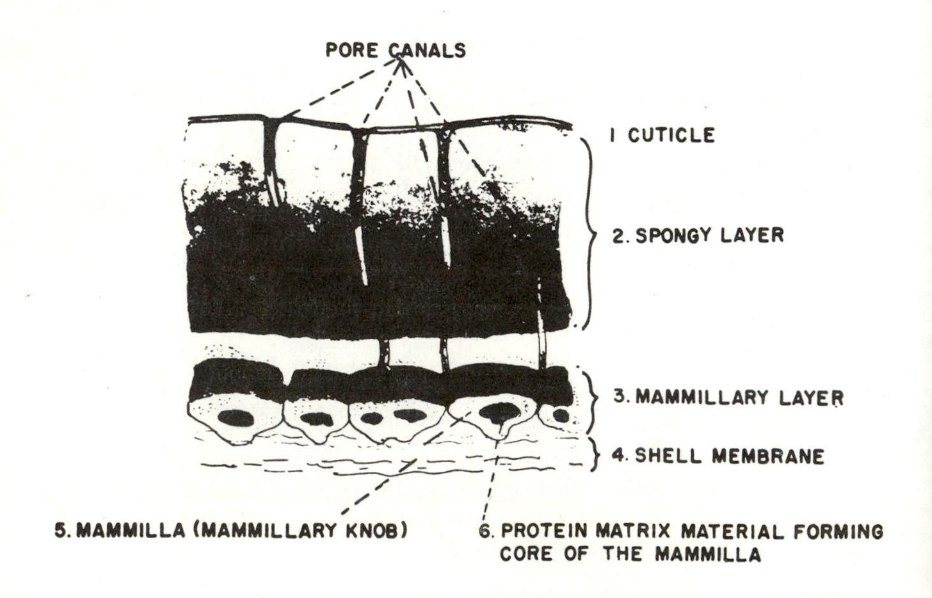

Courtesy of USDA

FIG. 4.6. MAGNIFIED RADIAL SECTION THROUGH AN EGG SHELL

Even if bacteria can penetrate these barriers, they are confronted with the antimicrobial mechanisms of the albumen before they can attack the nutrients in the yolk.

Barriers in Plant Products

Foods, such as fruits, vegetables, grains and nuts, contain a peel, skin or hard surface that covers the interior food and protects it from microbial attack. The effectiveness of the coverings is evidenced by the growth of mold primarily on the stem end of fruit, where injury to the cover may occur during picking. Also, when a worm hole or insect or bird bite is observed in an apple or other fruit, the microbial population in that area is high and rotting of the fruit occurs there before spreading to the rest of the fruit. Bruising of fruit or vegetables can disrupt the covering and allow deterioration to occur.

In the section on inhibition, the presence of essential oils in the peel of citrus fruit is discussed. There are inhibitors in the wax-like coating on many fruits. Cabbage leaves contain wax esters on their surface. The wax-like coating of apples increases during storage, the extent of the increase depending upon the cultivar.

The protective coverings are not only microbial barriers, but also help control diffusion of materials into and out of the fruits or nuts. The sale of nuts in the shell has remained fairly constant while purchases of shelled nuts have increased. Removing the nuts from the shell not only subjects them to microbial contamination, but also to oxidative rancidity of the fatty material and other deteriorative changes.

The seed coat and pericarp of grains which allow the long term storage of these foods were important in establishing the civilization of man.

TEMPERATURE

Since water, as the solvent, is necessary for microbial growth, the temperatures which will support microorganisms are limited to those at which water is a liquid. Pure water freezes at 0°C, but the addition of solutes lowers the freezing point. The exact minimum temperature for microbial growth is difficult to determine, since the solutes that are added to prevent freezing of water below −10°C may inhibit the microorganisms.

Water boils at 100°C and is no longer acceptable for growth. The boiling point is increased by adding solutes, but is decreased at pressures less than one atmosphere (above sea level). Organisms have been found to grow at temperatures approaching the boiling point, but generally a temperature somewhere around 90°C is considered the upper limit, although this estimate may be too high or too low. The upper temperature limit for bacteria may remain unknown, due to the difficulty in using high temperatures of incubation.

The biokinetic zone, for the present, is about −15°C to 90°C. No individual organism can grow over the entire potential temperature range. An organism is usually limited to growth in a range of some 25° to 40°C.

Not only is water a limiting factor for lower and upper temperatures, but also it has been suggested that there are phase transitions in water at temperatures near 15°, 30°, 45° and 60°C (Karmas 1973).

Temperature is one of the most important environmental factors that regulates the growth of microorganisms. Temperature is not only related to the ability of an organism to grow, but also to survive. The environmental temperature also has an effect on cell size, metabolic products such as pigments and toxins, nutritional requirements, enzymatic reactions and the chemical composition of cells.

Ranges for Growth

Each organism has a minimum, optimum and maximum temperature for growth. The minimum and maximum temperatures are those beyond which the organism ceases to grow. The optimum temperature is more difficult to describe, since it may be the optimum for total cell yield, rate of growth, rate of metabolism, respiration or the production of some metabolic product. Usually the optimum temperature is based on the rate of growth.

In general microbiology, the microorganisms are usually divided into three arbitrary classes: psychrophilic (low temperature), mesophilic (medium or middle temperature), and thermophilic (high temperature). Psychrophiles and thermophiles are divided into facultative and obligate types. Psychrotrophic is a term used to describe psychrophilic types of organisms. The temperature ranges for these types are listed in Table 4.14. There has never been agreement among microbiologists for the exact ranges of growth. The fact that some organisms grow at low, some at intermediate and others at high temperatures is more important than the exact temperatures for these ranges. The temperature ranges for the growth of several microorganisms are listed in Table 4.15. It should be remembered that the growth temperature depends upon the strain of a species and the chemical and physical properties of the substrate.

In general, organisms can survive and show some growth at temperatures considerably lower than the optimum temperature. When the temperature is increased above the optimum, there is usually a very rapid decline in the growth rate. The maximum temperature is usually only a few degrees (3°-10°C) above the optimum. Above the maximum temperature, not only does growth cease, but the life of the cell is in jeopardy.

TABLE 4.14

APPROXIMATE TEMPERATURE RANGES OF GROWTH FOR ARBITRARY CLASSES OF MICROORGANISMS

	Temperatures (°C)		
	Minimum	Optimum	Maximum
Psychrophilic	−15-5	10-30	20-40
Obligate	−15-0	10-20	20-22
Facultative	−5-5	20-30	30-40
Psychrotrophic	−5-5	25-30	30-40
Mesophilic	5-25	25-40	40-50
Thermophilic	35-45	45-65	60-90
Obligate	40-45	55-65	70-90
Facultative	35-40	45-55	60-80

Psychrophiles and/or Psychrotrophs.—There are many definitions of psychrophilic organisms. One of the simplest definitions is that psychrophiles grow well at 0°C. They should produce a visible colony at that temperature in 7, 10 or 14 days. With this definition there is no consideration of optimum, minimum or maximum temperatures. Perhaps some distinction should be made between those organisms with a maximum temperature of 20°C or less and those that can grow at a higher maximum. Morita (1975) suggested that psychrophiles have an optimal temperature for growth at about 15°C or lower, a maximum at about 20°C and a minimum at 0°C or lower. Organisms that grow at low temperatures, but do not meet the definition are called psychrotrophs. The term psychrotroph was suggested by Eddy (1960) to apply to microorganisms capable of multiplying at 5°C and below, regardless of the optimum temperature. This definition does not include the minimum or maximum temperature. Eddy's definition of psychrotroph is essentially that of a psychrophile, except that the temperature has been increased from 0°C to 5°C. There are few, if any, obligate psychrophiles in foods. Most microorganisms that grow in food at low temperature are psychrotrophic.

The pseudomonads are an important group of psychrotrophs. Some members of the genus can grow at 4°C while others grow at 43°C. Overall, the optimum temperature is somewhere between 20° and 30°C.

P. fluorescens grows at 4°C or less, but not at 41°C. The usual range for *P. aeruginosa* is 8° to 42°C. The optimum temperature for pigment production of *P. fluorescens* is 20°C. The diameter of colonies of *P. fluorescens* increases at a more rapid rate at 30°C than at 25°C or less (Johnson *et al.* 1970).

TABLE 4.15

TEMPERATURE RANGES OF SELECTED MICROORGANISMS

Microorganism	Minimum	Temperature (°C) Optimum	Maximum
Bacteria			
Acetobacter	5	–	42
Acinetobacter	5	–	50
Aeromonas	0-5	25-30	38-41
Bacillus cereus	10	–	–
Brevibacterium	5	–	42
Chromobacterium	2	–	44
C. violaceum	10-15	–	40-44
C. lividum	2	–	35
Clostridium	0-45	–	60
C. botulinum	3.3-10	30-40	–
C. perfringens	15-20	30-40	45-50
C. putrefaciens	0	20-25	30
C. thermosaccharolyticum	45	55	–
Cytophaga xantha	<0	15-20	25
Escherichia coli	5-10	37	–
Gluconobacter oxydans	7	25-30	41
Kurthia	5	25-30	45
Lactobacillus	5	30-40	53
Leuconostoc	10	20-30	40
Micrococcus	10	25-30	45
Moraxella	2	–	42
Propionibacterium	2-3	30-37	45
Proteus	10	–	43
Pseudomonas	4	20-30	43
P. aeruginosa	8	–	42
P. fluorescens	0-4	20-25	40
P. syringae	–	28	–
Salmonella	5-10	35-37	46
Staphylococcus	5-10	35-40	46-48
S. aureus	5-10	35-39	48
Streptococcus cremoris	–	25-30	–
S. faecalis	5-10	37	–
S. lactis	10-15	25-30	40
Vibrio	–	10-37	–
Xanthomonas	0-5	25-31	<40
Yersinia enterocolitica	0-4	–	37
Molds	−10	18-30	55
Aspergillus fumigatus	–	30-40	–
Botrytis cinerea	−1	20	30
Cladosporium	−5- −8	–	–
Mucor mucedo	0	–	25
M. pusilus	–	40-45	–

TABLE 4.15. (*Continued*)

Microorganism	Temperature (°C) Minimum	Optimum	Maximum
Penicillium rubrum	–	25-28	–
Rhizopus stolonifer	5	–	25
Yeasts	−5- −10	21-32	50-60
Candida	0	–	29-48
C. humicola	0-1	–	37
C. lipolytica	5	25	35-40
Hansenula	–	37-42	50
Kloeckera	–	30	–
Rhodotorula	10-1	–	>37
Saccharomyces	0-7	20-30	40
Torulopsis	0	17-25	30-35

Sources of Information: Buchanan and Gibbons (1974); Dennis and Cohen (1976); Inoue and Komagata (1976); Rosenberg (1975).

The spoilage of refrigerated food (meat, milk, eggs) is generally due to psychrotrophic growth. Psychrotrophs are found in many genera and include aerobes and anaerobes.

Mesophiles.—The mesophiles are organisms that grow in the middle temperature range. We can define mesophiles as those organisms with an optimum temperature of 25° -45°C. When attempting to set limits, there is overlapping between classes, since psychrotrophs or facultative psychrophiles are included in this definition.

There are two groups of organisms in the mesophilic range. The saprophytic organisms have an optimum temperature from 25° to 30°C, and potential pathogens have an optimum between 35° and 45°C.

Thermophiles.—These organisms have been defined as growing at high temperatures with an optimum of 45°C or higher. The information in Table 4.14 shows the minimum temperature for thermophiles somewhere between 35° and 45°C, the optimum between 45° and 65°C and the maximum between 60° and 90°C. Since the minimum temperatures for thermophiles overlap the higher mesophilic range, the thermophiles have been divided into facultative and obligate types. The facultative thermophiles can grow in the mesophilic range of 37°C or less, but the obligate thermophiles cannot grow at 37°C.

Enumeration

If the arbitrary temperature classes have any meaning, methods for determining the organisms should follow the definitions. The incubation condition for psychrotrophs is listed as 7°C for 10 days (APHA 1976).

Juffs (1972) compared the extremes of the British Standards Institute method of incubating at 5°-7°C for 7-10 days to determine psychrotrophs. The high extreme (7°C for 10 days) is the same as the APHA (1976) method. He found that 7°C for 7 days or 5°C for 10 days gave counts that were not significantly different. However, counts from plates incubated at 7°C for 10 days were significantly higher, and those using 5°C for 7 days were significantly lower.

The incubation period of 7 or 10 days becomes impractical as a control measure for perishable refrigerated foods such as milk. By the time the count is obtained, the product is consumed or may have developed defects. It is of value as a continuing control measure so that one is aware of potential problems. Oliveria and Parmelee (1976) suggested that incubation at 21°C for 25 hr gave comparable counts to the standard psychrotrophic method.

Comparing the incubation temperature of 7°C to the definitions of classes, it fits neither the 0°C given for psychrophiles nor the 5°C given for psychrotrophs. Since the term psychrotroph has been substituted for psychrophile, and using the APHA plating system, we might define psychrotrophs as those organisms that grow at 7°C and produce a visible colony in 10 days.

The mesophilic count can be related to the standard plate count, or aerobic plate count. Incubation temperatures of 25°-37°C have been used. The 37°C temperature indicates the potential pathogen population better than the lower temperature; however, since spoilage organisms are a problem, and higher counts are obtained, a temperature of 32° or 35°C is used (APHA 1976).

Thermophiles are enumerated by incubating the culture at 55°C ± 2°C for 48-72 hr (APHA 1976).

Relative Importance

The relative importance of each of the temperature classes (psychrophiles, psychrotrophs, mesophiles and thermophiles) depends upon the field of microbiology. Much of our food supply is held in cold storage which makes psychrotrophs important as potential spoilage organisms.

Psychrotrophic microorganisms are ubiquitous (widespread) in nature. Although they are favored by cold, they are not confined to temperatures of 5MC or less. Besides being found in ice and snow, psychrotrophs are in fresh as well as salt water, in soil and in or on nearly all raw food. Stokes (1968) reported that psychrophiles constituted 35 to 95% of the bacterial population of meats, 67% of the population on chicken and 17% on frozen fish sticks.

At low temperatures (0°-5°C), carbohydrate fermenters are generally inhibited and acids are not formed. The proteolytic and lipolytic organisms grow at low temperatures resulting in food defects.

Mesophilic species can be found in nearly every genus of organisms important in foods. Although spoilage of foods by mesophiles is significant, there is more concern with the mesophilic organisms that cause foodborne illness, such as certain species of *Salmonella, Staphylococcus, Clostridium, Shigella* and *Bacillus*.

Since the rate of growth of microorganisms and enzyme activity is higher in the mesophilic range than the psychrophilic range, spoilage of food occurs sooner.

Thermophiles can cause spoilage whenever foods are held at 50°-70°C. This may occur during heating of the food while cooking, processing or pasteurizing. Their growth in hot syrup may cause undesirable effects in sugar processing plants. In some tropical countries, thermophiles, not psychrotrophs, are the important spoilage organisms.

The generation time is shorter for thermophiles than for either psychrophiles or mesophiles when each is grown at its optimum temperature. This results in a more rapid spoilage of food by thermophiles.

The thermophiles may be more important than is presently realized. Since food is not normally held at temperatures conducive for growth of thermophiles, they have not been given the attention that has been allocated to the other classes.

Effect of Organisms on Temperature

Just as the environmental temperature alters the growth of microorganisms, the growth of microorganisms alters the temperature. When the organisms metabolize organic compounds, they not only grow and multiply, but also they produce heat as a by-product. In a compost pile, the temperature may rise to a level at which only thermophilic bacteria can grow. In the production of compounds by microbial fermentation, it is often necessary to use a cooling system in the fermenter to maintain the temperature so that the maximum amount of desired product is developed.

GASEOUS ATMOSPHERE

The type of gas in the atmosphere surrounding the food may determine the type(s) of organisms that become dominant. Oxygen in the atmosphere will favor the growth of aerobic types. The lack of oxygen, or a vacuum, will allow facultative anaerobes to become dominant.

Microorganisms vary widely in their tolerance to carbon dioxide. In a CO_2 atmosphere, the growth of some microorganisms is completely suppressed, while others are less affected. The lag period for growth of spoilage organisms on beef is increased in a 10% CO_2 atmosphere. Also, these organisms need a higher minimum water activity for growth in 10% CO_2 as compared to the normal atmosphere (0.033% CO_2). On the other hand, it is well documented that the presence of low concentrations (1% or less) of CO_2 stimulates oxygen uptake of bacteria. The amount of stimulation varies with bacterial species.

According to King and Nagel (1975), CO_2 inhibits the synthetic process of *Pseudomonas aeruginosa*, since resting cell metabolism is not affected.

Oxygen in the atmosphere was found to cause a loss in viability of freeze-dried and thawed *E. coli* and *Serratia marcesens* (Cox and Heckly 1973).

MICROBIAL INTERACTIONS

The environment affects the growth of microorganisms, and they can alter the environment. Also, microorganisms can inhibit or stimulate the growth of each other. Although pure cultures are usually used in microbial studies in the laboratory, mixed cultures are found in nature and on food products. Since the main goals of all living forms are self-preservation and perpetuation of the species, these organisms have to compete for food and other necessities. As a result, there are various actions and reactions (or coactions) which may be harmful or beneficial. Burkholder (1952) listed nine possible coactions when organisms are in a mixed population:

1. Predation—In this relationship, the strong predators damage the weak prey.
2. Parasitism—The weak parasite is benefited at the expense of the strong host.
3. Commensalism—The coaction results in the weak benefitting and the strong is unaffected.
4. Amensalism—The opposite of commensalism. The strong benefits and the weak is unaffected.
5. Allotrophy—In this relationship, the strong feeds the weak.
6. Allolimy—The strong starves the weak.
7. Symbiosis—In symbiosis, there is mutual aid, with both organisms benefitting from the relationship.
8. Synnecrosis—There is mutual conflict resulting in the death of both organisms.
9. Neutrality—Neither organism benefits or loses from the relationship. There is no coaction between them.

Food production involves various microorganism-host relationships such as found in the rumen of cattle and sheep. The interactions of microorganisms with animals and man result in potential sources of food contaminating microorganisms. A primary aspect is the association that can result in foodborne illness in man. This involves the interrelationships of intestinal microorganisms as well as their relationship with man.

The microbial relationships of most importance in foods are neutrality, symbiosis (cooperation) and allolimy (antagonism). The coactions of microorganisms can be due to competition for the necessities of life, parasitism or to products of metabolism.

Metabolism

The metabolism of an organism can alter the environmental factors that have been discussed (nutrients, a_w, pH, oxygen, temperature) so that the resultant environment may benefit or harm other organisms. A microorganism can produce antibiotics or antibiotic-like compounds that affect other organisms.

Nutrients.—Survival and growth of nutritionally dependent strains can be favored by the metabolic activities of another species in mixed culture. A strain of *Staphylococcus aureus* that required thymine and tryptophan was able to grow in mixed cultures or with culture filtrates of *Pseudomonas aeruginosa* (Gadbois *et al.* 1973). The *Pseudomonas* supplied the thymine and tryptophan essential for the growth of the *S. aureus.* With studies of infections with this mixture, they noted that the endotoxin of the *Pseudomonas* seemed to damage the animals' leucocytes which reduced phagocytosis of the *S. aureus* cells, allowing infection to occur. In this case, the *Pseudomonas* aided the *S. aureus* in obtaining needed nutrients and in protecting it from a defense mechanism of the animal.

Using a chemically defined medium that supported the growth of the yeast *Saccharomyces cerevisiae*, but not *Proteus vulgaris*, Shindala *et al.* (1965) found that both organisms grew in mixed culture, the yeast not being affected by the bacterial growth. A niacin-like factor, produced by the yeast, made it possible for the bacterium to grow. When niacin was added to the culture, the *Proteus* was no longer dependent upon the yeast and promptly outgrew it. This indicates that not only the organisms involved but also the environment, determine the coaction that occurs. The production of many fermented products involves the cooperation of microorganisms.

Although these citations show one organism aiding another organism, the dominance of one species can result in the utilization of the nutrients so there is little or none left for other species.

pH.—Microorganisms can alter the pH upward or downward, and thereby influence the growth of other microorganisms. The classical example of microorganisms influencing environmental pH and, in turn, influencing dominant organisms is that which occurs in raw milk. Such milk contains a variety of microorganisms (bacteria, yeasts, molds). As the lactose is converted to lactic acid by streptococci, other organisms become inhibited by the low pH and the streptococci dominate. As the pH continues to decrease, the streptococci are inhibited, but lactobacilli become dominant and continue the fermentation of lactose. The acid sours the milk and denatures the protein. The yeasts and molds are able to grow at a low pH. They become dominant and convert the lactic acid to nonacid products which raises the pH. When the pH is raised, the available proteins are utilized by proteolytic organisms such as *Bacillus* species. The putrefaction caused by the bacilli results in a clear and odorous product. Depending upon the environment (nutrients and pH), various organisms dominate during this sequence of events.

The lactic acid bacteria (streptococci and/or pediococci) inhibit several bacteria (both spoilage types and foodborne pathogens) partially by pH alteration and also by some undetermined factors (Mitchell and Kenworthy 1976).

Inhibitors.—The production of microbial inhibitors by microorganisms is well known. Besides the chemicals called antibiotics, there are other metabolic compounds that have antibiotic-like or inhibitory characteristics.

Goatcher and Westhoff (1975) isolated species of *Pseudomonas* from oysters. Some of these pseudomonads showed inhibition toward strains of *Vibrio parahaemolyticus*. The extent of inhibition varied with the strain of *Pseudomonas* and strain of *Vibrio*. Pseudomonads affect the salt tolerance and enterotoxin production of *S. aureus* (Collins-Thompson *et al.* 1973). An amino acid antimetabolite produced by *P. aeruginosa* inhibited a *Bacillus* species (Scannell *et al.* 1972).

Bacillus species interact with other microorganisms. *B. subtilis* produces an antibiotic called subtilin. This polypeptide has a marked action against a wide range of Gram positive, acid-fast and certain Gram negative bacteria. Depending on the concentration, subtilin is bacteriostatic or bactericidal. Barr (1975) found 10 antimicrobial metabolites produced by a strain of *B. subtilis*. Some of these were antibacterial while others were antifungal.

Serratia marcescens produces the red pigment, prodigiosin. Kalesperis *et al.* (1975) reported that certain fractions of the pigment showed antibiotic properties to *E. coli*, *Enterobacter aerogenes*, *S. aureus*, *B. subtilis* and *P. aeruginosa*.

Substances produced by *S. citrovorus* and *S. diacetilactis* inhibitory to *Pseudomonas fragi* were discussed by Pinheiro *et al.* (1968A,B). The inhibitory substances were possibly organic acids as well as acetoin and diacetyl.

Strains of species vary in their interactions with other organisms. In comparing the interactions of four strains of *Streptococcus cremoris* and four strains of *S. lactis*, it was found that two strains of *S. cremoris* dominated the four *S. lactis*, one strain of *S. cremoris* was compatible with the *S. lactis*, and one strain of *S. cremoris* was dominated by the four *S. lactis* strains (Reddy *et al.* 1971).

Inhibitory properties of pediococci on other microorganisms were reported by Fleming *et al.* (1975). A glycoprotein of *Saccharomyces cerevisiae* caused membrane damage to *Torulopsis glabrata* (Bussey and Skipper 1976).

Volatile metabolites of certain strains of bacteria were able to inhibit the growth, sporulation and mycotoxin production of *Penicillium* and *Aspergillus* species (Barr 1976).

Aflatoxin, a substance produced by the mold *Aspergillus flavus*, inhibits various *Bacillus* species (Lillehoj and Ciegler 1970).

There are several reports in the literature concerning interactions of organisms by some unnamed inhibitory substance. Just as some fungi produce substances inhibitory for *Bacillus* species, there are bacilli that produce toxic substances for fungi.

Besides producing microbial inhibitors, certain microorganisms may metabolize an antimicrobial agent so that other organisms can grow.

Fatty Acids.—These substances are discussed as inhibitors naturally present in foods. They also are formed by microorganisms during metabolism.

The growth of *Clostridium botulinum* is inhibited on surface-ripened cheese due to the formation of fatty acids by *Brevibacterium linens*. *Salmonella gallinarum* is inhibited by *Leuconostoc citrovorum* due to the acidic pH and production of acetic acid (Sorrells and Speck 1970).

Hydrogen Peroxide.—The mechanism of hydrogen peroxide (H_2O_2) formation by streptococci was described by Anders *et al.* (1970). The amount of H_2O_2 that accumulates varies among strains. A sufficient amount accumulated to inhibit growth, respiration and viability of these organisms. The addition of catalase or ferrous sulfate to milk prevents peroxide accumulation and results in an increased rate of acid production by lactic streptococci (Gilliland and Speck 1969). Lactobacilli produce sufficient H_2O_2 to inhibit *Pseudomonas, Bacillus* and *Proteus* species (Price and Lee 1970).

Organisms such as the micrococci produce catalase. It was reported by Nath and Wagner (1973) that in the presence of a *Micrococcus* species, the growth of, and acid production by, six cultures of lactic acid bacteria is stimulated. The stimulation is greater than observed with addition of catalase, indicating that other factors besides hydrogen peroxide removal are involved.

Hydrogen peroxide produced by *Streptococcus mitis* and *Lactobacillus acidophilus*, in combination with a peroxidase and a halide, has viricidal activity to polio virus (Klebanoff and Belding 1974).

Lytic Enzymes.—The lytic enzyme, lysozyme, was discussed in regard to natural inhibitors present in food. Microorganisms also produce enzymes that lyse cell walls.

Extracellular enzymes sometimes are excreted in large amounts by bacteria. Bacteriolytic enzymes have been isolated from cultures of many types of microorganisms. These enzymes are capable of causing lysis of a wide spectrum of Gram positive bacteria, although any one enzyme is specific for certain microorganisms.

Takahara *et al.* (1974 A,B) and Murao and Takahara (1974) described "B-enzyme" produced by a strain of *B. subtilis* that lysed cell walls of *Pseudomonas aeruginosa.* The enzyme is similar in composition to lysozyme. The enzyme also has a high rate of lysis for cell walls of *E. coli, Salmonella typhimurium, Klebsiella pneumoniae, P. fluorescens* and *B. megaterium.* Tsujisaka *et al.* (1975) described lytic enzymes from a strain of *Bacillus* that lyses cell walls of *Rhizopus* species.

Enzymes have been reported that act on *Vibrio parahaemolyticus* (Miyamoto *et al.* 1976) and others that degrade the coat proteins of enteroviruses (Herrmann *et al.* 1974).

Bacteriocins.—The bacteriocins are bactericidal substances produced by various species of bacteria. Although considered as antibiotics, they are not like the classical antibiotics since they are macromolecular, including or consisting of polypeptide or protein, and they generally act on strains of the same, or closely related, species.

There are several families of bacteriocins. Many strains of *P. vulgaris* and *P. mirabilis* produce phage tail-like bacteriocins.

The colicins have been studied the most extensively. They are found in *E. coli* and other Enterobacteriaceae. There are over 20 types of colicins. In some respects, they act similarly to phage. They can be induced by UV light, similarly to prophage. In some cases, they are held intracellularly, and when released, the cell is lysed. Under ordinary conditions, most of the cells in a culture do not produce colicins or phages, even though they have the capacity to do so.

Colicinogenicity can be transferred from one organism to another by plasmids called colicin factors or Col factors. Colicinogenic cells are resistant to the killing action of the colicin they produce.

When bacteriocins attach to the receptor sites of a sensitive cell, one or more things may occur, depending upon the bacteriocin and the cell. Some of the effects are the inhibition of macromolecule synthesis, such as proteins, RNA and DNA. Some colicins inhibit only protein synthesis, while others inhibit all three. Inhibition of respiration may occur with some colicins, while others may block the function of permeases. Interference with the formation of ATP, or enhancement of its breakdown, has been reported. Leakage of cellular constituents has been reported but this may be due to secondary reactions and not caused directly by the bacteriocin. Colicins disrupt transport mechanisms of amino acids which results in a lack of protein synthesis.

Serratia marcescens produces bacteriocins called marcescins. They affect not only other strains of *S. marcescens*, but also strains of *E. coli, Shigella, Klebsiella,* and *Enterobacter.* The mode of action is by inhibiting synthesis of DNA, RNA and protein (Eichenlaub and Winkler 1974).

Bacteriocins of Gram positive organisms generally have a wider spectrum of activity than those produced by Gram negative cells.

Pyocin is a bacteriocin produced by *P. aeruginosa.* It is inhibitory to not only various strains of these species, but also to strains of *P. putida* and *P. fluorescens* (Jones *et al.* 1974).

Parasites

Two of the parasites of bacteria are the bacterial viruses or bacteriophage and the bacterial genus, *Bdellovibrio.*

Some phages have a limited number of microbial species that are susceptible hosts. In a species of bacteria there may be susceptible and resistant strains of bacteria. For other phages, there may be several species and even different genera of bacteria which contain susceptible cells. For instance, some pasteurella phages also attack strains of *Salmonella* and *Shigella.* The limited hosts for phages makes it possible to use phage-typing to differentiate bacteria.

The *Bdellovibrio* resemble the virulent bacteriophages in their ability to lyse bacterial cells. However, unlike the phages, these microorganisms are actively motile (a single flagellum), small (0.25-0.4 μm wide and 0.8-1.2 μm long), vibroid, Gram negative bacteria. The formation of plaques (holes in a host lawn due to lysis of cells) is usually developed in 12-24 hr by phage, but *Bdellovibrio* plaques are visible only after 2-4 days, and they enlarge up to 6 days of incubation. Host-independent *Bdellovibrio* populations have been grown, which has not been possible with phage. About one in a million cells is host-independent, and these host independent cultures can revert to host-dependent varieties at about the same rate.

Although of academic interest, Neal and Banwart (1977) reported that *Bdellovibrio* cells do not reduce the count of host cells in foods.

STRESSES

Microorganisms may be subjected to various physical or chemical stresses during the processing of food. In many cases, the physical or chemical treatments are minimal in order to maintain quality attributes. Sometimes the effect that the treatment has on microorganisms is a secondary consideration. In other cases the treatment may be an attempt to control or regulate certain organisms, but other organisms are affected. The treatments that do not kill, but damage or injure the cells, are called sublethal.

The damages or injuries to the organism due to sublethal treatments are called lesions. As a result of the injuries, the growth capabilities of the organism may be altered, both in the food and on selective agars during enumeration. The stressed cells may have an extension of the lag phase, more exacting growth requirements or increased sensitivity to selective agents or inhibitors.

Sublethal Heating

In several processes (scalding, blanching, pasteurization, cooking and spray drying), the food and associated microbial flora are subjected to heat. Since none of these processes are designed to obtain a sterile product, some organisms may be killed, many damaged and others unaffected by the treatment.

Heat-induced lesions may be manifested as impairment of the cytoplasmic membrane with leakage of cellular components, alteration of the metabolic capabilities of the cell (which may appear as inactivation or stimulation with or without protein denaturation) and degradation of ribosomal RNA.

Yeasts heated in water near the maximum temperature for growth are essentially undamaged, but the presence of glucose induces leakage of cell contents (Hagler and Lewis 1974). Calcium ions temporarily protect the yeasts against glucose-induced leakage. This indicates damage to the cytoplasmic membrane.

Heat-induced damage results in loss of ability to grow in conditions where normal cells would grow. *S. aureus* is often enumerated on agar containing 7.5% NaCl; however, heat damaged *S. aureus* is sensitive to media containing over 4% salt (Smolka *et al.* 1974). The pH range for optimum growth is more narrow for heat-stressed as compared to normal *S. aureus.* This is also true for other organisms, including yeasts and molds. Yeasts are usually enumerated on acidified (to pH 3.5) media; however, when 12 species of yeast were heat-stressed, the lowest optimum pH for any of these was pH 6.8, and the maximum optimum pH was as high as pH 10 for two yeasts (Nelson 1972). The problem of using acidified media to enumerate sublethally treated yeast should be evident. Koburger (1972) found most retail food samples showed maximum counts of yeasts and molds when the medium was adjusted to pH 8.

Besides the more fastidious character of these injured cells, their growth has a prolonged lag phase. This longer than normal lag phase can be called the recovery period during which the damage to the cell is repaired. Although the cells usually can recover in a non-inhibitory growth medium, they can repair malfunctions in media in which they do not grow. In other words, growth is not needed for repair of the lesion.

Heat injury can occur in the spores of *Bacillus* and *Clostridium*. The germination system of *B. subtilis* is impaired by heating. Primary activation of spores at 90°C for 60 min, reduction of the incubation temperature by 10° to 15°C or addition of $CaCl_2$ and sodium dipicolinate, increases germination and colony formation of heat damaged spores. The addition of lysozyme to enumeration media improves the recovery of severely heated *Clostridium* spores.

Cold Effects

Freezing of cells can cause an apparent increase in not only nutritional requirements but also to sensitivity to selective agents used in media. Cells in the logarithmic growth phase are highly susceptible to cold shock and show an apparent loss in viability, which can be recovered by incubation with magnesium ions at 30°C. They gradually lose their capacity to recover if kept in cold magnesium-free buffer. Loss of viability due to cold shock is apparently due to damage of DNA (Sato and Takahashi 1970). The lethal effect of cold shock depends on the exposure time to low temperature, the growth phase of the cells, the concentration of bacteria, the diluent and the type of cell.

Drying

It is well established that organisms in dried foods may have metabolic injuries that impair the proliferation of the cells in selective media containing inhibitors in a concentration well tolerated by normal cells of the same species. These stressed cells can recover their tolerance to inhibitors if allowed to recuperate or repair in a nonselective medium.

Dried organisms are stressed by freezing (freeze drying or lyophilization), aerosolization (spray drying) or heat (roller, drum, tunnel or spray drying), as well as existing in an environment of very low water activity. Microorganisms in freeze-dried foods are subjected to several stresses, since the foods are processed, frozen, dried under vacuum at elevated temperatures, often packaged in an inert and dry gaseous atmosphere and stored at ambient temperatures.

The initial physiological state, the composition of the food, the location of the organisms in the food, the processing history, storage conditions and the method of rehydration, will influence the survival and the recovery of organisms from dried food.

Freeze-dried cells have altered permeability. Freeze-dried *E. coli* are susceptible to antibiotics that do not affect normal cells. Also, there is leakage of RNA from cells stressed by freeze drying. The alterations in the stressed cells are reversible, since during or immediately after recovery of cellular permeability, the damaged RNA, as well as metabolic damage, is repaired (Sinskey and Silverman 1970). After freeze drying, resynthesis of cell wall and membranes and the reestablishment of transport mechanisms are needed before cellular growth occurs. Many freeze dried cells require pyruvate, hematin and menadione for recovery. The effectiveness of the three compounds during repair of the cells is additive. It is believed these compounds may aid in reestablishing the permeability barrier of the cell.

Besides the need for nonselective nutrients in the recovery medium, due to the need for rehydration, the temperature at which this is performed is important. Ray *et al.* (1971A) recovered a higher number of cells at 15° to 25°C, but there was an earlier initiation of growth and more rapid growth if rehydration was done at 35°C. The rate of repair is reduced by lowering the temperature of the recovery medium from 35°C to 10°C and is extremely low at 1°C (Ray *et al.* 1971B).

Chemical Inhibitors

Since preservation processes cause injuries to microorganisms, perhaps chemical inhibitors may do likewise. Besides chemical preservatives, organisms on equipment surfaces are exposed to cleaning and sanitizing agents. If these organisms remain on the equipment, they may contaminate the food and, if injured, may be difficult to enumerate.

There is some evidence that injury and recovery due to chemical treatment can occur in a manner similar to heating, freezing or drying.

The action of physical factors can be readily altered and the microorganisms tested for injury, but it is more difficult to eliminate chemicals if they are sorbed to the cell or are inside the cell. In these cases, even dilution may not eliminate the chemical effect on the cell and death can result.

Injury to *E. coli* by acid was reported by Roth and Keenan (1971). *E. coli* exposed to hydrogen peroxide causes single-strand breaks in DNA. Repair of these breaks varies with the strain of cell (Ananthaswarmy and Eisenstark 1977).

The role of chemical preservatives on possible injury and recovery remains to be investigated.

Ultraviolet Light

The ultraviolet portion of the spectrum includes radiations from 150 to 390 nm. When living cells are irradiated with this light, some may be "killed" and others may be mutated. The UV wavelengths (around 260 nm) most active in producing either of these effects are those most readily absorbed by the nucleic acids. Both lethal and mutagenic effects of ultraviolet light can be partially reversed by repair or reactivation.

Although the ultraviolet light may be absorbed by many cellular components, absorption by the nucleic acids results in damage to cellular mechanisms for division. Repair of DNA in the light is called photoreactivation (PR). Reactivation of ultraviolet irradiated cells also can occur in the dark (dark repair, dark reactivation or excision repair).

Various compounds have been given credit for activation or protection of irradiated cells. One such, the enzyme catalase, breaks down the microbial inhibitor, peroxide, that is formed during irradiation. Another action of ultraviolet light is the inactivation of disulfide enzymes due to disruption of cystine. Inactivation of enzymes can affect microbial growth but not necessarily cause death.

Nalidixic acid (Eberle and Masker 1971), coumarin, pyronin Y, 6, 9-dimethyl 2-methylthiopurine and caffeine (Grigg 1972) inhibit or block repair of ultraviolet treated cells.

Other Repair Systems

Microorganisms can repair damage by gamma rays (Yatvin *et al.* 1972), fluorescent and photo light (Eisenstark and Ruff 1970), as well as X-rays and radiomimetic agents. The injury and repair of these sublethal treatments is similar to those discussed for the other treatments.

Summary

Various treatments as discussed in this section can cause injury to microorganisms and often these injuries can be repaired. It should be pointed out that the treatment rendered to the organism can be lethal depending upon the extent or amount of treatment. A truly lethal injury cannot be repaired, but microorganisms given a sublethal treatment, under certain conditions, can recover, grow and multiply. In some instances, the treatment appears to kill the cells since they are not able to multiply. However, they may still be alive, since they are metabolizing and even growing, such as the filamentous forms of *E. coli.*

The sublethal injuries are important since, without repair, the organisms are not able to grow on selective media used in enumeration of particular types. However, since they are metabolizing, they can cause food spoilage and, more importantly, they still maintain their potential for pathogenicity. Being in a stressed condition, the microorganisms are more susceptible to inhibition and death by other stress factors.

OTHER FACTORS

Environmental conditions, such as pressure, surface tension, surfaces, light and photosensitizers and pesticides, as well as colloidal characteristics, emulsion structure and concentration of nutrients, might help determine the microorganisms that are able to metabolize, grow, reproduce and spoil food products or result in a public health hazard. Compared to the other factors listed, these are of minor consequence in foods.

INTERACTIONS OF FACTORS

There are few instances when only one factor causes stresses on microorganisms. The interactions of these factors are important in foods.

Water Activity and Nutrition

The minimum a_w for growth of a microorganism depends upon the nutrients present in the growth medium. Christian (1955) showed that the range of a_w for growth of *Salmonella oranienburg* was appreciably smaller in a chemically defined glucose-salts medium than in a nutrient broth. The addition of small amounts of proline and four other amino acids to the glucose-salts medium stimulated the growth of *S. oranienburg* at 0.97 a_w, and permitted growth down to 0.96 a_w. Addition of vitamins showed further stimulation and growth occurred at 0.95 a_w. Thus, the minimum a_w levels for microorganisms should be determined in a medium with ample nutrients for the growth of the organism tested.

Since foods differ in their nutrient content, microorganisms may be able to grow in some foods at lower a_w than required in other foods.

Water Activity and Temperature

The greatest tolerance to low a_w occurs at the optimum temperature. As the temperature is changed from optimum, the range of a_w is reduced in which spore germination and growth occur. Studying the growth of fungi on jam, Horner and Anagnostopoulos (1973) found the interaction of a_w and temperature had a significant effect.

For *Salmonella* survival and multiplication, there was a close correlation between a_w and storage temperatures of meat and bone meal (Liu *et al.* 1969).

Water Activity and pH

As the pH is increased or decreased from the optimum, the minimum a_w needed for growth is increased. Decreasing the pH of the growth medium increases the minimum a_w at which *Clostridium botulinum* spores will germinate and initiate growth (Baird-Parker and Freame 1967).

pH and Temperature

Outgrowth of spores of *Clostridium botulinum* type E occurred at 15.6°C and a pH of 5.4-5.6. When the temperature was lowered to 5°C, a pH of 6.2-6.4 was needed for outgrowth to occur (Emodi and Lechowich 1969).

The optimum pH for *B. subtilis* spore germination is markedly changed by altering the incubation temperature (Ishida *et al.* 1976). The optimum pH was 7.4 at 37°C and 5.4 at 10°C.

Eh and Nutrients

The requirement for vitamin B_{12} by *Lactobacillus lactis* and *L. leichmannii* is dependent upon the OR potential of the medium. Anaerobiosis eliminates the requirement for this vitamin. In aerobic conditions, *L. lactis* requires pyrimidine, while anaerobically, it does not. The uracil requirement for *S. aureus* is reversed, requiring it anaerobically but not aerobically.

Under aerobic conditions, thiamin is essential and nicotinic acid stimulatory for *Sarcina lutea*, while under reduced pO_2, nicotinic acid is essential and thiamin is stimulatory.

Eh and pH

There is a definite relationship between the pH and the functional OR systems in bacterial cultures. Variations in the pH influence the Eh limits for growth of microorganisms.

Staphylococci are more acid-tolerant aerobically than anaerobically (Barber and Deibel 1972). Most staphylococcal strains initiate growth and produce detectable enterotoxin at pH 5.1 when grown aerobically. In anaerobic conditions, most strains fail to produce enterotoxin below pH 5.7.

Eh and a_w

The minimum a_w for growth of *Staphylococcus aureus* is reported to be

0.86 in aerobic conditions, but about 0.90 if the organism is grown anaerobically. Mead (1969) added salt to the medium of four strains of *Clostridium perfringens*, and found they all grew in 0.5% added salt at an Eh of +194 to +238 mv. When the salt level was increased to 5%, the maximum Eh at which growth was observed was +141 to +57 mv, depending upon the strain. Even with the lower Eh values, an extended lag phase was evident, during which the organisms became adjusted to the environment, and they adjusted the environment by lowering the Eh.

SUMMARY

Fresh foods contain a mixed population of microorganisms. However, during storage, one or a few species appear to become dominant. The continued presence of an organism in a particular environment suggests that it has some ecological advantage not possessed by other organisms. This advantage may be related to nutrition, a tolerance to toxic factors or a resistance to elimination by parasites. The survival and dominance of a species depends upon its fitness traits or characteristics for coping with the natural selection. Individuals lacking the fitness traits for survival in a specific environment will die and those that have the needed characteristics will survive, multiply and dominate.

Although gross environments are usually considered, it should be remembered that the microorganisms live in microenvironments. On a food product there may be many microenvironments and different organisms may be dominant in each of them. However, we usually determine the physical and chemical properties of the total environment and it is rare indeed if the microenvironments are specifically analyzed, even for the microorganisms that are present. Practically, it would be too time consuming and expensive to analyze anything but the gross product.

Many of the factors discussed in this chapter are used to control microorganisms in food as methods of preservation.

BIBLIOGRAPHY

ACTON J.C., and DICK, R.L. 1976. Composition of some commercial dry sausages. J. Food Sci. *41*, 971-972.

ALDERMAN, G.G., and MARTH, E.H. 1976. Inhibition of growth and aflatoxin production of *Aspergillus parasiticus* by citrus oils. Z. Lebensm. Unters.-Forsch. *160*, 353-358.

ANANTHASWAMY, H.N., and EISENSTARK, A. 1977. Repair of hydrogen peroxide-induced single-strand breaks in *Escherichia coli* deoxyribonucleic acid. J. Bacteriol. *130*, 187-191.

ANDERS, R.F., HOGG, D.M., and JAGO, G.R. 1970. Formation of hydrogen peroxide by group N streptococci and its effect on their growth and metabolism. Appl. Microbiol. *19*, 608-612.

APHA. 1976. Compendium of Methods for the Microbiological Examination of Foods. M.L. Speck (Editor). American Public Health Association, Washington, D.C.

ASHTON, D.H., BUSTA, F.F., and WARREN, J.A. 1968. Relief of casein inhibition of *Bacillus stearothermophilus* by iron, calcium, and magnesium. Appl. Microbiol. *16*, 628-635.

AUCLAIR, J.E., and HIRSCH, A. 1953. The inhibition of micro-organisms by raw milk. I. The occurrence of inhibitory and stimulatory phenomena. Methods of estimation. J. Dairy Res. *20*, 45-59.

BACHRACH, U., and WEINSTEIN, A. 1970. Effect of aliphatic polyamines on growth and macromolecular synthesis in bacteria. J. Gen. Microbiol. *60*, 159-165.

BAIRD-PARKER, A.C., and FREAME, B. 1967. Combined effect of water activity, pH and temperature on the growth of *Clostridium botulinum* from spore and vegetative cell inocula. J. Appl. Bacteriol. *30*, 420-429.

BARBER, L.E., and DEIBEL, R.H. 1972. Effect of pH and oxygen tension on staphylococcal growth and enterotoxin formation in fermented sausage. Appl. Microbiol. *24*, 891-898.

BARKAI-GOLAN, R., and AHARONI, Y. 1976. The sensitivity of food spoilage yeasts to acetaldehyde vapors. J. Food Sci. *41*, 717-718.

BARR, J.G. 1975. Changes in the extracellular accumulation of antibiotics during growth and sporulation of *Bacillus subtilis* in liquid culture. J. Appl. Bacteriol. *39*, 1-13.

BARR, J.G. 1976. Effects of volatile bacterial metabolites on the growth, sporulation and mycotoxin production of fungi. J. Sci. Food Agr. *27*, 324-330.

BISHOP, J.G., SCHANBACHER, F.L., FERGUSON, L.C., and SMITH, K.L. 1976. In vitro growth inhibition of mastitis-causing coliform bacteria by bovine apo-lactoferrin and reversal of inhibition by citrate and high concentrations of apo-lactoferrin. Infec. Immunity *14*, 911-918.

BOGIN, E., and ABRAMS, M. 1976. The effect of garlic extract on the activity of some enzymes. Food Cosmet. Toxicol. *14*, 417-419.

BROWN, A.D. 1976. Microbial water stress. Bacteriol. Rev. *40*, 803-846.

BUCHANAN, R.E., and GIBBONS, N.E. 1974. Bergey's Manual of Determinative Bacteriology, 8th Edition. William and Wilkins Co., Baltimore.

BULLOCK, D.H., and KENNEY, A.R. 1969. Effect of emulsion characteristics of a low-fat dairy spread on bacterial growth. J. Dairy Sci. *52*, 625-628.

BURKHOLDER, P.R. 1952. Cooperation and conflict among primitive organisms. Amer. Sci. *40*, 601-631.

BUSSEY, H., and SKIPPER, N. 1976. Killing of *Torulopsis glabrata* by *Saccharomyces cerevisiae* killer factor. Antimicrob. Agents Chemother. *9*, 352-354.

CARPENTER, J.A., WANG, Y.P., and POWERS, J.J. 1967. Effect of anthocyanin pigments on certain enzymes. Proc. Soc. Exp. Biol. Med. *124*, 702-706.

CHRISTIAN, J.H.B. 1955. The influence of nutrition on the water relations of *Salmonella oranienburg.* Aust. J. Biol. Sci. *8*, 75-82.

CHU, N.T., CLYDESDALE, F.M., and FRANCIS, F.J. 1973. Isolation and identification of some fluorescent phenolic compounds in cranberries. J. Food. Sci. *38*, 1038-1042.

COBB, B.F., VANDERZANT, C., THOMPSON, C.A., JR., and CUSTER, C.S. 1973. Chemical characteristics, bacterial counts, and potential shelf-life of shrimp from various locations on the northwestern Gulf of Mexico. J. Milk Food Technol. *36*, 463-468.

COLLINS-THOMPSON, D.L., ARIS, B., and HURST, A. 1973. Growth and enterotoxin B synthesis by *Staphylococcus aureus* S6 in associative growth with *Pseudomonas aeruginosa.* Can. J. Microbiol. *19*, 1197-1201.

COX, C.S., and HECKLY, R.J. 1973. Effects of oxygen upon freeze-dried and freeze-thawed bacteria: viability and free radical studies. Can. J. Microbiol. *19*, 189-194.

DALLYN, H., and EVERTON, J.R. 1970. Observations on the sporicidal action of vegetable oils used in fish canning. J. Appl. Bacteriol. *33*, 603-608.

DECALEYA, R.F., GONZALES-PASCUAL, B., GARCIA-OLMEDO, F., and CARBONERO, P. 1972. Susceptibility of phytopathogenic bacteria to wheat purathionins in vitro. Appl. Microbiol. *23*, 998-1000.

DENNIS, C. and COHEN, E. 1976. The effect of temperature on strains of soft fruit spoilage fungi. Ann. Appl. Biol. *82*, 51-56.

DEVORE, D.P., and SOLBERG, M. 1974. Oxygen uptake in postrigor bovine muscle. J. Food Sci. *39*, 22-28.

DOUGLAS, F., HAMBLETON, R., and RIGBY, G.J. 1973. An investigation of the oxidation-reduction potential and of the effect of oxygen on the germination and outgrowth of *Clostridium butyricum* spores using platinum electrodes. J. Appl. Bacteriol. *36*, 625-633.

DOUGLAS, F., and RIGBY, G.J. 1974. The effect of oxygen on the germination and outgrowth of *Clostridium butyricum* spores and changes in the oxidation-reduction potential of cultures. J. Appl. Bacteriol. *37*, 251-259.

EBERLE, H., and MASKER, W. 1971. Effect of nalidixic acid on semi-conservative replication and repair synthesis after ultraviolet irradiation in *Escherichia coli.* J. Bacteriol. *105*, 908-912.

EDDY, B.P. 1960. The use and meaning of the term 'psychrophilic'. J. Appl. Bacteriol. *23*, 189-190.

EICHENLAUB, R., and WINKLER, U. 1974. Purification and mode of action of two bacteriocins produced by *Serratia marcesens* HY. J. Gen. Microbiol. *83*, 83-94.

EISENSTARK, A., and RUFF, D. 1970. Repair in phage and bacteria inactivated by light from fluorescent and photo lamps. Biochem. Biophys. Res. Commun. *38*, 244-248.

EMEH, C.O., and MARTH, E.H. 1976. Cultural and nutritional factors that control rubratoxin formation. J. Milk Food Technol. *39*, 184-190.

EMODI, A.S., and LECHOWICH, R.V. 1969. Low temperature growth of type E *Clostridium botulinum* spores. 1. Effects of sodium chloride, sodium nitrite and pH. J. Food Sci. *34*, 78-81.

FAIR, W.R., and WEHNER, N. 1971. Antibacterial action of spermine: effect on urinary tract pathogens. Appl. Microbiol. *21*, 6-8.

FEENEY, R.E., and NAGY, D.A. 1952. The antibacterial activity of the egg white protein conalbumin. J. Bacteriol. *64*, 629-643.

FETT, H.M. 1973. Water activity determination in foods in the range of 0.80 to 0.99. J. Food Sci. *38*, 1097-1098.

FITZGERALD, T., and YOTIS, W.W. 1971. Interference with cellular incorporation of substrates into *Staphylococcus aureus* by hormones. J. Med. Microbiol. *4*, 97-106.

FLEMING, H.P., ETCHELLS, J.L., and COSTILOW, R.N. 1975. Microbial inhibition by an isolate of *Pediococcus* from cucumber brines. Appl. Microbiol. *30*, 1040-1042.

FLEMING, H.P., WALTER, W.M., JR., and ETCHELLS, J.L. 1973. Antimicrobial properties of oleuropein and products of its hydrolysis from green olives. Appl. Microbiol. *26*, 777-782.

FLORES, S.C., and CRAWFORD, D.L. 1973. Postmortem quality changes in iced Pacific shrimp (*Pandalus jordani*). J. Food Sci. *38*, 575-579.

FORSYTH, M.P., and KUSHNER, D.J. 1970. Nutrition and distribution of salt response in populations of moderately halophilic bacteria. Can. J. Microbiol. *16*, 253-261.

FOSSUM, K., and WHITAKER, J.R. 1968. Ficin and papain inhibitor from chicken egg white. Arch. Biochem. Biophys. *125*, 367-375.

FRAZIER, W.C. 1967. Food Microbiology. 2nd Edition. McGraw-Hill Book Co., New York, N.Y.

FREDEEN, H.T., MARTIN, A.H., and WEISS, G.M. 1974. Changes in tenderness of beef longissimus dorsi as related to muscle color and pH. J. Food Sci. *39*, 532-536.

FUKUI, H., EGAWA, H., KOSHIMIZU, K., and MITSUI, T. 1973. A new isoflavone with antifungal activity from immature fruits of *Lupinus luteus*. Agr. Biol. Chem. *37*, 417-421.

GADBOIS, T., DE REPENTIGNY, J., and MATHIEU, L.G. 1973. Favorable effects in vitro and in vivo of two clinical isolates of *Pseudomonas aeruginosa* on nutritionally deficient *Staphylococcus aureus* strains. Can. J. Microbiol. *19*, 973-981.

GALBRAITH, H., and MILLER, T.B. 1973. Effect of metal cations and pH on the antibacterial activity and uptake of long chain fatty acids. J. Appl. Bacteriol. *36*, 635-646.

GILLILAND, S.E., and SPECK, M.L. 1969. Biological response of lactic streptococci and lactobacilli to catalase. Appl. Microbiol. *17*, 797-800.

GOATCHER, L.J., and WESTHOFF, D.C. 1975. Repression of *Vibrio parahaemolyticus* by *Pseudomonas* species isolated from processed oysters. J. Food Sci. *40*, 533-536.

GRIGG, G.W. 1972. Effects of coumarin, pyronin Y, 6,9-dimethyl 2-methylthiopurine and caffeine on excision repair and recombination repair in *Escherichia coli*. J. Gen. Microbiol. *70*, 221-230.

HAGLER, A.N., and LEWIS, M.J. 1974. Effect of glucose on thermal injury of yeast that may define the maximum temperature of growth. J. Gen. Microbiol. *80*, 101-109.

HARMON, R.J., SCHANBACHER, F.L., FERGUSON, L.C., and SMITH, K.L. 1975. Concentration of lactoferrin in milk of normal lactating cows and changes occurring during mastitis. Amer. J. Vet. Res. *36*, 1001-1007.

HARRISON, D.E.F. 1972. Physiological effects of dissolved oxygen tension and redox potential on growing populations of micro-organisms. J. Appl. Chem. Biotechnol. *22*, 417-440.

HENTGES, D.J., and MAIER, B.R. 1972. Theoretical basis for anaerobic methodology. Amer. J. Clin. Nutr. *25*, 1299-1305.

HERRMANN, E.C., JR., and KUCERA, L.S. 1967. Antiviral substances in plants of the mint family (*Labiatae*). II. Nontannin polyphenol of *Melissa officinalis*. Proc. Soc. Exp. Biol. Med. *124*, 869-874.

HERRMANN, J.E., KOSTENBADER, K.D., JR., and CLIVER, D.O. 1974. Persistence of enteroviruses in lake water. Appl. Microbiol. *28*, 895-896.

HIBBITT, K.G., BROWNLIE, J., and COLE, C.B. 1971. The antimicrobial activity of cationic proteins isolated from the cells in bulk milk samples. J. Hyg. (Camb.) *69*, 61-68.

HORNER, K.J., and ANAGNOSTOPOULOS, G.D. 1973. Combined effects of water activity, pH and temperature on the growth and spoilage potential of fungi. J. Appl. Bacteriol. *36*, 427-436.

HUHTANEN, C.N., NAGHSKI, J., CUSTER, C.S., and RUSSELL, R.W. 1976. Growth and toxin production by *Clostridium botulinum* in moldy tomato juice. Appl. Environ. Microbiol. *32*, 711-715.

IDZIAK, E.S., and SUVANMONGKOL, P. 1972. Effect of pH on the pathogenic functions of *Salmonella typhimurium*. Can. J. Microbiol. *18*, 9-12.

IGLESIAS, H.A., and CHIRIFE, J. 1976. A model for describing the water sorption behavior of foods. J. Food Sci. *41*, 984-992.

INOUE, K., and KOMAGATA, K. 1976. Taxonomic study of obligately psychrophilic bacteria isolated from Antarctica. J. Gen. Appl. Microbiol. *22*, 165-176.

ISHIDA, Y., ISHIDO, T., and KADOTA, H. 1976. Temperature-pH effect upon germination of bacterial spores. Can. J. Microbiol. *22*, 322-323.

ITO, K.A. *et al.* 1976. Effect of acid and salt concentration in fresh-pack pickles on the growth of *Clostridum botulinum* spores. Appl. Environ. Microbiol. *32*, 121-124.

JOFFE, A.Z., and LISKER, N. 1969. Effects of light, temperature, and pH value on aflatoxin production in vitro. Appl. Microbiol. *18*, 517-518.

JOHNSON, M.G., PALUMBO, S.A., RIECK, V.T., and WITTER, L.D. 1970. Influence of temperature on steady-state growth of colonies of *Pseudomonas fluorescens.* J. Bacteriol. *103*, 267-268.

JONES, L.F. *et al.* 1974. Pyocin sensitivity of *Pseudomonas* species. Appl. Microbiol. *27*, 288-289.

JUFFS, H.S. 1972. Variation in psychrotroph counts obtained at the extremes of incubation prescribed by British standard 4285:1968. Aust. J. Dairy Technol. *27*, 26-27.

JUVEN, B., HENIS, Y., and JACOBY, B. 1972. Studies on the mechanism of the antimicrobial action of oleuropein. J. Appl. Bacteriol. *35*, 559-567.

KALESPERIS, G.S., PRAHLAD, K.V., and LYNCH, D.L. 1975. Toxigenic studies with the antibiotic pigments from *Serratia marcescens.* Can. J. Microbiol. *21*, 213-220.

KARMAS, E. 1973. Water in biosystems. J. Food Sci. *38*, 736-739.

KATO, F., OGATA, S., and HONGO, M. 1976. Killing activity of clostocin O. Agr. Biol. Chem. *40*, 1107-1111.

KAUFMANN, O.W., and MARSHALL, R.S. 1965. Factors affecting the development of *Clostridium botulinum* in whole milk. Appl. Microbiol. *13*, 521-526.

KEENEY, D.R. 1973. The nitrogen cycle in sediment-water systems. J. Environ. Quality *2*, 15-29.

KEMPSTER, A.J., and CUTHBERTSON, A. 1975. A national survey of muscle pH values in commercial pig carcasses. J. Food Technol. *10*, 73-80.

KING, A.D., JR., and NAGEL, C.W. 1975. Influence of carbon dioxide upon the metabolism of *Pseudomonas aeruginosa.* J. Food Sci. *40*, 362-366.

KLEBANOFF, S.J., and BELDING, M.E. 1974. Virucidal activity of H_2O_2-generating bacteria: requirement for peroxidase and a halide. J. Infec. Dis. *129*, 345-348.

KOBURGER, J.A. 1972. Fungi in foods. IV. Effect of plating medium pH on counts. J. Milk Food Technol. *35*, 659-660.

KONOWALCHUK, J., and SPEIRS, J.I. 1973. Identification of a viral inhibitor in ground beef. Can. J. Microbiol. *19*, 177-181.

KONOWALCHUK, J., and SPEIRS, J.I. 1978. Antiviral effect of commercial juices and beverages. Appl. Environ. Microbiol. *35*, 1219-1220.

KULSHRESTHA, D.C., and MARTH, E.H. 1975. Some volatile and nonvolatile compounds associated with milk and their effects on certain bacteria. A review. J. Milk Food Technol. *38*, 604-620.

LABUZA, T.P., CASSIL, S., and SINSKEY, A.J. 1972. Stability of intermediate moisture foods. 2. Microbiology. J. Food Sci. *37*, 160-162.

LABUZA, T.P. *et al.* 1976. Water activity determination: a collaborative study of different methods. J. Food Sci. *41*, 910-917.

LEE, Y.B., HARGUS, G.L., HAGBERG, E.C., and FORSYTHE, R.H. 1976. Effect of antemortem environmental temperatures on postmortem glycolysis and tenderness in excised broiler breast muscle. J. Food Sci. *41*, 1466-1469.

LILLEHOJ, E.B., and CIEGLER, A. 1970. Aflatoxin B_1 effect on enzyme biosynthesis in *Bacillus cereus* and *Bacillus licheniformis*. Can. J. Microbiol. *16*, 1059-1065.

LIU, T.S., SNOEYENBOS, G.H., and CARLSON, V.L. 1969. The effect of moisture and storage temperature on a *Salmonella senftenberg* 775W population in meat and bone meal. Poultry Sci. *48*, 1628-1633.

LOESCHE, W.J. 1969. Oxygen sensitivity of various anaerobic bacteria. Appl. Microbiol. *18*, 723-727.

LOVE, R.M., and HAQ, M.A. 1970. The connective tissues of fish. IV. Gaping of cod muscle under various conditions of freezing, cold storage and thawing. J. Food Technol. *5*, 249-260.

LOVE, R.M., HAQ, M.A., and SMITH, G.L. 1972. The connective tissues of fish. V. Gaping in cod of different sizes as influenced by a seasonal variation in ultimate pH. J. Food Technol. *7*, 281-290.

MACMILLAN, W.G., and HIBBITT, K.G. 1973. Surface properties and natural defence in mammals. Pestic. Sci. *4*, 863-870.

MATSUSHIMA, K. 1958. An undescribed trypsin inhibitor in egg white. Science *127*, 1178-1179.

MEAD, G.C. 1969. Combined effect of salt concentration and redox potential of the medium on the initiation of vegetative growth of *Clostridium welchii*. J. Appl. Bacteriol. *32*, 468-475.

MITCHELL, I.D.G., and KENWORTHY, R. 1976. Investigations on a metabolite from *Lactobacillus bulgaricus* which neutralizes the effect of enterotoxin from *Escherichia coli* pathogenic for pigs. J. Appl. Bacteriol. *41*, 163-174.

MIYAMOTO, S., KURODA, K., HANAOKA, M., and OKADA, Y. 1976. Isolation of a small rod with lytic activity against *Vibrio parahaemolyticus* from fresh sea water. Jap. J. Microbiol. *20*, 517-527.

MORITA, R.Y. 1975. Psychrophilic bacteria. Bacteriol. Rev. *39*, 144-167.

MOSSEL, D.A.A. 1971A. Physiological and metabolic attributes of microbial groups associated with foods. J. Appl. Bacteriol. *34*, 95-118.

MOSSEL, D.A.A. 1971B. Ecological essentials of antimicrobial food preservation. Symposia Soc. Gen. Microbiol. *21*, 177-195.

MOSSEL, D.A.A., and INGRAM, M. 1955. The physiology of the microbial spoilage of foods. J. Appl. Bacteriol. *18*, 232-268.

MOTOHIRO, T., and INOUE, N. 1970. pH of canned crab meat. I. Stages in the molting cycle in relation to pH. Food Technol. *24*, 1389-1391.

MURAO, S., and TAKAHARA, Y. 1974. Enzymes lytic against *Pseudomonas aeruginosa* produced by *Bacillus subtilis* YT-25. Agr. Biol. Chem. *38*, 2305-2316.

MURPHY, E.W., WATT, B.K., and RIZEK, R.L. 1973. Tables of food composition: Availability, uses, and limitations. Food Technol. *27*, 40-51.

NATH, K.R., and WAGNER, B.J. 1973. Stimulation of lactic acid bacteria by a *Micrococcus* isolate: evidence for multiple effects. Appl. Microbiol. *26*, 49-55.

NEAL, J.J., and BANWART, G.J. 1977. *Bdellovibrio* in foods. J. Food Sci. *42*, 555-556.

NELSON, F.E. 1972. Plating medium pH as a factor in apparent survival of sublethally stressed yeasts. Appl. Microbiol. *24*, 236-239.

OEGEMA, T.R., JR., and JOURDIAN, G.W. 1974. The physical and chemical properties of a chicken egg white glycoprotein purified by nondenaturing methodology. Arch. Biochem. Biophys. *160*, 26-39.

OLIVERIA, J.S., and PARMELEE, C.E. 1976. Rapid enumeration of psychrotrophic bacteria in raw and pasteurized milk. J. Milk Food Technol. *39*, 269-272.

ONDERDONK, A.B., JOHNSTON, J., MAYHEW, J.W., and GORBACH, S.L. 1976. Effect of dissolved oxygen and Eh on *Bacteroides fragilis* during continuous culture. Appl. Environ. Microbiol. *31*, 168-172.

ORR, M.L. 1969. Pantothenic acid, vitamin B_6 and vitamin B_{12} in foods. Home Econ. Res. Report No. *36*. U.S. Dept. Agr., Washington, D.C.

PINHEIRO, A.J.R., LISKA, B.J., and PARMELEE, C.E. 1968A. Properties of substances inhibitory to *Pseudomonas fragi* produced by *Streptococcus citrovorus* and *Streptococcus diacetilactis*. J. Dairy Sci. *51*, 183-187.

PINHEIRO, A.J.R., LISKA, B.J., and PARMELEE, C.E. 1968B. Inhibitory effect of selected organic chemicals on *Pseudomonas fragi*. J. Dairy Sci. *51*, 223-224.

PITT, J.I., and CHRISTIAN, J.H.B. 1968. Water relations of xerophilic fungi isolated from prunes. Appl. Microbiol. *16*, 1853-1858.

PRAGER, E.M.,, and WILSON, A.C. 1971. Multiple lysozymes of duck egg white. J. Biol. Chem. *246*, 523-530.

PRICE, R.J., and LEE, J.S. 1970. Inhibition of *Pseudomonas* species by hydrogen peroxide producing lactobacilli. J. Milk Food Technol. *33*, 13-18.

RAY, B., JEZESKI, J.J., and BUSTA, F.F. 1971A. Effect of rehydration on recovery, repair, and growth of injured freeze-dried *Salmonella anatum*. Appl. Microbiol. *22*, 184-189.

RAY, B., JEZESKI, J.J., and BUSTA, F.F. 1971B. Repair of injury in freeze-dried *Salmonella anatum*. Appl. Microbiol. *22*, 401-407.

REDDY, M.S., VEDAMUTHU, E.R., WASHAM, C.J., and REINBOLD, G.W. 1971. Associative growth relationships in two strain mixtures of *Streptococcus lactis* and *Streptococcus cremoris*. J. Milk Food Technol. *34*, 236-240.

REITER, B., MARSHALL, V.M.E., BJÖRCK, L., and ROSEN, C.G. 1976. Nonspecific bactericidal activity of the lactoperoxidase-thiocyanate-hydrogen peroxide system of milk against *Escherichia coli* and some Gram-negative pathogens. Infec. Immunity *13*, 800-807.

RHODES, M.B., BENNETT, N., and FEENEY, R.E. 1960. The trypsin and chymotrypsin inhibitors from avian egg whites. J. Biol. Chem. *235*, 1686-1693.

RIPPON, J.W. 1968. Monitored environment system to control cell growth, morphology, and metabolic rate in fungi by oxidation-reduction potentials. Appl. Microbiol. *16*, 114-121.

ROSENBERG, S.L. 1975. Temperature and pH optima for 21 species of thermophilic and thermotolerant fungi. Can. J. Microbiol. *21*, 1535-1540.

ROSS, E., BARTLETT, D.S., and HARD, M.M. 1953. An application of oxidation-reduction potentials to frozen fruits treated with ascorbic acid and ascorbic-citric acid mixtures. Food Technol. *7*, 153-156.

ROSS, K.D. 1975. Estimation of water activity in intermediate moisture foods. Food Technol. *29*, No. 3, 26-34.

ROTH, L.A., and KEENAN, D. 1971. Acid injury of *Escherichia coli.* Can. J. Microbiol. *17*, 1005-1008.

SATO, M., and TAKAHASHI, H. 1970. Cold shock of bacteria. IV. Involvement of DNA ligase reaction in recovery of *Escherichia coli* from cold shock. J. Gen. Appl. Microbiol. *16*, 279-290.

SCANNELL, J.P. *et al.* 1972. Antimetabolites produced by microorganisms. V. L-2-amino-4-methoxy-trans-3-butenoic acid. J. Antibiot. *25*, 122-127.

SEGNER, W.P., SCHMIDT, C.F., and BOLTZ, J.K. 1966. Effect of sodium chloride and pH on the outgrowth of spores of type E *Clostridium botulinum* at optimal and suboptimal temperatures. Appl. Microbiol. *14*, 49-54.

SEGNER, W.P., SCHMIDT, C.F., and BOLTZ, J.K. 1971. Minimal growth temperature, sodium chloride tolerance, pH sensitivity, and toxin production of marine and terrestrial strains of *Clostridium botulinum* type C. Appl. Microbiol. *22*, 1025-1029.

SHINDALA, A., BUNGAY, H.R., KRIEG, N.R., and CULBERT, K. 1965. Mixed-culture interactions. I. Commensalism of *Proteus vulgaris* with *Saccharomyces cerevisiae* in continuous culture. J. Bacteriol. *89*, 693-696.

SKIPPER, N., and BUSSEY, H. 1977. Mode of action of yeast toxins: energy requirement for *Saccharomyces cerevisiae* killer toxin. J. Bacteriol. *129*, 667-677.

SINSKY, T.J., and SILVERMAN, G.J. 1970. Characterization of injury incurred by *Escherichia coli* upon freeze-drying. J. Bacteriol. *101*, 429-437.

SMOLKA, L. R., NELSON, F.E., and KELLEY, L.M. 1974. Interaction of pH and NaCl on enumeration of heat-stressed *Staphylococcus aureus.* Appl. Microbiol. *27*, 443-447.

SORRELLS, K.M., and SPECK, M.L. 1970. Inhibition of *Salmonella gallinarum* culture filtrates of *Leuconostoc citrovorum.* J. Dairy Sci. *53*, 239-241.

STOKES, J.L. 1968. Nature of psychrophilic microorganisms. *In* Low Temperature Biology of Foodstuffs. J. Hawthorne (Editor). Pergamon Press, Oxford, England.

SUBBA, M.S., SOUMITHRI, T.C., and RAO, R.S., 1967. Antimicrobial action of citrus oils. J. Food Sci. *32*, 225-227.

SWAMINATHAN, B., and KOEHLER, P.E. 1976. Isolation of an inhibitor of *Aspergillus parasiticus* from white potatoes (*Solanum tuberosum*). J. Food Sci. *41*, 313-319.

TAKAHARA, Y., MACHIGAKI, E., and MURAO, S. 1974A. Lytic action of B-enzyme on *Pseudomonas aeruginosa*. Agr. Biol. Chem. *38*, 2349-2356.

TAKAHARA, Y., MACHIGAKI, E., and MURAO, S. 1974B. General properties of endo-N-acetylmuramidase of *Bacillus subtilis* YT-25. Agr. Biol. Chem. *38*, 2357-2365.

TEH, J.S. 1974. Toxicity of short-chain fatty acids and alcohols toward *Cladosporium resinae*. Appl. Microbiol. *28*, 840-844.

TEUBER, M. 1970. Low antibiotic potency of isohumulone. Appl. Microbiol. *19*, 871.

TOWNSEND, C.T., YEE, L., and MERCER, W.A. 1954. Inhibition of the growth of *Clostridium botulinum* by acidification. Food Res. *19*, 536-542.

TROLLER, J.A. 1971. Effect of water activity on enterotoxin B production and growth of *Staphylococcus aureus*. Appl. Microbiol. *21*, 435-439.

TSUJISAKA, Y., TOMINAGA, Y., and IWAI, M. 1975. Purification and some properties of the lytic enzyme from *Bacillus* R-4 which acts on *Rhizopus* cell wall. Agr. Biol. Chem. *39*, 145-152.

VANDERZANT, C., MATTHYS, A.W., and COBB, B.F. 1973. Microbiological, chemical, and organoleptic characteristics of frozen breaded raw shrimp. J. Milk Food Technol. *36*, 253-261.

VEDAMUTHU, E.R., WASHAM, C.J., and REINBOLD, G.W. 1971. Isolation of inhibitory factor in raw milk whey active against propionibacteria. Appl. Microbiol. *22*, 552-556.

WALDEN, W.C., and HENTGES, D.J. 1975. Differential effects of oxygen and oxidation-reduction potential on the multiplication of three species of anaerobic intestinal bacteria. Appl. Microbiol. *30*, 781-785.

WATT, B.K., and MERRILL, A.L. 1963. Composition of Foods—Raw, Processed and Prepared. Agr. Handbook No. 8, U.S. Government Printing Office, Washington, D.C.

WEDRAL, E.M., VADEHRA, D.V., and BAKER, R.C. 1971. Mechanism of bacterial penetration through eggs of *Gallus gallus*. 2. Effect of penetration and growth on permeability of inner shell membrane. J. Food Sci. *36*, 520-522.

WEISSBERGER, W., KAVANAGH, T.E., and KEENEY, P.G. 1971. Identification and quantitation of several nonvolatile organic acids of cocoa beans. J. Food Sci. *36*, 877-879.

YATVIN, M.B., WOOD, P.G., and BROWN, S.M. 1972. "Repair" of plasma membrane injury and DNA single strand breaks in γ-irradiated *Escherichia coli* B/r and B_{s-1}. Biochim. Biophys. Acta *287*, 390-403.

YOTIS, W.W., and BAMAN, S.I. 1969. An evaluation of diethylstilbestrol as an inhibitor of the growth of staphylococci. Yale J. Biol. Med. *41*, 311-322.

5

Sources of Microorganisms

The microbial flora of a food consists of the microorganisms associated with the raw material, those acquired during handling and processing and those surviving any preservation treatment and storage.

Since these microorganisms do not arise by spontaneous generation, they must contaminate the food at some stage of production, harvesting, handling, processing, storage, distribution and/or preparation for consumption. Most foods are subjected to many potential sources of microorganisms.

Why should we be concerned with sources of contamination? Primarily, so that we can control contamination and keep the microbial load on or in the food as low as possible. By doing this, we obtain a longer shelf life for the food and, hopefully, there is less chance of microbial foodborne illness when the food is ingested. By keeping the count low, it is easier to control or eliminate the microorganisms with food preservation techniques.

The potential sources of contamination are soil, water, air, plants, feed or fertilizer, animals, humans, sewage, processing equipment, ingredients, product to product and packaging materials.

Microorganisms can be exchanged between these sources. For example, animals contaminate the soil with fecal material. Then rain washes the microorganisms into the creeks or rivers. This water may be used for irrigation and contaminate plants used for food. Thus although water is the carrier, the microorganisms originally came from animals.

For some foods, it is difficult to determine how many of the organisms in the flora are contaminants and how many are the result of multiplication on or in the food.

SOIL

Soil is the natural habitat of many types of microorganisms which quite often are found in high numbers. The microbial density is greater near the soil surface and decreases in deeper soil. The types and numbers of microorganisms vary with the type of soil as well as the environmental conditions. These environmental conditions are constantly changing, especially the moisture and temperature.

Growth of Microorganisms

The microbial growth in soil is limited to areas of organic material. These areas include the roots of plants, plant debris falling onto the soil, dead animal carcasses, fecal deposition by animals and dead microorganisms. Besides the animal carcasses on the soil surface, dead earthworms, insects and other small animals are in the soil.

The growth of microorganisms is influenced by the chemical composition of the materials undergoing decomposition, the rate of decomposition of the chemical constituents of these substances and upon the environmental conditions. Fats, waxes and lignins may gradually accumulate in the soil due to their relative resistance to decomposition. This residual, called humus, is subject to slow and gradual attack by a variety of microorganisms.

Number and Types of Microorganisms

The microbial numbers can vary from a few organisms in sandy and desert soils to as many as 10^{10}/g in fertile soils. According to Gray and Williams (1971) many of the microorganisms in soil are in some form of resting stage. Endospore forming bacteria (*Bacillus* and *Clostridium*) are quite prevalent in some soils. Vegetative cells may exist in a reduced state of metabolic activity. The fungal flora is primarily some type of mold spore. Yeasts are found in soil and live on food washed off plants by rainwater. Yeasts are especially prevalent in soils of vineyards and orchards. There may be several thousand yeast cells per gram of soil. Bacteria outnumber other microorganisms in most soils. Although most of the different types of bacteria can be found in soil at some time, those that are common in soil and found in food include *Acinetobacter, Alcaligenes, Arthrobacter, Bacillus, Clostridium, Corynebacterium, Flavobacterium, Micrococcus, Pseudomonas* and *Streptomyces.*

The natural habitat of *Clostridium botulinum* types A and B is soil. These organisms enter foods as soil-borne contaminants. Thermophilic spore formers that can cause spoilage of canned foods are found in soil. Although usually in low numbers, as many as 10^8 spores/g of soil may be present.

Poliovirus survives in soil longer in winter than in summer (Tierney *et al.* 1977). Survival of poliovirus for 84 days was reported by Duboise *et al.* (1976).

Contamination of Food

Microorganisms in the soil can contaminate tubers or root crops by direct contact. Also, dirt is blown by the wind or is splashed by rain falling onto the soil, so that the dirt can contaminate crops such as strawberries, beans, cabbage or peas that grow near the ground level. The microbial numbers and types on crops are influenced by the degree of contamination of the soil in which they are grown.

Mechanical harvesting has increased the amount of soil contamination as well as breakage of fruits and vegetables. Cereal crops are contaminated mainly during harvesting.

Marine sediments have microbial counts in the range of 10^4 to 10^9/g. The numbers are usually higher near the shore than at deeper levels. The bacteria found in these sediments include *Aeromonas, Bacillus, Chromobacterium, Citrobacter, Escherichia, Pseudomonas* and *Vibrio*. These sediments serve as a source of microorganisms for water as well as for fish and shellfish. During trawling along the bottom with nets, the sediment is disturbed so that it contaminates the fish or shellfish that are caught in the nets.

WATER

Water is a potential source of microbial contamination of food. Rain contains microorganisms that are washed from the air. As the water lands on the ground, it is further contaminated by soil microorganisms. In the ocean, organisms are in greater abundance near the shore than in regions distant from land. The dumping of wastes such as sewage and the runoff from animal feedlots result in considerable microbial contamination of waterways with enteric types of bacteria.

Types of Microorganisms

The genera of bacteria which tend to be part of the "normal" flora of water include *Pseudomonas, Flavobacterium, Cytophaga, Acinetobacter, Moraxella, Aeromonas, Corynebacterium, Streptococcus, Klebsiella, Alcaligenes, Bacillus* and *Micrococcus*.

Escherichia coli is used as an indicator of fecal pollution of water in temperate climates. This organism supposedly dies quite rapidly in the relatively hostile environment of rivers due to low temperature, sunlight, toxic chemicals and a lack of nutrients. However, some reports indicate that the coliforms and *E. coli* seem to maintain their population quite well (Coleman *et al.* 1974; Goodrich *et al.* 1970). Water is still the main carrier of organisms that cause gastroenteritis in man.

Viruses are found in sewage and in water. The viruses can attach to suspended particles and remain infective (Moore *et al.* 1975).

Food Contamination

Water contacts food during production, harvesting and processing. If the water used for irrigation of various crops is contaminated or is sewage effluent, the fruits and vegetables can be potential health hazards (Tierney *et al.* 1977; Van Donsel and Larkin 1977; Tamminga *et al.* 1978).

Seafoods are harvested from water. The microorganisms in the water contaminate the surface, gills and intestinal tract of fish and shellfish. The occurrence of fecal coliforms in fish is a reflection of the pollution level of their water environment. The skin of Atlantic salmon was analyzed by Horsley (1973). The predominant genera were *Cytophaga, Flavobacterium, Moraxella* and *Pseudomonas.* Other organisms found were *Acinetobacter, Bacillus, Aeromonas, Vibrio,* coryneforms, members of the families Enterobacteriaceae and Micrococcaceae, as well as low numbers of *Chromobacterium* and fungi.

When bivalve mollusks feed, they filter large quantities of water and concentrate bacteria and viruses that are present in the water environment. Shellfish are normally found in water near the shore. This water is subject to contamination of run-off water carrying soil microorganisms and sewage outfalls. The accumulation and concentration of these microorganisms from the water is of particular concern when they are potential pathogens such as salmonellae and human enteric viruses.

If the drinking water of animals is contaminated by potential pathogens, they can be a health hazard to humans who handle the animals and can cause contamination of the carcass during slaughter.

During harvesting, water may be used for hydrocooling of vegetables. Irrigation or surface water which is polluted with sewage or animal wastes can serve as a source of inoculation with various organisms including potential pathogens. The water can redistribute organisms already on the vegetables from high levels of contamination or spoilage to other areas so that all of the vegetables are contaminated. Since many vegetables and fruits are eaten raw, the use of untreated water to wash these foods can serve as a vehicle of transmission of pathogenic organisms.

Ice is used to cool and maintain coolness in a variety of fresh foods. As the ice melts, the organisms associated with it are passed on to the foods and vice versa. The reuse of ice is neither a recommended nor an accepted procedure, since this ice is contaminated.

The use of water in a food processing plant can be a source of microorganisms for contaminating food. Water is the food industry's number one raw ingredient. Water enters into the processing of most foods for cleaning equipment and processing areas, for washing food, for conveying (fluming) food or as an ingredient.

The safety of water that is used in foods is of major importance. However, although considered to be safe for drinking (potable), municipal water supplies are not always acceptable for food processing. This is due to the presence of food spoilage types of microorganisms, as well as chemicals that may produce unsatisfactory odors and flavors in foods. The organisms that tend to be present in municipal water supplies are psychrotrophs such as pseudomonads which can grow in water with a limited nutrient content. Hence, when a food processor stores water in small tanks, the growth of the pseudomonads may result in microbial levels of 10^5 to 10^6 per ml.

In the dairy industry, these psychrotrophs are especially troublesome. The wash water may contain up to 10^3 psychrotrophs per ml and be one of the most important sources of microbial contamination. These psychrotrophs in the water cause spoilage in refrigerated milk and milk products.

In poultry processing plants, water is used in the scald tank, for washing of the carcass and for cleanup of the building and equipment. There has been much speculation that the scald tank water is a source of contamination of the carcass. This is due to the fact that the dirt from the head, feet and feathers, as well as feces (due to defecation at death) contaminate the scald water. The temperature of the water (53° to 61°C) can kill many pathogenic and spoilage types, but is not sufficient to kill all of the microorganisms that are present, and thermophilic types can grow. Bacterial counts of the scald water range from 10^2 to 10^6 per ml. When the birds enter the scald tank they may be taking their last gasp and the heart may be pumping. The contaminated water then can contaminate the internal portions of the bird through the lungs and vascular system.

In food canning operations, water is used to cool the hot cans of food after heat processing. Due to the heat and expansion of metal, the seams and seals on the cans are under stress and leakage can occur. *Leuconostoc mesenteroides* spoilage of canned tomato juice is due to contaminated cooling water and/or faulty can seams. In an outbreak of salmonellosis due to imported canned food, it was found that polluted river water was used to cool the processed cans of food. As a result, salmonellae in the cooling water gained access to the processed food. Since no further processing occurs after this point, contamination of the food by any microorganisms that are capable of growth in the canned product may cause spoilage or, if pathogenic, foodborne illness.

Water is used as an ingredient in many foods. In these cases the water is a direct source of microbial contamination.

With the present shortages and high cost of water, there is an effort to reduce water consumption and to recycle water in the food processing plant. Caution must be used so that contamination or high microbial loads on foods do not result from these practices.

AIR

Today, the main concern with air pollution is with chemicals (carbon monoxide, hydrocarbons, soot, fly ash) rather than with biological agents. Nature is the major contributor of not only the chemical pollutants, but also of biological agents such as plant cells, animal hair, pollen, algae, protozoa, bacteria, yeasts, mold spores and viruses. Food is subjected to airborne contamination until it is sealed in a package.

Types and Numbers of Microorganisms

There is no natural or normal microflora of air. It is contaminated from various sources. Generally, mold spores are more prevalent than are other microorganisms. The main source of microorganisms, especially mold spores, is decaying plant materials near the ground surface. Contamination of the air is caused by gusts of wind picking up the microorganisms or spores. Compared to mold spores, there are relatively few yeast cells in the atmosphere and these are mainly near ground level. One would expect yeasts to be prominent in the air of orchards and vineyards. However, even here, they may be only 25% of the fungal population of air.

There are many sources of microorganisms for contamination of the air. Trickling filters used in most sewage treatment plants produce aerosols by the spraying action and splashing liquid sewage. If not operated properly, incinerators can be a source of aerosols or infectious microorganisms. Liquids are aerosolized by splashing, spraying, the bursting of bubbles, by forcing through a small orifice or by vibrations.

The types of organisms in air are often associated with the type of activity in the area. Downwind from a sewage treatment plant, Pereira and Benjaminson (1975) found species of *Klebsiella*, *Bacillus*, *Flavobacterium*, *Streptococcus* and *Micrococcus*. Streptococci are near dairies and yeasts are near bakeries and breweries. The microbial flora of the air in a food processing plant reflects the sanitary condition of the plant unless the air is purified.

Humans shed organisms and also produce aerosols during talking, coughing and sneezing. The microbial load of the air is proportional to the numbers of persons in an enclosed space, their activity and the rate of air circulation. Although humans might not contribute significantly to the microorganisms in outdoor air, they can be a significant source for air in a processing room.

Gerba *et al.* (1975) reported that bacteria and viruses become airborne when a toilet is flushed. These organisms settle out on surfaces throughout the bathroom.

Thus, air is subjected to a number of sources of microorganisms (decaying matter, water, soil, humans, animals, sewage). The numbers and types of organisms present in the air or atmosphere depend upon many factors, such as tendency to settle out and maintenance of viability.

There is a considerable variation of the microbial load of air. It has been suggested that country air in the summer contains an average of 10,000 spores/m^3. Lacey (1973) suggested the level may reach $10^8/m^3$ during haymaking or harvesting. The handling of moldy grain or hay indoors can result in a count of 10^9 spores/m^3.

Lacey (1973) listed several genera of fungi as being present in rural air with *Cladosporium* predominating. This dominance also was reported by Ogunlana (1975). Spores of *Penicillium, Aspergillus* and *Fusarium* are quite prominent in air.

The air near the earth is more contaminated than at higher altitudes, although the microbial population at 2,000 to 3,000 meters is more stable than that at lower altitudes. Air over land is more contaminated than air over the ocean, although wind currents can carry microorganisms for several hundred miles out to sea. The upper atmosphere over the ocean is more contaminated than air near the ocean surface.

Clouds often contain high levels of bacteria and fungal spores. Turbulent air tends to carry high numbers of cells aloft. Temperature inversions can affect the microbial load at various altitudes. Rain and snow tend to wash the microflora from the air.

The atmosphere is more contaminated in the summer than in the winter.

Survival

Microorganisms cannot multiply in air. The stability of microorganisms in air is influenced by the relative humidity, oxygen, solar factors and chemical components. The ability of microorganisms to maintain their viability influences the flora of air. Spores of molds and bacteria retain their viability better than do vegetative cells. This is probably why mold spores generally dominate the fungal flora.

Aerosols of microorganisms have been produced in the laboratory and the viability determined. Aerosolization damages the cytoplasmic membrane and at least some of the associated transport mechanism (Benbough *et al.* 1972).

The effect of relative humidity (RH) on microbial survival has not been evaluated fully. Some research suggests that survival is usually greater at 30% RH or less than at 90% RH or over. However, the medium range of RH (40 to 80%) generally causes the highest rate of death. With aerosolized viruses, reportedly, the stability is lower at 50 to 60% RH than at either low or high RH values (Akers *et al.* 1973; Trouwborst and Kuyper 1974). There are variations to this generalization.

The RH has no influence on the stability of St. Louis encephalitis virus (Rabey *et al.* 1969). There is little difference in inactivation at various RH values at 21°C, but RH seemed to influence the viral stability at 32°C (Akers *et al.* 1973). Overall, the stability or the inactivation of viruses is influenced by the type of virus, the suspending medium, the method of aerosolization, the relative humidity, the temperature of the air, the method of aerosol collection and method of enumerating the virus. In all cases, there is a rapid inactivation immediately upon aerosolization, followed by a slower rate of inactivation.

The stability of bacteria in aerosols is similar to that of viruses. There is a rapid death during aerosolization, followed by a less rapid death rate. The method of recovery and enumeration influence the apparent effect of relative humidity and oxygen on survival of bacteria in aerosols.

In general, aerosolized bacteria survive better at low RH than at high RH, but they have the highest death rates at intermediate (40 to 80%) RH (Cox and Goldberg 1972; Stewart and Wright 1970).

When air is sampled downwind from a source of contamination there is a significant decrease in the microbial load as the distance from the source increases. This is due to the settling out of organisms onto various surfaces, the loss of viability of the organisms and the dilution of the contaminated air by less contaminated air.

Sunlight reduces the number of viable organisms in air. By analysis of air near a spray irrigation system, Teltsch and Katzenelson (1978) found higher counts at night than during the day.

Food Processing Operations

The contamination of air in food processing plants was reviewed by Heldman (1974). The effect that air has on the microflora of food depends upon the level of contamination of air and the time of contact of air with food.

Aerosols are produced in food processing plants by spray washing or spray cooling of food, by high pressure sprays used in cleaning, by flooding of floor drains, by mixers, motors and the operation of various other equipment. Workers in the area produce aerosols. The movement of equipment, supplies and people in a food plant causes turbulent wind currents that increase the microbial load of the air.

There is considerable variation in the microbial load of air in various areas of a processing plant. In clean areas there are very few organisms in the air. In areas in which live animals are handled or raw products are brought into the processing operation, the microbial load can be quite high.

One method that is used to control the microbial load in the air of a processing plant is to move air from clean areas to dirty areas or by using positive air pressure in clean areas. With positive air pressure, if a door is opened, air flows out of a room and outside air does not come in. Fresh air entering the clean areas is filtered to remove dirt as well as some microorganisms.

ANIMAL AND PLANT FOOD

Animal and plant food can serve as a source of potential chemical and biological contamination of food.

Microorganisms in animal feed can contaminate the feet, hide, hair or feathers of animals. Consumption of the feed adds organisms to the digestive tract. Those organisms that survive the rigors of digestion, upon elimination, can contaminate exterior portions of the animal. When the feed contains potential pathogens such as salmonellae, these may cause illness in the animal, invade the body and locate in lymph nodes, contaminate the carcass during slaughter, as well as contaminate external portions of the animal (MacKenzie and Bains 1976).

Plant food may contain organisms that can contaminate the surfaces of plants and associated human foods. When animal or human wastes are used as fertilizer, pathogens from these sources can contaminate the plant and associated foods. Since fruits and vegetables are eaten raw, in this case, they can be a source of foodborne illness. In countries that use night soil as a fertilizer for food crops, washing of the food is suggested. The decomposition of plant tissue can cause an increase of potential toxin-forming molds in fields (Griffin and Garren 1976).

Although pathogens may be present in manure or sewage used to fertilize crops, sunlight tends to eliminate microorganisms from the plant parts (Bell *et al.* 1976; Jones 1976).

PLANTS

Plants are contaminated by microorganisms from several sources (dirt, water, air, fertilizer, animals and humans). Once contaminated, certain microorganisms can grow on the plant surfaces and plant pathogens can attack their host plants. The microbial flora on plant surfaces varies with the kind of plant.

Goel *et al.* (1970) found a total count of 10^9/g including 10^7 coliforms, 10^8 psychrotrophs, 10^7 streptococci, 10^5 yeasts and 10^6 molds/g on wild rice.

Pseudomonas species are quite prevalent on vegetables. Kominos *et al.* (1972) isolated *P. aeruginosa* from tomatoes, radishes, celery, carrots, endive, cabbage, cucumbers, onions and lettuce. They believed that raw vegetables are a source of this organism contaminating patients in hospitals. Green *et al.* (1974) suggested that *P. aeruginosa* can colonize plants when the temperature and humidity are favorable.

The flowers of fruits are inhabited by many genera of yeasts including the ascospore formers *Saccharomyces* and *Hansenula* as well as the imperfect *Torulopsis, Candida, Rhodotorula* and *Kloeckera.* In some fruits such as grapes, the yeasts might not be present on the flowers but are found on the ripe fruit.

A *Pseudomonas* and an *Arthrobacter* introduced into open flowers of soybean plants were isolated from 24 of 177 resultant beans pods (Leben 1976).

There is evidence that not only the surface but also the interior tissue of plants can be contaminated. Vegetables can harbor fecal streptococci within unopened pods, heads and other structures. Meneley and Stanghellini (1974) found 44% of apparently healthy cucumbers contained bacteria internally. Enterobacteria (*Proteus, Citrobacter* and *Enterobacter*) were present in 10% of the cucumbers.

Thus, live plants can serve as a source of microorganisms. When the plant dies, the decaying vegetation becomes an important source of airborne, waterborne and soil microorganisms which then contaminate plants the following year.

ANIMALS

Animals have a normal or natural microflora that is established very early in life. Besides this microflora, they tend to harbor the types of organisms found in their environment, since they are contaminated by soil, water, air, feed and excreta.

Animals harbor organisms that can cause food spoilage or foodborne disease. Organisms are found in the animal's gastrointestinal tract, nasal passages, cutaneous lesions and on the skin, feet, hair or feathers on the outer surface. These organisms are readily transferred to the edible portions of the animal during processing.

Wild animals contaminate growing crops and stored products. One of the primary ways for plant viruses to get into the plant cells is by the stinging of insects. Insects, birds and rodents destroy the protective covering on foods so that not only do they contaminate the food, but they make the food more susceptible to spoilage. These animals also carry potential human pathogens which may be transferred to the foods they contaminate. Birds are a large reservoir of diseases that can be passed on to humans.

Flies, much more so than many other types of common food plant insects, are a serious potential carrier of disease producing organisms. Flies pick up organisms on their hairy legs and feet. They then contaminate man's food by walking or leaving their excreta on it. Flies have a part in the spreading of *Salmonella, Shigella, Vibrio, Escherichia* and other disease causing organisms as well as food spoilage types.

Insects destroy crops as well as contaminate the crops with microorganisms. Osmophilic yeasts that spoil honey and concentrated fruit products are mainly contaminants from insects.

Rodents (rats and mice) consume a tremendous amount of food each year. While eating, these animals also eliminate their wastes and further contaminate the food with microorganisms. Pests such as rodents are a source of *Salmonella* for contamination of food and feed. Kapperud (1975) studied the occurrence of *Yersinia* in rodents. *Y. enterocolitica* causes human infections and was widespread in rodents that were examined.

The muscle tissue of most animals is considered to be essentially sterile. In a diseased state, microbial invasion may occur. When the animal dies during slaughter (or harvesting in the case of fish), bacterial invasion of the tissue can occur. It is believed that most of the contamination of tissue is a result of surface or intestinal contamination, or is due to the processing operations. When found in the tissue of normally healthy animals, the numbers of microorganisms are usually very low. Vanderzant and Nickelson (1969) aseptically removed muscles from slaughtered hogs, ewes, bulls and steers. The maximum number of microorganisms per gram of meat was: ham, 700, lamb, 140 and beef, 1,400. Of 65 samples, 44 did not yield isolates when incubated at 37°C. Hence, it is evident that even after slaughter, a majority of samples of muscle tissue are sterile.

One of the sources of microorganisms in the interior portion of animals is the lymph nodes. These nodes tend to filter out bacteria from the lymphatic system. Microbial counts of 10^5/g have been obtained from the lymph nodes, with many types of organisms being present.

The number of organisms on external portions of fish vary from 10^2 to 10^5/cm^2, and the intestinal contents from 10^4 to 10^7 organisms/g. These microorganisms reflect the condition of the environment from which the fish are obtained.

The microbial flora that grows on refrigerated beef is composed mainly of soil organisms derived from the hide, hair and hoofs of the animals. The hides of animals may contain from 10^3 to over 10^9 microorganisms per cm^2. Although they do not penetrate the intact skin, these microorganisms are a source of contamination of the edible portions of the carcass during processing.

The digestive tract of animals contains numerous microorganisms and is a source of contaminants for soil, water, humans and food. During life, the intestinal or cecal wastes can contaminate the outer surface of the animal. During slaughtering and dressing, the sphincter muscle becomes nonfunctional and also the intestines may be punctured accidentally. In either case, the contents and associated organisms can contaminate the carcass (Howe *et al.* 1976; Linton *et al.* 1977). The presence of fecal material or ingesta on the internal or cut surface of poultry carcasses is cause for condemnation.

There is a possibility that microorganisms can pass through the intestinal barrier into the organs and tissues. In normal, healthy intestines, there is probably little passage of bacteria across the intestinal wall. Those that do get through the intestinal wall tend to concentrate in lymph nodes or some internal organ.

The microbial load of the intestine or cecum is very high. The total aerobic count varies from 10^6 to 10^{11}/g and about the same number for anaerobes. The species of microorganisms present in the digestive system are influenced by the environment present in the various portions of the gastrointestinal tract, the type and age of the animal, the diet and the husbandry. Coliforms, enterococci, lactobacilli and bacteroides are usually the dominant types of organisms in the fecal material of animals. The natural habitat of the human pathogen, *Salmonella,* is the intestinal tract of animals and man.

The circulatory system is sterile except for the occasional invader or during an infection.

The respiratory system acquires microorganisms from the air during breathing. The nose filters out many microorganisms and it tends to be the most contaminated part of the system. The respiratory system may serve as a passageway for microorganisms to contaminate internal tissues.

Milk in the healthy udder has few, if any, microorganisms. However, aseptically drawn milk usually ranges from less than 500 to 5,000 and occasionally, up to 10,000 organisms per ml (Thomas *et al.* 1971). The flora is predominantly coagulase negative staphylococci, micrococci and corynebacteria. Infection of the bovine mammary glands is called mastitis. With this condition, counts of 10^5 or more per ml of milk have been found. A variety of organisms can cause mastitis and these, such as coagulase positive staphylococci, coliforms or streptococci, will dominate in the milk. Aseptically drawn milk contains mesophiles, but few psychrotrophs or thermophiles. When milk is obtained normally, it is contaminated by bacteria in the teat canal and on the surface of the teat. The surfaces of the four teats, even after washing with disinfectant and drying, had a mean colony count of 10^6 bacteria (Thomas *et al.* 1971). Hibbitt *et al.* (1972) found an average of 22,500 viable microorganisms per teat canal. However, McKinnon *et al.* (1973) analyzed milk normally obtained from 25 cows and found the bacterial counts ranged from 44 to 11,400/ml, which is similar to the so-called aseptically drawn milk referred to by Thomas *et al.* (1971). Due to microorganisms in the teat canal, the first drawn milk contains more organisms than the last portion obtained from the particular quarter of the udder.

Some bacteria, such as salmonellae, can invade the ovaries of hens and contaminate the yolk before the egg is formed. However, only a very small proportion of freshly-laid eggs contain viable microorganisms and in these, the bacterial number is relatively low.

The egg is essentially sterile when it is laid. The bacterial contamination on the shells of normally laid eggs is due mainly to contact with nesting material, feather dust and the feet and body of the bird, as well as subsequent handling and storage. Eggs from flocks on wire have less bacterial contamination than eggs from flocks in houses with floor litter. This is due to less chance of nesting material, including feces, coming in contact with the shell.

Most of the contaminants on egg shells are of intestinal origin, indicating that the hen is a prime source of bacterial contamination of eggs.

HUMANS

A human embryo develops in a relatively sterile environment. When birth occurs, the baby is subjected to a massive invasion of microorganisms from the mother, other humans, bedding, air, food, water and other materials.

The skin is the most available part of the body for colonization after birth. The staphylococci are predominant on normal infant skin (Carr and Kloos 1977). The colonization is followed by microbial invasion of the nose, oral cavity, throat and the respiratory, digestive and urogenital tracts. Many microorganisms are merely transients, while others become established as a normal, perhaps permanent, flora. Besides the microbial population on or in a human, clothing can be contaminated by external sources or by the human, and then serve to pass the microorganisms on to the human or to food products.

The skin is never free of bacteria and the dirtier the skin, the greater the contamination. Normal human skin contains a relatively stable microflora. However, the numbers and types of microorganisms vary due to differences in the environment at various sites on the body. Some areas are too dry and in others the pH is not satisfactory for bacterial growth.

Washing the skin removes most of the transient microorganisms, but it is practically impossible to remove all of the normal flora. Some microorganisms associate with the sweat glands, sebaceous glands and hair follicles where they not only can grow, but also are difficult to remove.

S. aureus is found more often on the hands and face than other parts of the body. This is due to the association of this organism with the nose and the habit of people handling their nose and face, thus spreading the *S. aureus.* This organism is associated with infections such as acne, pimples and boils.

Sunga *et al.* (1970) studied organisms shed from the arms and hands and identified species of *Sarcina, Peptococcus, Staphylococcus, Micrococcus, Bacillus, Alcaligenes, Pseudomonas* and *Corynebacterium.*

The counts of organisms on the skin vary from 300 to 468/4 cm^2 on the back to 1,230 to 10,470/4 cm^2 on the thigh (Noble *et al.* 1976). McBride *et al.* (1977) and Noble (1975) reported higher counts on the skin.

The contact plate method yields lower results than the scrub count. Evans and Stevens (1976) found the swab method of sampling could be used to estimate surface flora while the scrub method also estimated the subsurface flora. The palm of the hand has primarily surface organisms whereas the flora of the forehead is both surface and subsurface, but mainly subsurface. The surface flora consists of transients and normal flora, while the subsurface flora is primarily normal flora.

Humans are a source of airborne microorganisms as well as an important source of food contamination through handling of food. The hands are subject to contamination by considerable numbers of bacteria, most of which are unable to multiply on the hands and usually die. However, these transient bacteria are passed on to food products when they are handled. Many of the transient bacteria result from the handling of foods (Seligmann and Rosenbluth 1975).

The hair covering the skin is a potential source of microorganisms. There is no normal flora of the hair, but hair acts as a carrier to retain and shed organisms. Beards, sideburns and mustaches may be especially bad if worn in a food handling area. Barbeito *et al.* (1967) found that, although washing with soap and water reduced the level of contamination of beards, it did not eliminate microorganisms and toxins from them. If a person grows a beard to hide acne or pimples or a nasal carrier has a mustache, these hairy growths will merely serve for continued dissemination of *S. aureus.* Besides the microorganisms that may be shed into foods, the hairs and hair clippings may contaminate the food.

Clothing is directly associated with humans and can serve as a carrier to contaminate foods with human microflora or can contaminate humans with environmental microorganisms. Laundering does not necessarily remove all of the microorganisms from clothing. The effectiveness of laundering depends upon the type of organism, fabric, temperature, detergent, antibacterial agents, bleaches, flushes and rinses, drying, ironing and final packaging. Viruses tend to adsorb to the fabric and are difficult to remove. Microorganisms tend to remain viable in wool longer than in cotton. Synthetic fabrics require less drastic laundering, allowing greater microbial survival. Microorganisms tend to survive storage for a shorter time on these fabrics than on either cotton or wool.

During laundering, microorganisms are transferred among the clothing in the same wash load and it is conceivable that microorganisms can carry over from one load to the next. Even dry cleaning does not assure the sterility of clothes.

The gastrointestinal tract may be considered as a hollow tube within the body but the internal portion is outside the body. The gastrointestinal tract is important as a potential source of organisms, as well as the primary site of action of foodborne organisms causing illness (gastroenteritis). Factors that can influence the flora of the gastrointestinal tract include the diet, antibiotics or other chemotherapeutic agents and certain diseases.

Although enormous numbers of microorganisms are ingested with food and with mucus from the respiratory tract, most do not survive in the digestive system.

In the normal stomach, the microbial flora increases and decreases. The ingestion of food or mucus causes a rise in numbers. The flora is reduced by acid and digestive enzymes and by emptying of the stomach into the duodenum. There are usually low numbers of organisms in the normal stomach, duodenum, jejunum and upper ileum. These organisms are mainly Gram positive facultative types and, at rare times, a few coliforms. The pH of the stomach contents varies among individuals. High pH values (low acidity) will allow the survival of more organisms. Foods tend to react with the acid and the pH is raised somewhat. Some microbial cells are imbedded in the food particles. This tends to protect the cells from the acid.

Normal peristalsis plus bile and immunoglobulin tend to maintain low counts in the upper small intestine. If any abnormal condition exists so that the effectiveness of the inhibitory conditions of the stomach and upper small intestine is reduced, then the microbial population in these areas may rise to 10^{10}/ml. Drugs, such as morphine, slow peristaltic action. This allows the establishment of certain organisms in these areas. The growth of organisms in the small intestine can lead to nutritional deficiencies due to malabsorption of nutrients (Neale *et al.* 1972), as well as enteric disturbances.

As the cells move down the intestines, the conditions improve for microbial growth. Active multiplication begins in the lower small intestine and continues throughout the large intestine. Bacterial levels of 10^{10} to 10^{12}/g of contents are attained in the colon and excreta. The redox potential varies from −150 mv the duodenum to −250 mv in the colon (Holdeman and Moore 1972). This favors facultative and anaerobic types.

In the breast-fed baby, 99% or more of the fecal flora consists of species of *Bifidobacterium.* The remaining 1% or less of the flora consists of coliforms, enterococci, aerobic lactobacilli and staphylococci. If the baby is not in a clean environment, other microorganisms, such as found in the adult, will be present in the fecal flora.

The fecal flora of the adult becomes putrefactive with the anaerobic bacteroides group surpassing the anaerobic bifidobacteria. These two groups range from 10^9 to 10^{11} cells/g of feces and account for about 95% of the flora (Haenel 1970). The remaining 5% includes coliforms, enterococci and lactobacilli, with small numbers of staphylococci, clostridia, bacilli, veillonellae, enterobacteria, pseudomonads, yeasts, molds and viruses. Even though clostridia are considered a minor constituent, and may not be present in all fecal samples, they may range from 10^3 to 10^9 cells/g.

The relationship of viruses to the fecal microflora has not been examined thoroughly. Enteric viruses are present in the intestines and feces of apparently healthy people.

In the abnormal stomach and small intestine, the microbial population often is dominated by aerobes and facultatives such as coliforms, lactobacilli and enterococci. The number of bifidobacteria is reduced, and staphylococci, yeasts and enteric organisms may increase.

Speck *et al.* (1970) tested the effect of diets on the fecal flora. Their results indicated that a change in diet influences the fecal microflora. The type of diet affected the aerobic population more than the anaerobic microflora. A protein diet supported a higher aerobic population than did either a carbohydrate or fat diet. Haenel (1970) found no dramatic differences in the intestinal microbial flora with meat-egg, milk-vegetable, vegetarian or raw vegetarian diets. Even the consumption of 1 kg of Bulgarian yogurt per day failed to dramatically alter the microbial flora in most individuals. Conn and Floch (1970) fed 8-20 capsules, each containing 5 (10^{10}) viable *Lactobacillus acidophilus* for 7-10 days. They found no significant increase in fecal lactobacilli, but did notice a decrease in total anaerobic bacteria. Similar results were reported by Paul and Hoskins (1972).

Only a few types of bacteria were at significantly different levels in the fecal flora with a Japanese or American diet (Finegold *et al.* 1974). Holdeman *et al.* (1976) found only two of the most common species were affected by a change in the diet.

Some strains of organisms produce enterotoxins that cause fluid secretion into the small intestine. As a result of the acute diarrhea, the microflora of the small intestine is radically disturbed. The enteric organisms may colonize this area and even move into the stomach.

People with acute diarrhea may show the following features of intestinal dysfunction: 1) bacterial colonization of the small intestine with defective microbial clearing systems; 2) malabsorption of fat, carbohydrate and vitamin B_{12}; 3) fluid accumulation due to alterations in electrolytes and water transport; and 4) alteration of the mucosal morphology. Those abnormalities are not necessarily manifested in all persons with acute diarrhea, or in all cases of diarrhea.

Mendes and Lynch (1976) found fecal organisms on various surfaces in restrooms (door handles, flush handles, tap handles, toilet seat, floor). The tap handles were more contaminated than the door handles. The wash basin overflow was more contaminated than the floor, or the area under the rim of the toilet. They believed their findings showed that a person with *Salmonella* or *Shigella* infection could contaminate various surfaces in a washroom or toilet. These areas could conceivably serve as a source of contamination for other people.

Since humans are carriers of pathogenic types of microorganisms, they are especially hazardous when handling processed (cooked or pasteurized) products that may be held for short periods and eaten with no further treatment.

In a review, Cliver (1971) discussed seven cases of infectious hepatitis. Five of these were the result of contamination of food by human carriers.

Both animals and humans are carriers of salmonellae and *C. perfringens* so it is difficult to assess the exact role of each in outbreaks of these foodborne illnesses. *S. aureus* food poisoning is primarily due to contamination of food by humans, and *Shigella,* a human pathogen, is rarely associated with animals.

The hands of workers have been cited as increasing the microbial load of many products. Poultry products show increased counts when the carcass is hand transferred from the picking to the eviscerating line, during eviscerating, cutting-up the carcass and deboning by hand. In the shellfish industry, the hand shucking of oysters and clams, shelling of shrimp and picking crabmeat from the shell are associated with increased bacterial counts. Slicing, weighing and hand packaging of meat products increase the contamination through human handling of the product. The unpacking, trimming, sorting and repackaging of fresh produce (fruit and vegetables) provide an opportunity for contamination of the food with human pathogens. Fresh produce may be eaten raw which makes contamination a potential hazard.

The carelessness of humans is an important cause of microbial contamination of foods. Failure to properly clean and sanitize equipment, failure to wash one's hands, working and handling food with an infection, poor personal hygiene, lack of care in handling of food, making unsatisfactory alterations of equipment and failure to keep foods at the proper temperature are some of the careless things that humans do that can increase the microfloral load of foods.

Some shoppers have been seen opening containers of foods. This is a deplorable practice that can cause contamination of the food with resultant spoilage or a health hazard.

During the analysis of foods, microbiologists can be a source of contamination (Denny 1972).

SEWAGE

Animal manure or, in some cases, human wastes are used as a fertilizer on crops. These biologically produced subtances contain microorganisms including human pathogens. When added to soil, these pathogens may survive for periods sufficient to contaminate the harvested crop.

In rural areas, septic tanks are used. Quite often these are not properly installed and do not operate effectively so that essentially raw sewage leaks into the soil.

Even in sewage disposal systems there is not adequate treatment to destroy all potential pathogens in the effluent.

In the United Kingdom, Evison and James (1973) found that raw domestic sewage contained 10^9 coliforms/g and effluent from a treatment facility had 10^6 coliforms/g. The fecal streptococci were at levels of 10^6/g and 10^4/g, respectively. Kampelmacher and van Noorle Jansen (1976) found salmonellae in 94% of the effluents from sewage treatment plants.

The survival of enteric viruses through treatment plants is well documented. Moore *et al.* (1975) discussed this aspect as well as the ability of viruses to remain infective when adsorbed to solid particles.

EQUIPMENT

During the course of the Industrial Revolution, machines were developed to do most of the work of humans. Hence, there is less contact with food by humans and more contact by machines and equipment.

There are thousands of small pieces of equipment, such as knives, cutting boards and bins which are used in food processing or handling establishments. Although most equipment is metal, some parts may be made of rubber or plastic. In some cases, cardboard is used in boxes used to hold harvested fruits or vegetables. Cardboard or paper is used as packaging material for bulk shipment and storage of various ingredients.

Metal processing equipment does not support the growth of microorganisms. It has no natural or normal microbial flora. Yet, food processing equipment is one of the major sources of contamination of foods.

Equipment may be cleaned and sanitized but this does not mean that it is sterile. Even on washed, visibly clean surfaces the survival and growth of bacteria are possible. This is due to the fact that even visibly clean surfaces may have food deposits or films that provide microenvironments acceptable for survival and growth of microorganisms. The potential for microbial buildup on equipment is enhanced when equipment is improperly cleaned and sanitized. If food residues are visible on cleaned equipment, one can be assured of a potentially high microbial level.

Pitted surfaces or poorly soldered joints are places in which foods can lodge on the equipment. During the course of the daily operation, bacterial growth will occur in these food films and deposits. This then serves as a source of contamination when food contacts these surfaces.

It is easier to remove soil and bacteria from stainless steel than from rubber or plastics. Hence, although stainless steel is more expensive than other metals, rubber or plastic, it pays for itself through savings in cleaning, durability and product quality. When flexibility is needed, such as in milking machines or in chicken picking machines, rubber or plastic becomes an essential part of the equipment.

Besides direct contact surfaces, a processing plant may include equipment or areas in which microorganisms can multiply and then contaminate food through an intermediary vehicle. One such case is floor drains. When the drains are flooded, bacteria in the drains become airborne.

Gilbert and Maurer (1968) surveyed equipment in shops and cafes. Swabbings from a meat slicing machine revealed bacterial loads of 10^6 to $10^7/cm^2$. The cleaning methods used were ineffective in reducing the bacterial loads to an acceptable level. When we consider that most outbreaks of foodborne illness are due to contamination of the food in food service units or in the home, it should be obvious that the equipment in the home and food service institutions needs more attention.

Meat

The contamination of meat by equipment begins with slaughtering the animal. Contaminants from the knife used to cut the artery may circulate through the body during the bleeding process. Humane slaughter by immobilizing the animal with CO_2 or electric stunning before bleeding helps reduce the heart action and hence, contamination.

During the slaughtering operation, the carcass contacts various equipment surfaces. The microflora on the outer surfaces of carcasses can be transferred to the cut surfaces by knives or saws.

When the carcass is cut up and packaged for retail, equipment such as band saws, knives, slicers and cutting blocks can transfer the microorganisms from the carcass surface to the freshly cut surface of the meat. These organisms may be psychrotrophs which are important in the spoilage of refrigerated meat. When the carcass is cut up, the surface area is increased and cut surfaces with their associated juices support large populations of bacteria.

The grinding of meat further increases the surface area, releases juices and distributes the original surface bacteria throughout the meat. Equipment may be involved in the distribution and increased counts of the product. Meat grinders are highly contaminated with millions of bacteria (Dempster 1973).

Poultry

During processing there may be a 10-fold increase in the microbial count of the skin of poultry carcasses. Some of this increase is due to contaminated scald water and to human factors. The equipment used in processing the birds plays a role in contaminating the carcasses.

Schuler and Badenhop (1972) sampled 17 equipment sites on two sampling days in 13 poultry processing plants. On only 5 of the pieces of equipment was the average bacterial count less than 1,000/cm^2 on both sampling days prior to start of the operation. Sites of contamination were the flexible fingers on the picking machine (11,000 to 99,000/cm^2), the hock cutter (7,000 to 97,000/cm^2), the transfer belt between dressing and eviscerating lines (5,000 to 56,000/cm^2), lung gun (3,500 to 73,000/cm^2) and cut-up belt (2,000 to 35,000/cm^2).

The picker fingers are difficult to clean properly and the contamination of these fingers increases significantly during the plant operation. There is increased contamination of poultry carcasses during feather removal, which can be due to leakage from the vent and the bacteria from the feathers contaminating the skin via the flexible rubber fingers.

Milk

Since milk is a liquid, it is in contact with some type of equipment from the time it is removed from the cow until it is packaged. Unsatisfactorily washed and sanitized milking and milk handling equipment constitute the main source of bacteria in milk at the farm level.

The bacteria in farm bulk milk tanks include many psychrotrophs and few thermodurics (MacKenzie 1973; Thomas *et al.* 1971). Thermodurics are more prevalent than psychrotrophs in pipeline milking plants (MacKenzie 1973).

Psychrotrophs can grow and cause spoilage of refrigerated milk, but few thermodurics grow below 10°C. Pasteurization of milk destroys most of the psychrotrophs. Contamination of pasteurized milk by psychrotrophs from poorly cleaned pipes and filling machines can result in spoilage of the refrigerated milk.

Seafood

When fish are caught they are subjected to contamination from fish boxes (plastic, wood or metal) and other fish holding systems on the ship. However, there are few, if any, bacteria in the flesh of fish before filleting and subsequent handling.

Filleting may be done manually or by filleting and skinning machines. When filleted by hand, the knives and cutting boards provide a source of microorganisms and filleting machines become contaminated by the surface microflora on the fish which is transferred to the flesh of the fillet. The bacterial count on the filleting equipment varies from 10^3 to 10^8 bacteria/cm^2. When the bacterial load on the surface of the fish is high, there is a higher bacterial count on the filleting boards. Gloves worn by the filleters often have a higher bacterial load than the incoming fish.

Fruits and Vegetables

The microbial load of fruits and vegetables increases during harvesting and processing. Peas aseptically removed from the pods are essentially sterile. When the peas are separated by a viner, the bacterial count may reach 10^6/g. The microbial load of most vegetables is reduced by washing and by blanching. After blanching, due to contamination by equipment, the bacterial load increases and may attain levels of 10^4-10^5/g. The equipment causing contamination includes conveyor belts, hoppers and filling machines.

Splittstoesser (1973) stated that the source of most organisms on frozen vegetables is contaminated equipment surfaces. His data showed that corn cutters, French bean slicers, spinach choppers and inspection belts for peas significantly increased the microbial level of these vegetables. In one instance, he found that moving peas to the freezer by an airlift (to conserve water) resulted in significantly more bacterial contamination of the peas than when they were flumed.

INGREDIENTS

The quality of a processed food is influenced by the quality of the ingredients. Although ingredients may constitute a small part of the total food, they may add a substantial number of microorganisms. Hence, specifications, including acceptable microbiological levels, are needed for the purchase or production of ingredients.

Spices or seasonings are often the source of high microbial numbers. Spices are parts of plants such as dried seeds, buds, fruit or flower parts, bark or roots, usually of tropical orgin. These seasonings and flavors may contain over 10^8 aerobic bacteria per gram. Also, they contain aerobic and anaerobic spores. A sample of ginger contained 48 million total bacteria, 12 million yeasts and molds, 26 million aerobic sporeformers and 0.72 million anaerobic sporeformers (Tjaberg *et al.* 1972). Pepper is usually highly contaminated with bacteria and fungi. Flannigan and Hui (1976) reported that 14 of 20 spices contained *Aspergillus flavus* and four of these spices supported growth of this mold and the production of aflatoxin. Mean standard plate counts of over 10^6 per gram were obtained for black pepper, ginger and paprika (Julseth and Deibel 1974). These spices also contained over 10^6 bacterial spores per gram. Psychrotrophic spores were absent in herbs and spices analyzed by Michels and Visser (1976).

Thermophilic bacteria, usually as spores, are added to foods with ingredients such as spices, starch, flour and sugar. Thermophilic spores are important when added to canned foods. The higher the level of heat resistant spores, the greater the chance that some may survive the heat treatment and be a potential spoilage problem.

The total aerobic count of flour sampled at the mills is generally in the range of 10^2 to 10^3/g (Graves *et al.* 1967). They reported levels of total aerobic thermophilic spores from zero to 1,200 per 10 g of flour, and although thermophilic flat-sour spores were not present in 10 g of most flour samples, one 10 g sample contained 730 flat-sour spores.

Since sugar is a potential source of spores, the National Canner's Association has set limits of aerobic thermophilic flat-sour and anaerobic thermophilic spores when this ingredient is to be added to canned foods.

Eskin *et al.* (1971) suggested that infections of confectionery items by osmophilic yeasts are due to contaminated nuts, fruits, chocolate, flour or flavoring syrup. Bakery products can be contaminated by the ingredients in the doughs or by adding materials to the baked food. Icing and nuts are added to already baked sweet rolls. The toasting of almonds and pecans is not sufficient to assure sterility. These products may contain 10^4-10^5 bacteria and 10^4 yeasts and molds/g.

Batters used in the breading of seafoods, onion rings or other products are a potential source of high levels of microorganisms. Batters used in the manufacture of breaded onion rings had a total bacterial count of 2,500 to 8,500,000/g (Maxcy and Arnold 1972). There were 110 to 32,000 coliforms/g of the batters they examined.

Since ingredients are an important source of microorganisms, food processors must establish acceptable microbiological limits for these substances. Then sufficient testing is needed to assure the processor that the ingredients are acceptable and, if they are used in the manufacture of a food, adjustments of the process are needed to take care of any excessive bacterial load.

The types of microorganisms in an ingredient or product are often more important than the total numbers that are present. Organisms that cause spoilage or are a potential health hazard are the important ones to consider in an ingredient or final product. However, it is assumed, although not always correctly, that an ingredient with a very high microbial load is more likely to contain spoilage or pathogenic organisms than one with a very low number of microorganisms. We should admit that we do not know everything. It may be that the ingestion of high levels of what are considered to be nonpathogenic microorganisms may cause some form of mild gastrointestinal distress. It may be that this occurs only in certain people

We do not know the ramifications of the ingestion of some types of microorganisms and the level that might or might not be acceptable. Therefore, it is essential for food microbiologists with the cooperation of people in other fields to make sensible, educated guesses concerning the potential effects of microorganisms and levels in foods, and determine approximate acceptable levels in foods by considering risk factors. It is neither possible nor wise to expect to eat foods with no microorganisms.

PRODUCT TO PRODUCT

There is an old saying that one bad apple can spoil the barrel of apples. This is very true since the spoilage organisms from the bad apple can contaminate and cause spoilage of surrounding apples until all of the apples are spoiled.

In this example, the transfer of the organisms can be disastrous since they are known to cause spoilage of apples. Although the transfer of microorganisms from one food to another can occur directly if the foods contact each other, more often it occurs through contamination of wash water, equipment or handling.

Mol *et al.* (1971) reported that the shelf life of vacuum-packed sliced cooked meats would be improved if they were handled and sliced in an area strictly separated from raw cured meats.

During World War II it was determined that dust from dried egg whites that contained salmonellae contaminated finished baked products. These contaminated products caused salmonellosis.

Cross contamination of food is one of the 10 main factors that contribute to foodborne illnesses.

People handling both cooked and raw foods can transfer microorganisms from the raw to the cooked product. With no further treatment, this is a potential health hazard. The housewife who cuts up a raw chicken on a cutting board which is then used for cutting up raw vegetables for a salad may transfer *Salmonella* from the raw chicken to the raw vegetables. The meat department in a retail store may use the same knife and block to cut fish, cold meat, chicken, beef and other types of food. Some of these foods may be eaten with no further treatment or only a minimal heating. It is evident that a potential health hazard can result.

PACKAGING

Packaging is a potential source of contamination. There are many studies concerning the effect of different types of packaging materials on the shelf life of food. The possibility that the packaging material might contribute to the flora is not mentioned. This is probably due to the fact that the flora on most foods is much greater than that on packaging material. In their survey of microorganisms on poultry processing equipment, Schuler and Badenhop (1972) reported the total bacterial count of packaging material. In this case, the packages were bulk containers for ice-packed poultry. The counts ranged from 290/4 in.^{2} (11/cm^2) to 9,770/4 in.^{2} (376/cm^2). They stated that the holding of boxes overnight led to contamination by dust, rodents and birds. In one case, the cardboard used to form the boxes was stored on a loading dock and birds roosted on the material. Although the material was contaminated with bird droppings, it was used for packing poultry.

Korab (1963) tested one-trip glass bottles used for carbonated beverages. He reported that a majority of bottles had no bacteria, yeasts or molds when they were received at the beverage company, but about 6.5% of the bottles had over 50 bacteria per bottle. The bacterial level of bottles received was about the same as after they were put through the bottle washer. Storage of uncovered bottles for one week increased the bacterial, yeast and mold contamination. Storage outside resulted in more contamination than storage inside, and the microbial level after holding for one month was greater than after one week. Water cleaning removed most of this storage contamination. Compared to returnable and reused bottles, the one trip bottles have very low microbial counts.

Plastic materials are used for packaging many food products. During manufacture and handling, electrostatic charges can occur on the plastic.

These charges attract airborne material such as dust and microorganisms (Baribo *et al.* 1966). Although plastics are essentially sterile when made into packaging material, they may become contaminated if not handled in an acceptable manner. The static charges add to the problem of maintaining plastic free of microbial contamination.

The reuse of containers can cause problems in the food industry. The reuse of damaged egg cases leads to increased loss due to breakage and higher microbial counts in the egg products.

There was a time when containers used for shipping poultry were reused to pack fresh vegetables for shipment. Due to the potential for contamination of the containers by salmonellae from the poultry and then the transfer of these pathogens from the container to the vegetables, this practice was halted.

Packages serve as a protective covering to limit or prevent microbial contamination; however, they do not prevent microbial growth. It is important that preparation of the food for packaging is designed to limit or prevent microbial contamination prior to packaging. Also, the package must be durable enough to maintain its integrity during storage and distribution.

BIBLIOGRAPHY

AKERS, T.G., PRATO, C.M., and DUBOVI, E.J. 1973. Airborne stability of simian virus 40. Appl. Microbiol. *26*, 146-148.

BARBEITO, M.S., MATHEWS, C.T. and TAYLOR, L.A. 1967. Microbiological laboratory hazard of bearded men. Appl. Microbiol. *15*, 899-906.

BARIBO, L.E., AVENS, J.S., and O'NEILL, R.D. 1966. Effect of electrostatic charge on the contamination of plastic food containers by airborne bacterial spores. Appl. Microbiol. *14*, 905-913.

BELL, R.G., WILSON, D.B., and DEW, E.J. 1976. Feedlot manure top dressing for irrigated pasture: good agriculture practice or a health hazard? Bull. Environ. Contam. Toxicol. *16*, 536-540.

BENBOUGH, J.E., HAMBLETON, P., MARTIN, K.L., and STRANGE, R.E. 1972. Effect of aerosolization on the transport of α-methyl glucoside and galactosides into *Escherichia coli*. J. Gen. Microbiol. *72*, 511-520.

CARR, D.L., and KLOOS, W.E. 1977. Temporal study of the staphylococci and micrococci of normal infant skin. Appl. Environ. Microbiol. *34*, 673-680.

CLIVER, D.O. 1971. Transmission of viruses through foods. Critical Rev. in Environ. Control *1*, 551-579.

COLEMAN, R.N., CAMPBELL, J.N., COOK, F.D., and WESTLAKE, D.W.S. 1974. Urbanization and microbial content of the North Saskatchewan River. Appl. Microbiol. 27, 93-101.

CONN, H.O., and FLOCH, M.H. 1970. Effects of lactulose and *Lactobacillus acidophilus* on the fecal flora. Amer. J. Clin. Nutr. *23*, 1588-1594.

COX, C.S., and GOLDBERG. L.J. 1972. Aerosol survival of *Pasteurella tularensis* and the influence of relative humidity. Appl. Microbiol. *23*, 1-3.

DEMPSTER, J.F. 1973. A note on the hygiene of meat mincing machines. J. Hyg. (Camb.) *71*, 739-744.

DENNY, C.B. 1972. Collaborative study of a method for the determination of commercial sterility of low-acid canned foods. J. Assoc. Offic. Anal. Chem. *55*, 613-616.

DUBOISE, S.M., MOORE, B,E., and SAGIK, B.P. 1976. Poliovirus survival and movement in a sandy forest soil. Appl. Environ. Microbiol. *31*, 536-543.

ESKIN, N.A.M., HENDERSON, H.M., and TOWNSEND, R.J. 1971. Biochemistry of foods. Academic Press, New York.

EVANS, C.A., and STEVENS, R.J. 1976. Differential quantitation of surface and subsurface bacteria of normal skin by the combined use of the cotton swab and the scrub methods. J. Clin. Microbiol. *3*, 576-581.

EVISON, L.M., and JAMES, A. 1973. A comparison of the distribution of intestinal bacteria in British and East African water sources. J. Appl. Bacteriol. *36*, 109-118.

FINEGOLD, S.M., ATTEBERY, H.R., and SUTTER, V.L. 1974. Effect of diet on human fecal flora: comparison of Japanese and American diets. Amer. J. Clin. Nutr. *27*, 1456-1469.

FLANNIGAN, B., and HUI, S.C. 1976. The occurrence of aflatoxin-producing strains of *Aspergillus flavus* in the mould floras of ground spices. J. Appl. Bacteriol. *41*, 411-418.

GERBA, C.P., WALLIS, C., and MELNICK, J.L. 1975. Microbiological hazards of household toilets: droplet production and the fate of residual organisms. Appl. Microbiol. *30*, 229-237.

GILBERT, R.J., and MAURER, I.M. 1968. The hygiene of slicing machines, carving knives and can-openers. J. Hyg. (Camb.) *66*, 439-450.

GOEL, M.C. *et al.* 1970. Microbiology of raw and processed wild rice. J. Milk Food Technol. *33*, 571-574.

GOODRICH, T.D., STUART, D.G., BISSONNETTE, G.K., and WALTER, W.G. 1970. A bacteriological study of the waters of Bozeman Creek's south fork drainage. Proc. Mont. Acad. Sci. *30*, 59-65.

GRAVES, R.R., ROGERS, R.F., LYONS, A.J., JR., and HESSELTINE, C.W. 1967. Bacterial and actinomycete flora of Kansas-Nebraska and Pacific Northwest wheat and wheat flour. Cereal Chem. *44*, 288-299.

GRAY, T.R.G., and WILLIAMS, S.T. 1971. Microbial productivity in soil. *In* Microbes and Biological Productivity. D.E. Hughes and A. H. Rose (Editors). Twenty-first Symposium of the Society for General Microbiology. The University Press, Cambridge, England.

GREEN, S.K. *et al.* 1974. Agricultural plants and soil as a reservoir for *Pseudomonas aeruginosa*. Appl. Microbiol. *28*, 987-991.

GRIFFIN, G.J., and GARREN, K.H. 1976. Colonization of rye green manure and peanut fruit debris by *Aspergillus flavus* and *Aspergillus niger* group in field soils. Appl. Environ. Microbiol. *32*, 28-32.

HAENEL, H. 1970. Human normal and abnormal gastrointestinal flora. Amer. J. Clin. Nutr. *23*, 1433-1439.

HELDMAN, D.R. 1974. Factors influencing air-borne contamination of foods. A review. J. Food Sci. *39*, 962-969.

HIBBITT, K.G., BENIANS, M., and ROWLANDS, G.J. 1972. The number and percentage viability of commensal microorganisms recovered from the teat canal of the cow. Brit. Vet. J. *128*, 270-274.

HOLDEMAN, L.V., GOOD, I.J., and MOORE, W.E.C. 1976. Human fecal flora: variation in bacterial composition within individuals and a possible effect of emotional stress. Appl. Environ. Microbiol. *31*, 359-375.

HOLDEMAN, L.V., and MOORE, W.E.C. 1972. Roll-tube techniques for anaerobic bacteria. Amer. J. Clin. Nutr. *25*, 1314-1317.

HORSLEY, R.W. 1973. The bacterial flora of the Alantic salmon (*Salmo salar L.*) in relation to its environment. J. Appl. Bacteriol. *36*, 377-386.

HOWE, K., LINTON, A.H., and OSBORNE, A.D. 1976. An investigation of calf carcass contamination by *Escherichia coli* from the gut contents at slaughter. J. Appl. Bacteriol. *41*, 37-45.

JONES, P.W. 1976. The effect of temperature, solids content and pH on the survival of salmonellas in cattle slurry. Brit. Vet. J. *132*, 284-293.

JULSETH, R.M., and DEIBEL, R.H. 1974. Microbial profile of selected spices and herbs at import. J. Milk Food Technol. *37*, 414-419.

KAMPELMACHER, E.H., and VAN NOORLE JANSEN, L.M. 1976. *Salmonella* in effluent from sewage treatment plants, wastepipe of butchers' shops and surface water in Walcheren. Zbl. Bakt. Hyg I. Abt. Orig. B *162*, 307-319.

KAPPERUD, G. 1975. *Yersinia enterocolitica* in small rodents from Norway, Sweden and Finland. Acta Pathol. Microbiol. Scand. B., *83*, 335-342.

KOMINOS., S.D., COPELAND, C.E., GROSIAK, B., and POSTIC, B. 1972. Introduction of *Pseudomonas aeruginosa* into a hospital via vegetables. Appl. Microbiol. *24*, 567-570.

KORAB, H.E. 1963. Microbiological aspects of one-trip glass bottles as used by the carbonated beverage industry. Food Technol. *17*, 108-109.

LACEY, J. 1973. Actinomycete and fungus spores in farm air. J. Agr. Labour. Sci. *1*, 61-78.

LEBEN, C. 1976. Soybean flower-to-seed movement of epiphytic bacteria. Can. J. Microbiol. *22*, 429-431.

LINTON, A.H. *et al.* 1977. Antibiotic resistance among *Escherichia coli* O-serotypes from the gut and carcasses of commercially slaughtered broiler chickens: a potential public health hazard. J. Appl. Bacteriol. *42*, 365-378.

MACKENZIE, E. 1973. Thermoduric and psychrotrophic organisms on poorly cleansed milking plants and farm bulk milk tanks. J. Appl. Bacteriol. *36*, 457-463.

MACKENZIE, M.A., and BAINS, B.S. 1976. Dissemination of *Salmonella* serotypes from raw feed ingredients to chicken carcasses. Poultry Sci. *55*, 957-960.

MAXCY, R.B., and ARNOLD, R.G. 1972. Microbiological quality of breaded onion rings. J. Milk Food Technol. *35*, 63-66.

MCBRIDE, M.E., DUNCAN, W.C., and KNOX, J.M. 1977. The environment and the microbial ecology of human skin. Appl. Environ. Microbiol. *33*, 603-608.

MCKINNON, C.H., COUSINS, C.M., and FULFORD, R.J. 1973. An in-line milk sampler for determining the numbers of bacteria derived from teat surfaces and udder infections of cows milked in recorder machines. J. Dairy Res. *40*, 47-52.

MENDES, M.F., and LYNCH, D.J. 1976. A bacteriological survey of washrooms and toilets. J. Hyg. (Camb.) *76*, 183-190.

MENELEY, J.C., and STANGHELLINI, M.E. 1974. Detection of enteric bacteria within locular tissue of healthy cucumbers. J. Food Sci. *39*, 1267-1268.

MICHELS, M.J.M., and VISSER, F.M.W. 1976. Occurrence and thermoresistance of spores of pyschrophilic and psychrotrophic aerobic sporeformers in soil and foods. J. Appl. Bacteriol. *41*, 1-11.

MOL, J.H.H., HIETBRINK, J.E.A., MOLLEN, H.W.M., and VAN TINTEREN, J. 1971. Observations on the microflora of vacuum packed sliced cooked meat products. J. Appl. Bacteriol. *34*, 377-397.

MOORE, B.E., SAGIK, B.P., and MALINA, J.F., JR. 1975. Viral association with suspended solids. Water Res. *9*, 197-203.

NEALE, G. *et al.* 1972. The metabolic and nutritional consequences of bacterial overgrowth in the small intestine. Amer. J. Clin. Nutr. *25*, 1409-1417.

NOBLE, W.C. 1975. Dispersal of skin microorganisms. Brit. J. Dermatol. *93*, 477-485.

NOBLE, W.C. *et al.* 1976. Quantitative studies on the dispersal of skin bacteria into the air. J. Med. Microbiol. *9*, 53-61.

OGUNLANA, E.O. 1975. Fungal air spora at Ibadan, Nigeria. Appl. Microbiol. *29*, 458-463.

PAUL, D., and HOSKINS, L.C. 1972. Effect of oral lactobacillus feedings on fecal lactobacillus counts. Amer. J. Clin. Nutr. *25*, 763-765.

PEREIRA, M.R., and BENJAMINSON, M.A. 1975. Broadcast of microbial aerosols by stacks of sewage treatment plants and effects of ozonation on bacteria in the gaseous effluent. Public Health Rep. *90*, 208-212.

RABEY, F., JANSSEN, R.J., and KELLEY, L.M. 1969. Stability of St. Louis encephalitis virus in the airborne state. Appl. Microbiol. *18*, 880-882.

SCHULER, G.A., and BADENHOP, A.F. 1972. Microbiology survey of equipment in selected poultry processing plants. Poultry Sci. *51*, 830-835.

SELIGMANN, R., and ROSENBLUTH, S. 1975. Comparison of bacterial flora on hands of personnel engaged in non-food and in food industries: a study of transient and resident bacteria. J. Milk Food Technol. *38*, 673-677.

SPECK, R.S., CALLOWAY, D.H., and HADLEY, W.K. 1970. Human fecal flora under controlled diet intake. Amer. J. Clin. Nutr. *23*, 1488-1494.

SPLITTSTOESSER, D.F. 1973. The microbiology of frozen vegetables. Food Technol. *27*, 54-56, 60.

STEWART, R.H., and WRIGHT, D.N. 1970. Recovery of airborne streptococcal L-forms at various relative humidities. Appl. Microbiol. *19*, 865-866.

SUNGA, F.C.A., HELDMAN, D.R., and HEDRICK, T.I. 1970. Microorganisms from arms and hands of dairy plant workers. J. Milk Food Technol. *33*, 178-181.

TAMMINGA, S.K., BEUMER, R.R., and KAMPELMACHER, E.H. 1978. The hygienic quality of vegetables grown in or imported into the Netherlands: a tentative survey. J. Hyg. (Camb.), *80*, 143-154.

TELTSCH, B., and KATZENELSON, E. 1978. Airborne enteric bacteria and viruses from spray irrigation with wastewater. Appl. Environ. Microbiol. *35*, 290-296

THOMAS, S.B., DRUCE, R.G., and JONES, M. 1971. Influence of production conditions on the bacteriological quality of refrigerated farm bulk tank milk—a review. J. Appl. Bacteriol. *34*, 659-677.

TIERNEY, J.T., SULLIVAN, R., and LARKIN, E.P. 1977. Persistence of poliovirus 1 in soil and on vegetables grown in soil previously flooded with inoculated sewage sludge or effluent. Appl. Environ. Microbiol. *33*, 109-113.

TJABERG, T.B., UNDERDAL, B., and LUNDE, G. 1972. The effect of ionizing radiation on the microbiological content and volatile constituents of spices. J. Appl. Bacteriol. *35*, 473-478.

TROUWBORST, T., and KUYPER, S. 1974. Inactivation of bacteriophage T_3 in aerosols: effect of prehumidification on survival after spraying from solutions of salt, peptone, and saliva. Appl. Microbiol. *27*, 834-837.

VANDERZANT, C., and NICKELSON, R. 1969. A microbiological examination of muscle tissue of beef, pork, and lamb carcasses. J. Milk Food Technol. *32*, 357-361.

VAN DONSEL, D.J., and LARKIN, E.P. 1977. Persistence of *Mycobacterium bovis* BCG in soil and on vegetables spray-irrigated with sewage effluent and sludge. J. Food Protect. *40*, 160-163.

6

Foodborne Illness

We are constantly subjected to risks or hazards during our lifetimes. There is no absolute safety in anything we do, including the consumption of food.

Foods should be safer today than in the "good old days," due to increased knowledge of bacteria and sanitation, as well as increased regulations. However, due to large-scale, high-speed food processing, alteration of traditional processing methods resulting in less control of microorganisms, proliferation of heat-and-eat convenience foods and nationwide distribution with increased potential for mishandling, there is concern that outbreaks of foodborne illness can occur that could involve very large numbers of people. The number of outbreaks and cases fluctuates from year to year but the number of cases per 100,000 people has tended to increase since 1967. Part of the increase may be due to better reporting of foodborne illnesses and part may be due to an actual increase in the prevalence of cases.

TYPES OF FOOD HAZARDS

The principal categories of food hazards considered in Chapter 1 were microbiological hazards, malnutrition, environmental contaminants, naturally occurring toxins, pesticides and conscious food additives.

Bryan (1973A) listed approximately 200 diseases that can be transmitted to man by foods. The etiological agents of these diseases are: bacteria, viruses, fungi, parasites, chemicals and plant and animal toxicants. For purposes of brevity, only bacterial, fungal and viral diseases are discussed in this text. It should not be construed that the other diseases are not important problems. We do not know all of the ramifications that might result from being exposed to certain chemical compounds or other hazards over a long time. It was suggested by Payne and Kotin (1969) that vague, chronic ill health may be due to multiple, minor and repeated environmental insults. Potential hazards of foods are included in the insults. Chemicals are involved in an increasing number of outbreaks and cases each year. In some areas of the USA and the world, parasites and protozoa are important agents of foodborne illness.

In 1961, the Communicable Disease Center (since renamed the Center for Disease Control, CDC), became responsible for maintaining records and reporting foodborne illnesses in the USA. It is well recognized that all outbreaks or cases of foodborne illness are not reported to the CDC. However, these are the best data that we have. Annual summaries of foodborne diseases have been published since 1966. The confirmed etiologies of foodborne outbreaks and cases for 1976 and 1972-1976 (CDC 1976 A,B,1977A) are listed in Tables 6.1 and 6.2. Bacterial and viral etiologies accounted for 68.5% of the outbreaks and 95.5% of the cases (1972-1976). Bacteria are the main causative agents of foodborne illness reported in the USA.

TABLE 6.1

CONFIRMED FOODBORNE DISEASE OUTBREAKS AND CASES BY ETIOLOGY, 1976

		Outbreaks		Cases	
		#	%	#	%
BACTERIAL					
B. cereus		2	1.5	63	1.8
C. botulinum		23	17.4	40	1.1
C. perfringens		6	4.5	509	14.2
Salmonella		28	21.2	1169	32.7
Shigella		6	4.5	273	7.6
Staphylococcus		26	19.7	930	26.0
Y. enterocolitica		1	0.8	286	8.0
	Total	92	69.6	3270	91.4
CHEMICAL					
Heavy metal		6	4.5	55	1.5
Ciguatoxin		6	4.5	19	0.5
Scombrotoxin		2	1.5	5	0.1
Paralytic shellfish poison		4	3.0	11	0.3
Monosodium glutamate		2	1.5	7	0.2
Mushroom poison		1	0.8	1	0.0
Other chemicals		7	5.3	59	1.6
	Total	28	21.1	157	4.2
PARASITIC					
T. spiralis		8	6.1	27	0.8
E. histolytica		1	0.8	9	0.3
	Total	9	6.9	36	1.1

TABLE 6.1. *(Continued)*

	Outbreaks #	%	Cases #	%
VIRAL				
Hepatitis A	2	1.5	37	1.0
Echo, type 4	1	0.8	80	2.2
Total	3	2.3	117	3.2
CONFIRMED TOTAL	132	99.9	3,580	99.9

Courtesy CDC (1977A)

TABLE 6.2

CONFIRMED FOODBORNE DISEASE OUTBREAKS AND CASES, 1972-1976

	Outbreaks #	%	Cases #	%
BACTERIAL				
Arizona hinshawii	1	0.1	15	0.1
Bacillus cereus	7	0.9	121	0.4
Brucella	1	0.1	4	0.1
Clostridium botulinum	72	9.2	146	0.4
C. perfringens	55	7.0	3,388	10.3
Salmonella	170	21.7	12,583	38.2
Shigella	23	2.9	2,372	7.2
Staphylococcus aureus	168	21.4	9,782	29.7
Streptococcus (group A)	3	0.4	610	1.9
Streptococcus (suspect group D)	4	0.5	138	0.4
Vibrio cholerae	1	0.1	6	0.1
V. parahaemolyticus	9	1.2	927	2.8
Yersinia enterocolitica	1	0.1	286	0.9
Total Bacterial	515	65.7	30,378	92.2
CHEMICAL	182	23.2	1,120	3.4
PARASITIC	65	8.3	373	1.1
VIRAL	22	2.8	1,087	3.3
Total	784		32,958	

Data from CDC Annual Summaries

Definition of an Outbreak

A foodborne disease outbreak is defined by CDC as an incident in which two or more persons experience a similiar illness, usually gastrointestinal, after ingestion of a common food, and epidemiological analysis implicates the food as the source of the illness. For botulism or chemical poisoning, one case constitutes an outbreak.

A microbial foodborne illness may result from ingesting a food containing either pathogenic microorganisms or a toxin or poison. When a pathogenic microorganism is the etiologic agent, the illness is called an infection. If a toxin or poison is the causative agent the illness is called a food intoxication or food poisoning.

Epidemiology

Epidemiology attempts to identify the cause and the mode of transmission of infections and to suggest and evaluate methods for control.

The diagnosis of the specific disease is important for treatment and control. With a known etiology, acceptable therapy can be prescribed, dangers from handling patients with infections can be avoided and the patient can be informed of the possible course of the illness.

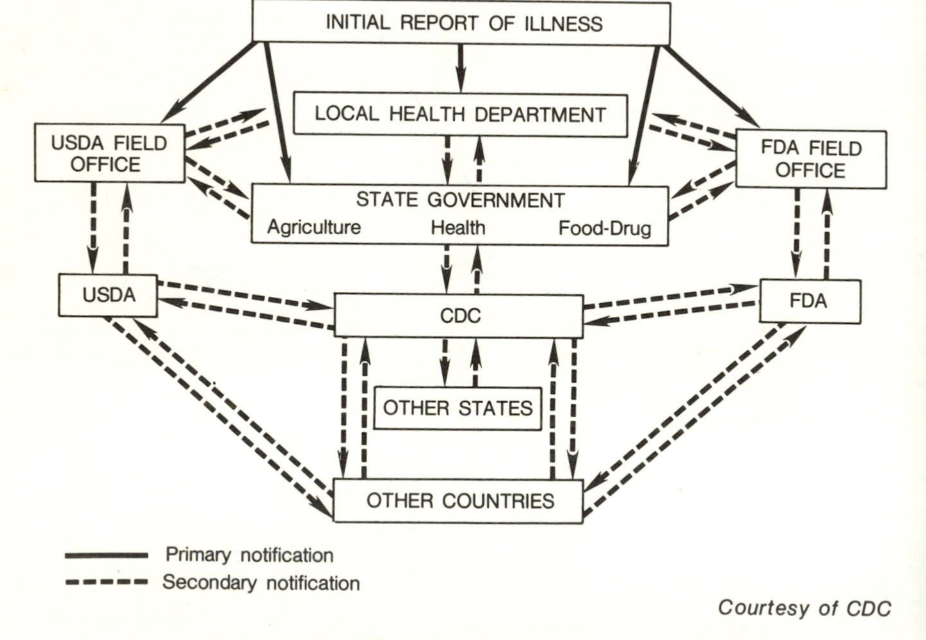

Courtesy of CDC

FIG. 6.1. FOODBORNE DISEASE SURVEILLANCE SYSTEM, UNITED STATES

Confirmed etiologies are those in which laboratory evidence is obtained and fulfills specific criteria established by the CDC. Although 301 outbreaks were reported in 1972, only 136 of these had confirmed etiologies. The present reporting system of illnesses involves many people and agencies (Fig. 6.1 and 6.2). These range from afflicted persons and doctors to local, state and federal offices and even foreign countries that might become concerned. The CDC not only collects data, but also relays information that is valuable to other agencies for controlling potential outbreaks. If there is a breakdown in the reporting system, then the outbreak is not recorded (Fig. 6.3). The afflicted person might not seek medical help or the doctor might not report the illness to a local authority. If there is a study, it might not be complete.

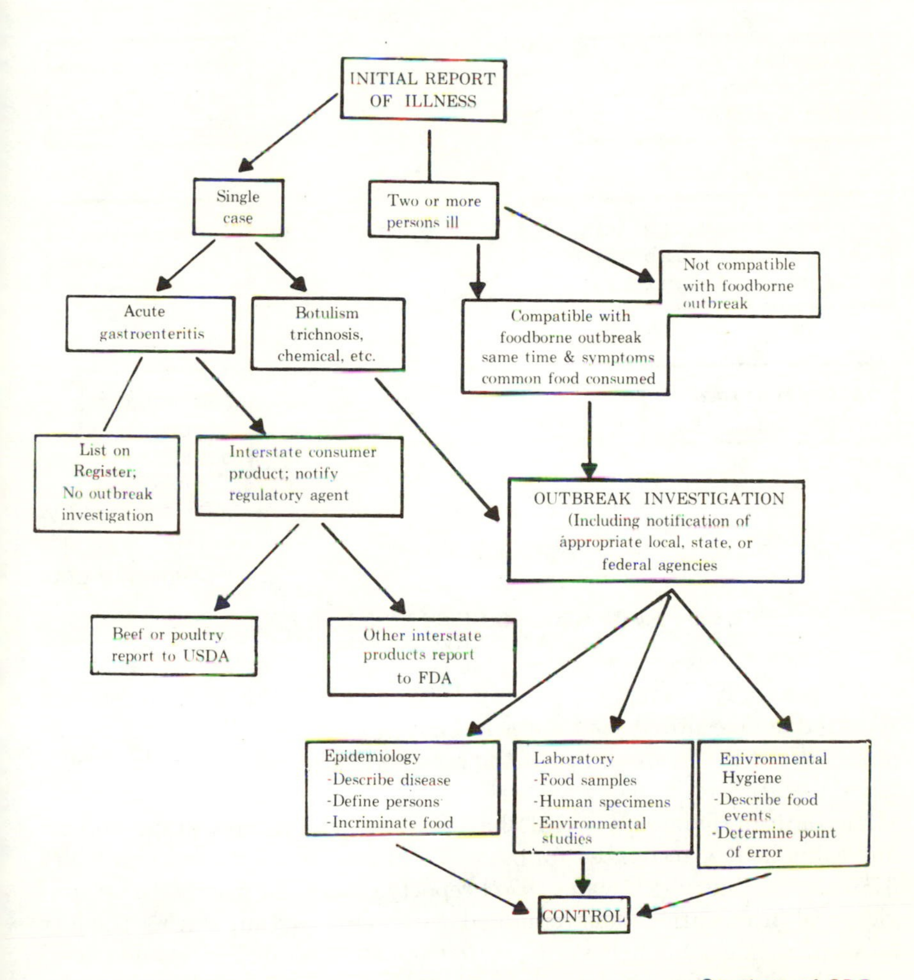

Courtesy of CDC

FIG. 6.2. A SCHEME FOR THE HANDLING OF FOODBORNE DISEASE COMPLAINTS BY STATE AND LOCAL HEALTH DEPARTMENTS

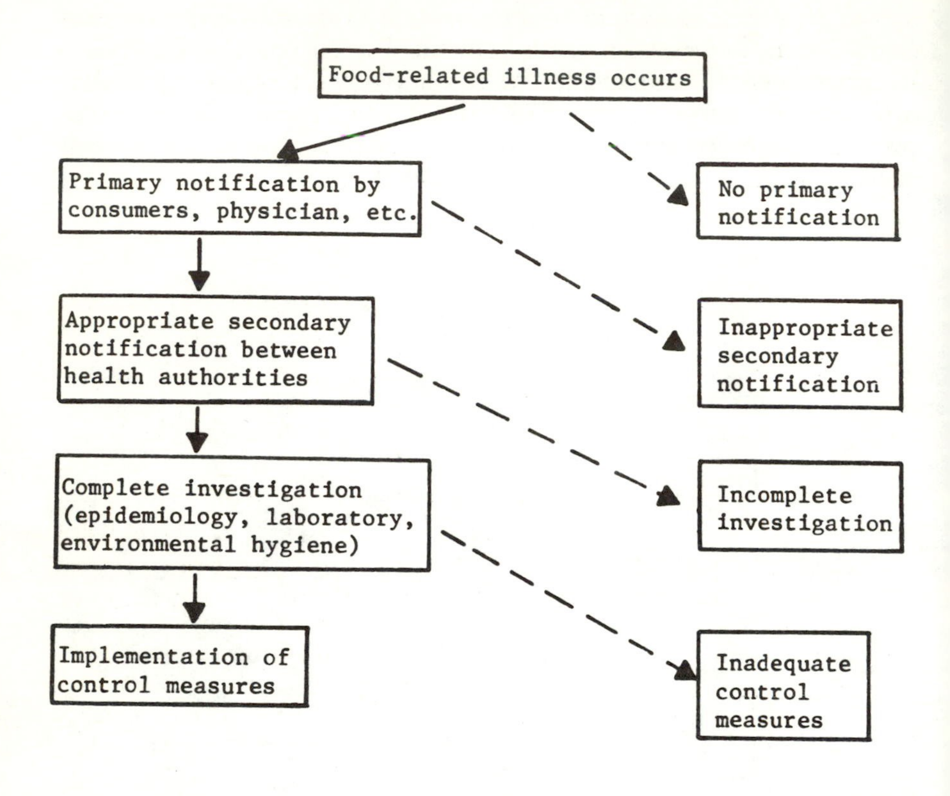

Courtesy of CDC

FIG. 6.3. CONTINGENCIES OF SUCCESSFUL FOODBORNE DISEASE SURVEILLANCE

It is generally assumed that only 10% or less of the actual cases are reported and recorded. Estimates as high as 10 or 20 million cases a year in the USA have been made. Of the reported outbreaks, only about 50% have a confirmed etiology.

The data in Table 6.2 reveal that two bacterial agents, staphylococci and salmonellae, accounted for over 43% of the total outbreaks and over 67% of the illnesses. Bryan (1972) reported that the number of staphylococcal intoxications has remained relatively constant during the last 10 years, while salmonellosis, *C. perfringens* foodborne illness and chemical poisoning outbreaks have increased. Various other etiologies such as infectious hepatitis, *Bacillus cereus* foodborne illness and *Vibrio parahaemolyticus* infection have emerged as foodborne problems. On a worldwide basis, cholera is an important illness spread by poor sanitation and contaminated food. There are diseases such as traveler's diarrhea in which the symptoms resemble several foodborne diseases, but no definite causative agent is listed. The CDC defined traveler's diarrhea as an acute intestinal illness which develops one or more days after arrival in a foreign country. Bacteria (*E. coli, Salmonella, Shigella*), parasites (*Giardia*) and viruses (parvoviruses, reovirus-like agents) may be responsible.

Foods Involved

Various types of foods are involved with foodborne illness as shown in Table 6.3. Pork, beef and their products are the vehicles for transmission of a major share of foodborne illnesses. Quite often, red meats and poultry products are contaminated by salmonellae and other pathogens. Guthertz *et al.* (1976, 1977) reported 28% of fresh and 38% of frozen samples of comminuted turkey meat contained salmonellae. They also found *C. perfringens, S. aureus* and *E. coli* in these samples. The public health aspects have been reported for poultry (Horwitz and Gangarosa 1976), fish and shellfish (Hughes *et al.* 1977) and cream filled pastries (Bryan 1976). Few foods are exempt as potential vehicles of infection or intoxication.

TABLE 6.3

FOODBORNE DISEASE OUTBREAKS BY VEHICLE OF TRANSMISSION, 1973-1976

Vehicle	Year 1973	1974	1975	1976	Total	%
Beef	28	22	54	28	132	7.8
Pork	12	11	11	3	37	2.2
Ham	15	20	23	8	66	3.8
Sausage	4	11	11	5	31	1.8
Other meat	3	22	21	6	52	3.1
Total meat					318	18.7
Poultry	19	15	27	18	79	4.7
Shellfish	4	9	9	8	30	1.8
Other fish	18	51	42	17	128	7.5
Total seafood					158	9.3
Milk	3	4	4	3	14	0.8
Ice cream	4	9	5	2	20	1.2
Other dairy	1	2	9	5	17	1.0
Total dairy					51	3.0
Total animal food					606	35.7
Baked foods	10	11	11	12	44	2.6
Fruit and vegetables	12	16	12	6	46	2.7

TABLE 6.3. (*Continued*)

Vehicle	Year 1973	1974	1975	1976	Total	%
Salads, miscellaneous	20	17	29	10	76	4.5
Mushrooms	11	8	5	1	25	1.5
Chinese food	7	7	22	19	55	3.2
Mexican food	9	11	15	5	40	2.4
Other food	31	64	68	100	263	15.5
Unknown	96	146	119	182	543	32.0
TOTAL	307	456	497	438	1,698	

Data from CDC Annual Summaries

Place of Mishandling

When a processor is involved in a foodborne disease outbreak, it is national news. This is due to the potential for many cases per outbreak. According to the data of the CDC, the processor is involved in relatively few outbreaks (Table 6.4). Most problems occur in food service establishments. Also, the home is an important place for mishandling of food. Unfortunately, the place of mishandling is unknown for over 40% of the outbreaks.

The outbreaks and cases resulting from mishandling in food processing establishments during 1973-1976 are shown in Table 6.5. There were 59 outbreaks and an average of about 65 cases per outbreak. Overall, the average number of cases per outbreak is about 42. It can be assumed that there are very few cases per outbreak for those occurring in the home.

TABLE 6.4

PLACE WHERE FOOD WAS MISHANDLED, 1968-1976

Place	No. of Outbreaks	%
Food processing establishment	163	4.9
Food service establishment	1,283	38.8
Home	504	15.3
Unknown or unspecified	1,354	41.0
Total	3,304	

Data from CDC Annual Summaries

TABLE 6.5

FOODBORNE DISEASE OUTBREAKS CAUSED BY MISHANDLING OF FOOD IN FOOD-PROCESSING ESTABLISHMENTS
1976

Etiology	Vehicle	Number of Cases
Salmonella heidelberg	Cheese	339
S. bovis-morbificans	Precooked roast beef	21
S. infantis	Nutrient supplement	4
Staphylococcus Enterotoxin D	Greek spaghetti	20
Staphylococcus	Beef ravioli	4
Staphylococcus Enterotoxin D	Custard-filled doughnuts	2
Staphylococcus	Pepperoni, sausage	3
Y. enterocolitica	Chocolate milk	286
Sodium nitrate	Table salt	2
Propyl paraben	Cake icing	9
Histamine	Cheese	38
Niacin	Hamburger	2
Niacin	Cubed steak	3
Unknown	Tuna extender	508
Unknown	Milk	42

Total Illnesses—1973-1976

	Year	Outbreaks	Cases
	1976	15	1,283
	1975	13	123
	1974	16	1,704
	1973	15	736
Total		59	3,846

Data form CDC Annual Summaries

Contributing Factors

Various factors contribute to outbreaks of foodborne illness. The main ones are: improper holding temperatures (failing to properly refrigerate food), inadequate cooking, contaminated equipment (failure to clean and disinfect kitchen or processing plant equipment) and poor personal hygiene (Table 6.6). Other factors that may contribute to foodborne illness include preparing food a day or more before serving with improper holding and reheating, cross contamination (from raw to cooked products) and adding contaminated ingredients to previously cooked food without reheating. After foods are contaminated, the main factor is letting them remain at a temperature that allows growth of potentially hazardous microorganisms.

Most of these problems could be controlled with only a little effort on the part of food handlers, whether in a processing plant, a restaurant, a cafeteria or in a home. The tremendous turnover of food workers makes effective training difficult. However, since many outbreaks occur due to carelessness at home, it might be suggested that everyone should be trained. Already trained microbiologists and food technologists should set an example for personal hygiene and food handling practices.

TABLE 6.6

CONTRIBUTING FACTORS CAUSING FOODBORNE ILLNESS, 1972–1976

	Year						
Factor	1972	1973	1974	1975	1976	Total	%
Improper holding temperature	117	109	131	214	160	731	42.7
Inadequate cooking	36	43	45	87	43	254	14.8
Contaminated equipment	38	34	31	62	54	265	15.5
Food from unsafe source	–	24	50	14	17	105	6.1
Poor personal hygiene	52	42	41	93	53	281	16.4
Other	–	10	9	14	44	77	4.5
TOTAL						1,713	

Data from CDC Annual Summaries

Symptoms and Severity

The symptoms of the illnesses are variable. However, diarrhea, nausea, vomiting and abdominal cramps are evident in most foodborne illnesses.

Fortunately, in most cases, the illness is not severe and the patient recovers in one or a few days. In some cases, especially botulism, paralysis and death can occur. The deaths caused by foodborne illnesses from 1973-1976 are listed in Table 6.7. Of the 32,958 cases (Table 6.2), there were 63 deaths. Botulism (22 deaths) and salmonellosis (17 deaths) accounted for almost 62% of the deaths. The mortality rate is much higher for botulism than for salmonellosis.

BACTERIAL DISEASES

Bryan (1973A) listed several foodborne diseases caused by bacteria as having contemporary importance. These are: staphylococcal intoxication, botulism, *Clostridium perfringens* foodborne illness, salmonellosis, typhoid fever, paratyphoid fever, enteritis necroticans, *Bacillus cereus* gastroenteritis, *Arizona* infection, *Vibrio parahaemolyticus* infection and bonkrek poisoning.

TABLE 6.7

DEATHS ASSOCIATED WITH FOODBORNE ILLNESSES, 1972–1976

	Year						
Etiology	1972	1973	1974	1975	1976	Total	%
Clostridium botulinum	4	4	7	2	5	22	34.9
C. perfringens	1	1	1	1	0	4	6.3
Salmonella	4	7	1	2	3	17	27.0
Shigella	0	0	0	0	1	1	1.6
Staphylococcus	0	0	0	0	0	0	0
Vibrio cholerae	0	0	1	0	0	1	1.6
Hepatitis A	0	0	1	0	0	1	1.6
Chemicals	4	1	2	2	0	9	14.3
Parasites	1	1	0	1	0	3	4.8
Unknown	—	1	1	2	1	5	7.9
TOTAL	14	15	14	10	10	63	

Data from CDC Annual Summaries

Diseases that sometimes are foodborne include shigellosis (bacillary dysentery), enteropathogenic *Escherichia coli* infection, beta hemolytic streptococcal infections, cholera, brucellosis, tuberculosis, diphtheria, tularemia, anthrax and haverhill fever.

Diseases in which transmission by foods is inconclusive include infections by enterococci, *Proteus, Klebsiella, Citrobacter, Enterobacter, Edwardsiella, Pseudomonas aeruginosa, Aeromonas, Bacillus subtilis, Vibrio, Campylobacter fetus* and *Actinomyces*.

Space does not permit a discussion of all of these foodborne illnesses. However, the more prominent ones are described.

Staphylococcal Intoxication

This illness accounted for 21.4% of the outbreaks and 29.7% of the cases of foodborne illness in the USA from 1972 to 1976. Individual cases are not reported and many afflicted people never seek medical help. Hence, the actual extent of this illness is not known.

Characteristics of the Intoxication.—The characteristics of an illness can aid in determining the causative agent in an outbreak of foodborne illness.

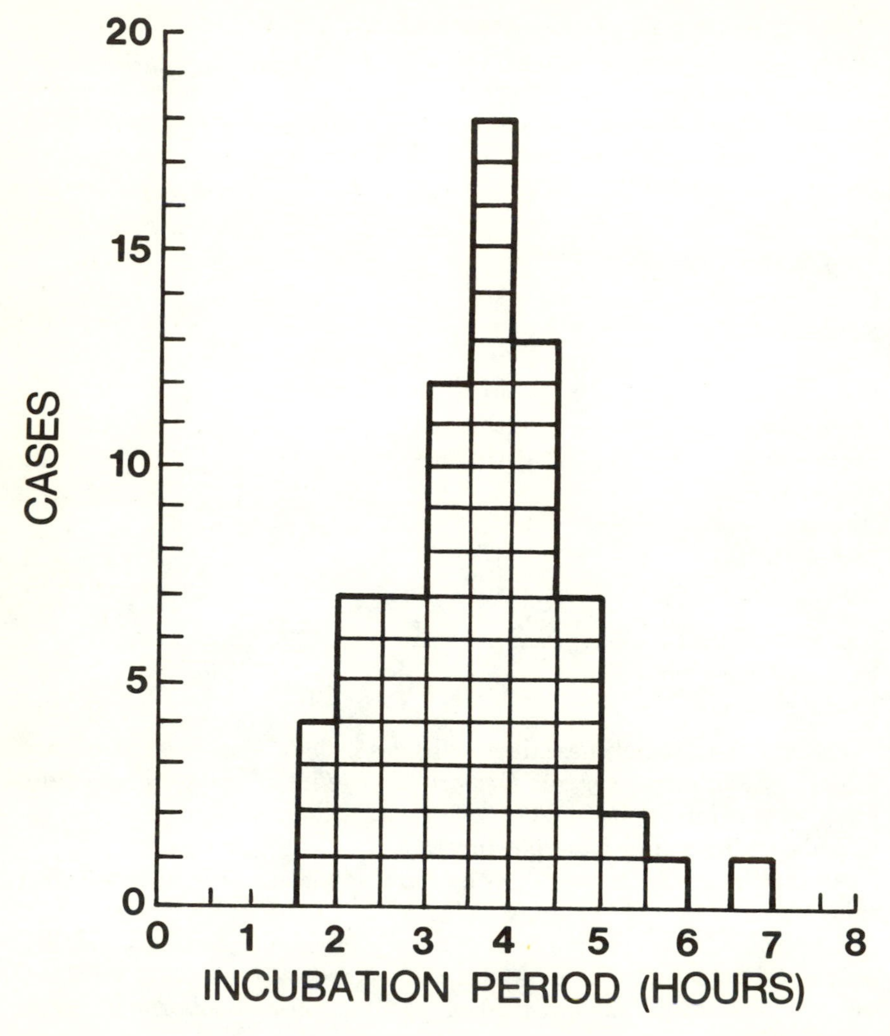

Courtesy of CDC

FIG. 6.4. DEPICTION OF AN OUTBREAK OF STAPHYLOCOCCAL INTOXICATION

Incubation period.

Incubation Period.—The incubation period of a typical outbreak usually ranges from 30 min to 8 hr with most illnesses occurring 2-4 hr after ingestion of the suspect food (Fig. 6.4). Longer than normal incubation periods may be attributed to the lack of vomiting, which is an early symptom.

Symptoms.—Not all of the people eating a suspect meal become ill, and not all ill people experience the same symptoms. The severity of the symptoms varies with the concentration of enterotoxin in the food, the amount of food consumed and the susceptibility of the individual. The principal symptoms listed in Table 6.8 are nausea, vomiting, abdominal cramps and diarrhea. Besides these symptoms, shock, anorexia, cyanosis, salivation, malaise, retching, sweating and dehydration have been expressed in various outbreaks.

TABLE 6.8

PERCENTAGE OF ILL PERSONS EXPERIENCING SPECIFIC SYMPTOMS DUE TO STAPHYLOCOCCAL INTOXICATION

Symptoms	Outbreaks					
	1	2	3	4	5	6-7
Nausea	—	76	100	50	41	45
Vomiting	70	44	100	77	76	76
Diarrhea	19	67	100	82	73	78
Abdominal cramps	71	71	58	15	16	16
Chills	—	25	46	—	—	—
Headache	42	—	6	41	35	38
Prostration	—	—	63	—	—	—
Weakness	—	—	—	68	73	70
Leg cramps	—	—	—	—	11	6
Muscle soreness	—	—	5	—	—	—
Collapse	—	9	—	—	—	—
Hypotension	—	1	—	—	—	—
Fever	—	25	7	21	24	23

—Not reported
Data from CDC Morbidity and Mortality Weekly Reports

Duration and Therapy.—Symptoms of the illness usually subside after one or two days. The disease is rarely fatal. Due to the sudden onset and short duration, treatment usually is not needed. However, hospitalization is required in cases in which shock, dehydration and extensive vomiting have occurred. In these cases, therapy includes replacing lost fluids and electrolytes.

Etiologic Agent.—This illness is called an intoxication because the etiologic agent is an enterotoxin. According to Payne and Wood (1974) there are six known enterotoxins: A, B, C, D, E and F. This classification is based on reactions of the enterotoxin with specific antibody. Although toxins C_1 and C_2 are different, they are similar serologically and usually are listed as enterotoxin C. Under certain conditions, other cross reactions may occur between antitoxins and enterotoxins. Some enterotoxins reportedly have multiple forms (Chesbro *et al.* 1976; Yamada *et al.* 1977).

Enterotoxin A is the most frequently encountered in food poisoning outbreaks in the USA, with enterotoxin D second in frequency. In New Zealand, Jarvis and Harding (1972) found enterotoxins C and D to be more prevalent than A. Enterotoxin B rarely is involved in food poisoning outbreaks.

Properties of the Enterotoxins.—Staphylococcal enterotoxins are simple proteins with a molecular weight between 25,000 and 35,000 daltons. They are readily soluble in water and salt solutions. The enterotoxins resist the action of proteolytic enzymes including papain, rennin, trypsin and chymotrypsin (Bergdoll 1972).

The enterotoxins are relatively heat stable. Enterotoxin B is more heat resistant than either A or D. The heat inactivation of enterotoxins is influenced by the pH and composition of the heating medium (Humber *et al.* 1975). According to Tatini (1976) enterotoxin A is more heat stable at pH 6.0 than at pH 4.5-5.5, while the reverse is true for enterotoxin D. Lee *et al.* (1977) reported the time needed to inactivate 90% of enterotoxin B in veronal buffer at 110°C was 18 min and in a beef broth was 60 min. There is a protective effect of proteins on enterotoxins during heating.

Reportedly there is more rapid inactivation at 80° than at 100°C (Jamlang *et al.* 1971; Satterlee and Kraft 1969). The presence of either myosin or metmyoglobin results in a rapid loss of enterotoxin (Satterlee and Kraft 1969). Reactivation during storage of the heated toxin has been reported (Reichert and Fung 1976). This reactivation appears to be greater for toxin heated at 70° or 80°C than that heated at 100° or 110°C.

In foods, the enterotoxins are not completely inactivated by normal cooking, pasteurization or other usual heat treatments. Tatini (1976) stated that thermal processing cannot be relied upon to inactivate these toxins. Further, he found that heated toxins had greater biological activity than unheated toxin when tested at the same dose level.

Amount of Enterotoxin Needed for Illness.—There is no definite data concerning the minimum amount of enterotoxin needed to cause symptoms in a human. Gilbert (1974) listed estimates that ranged from

0.015-0.357 μg of enterotoxin per kilogram of body weight. Besides body weight, individuals vary in their sensitivity to enterotoxins.

Action of Enterotoxins.—Using monkeys, Elwell *et al.* (1975) studied enterotoxin-induced emesis (vomiting). Their results indicate that enterotoxin is not absorbed from the intestine. Orally ingested enterotoxin causes emesis due to a local gastrointestinal initiated neural stimulus to the vomiting center.

The diarrheal symptom is defined as excessive fluid and electrolyte loss due to malabsorption or excessive secretion of these substances (Sidorov 1976).

In a normal average adult, between 9 and 10 liters of fluid enter the gastrointestinal (GI) tract each day. This fluid results from the ingestion of food and liquids (2-3 liters) and from secretions such as saliva, gastric juice, bile and pancreatic juice. Normally, through absorption, most of this fluid is removed from the GI tract. The absorption is influenced by the presence of electrolytes. Only about 100-200 ml of fluid passes out of the colon. Most of the excess fluid is removed from the body by the kidneys. Perspiration accounts for the remainder.

Phillips (1972) stated that only a few hundred milliliters of excess water in the GI tract can result in diarrhea.

The action of staphylococcal enterotoxins in the diarrheal syndrome is not known. Staphylococcal enterotoxin shows an affinity to the walls of the stomach and the small and large intestines. If sufficient enterotoxin is present in consumed food, it causes inflammation and irritation of the lining in the stomach and intestinal tract. Working with flounder intestine in vitro, the data of Huang *et al.* (1974) suggested that enterotoxin stimulates active sodium and chloride secretion. Kapral *et al.* (1976) found enterotoxin B did not interfere with water absorption in the guinea pig ileum, but staphylococcal delta toxin did inhibit absorption in the jejunum and ileum.

Foods Involved.—The foods involved in outbreaks of staphylococcal intoxication in recent years are listed in Table 6.9. Many types of food were the vehicle of the enterotoxin, but red meats (beef, pork and other meat) were involved in more outbreaks than were any other foods.

Corned beef, genoa salami, ham, bacon and barbecued pork have been the major vehicles of the enterotoxin. Fresh meats are involved less often since, being perishable, they are held under refrigeration at a temperature not conducive to growth or toxin production of *S. aureus*. Cured meats, on the other hand, often are mishandled by allowing them to remain at room temperature.

The poultry products involved are usually either barbecued or in salads. Other salads (potato, macaroni and tuna fish) have been implicated in outbreaks. Various bakery products have been involved but the primary offenders are products containing custard or cream, such as eclairs, coconut meringue pie, lemon chiffon pie and filled doughnuts.

TABLE 6.9

FOODS INVOLVED IN STAPHYLOCOCCAL INTOXICATIONS (1971–1976)

Food	Years 1971-1972	1973-1974	1975-1976	Total	%
Beef	11	2	4	17	6.6
Pork	52	19	21	92	35.7
Other meat	4	6	5	15	5.8
Poultry	10	4	4	18	7.0
Shellfish	2	0	0	2	0.8
Other fish	4	0	2	6	2.3
Eggs	3	2	0	5	1.9
Milk	0	1	0	1	0.4
Other dairy	1	0	0	1	0.4
Bakery products	6	4	3	13	5.0
Fruits and vegetables	2	2	1	5	1.9
Chinese food	1	0	0	1	0.4
Mexican food	1	2	1	4	1.6
Salads	0	5	12	17	6.6
Other foods	15	11	17	43	16.7
Unknown	14	3	1	18	7.0
TOTAL	126	61	71	258	

Data from CDC Annual Summaries

Miscellaneous other food products implicated in staphylococcal intoxication include hollandaise sauce, bread pudding, poultry dressing, sauces, gravies and fried rice.

The Organism (S. aureus).—The organisms in this species are Gram positive, nonsporeforming, nonmotile, spherical cells. The cells occur singly or in pairs and divide in more than one plane to form irregular (grape-like) clusters. Some strains possess a slime layer or capsule. *S. aureus* is facultatively anaerobic, but grows better in aerobic conditions. Some colonies are white and others have pigment ranging from yellow to orange. Both strain variations and growth conditions influence the color.

In both aerobic and anaerobic conditions, *S. aureus* produces acid from mannitol, glucose, lactose and maltose. Aerobically, many carbohydrates are metabolized with the production of acid, but acid is not produced from arabinose, cellobiose, dextrin, inositol, raffinose, rhamnose or xylose.

S. aureus is distinguishable from the other species in the genus by its ability to ferment glucose anaerobically and the production of α-toxin, heat resistant endonuclease and coagulase.

S. aureus is a relatively poor competitor and various bacteria can inhibit or outgrow it. The effect of streptococci on the growth of *S. aureus* is shown in Fig. 6.5. This inhibitory interrelationship of other bacteria with *S. aureus* is important in preventing toxin production in foods. It may be a primary reason for certain foods to be less involved than others in outbreaks of staphylococcal intoxication. In foods with a_w of 0.90-0.95 or with 5-10% salt, *S. aureus* can dominate because most other bacteria cannot grow. Sublethally injured *S. aureus* is inhibited by 5% salt, but this injury is reversible.

Sources.—Some reports call *S. aureus* ubiquitous since it is found widespread in nature (air, dust, clothing, floors, water, sewage and insects). The principal source of *S. aureus* is the human nose, although it is found on the skin, the hands, infected wounds, burns, boils, pimples, acne, in nose and throat discharges and in feces. The primary site on the hands is the fingertips, which relates to the habit of handling one's nose with the fingers. The extent of nasal carriers is difficult to determine, but surveys have shown the carrier rate to vary from 6% to over 60% of the population. People associated with hospitals tend to have a higher carrier rate than the normal population.

Much of the widespread distribution of *S. aureus* is due to contamination from human sources. Investigations of outbreaks reveal that most are the result of a food handler contaminating the food, usually after it is cooked. However, there are a few outbreaks in which contamination occurred before cooking. In either case, the intoxication developed by not refrigerating the food to prevent growth and toxin production by the *S. aureus*.

Animals are a source of *S. aureus*. Most of the strains of *S. aureus* isolated from animals tend to have biochemical characteristics different from those strains associated with humans.

Samples of ham, lamb, and beef were obtained aseptically immediately after slaughter (Vanderzant and Nickelson 1969). Most samples revealed no contamination, but of those containing bacteria, *Staphylococcus* was the predominating genus. Many of the staphylococci were coagulase positive (59.7% in ham, 92.3% in lamb and 45.9% in beef).

Growth.—*S. aureus* is found in many foods. It is generally believed that 10^5 to 10^6 cells of *S. aureus* per gram of food must be present be-

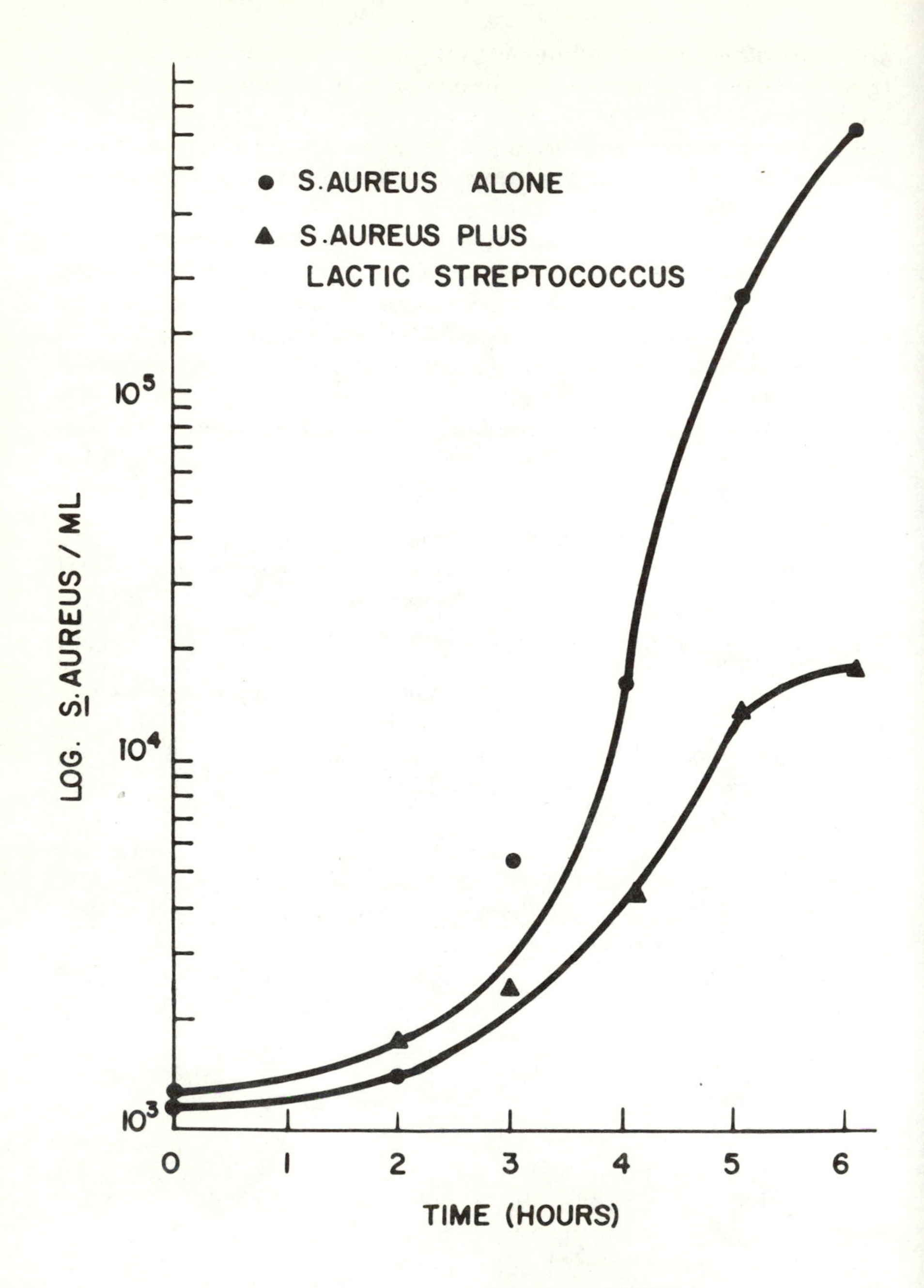

Courtesy of Gilliland and Speck (1972)

FIG. 6.5. THE EFFECT OF STREPTOCOCCI ON THE GROWTH OF *STAPHYLOCOCCUS AUREUS*

fore production of enterotoxin reaches a level that can cause intoxication. Due to the normally low numbers in food, multiplication must occur. By knowing and understanding the factors affecting the growth of *S. aureus,* we can control the growth, enterotoxin production and outbreaks of staphylococcal intoxication. The general factors affecting the growth of *S. aureus* are described in other sections of this text.

Competition by other organisms has been suggested as an important reason that some foods are involved in intoxications while other foods are not involved. Although cured meats, such as ham, are the main vehicles of the enterotoxin, fresh meats are rarely involved.

During the curing process, some organisms are killed and, due to the low a_w of the cured meat, others are inhibited. In cured meat, there is less competition from other organisms, and *S. aureus* is able to grow in this product.

Since lactic streptococci are used in cheesemaking and cheese has been a vehicle of staphylococcal enterotoxin, it is evident that microbial competition cannot always be relied upon to inhibit growth and toxin production of *S. aureus.*

Although many species and strains of microorganisms inhibit growth of *S. aureus,* others may stimulate growth. Mold growing on the surface of cheese altered the pH which allowed the growth of *S. aureus* (Duitschaever and Irvine 1971). Competitive organisms may interfere with toxin production and, since the enterotoxins are protein, it has been suggested that the proteolytic enzymes of some competitive organisms may degrade the toxin. In a study of toxin formation in meat, Venn *et al.* (1973) found that the proteolytic meat tenderizers seemed to stimulate enterotoxin production.

Other factors which act to control *S. aureus* and enterotoxin production are sublethal treatments and the interactions of all of the factors (pH, a_w, temperature, competing organisms, Eh) affecting growth (Table 6.10).

Obviously, *S. aureus* is able to grow on or in foods that have been involved in staphylococcal intoxication. Some outbreaks have occurred by holding the food at room temperature (22°-35°C) for less than 4 hr, although longer holding times increase the risk.

Although some foods are refrigerated, a long time may be needed for the food to cool sufficiently to prevent growth of the organism. Foods in gravies and sauces cool slower than those without, and the substrate is more readily available for growth of the organisms.

Protein foods, or mixtures of foods containing proteins, are often the vehicle of the toxin. Protein foods tend to have a pH in the range that allows *S. aureus* to grow readily.

Chocolate has not been involved in staphylococcal intoxication (Ostovar 1973). He found that during storage of chocolate at room temperature, the number of viable *S. aureus* cells decreased. This is presumably due to the low a_w or to some natural inhibitory agent in chocolate.

TABLE 6.10

FACTORS AFFECTING THE GROWTH OF *STAPHYLOCOCCUS AUREUS*

Factor	Minimum	Optimum	Maximum
a_w	0.83-0.86	0.99-1.00	0.99-1.00
pH	4.0	6.5-7.5	9.8
Temperature (°C)	6.5	30-37	45-47.8
Salt (NaCl) % w/w	0	0	18

Toxin Production.—*S. aureus* produces several toxins (Table 6.11). The main difference between exotoxins and endotoxins is not whether they are outside or inside the cell, but rather their structure. Exotoxins are proteins with little or no nonprotein residues. Endotoxins are primarily polysaccharide and lipid complexes (lipopolysaccharides). In food microbiology we are concerned primarily with the enterotoxins.

Not all strains of *S. aureus* are enterotoxigenic, although the toxin is produced by both pigmented and nonpigmented strains of the organism.

TABLE 6.11

SOME TOXINS OF *STAPHYLOCOCCUS AUREUS*

Toxin	Action
α-toxin	hemolytic, dermonecrotic, lethal
β-toxin (phospholipase C)	hemolytic
γ-toxin	hemolytic
δ-toxin	hemolytic, enteric
Hyaluronidase	spreading factor
Staphylococcal coagulase	coagulates plasma
Staphylokinase	fibrinolytic
Leukocidin	kills leucocytes
Epidermolytic	exfoliation
Enterotoxins (A, B, C, D, E, F)	emetic

It was estimated by Schroeder (1967) that only 4% of the staphylococcal strains in milk are capable of producing enterotoxin. In a survey of *S. aureus* in meat, dairy products and other foods, Payne and Wood (1974) found 75 of 200 strains produced enterotoxins. Wieneke (1974) found a somewhat higher ratio of enterotoxigenic strains in cooked food than in raw food. This may be the result of human strains contaminating the cooked food. Strains of human origin are more frequently enterotoxigenic than strains isolated from other sources.

Many years ago it was observed that only lysogenic cells of *Corynebacterium diphtheriae* produced toxin. Even then, not all phage-host relationships resulted in toxin production, but a specific phage invasion was necessary for the cell to become toxigenic. Zabriskie (1970) reported that lysogeny is common in *S. aureus* strains, and, although meager and not conclusive, the information that is available indicates that enterotoxin production by *S. aureus* is phage mediated. This is in conflict with the finding of Read and Pritchard (1963). They reported that prophage is not necessary for enterotoxin production by *S. aureus*. In experiments using streptomycin to inhibit toxin production, Rosenwald and Lincoln (1966) used phage-free *S. aureus* S-6 and noted the production of enterotoxin.

Casman (1965) found that a phage obtained from strain PS42D conferred the ability to produce enterotoxin A (SEA). Jarvis and Lawrence (1971) also found that two non-toxin producing strains of *S. aureus* became SEA producers when lysogenized with a phage from *S. aureus* strain PS42D. Lysogenic phage from other toxin producing strains of *S. aureus* did not confer enterotoxin production on non-producing strains.

When Dornbusch *et al.* (1969) treated four enterotoxigenic strains with acridine dyes, the cells lost the ability to produce enterotoxin B (SEB). They stated that SEB production did not appear to be the result of lysogenic conversion. Shalita *et al.* (1977) reported that a genetic determinant for SEB production is carried by a plasmid.

The relationship of growth of *S. aureus* and SEB production was reported by Markus and Silverman (1969) and shown in Fig. 6.6. In a review, Bergdoll *et al.* (1974) concluded that SEA and SEB are primary metabolites, being produced during all phases of growth. The rate of synthesis of SEB is greater than for SEA, so that higher concentrations of SEB are obtained. The presence of glucose in the substrate decreases the production of enterotoxin. Growth and SEB production can be either depressed or stimulated by the presence of fatty acids (Altenbern 1977). Czop and Bergdoll (1970) found SEA and SEC were produced by L-forms (without cell walls) of *S. aureus* but SEB was not produced. They suggested that SEB is produced in conjunction with the cell wall, and is located on the surface of the cell.

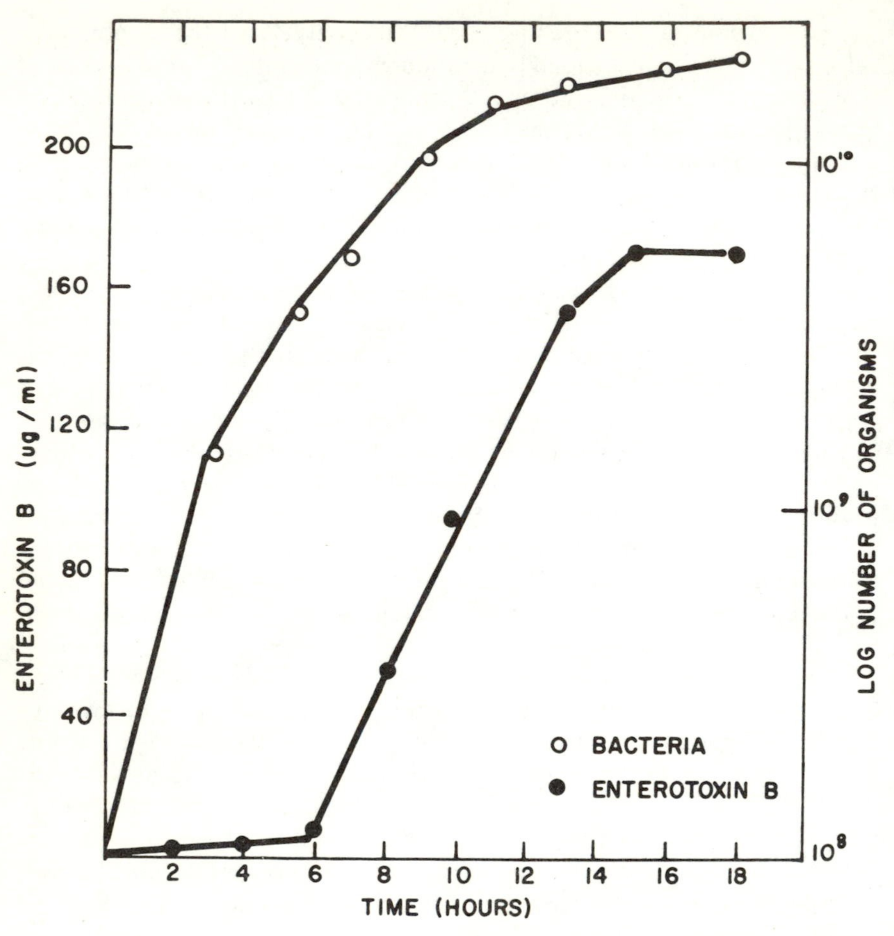

Courtesy of Markus and Silverman (1969)

FIG. 6.6. THE GROWTH AND ENTEROTOXIN B PRODUCTION OF *S. AUREUS*

Using controlled fermentor conditions, Carpenter and Silverman (1976) reported that constant aeration at the rate of 500 cm^2 per min and a pH of 6.5 and 7.0 yielded the highest titers of SEA. At high levels of oxygen, toxin production was reduced or inhibited. In a study of SEB production, Carpenter and Silverman (1974) reported a constant dissolved oxygen (DO) of 100% stimulated growth, but enterotoxin production was not observed. A DO of 50% yielded more enterotoxin than a DO of either 100% or 10%

Aerobically, certain strains of *S. aureus* produce enterotoxin at a pH of 4.8, but anaerobically, no enterotoxin is found at pH 5.4 (Barber and Deibel 1972). Their results indicate that biological acid production cannot be relied upon to inhibit *S. aureus* in fermented sausage, and they recommended chemical acidulation. Since the toxin is produced at a higher level aerobically than anaerobically, they suggested sampling aerobic portions of foods for the toxin. Mixing and cross sectional sampling merely dilute the toxin with anaerobic portions of the food which contain little or no toxin.

The conditions necessary for growth and enterotoxin production have been reviewed (Niskanen 1977). In general, toxin production occurs in a more narrow range of environmental characteristics than those listed for growth in Table 6.10.

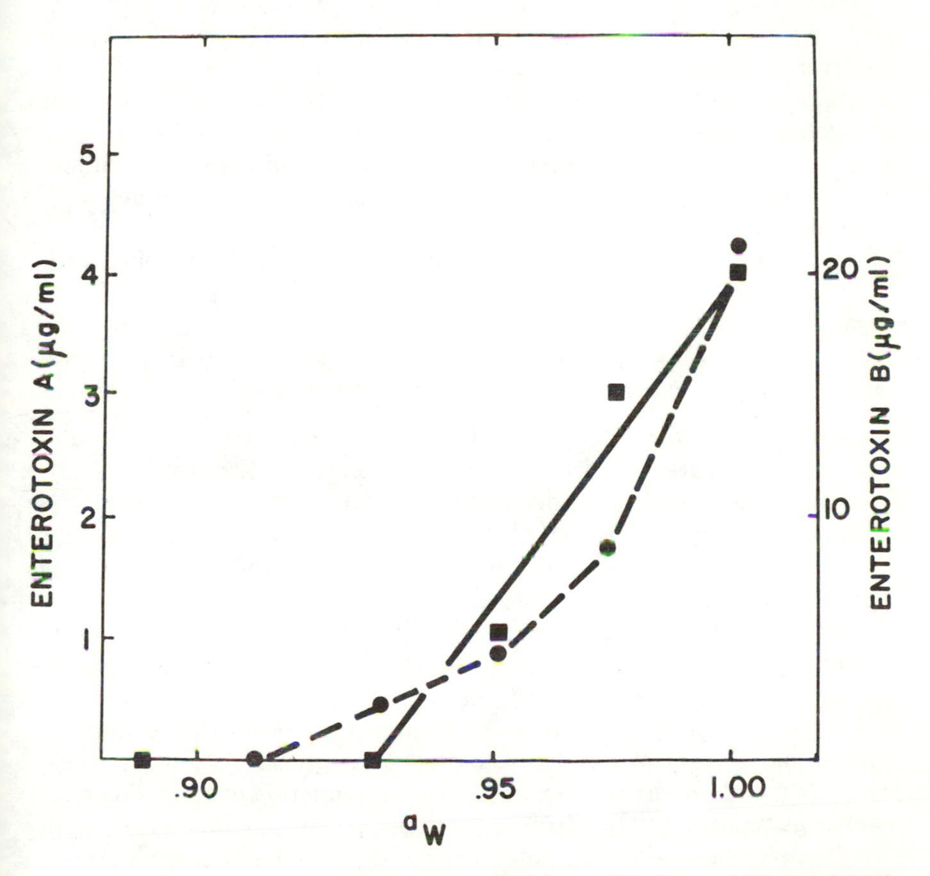

Courtesy of Troller and Stinson (1975)
Copyright Institute of Food Technologists

FIG. 6.7. THE EFFECT OF WATER ACTIVITY ON TOXIN PRODUCTION OF *S. AUREUS*

Lowering the a_w from 1.00 or 0.99 causes a significant reduction in the production of enterotoxin (Fig. 6.7) as reported by Troller and Stinson (1975). They found that *S. aureus* 196E produced enterotoxin in potato doughs at a_w 0.93 but not at 0.88. In a shrimp slurry, this strain produced enterotoxin at a_w 0.95 but not 0.93. Strain C-243 produced toxin at a lower a_w in shrimp slurry (0.93) than in potato dough (0.97).

Since cured meat has a water activity of 0.87-0.95, and ham is a common vehicle of enterotoxin, it is evident that toxin formation can occur in foods with the water activity below 0.95. The effect of sodium chloride on growth and enterotoxin production was reported by McLean *et al.* (1968) and is shown in Fig. 6.8.

Methodology.—Foods or other samples are examined for *S. aureus* and/or the enterotoxins to demonstrate post-processing contamination, to determine a potentially hazardous product or to confirm a causative agent in a foodborne illness. Since *S. aureus* is sensitive to heat and sanitizing agents, the presence of the organism or its toxins in processed food or on equipment generally indicates poor sanitation or handling. Post-processing contamination usually is due to human contact or to improperly sanitized food contact surfaces.

The methodology involved in conjunction with *S. aureus* includes the detection and enumeration of the organism; the characterization of the organism by testing for coagulase, nuclease and other enzymes; serological reactions; phage typing and the qualitative and quantitative analyses for the enterotoxins.

Detection and Enumeration of the Organism.—When attempting to detect and enumerate specific organisms, the media used are selective, differential or both selective and differential for the particular organism. The selective agents used are those which the particular organism can tolerate while other organisms are inhibited, and differential agents are those which can determine some characteristic(s) of the organism.

Many media have been developed and proposed for the direct plating and enumeration of *S. aureus.* The FDA (1976) method uses Baird-Parker agar. This medium contains potassium tellurite as a selective agent and egg yolk and tellurite as differential agents. On this medium, *S. aureus* colonies appear circular, smooth, convex, moist, gray to jet-black, frequently with an off-white margin, due to reduction of the tellurite to elemental tellurium. They are surrounded by an opaque zone and an outer clear zone due to the reaction on the emulsified egg yolk. When touched with an inoculating needle, the colonies have a buttery to gummy consistency. There may be variations to this description. Typical colonies are selected for further study.

A collaborative study by Baer *et al.* (1975) revealed that Baird-Parker agar is satisfactory for recovery of cells stressed or injured by heat or other processing conditions.

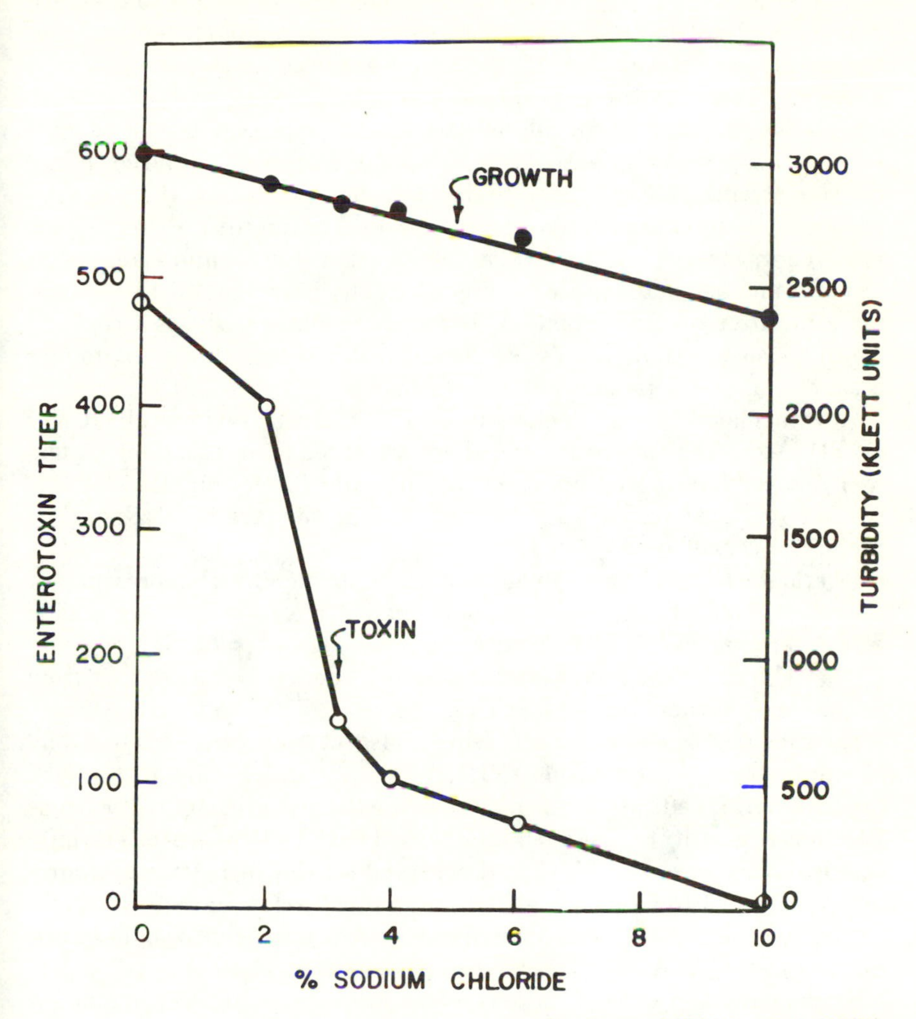

Courtesy of McLean et al. (1968)

FIG. 6.8. THE EFFECT OF SALT CONCENTRATIONS ON THE PRODUCTION OF ENTEROTOXIN BY *S. AUREUS*

Some of the drawbacks of Baird-Parker medium are the complexity of preparation, lability in storage, the similar appearance of *Proteus vulgaris* and *S. aureus* and the occasional occurrence of strains of *S. aureus* that do not clear egg yolk.

Characterization.—Colonies typical for *S. aureus* on an agar surface are selected for further testing and characterization. These tests may include various fermentations of carbohydrates, the presence of coagulase, heat stable nuclease or lysozyme and determining the resistance to chemical inhibitors or antibiotics. None of these tests or combinations of these tests is an absolutely reliable indicator of enterotoxin formation by the organism. Tests for the enterotoxins are available.

The coagulase test is considered the most reliable single test for differentiating potentially pathogenic *S. aureus*. However, not all coagulase positive strains produce enterotoxins. On the other hand, there are reports that coagulase negative strains produce enterotoxin or have been involved in cases of staphylococcal intoxication (Lotter and Genigeorgis 1975). Some enterotoxigenic strains of staphylococci found to be coagulase negative on first isolation, become coagulase positive after subculturing several times, and some strains may lose the ability to produce coagulase and still remain enterotoxigenic.

The coagulase reaction in clotting blood serum was reviewed by Tager (1974). According to this review, there are three main reactions ending with the conversion of fibrinogen to fibrin and formation of a clot. According to Baird-Parker (1972), over 90% of the strains of *S. aureus* produce a coagulase.

Several tests have been evaluated for determining the presence of coagulase. The tube test is used to determine free coagulase, and a slide test determines what is referred to as clumping factor or "bound coagulase." The clumping factor is generally regarded as being distinct from coagulase, but there are similarities.

The tube test consists of inoculating a suspect colony into 0.2 to 0.3 ml of sterile brain heart infusion (BHI) broth, and incubating at 35°-37°C for 18-24 hr. Dried rabbit plasma treated with an anticoagulant such as EDTA is reconstituted, and 0.5 ml is added to the BHI culture. The mixture is incubated at 35°-37°C, and observed for clot formation at hourly intervals from 1 to 6 hr. Only a firm and complete clot that stays in place when the tube is tilted or inverted is considered a positive coagulase test for *S. aureus* (FDA 1976).

The amount and type of anticoagulant are important. As the amount of anticoagulant is increased, a progressively softer clot is formed in the coagulase test. Although oxalate and heparin have been used in laboratory preparations, they have not been used widely in commercial plasma. The use of citrated plasma resulted in false-positive coagulase tests due to organisms utilizing the citrate anticoagulant. Therefore, EDTA is, at present, the desired additive in commercial preparations of plasma.

The production of a very heat stable nuclease is another characteristic used in the diagnosis of *S. aureus*. Although other organisms may produce nucleases, the heat stability of *S. aureus* nuclease is unique. It is generally agreed that the enzyme loses but little activity by boiling for 30 min. Erickson and Deibel (1973) reported that the enzyme retained 10% of its activity after heating at 100°C for 180 min, 120°C for 34 min or 130°C for 16.6 min.

The basic test for staphylococcal nuclease was described by Lachica *et al.* (1971). The test uses an agar diffusion procedure. The presently used agar consists of 1000 ml of 0.05 M tris buffer (pH 9.0) with DNA (0.3 g), $CaCl_2$ (1 ml of 0.01M), NaCl (10 g) and agar (10 g). This mixture is boiled and then 3 ml of 0.1M toluidine blue O is added. For use, the melted agar can be used as an overlay of colonies on a selective agar surface (Lachica 1976). Another system starts with adding the melted agar to a microslide or petri dish. After solidification, circular wells are cut into the agar and the agar plugs removed by aspiration. The wells are filled with a boiled (15 min) broth culture of the test organism. After preparation, the system is incubated at 37°C for 2-4 hr. A bright pink zone around the well is positive for nuclease. Kamman and Tatini (1977) suggested changes to optimize the reaction. They used 0.005M calcium chloride, 0.17M NaCl at pH 10.0 in preparing the toluidine blue test medium. The prepared system was incubated at 50°C rather than 37°C.

The heat stable nuclease test compares favorably with the coagulase test (Menzies 1977).

Rayman *et al.* 1975) suggested that the thermostable nuclease test should be made on any culture with a doubtful coagulase reaction. They found doubtful coagulase positive cultures generally were nuclease negative and not enterotoxigenic. Niskanen and Koiranen (1977) suggested that a culture that was either coagulase negative or thermonuclease negative is not likely enterotoxigenic.

The thermostable nuclease test might be used as a screening method for foods to detect the possible presence of high populations of *S. aureus* or enterotoxin (Tatini *et al.* 1975, 1976). This system seemed to be applicable to certain foods, but was of questionable value for others. More information on this procedure is needed.

Phosphatase, protease, lipase, lysozyme and various hemolysins have been suggested and tested as agents to determine enterotoxigenicity of *S. aureus*. Abramson (1974) suggested the possible relationship of isoenzymes to diseases caused by organisms. Resistotyping was suggested by Elek and Moryson (1974) as an epidemiological tool to amplify or clarify other findings. Toxin producing strains tend to lack or have a low resistance to methicillin. It has been inferred that these characteristics are associated with the same plasmid.

Serological Reactions.—Reviews have been published concerning the cellular antigens of *S. aureus* (Oeding 1974) and serotyping of staphylococci (Cohen 1974).

Although serological typing of *S. aureus* is not widely used at present, the system might be of assistance in determining the organisms in foods or in epidemiological investigations. Since not all *S. aureus* strains are phage typable, serology might help to differentiate those staphylococci.

Phage Typing.—Some bacteriophages are very specific for the cells they will lyse, while other phages have a wider spectrum of susceptible cells. Obviously, for phage typing of *S. aureus*, those phages that have a limited spectrum of susceptible cells are desirable.

The International Committee on Nomenclature of Bacteria established a Subcommittee on Phage Typing of Staphylococci in 1953. The Staphylococcus Reference Laboratory of the British Public Health Laboratory Service in London became the International Reference Center for the Subcommittee. In 1961, it became the World Health Organization Centre for Staphylococcus Phage-Typing.

TABLE 6.12

PHAGES FOR TYPING *STAPHYLOCOCCUS AUREUS*

Group	Phages
I	29, 52, 52A, 79, 80
II	3A, 3C, 55, 71
III	6, 42E, 47, 53, 54, 75, 77, 83A, 84, 85
IV	42D
Unassigned	187, 81, 94, 95, 96

More recently, phages 42D and 187 have been removed and 94, 95 and 96 added for a total of 23 phages for typing. Phage D11 is an unassigned experimental phage that may be used for typing. Phages 187, 42D or other phages may be used by some laboratories.

The Subcommittee established a set of phages for routine typing. In 1972, there were 22 phages in the set divided into groups as listed in Table 6.12. Since then, phages 94, 95 and 96 have been added while 42D and 187 were eliminated, so that the basic set now contains 23 phages.

Some phages are useful in certain countries but not in others. For example, phage 86 has been used in the USA for typing, but it is of little value in Europe. Enterotoxigenic strains of *S. aureus* usually are associated with phage group III, but also are found in other groups (Payne and Wood 1974; Wieneke 1974). This basic set of phages is not suitable for typing *S. aureus* strains isolated from animals.

The complete procedure for phage typing of *S. aureus* is too lengthy to repeat. However, briefly, the phages are propagated and harvested, then the routine test dilution (RTD) is determined. The RTD is the highest dilution of a phage suspension which just fails to produce confluent lysis of its homologous propagating strain, when a 0.02 ml drop is placed on a lawn of the organism. Then the determination of the lytic spectrum of the phage is tested to see if the new phage has the same host range as the parent phage. The specificity of each phage in the set must be maintained.

A surface of agar is seeded with a young broth culture by swabbing or spreading the cells uniformly over the surface. A template or guide is useful to divide the agar surface into sections and to number each area for a particular phage. One drop (0.02 ml) of each phage diluted to its RTD is placed in its respective area according to the template. After air drying, the prepared plate(s), along with control plates that are needed, are incubated overnight at 30°C. The incubated plates are observed and the lytic reactions are read semiquantitatively. A ± score means fewer than 20 plaques, a + is 20 to 50 plaques, a ++ indicates more than 50 plaques. Complete lysis may be recorded as +++, but a ++ is sufficient. For cultures showing no strong lysis, a phage concentration of RDT × 1000 is used on a second set of plates. Degre (1967) suggested using a single concentration of 100 RTD rather than using two concentrations of phage. Phages at high concentration (RTD × 1000) may inhibit the organism, which might be confused with lysis.

For a more complete description, the reader is referred to CDC (1976E) or Parker (1972).

Smith (1972) stated that there are only two good reasons for typing *S. aureus,* either by phages or antisera. The main reason is for epidemiological investigations, and the second reason is to give a "label" to the organism for research work.

Phage typing is of particular help in the epidemiological study of staphylococcal intoxications or infections. Various strains isolated during an outbreak from patients, foods and food handlers can be tested for their phage patterns. This information makes it possible to differentiate between a strain responsible for the outbreak and unrelated strains.

Enterotoxin Detection.—The detection of enterotoxin is important. If an outbreak occurs and isolates of *S. aureus* are obtained from a food, the enterotoxins produced can be compared to the enterotoxin causing the intoxication. In one outbreak involving dried milk, no *S. aureus* isolates were detected. In this case, the organisms multiplied, produced the toxin and then were killed by heat during the processing. In cases such as this, the food can be analyzed for the enterotoxin.

The principal methods for detecting enterotoxins can be separated into biological and serological systems.

A biological system can be some living entity or part thereof. The best living entity to determine enterotoxin activity is a human. However, humans are not always readily available. Also, the amount of enterotoxin cannot be determined since people vary in their sensitivity to enterotoxin.

Animals, such as monkeys, chimpanzees, dogs, pigs, cats and kittens have been tested as biological systems. Monkeys or chimpanzees can be given the enterotoxin orally, whereas the other animals are injected either intraperitoneally or intravenously. Because vomiting is the first symptom to occur and is the most readily observable reaction to enterotoxin, animals such as rodents that do not have a vomiting mechanism are not used. Pigeons, frogs, tropical fish, nematodes or various protozoa or bacteria show no obvious reaction with enterotoxins. Tissues including human intestinal cells, rabbit gut segments, chicken embryos, tissue cultures (HEp-2 and HeLa cells) and isolated enzyme systems have been tested for assay of enterotoxin. Although crude enterotoxins have shown some effect on some of these systems, the purified toxins have revealed little or no effect.

The kitten has been a valuable test animal, but injections are used. Other toxins, besides enterotoxins, can induce vomiting. Therefore, these other toxic substances must be inactivated. Digestion with trypsin removes the hemolytic toxins without appreciable destruction of the enterotoxin.

Since monkeys or chimpanzees can be fed the enterotoxin orally, they are the preferred test animal. The monkey is not as sensitive to enterotoxin as humans, and the sensitivity of chimpanzees is somewhere between humans and monkeys. Another disadvantage is that these animals are expensive to maintain and, as they are used to assay enterotoxin, they tend to develop a resistance (a limited immunity) to the toxins. Thus, their usefulness for enterotoxin detection is limited.

Tests using animals were necessary before the enterotoxins were purified and serological tests could be developed. Hence, it is evident that the animal tests are still important for the detection of any new types of enterotoxins, or testing the toxicity of isolated compounds, or a particular chemical portion of a toxin.

Serological methods have been used to detect and assay enterotoxins. These systems can be used only for those enterotoxins that have been identified and purified.

Serological systems that have been suggested for assaying enterotoxins include: the Ouchterlony double diffusion plates, Oudin single gel diffusion tube test, Oakley double gel diffusion tube test, microslide test, comparator cell, quantitative precipitin test, hemagglutination inhibition, reversed passive hemagglutination, immunofluorescence and radioimmunoassay.

In the Ouchterlony double diffusion method (Ouchterlony 1968), a petri dish poured with a semi-solid agar is used. Antiserum is added to a centrally located well in the agar, and antigens are added to peripheral wells. The antigens and the antibodies diffuse toward each other through the agar to form zones or lines where the reacting antigen and antibodies combine in optimal proportions.

The Ouchterlony system was adapted for use on a glass slide by Wadsworth. A modification of this microslide method was used by Casman and Bennett (1965) to detect enterotoxin in foods. Bennett (1971) stated that 0.005 μg of enterotoxin A per gram of cheese was detected.

According to a collaborative study reported by Bennett and McClure (1976), the microslide gel double diffusion test has a high degree of specificity, is simple and has good reproducibility in the identification of enterotoxins. This system has been adopted as official first action by the AOAC.

The other tests have been used with varying results. One procedure which has possibilities is the radioimmunoassay technique. These systems are based on the competition between toxin labeled with radioactive iodine (^{125}I) and unlabeled toxin for antigen binding sites on the antitoxin molecule. For solid-phase methods, the antitoxin is absorbed to bromacetyl cellulose (BAC) (Collins *et al.* 1973), or coated onto the inside wall of polystyrene tubes (Johnson *et al.* 1973). In the former test, a known amount of antibody on BAC is put into polycarbonate centrifuge tubes. The unknown amount of enterotoxin is added and allowed to react. Then, a known amount of the ^{125}I-labeled enterotoxin is allowed to react. The tube is centrifuged and 1 ml of the supernatant is counted with a gamma counter. This determines residual amounts of labeled enterotoxin that did not react, so one can determine the amount that did react, and also, the original amount of antibody that reacted with the enterotoxin being tested. The polystyrene tubes are treated in much the same way. The test enterotoxin is added to the prepared polystyrene tubes and the enterotoxin reacts with the antibody. Then, the labeled enterotoxin is added. It reacts with any residual antibody. The tube is emptied, and the amount of labeled enterotoxin that reacted with the excess antibody is determined by measuring the radioactivity of the polystyrene tube with a gamma counter.

In the double antibody system, the antigen-antibody complex is precipitated by a second antibody produced against the antigen-binding antibody (Robern *et al.* 1975). A double antibody solid-phase system was described by Lindroth and Niskanen (1977).

With a radioimmunoassay system, Pober and Silverman (1977) detected from 1 to 1.3 ng/g of food. The sensitivity depended upon the type of food being analyzed.

The radioimmunoassay systems are more expensive than other assays, due to the need for a radioactive level. It might not be adaptable for some laboratories. The decrease in the radioactivity of the labeled enterotoxin during storage, the strict regulation of the application of radioactive substances by various laws and the limited availability of highly purified enterotoxins needed for labeling are some problems and limitations of these systems.

Methods for the detection of enterotoxins have been reviewed by Bergdoll *et al.* (1976). An enzyme immunoassay for enterotoxins was described by Saunders and Bartlett (1977) to overcome some of the problems of the radioimmunoassay system.

Although the biological activity of the enterotoxins is not measured by these tests, Bergdoll (1970) stated that correlation between the two is adequate to justify the use of serological tests to analyze for enterotoxins. Chang and Dickie (1971) interpreted their observations with enterotoxin B to mean that antigenic sites and toxin sites are probably not the same. They did not speculate if this would invalidate the serological tests as methods for the analysis of the enterotoxins.

Control of Staphylococcal Intoxication.—It should be relatively simple and easy to prevent staphylococcal intoxication. The obvious control measure would be to keep the organisms out of foods. For those *S. aureus* that do invade food, methods to destroy them or to prevent their growth and enterotoxin production can be employed. If toxin is produced, then destruction of the toxin is needed.

Prevent Contamination.—It is impossible to keep all foods free from *S. aureus* due to the ubiquitous character of the organisms. However, it would seem possible to use common sense measures to keep the contamination low.

Humans are the main reservoir of the organism. The health, hygiene and work habits of food handlers can influence the level of contamination of foods. People that are sick should not be allowed to handle food. They not only contaminate the food with *S. aureus,* but also other disease organisms. People with colds, sinusitis or other respiratory disturbances are good sources of *S. aureus,* as are boils, pimples, acne and infected cuts. When people are so afflicted, they should not be allowed to handle or work with food. It is not possible to segregate all carriers of *S. aureus*, since a sizeable portion of the population may be involved. However, those people who are obviously contaminators of food should be removed from food handling areas.

The staphylococcal carrier state in a hospital was eliminated in 89% of the people treated with a nasal antibiotic spray (Hunter and Baker 1967). This practice might be practical or acceptable for food handlers.

Another source of contamination is milk from cows with mastitic udders. Food from diseased animals generally is not used for human food. However, cows have frequently harbored this organism without noticeable clinical manifestation. The prevention of mastitis is difficult, but not impossible. Without control, milk contaminated with *S. aureus* can become part of the general milk supply.

S. aureus has been isolated from knives and slicers during investigations of outbreaks of food poisoning. Food handling equipment should be designed so that it can be properly and effectively cleaned and sanitized. Cleaning and sanitizing should be accomplished as often as needed, but especially when work is completed with the equipment.

Prevent Growth and Toxin Production.—The main control of *S. aureus* is to hold the food at a temperature unsatisfactory for growth. There is little or no growth below 4°C or above 46°C.

Toxin production occurs in a more narrow temperature range than cellular growth. Troller (1976) suggested a range of 10°-45°C. Greater amounts of enterotoxin are produced in the range of 33° to 38°C than at either higher or lower temperatures.

Obviously, to warm the product for serving, or to cool leftover heated food, the temperature of the food must go through the growth range. This range must be traversed as fast as possible. Three hours has been suggested as the maximum time allowed. To facilitate cooling of leftover food, the food should be placed in shallow layers in shallow pans. The deeper the food, the longer it will take to cool. The warm food should not be held at room temperature to cool, but placed in the refrigerator. This means that the refrigeration system must be adequate to cool these potentially extra loads. The refrigerator cannot be jammed full of hot food, since no refrigerator can handle such a condition.

In fermented sausage products *S. aureus* can be controlled by lowering the pH with starter cultures or chemical acidulation (Daly *et al.* 1973). A combination of acidulation and starter cultures is more effective in inhibiting *S. aureus* than either treatment alone. The pH range for toxin production is about 5.0-9.0.

The minimum water activity for growth of *S. aureus* is about 0.83-0.86. Adding solutes or drying of foods to lower the a_w below 0.83 will prevent growth and enterotoxin production of *S. aureus*. Toxin production usually occurs at a higher a_w than that needed for growth.

A study of viability of *S. aureus* in intermediate moisture meats was reported by Plitman *et al.* (1973). *S. aureus* grew in pork adjusted to a_w of 0.88 by desorption, but the organism declined in numbers in pork adjusted to a_w 0.88 by adsorption. Troller (1976) found no enterotoxin production at a_w of 0.90 or less. The presence of NaCl inhibits enterotoxin production as shown in Fig. 6.8.

Destroy S. aureus, Inactivate Toxin.—Heating is the principal method used to kill *S. aureus* cells, as well as other organisms. Although heat may kill the cells, if enterotoxin has been produced, the toxin may persist since, in a food, the enterotoxin is more heat stable than the organism.

Not only is the temperature important, but also the time of exposure. The temperature and time needed to kill an organism depend on factors such as the heat resistance of the organism, the number of cells present, the age of the cells, the type of food or suspending medium, the ingredients added to the food, the pH, a_w and previous treatment of the food or organism (stress conditions). Thermal processing is described elsewhere in the text, so only a brief preview is given here.

To help describe the heat resistance of an organism, D and z values are used. The D value is used to describe the time-temperature relationship to the killing of an organism. A 1-D is the time needed at a specified temperature to reduce the number of cells by 90%, or, conversely, 10% of the cells survive this treatment. The z value is a measure of the effect of a change in temperature on the resistance of an organism. It is the °F or °C required for the thermal resistance to change by a factor of 10. Most of the thermal evaluations are reported in °F; however, when possible, the °F have been converted to °C.

Angelotti *et al.* (1961) determined the D values of various organisms in foods. In custard heated at 60°C, the D value for *S. aureus* 196E was 7.82 min, while for *S. aureus* Ms 149 it was 7.68 min. This means that if 100 cells of *S. aureus* Ms 149 were present per gram of custard, heating for 7.68 min at 60°C would reduce this number to 10. In another 7.68 min, there would be only 1 cell per g and after another 7.68 min, only 0.1 of a cell per g, or 1 cell/10 g of custard would be viable. The importance of aseptic methods to keep the original population as low as possible should be quite evident. In chicken a la king, the respective D values for *S. aureus* 196E and Ms 149 were 5.37 and 5.17 min. This indicates that the organisms are more susceptible to heat in chicken a la king than in custard, since the D values are lower for the organisms in chicken a la king.

The destruction of *S. aureus* is accomplished more readily in whole milk than in skim milk. In skim milk at 60°C, the D value for *S. aureus* strain 161-C was 1.30 min (Walker and Harmon 1966). In whole milk, the D value for strain 161-C was 0.75 min. Firstenberg-Eden *et al.* (1977) reported a D_{60} (D value at 60°C) in whole milk of 0.87 min for *S. aureus* if salt free media were used to detect heated cells. When salt was present in the recovery medium, the apparent D_{60} was 0.62 min. Sublethal heat treatments cause *S. aureus* to lose salt tolerance. These latter workers reported a z value of 9.46°C so that if D_{60} is 0.87 min, at 50.54°C, the D is 8.7, and at 69.46°C, it is only 0.087 min.

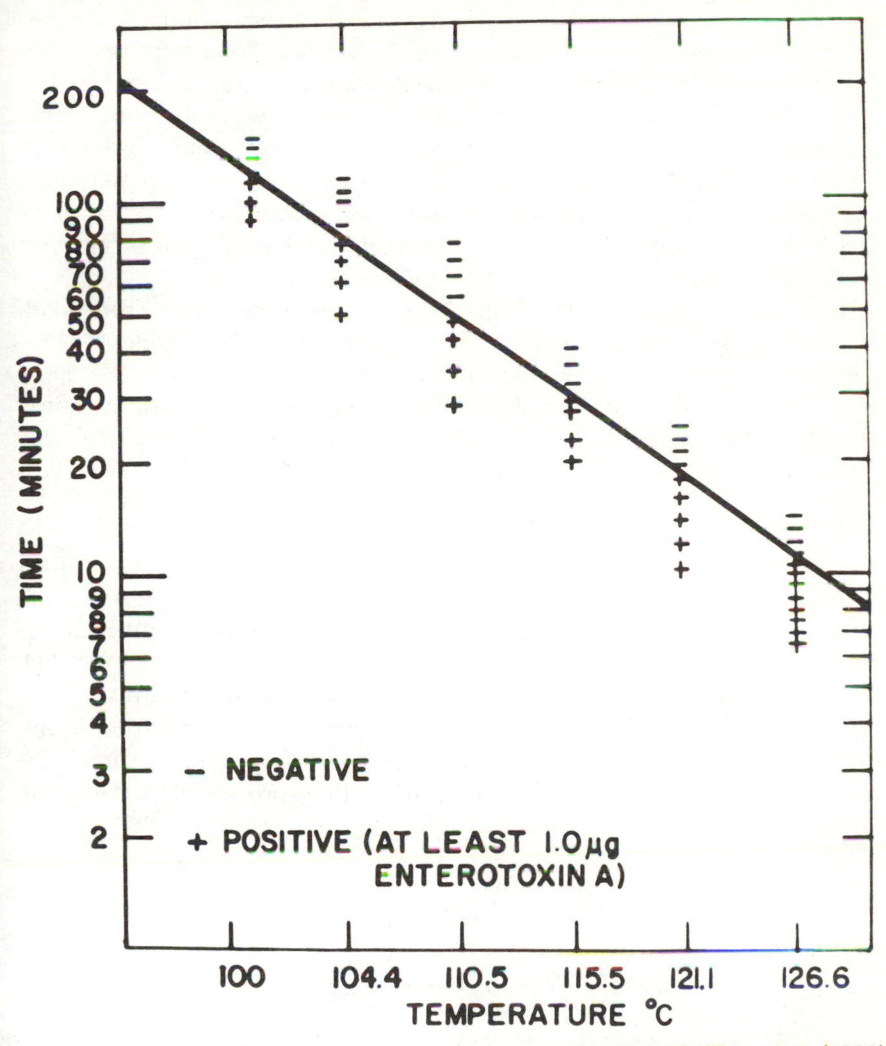

Courtesy of Hilker et al. (1968)

FIG. 6.9. THERMAL INACTIVATION OF STAPHYLOCOCCAL ENTEROTOXIN

According to Bean and Roberts (1975), levels of 8% salt (w/v) in the heating medium protects *S. aureus* from heat. They reported the D_{60} for a meat emulsion at pH 6.5 was 4.61-9.62 min. When 8-8.5% salt was added to the meat emulsion, the D_{60} increased to 18.62-26.38 min. The addition of sodium nitrite had little effect on the heat resistance of this organism.

It is relatively easy to kill *S. aureus* by normal pasteurization or cooking procedures. The problem lies in recontamination of the heated food with *S. aureus*. Deboning poultry or slicing of ham after cooking furnishes opportunities for recontamination from the hands of food handlers. With the destruction or reduction of the indigenous flora during heating, the recontaminating cells of *S. aureus* are not competitively inhibited. If this recontaminated food is allowed to remain at temperatures for growth of *S. aureus*, enterotoxin production may occur resulting in a potential outbreak of staphylococcal intoxication. To reduce the occurrence of staphylococcal intoxications, it is important that strict sanitation and hygiene be followed anywhere that food is handled.

Heating is not a practical means to eliminate enterotoxins from food, because of the heat stability of these agents (Fig. 6.9). Calcium hypochlorite at 50-200 μg/ml detoxified SEB contaminated water (Meyer *et al.* 1977). Using chlorine to detoxify food might not be acceptable, but it may be effective for surface decontamination.

Botulism

From 1899 through 1976, there were 1,875 reported cases of botulism in the USA. Although this number of cases is low when compared to some other foodborne illnesses, there were 992 deaths for a mortality rate of 52.9%. The mortality rate has decreased in recent years (Fig. 6.10, 6.11), but even the present rate is reason for concern about botulism. The average outbreak involves only two or three people. The largest outbreak reported in the USA occurred in Michigan in 1977. This outbreak involved 46 people, but there were no deaths. Home processed food was the source of the toxin.

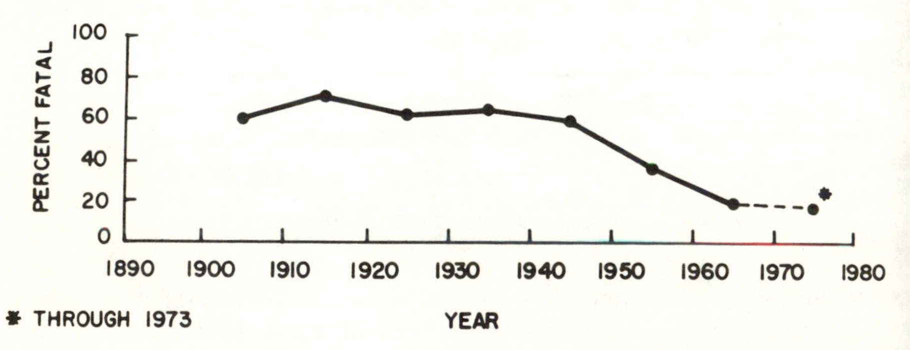

Courtesy of CDC

FIG. 6.10. FOODBORNE BOTULISM—DEATH TO CASE RATIOS BY 10 YEAR PERIODS, 1899-1973

BOTULISM – Reported Cases and Deaths by Year, United States, 1950–1976

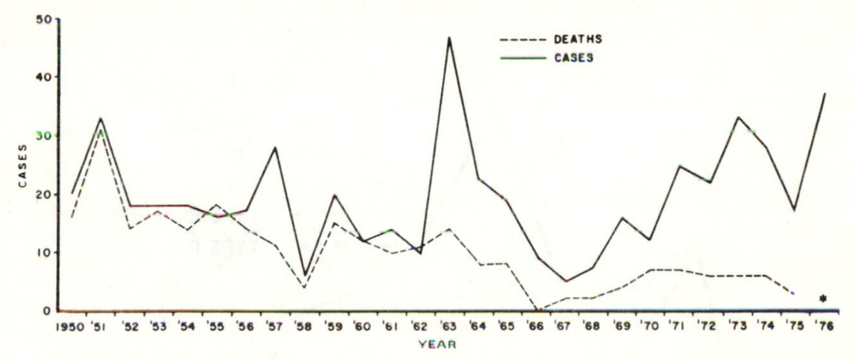

*Not available for 1976

Discrepancies between the number of cases and the number of deaths in 1955 and 1962 are believed to be due to different reporting mechanisms of the National Center for Health Statistics (deaths) and the National Morbidity and Mortality Statistical Activity (cases).

Courtesy of CDC

FIG. 6.11. CASES AND DEATHS DUE TO FOODBORNE BOTULISM, 1950–1976

Etiologic Agent.—Botulism is an intoxication caused by neurotoxins produced by toxigenic strains of *Clostridium botulinum.* Seven antigenically distinct toxin types (A, B, C, D, E, F and G) are recognized. Type C toxin appears as C_1 and C_2 toxins.

Types A, B, E and F cause botulism in humans. Although types C and D have been reported as causing illness in humans, the involvement is rare and uneventful.

Type F has been reported as the etiologic agent of very few cases of botulism. There is limited information about type G botulism. It was reported in 1970, and more recently was studied by Ciccarelli *et al.* (1977). It has not been the agent in any human illness.

The number of outbreaks caused by Type A, B, E and F in the USA from 1899 to 1973 is shown in Fig. 6.12. Type A toxin is the main cause of botulism in the USA, accounting for 164 of the outbreaks. Type B toxin was responsible for 48 outbreaks, type E for 22 and type F only 1. Most of the type A outbreaks occur in the western states, while type B is predominant in the east and central USA. This correlates with the finding that type A spores predominate in the soil in the west, while type B spores predominate in the east and central areas. In Europe, type B botulism is the predominant type, which correlates with the occurrence of type B spores in the soil. Although not prominent in the contiguous USA, type E botulism has accounted for over 40% of the outbreaks in

Japan, Canada, the Scandanavian countries and nearly all of the cases in Alaska. Outbreaks due to type A and type B toxins only recently have been reported in Alaska (Barrett *et al.* 1977).

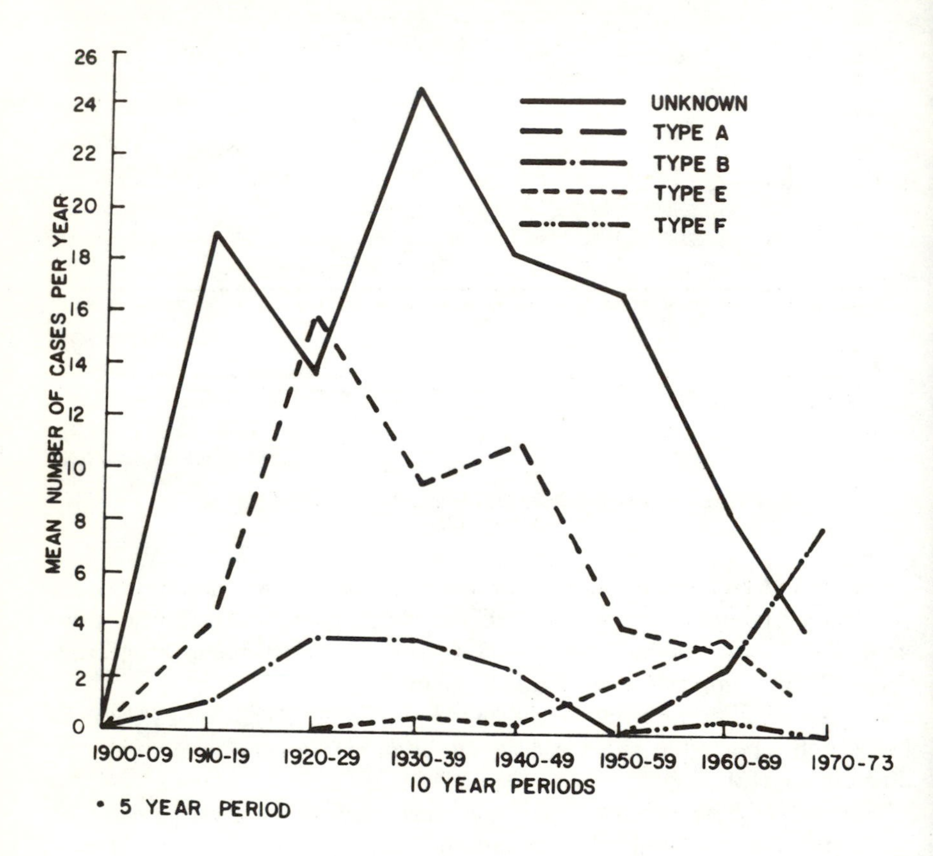

Courtesy of CDC

FIG. 6.12. THE NUMBER OF CASES OF BOTULISM BY THE TYPE OF TOXIN

Types C and D toxins cause botulism in lower animals. Some animals can be affected by other toxins. Swine are moderately to highly resistant to most of the toxins.

Although the toxins are ingested by the oral route, unlike the staphylococcal enterotoxins, they do not affect the alimentary tract, but do affect the nerves. Thus, they are designated as neurotoxins.

Besides oral toxicity, wound infection by *C. botulinum* with subsequent toxin production and symptoms of botulism have been reported for 15 cases since 1943 (CDC 1974C), with 5 cases reported in 1974. Three cases of wound botulism were reported in 1976.

Boroff and Das Gupta (1971) reviewed the theory that ingestion of the organisms or their spores can result in production of botulinum toxin in vivo. They present much evidence to support this proposition. The site of production and absorption of botulinum toxin in the chicken is the cecum (Miyazaki and Sakaguchi 1978). With wound botulism revealing that the toxin can be produced in vivo, we cannot completely rule out the possibility that ingestion of the organism might result in intoxication. According to Arnon *et al.* (1977), infant botulism may result from the apparent intraintestinal production of toxin by *C. botulinum.* This illness was not recognized until 1975. Since then, 15 cases have been reported in 1976 and 42 in 1977. Of the 58 known cases, 33 involved type A and 25 type B toxins. No known source of toxin has been found, but *C. botulinum* spores were found in house dust and/or soil outside the house in these cases. In three homes, opened jars of honey contained type B organisms and one unopened jar had type A organisms. Outbreaks have occurred in both breast-fed and formula-fed babies. An outbreak of botulism in a breast-feeding mother resulted in no toxin in her milk (CDC 1976C). The nursing infant did not develop botulism.

Almost all of the known cases of infant botulism recover. CDC (1978A) estimated at least 250 cases occur annually in the USA. How many unreported cases are fatal? We do not know.

Properties of the Toxins.—The toxins are simple proteins that are water soluble, heat sensitive and acid stable. On a molar basis, the toxins produced by *C. botulinum* are the most lethal natural products known. Since the toxins are protein, they are antigenic. Various molecular weights ranging from 5,000 to 900,000 daltons have been reported for these toxins. Most reports list the molecular weights of the toxins between 50,000 and 250,000 daltons. These variations indicate that the toxins are heterogeneous substances.

Generally it is accepted that these neurotoxins, as produced, require activation. The specific toxicities increase following activation by certain proteases. When the organisms producing the toxins are proteolytic, the endogenous enzymes of the organisms activate the toxin to a more toxic substance. All type A and some type B strains are proteolytic. When the strain is nonproteolytic, activation of the toxin requires treatment with an enzyme. Trypsin is the most effective of the enzymes that have been used. A short-term exposure to trypsin activates the toxin and increases the toxicity, but extended exposure to trypsin reduces the toxicity.

Since foods usually contain a mixed culture including proteolytic organisms, as well as proteolytic enzymes, or when taken orally, the toxins are exposed to trypsin in vivo; whether the toxin is activated before or after ingestion may seem to be immaterial. This need for activation is important to the analyst or epidemiologist.

The activation phenomenon is not clearly understood. One group believes that the toxin is fragmented into smaller toxin units, but other researchers believe this does not occur. Rather, it is thought that the enzyme breaks bonds with a change in the shape of the molecule. This either makes it easier for the toxin to penetrate the tissue to the site of toxic action, or it may increase the exposure of toxophoric groupings of amino acids.

During the logarithmic phase of growth, large amounts of toxin accumulate intracellularly, while smaller amounts can be found in the extracellular medium. Beyond the logarithmic growth phase, very large amounts of toxin are released to the extracellular medium, while the amount of intracellular toxin decreases.

The mechanism for releasing the toxin from the bacterial cell has not been fully explained. One explanation is that the toxin is released upon cell lysis.

The heat stability of botulinum toxin is considerably less than that of staphylococcal enterotoxin. The stability depends somewhat on the type of solution in which the toxin is heated.

In review articles, Boroff and Das Gupta (1971) reported that boiling for 1 min or heating at 75°-80°C for 5-10 min completely destroys the toxicity, while Jarvis (1972) stated that at 65°C, the toxins are inactivated in 90 min.

Freezing and frozen storage affect the heat stability of type E toxin (Yao *et al.* 1973), but not type A toxin (Woolford *et al.* 1978). The resistance to chemicals depends upon the type of toxin, temperature, pH and other substances in the medium. In water, the purified toxin is sensitive to chlorine, bromine or iodine. Free available chlorine with a residual of 1 mg/liter (1 ppm) will destroy at least 99.9% of all the types of botulinum toxin in 5 min or less. Type E toxin is the most resistant to destruction by chlorine.

Action of the Toxins.—Except for wound or infant botulism, the preformed toxin is considered to be ingested orally. To react with and affect the nerves, the toxin must traverse the barrier of the gastrointestinal tract and be transported to the susceptible nerves. The action on the nerves causes neuromuscular blockage, with paralysis of the muscles.

The type of food in the GI tract can affect the toxin and its action. The food might protect the toxins from the enzymes or other destructive actions, such as by stomach acids. Food increases or decreases the secretion of digestive juices. The food might combine with the toxin forming larger particles less able to penetrate the intestinal wall. Foods may affect the rate of peristalsis which increases or decreases the time the toxin is in an area of the intestine affording the greatest opportunity for penetration of the wall.

The major site for absorption of the toxin is the small intestine. The route from the intestine to the bloodstream is the lymphatic system. The toxin appears in the lymph as a protein, indicating that the small intestine is not an absolute barrier to the passage of protein. Perhaps only a small amount of the ingested toxin is absorbed, but not much is needed to cause botulism. The large intestine is also an absorption site since, according to Lamanna (1968), intrarectal instillation of toxin into monkeys and rabbits caused death, although the animals lived longer than when the toxin was given orally. The different toxins are absorbed at different rates (Sugii *et al.* 1977).

That the toxin is "wasted" in the alimentary tract is evident from work with animals. When type C toxin is administered to mink intraperitoneally, only 50 MLD (mouse lethal dose) causes death. Oral administration of 4,000 MLD causes death, but with 2,500 MLD, the mink survive. With pigs, the oral: intravenous ratio of amount of toxin needed to cause death is 16,700:1 (Smith *et al.* 1971). Data from various outbreaks indicate that not all of the toxin penetrates the intestinal wall, and that it can be present in the intestines for a prolonged period.

The toxin passes with the lymph through the thoracic duct and is dumped into the bloodstream. It is carried by the vascular system to the nerves. Lamanna (1968) reported that when the lymph system was prevented from emptying into the bloodstream, animals fed botulinum toxin did not develop symptoms of botulism.

At the nerves, the toxin attaches to the presynaptic terminals of cholinergic nerves, and interferes with the release of acetylcholine at the myoneural junctions.

The mechanism of the suppression of the release of acetylcholine is obscure (Boroff and Das Gupta 1971). The failure of impulses to be transmitted across the nerve fiber junctions results in paralysis of the muscles which the nerves control.

Characteristics of the Intoxication.—Since the toxins affect the nerves, this illness differs from that of staphylococcal enterotoxins, as well as other foodborne illnesses.

Incubation Period.—After ingestion of food contaminated with botulinum toxin, the usual time for symptoms or signs of botulism to appear is 12-48 hr (CDC 1977A). However, this time may vary from 2 hr to 8 days, depending upon the amount of toxin ingested, the type of toxin, the resistance of the individual and perhaps even the type of food. Generally, if the incubation period is less than 24 hr, the person will be more severely affected and death is more likely than if the incubation period is longer.

Since the toxin must traverse the intestinal barrier, the incubation period is longer than that for staphylococcal intoxication, which acts directly in the intestine.

Attack Rates.—Although botulinum toxin is the most potent natural poison known, not everyone consuming contaminated food will either acquire symptoms or succumb to the toxin. This has been related to the amount of contaminated food ingested. However, there have been reports that merely tasting a food and spitting it out resulted in intoxication and death. This led to the supposition that toxin could be absorbed directly from buccal exposure. The contention of Lamanna *et al.* (1967) is that the normal swallowing reflex transfers the toxin from the mouth to the GI tract.

Since the organisms grow in acceptable microenvironments and produce the toxin, there is undoubtedly unequal distribution of the toxin throughout the food. Hence, the food as a whole could be contaminated, but parts of it could be free of toxin.

There are differences among humans to susceptibility to the toxins. Also, a person may have had prior exposure to the toxin and developed an immunity. However, with a toxin such as botulinum toxin, the lethal dose is often smaller than the immunizing dose.

Symptoms and Signs.—The cardinal features of botulism are:

1. Fever is absent, but may develop if a complicating infection occurs.
2. Mental status is normal. Patients may be anxious or agitated and some are unusually drowsy; however, in the absence of secondary complications, patients are responsive.
3. The pulse rate is normal, or slow, but tachycardia may occur if hypotension develops.
4. Although vision may be blurred, numbness and paresthesias are absent, and a sensory deficit does not occur.
5. Neurological manifestations are symmetrical.

The symptoms and signs from cases reported to the Center for Disease Control are listed in Table 6.13. The first symptoms of illness are gastrointestinal. Nausea or vomiting, substernal burning or pain, abdominal distension and decreased bowel sounds may occur. Some cases have initial transitory diarrhea, but subsequently become constipated. These symptoms and signs may mislead physicians to diagnose the illness as appendicitis, bowel obstruction or even diaphragmatic myocardial infarction. The mucous membranes of the mouth, tongue and pharynx may be red, dry and painful, which might be diagnosed as pharyngitis.

Sometimes other illnesses are diagnosed as botulism. Investigation of 438 suspect botulism outbreaks by CDC (1974A) revealed that only 75 were actually botulism and the rest were due to other disorders.

TABLE 6.13

OUTBREAKS OF BOTULISM REPORTED TO CDC IN WHICH ONE OR MORE PERSONS WERE AFFECTED BY A GIVEN SYMPTOM OR SIGN, 1953–1973

	Type A	Type B	Type E	Type F	Und[1]	Total	
Outbreaks	34	15	10	1	44	104	
Cases	97	46	36	3	90	272	
Symptoms	Number of Outbreaks						% with Symptom or Sign
1. Blurred vision, diplopia, photophobia (double vision)	31	13	9	1	40	94	90.4
2. Dysphagia (difficulty in swallowing)	27	14	3		35	79	76.0
3. General weakness	22	12	4		22	60	57.7
4. Nausea and/or vomiting	15	13	10	1	19	58	55.8
5. Dysphonia (confused speech)	25	8	5		19	57	54.8
6. Dizziness or vertigo	8	4	5		15	32	30.8
7. Abdominal pain, cramps, fullness	5	6	3		7	21	20.2
8. Diarrhea	5	6			5	16	15.4
9. Urinary retention or incontinence	2	2	1		2	7	6.7
10. Sore throat	4	2	1			7	6.7
11. Constipation		2		1	3	6	5.8
12. Paresthesias	1					1	1.0
Signs							
1. Respiratory impairment	32	7	7		30	76	73.1
2. Specific muscle weakness or paralysis	23	9	3		13	48	46.2
3. Eye muscle involvement, including ptosis	16	9	3	1	17	46	44.2
4. Dry throat, mouth or tongue	7	6	2		7	22	21.2
5. Dilated, fixed pupils	3	4	2		8	16	15.4
6. Ataxia	3	1		1	4	9	8.7
7. Postural hypotension			1		2	3	2.9
8. Nystagmus	1		1		1	3	1.0
9. Somnolence		1				1	1.0

[1]Toxin type undetermined or unspecified
Data from CDC (1974A)

The neural symptoms usually begin with the eyes and face and paralysis progresses downward to the throat, chest and extremities. When the diaphragm and chest muscles become fully involved, respiration is not possible. Death is due to asphyxiation. This usually occurs in 3-6 days. In non-fatal cases, complete recovery may take several months. Mental processes are usually clear during the illness.

Therapy and Duration.—Prompt clinical, epidemiological and laboratory efforts are required since each hour is critical to the survival of a patient with botulism. The patient should be hospitalized for treatment.

The treatment of botulism can be separated into three parts: removing unabsorbed toxin from the alimentary tract, neutralizing toxin with anti-toxin and treating the symptoms, such as respiratory distress.

The sooner the antitoxin is administered, the better are the chances for recovery. There have been cases in which the patient died within 24-48 hr after ingestion of the toxic food. In these cases, botulism is often not treated due to the short time, and diagnosis of botulism is made postmortem, if at all. There are essentially no post-morten visual characteristics associated with botulism, but toxin can be detected by laboratory tests.

In some cases, guanidine hydrochloride is used as an adjunct. This compound seems to compensate for the neural effects of botulinum toxin. The use of germine in combination with guanidine was found to be beneficial in the treatment of botulism (Cherington *et al.* 1972).

The treatment of the symptoms is especially important with respiratory difficulty, or respiratory failure. Tracheotomy of the patient is used to assist in breathing. In more severe neuromuscular blockage, mechanical respirators are needed to maintain breathing.

Reduction of the mortality rate in recent years has been attributed to the prompt treatment of the patients and administration of antitoxin.

One reason given for the lower mortality rate in Europe compared to the USA is that type B is prevalent in Europe and type A in the USA. Type A toxin is known to bind rapidly to tissue. In one case, 24 hr between ingestion and treatment was sufficient for the type A toxin to adhere irreversibly to the patient's myoneural tissue (Dillon *et al.* 1969). The antitoxin reacts with free toxin. When the toxin is irreversibly bound, administration of antitoxin has little effect on the recovery of the patient.

Exposure to the toxin produces an immunity to that type of botulism. Even if a human does survive botulism and develops an immunity to one type of toxin, the person would still be subject to the effects of other toxin types. Although immunization could be accomplished, the rare occurrence of botulism makes widespread immunization impractical. Immunization is recommended for laboratory or other personnel who are exposed to the toxin.

The duration of the illness depends upon the severity. Death has occurred in 24 hr, while in other cases, 2 or 3 weeks of illness end suddenly with the death of a patient. With very little exposure to the toxin, the person may remain asymptomatic or develop symptoms with little distress that pass in a few days. With increased amounts of toxin and development of respiratory failure so that a respirator is needed, the patient may require several months to recover, and even then may experience some distress. Partial paralysis may persist for 6-8 months (Bryan 1973A).

Foods as Vehicles of the Toxin.—Botulism is usually associated with foods that have been given a preservation treatment, stored for some time and consumed without appropriate heating. The preservation treatment in these foods is inadequate to destroy the spores that were present in the food.

The foods involved in botulism outbreaks (1899-1976) in which the toxin type was determined, are listed in Table 6.14 and Fig. 6.13. In over 52% of the outbreaks, canned vegetables were the vehicle for the toxin. Of the remaining 48% of the outbreaks, fish, fruit and condiments (chili sauce, chili peppers, tomato relish and salad dressing) were the main vehicles of the toxins. Meat, poultry and dairy products rarely are involved in outbreaks of botulism. This is because these foods are consumed primarily as fresh, rather than canned, food.

TABLE 6.14

FOODS INVOLVED IN BOTULISM OUTBREAKS IN WHICH THE TYPE OF TOXIN WAS DETERMINED (1899-1976)

	Botulinum Toxin Type							
Food Group	A	B	D	E	F	A+B	Total	%
Vegetables	108	29	1	1	–	2	141	52.8
Fishery products	11	4	–	23	–	–	38	14.2
Fruit	22	6	–	–	–	–	27	10.1
Condiments	16	4	–	–	–	–	20	7.5
Beef	6	1	–	–	1	–	8	3.0
Dairy products	2	2	–	–	–	–	4	1.5
Pork	2	1	–	–	–	–	3	1.1
Poultry	2	2	–	–	–	–	4	1.5
Other	13	8	–	–	–	–	21	8.2
TOTAL	182	57	1	24	1	2	267	
%	68.2	21.3	0.4	9.0	0.4	0.7		

Sources: CDC (1974A, 1976A,B, 1977A)

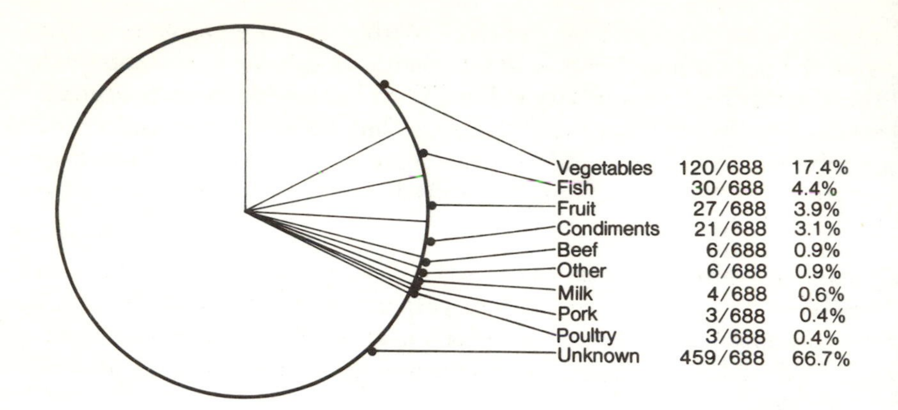

Courtesy of CDC

FIG. 6.13. FOODS INVOLVED IN OUTBREAKS OF BOTULISM IN THE USA, 1899–1973

Home processed foods accounted for the majority of the outbreaks (72%), while commercially processed foods were involved in less than 10% of the outbreaks. Unknown vehicles accounted for slightly less than 20% of the outbreaks of botulism (Table 6.15). Although commercially processed foods have been involved in fewer outbreaks than home processed foods, the incidents which involve commercial foods are publicized, while those caused by home packed foods rarely are mentioned. The reason is the widespread distribution of the commercial packs. If 500 cans of home-packed corn are contaminated, chances are only one of these will be involved in botulism in one family. If 500 cans of commercially canned corn are contaminated, they could cause botulism in 500 families. So far, this has not happened in the USA.

The outbreaks of botulism due to commercially packed foods since 1960 caused the health "experts" to suggest that families should process food at home. Data certainly do not support the theory that home canning would reduce the number of botulism outbreaks.

Due to the present economic situation, there has been more home canning, and an increase in the number of outbreaks of botulism. There were 23 outbreaks of botulism reported in 1976, which is the most for any one year since 1935. There was an economic problem in the 1930's and home canning was widespread.

The Organism.—The genus *Clostridium* is divided into four groups according to spore formation and gelatin liquefaction. *Clostridium botulinum* is in Group II. In this group, the spores are terminal and gelatin is hydrolyzed.

TABLE 6.15

OUTBREAKS OF FOODBORNE BOTULISM ATTRIBUTED TO COMMERCIALLY PROCESSED OR HOME PROCESSED FOODS 1899–1976

Source of Food	1899	1900-1909	1910-1919	1920-1929	1930-1939	1940-1949	1950-1959	1960-1969	1970-1976	Total	%
Home processed	1	1	48	77	135	120	50	42	60	534	71.6
Commercially processed	0	1	14	26	6	1	2	10	4	64	8.6
Unknown	0	0	8	13	13	13	51	26	24	148	19.8
TOTAL	1	2	70	116	154	134	103	78	88	746	

Source: CDC Annual Summaries

The species includes a heterogeneous group of strains that are divided into types A through G based on the antigenic neurotoxins that are produced. These seven types are divided into four groups according to their deoxyribonucleic acid homologies and biochemical, physiological and serological characteristics. The members of these groups are as follows: group 1, type A and proteolytic strains of types B, C, D and F; group II, type E and nonproteolytic strains of types B and F; group III, nonproteolytic strains of types C and D; and group IV, type G (Buchanan and Gibbons 1974).

The optimum temperature for growth is 30°-40°C for group I, 25°-37°C for group II and 30°-37°C for groups III and IV organisms. Group I organisms digest milk, group II coagulate milk with a soft curd but do not digest it, group III organisms do not change milk and those in group IV digest milk slowly.

Toxigenic Strains.—Apparently most strains of *C. botulinum* are capable of producing a toxin; however, this ability may be lost due to laboratory cultivation, and nontoxigenic clostridia have been isolated from both marine and terrestrial environments.

Inoue and Iida (1970) suggested that production of toxin by *C. botulinum* is influenced by prophages in the cells. By curing the cells of the phage, some toxigenic strains lose the ability to produce toxin. With reinfection by phage, the cells regain toxigenicity. It has been established that the phage type can determine the toxin type that is produced and not all lysogenic phages will confer toxigenicity on the infected cells (Hariharan and Mitchell 1976; Oguma *et al.* 1975). Most of the non-toxigenic strains that were converted to the toxigenic state reportedly were not stable to serial transfer through cooked meat medium (Oguma 1976).

Some strains of *C. botulinum* cured of their prophages by Eklund *et al.* (1970) remained toxigenic. They believed that the organisms might carry more than one prophage, or that not all *C. botulinum* toxins are induced by bacteriophage.

The studies of phage inducement of toxin formation primarily have used type C and D strains. The relationship of phage to toxigenicity of strains causing human botulism has not been evaluated sufficiently.

Although phage-host relationships may not be the entire answer for the toxigenic characteristic of all *C. botulinum* types and strains there appears to be some connection.

Sources.—The organism is widely distributed in nature. It occurs in soil apparently throughout the world. The type of *C. botulinum* depends upon the locality. The factors that affect the distribution of the different toxin types are not known. Healthy animals may be carriers of types of organisms to which they are immune, thus serving as reservoirs for replenishing the organisms in the soil and spreading the organisms to various places.

Bottom sediments of marshes, lakes and coastal ocean waters contain *C. botulinum.* The primary type in marine environments in northern areas appears to be type E; however, other toxin types of the organisms also are found. Type E is common in the intestines of fish taken from certain waters. The organism has been isolated from the environment of trout farms and from trout produced in these establishments. The organism does not multiply in living fish, but it does multiply in aquatic vegetation and bottom deposits. In marshes, the growth of algae reduces the oxygen level of the water to a level that allows *C. botulinum* to multiply.

Hayes *et al.* (1970) examined 240 samples of smoked fish products after holding for 1-3 weeks at 6°C. They found that 47 (19.6%) were toxic to mice and 11 (4.6%) contained type E toxin.

C. botulinum was detected in 11 of 263 samples of vacuum-packed bacon (Roberts and Smart 1976).

Since the organism is widely distributed in nature, it is evident that food can be contaminated during production, harvesting or processing. Fortunately, if present, the spores are usually in low numbers, and they can germinate only if the conditions of microenvironments of the food are favorable. After germination, sufficient time is needed for growth and toxin production.

Growth and Toxin Production.—The foods involved in outbreaks of botulism obviously present an environment or microenvironments that allow the growth of *C. botulinum.*

Various minimum and maximum temperatures have been reported for growth and toxin production. Eklund *et al.* (1967) reported growth and toxin production of a strain of type B *C. botulinum* at 3.3°C. The maximum temperature of growth for types A and B is about 48°C. Type E can grow in a range from 3.3° to 45°C. Growth and toxin production of type E is less pronounced at 37°C than at either 30° or 22°C (Ajmal 1968). A higher number of cells developed at 30° than at 22°C, but no difference was found for toxin production, indicating that the amount of toxin produced per cell was greater at 22° than at 30°C. When incubated at 4°C, growth and toxin production occurred in 4 to 5 weeks.

In controlled experiments, no growth of *C. botulinum* has been reported at pH 4.7 or less. There have been outbreaks in which foods supposedly lower than pH 4.7 were involved. In these instances, either the acid was not distributed homogeneously or other organisms were present, grew and raised the pH to a level favorable for *C. botulinum* spores and cells. Also, there is the possibility that the method of preparation of acid foods can neutralize part of the acid and bring the acidity within the pH growth range of *C. botulinum.*

The effect of salt on the growth and toxin production varies with the types and strains of *C. botulinum.* An aqueous phase level of 5.5% salt in homogenized cod was sporostatic to two strains of *C. botulinum* type E (Boyd and Southcott 1971). One strain produced toxin with low potency at 4.5% salt. With another strain, toxin formation was inhibited but not outgrowth or vegetative cell multiplication at a level of 3.5% salt. They found considerably less toxin produced in homogenates containing added salt than in the homogenates with no salt added. They suggested that a brine concentration of 5.5% would be required in certain substrates to inhibit toxin production of *C. botulinum* type E.

At pH 7.0, 30°C and using glycerol as the solute, the minimum a_w for *C. botulinum* types A and B is 0.93, while for type E it is 0.95 (Baird-Parker and Freame 1967). With lower pH or using other solutes, a higher a_w was needed for growth.

C. botulinum is an anaerobic organism. The clostridia lack cytochromes, cytochrome oxidase, catalase and peridoxase, so the oxidation-reduction potential (Eh) must be low for growth to occur.

C. botulinum type E inoculated into meat and fish, produced toxin under both aerobic and anaerobic conditions of incubation (Ajmal 1968). There are microenvironments in these foods which have low oxidation-reduction potentials favorable for the growth of this organism.

Sugiyama and Yang (1975) inoculated spores of *C. botulinum* into fresh mushrooms which were then packaged and stored at 20°C. The respiration of the mushrooms removed the free oxygen and allowed the spores to germinate and the cells to reproduce and form the toxin in three to four days. The mushrooms appeared to be edible. No toxin was detected in mushrooms held at 4°C.

Since vacuum packaging in gas-impermeable plastic produces anaerobic conditions, it was reasoned that this practice might encourage the growth of *C. botulinum* and create a toxic food product. Research shows that vacuum packaging has little if any effect on the ability of *C. botulinum* to grow on cured meats (Christiansen and Foster 1965). The type of package influenced spoilage, but not toxin production. Growth and toxin production occurred in fish or meat, regardless of the type of packaging material. In the past, vacuum packaged meats have not been an important vehicle for the toxin.

According to Pace and Krumbiegel (1973), the rate of toxin production is higher in vacuum packaged fish than fish packaged without vacuum. The slowest rate of toxin production is in unpackaged products. However, Johannsen (1965) stated that the formation of toxin in vacuum packed products is often lower than in those not packed in vacuum. According to him, the microaerophilic environment in vacuum packed products encourages the growth of lactobacilli. These organisms antagonize the growth and toxin formation of *C. botulinum*.

A *Moraxella* species produced a substance inhibitory to *C. botulinum* type E (Kwan and Lee 1974). The relationship of *Moraxella* to *C. botulinum* in nature is not known.

The interaction of microorganisms is both stimulatory and inhibitory to growth of *C. botulinum*. The growth of other organisms tends to reduce the Eh of the growth medium which may then allow growth of *C. botulinum*. Some organisms, such as yeasts, may produce growth factors favorable for *C. botulinum*.

Heat damaged spores, or spores heated in the presence of curing agents, are more sensitive to pH effects and salt concentration than are unheated spores. The original spore load may determine if outgrowth and toxin production will occur. The higher the level of spores the sooner will toxin be detected during the storage of a food product.

The growth of *C. botulinum* in foods can cause a foul, putrid odor which should serve as a warning to the consumer. In many of the outbreaks of botulism, either a patient or an asymptomatic participant stated that the food had an off-odor or flavor, but these warning signs were ignored by at least one person (the patient or patients). One problem is that some people have a high tolerance to off-odors or off-flavors. In some food products (especially those that are smoked, spiced or fermented), slight off-odors or off-flavors are difficult to detect.

There are reports that toxin is present in spores (spore-bound toxin) of *C. botulinum* (Booth *et al.* 1972). Phagocytosis of the cells releases the toxin. It has the same lethal effect as the free toxin.

Methodology.—The methodology of *C. botulinum* involves the detection and enumeration of the organism, the characterization of the organism and/or qualitative or quantitative determinations of the neurotoxins.

Detection and Enumeration.—Due to the hazards involved, before working with *C. botulinum,* a person should be protected by suitable toxoids.

The demonstration of botulinum toxin in a food implicated in a botulism outbreak is often all that is done. However, the detection and isolation of *C. botulinum* from the food furnishes confirmatory evidence. Also, methods of detection, isolation, identification and enumeration are basic in determining the distribution of the organism in nature.

The Bacteriological Analytical Manual (FDA 1976) described a system for detecting *C. botulinum.* Briefly, the food sample or toxic culture is treated with an equal volume of absolute ethanol for one hour at room temperature. Then, a loopful of the mixture is streaked onto anaerobic egg yolk agar and/or liver veal-egg yolk agar. The inoculated media are incubated anaerobically at 35°C for 48 hr. On egg yolk agar, *C. botulinum* colonies are white to light yellow, flat and irregular, 1-2 mm in diameter, surrounded by a yellow zone of precipitate about 2-4 mm in diameter. The precipitated zone is irridescent (pearly layer) to oblique light. Several other species of *Clostridium* produce the pearly layer. Several typical colonies are picked from the plates and inoculated into a broth. If type E toxin is suspected, trypticase peptone glucose yeast extract broth with sterile trypsin added (TPGYT) is used, and if type A or B, cooked meat medium is used. The TPGYT is incubated at 26°C and the cooked meat medium at 35°C each for 5 days. The type of organism is then determined by assaying the toxin produced. Trypsin is added to the medium, TPGYT, to inactivate boticin E. This bacteriocin is bacteriolytic for vegetative cells, and bacteriostatic for spores of *C. botulinum* type E, so it interferes with toxin production. It is extremely sensitive to proteolytic enzymes. The trypsin also activates the type E toxin or toxin produced by other nonproteolytic strains of *C. botulinum.* The addition of trypsin was shown to increase the level of toxin produced by *C. botulinum,* thus improving the possibility of detection of the toxin (Lilly *et al.* 1971).

For the enumeration or recovery of *C. botulinum* spores, Wynne agar, supplemented with egg yolk (Hauschild and Hilsheimer 1977) and yeast extract or pork infusion agars (Odlaug and Pflug 1977) have been suggested.

Characterization.—There is a widely held view that no organism isolated from a suspected case of botulism should be identified as *C. botulinum* unless it is toxigenic. Therefore, the outstanding characteristic of a strain of *C. botulinum* is the toxin that it produces.

Once an organism has been isolated and found to be a Gram positive rod that forms spores and is anaerobic, other tests can be made. These include digestion of gelatin, milk and meat; fermentation of various carbohydrates; indole production; hemolysis; and lecithinase and lipase activity. Various biochemical reactions of *C. botulinum* types are listed in Table 6.16, although not all of the literature sources agree on these reactions.

TABLE 6.16

REACTIONS OF *CLOSTRIDIUM BOTULINUM*

	Proteolytic Types A,B,C,D,F	Non-Proteolytic Types B,C,D,E,F	G
Digestion of			
Gelatin	+	+	+
Casein	+	−	+
Meat	+	−	N
Acids produced from			
Glucose	+	+	−
Fructose	+	+	−
Mannose	−	+	−
Lactose	−	−	−
Salicin	V	−	−
Other tests			
Indole	−	−	−
Urease	−	−	−
Nitrates reduced	−	−	−
Lipase	+	+	−
Hemolysis	+	+	+

+ Positive reaction
− No reaction
V Variable reactions
N Data lacking

It is evident from this information that biochemical tests can be considered reliable only to a limited extent in determining or distinguishing *C. botulinum*. This is because different toxin types of *C. botulinum*, or even strains within toxin types, represent quite distinct metabolic groups.

Chromatographic analysis can be used to measure the constituents or metabolic products of the organism (Mayhew and Gorbach 1975; Moss *et al.* 1970; Slifkin and Hercher 1974). This analysis can distinguish the species and some strains within the species of *C. botulinum*.

Serological Reactions.—Besides the toxins, the organism has three sources of antigenic substances: the flagellar antigens (H), the somatic antigens (O) and the spore antigens. The flagellar antigens show a narrow range of specificity, and the somatic antigens are often common to all members of a species or type. The spore antigens are type specific and are dissimilar to either the flagellar or somatic antigens of the vegetative cell (Solomon *et al.* 1969).

The fluorescent antibody (FA) system allows the differentiation of toxigenic types of *C. botulinum*. The direct or indirect FA test can be used. For the direct test, the specific antibody is conjugated with a fluorescent dye. When this conjugate is added to a solution containing homologous antigen, they will react, and the product will be fluorescent in ultraviolet light.

For the indirect method, if the antibody is produced in a rabbit, rabbit serum is injected into another animal, such as a goat, to produce anti-rabbit serum. This goat anti-rabbit serum is conjugated with the fluorescent dye. For the test, the antigen is reacted with the specific antiserum produced in the rabbit. Then the conjugated goat anti-rabbit serum is added, which will react with the antiserum produced in the rabbit.

The advantage of the indirect method is that, as long as the antibodies are produced in a rabbit, only one conjugated antiserum is needed (goat anti-rabbit). This conjugate will work in FA tests for clostridia, staphylococci, salmonellae or any organisms, as long as the specific antiserum to the antigens of the organism are produced in a rabbit.

Lynt *et al.* (1971) found the direct method had certain advantages over the indirect method when testing foods for *C. botulinum*. The direct test required less time to perform, produced less background and nonspecific staining and there were fewer cross reactions with other clostridia.

Toxin Detection.—The Bacteriological Analytical Manual (FDA 1976) described a procedure for detection of preformed toxin in food, and for detection of toxins produced by organisms in the food with an enrichment procedure.

An extract is analyzed for toxin by intraperitoneal injection of mice. The injected mice are observed over a period of 92 hr. Death indicates the presence of toxin.

Injection of crude preparations into mice has resulted in death in both protected and unprotected mice. Examination of the mice revealed that infections by various other organisms were the cause of death. It was suggested that the mice be observed for only 10 hr and the symptoms of botulism should be apparent. With botulism, mice have indrawn flanks and labored breathing, usually within a few hours after injection. These mice then enter a stage of frantic activity just before death. They jump wildly about the cage and then expire in a few seconds.

Elimination of some of the nonspecific toxicities, especially due to infectious organisms, can be accomplished by centrifuging and filtration procedures.

The mouse toxicity test has been accepted as official-first action by the AOAC.

Serological techniques, as decribed for detection of staphylococcal enterotoxin, have been tested for analysis of botulinum toxins. Due to the heterogeneous characteristics of the toxins, and the attachment of hemagglutinins, cross reactions and multiple precipitin bands are observed in gel diffusion tests between type A and B toxins and antitoxins.

No cross reactions have been noted between type A and E, or type B and E. The gel diffusion test is not nearly as sensitive as the mouse test in the detection of botulinum toxin. Mestrandrea (1974) described a capillary tube test that could detect type E toxin at a level of 100 MLD/ml. He suggested the procedure could be used as a rapid screening system.

Cross reactions in the serological tests can be reduced or eliminated by titration of antitoxins in the presence of the heterologous antigens. Reversed passive hemagglutination tests using purified systems have been reported as successful in detecting type A and type B toxins (Sakaguchi *et al.* 1974). The sensitivity of this test compares favorably with the mouse test.

The electroimmunodiffusion technique was used to detect type A toxin (Miller and Anderson 1971). They could detect as little as 14 mouse LD_{50} of type A toxin per 0.1 ml in 2 hr.

A radioimmunoassay procedure using ^{131}I as a label was reported by Boroff and Shu-Chen (1973). They suggested that the assay can be made sufficiently sensitive to detect as little as 100 MLD of type A toxin.

Control of Botulism.—For botulism to occur, the spores or cells of toxigenic *C. botulinum* must be present in the environment and gain access to the food; viable cells or spores must remain in the food after processing; and the food must have an environment favorable for germination and outgrowth of the spores and growth of the vegetative cells to produce the toxin; then the food is eaten cold or with insufficient heating to destroy the toxin.

Therefore, the principal methods of controlling staphylococcal intoxication also are important in the control of botulism. Quite simply, we can prevent contamination of the food, prevent growth and toxin production, destroy the organism or toxin or not eat the food.

Perhaps the best method of controlling botulism is to heat the food during processing to a temperature that will destroy the spores of toxigenic *C. botulinum*. However, there are many foods, such as cured meat products, that would suffer organoleptically if such heat treatments were used. For these foods, additives are needed that will inhibit toxin production by *C. botulinum* contaminants.

Prevent Contamination.—Generally, *C. botulinum* gains access to food as a soil borne or dust borne contaminant. Vegetables are harvested from the soil, or in close proximity to the soil, so they can be contaminated. Washing of vegetables will help remove soil and associated organisms, but it would be difficult to remove all organisms by this method.

Fish and marine animals feeding in waters that contain *C. botulinum* can be contaminated. Most of the contamination is in the intestines of fish. Proper cleaning and eviscerating procedures result in minimal contamination of the fish flesh.

High levels of contamination are more difficult to destroy than are low levels or, ideally, no spores. Therefore, sanitary practices are very important.

Prevent Growth and Toxin Production.—Foods that do not have an environment conducive to the growth and toxin production of *C. botulinum* will not become toxic.

One of the simplest methods to inhibit the growth of microorganisms is to hold the food at temperatures below which growth will occur. Since certain strains of *C. botulinum* can grow and produce toxins at 5°C or less, normal refrigerator temperatures cannot be relied upon to control toxin formation in foods. Freezing does not destroy the toxin or the spores of *C. botulinum,* but as long as the food is frozen, germination and growth of the vegetative cells with toxin production does not occur. Frozen foods should not be allowed to thaw and remain in that condition for extended periods. *C. botulinum* does not grow at an a_w of 0.93 or less. Foods dried below this level should not become toxic (there is the possibility that toxin could be formed prior to drying).

A chemical added to cured meats is sodium nitrite. This chemical is important in color fixation of cured meat, as well as inhibiting growth and toxin formation by *C. botulinum.*

To inhibit growth and toxin production in seafood cocktails, Lerke (1973) suggested the sauce should have a pH of 3.7 or less, and, after preparation, the product should be chilled to at least 10°C. After 24 hr at refrigerator temperature, a sauce having a pH of 3.7 is sufficient to acidify the seafood to a safe level to prevent toxigenesis.

During processing, the destruction of microbial competitors, the removal of air by heating (lowering the redox potential) and the destruction of cellular tissues with the release of cellular fluids, favors the growth of *C. botulinum.*

Destruction of C. botulinum.—Since this organism produces spores, it is more difficult to destroy than is *S. aureus.* The method used to destroy the spores is thermal processing. Radiation, or a combination of heat and radiation will destroy spores of *C. botulinum;* however, at the present time, radiation has not been approved in the USA as a method for food preservation.

Most of the outbreaks of botulism have been caused by consumption of toxic heat-processed food. Thus, it may seem odd that heating is the only presently acceptable method of destroying the spores of *C. botulinum.*

The spores of *C. botulinum* vary in their heat resistance. Type E spores are the least resistant, thus are most easily destroyed by heat. (They are also the least radiation resistant.)

Due to the outbreaks of botulism from eating contaminated fish, it was recommended that during smoking, the fish be held at 82.2°C for 30 min. Pace and Krumbiegel (1973) examined commercially smoked fish and found from 0.9% to 2.0% contained *C. botulinum* type E spores, although the fish had been subjected to 82.2°C for 30 min.

Various results of heat processing are undoubtedly obtained due to the difference in heat resistance of various strains of type E spores. Crisley *et al.* (1968) determined the thermal-death-time of five strains of type E spores in fish paste. The D values in this medium at 80°C varied from an average of 1.6 min to 4.3 min for the 5 strains.

Boiling (100°C) is not recommended to destroy spores of type A or B, since several hours would be required. Processing of low acid foods is accomplished at a temperature of 121.1°C. A 12D thermal process is considered essential for low acid canned foods. With this treatment, the probability of survival of *C. botulinum* spores is very remote.

Of the clostridial spores, those of *C. botulinum* are included in the most radiation resistant group. The spores vary in their resistance with type F generally considered to be the most resistant and type E spores the least resistant, with types A and B intermediate.

Proteinaceous solutions (foods or media) generally protect bacterial spores against radiation injury. Whole kernel corn is one of the most protective substances, followed by beef, pork, chicken, fish, phosphate buffer, green beans and carrots. Saline seems to be unusually effective in protecting spores against radiation.

Fernandez *et al.* (1969) reported that although beef can be effectively sterilized by gamma radiation, the damaged spores may be able to introduce toxin to the product during storage. They stated that the amount of toxin was relatively small and of theoretical interest only.

However, since we do not know the lower limit of acceptable ingestion of *C. botulinum* toxin, can we ignore even small amounts of toxin in a food product?

A combination of radiation and heat has a synergistic effect on destruction of spores of *C. botulinum*. Spores that are exposed to low-level radiation are more susceptible to heat inactivation than untreated spores. The heat is also of value in halting enzyme action and inactivating any toxin that might be present. Thus, pasteurization levels of radiation followed by heating at less than presently required values may be beneficial in controlling *C. botulinum*. Heating before radiation does not lower the radiation resistance, but radiation lowers the heat resistance of the spores.

Inactivation of Toxin.—Radiation pasteurization does not inactivate preformed botulinum toxin, whereas heat pasteurization does. The toxin is more stable at pH 6.5 and below, and is most stable at approximately pH 5.5.

Although the toxin is found to be heat labile, being destroyed in 2 min or less at 90°C, this heat treatment is not recommended for destroying botulinum toxin that might have been formed in canned foods.

It is recommended that, before serving, canned food should be heated to boiling (ca 100°C) and held for 5 min to 15 min (Bryan 1973A).

Canned food which is not normal should not be consumed. In most of the outbreaks of botulism involving canned food, one or more people noted that there was swelling of the can, there were bubbles in the food, off-odors were detected or the liquid was turbid. If the food is suspicious, it should be returned to the store, if purchased, or disposed of in a manner so that it is not consumed by animals or other humans.

Clostridium perfringens Foodborne Illness

This illness has been called a food poisoning, an intoxication, a foodborne illness, an infection and an infective food poisoning. These different designations are undoubtedly due to the fact that the release of the toxin is different from that of *S. aureus* or *C. botulinum*. Large numbers of the organism are associated with the illness. This is true for infections.

The actual number of outbreaks and cases in the USA is not known. Due to the rather mild symptoms and short duration of the illness, usually medical help is not needed. Hence, most cases probably are not reported to the CDC. Table 6.17 shows the number of confirmed outbreaks and cases reported in recent years. The majority of the reported outbreaks are due to food consumed in mass feeding establishments.

Outbreaks that occur in the home might not be reported due to the type of illness and the few people that would be involved.

TABLE 6.17

ILLNESS CAUSED BY *CLOSTRIDIUM PERFRINGENS*

Date	Outbreaks	Cases	Cases/Outbreak
1971	3	106	35
1972	9	973	108
1973	9	1,424	158
1974	15	863	57
1975	16	419	26
1976	6	509	84
TOTAL	58	4,294	74

Data from CDC Annual Summaries

Etiologic Agent.—The etiologic agent for this illness has been termed "*Clostridium perfringens* enterotoxin" (Hauschild 1971). Unlike the enterotoxin of *S. aureus,* or the neurotoxin of *C. botulinum,* which is present in toxic food, the enterotoxin of *C. perfringens* normally is not found in food. Instead, it is produced in vivo (in the intestine) by a sporulating culture of enterotoxigenic strains of *C. perfringens.* With media to induce sporulation, it has been possible to obtain the enterotoxin in vitro. This production of toxin might occur in foods if the organism sporulated. It was suggested that the enterotoxin protein is a structural part of the spore coat that is overproduced in certain strains of *C. perfringens* (Frieben and Duncan 1973).

Properties of the Enterotoxin.—Attempts to demonstrate the toxin in vegetative cells have failed. A direct correlation between sporulation and production of the toxin was demonstrated (Duncan *et al.* 1972). Enterotoxin is not produced in growth media in which sporulation is suppressed. However, in sporulation media, enterotoxin is produced. The enterotoxin is not released by the organisms except upon lysis of the sporulating cells. Synthesis of the spore coat is an early event in sporulation (Labbe and Duncan 1977). Niilo (1977) detected enterotoxin at 4 hr, and it accumulated up to the 10th or 11th hour. With lysis, the enterotoxin and mature spores were released. Not all sporulating cells produced the toxin.

He found no toxin in the mature spores. Frieben and Duncan (1975) reported an enterotoxin-like protein could be extracted from the spores.

The purified toxin is protein and has no nucleic acids, fatty acids, phosphatides or reducing sugars. The toxin yields 18 amino acids with aspartic acid, serine, leucine and glutamic acid being predominant (Hauschild *et al.* 1973). The molecular weight is between 33,000 and 40,000 daltons, with 36,000 daltons being the most probable value. The isoelectric point is pH 4.3 (Hauschild 1971).

The toxin is heat labile, with a D value at 60°C of 4 min (90% of the activity is lost in 4 min at 60°C). The toxin is not dialyzable. It is inactivated by pronase, but not by trypsin, lipase, chymotrypsin or papain (Duncan and Strong 1969A; Hauschild and Hilsheimer 1971).

The toxic factor was stable when stored for 5 days at 37°C, room temperature, 4°C, or −21°C (Duncan and Strong 1969A). Being protein it is an antigen. Only one antigenic type of enterotoxin has been detected, regardless of the producing strain of organism.

Enterotoxigenic Strains.—The toxins produced by *C. perfringens* were reviewed by Hauschild (1971). There are four major exotoxins produced (alpha, beta, epsilon and iota). On the basis of the production of these four toxins, strains of *C. perfringens* are divided into five groups, A to E (Table 6.18). This grouping makes it possible to correlate strains with illnesses of humans and animals.

TABLE 6.18

TYPES OF *CLOSTRIDIUM PERFRINGENS*

Types	Major Toxins Produced
A	alpha
B	alpha, beta, epsilon
C	alpha, beta
D	alpha, epsilon
E	alpha, iota

The main food poisoning strains are included in type A, although type C strains were involved with a severe gastroenteritis in Germany shortly after World War II, and in other countries since then (Hauschild 1973). He stated that the beta toxin was responsible for these outbreaks. Skjelkvale and Duncan (1975) isolated enterotoxin from type C strains and Uemura and Skjelkvale (1976) reportedly isolated enterotoxin from a strain of type D. These enterotoxins are identical serologically, whether from type A, C or D.

Not all strains of type A produce enterotoxin. Even organisms that have been isolated from foodborne outbreaks seem to lose the ability to produce the enterotoxin after transfer in laboratory media. Some organisms do not sporulate, and hence fail to produce the enterotoxin.

Strains of *C. perfringens* that fail to produce fluid accumulation in ligated rabbit ileum or overt diarrhea when injected into the rabbit ileum, do not cause illnesses when fed to human subjects (Strong *et al.* 1971). The results depend upon the conditions used to grow the organisms.

Characteristics of the Illness.—The illness is the result of a sequence of events. The food is contaminated by the organism. During cooking, the vegetative cells and the heat sensitive spores may be killed, but heat resistant spores will survive. The heat experienced during cooking may activate the spores to germinate.

If the food is held at a temperature allowing growth (10°-50°C), the vegetative cells will multiply. A mean germination time of 10-12 min was reported for the organism in cooked poultry at 37°C (Mead 1969). With this short generation time, the organism can increase a thousand-fold in two hours.

The exact number of *C. perfringens* needed to cause the illness has not been determined. However, over 10^6 to 10^8 cells per gram of food can be a potential health hazard. The organism must first pass through the low pH of the stomach to reach the intestines. Organisms in the early lag phase are least resistant, and resistance increases to a maximum at the end of the growth phase. The presence of proteins protects the organism from the acid in the stomach. It is well documented that the acid in the stomach combines with consumed proteins causing an increase in the overall pH. This effect of protein on the stomach acidity, and resultant protection of the bacterial cells, may be the factor in the involvement of protein foods in outbreaks of the illness.

When the remaining organisms reach the small intestine, they find an environment acceptable for multiplication and sporulation. The sporulating cells produce the enterotoxin, which is released during cell lysis. The organisms may sporulate in some foods, but usually, by the time the food is toxic, it also is not palatable.

Incubation Period.—The interval of time between the meal and onset of illness for an outbreak of *C. perfringens* food poisoning is shown in Fig. 6.14. In this example, the range was 2-29 hr, with a median of 13 hr. CDC (1977A) listed the incubation period as 9-15 hr. The sporulation of *C. perfringens* requires 7-12 hr. Incubation times less than this possibly can be explained by partial sporulation and enterotoxin production in the food prior to ingestion, a very high number of cells in the food, with some cells in various stages of the sporulation cycle or a very delicate digestive system of the individual ingesting the food.

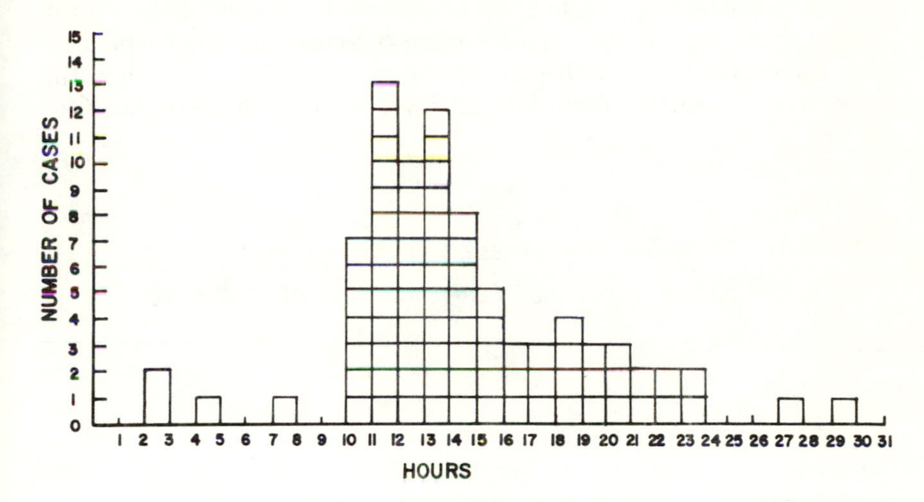

Courtesy of CDC

FIG. 6.14. A TYPICAL OUTBREAK OF *CLOSTRIDIUM PERFRINGENS* FOODBORNE ILNESS, SHOWING THE HOURS FROM INGESTION OF THE FOOD UNTIL THE APPEARANCE OF THE SYMPTOMS

In animal studies, rabbits develop the diarrheal symptom 12-18 hr after intraluminal injection of *C. perfringens* cells (Duncan and Strong 1969B), and the fluid volumes of ligated intestinal loops of rabbits are significantly larger than the fluid volume of control loops 90 min after injection of the toxin (Hauschild *et al.* 1971). It is difficult to compare the response of the rabbit to that of a human. However, these rabbit studies would indicate an incubation time of 13-20 hr.

Attack Rate.—Not everyone who eats the contaminated food becomes ill. The reasons for the escape of some people are similar to those described for staphylococcal intoxication and botulism. People with hyperacid stomachs may be protected, due to the death of *C. perfringens* cells. Previous contact, or any immunity, apparently has nothing to do with acquiring the illness, when contaminated food is ingested.

The disease is not prevented in parenterally immunized animals that possess neutralizing serum antibody against the toxin (Niilo 1971). In a survey of human sera, a majority contained antibody against *C. perfringens* enterotoxin (Uemura *et al.* 1974). They speculated that antibody could be induced by the acute *C. perfringens* food poisoning, or be due to the continuous presence of the organism and small amounts of toxin in asymptomatic people.

Symptoms.—The symptoms reported in four outbreaks of *C. perfringens* illness are listed in Table 6.19. The two prominent symptoms of the illness are diarrhea and abdominal cramps. Nausea and headache were fairly important in two outbreaks, but were of minor occurrence in the others. Fever, vomiting, dizziness and bloody stools may occur, but they are rare.

TABLE 6.19

SYMPTOMS OF *CLOSTRIDIUM PERFRINGENS* FOODBORNE ILLNESS

	Outbreak			
	1	2	3	4
Symptom	Percentage of People Reporting Symptoms			
Diarrhea	82	85	89	91
Abdominal cramps	75	70	86	72
Headache	40	4	—	44
Nausea	33	13	48	42
Fever	8	—	—	8
Bloody stools	7	—	—	1
Vomiting	6	—	16	11
Dizziness	—	2	—	—
Prostration	—	—	—	39
Chills	—	—	—	29

Data from CDC Morbidity and Mortality Weekly Reports

In young people, the symptoms are usually quite mild, but in elderly, ill or debilitated people, the illness can be severe. Although rare, death has occurred in cases complicated by other illness.

Duration and Therapy.—The illness is usually of short duration, from less than 12 hr to up to 24 hr. A few cases may persist for 48 hr. The illness is followed by a complete and uneventful recovery. Although weakness may be evident the day after symptoms begin, a person usually can return to work or normal activity. Due to the rather mild symptoms and short duration, usually no therapy is needed.

Action of the Enterotoxin.—Hauschild (1971) stated that the enterotoxin causes increased capillary permeability, vasodilation and excess fluid movement into the intestinal lumen, resulting in diarrhea. Also, there is an increase in intestinal motility. McDonel and Duncan (1977) reported that ileal loops responded to purified enterotoxin, with net secretion of fluid and sodium, inhibition of chloride and glucose uptake and substantial sloughing of epithelial cells. The action was greatest in the ileum, less in the jejunum and least in the duodenum.

Tests with ligated loops indicate that the toxin acts directly on the intestine, since fluid accumulates only in inoculated loops, and not in adjacent control loops (see Fig. 6.15). There does not seem to be any invasion of the intestinal mucosa by the cells of *C. perfringens,* which is suggestive of a toxic action on the intestine.

Foods Involved.—The foods involved in outbreaks of *C. perfringens* illness are usually protein-type foods that have been boiled, stewed or lightly roasted, or meat and poultry stews, sauces, gravies, pies, casseroles, salads and dressings. Usually the incriminated food is cooked a day or two in advance, refrigerated and then reheated prior to serving. When meat is cooked in bulk, the heat gain is slow, and subsequent cooling is slow. Heating lowers the oxygen content of the food, providing a more anaerobic environment for the clostridia to grow.

The foods involved in 46 outbreaks from 1973-1976 are listed in Table 6.20. All of the foods were protein (meat, poultry or fish). Beans were involved in an outbreak in 1978 (CDC 1978C).

TABLE 6.20

FOODS INVOLVED IN OUTBREAKS OF *CLOSTRIDIUM PERFRINGENS* FOODBORNE ILLNESS, 1973–1976

Food	Number	%
Beef	19	41.3
Pork	2	4.3
Other meat	2	4.3
Poultry	8	17.4
Fishery products	2	4.3
Other, multiple, unknown	13	28.3
TOTAL	46	

Data from CDC Annual Summaries

The Organism.—The cells in this species are anaerobic, Gram positive, sporeforming, straight rods. They occur singly, in pairs and, rarely, in short chains, are not motile, produce capsules and liquify gelatin. This species ferments many of the common carbohydrates, Acid is produced in litmus milk (fermentation of lactose), but the clot is broken up by the production of large amounts of gas, resulting in a stormy fermentation.

C. perfringens requires 13-14 amino acids and 5-6 vitamins for growth. This requirement may play a role in the type of foods associated with foodborne illness, since protein foods usually are involved.

Sources.—*C. perfringens* has been called ubiquitous, due to its widespread distribution in nature. It is found in soil, dust, air, water, sewage, human and animal feces and on many food products.

The presence of *C. perfringens* in the intestines of humans and animals is well established. In a review, Bryan (1969) reported that from 2% to 100% of human feces contain *C. perfringens* type A. These variations are probably due to the population sampled, as well as the method used for detection of the organism. People with poor hygiene, or who partake of communal meals tend to have a higher incidence of *C. perfringens* than do people using good hygienic practices. The level of *C. perfringens* contamination normally is 10^2-10^4 organisms per gram of fecal material. However, Nakagawa and Nishida (1969) reported about ⅓ of the positive samples had 10^3 or less, ⅓ had 10^4-10^5 and ⅓ had 10^6 or more *C. perfringens* per gram from normal human intestines. Yamagishi *et al.* (1976) found some healthy aged adults persistently excreted from 10^7-10^9 *C. perfringens* per gram of feces. As long as humans are intimately involved in food handling, we can expect to encounter foodborne pathogens of human origin.

Animal feces may contain from 10^3 to 10^9 *C. perfringens* per gram. Lozano *et al.* (1970) reported that most of the strains of *C. perfringens* isolated from cows and calves were type C, and only about 10% were type A.

With the widespread distribution of *C. perfringens* in the environment, and the fact that the organism can produce spores which are resistant to adverse conditions, it is logical that the organism should be a common contaminant of foods.

Growth.—Although *C. perfringens* is an anaerobe, it is aerotolerant. Therefore, strict anaerobic conditions are not needed for growth. The redox potentials of meat products are favorable for growth of the microorganism.

Good growth occurs between pH 5.5 and 8.0, and no growth occurs below pH 5.0 or above pH 9.0. At pH 5.5, vegetative growth occurred, but not sporulation and enterotoxin production (Labbe and Duncan 1974). They reported the optimum range for enterotoxin production as pH 6.5-7.3.

C. perfringens grows readily at temperatures between 20° and 50°C, with maximum growth between 37° and 47°C. Mead (1969) found no growth at either 15° or 52°C. Reportedly, some strains show limited growth at 15°C. A long lag phase is a characteristic of growth at low temperatures. Sporulation and toxin production were optimum at 37°C (Labbe and Duncan 1974). No sporulation or toxin production was noted at 46°C.

At a_w values less than 0.995, the rate of growth is reduced. The minimum a_w for growth depends upon the solute, temperature, pH and other factors. The absolute minimum a_w for growth appears to be 0.93 (Kang *et al.* 1969). In this case, glycerol was used to lower the a_w.

C. perfringens can grow in concentrations of curing salts (sodium nitrite and nitrate) considerably higher than those used in normal curing operations (Gough and Alford 1965). Ando (1975) reported that with 2% NaCl, a level of 0.05% (500 ppm) $NaNO_2$ was needed to completely prevent outgrowth of germinated spores of *C. perfringens.* Lower concentrations permitted some outgrowth. Certain ions (Ca, Mg, Fe, Na, K) are needed in trace amounts for growth of *C. perfringens.*

Methodology.—The clinical and epidemiological pattern of *C. perfringens* foodborne illness is so characteristic it almost is diagnostic. However, it is necessary to analyze appropriate samples in order to confirm the presence of the organism. Since the enterotoxin is released during sporulation, and sporulation usually occurs in vivo, the analysis of food samples for enterotoxin is of little value. Although alpha toxin is apparently not involved in gastroenteritis, it is produced by the organism, and is more stable than the vegetative cells. Thus, the presence of alpha toxin in a food has been related to the presence of the organism.

To help establish *C. perfringens* as the cause of a foodborne outbreak, three laboratory tests are suggested: a) demonstration of greater than 10^6 *C. perfringens* per gram in the implicated food and/or in the stools of affected individuals; b) demonstration of the same serotype of isolates from the stools of ill individuals and from the implicated food; and c) demonstration of elevated levels of alpha toxin in the implicated food.

Detection and Enumeration of C. perfringens.—Food should always be analyzed promptly. This is especially true for *C. perfringens* detection. Refrigeration of the food may result in a 90% reduction, and freezing in loss of 99% of the vegetative cells of this microorganism.

Another problem arises when the sample is blended with diluent. Although the organism is aerotolerant, during high speed blending, considerable oxygen becomes mixed into the sample and is harmful to the microorganism. The recommended method is to macerate the sample in the diluent in a mortar with sterile sand. Blending with the diluent for one minute at low speed is an alternative.

For the enumeration of *C. perfringens,* tryptose sulfite cycloserine (TSC) agar without egg yolk (Harmon *et al.* 1971) is recommended (FDA 1976). The medium is differential due to the presence of sodium bisulfite and ferric ammonium citrate. Bacteria that reduce sulfite to sulfide produce black colonies due to the formation of iron sulfide. Since many organisms can reduce sulfite to sulfide, D-cycloserine is added to select for clostridia.

Confirmational Tests.—*C. perfringens* is the only clostridial species that reduces sulfite, is nonmotile, reduces nitrate and produces a stormy fermentation in milk (Buchanan and Gibbons 1974). *C. perfringens* is provisionally identified as a nonmotile, Gram positive bacillus which produces black colonies on TSC agar, reduces nitrates to nitrites, produces acid and gas from lactose and liquifies gelatin in 48 hr (FDA 1976).

Nitrate reduction and motility are determined by stab inoculation of nitrate motility medium. It is recognized that all strains do not reduce nitrate in nitrate motility media, but when the medium is modified by adding 0.5% of both galactose and glycerol, then good nitrate reduction is obtained by nearly all strains (Hauschild and Hilsheimer 1974).

Lactose gelatin medium can be used to determine lactose fermentation and gelatin liquefaction. Salicin and raffinose fermentations also are suggested (FDA 1976).

These tests have been adopted by the AOAC as official-first action.

Serology.—Not all of the strains are typable with the presently available antisera. Using 57 antisera, Hughes *et al.* (1976) could type 65% of the strains isolated from outbreaks of foodborne illness.

The serological reaction is determined by the slide agglutination test. A drop of formalinized suspension of a pure culture is mixed with a drop of pooled or specific antisera on a slide. When the cells and sera are homologous, clumping occurs.

Although the organism is common, it is rare to find the same serotype in a significant number of people, except if they have been involved in the same foodborne outbreak. Even in an outbreak, the feces of patients may contain more than one serotype, due to the presence of normal resident and transient strains, as well as those involved in the outbreak. Therefore, several cultures from fecal samples must be tested serologically to correlate them with the serotype isolated from the suspect food.

If a homologous serotype is isolated from a food handler, one cannot assume that the person is a carrier or source. Rather than being the source, the person might be a victim.

Bacteriocins.—The specificity of bacteriocins can be used to type microorganisms. Mahony (1974) proposed a system for bacteriocin typing of *C. perfringens.* Although several bacteriocins were produced by strains of the organism, 10 of these were selected for the typing procedure. With these 10 bacteriocins, 274 cultures of *C. perfringens* were divided into 50 types on the basis of bacteriocin sensitivity.

As with any scheme, some adjustments may need to be made, especially since cross reactions between bacteriocins from type A, C and D were observed. However, it does present another means of distinguishing strains of the organism. Mahony (1974) found the method to be simple and, with few exceptions, the results are easy to read.

Toxin Detection—Enterotoxin.—Although the determination of enterotoxin may not be important epidemiologically, it is important when studying the role of toxin in pathogenesis, and the mechanisms of enterotoxic action. Various systems have been used for the assay of the enterotoxin.

Ligated intestinal loops have been used to assay for the enterotoxin (Hauschild *et al.* 1971; Niilo 1974). When toxin is injected into a ligated loop, it causes a fluid accumulation (Fig. 6.15). Measurement of this fluid, as compared to control loops, is an indication of the amount of toxin that is injected. Since it is inconvenient to assay toxin in this manner, other methods have been devised.

Intradermal injection of the enterotoxin into rabbits and guinea-pigs

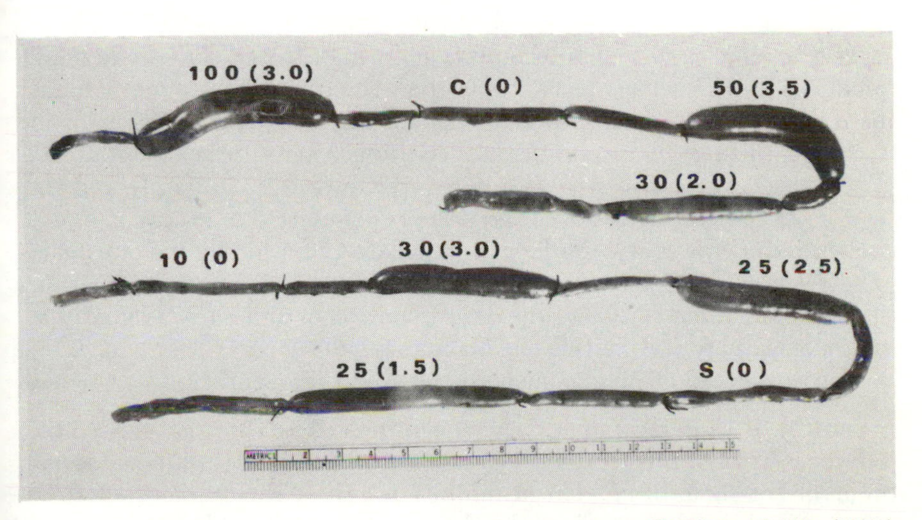

Courtesy of Niilo (1974)

FIG. 6.15. REACTIONS IN LIGATED INTESTINAL LOOPS OF TWO CHICKENS

Symbols: micrograms of enterotoxin injected per fluid volume in milliliters (in brackets); (C) enterotoxin-free cell extract; (S) saline.

causes erythema around the injection site in 1-2 hr, and reaches a maximum in 18-24 hr. The diameter of this reddened area is related to the concentration of enterotoxin (Niilo 1975). The reaction is distinctive for the enterotoxins. According to Hauschild (1970), this system is 1,000 times as sensitive as the ligated intestinal loop, as well as being more rapid and accurate.

The gel immunodiffusion methods (described for *S. aureus* enterotoxin) are used for assay of *C. perfringens* enterotoxin. Methods are available for purification of the enterotoxin (Hauschild *et al.* 1973; Sakaguchi *et al.* 1973). When antisera and enterotoxin are reacted, a precipitin line of identity is obtained, regardless of the strain source of the enterotoxin. This indicates the homogeneous nature of the enterotoxin.

The assay methods for enterotoxin were compared by Genigeorgis *et al.* (1973). They reported that reversed passive hemagglutination is by far the most sensitive test, followed by the microslide diffusion, single gel diffusion, electroimmunodiffusion, guinea pig skin test, mouse lethality test and the rabbit ileal loop test. Naik and Duncan (1977) detected as little as 0.2 μg/ml of enterotoxin with a counterimmunoelectrophoresis technique. Both the reverse passive hemagglutination and counterimmunoelectrophoresis tests reportedly are rapid and reliable assay methods (Skjelkvale and Uemura 1977).

Toxin Detection—Alpha Toxin.—In food samples, the vegetative cells of *C. perfringens* lose viability when refrigerated or frozen. The alpha toxin produced by the cells is more stable than the cells. Thus, a method has been developed for estimating the population level of *C. perfringens* in food by utilizing the hemolytic and lecithinase activities of alpha toxin. These tests are described by Harmon and Kautter (1970, 1974), and are useful for samples with the *C. perfringens* population in excess of 10^6/g, which is the case when the food is implicated in a foodborne outbreak. The results of a collaborative study (Harmon and Kautter 1974) indicated that the precision of the determination of alpha toxin makes the method useful for estimating the maxium population of *C. perfringens* in a food suspected of involvement in a foodborne outbreak.

Control of C. perfringens Food Poisoning.—The three general methods for control of this foodborne illness are: a) limit or prevent contamination of the food, b) prevent or inhibit growth of the organism, and/or c) destroy the organism. Since the *C. perfringens* enterotoxin is produced in vivo, systems to prevent toxin production or to destroy the toxin are not applicable.

Limit or Prevent Contamination.—Due to the ubiquitous character of the organism, preventing contamination of food cannot be relied upon as a means of control. However, this does not mean that we should ignore good sanitary practices. If these are followed, the amount of contamination can be limited. If conditions exist for growth, a larger original number will reach the hazardous level of 10^6-10^8 organisms per g of food sooner than if the original contamination is low.

Prevent Growth.—The presence of a few *C. perfringens* cells in a food product has not been shown to cause foodborne illness. The prevention of germination of spores, or multiplication of the vegetative cells is the only practical method to control *C. perfringens* food poisoning.

One simple method of control is to cook and serve the food without an extended holding period. If this is not possible, refrigeration of the food in small quantities for quick cooling will inhibit growth. The minimum temperature of growth is about 15°C. Any acceptable refrigerator should be able to maintain this temperature, unless it is overloaded. The lower the temperature of the refrigerator, the sooner the food will be cooled. As the temperature approaches 0°C, the population tends to decline, since the vegetative cells are not stable at such low temperatures. Even the spores appear to be damaged at low refrigerator or freezer temperatures.

The maximum temperature for growth is 50°-52°C. Above 52°C, no growth should occur, and the vegetative cells may die. There was greater than a 99% reduction in the number of *C. perfringens* on cooked beef cubes when held at 53.3°C (Brown and Twedt 1972).

Destroy the Organism.—Vegetative cells of *C. perfringens* are destroyed by thorough cooking, but heat resistant spores can survive. Subjecting the spores to sublethal temperatures stimulates germination. The heat resistance of the spores varies from strain to strain; hence, it is not possible to state that heating to a certain temperature is required, since the killing effect is dependent upon the strain of organism that is present.

If heat resistant spores are present, it is not possible to heat foods sufficiently to inactivate all of them without damaging the organoleptic properties of the food. One should assume that there are surviving spores in the cooked food, and take the necessary precautions by keeping the food hot or cooling it below 15°C.

If any food is cooked one day for serving the next, and even if it is refrigerated, it should be reheated prior to serving to kill the vegetative cells resulting from germination of spores and possible further multiplication. To be safe, foods, such as gravies, should be reheated by boiling for 10-15 min, and roasts or poultry carcasses should be reheated to a temperature of 80°C to assure destruction of the vegetative cells.

A sudden reduction of temperature (cold shock) kills the cells. When cells in the exponential growth phase were cold shocked by suddenly lowering the temperature from 37° to 4°C, the death rate was 96%. Another 3.8% were killed during a holding period of 90 min at 4°C, for a total kill of 99.8% of the cells (Traci and Duncan 1974).

Gamma radiation will destroy spores of *C. perfringens.* Clifford and Anellis (1975) calculated 12D values for eight strains of *C. perfringens* spores. These values, divided by 12, revealed D values ranging from 0.13 to 0.35 Mrad. Radiation is not permitted in the USA for the destruction of microorganisms in foods.

Salmonellosis

All members of the genus *Salmonella* are potentially pathogenic for humans as well as vertebrate animals. The transmission of the disease is usually from animals to humans by the ingestion of food of animal origin. Also, there is direct transmission from human to human, from human to animal and from animal to human (See Fig. 6.16). Diseases or infections naturally transmitted between vertebrate animals and humans are called zoonoses.

The illness caused by salmonellae can be divided into four syndromes which may occur individually, simultaneously or consecutively in the course of an infection. These syndromes are the carrier state (convalescent or asymptomatic), enteric fever (typhoid or paratyphoid fever), gastroenteritis (food infection) and septicemia (characterized by a brief febrile illness or a prolonged or relapsing illness with localized lesions).

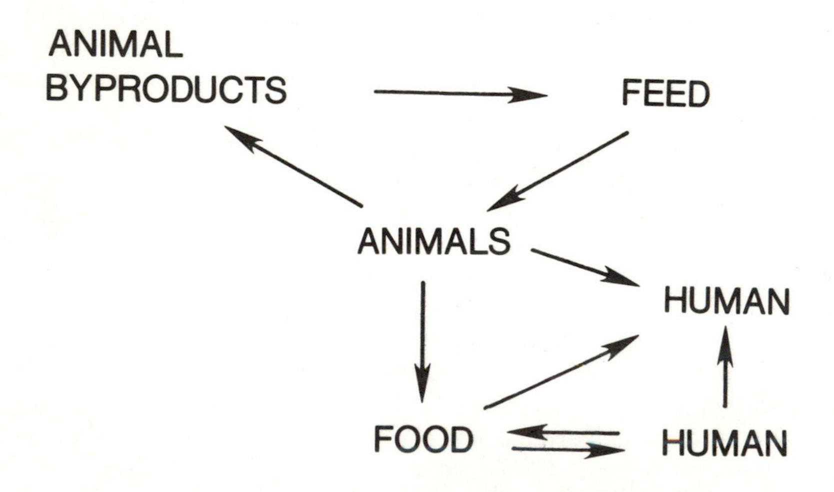

FIG. 6.16. CYCLES OF INFECTION FOR SALMONELLOSIS

Acute gastroenteritis, caused by nearly all serotypes of *Salmonella*, is the most frequent syndrome encountered. It is of primary importance to food microbiologists.

According to CDC (1977A), salmonellosis accounted for 21.2% of the outbreaks and 32.7% of the cases of foodborne illness in 1976. No other foodborne illness surpassed these percentages. The reported number of isolations of *Salmonella* per month from humans from 1966 to 1973 is shown in Fig. 6.17. There are more isolations during warm than cold ambient temperatures. It appears that there is a continual increase in isolations each year.

The increase in the number of cases has been attributed to better diagnosis, laboratory procedures and reporting. However, in those areas in which diagnosis and reporting have always been adequate, increases have occurred in the number of outbreaks and cases of salmonellosis. The prevalence of salmonellosis is increasing, even with all the efforts for controlling the organisms. The increase in *Salmonella* isolations from both human and animal sources is associated with national and international commerce in animals, feeds and foods; with large scale, intensive animal raising; and with increased use of ready-to-serve or heat-and-eat foods.

It has been estimated that only 1 to 10% of the actual number of cases is reported. There may be over 2,000,000 cases of salmonellosis each year in the USA. Many cases are mild and no professional care is solicited. Perhaps salmonellae are the major cause of bacterial gastroenteritis and foodborne illness in developed countries.

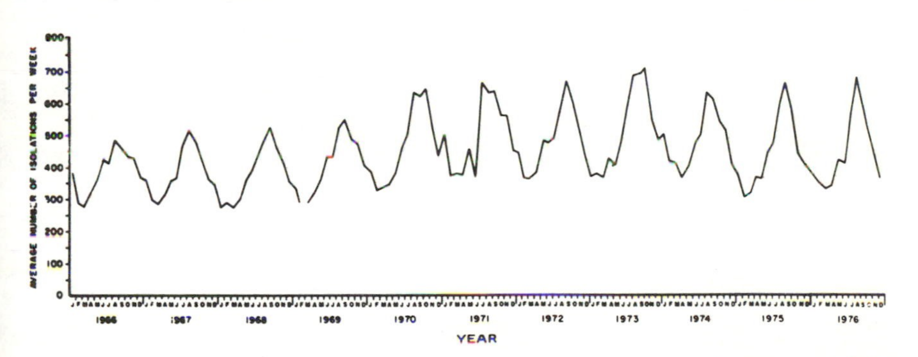

Courtesy of CDC

FIG. 6.17. THE REPORTED NUMBER OF ISOLATIONS OF SALMONELLAE FROM HUMANS, BY MONTH, 1966–1976

The economic loss due to *Salmonella* was estimated to be at least $300 million annually in the USA (Foster *et al.* 1970). The cost includes medical care, hospitalization, lost time and income through absence from work, death of animals, decreased production of animals, loss or reduced value of contaminated products, testing and control procedures and recall of products from market channels. A more recent estimate (CDC 1978B) is that the annual cost for salmonellosis is $1.5 billion for medical expenses alone.

Almost as many deaths are caused by salmonellosis as by botulism (Table 6.7). The death to case ratio is rather low, with a reported case fatality ratio of 0.26% from 1962 to 1968 (CDC 1971) and 0.15% for 1974 (CDC 1975B). Including typhoid and paratyphoid fevers, more deaths occur due to salmonellae than to *C. botulinum*. Most deaths occur in the very young or old people.

Characteristics of the Illness.—Salmonellosis has been considered to be an infection caused by the action of the organism in the intestine. However, there is evidence that a toxin or toxins may be involved.

Incubation Period.—In 33 random outbreaks reported by the CDC, the range for the incubation period varied from 1 hr to 168 hr. The usual incubation period is reported as 6-48 hr (CDC 1977A).

Symptoms.—The symptoms of nine random outbreaks reported by CDC are listed in Table 6.21. The symptoms and their severity depend on the number of organisms and the serotype of *Salmonella,* as well as the resistance of the host. The most reported symptom is diarrhea, followed by abdominal cramps, fever, nausea, vomiting, chills and headache. Dizziness and muscle aches were each reported in other outbreaks, but they were at a low rate.

TABLE 6.21

SYMPTOMS OF SALMONELLOSIS:
PERCENTAGE OF PEOPLE EXPERIENCING SYMPTOMS IN RANDOM OUTBREAKS

Symptom	Outbreak 1	2	3	4	5	6	7	8	9
Diarrhea	87	100	100	96	93	75	93	96	95
Diarrhea (bloody)	4	—*	—	—	5	—	5	—	—
Abdominal cramps	70	47	70	81	86	82	76	66	57
Fever	68	82	—	85	62	80	48	97	43
Nausea	53	69	80	62	69	70	52	—	38
Vomiting	53	82	80	40	40	62	26	54	19
Chills	38	54	70	79	—	—	52	86	—
Headache	36	66	60	—	65	—	63	—	29
Dizziness	—	—	—	—	42	—	—	—	—
Muscle aches	—	—	70	—	—	—	—	95	—

*Not reported; however, the symptom might have existed.
Data from CDC Morbidity and Mortality Weekly Reports

Attack Rate.—The attack rate of a food contaminated with *Salmonella* may be rather low, or it may reach 100% of the participants. Part of this is presumably due to individual resistance. Included in the resistance factor (besides illness or age) are genetic traits of the individual and partial immunity due to previous exposure to the particular serotype or closely related serotype. Unequal distribution of the organisms in the food may play a role in the attack rate.

The number of *Salmonella* apparently influences the attack rate. In an outbreak involving 18 people, reported by Janeway *et al.* (1971), over a billion *S. enteritidis* per gram were found in left-over turkey stuffing. This outbreak had an attack rate of 95%, and a case-fatality rate of 14%. Although 95% had symptoms, *S. enteritidis* was isolated from 100% of the people attending the dinner.

Duration.—In normal, healthy adults, the gastroenteritis usually lasts for only 2 to 3 days, but if further infection occurs, the illness may persist for months or years, and it may result in death. Thomas and Mogford (1970) stated that the duration of infection was longer than two months in nearly 25% of the cases, and intermittent excretion of the organisms was observed in 20% of the cases. Children are more susceptible than adults to prolonged excretion of the organism.

Therapy.—The majority of the cases need no therapy. When hospitalization is required, an attempt is made to stabilize loss of fluids and prevent dehydration, as well as to maintain the electrolyte balance.

The administration of broad-spectrum antibiotics usually gives no appreciable benefit. The salmonellae become resistant to antibiotics. Attempts to hasten clearance of *Salmonella* with antibiotic treatment may be counterproductive, according to a review by Baine *et al.* (1973). Treatment with antibiotics can cause the asymptomatic carrier state to become an active case. If the organism is resistant to the prescribed drug, it may aggravate the illness, and it may prolong the carrier state. Smith and Badley (1971) found no statistically significant benefit from treatment with antibiotics in relation to duration and severity of symptoms or duration of the carrier state of patients.

Foods Involved.—Various foods have been the vehicle for transmission of salmonellae. Most of these foods are of animal origin or contaminated by foods of animal origin, as shown in Table 6.22.

The involvement of meat and meat products as vehicles of *Salmonella* appears to have increased in the last 10 years, and accounted for almost 50% of the known foods from 1972 to 1976. Eggs and egg products were an important source of salmonellae, but pasteurization of egg products has eliminated this food as a vehicle. Ice cream was, in most cases, contaminated by infected shell eggs. Raw milk has been the carrier in outbreaks throughout the world. One outbreak in the USA was due to ingestion of pasteurized milk.

TABLE 6.22

FOODS INVOLVED IN SALMONELLOSIS, 1972–1976
NUMBER OF OUTBREAKS

Food	1972	1973	Year 1974	1975	1976	Total	%
Beef	6	7	1	4	4	22	12.9
Pork	3	–	1	2	–	6	3.5
Other meat	–	–	1	3	–	4	2.4
Poultry	3	3	2	3	2	13	7.6
Eggs	1	1	–	–	–	2	1.2
Fishery products	2	1	1	–	–	4	2.4
Dairy products	5	5	7	4	2	23	13.5
Bakery products	2	3	–	2	–	7	4.1
Salads	–	1	3	3	1	8	4.7
Other, unknown	14	12	19	17	19	81	47.6
TOTAL	36	33	35	38	28	170	

Source: CDC Annual Reports

Strawberries were contaminated by using animal waste as fertilizer for the strawberry patch. *S. eastbourne* was isolated from the environment in which cocoa beans were processed (CDC 1974B), apparently causing the contamination of chocolate candy which was involved in a widespread outbreak. Plant products generally are not vehicles of salmonellae unless there is cross contamination by animal products or by humans.

During 1977, there were several outbreaks and over 180 cases of salmonellosis related to the consumption of precooked roast beef. Seven serotypes of salmonellae were associated with one outbreak (CDC 1977B). Outbreaks due to precooked roast beef also occurred in 1975 and 1976. These outbreaks have stimulated action by the USDA to help assure the safety of this product by requiring the heating of beef roasts to at least 63°C.

Etiologic Agent.—The mechanism by which the salmonellae cause gastroenteritis is not fully explained. Perhaps more than one system is involved. Usually high numbers of salmonellae are needed to cause gastroenteritis. The heating of food to destroy salmonellosis results in a safe product. Hence, it was thought that salmonellosis is an infection. However, heating also could inactivate heat labile toxins.

The number of ingested *Salmonella* needed to cause an infection varies with the strain of the organism and the characteristics of the individual ingesting the organisms. Healthy, adult males can ingest 10^5 to 10^7 cells (depending upon the strain of the organism) before symptoms of the illness occur. However, human to human transmission in hospitals indicates that only a few organisms may be needed to cause the illness. The very young, the aged, debilitated, undernourished and individuals with other illness are more susceptible than normal healthy individuals to infection by *Salmonella*.

The pH of the gastric juice can have a marked effect on salmonellae (Gray and Trueman 1971; Giannella *et al.* 1971). Healthy gastric juice with low pH (1-2) kills small numbers of salmonellae, but when it is deficient (pH 3.0 or higher), an otherwise harmless number of cells may cause serious illness. The destruction of organisms in the stomach accounts for the lower number of infectious agents needed to cause illness by the respiratory, intraperitoneal or intravenous routes.

When salmonellae reach the small intestine, they can live and multiply. The extent of multiplication depends on factors such as the peristaltic rate, if they can become attached to the mucosa, the composition of the intestinal flora, the ingestion of various therapeutic agents (such as antibiotics), exposure to radiation and iron deficiency diseases.

To explain salmonellosis, Bryan (1968) simply stated that salmonellae grow and cause inflammation in the small intestine, resulting in the gastroenteritis syndrome.

Mice given a sublethal dose of *S. enteritidis,* eliminated 90% of the organism in the first 6 hr (Collins and Carter 1974). By this time, small numbers of the organism had established themselves within the wall of the small intestine, and some were detected in the mesenteric lymph nodes. Salmonellae have been seen within the intestinal epithelial cells. This penetration and invasion may be influenced by the lipopolysaccharide units of the bacterial cell envelope. A review by Smith (1977) inferred that an extracellular product is involved, since the microvilli degenerate before being touched by the bacteria. The mechanism of penetration and passage of salmonellae from the gut lumen into the intestinal epithelium was discussed by Takeuchi (1975).

In the past it was believed that the endotoxins associated with the cell wall were the agents responsible for the illness. It is not known with certainty that ingestion of endotoxin produces any effect in man, and the bulk of evidence indicates that it does not. Emody *et al.* (1974) fed endotoxin extracted from a strain of *Salmonella* to healthy adult humans with no adverse effects.

Although penetration of the epithelium apparently is essential for occurrence of the syndrome (Fromm *et al.* 1974), it is not sufficient, by itself, to cause overt disease (Formal and Gemski 1976).

A study of ion transport during *Salmonella* gastroenteritis was made by Fromm *et al.* (1974). Their data suggested that stimulation of active chloride ion secretion combined with inhibition of spontaneous sodium ion absorption may account for the diarrhea of *Salmonella* enteritis. Giannella *et al.* (1975) suggested that *S. typhimurium* does not secrete an enterotoxin, but causes fluid secretion by altering the sodium and chloride transport mechanisms. When infected, the mucosa showed an increase in adenylate cyclase activity and in cyclic adenosine monophosphate concentration. They suggested the stimulation of adenyl cyclase causes ileal secretion. They stated the mechanism of salmonellae activation of adenyl cyclase is unclear, but is different from the action of cholera toxin.

Only 4 of 13 strains of *Salmonella* tested produced a positive reaction in the rabbit ligated gut loop, when living organisms were used (Sakazaki *et al.* 1974B). When culture filtrates were tested in the ligated loops, 11 of 13 strains of *Salmonella* gave a positive reaction. It was their contention that the culture filtrates contained enterotoxic activities.

Koupal and Deibel (1975) reported isolation of an enterotoxic factor from *S. enteritidis*. It was protein associated with the cell wall, or outer membrane of the organism, and difficult to separate from other cell wall constituents. The enterotoxin maintained its activity when heated for 30 min at 70°C, but was inactivated at 80°C. As is typical of enterotoxins, it was inactivated by the enzyme pronase, but not by trypsin, lysozyme, amylase or phospholipases A, C or D. They found that the enterotoxin activity was associated with, but was not, endotoxin per se.

The rabbit ileal loop model was used to detect enterotoxin produced by salmonellae (Sedlock and Deibel 1978). The serotypes that were tested varied in their ability to produce enterotoxin. Stock cultures maintained this trait.

Giannella *et al.* (1977) reported that indomethocin (a non-steroidal anti-inflammatory agent) if given prior to infection, caused a change in fluid from secretion to absorption. The mechanism for this was unknown.

A factor isolated from *S. typhimurium*, caused Chinese hamster ovary cells to elongate (Sandefur and Peterson 1977). This action is indicative of an enterotoxin. They found this effect was blocked by antisera for cholera toxin, and the B fragment of cholera toxin.

There have been many suggestions concerning the action of salmonellae in the GI tract to cause enteritis. Endotoxin is not the causative agent, although it may play a role in the ability of the organisms to penetrate the intestinal epithelium cells. It seems likely that an enterotoxin is involved in the syndrome. Perhaps it is produced in response to a plasmid that can be transferred.

The Organisms.—The organisms in this genus are separated on the basis of their somatic (O) and flagellar (H) antigens. The antigenic classification of *Salmonella* is known as the Kauffmann-White scheme. New species or serotypes are continually being found, and added to the many already classified. In 1964, there were about 900 known serotypes and, by 1974, over 1,700 serotypes.

Even though there are more than 1,700 known serotypes, only 50 or so cause most of the outbreaks and cases of salmonellosis. Each year, some 10 serotypes account for over 60% of the isolations of *Salmonella*. The 10 most prominent serotypes isolated from humans, for the years 1971-1975, are listed in Table 6.23. *S. typhimurium* is the most frequently isolated and reported serotype from both human and nonhuman sources (Table 6.24). *S. typhimurium* is also the most frequently reported serotype in many other countries.

TABLE 6.23

THE MOST FREQUENTLY REPORTED *SALMONELLA* SEROTYPES FROM HUMAN SOURCES (1971-1975)

	Ranking for the Year (Top 10)				
Serotype	1975	1974	1973	1972	1971
*S. typhimurium**	1	1	1	1	1
S. newport	2	2	2	2	3
S. enteritidis	3	3	3	3	2
S. heidelberg	4	5	6	5	4
S. agona	5	6	7	-	-
S. infantis	6	4	4	4	5
S. saint-paul	7	7	5	6	6
S. typhi	8	8	8	-	10
S. oranienburg	9	10	-	9	-
S. javiana	10	-	10	10	-
S. derby	-	9	9	8	-
S. thompson	-	-	-	7	7
S. blockley	-	-	-	-	8
S. java	-	-	-	-	9

*Includes *S. typhimurium* var. *copenhagen*

Growth Factors.—Most strains of salmonellae can grow in a simple medium consisting of ammonium nitrogen, mineral salts and glucose. A few strains need essential growth factors, primarily vitamins. The minimum water activity for growth is between 0.93 and 0.96. The salt concentration needed to inhibit growth of salmonellae depends on the temperature and other factors (Alford and Palumbo 1969). In ground pork, at pH 5.0, stored at 10°C, a salmonellae grew with 3.5% salt, but did not grow at 5% salt. At 20°C or 30°C, salmonellae grew at the 5% salt level. With ground pork at pH 6.5, growth occurred with 8% salt added at 20°C, or 30°C, but no growth occurred at 10°C.

TABLE 6.24

THE MOST FREQUENTLY REPORTED *SALMONELLA* SEROTYPES FROM NONHUMAN SOURCES (1969–1973)

Serotype	Ranking for the Year (Top 10) 1973	1972	1971	1970	1969
*S. typhimurium**	1	1	1	1	1
S. senftenberg	2	3	8	9	9
S. newport	3	5	9	-	-
S. oranienberg	4	2	-	-	-
S. litchfield	5	-	-	-	-
S. infantis	6	-	5	6	8
S. saint-paul	7	4	3	5	5
S. anatum	8	6	4	2	4
S. heidelberg	9	10	2	4	2
S. montevideo	10	7	6	7	7
S. eimsbuettel	-	8	-	-	-
S. derby	-	9	-	3	10
S. reading	-	-	7	-	-
S. cholerae-suis	-	-	10	-	3
S. worthington	-	-	-	8	-
S. thompson	-	-	-	10	6

*Includes *S. typhimurium* var. *copenhagen*
From CDC Annual Summaries

The minimum pH for growth of salmonellae has been reported as 4.0 to 5.5. The variations in minimum pH values for growth may be partly due to the strains or serotypes tested, and partly due to the acids used to lower the pH of the substrate, as well as to other conditions or factors that might affect the growth of the organisms. The maximum pH for growth is somewhere between 9.0 and 11.0. The optimum pH range is from 6.5-7.5.

Since salmonellae are potential pathogens, they generally are considered to be mesophilic. The temperature range for growth is 5°C to 45°-47°C. The serotypes that grow at 5°C are psychrotrophic. The optimum temperature of growth is 35°-37°C. The minimum temperatures of growth vary due to the interrelationships with other growth factors. In general, the organisms require a temperature of 2°-4°C above the minimum reported in order to grow in food products. Even then, at these low temperatures, the more psychrophilic organisms tend to overgrow the salmonellae.

The interaction of other microorganisms naturally present in the food with salmonellae can result in the inhibition of these potential pathogens. There are numerous reports on the observed inhibition of *Salmonella* by lactic acid bacteria. Gilliland and Speck (1972) believed the antagonistic action was caused by factors in addition to the acidic environment created by the bacterial fermentation.

To overcome the antagonism and inhibition during analysis of food for *Salmonella*, selective media are used that inhibit the non-salmonellae, allowing the salmonellae to grow.

Biochemical Reactions.—The main biochemical reactions of the salmonellae are listed in Table 6.25. Even though a + or − reaction is listed, this does not mean that all of the 1,700 serotypes and the potential strains,

TABLE 6.25

THE MAIN BIOCHEMICAL REACTIONS OF THE SALMONELLAE

Test or Substrate	Subgenus I	Subgenus III
Gas from glucose	+	+
Methyl red test	+	+
Indole production	−	−
Voges-Proskauer test	−	−
H_2S production	+	+
Growth in Simmons citrate	+	+
Urease production	−	−
Gelatin liquefaction	−	(+)
Phenylalanine deamination	−	−
Growth in KCN medium (D)	−	−
Growth in malonate	−	+
Motility	+	+
Reduction of nitrate	+	+
Fermentation of:		
Adonitol	−	−
Dulcitol	d	−
Inositol	d	−
Lactose	−	+ or X
Maltose	+	+
Mannitol	+	+
Salicin	−	−
Sucrose	−	−

+ = prompt, positive; X = late and irregularly positive
(+) = delayed postive; d = different biochemical types
− = negative, ? various

types or variants give this reaction. Subgenus I includes the more prominent salmonellae serotypes, and subgenus III consists of the organisms commonly referred to as the Arizona group, now designated as *S. arizonae.*

One of the main reactions for detection of *Salmonella* is the inability to ferment lactose. About 95-99% of the organisms in subgenus I do not ferment lactose, while over 60% of the *S. arizonae* do ferment lactose.

Motility.—Although the organisms are said to be motile, there are mutants that lose this ability, and two serotypes, *S. pullorum* and *S. gallinarum* are not motile and do not possess flagella.

Genetic Exchange.—There is much information concerning the genetics and genetic exchange between members of the family Enterobacteriaceae. Nutritional, fermentative, drug resistance and antigenic characters have been transduced by phage. Sexual mating (conjugation) also occurs. The sex factor (F) is transferred from *E. coli* to various *Salmonella* strains, and from these to other strains of *Salmonella*. The F factor (named F for fertility) provides the cell with a mechanism for conjugation, allowing it to act as a genetic donor. Plasmids for colicins or drug resistance (R) can be transferred to other related organisms.

Sources.—As a genus, the salmonellae are ubiquitous, being everywhere. They are worldwide and found in or on soil, air, water, sewage, animals, humans, food, feed, processing equipment and some plant products. There are some serotypes that tend to be localized in a region or a country, but with national and international travel and trade, the organisms are easily disseminated.

The natural habitat of the organisms is the intestinal tract of humans and animals. Thus, it is logical that humans, animals and their environments are the primary sources of *Salmonella*.

The reported isolations from nonhuman sources (CDC 1973B) revealed that animal feeds accounted for 32.6%, cattle, swine and their products for 14.9%, and poultry and poultry products for 8.2%. The remaining 44.3% were from various sources.

Few, if any, animals are born with intestinal contamination by *Salmonella.* Just as human infants are highly susceptible to salmonellosis, so are very young animals. Some young animals may survive salmonellosis and become temporary or long-term shedders of the organisms. However, it is more likely that the animal is continually infected by some source of *Salmonella.*

The sources of salmonellae for infection of domestic animals include the parent stock, feed, water, rodents, wildlife, pets, humans, insects, arachnids, soil and vegetation. The salmonellae can spread from animal to animal during production and processing. They spread from animals to humans by direct interactions of man and livestock, wild animals and pets, as well as through poor food handling practices and the consumption of raw or undercooked foods of animal origin.

The relationship of *Salmonella* in feed to human salmonellosis is apparent with *S. agona*. Prior to 1971, this serotype was reported from humans in the USA only six times. It was isolated from imported fishmeal on several occasions in 1970-1972. Since these isolations, it has been found in domestic animals and, since 1973, it has ranked as one of the ten most prevalent salmonellae from humans (Table 6.23).

MacKenzie and Bains (1976) found a significant correlation between *Salmonella* serotypes in feed ingredients and those on poultry carcasses. This suggests the involvement of feed in a chain or cycle of infection.

The role of free-living (wild) animals as a source of *Salmonella* was thought to be quite important. Now it is thought that rodents, birds, and other wild animals may be victims of their environment, rather than the source of the organisms. Even so, these free-living animals can transfer the organism from infected areas to clean areas. In various surveys throughout the world, reptiles are found to be contaminated with salmonellae.

The introduction of *Salmonella* to uninfected farms can occur by purchase of outside animals. When an animal excretes the organism, the environment becomes contaminated and this can serve to infect other animals.

Since the waste materials from infected farms contain salmonellae, the use of this material as fertilizer for crops or pastures can spread the potentially infectious material to contaminate future animals. Rain falling on the farms, feed lots or fields containing contaminated wastes can wash the organism into streams or lakes. Besides the run-off from animal quarters, sewage sludge and effluents from abattoirs and poultry processing plants contain salmonellae. These polluted waters can cause infection of farm animals or wild animals that drink from them.

Various pets are potential sources of salmonellae. They may contaminate food or directly transmit these organisms to humans. Pet turtles are an important source of *Salmonella* for transmission primarily to children. It was estimated by Lamm *et al.* (1972) that there were 280,000 cases of turtle-associated salmonellosis in the USA in 1970 and 1971. The interstate shipment of pet turtles has been banned. Salmonellae were isolated from 21% and 57% of freshwater aquarium frogs and snails, respectively (Bartlett and Trust 1976; Bartlett *et al.* 1977).

The direct transmission of *Salmonella* from animals to man poses a problem in zoos that have children's petting areas or other open areas such as the aviary (Komorowski and Hensley 1973).

Humans may be one of the main reservoirs of *Salmonella*. If not a source, humans can act as the vehicle for transfer of the organisms from one area to another.

Eggs with checked or cracked shells have been the vehicle of *Salmonella* in several outbreaks of salmonellosis. The *Salmonella* infection of shell eggs can carry over into egg products processing plants. The equipment, environment and people can become contaminated and pass the organisms into the liquid egg and other products. Since the liquid eggs are accumulated in large vats, one contaminated egg can contaminate the entire vat of egg product. Although this will dilute the contamination, any faulty handling of the product can allow the organisms to multiply to a level that can cause salmonellosis.

The sources of *Salmonella* at the slaughterhouse or abattoir include the animal brought to the plant, the workers and the plant environment (including the equipment). When animals are stressed, such as by transport to the slaughterhouse, the rate of excretion of salmonellae is increased.

Many surveys have been made of poultry processing operations. It was the contention of Dougherty (1974) that there were only low levels of contamination of poultry received at the processor. However, during processing, this contamination increased so that from 47 to 90% of the final product shipped from the plant contained *Salmonella*. All of the seven samples he obtained at a retail store revealed the presence of *Salmonella.*

In a survey of 25 turkey flocks, the incidence of salmonellae varied from 0 to 72% (McBride *et al.* 1978). The flocks with a high incidence may contribute to contamination of the processing plant. If birds entering the plant are *Salmonella*-negative, generally the plant environment also is *Salmonella*-negative. When salmonellae enter the plant with the live birds, the equipment becomes contaminated.

Surveys of poultry carcasses at retail stores indicate from 0% to over 50% are contaminated with salmonellae. The extent of sampling (swabbing a small area or washing the entire carcass), the location on the carcass that was sampled, and the system for detecting the organisms can influence the results of any survey.

Kampelmacher (1963) reported that 25% of normal fattening pigs arriving at slaughter without any lesions, harbor *Salmonella* in their mesenteric and/or portal lymph nodes and/or feces. Usually, the environment and equipment of the abattoir is contaminated with salmonellae. Thus, contamination of the carcass begins at the time of slaughter, and continues during processing and handling.

Various levels of contamination have been reported for hog carcasses and pork products. The overall average may be expected to range from 30 to 40%.

Most reports claim a very low incidence of salmonellae in beef and beef products. Reportedly, beef is not important as a source of salmonellae, when compared to poultry or pork products. This assessment does not agree with the data of CDC (Table 6.22) in which beef is involved in more outbreaks than poultry and pork combined. This might be due to the tendency to eat beef less well cooked than either pork and poultry. However, this should be considered when comparing the potential health hazard of these products.

Although the cooking of raw meat and poultry will destroy salmonellae, the raw foods are a source for contaminating the food preparer, the kitchen equipment and cross contamination of cooked foods, or other foods that are eaten raw. At the present time, we have no approved method to eliminate salmonellae from raw animal products.

Fish and shellfish are contaminated by water that contains waste products from humans and animals. According to Metcalf *et al.* (1973), the consumption of raw or improperly cooked shellfish harvested from sewage-polluted water is important in the transmission of hepatitis and typhoid fever. They obtained shellfish from 5 collecting stations, and found from 7 to 20% of the samples were contaminated with *Salmonella.*

Mussels were examined by Thomas and Jones (1971). They found several serotypes of *Salmonella* with 40% of the unpurified samples contaminated by these organisms.

Fish are not considered an important source of *Salmonella,* since they do not carry these organisms as indigenous flora, but rather as contaminants which are probably quickly eliminated.

In a review by Janssen (1970), literature is cited that indicates salmonellae actively infect fish in nature. Bottom feeding fish are considered especially prone to *Salmonella* contamination. The review by Morse and Duncan (1974) indicated that infection could be induced in fish, such as catfish.

Fruits and vegetables are rarely involved in outbreaks of salmonellosis. However, vegetables can be contaminated by organisms present in animal wastes, or polluted irrigation water used on crops. Velaudapillai *et al.* (1969) found 1.3% of the vegetables sampled were infected with *Salmonella, Shigella* or enteropathogenic *E. coli. Salmonella* serotypes were found on celery, leeks, spinach, green beans and carrots. Wild olives contained *S. sandiego.* Samples of lettuce (68.3%) and fennel (71.9%) yielded one or more serotypes of salmonellae (Ercolani 1976). Salmonellae were isolated from over 22% of the vegetable samples analyzed by Tamminga *et al.* (1978).

Chocolate has been involved in outbreaks of salmonellosis. The cocoa beans are thought to be contaminated in the producing country, and serve as a source of salmonellae in the processing plant.

Coconut products have been a source of salmonellae. The coconut is believed to become contaminated by salmonellae in the soil. This contamination is transferred to the inner edible portion during processing.

Although the main route of salmonellosis is considered to be contaminated food, there are other routes of infection. In hospital outbreaks, human to human contact is apparent.

Food handlers or workers in animal by-product plants often have a high rate of carriage of *Salmonella*. Although they might not have an active infection, it has been observed that the children of these workers often have a higher rate of gastroenteritis than children of other workers. This indicates that a worker who handles products that contain *Salmonella* can bring the organisms home to the other members of the family.

Virulence.—The various serotypes, as well as strains of the serotypes, vary in their ability to cause gastrointestinal distress. Part of this may be due to host-specific types, and part may be due to previous encounters with a specific organism with resultant partial immunity.

The change in colony morphology of the salmonellae from smooth to rough, with accompanying loss of O agglutinating ability, reduces the virulence of the organisms.

Unfortunately, we do not know how many organisms are needed to cause salmonellosis. The studies that have been done used healthy adults. The young, old and debilitated are more susceptible than healthy adults. Stressed animals show an increased tendency toward infection. It might be assumed that humans in a stressed condition also are more susceptible. Infection of animals by the respiratory route requires fewer organisms than by the oral route.

Cases due to contaminated carmine dye indicate that 15,000 to 30,000 salmonellae will cause a reaction in infants and adults, respectively. Dried milk involved in an outbreak of salmonellosis, contained *Salmonella* at the level of 9/100 g (Julseth and Deibel 1969). In this case, growth could have occurred after reconstitution of the product.

Survival in Nature.—The natural habitat is the intestinal tract of animals. When the organisms are excreted, they are found in fecal material, food and various parts of the environment. To cause salmonellosis, they must be able to survive in these unnatural surroundings.

Salmonellae in slurry (liquid manure) sprayed onto grass, may survive from 18 days (Taylor and Burrows 1971) to 33 weeks (Findlay 1972). Jones *et al.* (1977) found the survival of salmonellae in cattle slurry was influenced by the normal microbial flora. In sterilized slurry, *S. dublin* multiplied and survived for more than 370 days, while in natural slurry, there was no multiplication and shorter survival times. Survival time in a pasture is influenced by grass cover, sunlight, temperature and rainfall. The freeze-thaw cycles, such as expected in the winter, rapidly kill the organisms.

Liu *et al.* (1969) found the survival of salmonellae in meat and bone meal was related to water activity and holding temperature. As water activity and temperature increased from low levels, the survival decreased. At an a_w of 0.82 and temperature of 50°C, a 5 log reduction occurred in 72 hr. However, at 10% moisture (a_w about 0.55) at 4°C, there was essentially no reduction in viable cells of *S. senftenberg* 775W for 20 days. Elevated temperatures, oxygen and unsaturated fatty acids accelerate the death of *S. oranienburg* in fish meal (Lamprecht and Elliot 1974).

When fingertips were contaminated with 500 to 2,000 cells of *S. anatum,* the cells could be recovered 3 hr later. Contamination, followed by washing, eliminated levels of less than 1,000, but when 6,400 were inoculated, they could be recovered 10 min after washing. Even fewer than 100 cells of *S. anatum* inoculated onto fingertips infected samples of corned beef and ham handled 10 min after exposure (Pether and Gilbert 1971). This shows how *Salmonella* can be spread by workers handling contaminated products and then clean products.

Survival in Food.—There is less destruction of inoculated salmonellae during pan drying than during spray drying of egg white. Pan drying is more similar to the natural drying of materials. Once dried, the organisms can survive for long periods at room temperature or below. Salmonellae were isolated from dried eggs after 41 weeks at 20°C, and at 65 weeks when held at 2°C.

Salmonellae can survive up to 11 weeks on vegetables held at 2°-4°C, and 7 weeks at room temperature. In acid foods, survival is relatively short, but if salmonellae survive freezing, such as in certain fruits, they may remain viable for considerable periods. *Salmonella* can survive at least 13 months on frozen poultry held at −21°C. In ice cream, the organisms can survive for several years.

Salmonellae inoculated into cold-pack cheese food survived in excess of 27 weeks at 4.4°C, when no preservative (potassium sorbate) or acids were added (Park *et al.* 1970).

When inoculated into a beef-pork mixture at a level of 10^4 cells per gram, *S. dublin* survived pepperoni processes and persisted after 42-43 days of drying (Smith *et al.* 1975A). *S. typhimurium* was more sensitive than *S. dublin* to the acid condition of Lebanon bologna during processing and aging (Smith *et al.* 1975B).

Salmonellae inoculated into chocolate bars at a level of about 10^6 cells/g were viable after storage for 9 months (Tamminga *et al.* 1976). The reduction was greater in bitter chocolate than in milk chocolate. *S. typhimurium* inoculated at a level of about 1,000/g of chocolate was not detected in 55 g after 15 months (Tamminga *et al.* 1977). *S. eastbourne* survived better than *S. typhimurium.*

In a review, Bryan (1968) summarized the survival of salmonellae.

These organisms can survive in the low range of temperature and relative humidity. According to references cited by him, they can survive in contaminated earth and pasture over 200 days, cloth for 228 days, plastic cover slips for 93 days, sweeper dust for 10 months, rodent feces for 148 days, roach pellets for 199 days, poultry feces for more than 9 days, dried cattle feces for over 1,000 days, on egg shells from 21 to 350 days, in dried whole eggs for over 4 years and in meat salad for 77 days.

Basic Methodology.—It is not the intent of this text to precisely describe the procedures needed to obtain samples, or to detect or enumerate salmonellae in the samples. For the procedures, the reader is referred to FDA (1976), AOAC (1975), APHA (1976) and/or USDA (1971).

We are primarily concerned with the analysis of food samples, swabs from food surfaces and water and environmental swabs, although the analysis of other samples may be required, even in a food microbiology laboratory.

In many foods, the organisms have been subjected to debilitating or sublethal processes, such as freezing, desiccation, extremes of pH, heat, osmotic pressures or curing ingredients during the manufacture or storage of the product. When present, salmonellae usually comprise a very small component of the total population. For most foods, there is no tolerance for *Salmonella*. Even one *Salmonella* detected in a 25 g sample is considered to be adulteration of the food (exceptions are raw chicken and meat). Hence, it is necessary to detect very low levels of these organisms. This small number of salmonellae, especially in comparison to the larger number of other organisms, means that the presence of salmonellae is unlikely to be demonstrated by direct plating on selective media, and that an enrichment procedure in broth is needed.

The basic procedure for the examination of foods for *Salmonella* consists of preenrichment and enrichment in broths, streaking and detection on selective-differential agars and then characterizing typical colonies to confirm that they are salmonellae. Confirmation is by means of biochemical tests, serological typing and, for a few serotypes, phage typing. It is evident that a long time period is needed before results of any kind are obtained. Hence, many attempts have been made to develop rapid procedures. Besides the basic procedure, some of these rapid methods are discussed.

Preenrichment.—The preenrichment phase is accomplished by incubating the sample in a non-inhibitory broth (usually in the ratio of one gram of sample to nine of broth) for 18-24 hr at 37°C.

Generally, processed foods that may contain sublethally injured salmonellae are preenriched. Raw or heavily contaminated products are put directly into a selective enrichment broth.

Enrichment.—The enrichment process is the creation of a special environment which permits the growth or selection of the desired species of a group of microorganisms from a mixture of microorganisms with which it is likely to be found. The procedure usually functions by permitting the desired microorganisms to outgrow all others. Ideally, an enrichment culture for salmonellae should allow these organisms to multiply and to inhibit all others.

It has been difficult to develop the ideal enrichment medium. Since there are some 1,700 serotypes of *Salmonella,* any selective agent is likely to be inhibitory to one or more serotypes, or to be not effective in inhibiting the non-salmonellae that are present.

Foods vary in their nutrient content, pH, inhibitory substances, and other factors that may influence the growth of *Salmonella.* Therefore, when a food is added to an enrichment medium, usually in a ratio of 1:9, the characteristics of the medium are changed. Each food will alter the medium in its own peculiar way. Therefore, to be effective, a different enrichment medium should be designed for each food.

The enrichment broths commonly used in the USA are selenite cystine (SC), tetrathionate (T) and tetrathionate brilliant green (TBG).

The enrichment broths are used as secondary enrichment of samples that have been preenriched (usually 1 ml of sample from the preenriched broth to 10 ml of enrichment broth), and as direct enrichment for raw or heavily contaminated food. Most analytical procedures suggest adding 25 g of food sample to 225 ml of broth. For feed samples, 30 g is added to 100 ml of broth. Due to multiple sampling of a food lot to assure the absence of *Salmonella,* the analysis of larger quantities of pooled material from these samples has been suggested.

Some foods have certain peculiarities so that special procedures are needed for better detection of *Salmonella.* When gelatin is added in enrichment media, the viscous suspension that results can interfere with *Salmonella* analysis. Rose (1972) suggested adding gelatinase to the enrichment medium. The enzyme was not inhibitory to salmonellae, and better recovery was made from gelatin artificially inoculated with *S. typhimurium.*

When milk is analyzed, the acid produced causes formation of curds. *Salmonella* trapped inside these clots of milk are not detected. The addition of trypsin aids detection by digesting the protein of the milk, but apparently it does not injure the salmonellae.

The inoculated enrichment broth is incubated at either 37° or 43°C for 18-24 hr and then a loopful is streaked onto an agar surface that is selective and differential.

Isolation.—After incubation of the preenrichment and/or enrichment broths containing the sample, a loopful of the broth is streaked onto an agar surface in a manner that will reveal well isolated colonies. The agars are selective and differential for salmonellae.

In studies comparing isolation agars, it is usually evident that no one agar is acceptable. Many more *Salmonella* positive samples are found when two or three agars are used. The agars generally used in the USA for analysis of food for *Salmonella* are bismuth sulfite (BS), salmonella-shigella (SS) and either brilliant green (BG) or BG sulfa. Hektoen enteric (HE) agar has been recommended for evaluation for possible incorporation into the AOAC method. *Proteus* has a tendency to spread on BG, but not on BG sulfa. The *Salmonella* colonies are generally smaller on BG sulfa than on BG agar.

BG, BG sulfa and SS agars use lactose fermentation as a differential agent, while BS agar uses the production of bismuth sulfide.

It is recommended that BS plates be observed after 24 hr and, if no typical salmonellae colonies are observed, the plates should be incubated for another 24 hr before they are called negative.

With preenrichment for 24 hr, enrichment for 24 hr and BS incubation for 48 hr, a total of 96 hr (4 days) is needed even for samples with no *Salmonella*, before any decision can be made.

The procedure to be used (type of preenrichment medium or no preenrichment, types of enrichment and plating media, as well as incubation temperatures) is dictated somewhat by the type of sample being analyzed, as well as the types of organisms and serotypes of *Salmonella* that might be present.

There is a desire to have standard methods for the analysis of foods for salmonellae that everyone will use. Designing a method is not the difficult part, but getting everyone to use it is not easy. Erdman (1974) reported a collaborative study in which 13 laboratories in 8 countries used a specified method and their own method to analyze ground meat. Best results from selective agars were obtained when the laboratories used their own, indicating that experience or familiarity with a particular method or medium is important. Although a standardized method makes it possible to compare results from different laboratories, it would tend to stifle change.

Biochemical Characterization.—If no typical colonies are found on the isolation medium, it is satisfactory evidence that the sample is *Salmonella*-negative. If typical colonies are detected, this is presumptive evidence that the sample may be *Salmonella*-positive. For further characterization of the organisms in these typical colonies, both biochemical and serological tests are needed. Neither the biochemical nor the serological tests are specific for *Salmonella*, and neither system will detect all of the serotypes of the genus.

The biochemical reactions listed as + or − usually mean that 90% or more of the strains of the organism are either positive or negative. This means there are strains of the organism that will not give a typical + or − reaction. Thus, many different biochemical tests are used in an effort to characterize the *Salmonella* and eliminate the non-salmonellae.

With the number of biochemical tests that need to be run, considerable time and material are required to prepare the media, innoculate, incubate, determine the reactions and classify the cultures. To cut down on the time and cost, rapid methods, composite media and simplified test kits have been developed.

Rapid and easily run methods have been the desire of microbiologists for a long time. One method employed to get a quick reaction is to use a large inoculum in a small amount of medium.

Various commercial systems or test kits for differentiation and identification of the Enterobacteriaceae are available. Commercial preparations have the potential benefit of providing many laboratories with standard materials for use in assessing bacterial reactions. These commercial systems include the API Analytab, Auxotab, Enterotube, Inolex Enteric 20, Minitek, Patho-tec and r/b enteric differential systems.

These systems have certain things in common. They are all miniaturized and rapid tests based primarily on reports in the literature by various workers. None of the systems agrees 100% with the conventional tube method for determining reactions. Most of the comparisons have been done in clinical laboratories, rather than food laboratories. The general opinion seems to be that these tests are acceptable for a small laboratory that analyzes relatively few cultures

At the present time, the genus of Enterobacteriaceae of most concern to food processors is *Salmonella*. Some of these systems have superfluous tests not needed for salmonellae characterization, and they omit other tests that are desirable. Thus, these tests probably are not as applicable for a food microbiology laboratory as a clinical laboratory.

Serology.—Serological considerations are used in the immunization of humans or animals for certain diseases caused by salmonellae, for testing the serum of humans or animals for potential carriers of the organisms and for determining the serotype of an isolated test culture. The serological reactions of the test culture are needed to supplement the information obtained from biochemical tests.

The salmonellae possess O, H and K antigens (Kauffmann 1972). An antigen derived from filamentous appendages called fimbriae, and designated as F antigen was described by Old and Payne (1971).

The K (Kapsel) antigens are either envelope or capsular antigens. So far, three K antigens have been demonstrated in the salmonellae. The K antigens are heat labile.

The O (somatic) antigens are considered to be a constitutive part of the cell wall, but may extend beyond this barrier. They are a complex of lipopolysaccharide (LPS) and protein. The detailed structure of the O antigen varies from strain to strain. This variation forms the basis for the serological differences of the organisms. The O antigens are considered to be heat stable, resisting boiling for 2½ hr.

Salmonella can change in colony morphology from smooth to rough (S→R variation). These rough forms are caused by mutations that block the synthesis of polysaccharides. Thus, they have no O side chains, lose O agglutinating ability and have a reduced virulence.

The H antigens, or flagellar antigens, are found only in motile cultures. They are destroyed at 100°C (heat labile), and by dilute alcohol or acid. Flagella are composed of the protein flagellin, with a molecular weight of about 40,000 daltons. *Salmonella* flagella exhibit a wide variety of antigenic specificities. Many species have two types of flagellar antigen (phase 1 and phase 2). A single bacterium manifests only one of these phases, and its descendants in a culture primarily will be in the same phase, but a minority may be in the other phase apparently by a random mutation-like process.

Phase 1 antigens are specific and are shared by only a few serotypes of salmonellae; while phase 2 antigens are non-specific, and are common to many serotypes.

Usually, only the O and H antigens are determined when typing a culture, although the V_i antigen is sometimes of interest. The cultures are typed according to the Kauffmann-White scheme. This system is outlined by Kauffmann (1972) and Buchanan and Gibbons (1974).

The specific H or O antigens are determined by the genetics of the cell. Due to mutations, or phage conversions, the antigens can be altered (Bartlett *et al.* 1978). This may account for the multiplicity of serotypes and the overlapping patterns of antigenicity.

Usually the specific typing of a culture is not necessary, since any serotype in a food can be potentially pathogenic for humans. However, in cases in which the source of contamination is being sought, or the organism that caused an illness must be determined, complete serotyping is needed.

If there is an agglutination with the O antisera, it is not proof that the culture is a salmonella. There are cross reactions with the *Salmonella* O antisera and non-salmonellae. Some strains of closely related organisms, especially strains of *E. coli* and *Citrobacter, Shigella* and *Enterobacter,* either share certain somatic antigens or have antigens similar to those of salmonellae (Refai and Rohde 1975). Other unrelated microorganisms reportedly react with *Salmonella* O antisera (Aksoycan and Saganak 1977; Corbel 1975).

The H antigens are determined by the tube test. A motile organism is necessary for reaction with H antigens. Since *S. pullorum* and *S. gallinarum* are not motile, they will not react. For the exact procedure for detecting H antigens, the reader is referred to AOAC (1975) or FDA (1976).

The medium in which the organism is grown can influence the reaction with H antiserum (Banwart and Kreitzer 1972; Stamper and Banwart 1974).

Other Methods.—It is evident that the conventional procedure requires considerable material and time in order to detect *Salmonella*. Other tests have been devised in order to obtain results quicker and easier. Usually rapid tests result in a loss of accuracy or precision. However, even with the conventional test, not all of the *Salmonella* serotypes are detected. It is often stated that only 50 serotypes account for over 95% of the isolations of *Salmonella*. It should be admitted that the conventional test, as run by many people in the past, overlooked such things as lactose-fermenters or unusual serotypes. The rapid test should at least compare favorably with the conventional test.

Fluorescent Antibody (FA).—The fluorescent antibody (FA) technique is regarded as an alternative "rapid" procedure to the conventional preenrichment, enrichment and plating procedure. Even with the FA method, the sample is preenriched and/or enriched so that the salmonellae will multiply. The FA system described by Fantasia *et al.* (1975B) requires the transfer of the sample through three broths, and a total of 40-52 hr of incubation before slide smears are made for the FA reaction.

The FA system was adopted as official first action as a method for screening food for salmonellae (Sanders 1975).

Besides the requirement for a good fluorescent (ultraviolet) microscope and well trained personnel, the diagnostic sera used to detect salmonellae by the FA is expensive. The serum should react with all of the serotypes of *Salmonella* and not cross react with the non-salmonellae. At the present time, this type of serum is not available. Since not all *Salmonella* have flagella, a polyvalent mixture of H and O antibodies is needed. These mixtures are commercially available. Thomason and Hebert (1974) evaluated four commercially prepared conjugated salmonellae antisera. The results of 769 food samples, using one polyvalent antiserum, agreed 90% with those using the culturing system. The other antisera yielded poorer results. The authors stated that all four of these antisera stained the homologous strains of *Salmonella*. Cross reactions occurred with *Citrobacter* and *E. coli* strains. Tharrington *et al.* (1978) found a *Lactobacillus plantarum* strain that was FA positive. The bacterial cell possessed an affinity for the nonantigenic region of the IgG molecule in the FA antisera.

In many surveys, the FA procedure reveals more *Salmonella* positive samples than the conventional culture test. Various reasons have been suggested. One proposal is that the FA is more sensitive than culturing. A more logical explanation is that cross reactions with non-salmonellae by the FA antisera result in more apparently *Salmonella* positive samples. Thus, the samples that appear positive with the FA technique must be run through the conventional test for confirmation.

Even more important is the ability or inability of the FA system to detect all of the samples that contain *Salmonella.* Thomason and Hebert (1974) found one swab sample and one meat meal sample that were negative by the FA technique, but positive by the conventional method. This was with nearly 1,000 samples, which could be considered very good. The conventional culture method may not be any better.

Phage Systems.—The use of bacteriophage has been suggested for detection, characterization and the typing of certain serotypes of salmonellae.

To detect *S. typhosa (S. typhi)* in soil samples, Kandelaki (1964) added typhoid indicator phage to the sample. An increase in the phage titer during incubation was evidence that *S. typhi* was present in the sample. The phage titer increase detected more positive samples than did the bacteriological plating system he used.

For determining the character of colonies on selective agars, Cherry *et al.* (1954) suggested using bacteriophages. According to them, the method is rapid, simple and quite specific. They used the O-1 phage of Felix and Callow. This system was reevaluated by Welkos *et al.* (1974). They tested 652 strains of *Salmonella,* and found the phage (O-1) reacted with 640 (98.2%). Of 1,463 non-salmonellae strains, only *E. coli* strains were lysed. Of 239 strains of *E. coli* tested with 10^{12} plaque forming units/ml (PFU/ml), 14 (5.9%) were susceptible to lysis. Gunnarsson *et al.* (1977) found that 98% of 5,287 strains of salmonellae were lysed by O-1 phage. They warned that monophasic strains of subgenus III and strains of subgenus IV usually are not sensitive to the O-1 phage.

Farmer and Sikes (1974) stated that bacteriophage typing is the most useful laboratory method for evaluating a possible epidemiological association among strains of *S. typhi.*

Phage typing schemes have been reported for certain groups or individual serovars. Gershmann (1977) used a set of 50 phages to differentiate 735 salmonellae cultures on the basis of 347 phage patterns.

Motility Systems.—Except for *S. pullorum* and *S. gallinarum,* all serotypes of salmonelae are motile. By using semi-solid media, these motile organisms can be separated from nonmotile types. These systems required special media or apparatus and there was lack of interest by officials in the FDA.

Immunoimmobilization.—A system combining the motility, growth in selective media and reaction with *Salmonella* H antisera, was reported by Mohit *et al.* (1975). They suggested the system could detect salmonellae in fecal samples.

Swaminathan *et al.* (1978) modified the system to detect salmonellae in food. Essentially, the food is enriched for eight hours in selenite cystine. A portion of broth and growth is filtered with a membrane filter. The filter disc is placed on the surface of a selective motility agar. About 6.5 cm from the filter disc, another disc containing polyvalent H antiserum also is placed on the agar surface. As the motile salmonellae migrate through the agar, they contact the diffusing H antibodies and form a line of precipitation in the agar.

Control of Salmonella.—The measures needed to control *Salmonella* and salmonellosis are essentially the same as those needed for the control of *S. aureus* and the clostridia (prevent contamination, prevent growth, destroy the organisms). The number of cases of typhoid fever due to *S. typhi* has been drastically reduced. Proper sewage disposal, protection and chlorination of communal water supplies, restricting the harvesting of aquatic food from contaminated water, pasteurization of certain products, sanitary control of food processing and sales establishments, exclusion of typhoid carriers from food handling occupations and immunization programs, have reduced the incidence of typhoid fever in the USA.

Since the prevalance of salmonellosis is increasing, it is evident that these procedures are not effective in controlling this illness and other measures need to be incorporated.

There is an incentive for the food industry to market products with no salmonellae. Section 402 of the Federal Food, Drug and Cosmetic Act defines a food to be adulterated if it bears or contains any poisonous or deleterious substance which may render it injurious to health, and if it has been prepared, packed or held under unsanitary conditions whereby it may become contaminated with filth or whereby it may be rendered injurious to health. Foods containing *Salmonella* or other pathogens fall within this definition.

The control of salmonellae in foods includes the acquisition of *Salmonella*-free raw materials, with processing, storage and distribution under conditions which prevent the increase of *Salmonella* and, ideally, a terminal treatment of the food to destroy any salmonellae that may be present.

Prevent Contamination.—Due to the ubiquitous nature of the salmonellae, it might seem to be a hopeless task to keep them out of our food. However, it is generally believed that, with a sincere effort by all factions of feed and food production, handling, processing and preparation, the contamination of food by salmonellae can be prevented or reduced.

There are cycles of infection. The major one is animal by-product to feed, feed to animals and animals to food to humans (Fig. 6.16). There are other cycles, such as human to food to human, or even human to human.

It is thought that if the cycle of feed to animal to food to humans could be broken, the incidence of salmonellosis would be reduced significantly. The cycle could be broken by eradicating salmonellae from domestic animals. Surveys of animals at the farm indicate a low level of infection with *Salmonella*. The organisms may not be detected easily, so that the contamination could be greater. Regardless of the present level of infection, or the feasibility of complete elimination of salmonellae from animals, there are many things that could minimize *Salmonella* at the farm level. All of the suggestions cannot be listed here, but briefly, there should be *Salmonella*-free breeding stock, strict sanitary practices, *Salmonella*-free water and feed, segregation of sick animals and sale of only healthy animals. Sanitation includes many facets, such as pest control, cleaning and disinfecting housing, removal and disposal of waste, storing feed so that it is not contaminated, preventing workers from carrying diseases with them from one farm area to another, and keeping out visitors. There are sanitary requirements for dairy farms, but other farming practices are usually left to the farmer.

In order to provide the animals with *Salmonella*-free feed, the renderer of animal by-products must cooperate by maintaining sanitation in the plant. The heat treatment given the product is usually sufficient to kill salmonellae, but the product is recontaminated. By producing a *Salmonella*-free feed supplement, and through the efforts of the farmer, the raw material that the renderer gets may have a lower incidence of salmonellae, so that producing a *Salmonella*-free feed will be easier.

For seafoods, the habitat in which these animals grow must be free of *Salmonella*. The first step is to prevent the access of untreated sewage or water to streams, lakes or oceans.

Shellfish, such as oysters, are believed to cleanse themselves of salmonellae and *E. coli* in 48 to 72 hr if placed in clean water. However, experimentally contaminated oysters retained *S. typhimurium* for at least 49 days (Janssen 1974). This indicates that the depuration process cannot be relied upon, and clean beds are needed for the production and harvesting of shellfish.

The food processor should maintain strict sanitary practices. Prior to the requirement that all processed poultry must be inspected, the incidence of *Salmonella* was usually lower in USDA inspected plants than in uninspected plants. Two reasons could account for this. The plant facilities, equipment and the sanitary requirements for the inspected plants could help lower the incidence. Questionable flocks would more likely be taken to an uninspected plant to reduce the losses due to condemnation.

Thorough daily cleaning and sanitizing of the processing plant and equipment can reduce the dissemination of salmonellae in the environment and on the poultry or meat from these processing plants.

Nickelson *et al.* (1975) found 35% of the commercially prepared frog legs contained *Salmonella.* Aseptically dissected tissue was not contaminated. By using sanitary measures and reducing contamination of the legs with contents of the intestinal tract, the incidence of *Salmonella* was reduced to 5.1%. Although the organisms were not eliminated completely, at least the number of people put in jeopardy by *Salmonella* contaminated food was lowered.

Simple sanitary practices reduced the contamination on hog carcasses by 50% to 75% (Childers *et al.* 1977).

There are many reports stating that regulations (administered by government agencies) are necessary to assure that food is not contaminated by *Salmonella,* or other pathogenic organisms. The laws of man will not eliminate *Salmonella* from natural products. The quality of sanitation in a processing plant is very often an indication of the type of people who work there. It is sad when government regulations are the incentive for people to do a better job.

It was pointed out by Foster (1969) that elimination of salmonellae from foods would not necessarily eliminate human salmonellosis. We also need to consider the human carrier as well as household pets. The human carrier is important in the processing plant, in the food preparation and serving industry, and in the home.

Testing for salmonellae should be part of the control effort. If the organisms are in the plant environment, there is a good chance they can be in the finished product.

In the preparation of food, one of the important factors for spreading *Salmonella* is the cross-contamination from raw to cooked food. Until such time as raw poultry or meat products can be produced with no *Salmonella,* we must assume that the organisms are present. Handling of these products and then handling ready-to-eat products transfers the bacteria from the raw to the ready-to-eat products.

In order to control the transmission of salmonellae by pet turtles, the FDA banned the interstate shipment of these animals. The ban applies to turtles with shells less than four inches across. Turtles for educational or scientific purposes, or exhibition at zoos are not included in this regulation.

Foods have been placed into five categories in terms of hazard to human health (Foster 1971). No safe level of salmonellae has been established, but it is evident that illness is more likely when the number ingested is increased. Three hazard characteristics were established for a particular food. These are: a) the food or an ingredient is a significant source of *Salmonella*; b) there is no control step (heating) in the process to destroy the organism; and c) if the product is mishandled, growth can occur, resulting in an increased hazard. The higher the number of hazard characterisitics of a food, the more potentially dangerous it is. A food with no hazard characterisitics is not considered dangerous, foods with one or two are intermediate and foods with all three are hazardous. The susceptibility of people varies, with infants, the aged and infirm the most susceptible to *Salmonella* infection. With this aspect included, the five categories are established.

Category I contains foods that possess any of the hazard characteristics, and are intended for use by infants, the aged or infirm.
Categories II, III, IV and V include foods for general uses.
Category II includes foods that possess all three hazard characteristics.
Category III includes foods that have two of the hazard characteristics.
Category IV is those foods that have one hazard characteristic.
Category V foods have no hazard characteristics.

With these hazard categories, it is evident that fewer samples of foods in category V need to be tested (if negative results are obtained), than those in category I or II. Considering 25 g samples, it was proposed that the food should be accepted if there are no positive samples in 15 samples (Category III, IV or V), 30 samples (Category II) and 60 samples (Category I) or, if no more than one positive sample is detected in 24 samples (III, IV, V), 48 samples (Category II) or 95 samples (Category I). Thus, the lower the risk category number and the greater the hazard, the greater the number of samples that need to be analyzed and the lower the percentage of positive samples that are allowed.

Prevent Growth.—Although refrigeration at 4°-5°C will control the multiplication of *Salmonella,* it will not improve the hazard characteristic, since the food can be abused in market channels or during preparation and serving. Foods should not be held between 10° and 50°C. These temperatures should be traversed as rapidly as possible during heating and cooling.

The salmonellae do not grow in foods with naturally low pH (below pH 4.0). Acids can be added to some foods or feeds to inhibit growth of the organisms. Although salmonellae might not grow, they may survive in some foods at a low pH.

Destroy the Organisms.—The destruction of the organisms is the best way to make certain that they do not present a health hazard. It is preferable if the destruction can be accomplished in the final package, so that recontamination does not occur during further handling.

Liquid propylene oxide was effective in destroying salmonellae in meat and bone meal (Tompkin and Stozek 1974). The concentration of propylene oxide needed for destruction of salmonellae depends on the relative humidity, level of contamination and time of exposure. There is a residual limit of 300 $\mu g/g$ in certain food ingredients.

Fumigation of feed (fish, meat or bone meal) with formaldehyde gas is an effective means of destroying salmonellae (Duncan and Adams 1972). The gas eliminates salmonellae to a depth of 1.91 cm of feed.

The chlorination of water is used to destroy *S. typhi.* This treatment can aid in the control of salmonellae in food processing plants whether chlorine is used in wash water, equipment rinse water or cooling water.

Heat is the principal method of destroying salmonellae.

The D value of *S. manhattan* at 60°C in chicken a la king was 0.40 min and in custard, 2.44 min. For *S. senftenberg* 775W, the D values in these foods were 9.61 and 11.32 min, respectively (Angelotti *et al.* 1961).

The D values for various *Salmonella* serotypes in whole milk were determined by Read *et al.* (1968). The values ranged from 3.6 to 5.7 sec at 62.8°C, 1.1 to 1.8 sec at 65.6°C, or from 0.28 to 0.52 sec at 68.3°C. *S. senftenberg* 775W is the most heat resistant strain isolated thus far. For this organism, they reported D values of 34.0 sec (65.5°C), 10.0 sec (68.3°C), 1.2 sec (71.7°C) and 0.55 sec (73.9°C), which are much greater than the D values for other serotypes. If we consider a 12D for serotypes other than *S. senftenberg* 775W, 68.4 sec at 62.8°C should be sufficient to destroy salmonellae. Read *et al.* (1968) stated that the present milk pasteurization process (63.33°C for 30 min or 71.67°C for 16 sec) will inactivate salmonellae, as long as the number does not exceed 3×10^{12} salmonellae per ml of milk. It would be extremely unlikely that this level of salmonellae would be attained without spoilage, due to other bacteria.

Pasteurization is used to control salmonellae in liquid egg products. The heat treatment given depends upon the type of product, due to the effect of heat on the functional quality of the product and the effect of the egg product on the heat resistance of the organisms.

Storage of dried albumen at elevated temperatures has effectively eliminated salmonellae from the product. As the moisture content is lowered, increased time or temperature is needed to destroy the organisms.

Since bacteria show increased heat resistance in dry products, it is difficult, but not impossible, to eliminate salmonellae from dry animal feed by a heat treatment. Carroll and Ward (1967) found a combination of 87.8°C for 10 min consistently reduced salmonellae to a nondetectable level in fish meal.

Pasteurization of coconut at 80°C for 8 to 10 min effectively killed salmonellae without affecting the quality of the product (Schaffner *et al.* 1967).

Cooking of food usually destroys the salmonellae that may be present. Baked goods reaching a temperature of 71.1°C or higher in the slowest heating region can be considered safe from salmonellae. *S. senftenberg* 775W is killed during normal processing of frankfurters in the smoke house when the internal temperature of the product reaches 71.1°C (Palumbo *et al.* 1974). Heating pepperoni to 60°C or Lebanon bologna to 51.7°C, eliminated 10^4 inoculated cells of salmonellae (Smith *et al.* 1975 A,B).

Salmonellae are only moderately resistant to radiation treatment. Although the needed radiation dose varies with the type of product, conditions of radiation and level of contamination, Thornley (1963) suggested that 0.5 Mrad would reduce the number of salmonellae by a factor of 10^7. Ostovar *et al.* (1971) found that 0.5 Mrad was sufficient to destroy salmonellae in smoked whitefish. Ley *et al.* (1970) reported that for frozen meat, 0.6 Mrad would reduce *Salmonella* numbers by a factor of at least 10^5. For decontamination of herring meal, radiation doses of 0.8 to 1.3 Mrad were recommended (Underdal and Rossebo 1972). Mulder *et al.* (1977) suggested treating poultry carcasses with 0.25 Mrad to eliminate salmonellae.

The low doses of radiation needed to eliminate salmonellae have no apparent adverse effect on the flavor or nutritional quality of the food or feed. It would seem to be a very useful process for animal feed, since gamma irradiation can be accomplished after the feed has been packed into impervious bags.

Shigellosis (Bacillary Dysentery)

This illness accounts for less than 10% of the reported outbreaks of foodborne illness in the USA; however, the number of cases per outbreak is relatively high. From 1973-1976, the average was about 114 cases/outbreak. Since few states report this illness to CDC, it is probably more widespread than the data indicate.

The predominant mode of transmission of shigellosis is by person-to-person. In 1969, there were 16 outbreaks, but only 1 was attributed to a common food source, 3 were due to contaminated water and, for the other 12 outbreaks, person-to-person spread was thought to be the mode of transmission. The isolations from humans (1968-1973) are depicted in Fig. 6.18.

Sometimes the illness is called the filth disease, since it is associated with poor personal hygiene and sanitation. The illness is often prominent in areas where there are large groups of people. Summer camps, mental institutions, Indian reservations and urban low socioeconomic communities are the high-risk population areas. With the increase in infant day care centers, they are becoming important in the spread of shigellosis.

SHIGELLA—Surveillance Program Reported Isolations from Humans by Month, United States, 1968-1973

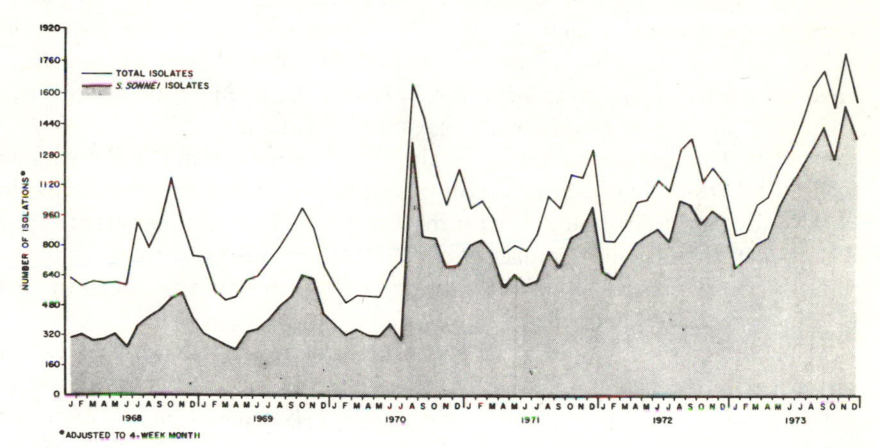

Courtesy of CDC

FIG. 6.18. REPORTED ISOLATIONS OF SHIGELLAE FROM HUMANS BY MONTH, 1968–1973

The mortality rate appears to be greatest in infants and in adults over 50 years of age. Most of the cases are children under 10 years, and the highest attack rate is children in the 1-4 age group. This is probably due to their lack of training in personal hygiene at this young age.

In the 20-40 age group, there is a higher rate of illness in females than males. This is believed due to women having closer contact with children, especially sick children, than do men of this age group.

Although considered a disease of the poor, the more affluent have acquired the illness through travel to foreign areas, such as Mexico, Central America and the Far East.

Shigellosis is primarily an illness of humans. Other primates, such as monkeys and chimpanzees, can acquire the disease (Rout *et al.* 1975). By using procedures such as starvation, folic acid deficiency, control of normal intestinal flora and administration of opium to slow peristaltic action, experimental shigellosis has been established in mice and guinea pigs (Nelson and Haltalin 1972; Maier and Hentges 1972). The ligated small intestine of rabbits serves as an animal model for studying shigellosis.

Pathogenesis.—Virulent shigellae attach to and penetrate the epithelial cells of the intestinal mucosa. After invasion, they multiply and cause the destruction of tissues (Levine *et al.* 1973). The severity of the disease is influenced by the extent of multiplication in the epithelium, and the formation of ulcerative lesions (Formal and Gemski 1976).

Rout *et al.* (1975) suggested that, unlike the toxigenic diarrheas, shigellosis is a disease of both the small and large intestine. They stated that dysentery results from a transport defect of the colon, while diarrhea is secondary to jejunal secretion superimposed on the defect in colonic absorption. Also, Kinsey *et al.* (1976) believed that watery diarrhea may result from an interaction between the jejunal mucosa and the shigellae, as they move through the small intestine.

As with other Enterobacteriacae, the shigellae have somatic O antigens and endotoxins. The chemical composition and structure of the O side chain may represent one determining factor for bacterial penetration of mucosal epithelial cells (Gemski *et al.* 1972). They also suggested that the O antigens may serve as protection against the destructive mechanisms of the host after penetration of the cells.

S. dysenteriae 1, which is the most virulent shigella, produces an enterotoxin (Keusch *et al.* 1972A). Although this enterotoxin induces the rabbit ileum to secrete water and electrolytes (Keusch *et al.* 1972A), and causes cytotoxic changes in the intestinal epithelium (Keusch *et al.* 1972B), the role of this toxin in shigellosis is not completely understood (Donowitz *et al.* 1975). Phillips (1975) inferred that no enterotoxin is involved in shigellosis.

The results of Formal *et al.* (1972) revealed that penetrating strains, regardless of whether or not they produced toxin, caused an inflammatory response and structural changes in the small and large intestine of guinea pigs. A toxigenic strain caused more severe damage than did a nontoxigenic strain. A toxigenic noninvasive strain caused no inflammation or mucosal alteration in guinea pigs. Flores *et al.* (1974) and Donowitz *et al.* (1975) reported the enterotoxin did not increase adenylate cyclase or cyclic AMP, as is the case with cholera enterotoxin. However, the results of Charney *et al.* (1976) suggested that part of the alteration of fluids and electrolytes may be induced by the adenylate cyclase-cyclic AMP system.

Exotoxins with cytotoxic or enterotoxic activity reportedly are produced by strains of *S. flexneri* and *S. sonnei* (Keusch and Jacewicz 1977; O'Brien *et al.* 1977). The toxin produced by *S. flexneri* 2a has biological activities similar to the *S. flexneri* 1 enterotoxin (O'Brien *et al.* 1977).

Thus, there are differing views on the action of shigellae in causing shigellosis or bacillary dysentery. It may be that enterotoxin merely increases the severity of the illness. Perhaps penetration of the organism is important for the activity of the enterotoxin. The inability to detect enterotoxin in cultures is not absolute proof that none is present. Keusch *et al.* (1976) fed human volunteers a reported nontoxigenic strain. The sera from these subjects contained antitoxin indicating that this strain did produce a toxin.

The mechanism of the pathogenic action of shigellae and/or their toxins is not known exactly. The illness may be an infection due to the organism, some form of intoxication or a combination of these actions.

Characteristics of the Illness.—The number of organisms needed to cause the infection may be rather low, since human-to-human transfer is the main mode of transmission. Levine *et al.* (1973) reported that as few as 10 organisms of virulent strains produced illness in human volunteers.

In most outbreaks of shigellosis there is an index case. One person develops symptoms of the disease and then one or two days later, others became ill. If the index case is a food handler, a foodborne outbreak can occur. An example of an outbreak is shown in Fig. 6.19.

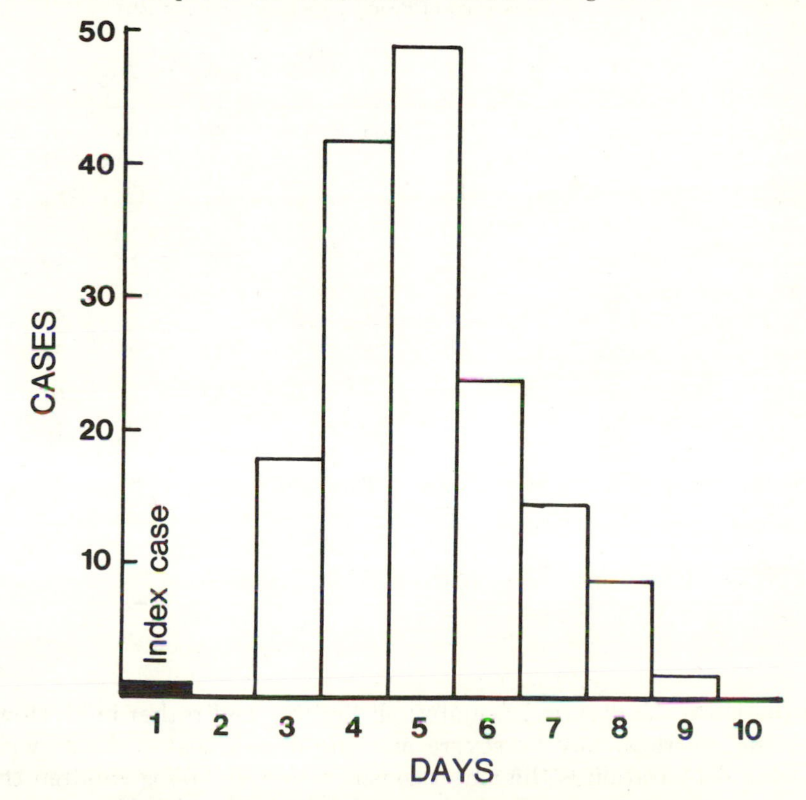

FIG. 6.19. THE PROGRESS OF A TYPICAL OUTBREAK OF SHIGELLOSIS

Incubation Period.—The time from ingestion of the organisms until symptoms appear varies with the individual (health, age), number of organisms ingested and virulence of the organism. CDC (1977A) listed the incubation period for shigellosis as 12-50 hr. Due to person-to-person spread, it sometimes is difficult to determine the exact times for the infections.

Symptoms.—The symptoms of shigellosis as listed in several outbreaks are shown in Table 6.26. The main symptom is diarrhea. Infections associated with mucosal ulceration usually cause more fever, abdominal pain and rectal bleeding than do toxin induced diarrheas.

TABLE 6.26

SYMPTOMS OF SHIGELLOSIS

	Percentage of People Experiencing Symptoms							
	Outbreaks							
Symptom	1	2	3	4	5	6	7	8
Diarrhea	100	98	91	92	100	98	100	91
Bloody	6	6	11	9	23	6	5	24
Mucus	—	19	19	—	—	19	—	—
Tenesmus	—	31	54	17	—	31	—	54
Abdominal pain or cramps	94	85	64	76	79	85	16	69
Chills	56	54	57	27	51	54	—	—
Fever	53	47	57	78	95	47	27	76
Nausea	41	59	47	56	—	59	35	57
Vomiting	33	27	21	53	49	27	13	48
Headache	—	66	43	29	51	66	—	57
Muscle ache	—	55	—	—	—	55	—	—
Weakness	—	—	71	—	—	—	—	—

Source: CDC Reports

In the USA, the disease is usually self limiting and rather mild. However, the diarrhea may be severe and the drastic loss of fluids with resulting dehydration is the major consideration. In young children the disease may be very severe, or even fatal, due to dehydration and extraintestinal manifestations.

Duration and Therapy.—The duration of the illness may range from 12 hr to 3 weeks, with the average illness lasting from 5 to 6 days. The infection usually persists beyond clinical recovery, and the duration of the convalescent carrier state is commonly three to four times that of the duration of symptoms.

Since very young patients may not understand the need for strict personal hygiene, drug therapy is often warranted to prevent secondary spread. Food handlers should be considered as potential spreaders of the organism, and should be given therapy. The type of drug used is dependent on the sensitivity of the cultured strain, as well as the patient, to various antibiotics.

In toxin induced diarrheas, parenteral fluid replacement and glucose-electrolyte solutions are effective, but there is less response to this treatment for invasive dysenteries, such as shigellosis.

After recovery from shigellosis, an apparent state of immunity often develops, although repeated infections may occur. Immunization of a population would not be practical, due to the low incidence of shigellosis. In institutions or other areas of high incidence, immunization would be beneficial.

Foods Involved.—As might be expected, the foods that have been involved in most outbreaks are those which are handled the most. These are salads (potato, tuna, shrimp, macaroni and chicken). The ingredients may be clean, but during the preparation, the salad is contaminated by hand manipulation or mixing. The organisms can multiply readily in moist foods held at room temperature, so even a small inoculum can cause a large-scale outbreak of shigellosis.

The Organisms.—The genus *Shigella* contains four species: *S. dysenteriae, S. flexneri, S. boydii* and *S. sonnei.* The relative frequency of isolation of these serotypes from humans in the USA is shown in Table 6.27. The number of shigellae isolated in 1976 was less than 50% of that in 1973. On a relative basis, *S. sonnei* is decreasing and *S. flexneri* is increasing in frequency of isolation.

Sources.—The normal habitat is the intestinal tract of humans and other primates. It is rarely isolated from other animals.

In a survey of market vegetables in Ceylon, Velaudapillai *et al.* (1969) found cabbage, leeks, spinach, beans, carrots and pumpkins contaminated with shigellae. They believed the contamination may have originated from the water used for irrigation, fertilizer (manure) or from handling the vegetables at the market.

The main source of shigellae involved in outbreaks is humans who are symptomless carriers, or ambulant cases. The organisms can be isolated

TABLE 6.27

RELATIVE FREQUENCIES OF *SHIGELLA* SEROTYPES, 1976

Serotype	No. Reported	Calculated No.	Calculated %
A. *S. dysenteriae*			
Unspecified	14		
1	10	12	0.1
2	39	46	0.6
3	18	21	0.3
4	4	5	0.1
6	1	1	0.0
9	1	1	0.0
10	1	1	0.0
B. *S. flexneri*			
Unspecified	538		
1 unspecified	139		
1A	183	319	4.0
1B	196	338	4.3
2 unspecified	158		
2A	479	771	9.7
2B	89	143	1.8
3 unspecified	184		
3A	309	584	7.4
3B	52	98	1.2
3C	7	13	0.2
4 unspecified	77		
4A	81	195	2.5
4B	4	10	0.1
5	26	33	0.4
6	205	258	3.3
Variant X	6	8	0.1
C. *S. boydii*			
Unspecified	19		
1	8	9	0.1
2	52	62	0.8
4	7	8	0.1
5	13	15	0.2
9	1	1	0.0
10	20	24	0.3
11	1	1	0.0
12	1	1	0.0
14	4	5	0.1
15	1	1	0.0
D. *S. sonnei*	4866	4924	62.3
Unknown	93		
Total	7907	7908	

Source: CDC (1977C)

for several weeks after recovery from the illness. Testing of healthy food handlers for these organisms would be meaningless; however, a person experiencing diarrhea should not be allowed to handle food.

Methodology.—The methodology involved with the shigellae includes enrichment, plating and differentiation to detect the organism. Serological reactions and colicin typing are aids to further separation of the shigellae.

Virulent and avirulent strains can be differentiated by feeding human volunteers. A simpler method is based on the fact that keratoconjunctivitis develops in the eyes of rabbits and guinea pigs when they are inoculated with virulent strains of shigellae (Cross and Nakamura 1970).

Control.—Foodborne outbreaks of shigellosis are due to mishandling of food. A high standard of personal hygiene by food handlers (washing hands after using the toilet, not handling food during illness or a diarrheal symptom) and sanitation of the premises, with proper cooking and refrigeration of foods, should help control shigellosis.

Escherichia coli Enteritis

Certain strains of *E. coli* cause enteric disease syndromes in the young of humans and other vertebrate animals. There is ample evidence that *E. coli* is involved in traveler's diarrhea of adults. In most of the outbreaks of *E. coli* enteritis, the organisms are transmitted by water or person-to-person. Some outbreaks are foodborne, although *E. coli* is not listed in Tables 6.1 and 6.2.

In 1971, an outbreak due to enteropathogenic *E. coli* (EEC) in imported cheese resulted in 387 reported cases. This was the first well documented outbreak in the USA in which food containing EEC affected adults.

In a survey of infants and children with diarrhea, salmonellae or shigellae were isolated from 42% of the patients, while enterotoxin producing strains of *E. coli* were found in 86% of the diarrhea group and 41% of the control group (Rudoy and Nelson 1975). This indicates that in enteric diseases of young people, *E. coli* may be as important as salmonellae or shigellae.

Etiologies.—The role of *E. coli* in enteric disease, especially the diarrheal syndrome, is quite complex. There are invasive strains and noninvasive strains that may or may not produce an enterotoxin. Reportedly, the *E. coli* can produce the diarrheal syndrome by invasion of the intestinal mucosa or by action of enterotoxin. Sakazaki *et al.* (1974A,D) found noninvasive strains which apparently did not produce an enterotoxin, but caused water accumulation in the rabbit ileal loop test. It led them to state that the mechanism of pathogenicity of noninvasive *E. coli* has not been fully explained.

The ability to produce enterotoxin is determined by an episome or plasmid, which can be transferred to various serotypes of *E. coli* and perhaps other Enterobacteriaceae (Gyles *et al.* 1974).

The invasive strains of *E. coli* cause a dysentery (fever, cramps, and diarrhea containing blood and mucus) similar to that caused by shigellae (Sakazaki *et al.* 1974A,B). The invasive strains can be distinguished by the development of keratoconjunctivitis when placed on the guinea pig eye, or by penetration of HeLa or HEp-2 cells (Mehlman *et al.* 1977).

Two types of enterotoxin have been described for *E. coli* strains. One toxin has a low molecular weight, is not antigenic and is heat stable (ST). The other enterotoxin has a higher molecular weight, is antigenic and is heat labile (LT). The heat stability of these toxins is somewhat arbitrary. Most LT preparations are inactivated in 30 min at 60°C. According to Smith and Gyles (1970), the ST retains most of its activity after 30 min at 100°C, but is inactivated after 30 min at 121°C.

Jacks and Wu (1974) reported the molecular weight of ST is between 1,000 and 10,000 daltons, and its activity is resistant to acid, trypsin and pronase. Molecular weights ranging from 14,000 to 200,000 daltons have been reported for LT (Schenkein *et al.* 1976). Most enterotoxigenic strains produce LT either alone or in combination with ST. Some strains produce only ST.

The two toxins are distinctly different enterotoxins. Gyles *et al.* (1974) found the plasmids that determine the production of both ST and LT enterotoxins (Ent plasmids) are homogeneous and consist of a single DNA species, while the Ent plasmids that code for ST enterotoxins only are heterogeneous.

When the enterotoxins were inoculated into ligated gut loops of rabbits, the onset of net fluid accumulation in response to the ST appeared to be immediate, even at the lowest dose tested (Evans *et al.* 1973). The reaction to the LT was rapid at high doses, but was delayed at low doses. The response to this enterotoxin was sustained for at least 18 hr. The delayed action of LT is similar to that of cholera enterotoxin. The LT has an immunological relationship to cholera enterotoxin. The cholera enterotoxin activates an enzyme, adenyl cyclase, in the epithelial cells of the small intestine. This causes an increase in cyclic adenosine 3′, 5′-monophosphate (cyclic AMP), which mediates the fluid transport system and causes an enhanced secretion of fluid and electrolytes by the cells in the intestinal mucosa (Guerrant *et al.* 1974).

The mechanism causing the enteritis due to ST is not known at present. There is evidence that noninvasive strains that produce only ST have been involved in diarrhea in infants and adults. The feeding of 10^8 cells caused mild illness and 10^{10} cells caused a cholera-like enteritis in human adults (Levine *et al.* 1977).

It appears there are several steps in the development of *E. coli* gastroenteritis. Normally, when a person ingests cells of *E. coli,* they pass through the stomach and small intestine and may colonize in the large intestine. If the cells can adhere to the epithelial cells in the small intestine, they may remain for a sufficient period to multiply and, if toxigenic, to produce enterotoxin. Wilson and Hohman (1974) believed that adherence is not an essential factor of pathogenesis, but it could enhance the pathogenicity of some *E. coli.*

Evans *et al.* (1975) found a surface-associated colonizing factor on a toxigenic strain of *E. coli,* which was mediated by a plasmid of 6×10^7 daltons. They further described the colonization factor as a pilus-like antigen (Evans *et al.* 1977).

Reportedly, the enterotoxins are produced by the organisms in the small intestine. However, they have been produced in laboratory media with the culture filtrates showing fluid accumulation in ileal loops. Hence, it is possible that some enterotoxin is preformed in foods. No one has reported finding *E. coli* enterotoxin in foods.

Characteristics of the Illness.—There are two types of illness due to *E. coli.* The invasive strains cause a dysentery similar to shigellosis, while the enterotoxin producing strains cause an illness similar to cholera.

Incubation Period.—The incubation period is 6-36 hr (CDC 1977A). In an outbreak involving imported cheese (Marier *et al.* 1973), the mean incubation time was 18 hr, and the average duration of symptoms was two days. In a similar outbreak, the incubation period ranged from less than 24 hr to over 72 hr, with most cases between 24 and 47 hr (Tullock *et al.* 1973). These outbreaks were due to invasive strains of *E. coli.*

In an outbreak due to heat labile enterotoxin, symptoms began by 12 hr and peaked at 36-48 hr (CDC 1976D).

Symptoms.—The main symptom is diarrhea. In a foodborne outbreak due to imported cheese (Marier *et al.* 1973), 88% of the people reported this symptom. Other symptoms (with the percentage of people experiencing the symptom) were: fever (72%), nausea (71%), cramps (66%), chills (38%), vomiting (35%), malaise (34%), aches (28%) and headache (7%). Since this was an invasive strain, the symptoms are similar to shigellosis. With severe diarrhea, dehydration can occur. Newborn infants are especially susceptible to enteropathogenic *E. coli.* In an outbreak, 100% of the patients had diarrhea and 7% had bloody diarrhea (Tullock *et al.* 1973). Tenesmus was a symptom of 96% of the patients.

These symptoms are essentially the same as those due to enterotoxins, except bloody diarrhea does not occur. The illness due to LT includes the symptoms of copious rice-water diarrhea, similar to cholera.

Duration and Therapy.—The duration of the illness depends upon the severity of the disease and the type of individual. The average illness persists for 3 to 4 days, but in some cases, the diarrhea may persist for 14 days. Usually, coliforms are removed from the upper intestine in 7-10 days, but in rare cases, they may persist for several weeks.

With severe watery diarrhea, fluids and electrolytes must be replaced. Certain antibiotics, such as nalidixic acid, have been effective in controlling *E. coli.*

Foods Involved.—The only well documented foodborne outbreak in the USA involved soft cheese. In other countries, outbreaks have been associated with consumption of dairy products and salads containing raw vegetables.

Number of Organisms Needed.—Although newborn humans and animals are highly susceptible, adults are more resistant. In a survey by Rudoy and Nelson (1975), EEC was isolated from 41% of a control group of children without diarrhea. This indicates that, even if a few organisms are present, they do not cause illness if the numbers are kept under control by the body defenses or by interactions with other organisms in the intestine.

Donta *et al.* (1974) found that all of their human volunteers developed diarrhea 11 to 48 hr after ingestion of 10^6 to 10^{10} colony forming units of *E. coli.* The type and number of cells of *E. coli,* and the age and physical condition of the host determine the occurrence and the severity of the illness.

Nature of the Organism.—*Escherichia coli* is an important organism in the microbiology of foods. Besides being involved in gastroenteritis, the organisms in this species are considered to be indicators of possible fecal contamination and can cause spoilage of some foods. There are over 160 serotypes of *E. coli.*

Sources.—The main habitat of *E. coli* is the intestinal tract of humans and animals. However, due to fecal contamination, it is found in other environments, especially soil and water.

In most surveys for the presence of *E. coli,* the organisms are not serotyped or tested for pathogenicity or enterotoxigenicity. What is the relationship of the presence of *E. coli* to that of EEC?

Schiff *et al.* (1972) surveyed 866 laboratory animals and isolated EEC from 401 (46.3%). About 75% of these animals were healthy, with no signs of overt disease.

Yang and Jones (1969) reported the isolation of EEC from pasteurized dairy products. The contamination was determined to be post-pasteurization, since the organisms were destroyed by pasteurization.

In an analysis of food samples, EEC was found in only 3 (0.6%) of 490 (Hall *et al.* 1967). They concluded that a foodborne EEC outbreak was likely to be due to contamination by food handlers.

Sack *et al.* (1977) analyzed various foods of animal origin and found enterotoxigenic *E. coli* in each type. They believed these foods are potentially important vehicles for these toxigenic organisms, especially if the food is not handled properly.

EEC was detected in about 10% of 2,000 samples of cheese (Fantasia *et al.* 1975A). Park *et al.* (1973) reported that inoculated EEC increased from 10^2 up to 10^5 cells per ml during the first 5 hr of cheesemaking. As the pH dropped during fermentation of lactose, *E. coli* declined in number. Even when the pH increased to near pH 7.0 due to protein decomposition, the number of *E. coli* continued to decline toward zero. If the milk used to make cheese contained penicillin, the EEC increased to 10^9/ml by 24 hr, and by the time the cheese was 9 weeks old, the number of EEC declined only to 10^7/g of cheese.

Methodology.—The methodology for recovery and identification of enteropathogenic *E. coli* was discussed by Mehlman *et al.* (1975). Simple culture techniques fail to differentiate either enteropathogenic or enterotoxigenic strains. The proposed method involves double enrichment, isolation on selective agar and characterization with biochemical and serological tests.

The *E. coli* possess O, H and K antigens. The serological scheme described by Orskov *et al.* (1977) utilized 164 O groups. There are some 100 K antigens. Although commercial sera are available for most of the organisms classed as EEC, there are other serotypes that are toxigenic.

Invasiveness and Enterotoxigenicity.—Laboratory models are available to detect and assay invasiveness or enterotoxigenicity of *E. coli* (Table 6.28).

Invasive strains are detected by their ability to penetrate the superficial layers of the cornea, and cause keratoconjunctivitis in the eye of the guinea pig. The invasion of rabbit intestinal mucosa can be tested if the culture causes keratoconjunctivitis.

The ligated intestinal loop test can be used to determine pathogenicity of *E. coli.* By using bacteria-free inocula, the reactions of enterotoxins, assay of enterotoxins, as well as the effect of antibodies on the reaction, can be obtained. The rabbit intestine is generally used, but other animal models are acceptable. Ligated intestinal loops of infant mice were found suitable for assay of the heat stable enterotoxin (Dobrescu and Huygelen 1973). The heat labile enterotoxin failed to react in the mouse.

In rabbit ileal loops, the LT causes maximum fluid accumulation in about 18 hr, with the ratio of volume (ml) to length (cm) about 2.0-2.5.

To detect ST, the loops are examined at 6-8 hr, although some effect of LT is evident at the time. To determine both LT and ST, the test material is introduced through the stomach with a tube or injected into the intestine of a 7-9 day old rabbit. After 6-7 hr, the animal is sacrificed, the entire GI tract is removed and weighed, the fluid is removed and measured and the ratio of volume to fluid is determined (Gorbach and Khurana 1972).

TABLE 6.28

MODELS FOR DETECTION AND ASSAY OF INVASICN OF ENTEROTOXIN OF *ESCHERICHIA COLI*

Property	Test
Invasiveness	Guinea pig eye-keratoconjunctivitis test (Sereney test)
	Invasion of rabbit (or other animal) intestinal mucosa
Heat labile enterotoxin	Rabbit ileal loop test (18 hr) Skin permeability test Chinese hamster ovary tissue culture Y-1 mouse adrenal tumor cells
Heat stable enterotoxin	Rabbit ileal loop test (6 hr) Infant mouse test
Stable and labile enterotoxins	Infant rabbit test (6 hr)

Certain cell lines can be used to assay for these enterotoxins. The LT causes elongation of Chinese hamster ovary cells. This is related to the ability of the toxin to activate adenylate cyclase (Guerrant and Brunton 1977).

Originally flat Y-1 mouse adrenal tumor cells become rounded when exposed to the enterotoxins (Gurwith 1977). Vero (African green monkey) cells appear enlarged, thick-walled and refractile with several filamentous tendrils, in response to *E. coli* enterotoxin (Speirs *et al.* 1977).

Rabbit skin tests have been used to detect LT. The preparation is injected intracutaneously, followed 24 hr later by an injection of Evans blue dye. Areas of induration and bluing are read 1-2 hr later. Young rabbits show a better reaction than do old rabbits.

Klipstein *et al.* (1976) suggested detection of both LT and ST by perfusion of the rat jejunum. A system using serological methods was described by Evans and Evans (1977A,B).

These tests require biological systems that are not adaptable to all laboratories. Also, there is some question of the ability of these tests to detect the small quantities of enterotoxin which might be present in food. Immunological tests, similar to those for *S. aureus* enterotoxin, have been described by Ceska *et al.* (1978) and Yolken *et al.* (1977).

Control.—The control of the illness is similar to that for other enteric diseases. As far as foodborne illness is concerned, the number of cases would not warrant vaccination or immunization of humans.

To prevent contamination, Bryan (1973A) simply stated that one should practice personal hygiene, prepare foods in a sanitary manner, protect and treat (chlorinate) water and dispose of sewage in a sanitary manner.

The easiest way to prevent growth of the organism is to provide an unacceptable environment for *E. coli*.

Since apparently large numbers of cells are needed to cause illness in other than newborn babies, simply cooking the food sufficiently provides a safe product.

Bacillus cereus Gastroenteritis

Under normal circumstances, *B. cereus* is not considered to be a pathogen. Foodborne illness due to *B. cereus* is rarely reported in the USA. The number of reported outbreaks is much higher in European countries, and is one of the major causes of foodborne illness in Hungary.

The incubation period, from ingestion to symptoms, usually ranges from 1 to 16 hr. The symptoms include mild to profuse diarrhea, stomach cramps or abdominal pains, moderate nausea and, rarely, vomiting or fever. However, an outbreak in Finland (Raevuori *et al.* 1976) also had a short incubation period, and symptoms of nausea and vomiting. Due to these apparently different characterisitics of the illness, Melling *et al.* (1976) proposed that at least two enterotoxins are involved. Turnbull (1976) reported three distinct enterotoxins, and suggested there may be four enterotoxins. Portnoy *et al.* (1976) believed the amount of enterotoxin ingested could account for the different symptoms.

The symptoms are usually mild with a duration of 10-12 hr. With mild symptoms and a short duration, people usually do not seek medical aid. Therefore, most cases can go undetected and unreported. Although the symptoms are usually mild, diarrhea may become severe, and young children seem to be affected more than adults.

High numbers of *B. cereus* (10^6-10^9/g) have been associated with foods involved in the illness. Due to multiplication of the cells in the food before analysis, these numbers may be higher than when the food was ingested. However, the large number of cells, short incubation period and duration with no fever, indicate that an enterotoxin is the etiological agent of the illness.

The Organism.—*B. cereus* is a Gram positive, sporeforming, motile rod. The organism is aerobic, but is capable of growth anaerobically in complex media. Anaerobic growth is promoted by the presence of glucose or nitrate.

The minimum temperature for growth is 10°-20°C, the optimum, 30°-35°C, and the maximum is 35°-45°C (Buchanan and Gibbons 1974). Growth at 49°-50°C has been reported (Goepfert *et al.* 1972). For growth of *B. cereus,* the pH range is approximately pH 4.9 to 9.3 and the minimum a_w is 0.95. The organism can grow in 7.5%, but not in 10% salt.

Since large numbers of cells are needed to cause foodborne illness, the food is abused sometime during preparation. In outbreaks in England due to rice, the rice was boiled and then allowed to sit at room temperature for from 12 hr to 3 days. The rice was either boiled or fried before being served. The original heating was not sufficient to kill the spores of *B. cereus,* but could activate the spores to germinate. Holding the rice at room temperature allowed the vegetative cells to multiply to levels that caused illness. There is reluctance to refrigerate boiled rice because the grains then stick together, making it difficult to toss them into beaten egg during frying, prior to serving.

Source.—*B. cereus* is common in soil and dust, so it is logical that foods that are readily contaminated by soil and dust will likely contain the organism. Plant products (cereals, flour, starch, bakery products, spices), animal products and mixtures of ingredients (spaghetti sauce, pudding, soup mixes, gravy mixes) can contain a few or many cells or spores of *B. cereus.*

There is a similarity between foods involved in illness due to *B. cereus* and *C. perfringens.* In either case, the food is prepared ahead of time, in large batches, which are not properly cooled prior to reheating (if needed) and serving. The reheating is not sufficient to destroy the cells.

Methodology.—To detect *B. cereus,* a spread plate system using phenol red-egg yolk-polymyxin agar is used (FDA 1976). On this medium, *B. cereus* produces colonies surrounded by a dense precipitate (lecithinase activity) with a distinct violet-red background.

Selected colonies are isolated and the organisms tested for the Gram reaction and biochemical tests. Methods to effectively detect and differentiate *B. cereus* from foods have not been developed to the extent that they have for most other organisms causing enteric disease.

Enterotoxin.—The enterotoxin is believed to be produced during the logarithmic phase, and is released upon lysis of the cells.

Turnbull (1976) suggested the possibility of four enterotoxins. One stimulates the adenylate cyclase-cyclic AMP system causing fluid accumulation. A second toxin causes fluid accumulation, but by a different mechanism than the adenylate cyclase system. A third toxin causes tissue damage to the mucosa, and a fourth toxin causes the emetic response.

The enterotoxin can be detected by the ileal loop test. Gorina *et al.* (1975) suggested a hemagglutination system to quantitate the enterotoxin. Reportedly, they could detect 0.004 μg of enterotoxin/ml of food or culture medium.

Control.—The methods of control are the same as those for *C. perfringens.* Keeping *B. cereus* out of the food would appear to be a difficult task. However, proper holding temperatures (55°C or higher, or below 10°C) for the food would prevent growth of the organisms. Multiplication of the cells is needed to attain sufficient numbers to cause illness.

Streptococcal Infections

Two groups of streptococci are listed as causing foodborne disease. Group A (*Streptococcus pyogenes*) may be present in food. This organism causes scarlet fever or septic sore throat. An outbreak occurred in Florida due to contaminated egg salad (Saslaw *et al.* 1974). Besides a sore throat, 18% experienced vomiting and 20% diarrhea.

Egg salad contaminated with a group G streptococcus was believed to be the cause of an outbreak of pharyngitis (Leaven *et al.* 1968). Included in the symptoms were nausea and vomiting (55%) and diarrhea (25%).

When streptococcal foodborne infection is mentioned, the organisms in group D, often referred to as enterococci, or fecal streptococci, are considered to be the etiologic agent of the illness.

The reported illness due to the enterococci is not significant. There were four outbreaks reported from 1971-1976. In those cases in which enterococci were thought to be involved, the incubation period ranged from 2 hr to 30 hr. The symptoms were similar to those produced by *B. cereus* and *C. perfringens.* They included nausea, vomiting and diarrhea, but were usually milder than those caused by other types of foodborne illnesses. The duration was usually only a few hours.

Some people believe that the enterococci do not cause foodborne illness, and some are convinced that these organisms are involved in outbreaks. Other workers are unsure, but they realize that more convincing proof is needed to relate enterococci to foodborne illness.

Illness Due to Vibrios

Vibrio cholerae causes a gastrointestinal illness in humans called cholera. The disease is of worldwide significance, but it is rarely reported in the USA. Cholera is associated with crowded conditions, with poor sewage disposal and inadequate treatment of drinking water. The diarrhea is severe (rice water stools) with the patient often becoming dehydrated due to loss of fluids.

An organism named *Campylobacter fetus* (formerly called *Vibrio fetus*) was isolated from poultry carcasses in retail stores (Smith and Muldoon 1974). This organism causes gastroenteritis in humans (vibriosis). According to them, the incidence of the illness is highly underrated.

Vibrio parahaemolyticus causes gastroenteritis in humans. Although perhaps not as important as cholera on a worldwide basis, the prevalence of this illness seems to have increased. In 1968, *V. parahaemolyticus* was not listed as an etiological agent of foodborne illnesses by CDC. In 1971, there were 3 outbreaks with 370 cases and in 1972, 6 outbreaks with 701 cases. However, from 1973 to 1976, only three outbreaks were reported.

Among the enteric pathogens, *V. parahaemolyticus* is a recent addition. The organism was recognized as a cause of foodborne illness in Japan during the early 1950's. During the summer months, the organism accounts for about 70% of the reported foodborne illnesses in Japan.

The illness was thought to be unique to Japan. However, due to a spreading epidemic of cholera in the early 1970's, techniques to isolate vibrios were improved, and diarrheal stools were examined for vibrios. It was discovered that *V. parahaemolyticus* was an important cause of foodborne illness, not only in Japan, but also in the USA and other countries.

The Illness.—This is a typical gastroenteritis, with diarrhea as the main symptom. Other symptoms include: abdominal cramps, nausea, vomiting, headache, chills and fever (Table 6.29). In severe cases, rice-water diarrhea, resembling cholera, may develop. In very severe cases, death may occur, but it is rare.

The incubation period usually is 12-24 hr. The symptoms persist from a few hours to 10 days with the usual duration of 2-3 days.

If treatment is needed, both supportive, to control fluid loss, and antibiotic medication may be used. The organism is reported to be sensitive to several antibiotics, including tetracycline. Antibiotic treatment decreases the length of time that the organisms are shed.

Foods that have been incriminated in outbreaks in the USA include steamed crabs, crab salad (made from canned crabmeat), raw crab, processed lobster, boiled shrimp, roasted oysters and raw oysters. The raw products were inadequately refrigerated. In two outbreaks, the "cooked" seafoods were inadequately heated, followed by inadequate refrigeration. In nine outbreaks, there was cross contamination between cooked and raw products.

TABLE 6.29

SYMPTOMS OF *VIBRIO PARAHAEMOLYTICUS* GASTROENTERITIS

	% of Patients with Symptom			
	Outbreak			
Symptom	1	2	3	4
Diarrhea	100	100	10	98
Abdominal cramps	85	96	86	78
Nausea	46	63	51	76
Vomiting	33	59	38	74
Headache	33	46	32	25
Chills	45	71	37	10
Fever	28	34	17	26
Bloody diarrhea	3	5	1	—

From CDC Reports

Besides these foods, in other countries, the consumption of raw fish, as well as such diverse foods as meat, eggs, cereal products and vegetables have been involved in outbreaks.

Etiologic Agent.—The exact mechanism whereby *V. parahaemolyticus* causes gastroenteritis has not been determined. When the organism is grown on Wagatsuma blood agar (Miyamoto *et al.* 1969), those strains isolated from human cases exhibit a beta hemolysis, while most of the strains from natural environments or seafoods are not hemolytic. This hemolytic activity is called the Kanagawa phenomenon, with hemolytic strains being Kanagawa (K) positive and non-hemolytic strains, K-negative. In human feeding trials, K-positive strains cause illness and K-negative strains do not. Thus, there is a close correlation of hemolysis and human pathogenicity.

Using the ligated intestinal loop of rabbits, Sakazaki *et al.* (1974D) tested living cultures of 16 K-positive strains and 32 K-negative strains. With these cultures, 14 K-positive and 7 K-negative strains gave a positive reaction in the ileal loop. The purified Kanagawa hemolysin failed to produce pathological changes in the gut loop. They believed the ability to grow in the intestine was important in the pathogenic character of the strains, and K-positive strains outgrow K-negative strains.

It is not known how many cells need to be ingested to cause the illness.

Smith (1971) stated that 10^6 to 10^9 cells were needed, depending on the condition of the stomach, type and quantity of food eaten and buffering capability of the ingested food. Accidental ingestion of 10^5 cells caused symptoms in a laboratory worker. Person to person transmission of the organism has not been reported.

Craig (1972) suggested that the major manifestations of the illness are caused by an enterotoxic elaborated by the organism in the small intestine, as in cholera. He stated that it seemed unlikely that hemolytic and enterotoxic properties would reside in the same molecule. Sakurai *et al.* (1974) found a pathogenic strain that was not hemolytic. This led them to believe that there was a factor or factors other than a hemolysin responsible for the illness.

Although there is evidence that the K-hemolysin is not the pathogenic factor, Miyamoto *et al.* (1975) purified a factor that hemolyzed human, sheep and rabbit red blood cells. They called this substance Fr 5 and reported that it caused a reaction in the rabbit ileal loop. The Fr 5 caused diarrhea when administered to suckling mice. They stated that their findings strongly support the hypothesis that the hemolytic factor plays a role in causing disease symptoms in humans.

Carruthers (1977) reported the adherence to intestinal cells was greater for living K-positive strains than for K-negative strains of *V. parahaemolyticus.* The organism is cytotoxic to epithelium cells, but this effect apparently is not due to invasion (Carruthers 1975).

Thus, it is evident that the factor or factors causing the illness have not been completely determined. There is a need for more information concerning the toxin, or other pathogenic characters of the organism.

Nature of the Organism.— *Vibrio parahaemolyticus* is a Gram negative, straight or curved, facultatively anaerobic rod. It is halophilic, requiring 1-3% salt for growth. The cells grow in broth with 8%, but not 10% salt (Kourany and Vasquez 1975). The minimum a_w is about 0.94. Usually the pH range for growth is 5 to 11 (Twedt *et al.* 1969). The optimum is pH 7.5 to 8.5.

The organism is sensitive to cold temperatures, so the minimum for growth is usually listed as 10° to 13°C. The upper temperature is about 42° or 43°C, with an optimum between 30° and 37°C. The effect of low temperature on the cells is related to the type of product and the temperature of storage. When exposed to 2°C, the membrane is damaged (Van den Broek and Mossel 1977). The presence of salt tends to protect the cells from low temperature damage.

Smith (1971) stated that the organism has a very short generation time of 9 to 11 min. The average generation time of four strains was 13.6 min (Lee 1973).

Sources.—The natural habitat of *V. parahaemolyticus* is coastal and estuarine ocean water throughout the world. It lives in sediment during the cold winter months, and, as the temperature rises in late spring, the organism colonizes the water and animal life. Direct relationships have been demonstrated between the water temperature, abundance of the organism in the water and outbreaks of vibrio foodborne illness.

Since the organism is associated with the coastal and estuarine marine environment, it has been recovered from a variety of seafoods.

Methodology.—*V. parahaemolyticus*, if present in a food, is usually in low numbers. Therefore, an enrichment procedure, using glucose salt teepol broth (GSTB) is employed. Media used for analysis of the organism contain 2-3% salt. A loopful of the enriched sample is streaked onto thiosulfate citrate bile salts sucrose (TCBS) agar (FDA 1976). On this medium, the organism appears as round colonies, 2-3 mm in diameter, with green or blue centers. The salt concentration (1%) and pH (8.4), plus other inhibitors, suppress the growth of most organisms except halophiles, such as *V. parahaemolyticus*. Sucrose fermenting organisms produce yellow colonies. These include *V. cholerae*, some enterics and *S. aureus*. Organisms besides *V. parahaemolyticus* that can form green colonies include some vibrios, pseudomonads and enterics.

A salt teepol buffer (STB) medium was reported to be a better enrichment broth than GSTB (Chun *et al.* 1974). With samples grossly contaminated with *V. parahaemolyticus*, enrichment is not necessary; better results are obtained by direct streaking in the isolation agar (Earle and Crisley 1975).

Since organisms besides *V. parahaemolyticus* can grow on the isolation media, biochemical tests are needed to differentiate Gram negative organisms (Table 6.30). *V. parahaemolyticus* can be distinguished from members of the Enterobacteriaceae by a positive oxidase reaction, and from pseudomonads by its fermentative metabolism, using Hugh-Liefson glucose broth. Differentiation of anaerogenic *Aeromonas* strains is by sensitivity to the vibriostatic agent pteridine (*Aeromonas* is resistant).

Serology.—Strains of *Vibrio parahaemolyticus* possess O, H and K antigens. A major problem is that from 10 to 60% of the isolates are not typable. Although serological typing has provided epidemiological data in foodborne outbreaks, at present, it appears to be of little value for identification of this organism due to the prevalence of nontypable strains. More work is needed before the serotyping of *V. parahaemolyticus* can become a routine procedure.

Control.—The control of *V. parahaemolyticus* should be relatively simple. It is found in marine environments and contaminates seafoods during the warm summer months. Recontamination by raw seafoods of cooked or ready-to-eat foods must be avoided. The application of good sanitary practices and personal hygiene would help in preventing this cross contamination.

TABLE 6.30

IDENTIFYING CHARACTERISTICS OF *VIBRIO PARAHAEMOLYTICUS*

Test	Result
Oxidase	+
Catalase	+
Lipase	+
Lysine decarboxylase	+
Ornithine decarboxylase	+
Voges-Proskauer	−
Hydrogen sulfide	−
Indole	+
Citrate utilization	+
Gelatin liquefaction	+
Growth in 3-8% salt	+
Growth in 0 or 10% salt	−
Motility	+
Casein hydrolysis	+
Starch hydrolysis	+
Nitrate reduction	+
Growth in KCN broth	+
Fermentation (acid)	Fermentation (−)
Glucose	Cellobiose
Fructose	Sucrose
Maltose	Lactose
Mannitol	Salicin
Trehalose	Adonitol
Arabinose	Inositol
Mannose	
Galactose	

The organism is readily killed by heat. When inoculated at a level of 10^5/ml in shrimp homogenate, no survivors were found after heating at 60° or 80°C for 15 min, or at 100°C for 1 min (Vanderzant and Nickelson 1972). Levels of 200 per ml were destroyed when heated to 60°C for 1 min.

Although cooking of seafoods will destroy the organisms, it is difficult to change the habit of eating raw seafoods by some cultures. The use of radiation treatment of seafoods might be a solution.

Histamine

Biologically active amines, such as histamine, have been involved in foodborne illness. The ingestion of 70-1,000 mg of histamine will usually cause intoxication. Many fermented foods (wine, cheese, sauerkraut, sausage) contain histamine, but not at a level that will cause illness.

Fish of the suborder *Scombroidei* contain relatively large amounts of free histidine. During spoilage, this is converted to histamine. Foo (1977) analyzed canned skipjack tuna implicated in an outbreak of food poisoning. The histamine content varied from 153 to 1,149 mg/kg of fish.

Various microorganisms (*Proteus morganii, C. perfringens, E. aerogenes, E. coli,* salmonellae, shigellae and pseudomonads) contain histidine decarboxylase. Foo (1977) found free histidine at high levels in skipjack (6,900 mg/kg) and mackerel (5,000 mg/kg). That toxic amounts of histamine could be produced before spoilage is evident.

Some controversy exists concerning the role of histamine and other amines in foodborne illness of humans.

Aeromonas

This genus is associated with water and animals that live in the water. Certain species have been isolated from humans. Strains of *A. hydrophilia* and other species are enterotoxigenic (Boulanger *et al.* 1977; Wadstrom *et al.* 1976A). The extent of enteric infections and the role of food, such as fish, as a vehicle of infection is not known at the present time. Swimming in, and ingestion of contaminated water is one of the means of infection (Fritsche *et al.* 1975).

Brucellosis

The number of cases of brucellosis has generally decreased in the USA, although since 1973, there has been an increased prevalence. It has become a disease of people who handle cattle and swine. This includes veterinarians, farmers and packinghouse workers. The workers in the kill department are most subject to contamination, although workers handling the carcasses and meat also have become infected.

In 1975, there were 24 cases caused by the ingestion of unpasteurized dairy products, such as raw domestic milk and imported cheese.

Citrobacter

Although this organism is not listed as a major cause of gastroenteritis, it is found in foods and certain strains reportedly produce enterotoxins (Wadstrom *et al.* 1976B).

Q Fever

Coxiella burnetii causes Q fever in humans. Reportedly, Q fever is usually transmitted by the inhalation of aerosols from contaminated animal products. The consumption of raw milk has been associated with cases of Q fever. The pasteurization process of milk was adjusted so the treatment would inactivate this organism. *C. burnetii* is a potential hazard to people who work in meat packing plants and handle raw meat.

Edwardsiella

E. tarda is found in the intestinal tract of reptiles and marine animals. Most isolates of this organism are from humans with intestinal disorders. It can cause an illness similar to salmonellosis, as well as septicemia and abscesses. Quite often, *E. tarda* infections are not recognized because the organism is overlooked, and proper tests for identification are not made.

Enterobacter

Some strains of *E. cloacae* are enterotoxigenic (Klipstein *et al.* 1977). This organism has been isolated from the intestinal tract of people with tropical sprue or acute diarrhea. Of 173 enterotoxigenic strains of bacteria isolated from children, 12% were *Enterobacter* (Wadstrom *et al.* 1976B). The role of *Enterobacter* as a foodborne gastrointestinal disease organism has not been established.

Klebsiella

K. pneumoniae produces a heat stable and heat labile enterotoxin similar to *E. coli* (Klipstein *et al.* 1977). This organism is quite prevalent in the gastrointestinal illness, tropical sprue.

Yersinia

Y. enterocolitica produces several disease syndromes in humans. Various sources of the organism for human illness have been suspected. An outbreak affecting 218 children was caused by contaminated chocolate milk (Black *et al.* 1978). The source of the contamination was not determined, but it was thought to be after pasteurization, during addition of the chocolate syrup.

The gastrointestinal illness is characterized by abdominal pain and fever. Other symptoms may include vomiting and diarrhea. The abdominal pain and fever resemble symptoms of appendicitis. Une (1977) considered *Y. enterocolitica* to be invasive. The organism produces an enterotoxin (Pai and Mor 1978). The prevalence of this organism and the extent of its involvement in human illness is not known.

FUNGAL ILLNESSES

Fungi are a diverse group of organisms which include molds, yeasts, rusts, smuts and mushrooms. These organisms can cause many diseases in plants, animals and humans.

In humans, the fungi can cause mycoses, allergies and/or toxicoses. Mycoses are diseases resulting from the invasion of living cells by the fungi. Allergies are diseases resulting from the development of hypersensitivity to fungal antigens. Toxicoses consist of illnesses due to ingesting toxic fungal metabolites formed in the food (mycotoxicoses) and the mycetisms caused by ingesting toxic fungal fruiting bodies.

Besides producing toxins, the fungi can induce host plants to produce substances which are toxic. Also, they can degrade the food so that it is deficient in certain nutrients, or the mycosis can upset the animal metabolism so that a nutrient deficiency exists.

Our concern is primarily with mycotoxicoses.

Mycotoxins

Mycotoxins are secondary metabolites produced by fungi and can cause unnatural or deleterious biological changes in plants, animals, humans or microorganisms. The mycotoxins may be contained within the spore or fungal thallus, or they may be secreted into the growth substrate of the fungus.

Mushroom poisoning was experienced by humans in ancient times. Every year, there are outbreaks in the USA due to mushroom poisoning.

Ergotism, or St. Anthony's Fire, was reported in the Middle Ages. This illness is caused by eating cereals infected with *Claviceps purpurea*. This fungus can grow and produce toxic alkaloids on wheat, barley, rye, oats and wild grasses. About 40 alkaloids have been isolated from the fungus. A tetracyclic ring called lysergic acid is associated with these alkaloids. The symptoms of the illness can include gangrene of the limbs, violent pains, convulsions, chills and hallucinations, as well as abortions of pregnant women. The last major outbreak of ergotism was in France in 1951.

Alimentary toxic aleukia (ATA), or septic angina, has occurred in Russia at infrequent intervals. Symptoms of this illness include a burning sensation in the mouth, the tongue feels stiff, with diarrhea, nausea, vomiting and perspiration. After these symptoms, there is a quiescent period, followed sometime later by leukopenia, weakness, hemorrhages of the skin and mucous membranes, necrotic areas in the mouth, throat and skin, gangrenous pharyngitis, fever, and recovery or death. The fatality rate varies from 2 to 80%.

This illness (ATA) was especially severe during World War II. The Russians did not have farm labor to harvest the grain in the fall. When the grain that was harvested during the following spring was consumed, it caused outbreaks of ATA. Investigations revealed several toxin producing fungi on overwintered grain, with *Fusarium* predominating.

Also, during the period of World War II, rice that Japan imported from other Asian countries was responsible for outbreaks of an illness called "yellow rice disease," which caused several deaths. The illness was associated with the invasion of rice by *Penicillium islandicum, P. citrinum*, and *P. citreoviride.*

In 1929, Fleming described a substance from a *Penicillium* mold that he called penicillin. This, and other antibiotics, affect microorganisms more than animals or humans. During the search for antibiotics, several substances were found that are more toxic to animals or humans than to microorganisms. These are mycotoxins. Even with the finding of toxic substances, there was little activity in mycotoxin research until 1960, when over 100,000 turkeys died in England. An investigation revealed that the deaths were due to peanut meal, in which *Aspergillus flavus* had produced mycotoxins called aflatoxins.

With an increased interest in fungal toxins, over 150 species of fungi have been reported to be capable of producing substances that are toxic. Some of those that might affect humans are listed in Table 6.31. Each year, several new mycotoxins are found.

One reason for the interest in mycotoxins is that some, such as aflatoxins, luteskyrin, patulin, penicillic acid and sterigmatocystin, are carcinogens.

Many genera of molds can produce mycotoxins; however, most molds produce no apparent toxins. Not all species of a mycotoxin producing genus, and not all of the strains of a mycotoxin producing species, are toxigenic.

There are certain conditions needed for the contamination of foods with mycotoxins. First, a mycotoxin producing fungus must be present. These fungi are widespread, and have been found in many foods or feeds. Next, the food must be suitable as a substrate, and environmental factors must be acceptable for growth of the fungus and elaboration of the mycotoxin. Growth is not synonymous with the presence of mycotoxin, since some substrates suitable for growth are not suitable for production of the secondary metabolite.

Molds can grow over a wide range of temperatures, as compared to other organisms. Even cold storage does not prevent the formation of mycotoxins. Frequently, toxins are produced at sub-optimal growth temperatures, and sometimes they are not produced at optimum growth temperatures or above. Since the mycotoxins are secondary metabolites, they do not appear during the early stages of growth. Hence, growth may be evident before mycotoxins are produced in a quantity sufficient for detection.

TABLE 6.31

SOME MYCOTOXINS THAT MIGHT CAUSE HUMAN ILLNESS

Mycotoxin	Some Producing Organisms
Aflatoxins	*Aspergillus flavus, Aspergillus parasiticus*
Citreoviridin	*Penicillium citreoviride*
Citrinin	*Penicillium citrinum, Penicillium viridicatum*
Cyclopiazonic acid	*Penicillium cyclopium*
Ergotoxins (alkaloids)	*Claviceps purpurea*
Luteoskyrin	*Penicillium islandicum*
Ochratoxins	*Aspergillus ochraceus, P. viridicatum*
Patulin	*Penicillium expansum, Penicillium patulum, Penicillium urticae, Aspergillus clavatus, Byssochlamys nivea*
Penicillic acid	*Penicillium martensii, P. viridicatum, P. cyclopium, Penicillium puberulum, Penicillium palitans*
Roquefortine	*Penicillium roqueforti*
Rubratoxin	*Penicillium rubrum*
Rugulosin	*Penicillium rugulosum*
Stachybotrytoxin	*Stachybotrys atra*
Sterigmatocystin	*Aspergillus versicolor, Aspergillus nidulans*
Tenuazonic acid	*Alternaria tenuissima, Alternaria alterata*

TABLE 6.31. *(Continued)*

Mycotoxin	Some Producing Organisms
Trichothecenes	
Deoxynivalenol (vomitoxin)	*Fusarium graminearum* (*Gibberella zeae*)
Fusarenon-x	*Fusarium nivale*
Neosolaniol	*Fusarium solani*
Nivalenol	*F. nivale, Fusarium roseum*
Fusarotoxin T-2	*Fusarium tricinctum, F. solani, F. roseum*
Trichothecin	*Trichothecium roseum*
Zearalenone	*F. graminearum* (*G. zeae*), *F. tricinctum, Fusarium culmorum, Fusarium oxysporum*

Moisture satisfactory for growth and toxin production may be present in foods prior to or during harvesting, before they are dried adequately. Improper storage of dried foods can result in a moisture level adequate for mold growth. During harvesting, damage to the protective covering of foods increases the susceptibility of the food to invasion by fungi. High moisture corn is more readily damaged during harvesting than is low moisture corn. Damage to the food by rodents and insects aids in the penetration and growth of mold.

Besides temperature and moisture, other factors influencing the production of mycotoxins include light, aeration, nutritional factors, presence of fungal inhibitors and competitive growth of other microorganisms (Jarvis 1971).

A discussion of all of the individual mycotoxins is not possible in this text; however, since the aflatoxins have attained an important position, it seems desirable to investigate this group of fungal toxins.

Aflatoxins.—More than 100,000 turkey poults died within a few months in England during 1960. Investigations revealed that ducklings, young pheasants and partridges also had a high mortality. About this same time, a trout hatchery in California experienced severe losses due to hepatomas. Reports from Africa also revealed heavy losses of ducklings. When the information of these incidents was accumulated and examined, all of these isolated losses were due to the presence of aflatoxins in the animal feed.

In the turkey poult outbreak, peanut meal was found to be the suspect feed ingredient. Peanut meal and extracted fractions showed similar toxicity. Molds grown from other meals included *Aspergillus flavus.* Toxic fractions were extracted from this fungus. Since the toxin was produced by *A. flavus,* it was called aflatoxin.

A chromatograph of the extract of moldy feed was prepared and observed with ultraviolet light. Four fractions appeared. Two were blue and two were green. The toxic substances were called aflatoxin B_1, B_2, G_1 and G_2, the letters corresponding to the fluorescent color, and the numbers denoting the relative mobility. At the present time, 13 of these compounds occur in nature (Ciegler 1975). They all possess a coumarin nucleus fused to a bifuran moiety. Eight aflatoxins (B_1, B_2, B_{2a}, M_1, M_2, P_1, R_0 and Q_1) have a pentene ring, while four (G_1, G_2, G_{2a} and G_{ml}) have a 6-membered lactone ring. The other toxin was referred to as parasiticol (B_3) (Ciegler 1975) and is similar to the B series, except that the pentene ring is opened. Eight of these aflatoxins are shown in Fig. 6.20.

Biological Effects.—Aflatoxins can cause a response in microorganisms, cell cultures, plants and animals. There may be acute or chronic effects, depending upon the dosage and the frequency of exposure to the toxins. The effects of aflatoxins can be toxigenic, mutagenic, teratogenic or carcinogenic.

Several species of *Bacillus* and certain *Flavobacterium* are inhibited by aflatoxins. Other microorganisms, including *Actinomycetes,* and various yeasts and molds are altered by the presence of aflatoxin in the substrate. Aflatoxins can interfere with plants by inhibiting the germination of seeds and/or chlorophyll synthesis.

In cell cultures, aflatoxin causes a decrease in cell numbers, protein level, RNA and DNA. There is an inhibition of mitosis, the formation of giant cells and accumulation of cellular debris.

The very young animals are more susceptible to aflatoxicosis than are mature animals. Species of animals vary in their resistance. For farm poultry, ducklings are the most susceptible, followed by turkey poults, pheasant chicks, goslings, chickens and quail. For farm animals, the order from high susceptibility to resistance is young pigs, pregnant sows, calves, fattening pigs, mature cattle and sheep.

Ducklings are highly sensitive to the toxins. Within a few days after ingesting toxic feed, the proliferation of bile-duct epithelial cells is clearly visible. Ducklings are a biological system for assaying of aflatoxins.

Most laboratory animals are susceptible to the effects of aflatoxins. The acute lethality of aflatoxin is very evident in rainbow trout (LD_{50} about 0.5 mg/kg). Coho salmon are about 10 to 20 times more resistant and channel catfish are about 30 times more resistant (Halver 1969).

Aflatoxin B_1 is the most abundant and considered to be the most toxic of the aflatoxins. When given orally to ducklings, the LD_{50} in mg/kg for some of the aflatoxins is 0.36 (B_1), 0.33 (M_1), 0.78 (G_1), 1.7 (B_2) and 3.5 (G_2) (deWaart 1973). B_{2a} and G_{2a} are not toxic to ducklings, even when

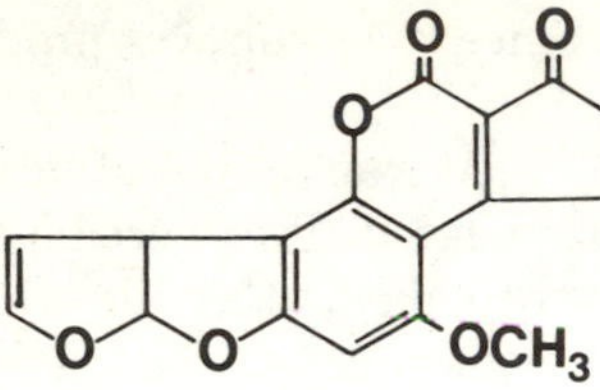

O O O O O OCH_3

B_1

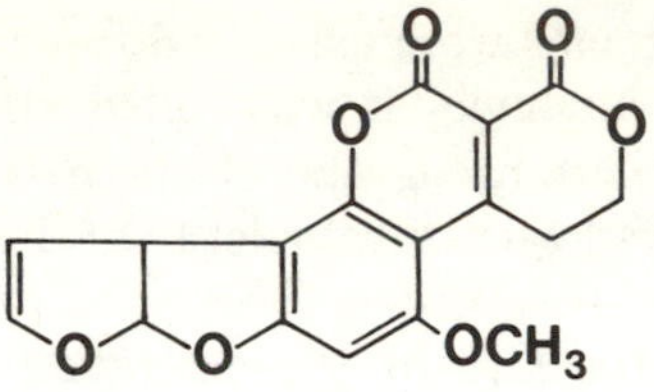

O O O O O O OCH_3

G_1

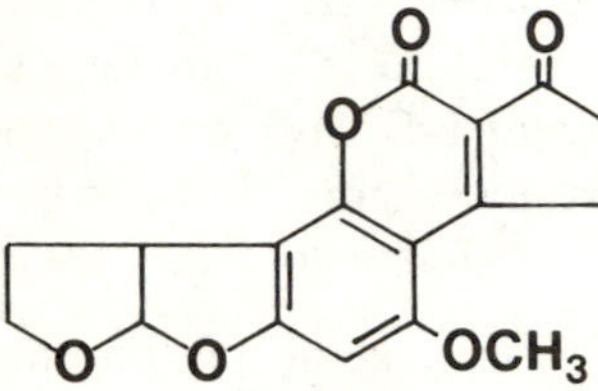

O O O O O OCH_3

B_2

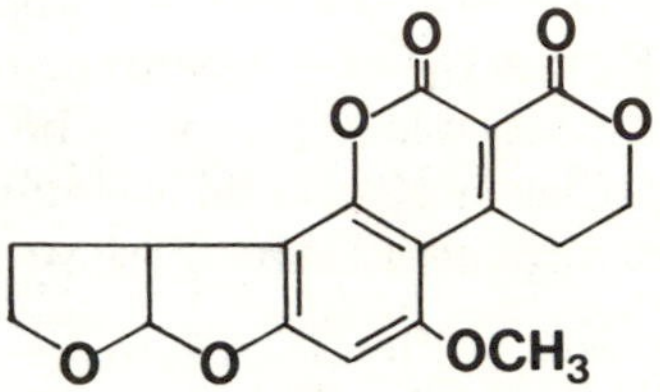

O O O O O O OCH_3

G_2

O

Furan

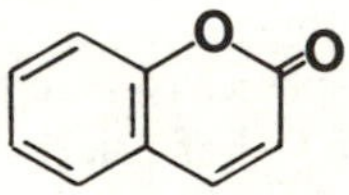

O O

Coumarin

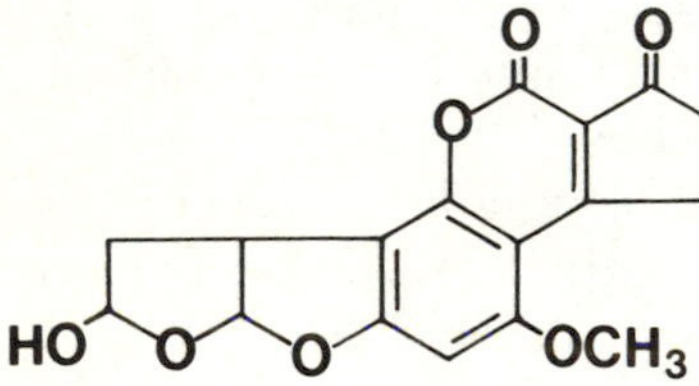

O O O HO O O OCH_3

B_{2a}

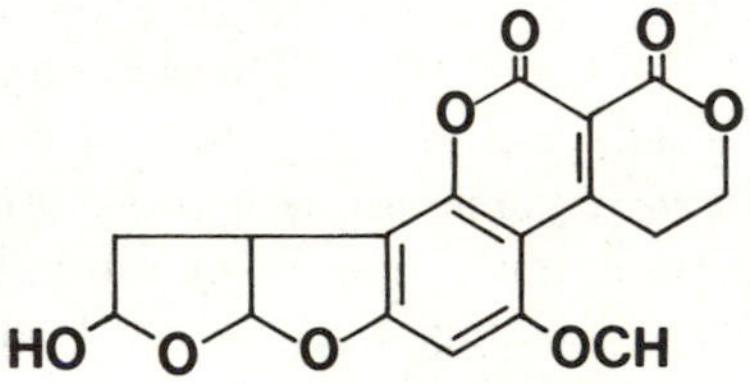

O O O O HO O O OCH

G_{2a}

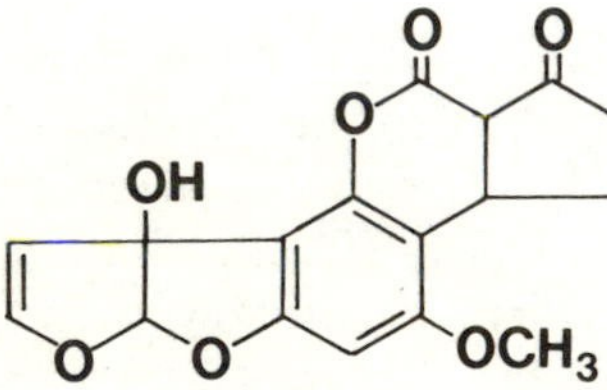

O O O OH O O OCH_3

M_1

O O O OH O O OCH_3

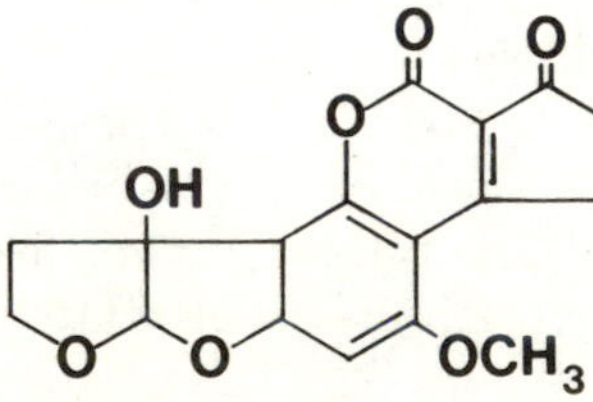

M_2

FIG. 6.20. FORMULAE OF AFLATOXINS

They contain a coumarin molecule and bifuran ring.

at very high levels. Thus, there may be justification for not calling these substances toxins.

Acute aflatoxicosis has been characterized by hemorrhage in tissues, anorexia, hepatitis and death of animals. The liver is the primary tissue that is affected. However, the spleen, pancreas, kidneys, as well as other tissues may be involved. Aflatoxin also causes fragility of the capillaries, so there is a greater tendency for bruising.

The exact biochemical mechanism of aflatoxicosis is not known. However, there is inhibition of DNA synthesis, nuclear RNA synthesis and alteration of gene transcription (Fujimoto and Ohba 1975). It is thought that aflatoxin or a metabolic product of aflatoxin, can bind to deoxyribonucleic acid (DNA), and its primary action is to inhibit RNA polymerase. Mutagenic effects of aflatoxin were described by Ong (1975).

The effect of aflatoxins on animals is influenced by the type of aflatoxin, species of animal, age, weight, health and diet. Environmental stresses may alter the susceptibility of the animal.

Aflatoxin is carcinogenic in ducklings, rainbow trout, rats and ferrets (Detroy *et al.* 1971). They stated that some reports indicate that aflatoxin induces carcinomas in pigs, hamsters, guinea pigs, mice and sheep. A monkey developed a cancer due to aflatoxin ingestion (Gopalan *et al.* 1972). Although the carcinogenic activity is primarily in the liver, aflatoxin can induce tumor formation in other organs, especially the kidney.

There is ample evidence that aflatoxins produce acute and chronic illness in animals. What is their effect when ingested by humans? The deliberate feeding of aflatoxins to human subjects has not been reported. The only information concerning humans is from instances in which the food, naturally contaminated with aflatoxins, has been consumed. Thus, evidence for human aflatoxicosis is not based on scientific experiments. Information accumulated after an incident is often incomplete.

Aflatoxin B_1 was detected in specimens of stools, urine, liver, brain and kidney obtained from Thai children who died of encephalopathy and fatty degeneration of the liver (Shank *et al.* 1971). Krishnamachari *et al.* (1975) described an outbreak caused by consuming moldy maize, in which 106 people died and 291 showed signs of hepatic dysfunction.

Attempts have been made to associate the high incidence of liver cancer in humans in certain areas to the presence of toxigenic molds (Peers and Linsell 1973; Van Rensberg *et al.* 1975). Newberne (1974) listed evidence associating aflatoxins with cirrhosis of the liver, as well as hepatoma, hepatitis and Reye's syndrome. The implication of aflatoxins with various illnesses of humans was reviewed by Ciegler (1975) and Steyn (1977).

Apparently, the exposure to below a certain level of aflatoxin is not harmful, but high levels may be responsible for carcinoma of the liver. Until it is possible to determine the exact relationship of aflatoxin to liver cancer, precautions will be needed to reduce the level of these agents in our foods and animal feeds.

Conditions for Toxin Production.—The ability of a fungus to produce and accumulate aflatoxin is dependent upon such factors as genetic potential, environmental conditions (substrate, moisture, temperature, pH) and the time of contact between the fungus and the substrate. The production of aflatoxin was reviewed by Maggon *et al.* (1977).

Aflatoxins have been reported to be secondary metabolites of various species of *Aspergillus,* as well as *Penicillium, Rhizopus* and *Streptomyces.* However, the screening of 121 fungal isolates, including 29 species, showed that only *A. flavus* and *A. parasiticus* produce aflatoxin (Wilson *et al.* 1968). Not all of the strains of these aspergilli produce aflatoxins. Generally, the toxin producing strains produce B_1 and lesser amounts of G_1, and the other aflatoxins. The composition of the toxin complex can be quite variable, depending upon the strain of the mold, as well as substrate and environmental conditions. Schroeder and Carlton (1973) described an unusual strain of *A. flavus* that produced only B_2 aflatoxin.

Although resting cells of *A. flavus* produce aflatoxin under laboratory conditions, growth of the microorganism is considered to be needed to produce toxin under natural conditions.

The minimum relative humidity (RH) for growth of *A. flavus* is 80% (a_w of 0.80). At other than optimum temperature, pH and other growing conditions, the minimum a_w for growth is increased. At 30°C, the minimum RH for aflatoxin production is 83% (Davis and Diener 1970). The minimum RH for growth varies with the condition of peanut kernels (Diener and Davis 1970). The limiting RH was 83% (broken immature kernels), 84% (sound mature kernels) or 86% (kernels from unshelled peanuts). Northolt *et al.* (1976) found the optimum a_w to be 0.99, with little aflatoxin production at a_w 0.87 with *A. parasiticus.*

Aspergillus is mesophilic, with a temperature growth range of 6°-8°C to 44°-60°C, and an optimum of 35°-38°C. The range of temperatures for aflatoxin production is more limited, being about 11° to 41°C. Maximum aflatoxin is produced at temperatures below those for optimum growth, or about 24°-30°C.

Incubation at 15°C for 24 hr, then 21°C for 24 hr and finally 28°C for 4 days, resulted in 4 times as much aflatoxin as when incubation was at a constant 28°C for 6 days (West *et al.* 1973).

The effect of pH of the medium on aflatoxin production is related to the type of substrate, the acids or bases used to alter the pH and other environmental factors. Jarvis (1971) observed that the fungus did not grow well when the pH was below 4.0. Maximum production of aflatoxin occurs at pH 5.5-7.0 (Buchanan and Ayres 1975).

Substrates with high carbohydrate concentration favor aflatoxin production (Diener and Davis 1969). Maximum growth of *A. parasiticus* occurred when the medium contained 10% glucose, but maximum toxin production required 30% glucose (Shih and Marth 1974). Glucose, galactose and sucrose seem to be acceptable carbohydrates, maltose and lactose are inferior to sucrose, and sorbitol and mannitol fail to support aflatoxin production.

Aflatoxin production was enhanced by 1-3% NaCl (Shih and Marth 1972). At 8% NaCl, growth and toxin production were inhibited at 21°C, but not at 28° or 35°C. No growth occurred at 14% NaCl.

Trace metals influence aflatoxin production. Zinc stimulates the formation of aflatoxin. Increasing the Zn level from 0 to 10 μg/ml increased the amount of aflatoxin over 1,000 fold (Marsh *et al.* 1975). Less aflatoxin was produced at 25 μg/ml than at 10 μg/ml of Zn. Low levels of Mn increase the amount of aflatoxin, but copper tends to decrease the production of aflatoxin. Iron has been reported to both increase and decrease aflatoxin production. Barium inhibits production of aflatoxin. Growth in a barium-containing medium has resulted in the development of an atoxigenic mutant.

Generally, molds are aerobic, so it might be expected that oxygen is needed for growth and aflatoxin production. Reducing the oxygen concentration or increasing the carbon dioxide or nitrogen concentration reduces aflatoxin production (Wilson and Jay 1975). The inhibitory effect of CO_2 is enhanced as the temperature and RH are lowered.

The time needed for aflatoxins to appear varies with the environmental factors. The toxins may appear in 24 hr, peak production may require one or two weeks or longer and, with adverse conditions, it may not be produced.

The toxin is produced in the fungal mycelium, and is released into the substrate.

Competitive growth of fungi can result in inhibition of aflatoxin production. *Scopulariopsis brevicaulis, Nocardia* and *Aspergillus niger* detoxify aflatoxins. *Rhizopus oryzae* metabolizes aflatoxin (Jarvis 1971). *A. chevalieri* and *A. candidus* reduced or prevented aflatoxin formation by *A. parasiticus* (Boller and Schroeder 1973, 1974).

Even in pure cultures of *A. flavus* or *A. parasiticus,* the aflatoxin level falls after reaching a maximum. This seems to be true, even though some strains appear to be nondegraders of the aflatoxin. Perhaps there is some nonspecific chemical mechanism for degradation of the toxin.

Sources of Aspergillus and Aflatoxin.—The mold is found in soil, air and on or in living or dead animals and plants. Aflatoxins have been found in foods at the time of harvest, indicating naturally occurring contamination. However, poor harvesting and storage systems may increase the proliferation of the mold and production of aflatoxins.

Any food that becomes contaminated with spores of toxigenic *A. flavus* or *A. parasiticus,* is capable of supporting growth of the mold, and is held in an environment favorable for growth and toxin production, can become a source of aflatoxin. Foods found to contain aflatoxins are listed in Table 6.32. In the USA, the main concern has been with aflatoxins on peanuts, cottonseed and corn, but now other foods are receiving attention.

TABLE 6.32

FOODS AND FEEDS FOUND TO CONTAIN AFLATOXIN

almonds	noodles
almond paste	nutmeg
bakery products	oats
barley	palm kernel
beans	pea
beer	peach seed paste
black pepper	peanuts
Brazil nut	peanut butter
bread	peanut meal
capsicum pepper	pecan
cayenne pepper	pepper corns
chili powder	pistachio nut
cocoa bean	raisin
cocoa	rapeseed
coconut	rice
coconut oil cake	rye
copra	sesame
corn	sorghum
corn grits	soybean
cottonseed	soybean meal
cottonseed meal	spaghetti
cowpea	sunflower flour
dried chili pepper	sunflower meal
dried fish	sweet potato
flour	various leafy foods
garlic	wheat
grape wine	wheat flour
hazel nut	
locust bean	
milk	
millet	

The examination of peanut meal samples from eight states in 1970 revealed that 83 samples had less than 10 $\mu g/kg$ aflatoxin, while 223 samples had from 31 to 100 $\mu g/kg$ and 70 samples contained over 100 $\mu g/kg$ of aflatoxin (Wessel and Stoloff 1973).

Peanuts can become contaminated with mold and aflatoxin during growth, harvesting or storage. Peanuts grown in soil used consistently for peanuts tend to have a higher accumulation than when crops are rotated. Infection of peanuts with toxigenic fungi usually is associated with insect or mechanical damage to the shells before, during or after harvesting.

Mechanically harvested peanuts are more apt to become contaminated than those that are hand picked, due to the damage done by the mechanical harvesters. After harvesting, it is essential that the peanuts be rapidly dried to 8% or less moisture content, and stored under conditions so that the moisture content will not increase.

When contaminated peanuts are pressed to remove the oil, most of the toxin remains with the meal cake, and only small amounts remain with the oil. Clarification of the oil removes much of the residual toxin. During refining, a hot alkali wash and bleaching treatment given the oil removes any remaining aflatoxin. Thus, peanut oil is considered to be safe. However, the peanut cake or meal can be highly contaminated.

Aflatoxin was found in 9% of the corn samples examined in 1971 (Wessel and Stoloff 1973). The greatest potential for contaminated samples occurs in the Southeast USA. This is due to the more conducive climate for mold growth during the late fall harvesting season than is found in the Midwest USA. In 1977, nearly 50% of the corn in the southeastern states contained aflatoxin above the allowable level of 20 $\mu g/kg$. In some tests, the samples had 200 $\mu g/kg$.

Insect damaged corn was detected in 90% of the aflatoxin containing samples of corn in the field (Anderson *et al.* 1975). Hesseltine *et al.* (1976) reported a direct correlation of insect damage to aflatoxin level and mold contamination of corn kernels.

From the list of foods found to contain aflatoxin, it is evident that there can be a carryover of aflatoxin from the field crop through processing to human food. An example of carryover is wheat to wheat flour and then to bread, spaghetti and noodles.

For corn, 90% is used as animal feed, and the remainder is used for breakfast food, grits, corn meal, starch, sugar, syrup and alcohol. There is no carryover of aflatoxin into the main edible products of the wet milling process (starch, sugar, syrup). However, the residuals of the process (steepwater, gluten and germ meal) retain the aflatoxin. The aflatoxin is not carried through the distillation process for alcohol production.

The feeding of aflatoxin contaminated corn to animals may result in the meat, eggs or milk having aflatoxin. Aflatoxins have been isolated from the liver and muscle tissue of poultry and swine. The B_1 aflatoxin in the rations of dairy cows is transformed into M_1 aflatoxin and found in the milk. Aflatoxin M_1 is about as potent as B_1. The amount of M_1 deposited in milk is proportional to the intake of B_1 in the diet. Dietary levels as low as 0.05 to 0.10 $\mu g/kg$ can give detectable amounts of aflatoxin in milk. Since milk is the main dietary item of infants and young animals, the presence of aflatoxin in this food can be especially hazardous. Measureable amounts of aflatoxin B_1 are in eggs from hens with as little aflatoxin as 0.05 to 0.10 $\mu g/kg$ in their diet. Smith *et al.*

(1976) believed the incidence and severity of aflatoxicosis in farm animals is greater than realized. With animals fed aflatoxins, levels of 6 μg/kg in the liver and 0.1 μg/kg in swine muscle were found (Murthy *et al.* 1975).

Methodology.—It would be an endless task to discuss all of the existing methods for determining aflatoxins in food or feed.

Although the presence of toxigenic mold in food can be a danger signal, it is possible to have toxin present in a food product with no visible mold. Therefore, it is necessary to analyze the product for the toxin.

The first problem is to obtain a representative sample. Aflatoxin is not homogeneously distributed throughout a product. If only one highly contaminated kernel or seed is included in the analyzed sample, the apparent aflatoxin content may be significantly higher than that of the lot being sampled. Because of the unequal distribution, it is difficult to obtain a representative sample, variation between replicates may be great and the aflatoxin content of any lot of product is difficult to estimate with any accuracy.

The simplest test, which can be used for some commodities, is merely the scanning of grain or seed with long wave ultraviolet light (365 nm). A bright greenish yellow or greenish gold fluorescence is correlated with the presence of aflatoxin. Infrequently, nonfluorescent samples contain aflatoxin, but usually only in small amounts (less than 30 μg/kg). Fennell *et al.* (1973) stated that good correlation was observed between the presence of *A. flavus,* a bright greenish yellow fluorescence, and aflatoxin in naturally contaminated white corn. Some nonfluorescent undamaged kernels assayed at less than 20 μg/kg, and some damaged kernels and fragments that contained 3,000 μg/kg did not show fluorescence. The fluorescence is not due to the presence of aflatoxin, but rather to the fungal produced kojic acid. Over 90% of the strains of aflatoxin producing aspergilli also form kojic acid.

Bothost and Hesseltine (1975) tested the bright greenish yellow fluorescence for screening several commodities for aflatoxin. They reported that it could be used as a presumptive test for aflatoxin in wheat, oats, barley, corn and sorghum. It was not satisfactory for peanuts, rice or soybeans.

To determine aflatoxins, they are extracted from the sample with an organic solvent, separated by thin layer chromatography and detected with long wave UV light (Fig. 6.21). The aflatoxins are soluble in methanol, acetone, chloroform and various other polar solvents and are sparingly (10-30 μg/ml) soluble in water. The solvents used for extraction depend upon the food or feed being analyzed.

Confirmation of the presence of aflatoxin can be accomplished chemically, or by mass spectral methods (Haddon *et al.* 1977). The extraction and thin layer chromatography method is sensitive to 2-4 μg/kg of aflatoxin (Romer 1973).

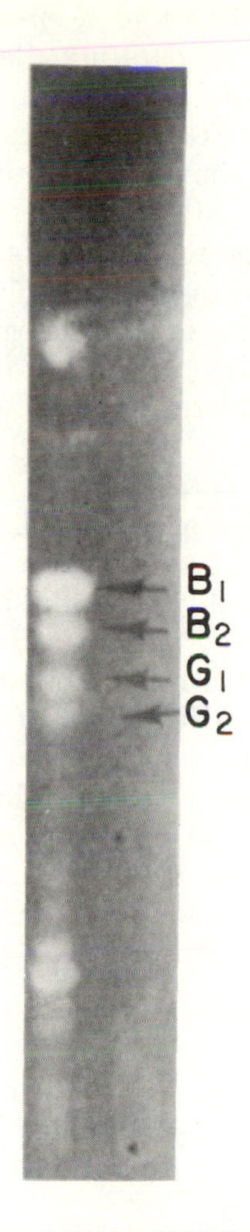

Courtesy of Wilson (1968)

FIG. 6.21. THIN-LAYER CHROMATOGRAM SHOWING FOUR AFLATOXINS ALONG WITH OTHER FLUORESCENT SUBSTANCES

Many variations of the physiochemical assay of aflatoxins have been reported.

In one system, a minicolumn or florisil tube is used for separation. A direct readout of aflatoxin concentration in the tube can be obtained with a Velasco fluorotoxin meter. Alternately, the column can be observed with longwave UV light (Romer 1975; Shotwell *et al.* 1977; Velasco 1972).

According to Stoloff (1976) high pressure liquid chromatography (HPLC) shows promise for separating aflatoxins. This system has been used not only for aflatoxins, but to separate other mycotoxins (Engstrom *et al.* 1977; Panalaks and Scott 1977; Stubblefield and Shotwell 1977). This may be adopted as the official method for aflatoxin.

Radioimmunoassays for aflatoxins were described by Langone and Van Vunakis (1976). They found the system was sufficiently sensitive and specific for aflatoxins in food or biological samples.

There are several systems for the biological assay of aflatoxin. In the chicken embryo test, specific amounts of aflatoxin deposited in a fertile egg will destroy the life of the embryo. Typical lesions develop in the embryo with subacute levels (less than 0.1 μg/kg) of aflatoxin B_1.

The effect of aflatoxin on ducklings can be used for assay purposes. Aflatoxins inhibit cell cleavage in fertilized mollusk (*Bankia setacea*) eggs without preventing nuclear division. The resultant cells are multinuclear. Over 60% mortality of brine shrimp (*Artemia salina*) is obtained in 24 hr by 0.5 μg/ml (aflatoxin B_1). With 1 μg/ml, greater than 90% mortality occurs. At a concentration of 1 μg/ml, aflatoxin B_1 is lethal to zebra fish larvae. With a concentration of 0.15 μg/ml, the water flea (*Daphnia*) is killed in 40 hr. Schuller *et al.* (1976) stated the chick embryo bioassay is the most useful of these biological systems.

The USFDA insists on a positive presumptive physiochemical test, a positive chemical derivative test and a positive biological toxicity test before taking regulatory action for aflatoxin.

Control of Aflatoxins.—The systems for controlling aflatoxins in foods or feeds are essentially the same as for controlling any microbial toxin. We might prevent contamination of the food by toxigenic *A. flavus* or *A. parasiticus* and inhibit their growth and toxin production. We can analyze food for aflatoxin and, if present, remove, destroy or detoxify the toxin.

Prevent Contamination.—Some reports infer that many crops are free of the toxigenic fungal spores until harvested. The ubiquitous nature of these fungi makes it difficult to grow crops without their being subject to fungal contamination. If we assume that crops are contaminated with the fungi, we can plan remedies to prevent the presence of aflatoxins in food or feed.

Prevent Fungal Growth.—The fungi attack damaged seeds more readily than they attack sound seeds. Control of insects and using care in harvesting can reduce the number of damaged seeds. The removal of damaged seeds and foreign material will help control growth and aflatoxin production.

Present evidence suggests that most of the production of and contamination with aflatoxin occurs after the seeds have begun to dry, but before they attain a moisture level to prevent growth of the mold.

The development of a low moisture content in the seeds, and storage at low RH is perhaps the simplest and best method for controlling growth and aflatoxin production. Rapid drying of the crop does not necessarily result in lower mold contamination. The rapid drying of peanuts by artificial methods can produce small cracks or fissures in the pods. This can result in increased mold contamination, but the low moisture level will prevent growth and toxin production.

Even in low moisture products, if there is improper ventilation, localized areas can develop with sufficient moisture for fungal growth. These localized areas can be caused by sweating, movement of moisture through the food and biological activity of the seeds.

Cold storage will prevent growth of the aspergilli and production of aflatoxin, but other toxigenic molds can produce mycotoxins at low temperatures (0°-5°C). Hence, cold storage alone is not beneficial from an overall viewpoint.

Lowering the O_2 and increasing the CO_2 content of the air will reduce or prevent aflatoxin production. With a small amount of food, this might be practical, but it would hardly be worthwhile for the large supplies of grain that require storage.

The addition of certain chemicals will inhibit or reduce the production of aflatoxin. Davis and Diener (1967) found that soaking peanuts in a solution of *p*-aminobenzoic acid reduced the production of aflatoxin by 50%. Potassium sulfite and postassium fluoride also inhibited aflatoxin production by the molds.

Soaking peanuts in a 20% solution of dimethyl sulfoxide with subsequent inoculation with toxigenic mold, reduced the production of aflatoxin by over 90% from untreated control peanuts (Bean and Rambo 1975). These, and other reports, indicate that chemical methods can be helpful in preventing aflatoxin biosynthesis in certain situations.

Destroy Organism.—Spores of *A. flavus* do not survive a 45 sec treatment with ultraviolet light (Bean and Rambo 1975). Gamma radiation at levels of 0.25 to 1.0 Mrad inactivate fungi in stored products.

According to Doyle and Marth (1975), the conidia of *A. flavus* and *A. parasiticus* strains have a D_{55} of 3 to 29 min and D_{60} of 8 to 59 sec at pH 7.0. Hence, it might be possible to heat certain foods prior to storage to destroy the organisms.

Removal of Aflatoxin.—Aflatoxins can be removed from food by segregating obviously contaminated grains or kernels, or by extraction of the food with solvents. These procedures can be used for only certain foods or feeds.

The aflatoxin content of a batch of peanuts may be confined to a few highly contaminated kernels. When peanuts are removed from the shell, a sorting procedure is used to separate any moldy, discolored, shriveled or damaged raw peanuts. Hand sorting is effective, but also expensive, so electronic sorters are used. Results show that the aflatoxin content of peanut lots is significantly reduced by these segregation procedures.

The removal of mold growth from products, such as cheese, does not remove all of the aflatoxin, since aflatoxins tend to diffuse away from the mold mycelia. Hence, not only must the mold be removed, but also any food material that may contain diffused aflatoxin.

For rice, 60-80% of the aflatoxin appears in the bran portion with less aflatoxin in the milled rice.

Examination of seeds with UV light for greenish yellow fluorescence, and segregation of the fluorescent seeds, should aid in reducing the aflatoxin content of the remaining seeds.

For some types of foods, solvents have been used to extract the aflatoxins. Due to the distribution of toxin throughout the peanut, solvent extraction of whole kernels is ineffective. However, with crushed peanuts or peanut meal, extraction can be used. Besides peanut meal, the extraction can be useful for other oilseed meals. Detroy *et al.* (1971) reviewed the systems, solvents and results of the extraction procedures for removing aflatoxins.

Biological systems might be used to degrade aflatoxins (Mann and Rehm 1976). *Corynebacterium rubrum, Aspergillus niger, Trichoderma viride* and *Mucor ambiguus* degrade aflatoxin B_1 to aflatoxin R_0. Before such systems can be used in industry, much more research is needed, as well as approval by the FDA. Aflatoxin R_0 reportedly is less toxic than B_1 to ducklings, but is carcinogenic to trout.

Inactivation of Aflatoxin.—Either chemical or physical treatments can be used to inactivate or detoxify aflatoxin. Methods for detoxification were reviewed by Dollear (1969).

Chemical treatments that have been reported for inactivation include hydrogen peroxide, chlorine gas, sodium hypochlorite, benzoyl peroxide, sodium perborate, sodium hydroxide, ammonia, methylamine, chlorine dioxide, formaldehyde, nitrogen dioxide and ozone.

Ammonia is effective in inactivation of aflatoxins in cottonseed and peanut meals. The major reaction product of aflatoxin B_1 and NH_4OH at 100°C under pressure has been isolated and characterized by Lee *et al.* (1974). The product is not fluorescent, exhibits phenolic properties and lacks the lactone group characteristic of B_1. The new compound is a result of opening of the B_1 lactone ring, formation of the ammonium salt of the hydroxy acid, and decarboxylation of this beta-keto acid. Ammoniation inactivates the aflatoxin, reduces the carcinogenicity to rainbow trout and does not reduce the nutritive value of treated corn (Brekke *et al.* 1977).

Treatment of peanut meal with NaOH reduces the aflatoxin content. A combination of calcium hydroxide and formaldehyde reacts with aflatoxin in highly contaminated meal, so that it is acceptable for feed use (Codifer *et al.* 1976).

Sodium hypochlorite (NaOCl), or bleach, has been effective in inactivation of aflatoxin. According to Dollear (1969), the most promising reagents for detoxifying aflatoxin are ammonia, methylamine, sodium hydroxide and ozone.

Oil, as extracted from oil seeds (peanuts and cottonseed) may contain small amounts of aflatoxins. The oil is treated with alkali and bleach to purify it. This processing reduces the aflatoxin content from 812 $\mu g/kg$ to less than 1 $\mu g/kg$ (Kensler and Natoli 1969).

The primary physical treatment used to reduce toxicity of aflatoxins is heat. Roasting peanuts at 150°C for ½ hr reduces aflatoxin B_1 about 80% and B_2 about 60% (Lee *et al.* 1968). Waltking (1971) roasted peanuts at 204°C and found an average loss of 40-50% aflatoxin B_1 and G_1 and 20-40% of B_2 and G_2. Oil roasting of peanuts at 163°C for 3 min reduces B_1 and G_1 by 65% (Lee *et al.* 1969).

Dollear (1969) reviewed the effect of heat treatment of cottonseed meal to reduce the aflatoxin content. Moisture is required for aflatoxin destruction by heat. Cooking of cottonseed meal, even at a high moisture level, did not reduce aflatoxin to an acceptable level. Aflatoxins are considered to be heat stable, at least at normal cooking temperatures.

Control by Regulation.—The production of aflatoxin by toxigenic molds cannot be controlled by a regulation of a government agency. However, it is possible to regulate against the use of food or feed if it contains aflatoxin.

Aflatoxins are considered to be carcinogens. The Delaney clause to Section 409 of the Federal Food, Drug and Cosmetic Act prohibits establishment of a tolerance for additives which "induce cancer when ingested by man or animals."

Instead of tolerances, the FDA has used administrative guidelines, the acceptable level depending upon the ability to detect aflatoxin. In 1964, a control program was established requiring the analysis of shelled peanuts by a certified laboratory. Peanuts with over 40 $\mu g/kg$ (ppb) could not be used in feed. In 1965, this was lowered to 30 $\mu g/kg$, and in 1969, to 20 $\mu g/kg$. Present methods enable the detection of aflatoxin at levels of 2.5 μg/kg or less. Ideally, there should be no aflatoxin, but realistically this would eliminate a large amount of food from an already overtaxed food supply. In 1975, the FDA proposed a formal tolerance of 15 μg/kg for aflatoxin in peanuts and peanut products.

Besides peanuts and peanut products, FDA surveillance activity is directed at any food that might contain aflatoxin and action is taken against any food or animal feed with 20 μg or more aflatoxin/kg. This level presumably will be lowered to 15 μg/kg. Animal feed is included, since aflatoxins consumed by animals can be found in the food produced by the animals (milk, meat, eggs).

The continual surveillance and the elimination of aflatoxin contaminated products from the food or feed supply constitutes a system of controlling these toxins.

VIRAL HEALTH HAZARDS

For most foodborne illnesses caused by bacteria, the organisms must proliferate in the food for illness to occur. Viruses that cause human illness cannot multiply in food. Thus, a virus that contaminates a food either survives or becomes inactivated. This means that viruses tend to be present only in very low concentration. This has made it difficult to detect viruses in foods. Since viruses do not multiply in food, and generally become inactivated, their possible importance in foodborne diseases often is overlooked.

Bryan (1973A) listed four viral diseases in which there is evidence of foodborne transmission. These diseases are infectious hepatitis, poliomyelitis, Bolivian hemorrhagic fever and Russian spring-summer encephalitis. He also listed eight viral diseases which might be foodborne, but proof is lacking. These include reovirus infections, serum hepatitis and nonbacterial gastroenteritis.

There were 22 outbreaks of confirmed foodborne viral illness from 1972 to 1976. This represents only a small part of the total outbreaks of foodborne illness. Many viral diseases are neither reported nor confirmed. One problem is that the isolation and identification of viruses require special techniques.

Humans are host to over 100 members of the enteric virus group (polioviruses, adenoviruses, echoviruses, reoviruses and coxsackieviruses). Ingested viruses must be able to survive the digestive enzymes, the acidity of the stomach and the bile in the duodenum. Some viruses survive these barriers. Certain viruses, such as those in the enteric group, can initiate a primary infection in the cells of the intestines. The ease with which viruses pass across the intestinal epithelium and cause infection may depend upon the type of food eaten, the stage of digestion and the presence of other microorganisms.

Studies of infectious doses indicate that only a few viral particles can cause infection. It may be that only one virus can produce an infection, especially in a hypersusceptible host. In most cases, little is known as to which host cells are involved, how the viruses pass through the mucous membranes or viral factors that influence the initiation or spread of the infections.

Food animals are susceptible to a wide variety of viral agents. Systemic infections of these animals are important since the virus spreads throughout the tissues and organs of the animal. Although viruses are usually species specific, a few can cause infection in more than one species of animal, including humans.

Viruses causing human diseases have been isolated from various domestic animals. In some cases, the same virus was isolated from animals and from humans who had close contact with the animals.

It has been difficult to implicate food in viral diseases, due to the long incubation periods and the difficulty of isolating viruses from foods.

Human and animal viruses have been recovered from raw and heated milk, dairy products, meat, eggs, oysters, mussels, clams and crabs. Raw or only partially cooked foods are those primarily implicated as vehicles for virus transmission.

The first viral illness reported to be transmitted by foods was poliomyelitis. Raw milk was the predominant food vehicle for this illness. Pasteurization of milk has been an effective control, since no food-associated outbreak of poliomyelitis has been reported in nearly 25 years.

Viruses also are found in sewage and polluted water. They can contaminate various inanimate objects, which can act as sources to contaminate other items, such as food.

Survival of Viruses

Mahl and Sadler (1975) studied the survival of various viruses on hard-surfaced inanimate objects (glass, stainless steel, tile). Although inactivation was observed, some viruses persisted for eight weeks at room temperature (25°C) and low RH (3-7%). At 37°C and 55% or 93-96% RH, survival time varied from one day to eight weeks, depending upon the virus type.

The action of time, sunlight, sewage, water treatment processes, physical and chemical inactivation by suspended and dissolved materials, bacterial antagonism, as well as simple dilution serve to decrease the viral concentration in water. In a rather extensive survey, viruses were not detected in chlorinated drinking water (Clarke *et al.* 1975). Enteroviruses are inactivated less rapidly in sterile lake water than in a lake (Herrmann *et al.* 1974). Proteolytic enzymes, including those of microbial origin, can degrade the protein coat of certain viruses, and cause inactivation, which accounts for greater survival in the sterile water.

The stability of viruses in foods is determined by the type of virus, the moisture and pH of the food and the temperature of storage. Enteric viruses survive longer than influenza viruses (Cliver *et al.* 1970). At low temperatures, poliovirus is extremely stable in food with pH 7.0 or greater. As the temperature of storage is increased from 4°C, the inactivation also is increased. They found that enteroviruses persisted in low moisture food for over 2 weeks at room temperature, and over 2 months in the refrigerator.

Herrmann and Cliver (1973A) inoculated ground beef and Thuringer sausage with a coxsackie virus. When ground beef was held at 4°C, about 85% of the viruses survived for 8 days, but less than 1% survived for 14 days. During fermentation of Thuringer sausage, (30°C for 24 hr), 84% of the inoculated viruses were inactivated. Holding the sausage at 49°C for 6 hr resulted in a 99.9% reduction of infective viruses.

Kantor and Potter (1975) reported that a poliovirus and echovirus persisted in high numbers and were virtually unaffected during the commercial production of salami and cervelat sausages.

McKercher *et al.* (1978) processed hogs infected with hog cholera virus and African swine fever virus. They recovered viruses from hams before heating, but not from the partly cooked hams. Both viruses were recovered from prepared salami and pepperoni sausages, but not after curing.

Larkin *et al.* (1976) irrigated lettuce and radish crops with sewage sludge and sewage effluent inoculated with poliovirus. This virus persisted for up to 36 days on these crops. They suggested that a cycle of infection could be established if contaminated effluent or sludge were used on crops in the food chain.

Enteric viruses survived at least 5 months in food stored at −20°C.

Inactivation

Viruses can be removed from materials or be inactivated, by essentially the same systems used to control bacteria. The viruses in sewage were reduced by means of a rotary-tube type of trickling filter (Clarke and Chang 1975). The reduction in virus content depended upon the type of virus, as well as the flow rate through the filter. To assure a virus-free effluent, further treatment is needed.

Shellfish tend to rid themselves of contaminants when removed from polluted water and placed in clean water. The rate of cleansing depends upon the initial uptake and contamination. After 48-72 hr of depuration, a few shellfish still harbored virus (Seraichekas *et al.* 1968). Canzonier

(1971) found that most of the virus contaminants were eliminated in 24 to 48 hr, but some viruses persisted for days or even weeks. Thus, depuration cannot be relied upon to eliminate all viruses from shellfish.

Just as viruses (bacteriophage) can affect bacteria, certain bacteria may inhibit some viruses. Hydrogen peroxide generating bacteria are antagonistic to poliovirus (Klebanoff and Belding 1974), culture filtrates of *Proteus mirabilis* contain a substance inhibitory to Sindbis and vesicular stomatitis viruses (Mahdy and Bansen 1974), and proteolytic bacteria inactivate coxsackie virus type 9 (Herrmann and Cliver 1973B).

Viruses can be inhibited or inactivated by various chemical agents, such as β-propiolactone, guanidine hydrochloride, quaternary ammonium, phenolic and iodophor compounds (Gaustad *et al.* 1974), ozone (Burleson *et al.* 1975), glutaraldehyde (Saitanu and Lund 1975) and calcium elenolate (Heinze *et al.* 1975).

Viruses are more resistant to gamma radiation than are other types of microorganisms. Considering the high radiation doses necessary to inactivate viruses, the use of radiation to inactivate viruses is not practical. Dugan and Trujillo (1975) suggested radiation at 45°C, since there is a synergistic action of radiation and heat in the inactivation of viruses.

Thermal inactivation seems to be an acceptable and practical method of controlling viruses in food. Many viruses are inactivated in 5 min when heated to 65°C. Therefore, it is sometimes assumed that properly cooked or pasteurized foods are not a concern of public health unless the product is recontaminated. However, DiGirolamo *et al.* (1972) reported that the time used to cook crabs may not be sufficient to inactivate viruses. Filppi and Banwart (1974) inoculated ground beef with poliovirus type 1 and heated the mixture at various temperatures. From their data, it is evident that meat, when highly contaminated with a potentially pathogenic virus, might be a source of infection, even if heated to 70°C. Sullivan *et al.* (1975) found no surviving poliovirus 1 or coxsackievirus B-2 in hamburgers broiled to either 71°C or 76.7°C.

Cliver (1971) reported that inoculated polio 1 virus was not completely inactivated in oysters by stewing, frying, baking or steaming. Pasteurization treatments, such as in egg products or milk, cannot be relied upon to destroy large numbers of heat resistant viruses.

Methodology

The analysis of food for viruses requires techniques different from those used for bacterial enumeration. Viruses are determined by inoculation into host cells for enumeration, by observation with an electron microscope, or by immunological methods, either by determining viruses directly or the antibody produced by a host.

When viruses are determined by using cells, such as tissue cultures, liquid foods, or those made into a liquid suspension, can be inoculated directly. However, if present, the viruses are usually low in numbers. The viruses cannot be enriched like bacteria, since they do not multiply in the food or a simple medium. Therefore, concentration of the viruses from a large food sample is needed.

For concentration, either the food solids or viruses, or both, must be removed from the suspension. Food solids can be removed by various systems. For ground beef, Tierney *et al.* (1973) used glass wool, woven fiber glass, cheese cloth or low-speed centrifugation to remove the food particles. The glass wool and fiber glass were more effective than the other two systems for the recovery of inoculated viruses. Even then there was only a 50% recovery.

The viruses in the supernate are concentrated. In a review of methods of detecting viruses in water, Hill *et al.* (1971) reported that the most promising methods for concentrating viruses are membrane-adsorption, adsorption to precipitable salts, iron oxide or electrolytes, aqueous polymer two-phase separation and a soluble alginate filter technique. Less effective are ultracentrifugation, electrophoresis or electro-osmosis and hydroextraction (dialysis).

Bacteria that are present in food can interfere with the growth of cells in a tissue culture. Therefore, if not removed, they must be inhibited. This is accomplished by adding antibiotics that inactivate the bacteria, but not the tissue culture or viruses. Sullivan *et al.* (1970) used chloroform to destroy the contaminating bacteria. After reacting with the bacteria, the chloroform must be removed prior to analysis for viruses on the tissue culture, since residual chloroform can affect the cells.

To determine viruses in the concentrate, the number of plaque forming units (PFU) is determined, using a susceptible tissue culture. A measured amount of viral concentrate (0.1 ml) is added to an acceptable cell sheet and, after adsorption of the viruses to the cell sheet (about 30 min), 2 ml of an agar with nutrients is used to overlay the cell-virus sheet. An overlay of a liquid nutrient may then be added, and the prepared plate is incubated at a time and temperature dependent upon the virus being determined. Afer incubation, the tissue culture is observed for plaques caused by cell destruction. Plaque counting can be facilitated by adding a stain, such as crystal violet.

Due to the cell specificity of viruses, the more types of tissue culture cells that are used, the more likely that most types of viruses can be detected and isolated.

Viruses causing illnesses, such as winter vomiting disease and hepatitis, have not been cultured. In some cases, procedures using the electron microscope have been used to detect particles, which apparently are viruses. Fong *et al.* (1975) pointed out that when specific antibodies to the viruses are present, such as in illnesses, the viruses cannot be detected by routine isolation methods. In these instances, the electron microscope can be utilized to visualize the virus-antibody complex.

The detection of viral antibodies is of value only when the virus is in a living host, which produces the antibody. These systems are of value to a clinical microbiologist, but not to a food microbiologist.

Serological techniques can be used to aid in the identification of viruses that are isolated from plaques formed in tissue cultures. Serological systems include fluorescent techniques, immunoperoxidase, radioimmunoassay, hemagglutination inhibition, and immune electron microscopy.

Diseases

Although several human illnesses are caused by viruses, only those that may be considered a foodborne illness are discussed.

Infectious Hepatitis.—This illness also is known as hepatitis A, with the etiologic agent, hepatitis virus A. Hepatitis B, or serum hepatitis, caused by hepatitis virus B, is similar to infectious hepatitis, but there are enough differences so that the illnesses can be differentiated. There is no proof that serum hepatitis is transmitted through food.

Person to person contact is perhaps the most common means of transmitting infectious hepatitis. Since the illness is spread by the fecal-oral route, waterborne and foodborne outbreaks can occur. The data of CDC (1976B) indicate only one death due to foodborne hepatitis from 1973 to 1975. However, according to CDC (1977D), there were 6,135 cases of hepatitis and 32 deaths due to the ingestion of raw shellfish from 1973 to 1975.

The incubation period for viral hepatitis varies from 10 to 50 days, with an average of 30 days. It is evident that tracing the source of infection back to a specific food is difficult with this long incubation period.

The symptoms and signs of the illness may include nausea, vomiting, lethargy, abdominal pain, diarrhea, fever, chills, anorexia, lassitude, sore throat, bile in urine (dark urine) and jaundice. Also, the level of serum glutamic-oxalacetic transaminase (SGOT) and/or serum glutamic-pyruvic transaminase (SGPT) is at least twice as high as the normal laboratory standard, and there is a total level of serum bilirubin over 2 mg/100 ml. The illness may be rather mild, allowing cases to be ambulatory. In other cases, liver injury occurs, resulting in cirrhosis.

Various foods have been implicated as the source of the virus in outbreaks of infectious hepatitis. The eating of raw shellfish (oysters, mussels or clams) is an important source of hepatitis. This is due to the potential contamination of waters with raw sewage and the collection and concentration of the viruses by the shellfish. However, only one of the five reported foodborne outbreaks in 1973 was due to shellfish (oysters). In the other four outbreaks, the implicated foods were tossed salad, garnished hamburgers, spaghetti and sandwiches. In these outbreaks, infected food handlers, while shedding the virus, contaminated the food during preparation, and there was no further cooking of the food prior to serving.

Other foods that have been implicated in outbreaks include raw milk, unclean fruit, potato salad, avocado salad, roast pork, orange juice, cold cuts, custard, frozen strawberries, glazed doughnuts, whipped cream cakes and freeze-balls. Most of these foods were contaminated through the carelessness of an infected food handler. Due to this source of contamination, almost any type of food may be involved. In one outbreak, raw watercress obtained from a contaminated stream, was the source of the virus.

Due to the limitations of hosts (human and non-human primates) and failure to multiply in tissue culture, it has been difficult to study the virus. The illness is diagnosed by clinical manifestations (symptoms and/or an increase in SGOT or SGPT). By the use of immune electron microscopy, the virus has been detected (Maynard *et al.* 1975). Inoculation of non-human primates with stool filtrates containing these inoculated virus-like particles (27 nm) resulted in hepatitis (Maynard *et al.* 1975). Krugman *et al.* (1975) reported the detection of hepatitis A antibody by specific immune adherence and complement-fixation tests. They stated the immune adherence test should be a valuable aid in diagnosis, for epidemiological surveys, for identification of susceptible and immune persons, for quantitative assays of gamma globulin and for identification of hepatitis A virus in attempts to propagate the virus in cell or organ culture.

A microtiter solid-phase immunoradiometric assay to detect hepatitis A antigen and antibody was described by Hollinger *et al.* (1976).

The disease can be attacked from four areas. First, people who have been exposed or might be exposed to the virus can be given immune serum globulin. This has been an effective deterrent for the secondary spread of the illness. Secondly, the beds from which seafoods, such as shellfish, are gathered should be clean and not polluted with raw sewage. Harvested shellfish should not be eaten raw. However, there are some people who insist on consuming these animals in the raw state.

The third means of control is to inactivate the virus in a food product by heating, radiation, drying or by chemicals. Unfortunately, little is known about the exact treatment to inactivate hepatitis virus A. Clams added to a pot of boiling water, heated until they opened and then eaten, were incriminated in an outbreak of hepatitis. This virus evidently is not inactivated by "limited heating." However, adequate cooking should inactivate the virus.

Inasmuch as most of the foodborne outbreaks seem to be caused by a food handler contaminating a food with no further cooking, the fourth area of control is the most important. The infected food handler must be controlled. This is not easily accomplished, since the virus may be excreted in stools during the incubation period, from 7-10 days prior to the onset of symptoms and by asymptomatic persons. However, those clinically ill persons should not be allowed to work with food until they are clearly convalescent. Employees can be screened by the analysis of the SGOT or SGPT levels of their serum. If the levels are over twice the normal values, then the employee should not work with food. Better personal hygiene of food handlers including more frequent handwashing and less handling of food and ingredients, would help prevent the spread of not only infectious hepatitis, but also other illnesses.

Nonbacterial Gastroenteritis.—This illness is one of the most prevalent of the gastroenteritis diseases, Nonbacterial gastroenteritis has been given several names, such as viral diarrhea, winter vomiting disease, epidemic collapse and epidemic nausea and vomiting. Acute nonbacterial gastroenteritis is thought to be transmitted from person to person, or from a common source with subsequent person to person transmission. The illness is self-limited and is characterized by such symptoms as diarrhea, vomiting, abdominal pain and malaise. Some cases have either vomiting or diarrhea, and some have both symptoms. The illness lasts 24-48 hr, and usually there are no complications.

The etiologic agents of these illnesses have been detected through immune electron microscopy. Also, using bacteria-free diarrheal stool filtrates, the syndrome can be transferred to other humans through the oral route. It can be transmitted through a series of human volunteers, indicating that the agent is not a toxin, but increases in amount in the human intestinal tract. Simple dilution of a nonmultiplying chemical agent would not cause illness through a series of human transfers. There is some immunity developed, since refeeding of previously ill volunteers does not produce the symptoms. Various histopathological, serological and biochemical responses to the ingested Norwalk and Hawaii agent were reported by Dolin *et al.* (1975) and Levy *et al.* (1976).

Since no etiologic agent has been isolated or grown outside a living human, it is difficult to determine if there is a common source for the virus. If there is, it could likely be a food that is contaminated by an infected food handler. As techniques are developed and improved, it may be possible in the future to determine if food is involved with these illnesses and if so, to what extent.

Infantile Gastroenteritis.—This is a sporadic and epidemic acute diarrheal illness in neonates. The virus associated with this illness has been called reovirus, reovirus-like, orbivirus-like, rotavirus or duovirus.

Using electron microscopy, Kapikian *et al.* (1974) described reovirus-like particles in stool specimens of infants and young children with severe gastroenteritis. The particles were related antigenically to the epizootic diarrhea of infant mice virus and the Nebraska calf diarrhea virus.

The virus and illness were discussed by Kimura and Murakami (1977), Lecce *et al.* (1978), Palmer *et al.* (1977) and Schoub *et al.* (1977).

Bryan (1973A) listed reovirus infections as possibly transmitted by foods.

Control

The control of viral infections involves essentially the same effort as other foodborne illnesses. There is a need for harvesting shellfish only from satisfactory waters and practicing good personal hygiene.

The viruses do not multiply in food and tend to become inactivated during storage. They can be inactivated with chemical and physical agents in a manner similar to bacteria.

BIBLIOGRAPHY

ABRAMSON, C. 1974. Staphyloccoccal hyaluronidase isoenzyme profiles related to staphylococcal disease. Ann. N.Y. Acad. Sci. *236*, 495-506.

AJMAL, M. 1968. Growth and toxin production of *Clostridium botulinum* type E. J. Appl. Bacteriol. *31*, 120-123.

AKSOYCAN, N., and SAGANAK, I. 1977. On the similarity of antigen between torulopsis and salmonella. 2. Communication. Zbl. Bakt. Hyg. I. Abt. Orig. *A238*, 489-493.

ALFORD, J.A., and PALUMBO, S.A. 1969. Interaction of salt, pH, and temperature on the growth and survival of salmonellae in ground pork. Appl. Microbiol. *17*, 528-532.

ALTENBERN, R.A. 1977. Effects of exogenous fatty acids on growth and enterotoxin B formation by *Staphylococcus aureus* 14458 and its membrane mutant. Can. J. Microbiol. *23*, 389-397.

ANDERSON, H.W., NEHRING, E.W., and WICHSER, W.R. 1975. Aflatoxin contamination of corn in the field. J. Agr. Food Chem. *23*, 775-782.

ANDO, Y. 1975. Studies on germination of spores of clostridial species capable of causing food poisoning (VIII). Effect of some chemicals on the growth from spores of a heat-resistant strain of *Clostridium perfringens* Type A. J. Food Hyg. Soc. Jap. *16*, 30-33.

ANGELOTTI, R., FOTER, M.J., and LEWIS, K.H. 1961. Time-temperature effects on salmonellae and staphylococci in foods. III. Thermal death time studies. Appl. Microbiol. *9*, 308-315.

AOAC. 1975. Official Methods of Analysis, 12th Edition. Association of Official Analytical Chemists, Washington, D.C.

APHA. 1976. Compendium of Methods for the Microbiological Examination of Foods. American Public Health Association, Washington, D. C.

ARNON, S.S. *et al.* 1977. Infant botulism. Epidemiological, clinical, and laboratory aspects. J. Amer. Med. Assoc. *237,* 1946-1951.

BAER, E.F., MESSER, J.W., LESLIE, J.E., and PEELER, J.T. 1975. Direct plating method for enumeration of *Staphylococcus aureus:* collaborative study. J. Assoc. Offic. Anal. Chem. *58,* 1154-1158.

BAINE, W.B., GANGAROSA, E.J., BENNETT, J.V., and BARKER, W.H. JR. 1973. Institutional salmonellosis. J. Infec. Dis. *128,* 357-360.

BAIRD-PARKER, A.C. 1972. Classification and identification of staphylococci and their resistance to physical agents. *In* The Staphylococci. J.O. Cohen (Editor). Wiley-Interscience, New York.

BAIRD-PARKER, A.C., and FREAME, B. 1967. Combined effect of water activity, pH and temperature on the growth of *Clostridium botulinum* from spore and vegetative cell inocula. J. Appl. Bacteriol. *30,* 420-429.

BANWART, G.J., and KREITZER, M.J. 1972. Effects of carbohydrates in growth medium on agglutination of several species of *Salmonella* with polyvalent H antiserum. Appl. Microbiol. *23,* 62-65.

BARBER, L.E., and DEIBEL, R.H. 1972. Effect of pH and oxygen tension on staphylococcal growth and enterotoxin formation in fermented sausage. Appl. Microbiol. *24,* 891-898.

BARRETT, D.H. *et al.* 1977. Type A and type B botulism in the north: first reported cases due to toxin other than type E in Alaskan Inuit. Can. Med. Assoc. J. *117,* 483-485.

BARTLETT, K.H., and TRUST, T.J. 1976. Isolation of salmonellae and other potential pathogens from the freshwater aquarium snail *Ampullaria.* Appl. Environ. Microbiol. *31,* 635-639.

BARTLETT, K.H., TRUST, T.J., and LIOR, H. 1977. Small pet aquarium frogs as a source of *Salmonella.* Appl. Environ. Microbiol. *33,* 1026-1029.

BARTLETT, K.H., TRUST, T.J., and LIOR, H. 1978. Isolation of bacteriophage 14-lysogenized *Salmonella* from the freshwater aquarium snail *Ampullaria.* Appl. Environ. Microbiol. *35,* 202-203.

BEAN, G.A., and RAMBO, G.W. 1975. Use of dimethyl sulfoxide to control aflatoxin production. Ann. N.Y. Acad. Sci. *243,* 238-245.

BEAN, P.G., and ROBERTS, T.A. 1975. Effect of sodium chloride and sodium nitrite on the heat resistance of *Staphylococcus aureus* NCTC 10652 in buffer and meat macerate. J. Food Technol. *10,* 327-332.

BENNETT, R.W. 1971. A review of collaborative studies of the microslide gel diffusion test for the detection of serologically identifiable staphylococcal enterotoxins. J. Assoc. Offic. Anal. Chem. *54,* 1037-1038.

BENNETT, R.W., and MCCLURE, F. 1976. Collaborative study of the serological identification of staphylococcal enterotoxins by the microslide gel double diffusion test. J. Assoc. Offic. Anal. Chem. *59,* 594-601.

BERGDOLL, M.S. 1970. Enterotoxins. *In* Microbial Toxins. Vol. III. Bacterial Protein Toxins. T.C. Montie, S. Kadis, and S.J. Ajl (Editors). Academic Press, New York.

BERGDOLL, M.S. 1972. The enterotoxins. *In* The Staphylococci. J.O. Cohen (Editor). Wiley-Interscience, New York.

BERGDOLL, M.S., CZOP, J.K., and GOULD, S.S. 1974. Enterotoxin synthesis by the staphylococci. Ann. N.Y. Acad. Sci. *236,* 307-316.

BERGDOLL, M.S., REISER, R., and SPITZ, J. 1976. Staphylococcal enterotoxins—detection in food. Food Technol. *30,* No. 5, 80-84.

BLACK, R.E. *et al.* 1978. Epidemic *Yersinia enterocolitica* infection due to contaminated chocolate milk. N. Engl. J. Med. *298,* 76-79.

BOLLER, R.A., and SCHROEDER, H.W. 1973. Influence of *Aspergillus chevalieri* on production of aflatoxin in rice by *Aspergillus parasiticus.* Phytopathology *63,* 1507-1510.

BOLLER, R.A., and SCHROEDER, H.W. 1974. Influence of *Aspergillus candidus* on production of aflatoxin in rice by *Aspergillus parasiticus.* Phytopathology *64,* 121-123.

BOOTH, R., SUZUKI, J.B., BERG, P.E., and GRECZ, N. 1972. Influence of weight and sex of mice in assaying spore-bound *Clostridium botulinum* type A toxin. J. Food Sci. *37,* 183-184.

BOROFF, D.A., and DASGUPTA, B.R. 1971. Botulinum toxin. *In* Microbial Toxins. Vol. IIA. Bacterial Protein Toxins. S. Kadis, T.C. Montie, and S.J. Ajl (Editors). Academic Press, New York.

BOROFF, D. A., and SHU-CHEN, G. 1973. Radioimmunoassay for type A toxin of *Clostridium botulinum.* Appl. Microbiol, *25,* 545-549.

BOTHAST, R.J., and HESSELTINE, C.W. 1975. Bright greenish-yellow fluorescence and aflatoxin in agricultural commodities. Appl. Microbiol, *30,* 337-338.

BOULANGER, Y., LALLIER, R., and COUSINEAU, G. 1977. Isolation of enterotoxigenic *Aeromonas* from fish. Can. J. Microbiol. *23,* 1161-1164.

BOYD, J.W., and SOUTHCOTT, B.A. 1971. Effects of sodium chloride on outgrowth and toxin production of *Clostridium botulinum* type E in cod homogenates. J. Fish. Res. Board. Can. *28,* 1071-1075.

BREKKE, O.L. *et al.* 1977. Aflatoxin in corn: ammonia inactivation and bioassay with rainbow trout. Appl. Environ. Microbiol. *34,* 34-37.

BROWN, D.F., and TWEDT, R.M. 1972. Assessment of the sanitary effectiveness of holding temperatures on beef cooked at low temperature. Appl. Microbiol. *24,* 599-603.

BRYAN, F.L. 1968. What the sanitarian should know about staphylococci and salmonellae in non-dairy products. II. Salmonellae. J. Milk Food Technol. *31,* 131-140.

BRYAN, F.L. 1969. What the food sanitarian should know about *Clostridium perfringens* foodborne illness. J. Milk Food Technol. *32,* 381-389.

BRYAN, F.L. 1972. Emerging foodborne diseases. I. Their surveillance and epidemiology. J. Milk Food Technol. *35,* 618-625.

BRYAN, F.L. 1973A. Diseases Transmitted by Foods (a classification and summary). Publication No. (HSM) 73-8237. U.S. Department Health, Education, and Welfare, Atlanta, Georgia.

BRYAN, F.L. 1973B. Activities of the Center for Disease Control in public health problems related to the consumption of fish and fishery products. *In* Microbial Safety of Fishery Products. C.O. Chichester and H.D. Graham (Editors). Academic Press, New York.

BRYAN, F.L. 1976. Public health aspects of cream-filled pastries. A review. J. Milk Food Technol. *39,* 289-296.

BUCHANAN, R.E., and GIBBONS, N.E. 1974. Bergey's Manual of Determinative Bacteriology, 8th Edition. Williams & Wilkins Co., Baltimore.

BUCHANAN, R.L., JR., and AYRES, J.C. 1975. Effect of initial pH on aflatoxin production. Appl. Microbiol. *30,* 1050-1051.

BURLESON, G.R., MURRAY, T.M., and POLLARD, M. 1975. Inactivation of viruses and bacteria by ozone, with and without sonication. Appl. Microbiol. *29,* 340-344.

CANZONIER, W.J. 1971. Accumulation and elimination of coliphage S-13 by the hard clam, *Mercenaria mercenaria.* Appl. Microbiol. *21,* 1024-1031.

CARPENTER, D.F., and SILVERMAN, G.J. 1974. Staphylococcal enterotoxin B and nuclease production under controlled dissolved oxygen conditions. Appl. Microbiol. *28,* 628-637.

CARPENTER, D.F., and SILVERMAN, G.J. 1976. Synthesis of staphylococcal enterotoxin A and nuclease under controlled fermentor conditions. Appl. Environ. Microbiol. *31,* 243-248.

CARROLL, B.J., and WARD, B.Q. 1967. Control of salmonellae in fish meal. Fish. Ind. Res. *4,* No. 1, 29-36.

CARRUTHERS, M.M. 1975. Cytotoxicity of *Vibrio parahaemolyticus* in HeLa cell culture. J. Infec. Dis. *132,* 555-560.

CARRUTHERS, M.M. 1977. In vitro adherence of Kanagawa-positive *Vibrio parahaemolyticus* to epithelial cells. J. Infec. Dis. *136,* 588-592.

CASMAN, E.P. 1965. Staphylococcal enterotoxin. Ann. N.Y. Acad. Sci. *128,* 124-131.

CASMAN, E.P., and BENNETT, R.W. 1965. Detection of staphylococcal enterotoxin in food. Appl. Microbiol. *13,* 181-189.

CDC. 1971. *Salmonella* Surveillance. Annual summary, 1970. Center for Disease Control, Atlanta, Ga.

CDC. 1973A. *Salmonella* Surveillance Report No. 115. Center for Disease Control, Atlanta, Ga.

CDC. 1973B. *Salmonella* Surveillance Annual Summary 1972. Center for Disease Control, Atlanta, Ga.

CDC. 1974A. Botulism in the United States, 1899-1973. Handbook for Epidemiologists, Clinicians, and Laboratory Workers. Center for Disease Control, Atlanta, Ga.

CDC. 1974B. Follow-up on *Salmonella eastbourne* outbreak. Morbidity Mortality *23,* 85-86.

CDC. 1974C. Wound botulism—California. Morbidity Mortality *23,* 355-356.

CDC. 1975A. Surveillance Summary. Botulism—United States, 1974. Morbidity Mortality *24,* 39.

CDC. 1975B. *Salmonella* Surveillance. Report No. 125. Center for Disease Control, Altanta, Ga.

CDC. 1976A. Foodborne and Waterborne Disease Outbreaks. Annual Summary 1974. Center for Disease Control, Atlanta, Ga.

CDC. 1976B. Foodborne and Waterborne Disease Outbreaks. Annual summary 1975. Center for Disease Control, Atlanta, Ga.

CDC. 1976C. Botulism—Alaska. Morbidity Mortality *25,* 399-400.

CDC. 1976D. Diarrheal illness on a cruise ship caused by enterotoxigenic *Escherichia coli.* Morbidity Mortality *25,* 229-230.

CDC. 1976E. Staphylococcus Typing Phages. Center for Disease Control, Atlanta, Ga.

CDC. 1977A. Foodborne and Waterborne Disease Outbreaks. Annual Summary 1976. Center for Disease Control, Atlanta, Ga.

CDC. 1977B. Salmonellae in precooked beef. Morbidity Mortality *26,* 310.

CDC. 1977C. *Shigella* Surveillance. Report No. 39. Annual Summary 1976. Center for Disease Control, Atlanta, Ga.

CDC. 1977D. Hepatitis Surveillance. Report No. 39. Center for Disease Control, Atlanta, Ga.

CDC. 1978A. Follow-up on infant botulism—United States. Morbidity Mortality *27,* 21-23.

CDC. 1978B. *Salmonella* seminar, Washington. CDC Veterinary Public Health Notes, Jan., 7-8.

CDC. 1978C. *Clostridium perfringens* food poisoning-California. Morbidity Mortality *27,* 164-165.

CESKA, M., GROSSMULLER, F., and EFFENBERGER, F. 1978. Solid-phase radioimmunoassay method for determination of *Escherichia coli* enterotoxin. Infec. Immunity. *19,* 347-352.

CHANG, P.C., and DICKIE, N. 1971. Heterogeneity of staphylococcal enterotoxin B. Can. J. Microbiol. *17,* 1479-1481.

CHARNEY, A.N., GOTS, R.E., FORMAL, S.B., and GIANNELLA, R.A. 1976. Activation of intestinal mucosal adenylate cyclase by *Shigella dysenteriae* 1 enterotoxin. Gastroenterology *70,* 1085-1090.

CHERINGTON, M., SOYER, A., and GREENBERG, H. 1972. Effect of guanidine and germine on the neuromuscular block of botulism. Curr. Ther. Res. *14,* 91-94.

CHERRY, W.B., DAVIS, B.R., EDWARDS, P.R., and HOGAN, R.B. 1954. A simple procedure for identification of the genus *Salmonella* by means of a specific bacteriophage. J. Lab. Clin. Med. *44,* 51-55.

CHESBRO, W., CARPENTER, D., and SILVERMAN, G.J. 1976. Heterogeneity of *Staphylococcus aureus* enterotoxin B as a function of growth stage: implication for surveillance of foods. Appl. Environ. Microbiol. *31,* 581-589.

CHILDERS, A.B., KEAHEY, E.E., and KOTULA, A.W. 1977. Reduction of *Salmonella* and fecal contamination of pork during swine slaughter. J. Amer. Vet. Med. Assoc. *171,* 1161-1164.

CHRISTIANSEN, L.N. and FOSTER, E.M. 1965. Effect of vacuum packaging on growth of *Clostridium botulinum* and *Staphylococcus aureus* in cured meats. Appl. Microbiol. *13,* 1023-1025.

CHUN, D., CHUNG, J.K., and SEOL, S.Y. 1974. Enrichment of *Vibrio parahaemolyticus* in a simple medium. Appl. Microbiol. *27,* 1124-1126.

CICCARELLI, A.S. *et al.* 1977. Cultural and physiological characteristics of *Clostridium botulinum* type G and the susceptibility of certain animals to its toxin. Appl. Environ. Microbiol. *34,* 843-848.

CIEGLER, A. 1975. Mycotoxins: occurrence, chemistry, biological activity. Lloydia *38,* 21-35.

CLARKE, N.A., and CHANG, S.L. 1975. Removal of enteroviruses from sewage by bench-scale rotary-tube trickling filters. Appl. Microbiol. *30,* 223-228.

CLARKE, N.A. *et al.* 1975. Virus study for drinking-water supplies. J. Amer. Water Works Assoc. *67,* 192-197.

CLIFFORD, W.J., and ANELLIS, A. 1975. Radiation resistance of spores of some *Clostridium perfringens* strains. Appl. Microbiol. *29,* 861-863.

CLIVER, D.O. 1971. Transmission of viruses through foods. Critical Reviews in Environmental Control. *1,* No. 4, 551-579.

CLIVER, D.O., KOSTENBADER, K.D. JR., and VALLENAS, M.R. 1970. Stability of viruses in low moisture foods. J. Milk Food Technol. *33,* 484-491.

CODIFER, L.P., JR., MANN, G.E., and DOLLEAR, F.G. 1976. Aflatoxin inactivation: treatment of peanut meal with formaldehyde and calcium hydroxide. J. Amer. Oil Chem. Soc. *53,* 204-206.

COHEN, J.O. 1974. Serological typing of staphylococci for epidemiological studies. Ann. N.Y. Acad. Sci. *236,* 485-494.

COLLINS, F.M., and CARTER, P.B. 1974. Cellular immunity in enteric disease. Amer. J. Clin. Nutr. *27,* 1424-1433.

COLLINS, W.S., JOHNSON, A.D., METZGER, J. F., and BENNETT, R.W. 1973. Rapid solid-phase radioimmunoassay for staphylococcal enterotoxin A. Appl. Microbiol. *25,* 774-777.

CORBEL, M.J. 1975. The serological relationship between *Brucella* spp., *Yersina enterocolitica* serotype IX and *Salmonella* serotypes of Kauffmann-White group N. J. Hyg. Camb. *75,* 151-171.

CRAIG, J.P. 1972. The enterotoxic enteropathies. Symposia Soc. Gen. Microbiol. *22,* 129-155.

CRISLEY, F.D., PEELER, J.T., ANGELOTTI, R., and HALL H.E. 1968. Thermal resistance of spores of five strains of *Clostridium botulinum* type E in ground whitefish chubs. J. Food Sci. *33,* 411-416.

CROSS, W.R., and NAKAMURA, M. 1970. Analysis of the virulence of *Shigella flexneri* by experimental infection of the rabbit eye. J. Infec. Dis. *122,* 394-400.

CZOP, J.K., and BERGDOLL, M.S. 1970. Synthesis of enterotoxin by L-forms of *Staphylococcus aureus.* Infec. Immunity *1,* 169-173.

DALY, C., LACHANCE, M., SANDINE, W.E., and ELLIKER, P.R. 1973. Control of *Staphylococcus aureus* in sausage by starter cultures and chemical acidulation. J. Food Sci. *38,* 426-430.

DAVIS, N.D., and DIENER, U.L. 1967. Inhibition of aflatoxin synthesis by *p*-aminobenzoic acid, potassium sulfite, and potassium fluoride. Appl. Microbiol. *15,* 1517-1518.

DAVIS, N.D., and DIENER, U.L. 1970. Environmental factors affecting the production of aflatoxin. *In* Proceedings of the First U.S.-Japan Conference on Toxic Micro-organisms. M. Herzberg (Editor). U.S. Govt. Printing Office, Washington, D.C.

DEGRE, M. 1967. Reproducibility of staphylococcal bacteriophage types by use of different concentrations of bacteriophage. Appl. Microbiol. *15,* 907-911.

DETROY, R.W., LILLEHOJ, E.B., and CIEGLER, A. 1971. Aflatoxin and related compounds. *In* Microbial Toxins. Vol. VI. Fungal Toxins. A. Ciegler, S. Kadis, and S.J. Ajl (Editors). Academic Press, New York.

DEWAART, J. 1973. Mycotoxins. Antonie van Leeuwenhoek J. Microbiol. Serol. *39,* 361-366.

DIENER, U.L., and DAVIS, N.D. 1969. Aflatoxin formation by *Aspergillus flavus. In* Aflatoxin Scientific Background, Control, and Implications. L.A. Goldblatt (Editor). Academic Press, New York.

DIENER, U.L., and DAVIS, N.D. 1970. Limiting temperature and relative humidity for aflatoxin production by *Aspergillus flavus* in stored peanuts. J. Amer. Oil Chem. Soc. *47,* 347-351.

DIGIROLAMO, R. *et al.* 1972. Uptake of bacteriophage and their subsequent survival in edible west coast crabs after processing. · Appl. Microbiol. *23,* 1073-1076.

DILLON, C.E. *et al.* 1969. Botulism—Pendleton, Oregon. Morbidity Mortality *18,* 69-70.

DOBRESCU, L., and HUYGELEN, C. 1973. Susceptibility of the mouse intestine to heat-stable enterotoxin produced by enteropathogenic *Escherichia coli* of porcine origin. Appl. Microbiol. *26,* 450-451.

DOLIN, R. *et al.* 1975. Viral gastroenteritis induced by the Hawaii agent. Jejunal histopathology and serological response. Amer. J. Med. *59,* 761-768.

DOLLEAR, F.G. 1969. Detoxification of aflatoxins in foods and feeds. *In* Aflatoxin Scientific Background, Control and Implications. L.A. Goldblatt (Editor). Academic Press, New York.

DONOWITZ, M., KEUSCH. G.T., and BINDER, H.J. 1975. Effect of shigella enterotoxin on electrolyte transport in rabbit ileum. Gastroenterology *69,* 1230-1237.

DONTA, S.T. *et al.* 1974. Tissue-culture assay of antibodies to heat-labile *Escherichia coli* enterotoxins. N. Engl. J. Med. *291,* 117-121.

DORNBUSCH, K., HALLANDER, H.O., and LÖFQUIST, F. 1969. Extrachromosomal control of methicillin resistance and toxin production in *Staphylococcus aureus.* J. Bacteriol. *98,* 351-358.

DOUGHERTY, T.J. 1974. *Salmonella* contamination in a commercial poultry (broiler) processing operation. Poultry Sci. *53,* 814-821.

DOYLE, M.P., and MARTH, E.H. 1975. Thermal inactivation of conidia from *Aspergillus flavus* and *Aspergillus parasiticus.* 1. Effects of moist heat, age of conidia, and sporulation medium. J. Milk Food Technol. *38,* 678-682.

DUGAN, V.L., and TRUJILLO, R. 1975. Heat-accelerated radioinactivation of attenuated poliovirus. Radiat. Environ. Biophys. *12,* 187-195.

DUITSCHAEVER, C.L., and IRVINE, D.M. 1971. A case study: Effect of mold on growth of coagulase-positive staphylococci in cheddar cheese. J. Milk Food Technol. *34,* 583.

DUNCAN, C.L., and STRONG, D.H. 1969A. Ileal loop fluid accumulation and production of diarrhea in rabbits by cell-free products of *Clostridium perfringens.* J. Bacteriol. *100,* 86-94.

DUNCAN, C.L., and STRONG, D.H. 1969B. Experimental production of diarrhea in rabbits with *Clostridium perfringens.* Can. J. Microbiol. *15,* 765-770.

DUNCAN, C.L., STRONG, D.H., and SEBALD, M. 1972. Sporulation and enterotoxin production by mutants of *Clostridium perfringens.* J. Bacteriol. *110,* 378-391.

DUNCAN, M.S. and ADAMS, A.W. 1972. Effects of a chemical additive and of formaldehyde-gas fumigation on *Salmonella* in poultry feeds. Poultry Sci. *51,* 797-802.

EARLE, P.M., and CRISLEY, F.D. 1975. Isolation and characterization of *Vibrio parahaemolyticus* from Cape Cod soft-shell clams *(Mya arenaria).* Appl. Microbiol. *29,* 635-640

EKLUND, M.W., POYSKY, F.T., REED, S.M., and SMITH, C.A. 1970. Bacteriophage and the toxigenicity of *Clostridium botulinum* type C. Science *172,* 480-482.

EKLUND, M.W., WIELER, D.I., and POYSKY, F.T. 1967. Outgrowth and toxin production of nonproteolytic type B *Clostridium botulinum* at 3.3° to 5.6°C. J. Bacteriol. *93,* 1461-1462.

ELEK, S.D., and MORYSON, C. 1974. Resistotyping of *Staphylococcus aureus.* J. Med. Microbiol. *7,* 237-249.

ELWELL, M.R., LIU, C.T., SPERTZEL, R.O., and BEISEL, W.R. 1975. Mechanisms of oral staphylococcal enterotoxin B-induced emesis in the monkey. Proc. Soc. Exp. Biol. Med. *148,* 424-427.

EMODY. L., *et al.* 1974. Physiological effect of orally administered endotoxin to man. J. Hyg. Epidemiol. Microbiol. Immunol. *18,* 454-458.

ENGSTROM, G.W., RICHARD, J.L., and CYSEWSKI, S.J. 1977. High-pressure liquid chromatographic method for detection and resolution of rubratoxin, aflatoxin, and other mycotoxins. J. Agr. Food Chem. *25,* 833-836.

ERCOLANI, G.L. 1976. Bacteriological quality assessment of fresh marketed lettuce and fennel. Appl. Environ. Microbiol. *31,* 847-852.

ERDMAN, I.E. 1974. ICMSF methods studies. IV. International collaborative assay for the detection of *Salmonella* in raw meat. Can. J. Microbiol. *20,* 715-720.

ERICKSON, A., and DEIBEL, R.H. 1973. Production and heat stability of staphylococcal nuclease. Appl. Microbiol. *25,* 332-336.

EVANS, D.G., EVANS, D.J., JR., and DUPONT, H.L. 1977. Virulence factors of enterotoxigenic *Escherichia coli.* J. Infec. Dis. *136, S118-S123.*

EVANS, D.G., EVANS, D.J., JR., and PIERCE, N.F. 1973. Differences in the response of rabbit small intestine to heat-labile and heat-stable enterotoxins of *Escherichia coli.* Infec. Immunity *7,* 873-880.

EVANS, D.G. *et al.* 1975. Plasmid-controlled colonization factor associated with virulence in *Escherichia coli* enterotoxigenic for humans. Infec. Immunity *12,* 656-667.

EVANS, D.J., JR., and EVANS, D.G. 1977A. Inhibition of immune hemolysis: serological assay for the heat-labile enterotoxin of *Escherichia coli.* J. Clin. Microbiol. *5,* 100-105.

EVANS, D.J., JR., and EVANS, D.G. 1977B. Direct serological assay for the heat-labile enterotoxin of *Escherichia coli,* using passive immune hemolysis. Infec. Immunity *16,* 604-609.

FANTASIA, L.D., MESTRANDREA, L., SCHRADE, J.P., and YAGER, J. 1975A. Detection and growth of enteropathogenic *Escherichia coli* in soft ripened cheese. Appl. Microbiol. *29,* 179-185.

FANTASIA, L.D., SCHRADE, J.P., YAGER, J.F., and DEBLER, D. 1975B. Fluorescent antibody method for the detection of *Salmonella:* development, evaluation, and collaborative study. J. Assoc. Offic. Anal. Chem. *58,* 828-844.

FARMER, J.J., and SIKES, J.V. 1974. Bacteriophage typing of *Salmonella typhi* in the United States: 1966-1973. *Salmonella* Surveillance Report No. 119. Center for Disease Control, Atlanta, Ga.

FDA. 1976. Bacteriological Analytical Manual for Foods, 4th Edition. U.S. Food and Drug Administration, Washington, D. C.

FENNELL, D.I., BOTHAST, R.J., LILLEHOJ, E.B., and PETERSON, R.E. 1973. Bright greenish-yellow fluorescence and associated fungi in white corn naturally contaminated with aflatoxin. Cereal Chem. *50,* 404-414.

FERNANDEZ, E., TANG, T., and GRECZ, N. 1969. Toxicity of spores of *Clostridium botulinum* strain 33A in irradiated ground beef. J. Gen. Microbiol. *56,* 15-21.

FILPPI, J.A., and BANWART, G.J. 1974. Effect of the fat content of ground beef on the heat inactivation of poliovirus. J. Food Sci. *39,* 865-868.

FINDLAY, C.R. 1972. The persistence of *Salmonella dublin* in slurry in tanks and on pasture. Vet. Rec. *91,* 233-235.

FIRSTENBERG-EDEN, R., ROSEN, B., and MANNHEIM, C.H. 1977. Death and injury of *Staphylococcus aureus* during thermal treament of milk. Can. J. Microbiol. *23,* 1034-1037.

FLORES, J. *et al.* 1974. Comparison of the effects of enterotoxins of *Shigella dysenteriae* and *Vibrio cholerae* on the adenylate system of the rabbit intestine. J. Infec. Dis. *130,* 374-379.

FONG, C.K.Y., GROSS, P.A., HSIUNG, G.D., and SWACK, N.S. 1975. Use of electron microscopy for detection of viral and other microbial contaminants in bovine sera. J. Clin. Microbiol. *1,* 219-224.

FOO, L.Y. 1977. Scombroid poisoning-recapitulation on the role of histamine. N.Z. J. Med. *85,* 425-427.

FORMAL, S.B., and GEMSKI, P., JR. 1976. Studies on the pathogenesis of intestinal disease caused by invasive bacteria. Zbl. Bakt. Hyg. I. Abt. Orig *A235,* 9-12.

FORMAL, S.B., GEMSKI, P., JR., GIANNELLA, R.A., and AUSTIN, S. 1972. Mechanisms of *Shigella* pathogenesis. Amer. J. Clin. Nutr. *25,* 1427-1432.

FOSTER, E.M. 1969. The problem of salmonellae in foods. Food Technol. *23,* No. 9, 74-78.

FOSTER, E.M. 1971. The control of salmonellae in processed foods: a classification system and sampling plan. J. Assoc. Offic. Anal. Chem. *54,* 259-266.

FOSTER, E.M. *et al.* 1970. An evaluation of the *Salmonella* problem: summary and recommendations. J. Milk Food Technol. *33,* 42-51.

FRIEBEN W.R., and DUNCAN, C.L. 1973. Homology between enterotoxin protein and spore structural protein in *C. perfringens* type A. Eur. J. Biochem. *39,* 393-401.

FRIEBEN, W.R., and DUNCAN, C.L. 1975. Heterogeneity of enterotoxin-like protein extracted from spores of *Clostridium perfringens* type A. Eur. J. Biochem. *55,* 455-463.

FRITSCHE, D., DAHN, R., and HOFFMANN, G. 1975. *Aeromonas punctata* subsp. *caviae* as the causative agent of acute gastroenteritis. Zbl. Bakt. Hyg. I. Abt. Orig. *A233,* 232-235.

FROMM, D. *et al.* 1974. Ion transport across isolated ileal mucosa invaded by *Salmonella.* Gastroenterology *66,* 215-225.

FUJIMOTO, S., and OHBA, Y. 1975. Interaction of aflatoxin B_1 with DNA. J. Biochem. *77,* 187-195.

GAUSTAD, J.W., MCDUFF, C.R., and HATCHER, H.J. 1974. Test method for the evaluation of virucidal efficacy of three common liquid surface disinfectants on a simulated environmental surface. Appl. Microbiol. *28,* 748-752.

GEMSKI, P., JR., SHEAHAN, D.G., WASHINGTON, O., and FORMAL, S.B. 1972. Virulence of *Shigella flexneri* hybrids expressing *Escherichia coli* somatic antigens. Infec. Immunity *6,* 104-111.

GENIGEORGIS, C., SAKAGUCHI, G., and RIEMANN, H. 1973. Assay methods for *Clostridium perfringens* type A enterotoxin. Appl. Microbiol. *26,* 111-115.

GERSHMAN, M. 1977. Single phage-typing set for differentiating salmonellae. J. Clin. Microbiol. *5,* 302-314.

GIANNELLA, R.A., BROITMAN, S.A., and ZAMCHECK, N. 1971. *Salmonella enteritis.* I. Role of reduced gastric secretion in pathogenesis. Amer. J. Dig. Dis. *16,* 1000-1006.

GIANNELLA, R.A., ROUT, W.R., and FORMAL, S.B. 1977. Effect of indomethacin on intestinal water transport in *Salmonella*-infected Rhesus monkeys. Infec. Immunity. *17,* 136-139.

GIANNELLA, R.A. *et al.* 1975. Pathogenesis of salmonella-mediated intestinal fluid secretion. Activation of adenylate cyclase and inhibition by indomethacin. Gastroenterology *69,* 1238-1245.

GILBERT, R.J. 1974. Staphylococcal food poisoning and botulism. Postgrad. Med. J. *50,* 603-611.

GILLILAND, S.E., and SPECK, M.L. 1972. Interactions of food starter cultures and food-borne pathogens: lactic streptococci versus staphylococci and salmonellae. J. Milk Food Technol. *35,* 307-310.

GOEPFERT, J.M., SPIRA, W.M., and KIM, H.U. 1972. *Bacillus cereus:* food poisoning organism. A review. J. Milk Food Technol. *35,* 213-227.

GOPALAN, C., TULPULE, P.G., and KRISHNAMURTHI, D. 1972. Induction of hepatic carcinoma with aflatoxin in the Rhesus monkey. Food Cosmet. Toxicol. *10,* 519-521.

GORBACH, S.L., and KHURANA, C.M. 1972. Toxigenic *Escherichia coli.* A cause of infantile diarrhea in Chicago. N. Engl. J. Med. *287,* 791-795.

GORINA, L.G., FLUER, F.S., OLOVNIKOV, A.M., and EZEPCUK, Y.V. 1975. Use of the aggregate-hemagglutination technique for determining exoenterotoxin of *Bacillus cereus.* Appl. Microbiol. *29,* 201-204.

GOUGH, B.J., and ALFORD, J.A. 1965. Effect of curing agents on the growth and survival of food-poisoning strains of *Clostridium perfringens.* J. Food Sci. *30,* 1025-1028.

GRAY, J.A., and TRUEMAN, A.M. 1971. Severe *Salmonella* gastroenteritis associated with hypochlorhydria. Scot. Med. J. *16,* 255-258.

GUERRANT, R.L., and BRUNTON, L.L. 1977. Characterization of the Chinese hamster ovary cell assay for the enterotoxins of *Vibrio cholerae* and *Escherichia coli* and for antitoxin: differential inhibition by gangliosides, specific antisera, and toxoid. J. Infec. Dis. *135,* 720-728.

GUERRANT, R.L. *et al.* 1974. Cyclic adenosine monophosphate and alteration of Chinese hamster ovary cell morphology: a rapid, sensitive in vitro assay for the enterotoxins of *Vibrio cholerae* and *Escherichia coli.* Infec. Immunity *10,* 320-327.

GUNNARSSON, A., HURVELL, B., and THAL, E. 1977. Recent experiences with the salmonella-O-1-phage in routine diagnostic work. Zbl. Bakt. Hyg. I. Abt. Orig. *A237,* 222-227.

GURWITH, M. 1977. Rapid screening method for enterotoxigenic *Escherichia coli.* J. Clin. Microbiol. *6,* 314-316.

GUTHERTZ, L.S., FRUIN, J.T., OKOLUK, R.L., and FOWLER, J.L. 1977. Microbial quality of frozen comminuted turkey meat. J. Food Sci. *42,* 1344-1346, 1348.

GUTHERTZ, L.S., FRUIN, J.T., SPICER, D., and FOWLER, J.L. 1976. Microbiology of fresh comminuted turkey meat. J. Milk Food Technol. *39,* 823-829.

GYLES, C., SO, M., and FALKOW, S. 1974. The enterotoxin plasmids of *Escherichia coli.* J. Infec. Dis. *130,* 40-49.

HADDON, W.F. *et al.* 1977. Mass spectral confirmation of aflatoxin. J. Assoc. Offic. Anal. Chem. *60,* 107-113.

HALL, H.E., BROWN, D.F., and LEWIS, K.H. 1967. Examination of market foods for coliform organisms. Appl. Microbiol. *15,* 1062-1069.

HALVER, J.E. 1969. Aflatoxicosis and trout hepatoma. *In* Aflatoxin. Scientific Background, Control, and Implications. L. Goldblatt (Editor). Academic Press, New York.

HARIHARAN, H., and MITCHELL, W.R. 1976. Observation on bacteriophages of *Clostridium botulinum* type C isolates from different sources and the role of certain phages in toxigenicity. Appl. Environ. Microbiol. *32,* 145-158.

HARMON, S.M., and KAUTTER, D.A. 1970. Method for estimating the presence of *Clostridium perfringens* in food. Appl. Microbiol. *20,* 913-918.

HARMON, S.M., and KAUTTER, D.A. 1974. Collaborative study of the α-toxin method for estimating population levels of *Clostridium perfringens* in food. J. Assoc. Offic. Anal. Chem. *57,* 91-94.

HARMON, S.M., KAUTTER, D.A., and PEELER, J.T. 1971. Improved medium for enumeration of *Clostridium perfringens.* Appl. Microbiol. *22,* 688-692.

HAUSCHILD, A.H.W. 1970. Erythemal activity of the cellular enteropathogenic factor of *Clostridium perfringens* type A. Can. J. Microbiol. *16,* 651-654.

HAUSCHILD, A.H.W. 1971. *Clostridium perfringens* enterotoxin. J. Milk Food Technol. *34,* 596-599.

HAUSCHILD, A.H.W. 1973. Food poisoning by *Clostridium perfringens.* Can. Inst. Food Sci. Technol. *J.6,* 106-110.

HAUSCHILD, A.H.W., and HILSHEIMER, R. 1971. Purification and characteristics of the enterotoxin of *Clostridium perfringens* type A. Can. J. Microbiol. *17,* 1425-1433.

HAUSCHILD, A.H.W., and HILSHEIMER, R. 1974. Evaluation and modification of media for enumeration of *Clostridium perfringens.* Appl. Microbiol. *27,* 78-82.

HAUSCHILD, A.H.W., and HILSHEIMER, R. 1977. Enumeration of *Clostridium botulinum* spores in meats by a pour-plate procedure. Can. J. Microbiol. *23,* 829-832.

HAUSCHILD, A.H.W., HILSHEIMER, R., and MARTIN, W.G. 1973. Improved purification and further characterization of *Clostridium perfringens* type A enterotoxin. Can. J. Microbiol. *19,* 1379-1382.

HAUSCHILD, A.H.W., HILSHEIMER, R., and ROGERS, C.G. 1971. Rapid detection of *Clostridium perfringens* enterotoxin by a modified ligated intestinal loop technique in rabbits. Can. J. Microbiol. *17,* 1475-1476.

HAYES, S., CRAIG, J.M., and PILCHER, K.S. 1970. The detection of *Clostridium botulinum* type E in smoked fish products in the Pacific Northwest. Can. J. Microbiol. *16,* 207-209.

HEINZ, J.E., HALE, A.H., and CARL, P.L. 1975. Specificity of the antiviral agent calcium elenolate. Antimicrob. Agents Chemother. *8,* 421-425.

HERRMANN, J.E., and CLIVER, D.O. 1973A. Enterovirus persistence in sausage and ground beef. J. Milk Food Technol. *36,* 426-428.

HERRMANN, J.E., and CLIVER, D.O. 1973B. Degradation of coxsackievirus type A9 by proteolytic enzymes. Infec. Immunity *7,* 513-517.

HERRMANN, J.E., KOSTENBADER, K.D., JR., and CLIVER, D.O. 1974A. Persistence of enteroviruses in lake water. Appl. Microbiol. *28,* 895-896.

HESSELTINE, C.W. *et al.* 1976. Aflatoxin occurrence in 1973 corn at harvest. II. Mycological studies. Mycologia *68,* 341-353.

HILKER, J.S., *et al.* 1968. Heat inactivation of enterotoxin A from *Staphylococcus aureus* in veronal buffer. Appl. Microbiol. *16,* 308-310.

HILL, W.F., JR., AKIN, E.W., and BENTON, W.H. 1971. Detection of viruses in water: a review of methods and application. Water Res. *5,* 967-995.

HOLLINGER, F.B., BRADLEY, D.W., DREESMAN, G.R., and MELNICK, J.L. 1976. Detection of viral hepatitis type A. Amer. J. Clin. Pathol. *65,* 854-865.

HORWITZ, M.A., and GANGAROSA, E.J. 1976. Foodborne disease outbreaks traced to poultry, United States, 1966-1974. J. Milk Food Technol. *39,* 859-863.

HUANG, K.C., CHEN, T.S.T., and ROUT, W.R. 1974. Effect of staphylococcal enterotoxins A, B, and C, on ion transport and permeability across the flounder intestine. Proc. Soc. Exp. Biol. Med. *147,* 250-254.

HUGHES, J.A., TURNBULL, P.C.B., and STRINGER, M.F. 1976. A serotyping system for *Clostridium welchii (C. perfringens)* type A, and studies on the type-specific antigens. J. Med. Microbiol. *9,* 475-485.

HUGHES, J.M., MERSON, M.H., and GANGAROSA, E.J. 1977. The safety of eating shellfish. J. Amer. Med. Assoc. *237,* 1980-1981.

HUMBER, J.Y., DENNY, C.B., and BOHRER, C.W. 1975. Influence of pH on the heat inactivation of staphylococcal enterotoxin A as determined by monkey feeding and serological assay. Appl. Microbiol. *30,* 755-758.

HUNTER, D.T., and BAKER, C.E. 1967. Control of staphylococcal carriers in three hospitals. Public Health Rep. *82,* 329-333.

INOUE, K., and IIDA, H. 1970. Conversion of toxigenicity in *Clostridium botulinum* type C. Jap. J. Microbiol. *14,* 87-89.

JACKS, T.M., and WU, B.J. 1974. Biochemical properties of *Escherichia coli* low-molecular-weight, heat-stable enterotoxin. Infec. Immunity *9,* 342-347.

JAMLANG, E.M., BARTLETT, M.L., and SNYDER, H.E. 1971. Effect of pH, protein concentration, and ionic strength on heat inactivation of staphylococcal enterotoxin B. Appl. Microbiol. *22,* 1034-1040.

JANEWAY, C.M. *et al.* 1971. Foodborne outbreak of gastroenteritis possibly of multiple bacterial etiology. Amer. J. Epidemiol. *94,* 135-141.

JANSSEN, W.A. 1970. Fish as potential vectors of human bacterial diseases. *In* A Symposium on Diseases of Fishes and Shellfishes. S.F. Snieszko (Editor). Special Publication No. 5. American Fisheries Society, Washington, D.C.

JANSSEN, W.A. 1974. Oysters: retention and excretion of three types of human waterborne disease bacteria. Health Lab. Sci. *11,* 20-24.

JANSSEN, W.A., and MEYERS, C.D. 1968. Fish: serologic evidence of infection with human pathogens. Science *159,* 547-548.

JARVIS, A.W., and HARDING, S. 1972. Staphylococcal food-poisoning in New Zealand. N. Z. Med. J. *76,* 182-183.

JARVIS, A.W., and LAWRENCE, R.C. 1971. Production of extracellular enzymes and enterotoxins A, B, and C by *Staphylococcus aureus.* Infec. Immunity *4,* 110-115.

JARVIS, B. 1971. Factors affecting the production of mycotoxins. J. Appl. Bacteriol. *34,* 199-213.

JARVIS, B. 1972. The significance of microbial toxins in foods with respect to the health of the consumer. *In* Health and Food, G.G. Birch *et al.* (Editors). Applied Science Publishers, London.

JOHANNSEN, A. 1965. *Clostridium botulinum* type E in foods and the environment generally. J. Appl. Bacteriol. *28,* 90-94.

JOHNSON, H.M., BUKOVIC, J.A., and KAUFFMAN, P.E. 1973. Staphylococcal enterotoxins A and B: solid-phase radioimmunoassay in food. Appl. Microbiol. *26,* 309-313.

JONES, P.W., SMITH, G.S., and BEW, J. 1977. The effect of the microflora in cattle slurry on the survival of *Salmonella dublin* Brit. Vet. J. *133,* 1-8.

JULSETH, R.M., and DEIBEL, R.H. 1969. Effect of temperature on growth of *Salmonella* in rehydrated skim milk from a food-poisoning outbreak. Appl. Microbiol. *17,* 767-768.

KAMMAN, J.F., and TATINI, S.R. 1977. Optimal conditions for assay of staphylococcal nuclease. J. Food Sci. *42,* 421-424.

KAMPELMACHER, E.H. 1963. Salmonellosis in the Netherlands. Inst. Pasteur Ann. *104,* 647-659.

KANDELAKI, N.V. 1964. Determination of dysentery and typhoid bacteria in soil by the phage titer increase test. Hyg. Sanitation *29,* No. 11, 135-136.

KANG, C.K., WOODBURN, M., PAGENKOPF, A., and CHENEY, R. 1969. Growth, sporulation, and germination of *Clostridium perfringens* in media of controlled water activity. Appl. Microbiol. *18,* 798-805.

KANTOR, M.A., and POTTER, N.N. 1975. Persistence of echovirus and poliovirus in fermented sausages. Effects of sodium nitrite and processing variables. J. Food Sci. *40,* 968-972.

KAPIKIAN, A.Z. *et al.* 1974. Reoviruslike agent in stools: association with infantile diarrhea and development of serological tests. Science *185,* 1049-1053.

KAPRAL, F.A., O'BRIEN, A.D., RUFF, P.D., and DRUGAN, W.J., JR. 1976. Inhibition of water absorption in the intestine by *Staphylococcus aureus* delta-toxin. Infec. Immunity *13,* 140-145.

KAUFFMANN, F. 1972. Serological Diagnosis of *Salmonella*-species Kauffmann-White-Schema. The Williams and Wilkins Co., Baltimore.

KENSLER, C.J., and NATOLI, D.J. 1969. Processing to ensure wholesome products. *In* Aflatoxin. Scientific Background, Control, and Implications. L.A. Goldblatt (Editor). Academic Press, New York.

KEUSCH, G.T., GRADY, G.F., MATA, L.J., and MCIVER, J. 1972A. The pathogenesis of *Shigella* diarrhea. I. Enterotoxin production by *Shigella dysenteriae* 1. J. Clin. Invest. *51,* 1212-1218.

KEUSCH, G.T., GRADY, G.F., TAKEUCHI, A., and SPRINZ, H. 1972B. The pathogenesis of shigella diarrhea. II. Enterotoxin-induced acute enteritis in the rabbit ileum. J. Infec. Dis. *126,* 92-95.

KEUSCH, G.T., and JACEWICZ, M. 1977. The pathogenesis of shigella diarrhea. VI. Toxin and antitoxin in *Shigella flexneri* and *Shigella sonnei* infections in humans. J. Infec. Dis. *135,* 552-556.

KEUSCH, G.T. *et al.* 1976. Pathogenesis of shigella diarrhea. Serum anticytotoxin antibody response produced by toxigenic and nontoxigenic *Shigella dysenteriae* 1. J. Clin. Invest. *57,* 194-202.

KIMURA, T., and MURAKAMI, T. 1977. Tubular structures associated with acute nonbacterial gastroenteritis in young children. Infec. Immunity *17,* 157-160.

KINSEY, M.D., FORMAL, S.B., DAMMIN, G.J., and GIANNELLA, R.A. 1976. Fluid and electrolyte transport in Rhesus monkeys challenged intracecally with *Shigella flexneri* 2a. Infec. Immunity *14,* 368-371.

KLEBANOFF, S.J., and BELDING, M.E. 1974. Virucidal activity of H_2O_2-generating bacteria: requirement for peroxidase and a halide. J. Infec. Dis. *129,* 345-348.

KLIPSTEIN, F.A., ENGERT, R.F., and SHORT, H.B. 1977. Relative enterotoxigenicity of coliform bacteria. J. Infec. Dis. *136,* 205-215.

KLIPSTEIN, F.A., LEE, C.S., and ENGERT, R.F. 1976. Assay of *Escherichia coli* enterotoxins by in vivo perfusion in the rat jejunum. Infec. Immunity *14,* 1004-1010.

KOMOROWSKI, R.A., and HENSLEY, G.T. 1973. Epizootic salmonellosis in an open zoo aviary. Arch. Environ. Health *27,* 110-111.

KOUPAL, L.R., and DEIBEL, R.H. 1975. Assay characterization, and localization of an enterotoxin produced by *Salmonella.* Infec. Immunity *11,* 14-22.

KOURANY, M., and VASQUEZ, M.A. 1975. The first reported case from Panama of acute gastroenteritis caused by *Vibrio parahaemolyticus.* Amer. J. Trop. Med. Hyg. *24,* 638-640.

KRISHNAMACHARI, K.A.V.R., BHAT, R.V., NAGARAJAN, V., and TILAK, T.B.G. 1975. Hepatitis due to aflatoxicosis. Lancet *1,* (May 10), 1061-1063.

KRUGMAN, S., FRIEDMAN, H., and LATTIMER, C. 1975. Viral hepatitis, type A. Identification by specific complement fixation and immune adherence tests. N. Engl. J. Med. *292,* 1141-1143.

KURTZMAN, R.H., JR., and HALBROOK, W.U. 1970. Polysaccharide from dry navy beans, *Phaseolus vulgaris:* its isolation and stimulation of *Clostridium perfringens.* Appl. Microbiol. *20,* 715-719.

KWAN, P.L., and LEE, J.S. 1974. Compound inhibitory to *Clostridium botulinum* type E produced by a *Moraxella* species. Appl. Microbiol. *27,* 329-332.

LABBE, R.G., and DUNCAN, C.L. 1974. Sporulation and enterotoxin production by *Clostridium perfringens* type A under conditions of controlled pH and temperature. Can. J. Microbiol. *20,* 1493-1501.

LABBE, R.G., and DUNCAN, C.L. 1977. Spore coat protein and enterotoxin synthesis in *Clostridium perfringens.* J. Bacteriol. *131,* 713-715.

LACHICA, R.V.F. 1976. Simplified thermonuclease test for rapid identification of *Staphylococcus aureus* recovered on agar media. Appl. Environ. Microbiol. *32,* 633-634.

LACHICA, R.V.F., GENIGEORGIS, C., and HOEPRICH, P.D. 1971. Metachromatic agar-diffusion methods for detecting staphylococcal nuclease activity. Appl. Microbiol. *21,* 585-587.

LAMANNA, C. 1968. The concept of toxins. *In* The Anaerobic Bacteria. V. Fredette (Editor). Institute of Microbiology and Hygiene, Montreal, Canada.

LAMANNA, C., HILLOWALLA, R.R., and ALLING, C.C. 1967. Buccal exposure to botulinal toxin. J. Infec. Dis. *117,* 327-331.

LAMM, S.H. *et al.* 1972. Turtle-associated salmonellosis. I. An estimation of the magnitude of the problem in the United States, 1970-1971. Amer. J. Epidemiol. *95,* 511-517.

LAMPRECHT, E.C., and ELLIOT, M.C. 1974. Death rate of *Salmonella oranienburg* in fish meals as influenced by autoxidation treatment. J. Sci. Food Agr. *25,* 1329-1338.

LANGONE, J.J., and VAN VUNAKIS, H. 1976. Aflatoxin B_1: specific antibodies and their use in radioimmunoassay. J. Nat. Cancer Inst. *56,* 591-595.

LARKIN, E.P., TIERNEY, J.T., and SULLIVAN, R. 1976. Persistence of virus on sewage-irrigated vegetables. J. Environ. Eng. Div. ASCE. *102,* No. EE1, 29-35.

LEAVEN, L.J. *et al.* 1968. Foodborne epidemic of group G, streptococcal pharyngitis—Vermont. Morbidity Mortality *17,* 406-407.

LECCE, J.G., KING, M.W., and DORSEY, W.E. 1978. Rearing regimen producing piglet diarrhea (rotavirus) and its relevance to acute infantile diarrhea. Science *199,* 776-778.

LEE, I.C., STEVENSON, K.E., and HARMON, L.G. 1977. Effect of beef broth protein on the thermal inactivation of staphylococcal enterotoxin B. Appl. Environ. Microbiol. *33,* 341-344.

LEE, J.S. 1973. What seafood processors should know about *Vibrio parahaemolyticus.* J. Milk Food Technol. *36.* 405-408.

LEE, L.S., CUCULLU, A.F., FRANZ, A.O., JR., and PONS, W.A., JR. 1969. Destruction of aflatoxins in peanuts during dry and oil roasting J. Agr. Food Chem. *17,* 451.

LEE, L.S., CUCULLU, A.F., and GOLDBLATT, L.A. 1968. Appearance and aflatoxin content of raw and dry roasted peanut kernels. Food Technol. *22,* 1131-1134.

LEE, L.S. *et al.* 1974. Ammoniation of aflatoxin B_1: isolation and identification of the major reaction product. J. Assoc. Offic. Anal. Chem. *57,* 626-631.

LERKE, P. 1973. Evaluation of potential risk of botulism from seafood cocktails. Appl. Microbiol. *25,* 807-810.

LEVINE, M.M. *et al.* 1973. Pathogenesis of *Shigella dysenteriae* 1 (Shiga) dysentery. J. Infec. Dis. *127,* 261-270.

LEVINE, M.M. *et al.* 1977. Diarrhea caused by *Escherichia coli* that produce only heat-stable enterotoxin. Infec. Immunity *17,* 78-82.

LEVY, A.G. *et al.* 1976. Jejunal adenylate cyclase activity in human subjects during viral gastroenteritis. Gastroenterology *70,* 321-325.

LEY, F.J. *et al.* 1970. The use of gamma radiation for the elimination of *Salmonella* from frozen meat. J. Hyg. *68,* 293-311.

LILLY, T., JR. *et al.* 1971. An improved medium for detection of *Clostridium botulinum* type E. J. Milk Food Technol. *34,* 492-497.

LINDROTH, S., and NISKANEN, A. 1977. Double antibody solid-phase radioimmunoassay for staphylococcal enterotoxin A. Eur. J. Appl. Microbiol. *4,* 137-143.

LIU, T.S., SNOEYENBOS, G.H., and CARLSON, V.L. 1969. The effect of moisture and storage temperature on a *Salmonella senftenberg* 775W population in meat and bone meal. Poultry Sci. *48,* 1628-1633.

LOTTER, L.P., and GENIGEORGIS, C.A. 1975. Deoxyribonucleic acid base composition and biochemical properties of certain coagulase-negative enterotoxigenic cocci. Appl. Microbiol. *29,* 152-158.

LOZANO, E.A., CATLIN, J.E., and HAWKINS, W.W. 1970. Incidence of *Clostridium perfringens* in neonatal enteritis of Montana calves. Cornell Vet. *60,* 347-359.

LYNT, R.K., JR., SOLOMON, H.M., and KAUTTER, D.A. 1971. Immunofluorescence among strains of *Clostridium botulinum* and other clostridia by direct and indirect methods. J. Food Sci. *36,* 594-599.

MACKENZIE, M.A., and BAINS, B.S. 1976. Dissemination of *Salmonella* serotypes from raw feed ingredients to chicken carcasses. Poultry Sci. *55,* 957-960.

MAGGON, K.K., GUPTA, S.K., and VENKITASUBRAMANIAN, T.A. 1977. Biosynthesis of aflatoxins. Bacteriol. Rev. *41,* 822-855.

MAHDY, M.S., and BANSEN, E. 1974. A viral inhibitory substance in culture filtrates of *Proteus mirabilis.* I. Inhibition of Sindbis virus plaque formation without a parallel antiviral action in fluid medium. Can. J. Microbiol. *20,* 1195-1203.

MAHL, M.C., and SADLER, C. 1975. Virus survival on inanimate surfaces. Can. J. Microbiol. *21,* 819-823.

MAHONY, D.E. 1974. Bacteriocin susceptibility of *Clostridium perfringens:* a provisional typing schema. Appl. Microbiol. *28,* 172-176.

MAIER, B.R. and HENTGES, D.J. 1972. Experimental *Shigella* infections in laboratory animals. I. Antagonism by human normal flora components in gnotobiotic mice. Infec. Immunity *6,* 168-173.

MANN, R., and REHM, H.J. 1976. Degradation products from aflatoxin B_1 by *Corynebacterium rubrum, Aspergillus niger, Trichoderma viride* and *Mucor ambiguus.* Eur. J. Appl. Microbiol. *2,* 297-306.

MARIER, R., *et al.* 1973. An outbreak of enteropathogenic *Escherichia coli* foodborne disease traced to imported French cheese. Lancet *2,* 1376-1378.

MARKUS, Z., and SILVERMAN, G.J. 1969. Enterotoxin B synthesis by replicating and nonreplicating cells of *Staphylococcus aureus.* J. Bacteriol. *97,* 506-512

MARSH, P.B., SIMPSON, M.E., and TRUCKSESS, M.W. 1975. Effects of trace metals on the production of aflatoxins by *Aspergillus parasiticus.* Appl. Microbiol. *30,* 52-57.

MAYHEW, J.W., and GORBACH, S.L. 1975. Rapid gas chromatographic technique for presumptive detection of *Clostridium botulinum* in contaminated food. Appl. Microbiol. *29,* 297-299.

MAYNARD, J.E. *et al.* 1975. Review of infectivity studies in nonhuman primates with virus-like particles associated with MS-1 hepatitis. Amer. J. Med. Sci. *270,* 81-85.

MCBRIDE, G.B., BROWN, B., and SKURA, B.J. 1978. Effect of bird type, growers and season in the incidence of salmonellae in turkeys. J. Food Sci. *43,* 323-326.

MCDONEL, J.L., and DUNCAN, C.L. 1977. Regional localization of activity of *Clostridium perfringens* type A enterotoxin in the rabbit ileum, jejunum, and duodenum. J. Infec. Dis. *136,* 661-666.

MCKERCHER, P.D., HESSE, W.R., and HAMDY, F. 1978. Residual viruses in pork products. Appl. Environ. Microbiol. *35,* 142-145.

MCLEAN, R.A., LILLY, H.D., and ALFORD, J.A. 1968. Effects of meat-curing salts and temperature on production of staphylococcal enterotoxin B. J. Bacteriol. *95,* 1207-1211.

MEAD, G.C. 1969. Growth and sporulation of *Clostridium welchii* in breast and leg muscle of poultry. J. Appl. Bacteriol. *32,* 86-95.

MEHLMAN, I.J. *et al.* 1975. Methodology for enteropathogenic *Escherichia coli.* J. Assoc. Offic. Anal. Chem. *58,* 283-292.

MEHLMAN, I.J. *et al.* 1977. Methodology for recognition of invasive potential of *Escherichia coli.* J. Assoc. Offic. Anal. Chem. *60,* 546-562.

MELLING, J., CAPEL, B.J., TURNBULL, P.C.B., and GILBERT, R.J. 1976. Identification of a novel enterotoxigenic activity associated with *Bacillus cereus.* J. Clin. Pathol. *29,* 938-940.

MENZIES, R.E. 1977. Comparison of coagulase, deoxyribonuclease (DNase), and heat-stable nuclease tests for identification of *Staphylococcus aureus.* J. Clin. Pathol. *30,* 606-608.

MESTRANDREA. L.W. 1974. Rapid detection of *Clostridium botulinum* toxin by capillary tube diffusion. Appl. Microbiol. *27,* 1017-1022.

METCALF, T.G., SLANETZ, L.W., and BARTLEY, C.H. 1973. Enteric pathogens in estuary waters and shellfish. *In* Microbial Safety of Fishery Products. C.O. Chichester and H.D. Graham (Editors). Academic Press, New York.

MEYER, G., HINTERBERGER, J., and KORTE, R. 1977. Detoxification of staphylococcal-enterotoxin B in water. Zbl. Bakt. Hyg., I Abt. Orig. *B164,* 352-359.

MILLER, C.A. and ANDERSON, A.W. 1971. Rapid detection and quantitative estimation of type A botulinum toxin by electroimmunodiffusion. Infec. Immunity *4,* 126-129.

MIYAMOTO, Y. *et al.* 1969. *In vitro* hemolytic characteristics of *Vibrio parahaemolyticus:* its close correlation with human pathogenicity. J. Bacteriol. *100,* 1147-1149.

MIYAMOTO, Y. *et al.* 1975. Extraction, purification, and biophysicochemical characteristics of a "Kanagawa phenomenon"-associated hemolytic factor of *Vibrio parahaemolyticus.* Jap. J. Med. Sci. Biol. *28,* 87-90.

MIYAZAKI, S., and SAKAGUCHI, G. 1978. Experimental botulism in chickens: the cecum as the site of production and absorption of botulinum toxin. Jap. J. Med. Sci. Biol. *31,* 1-15.

MOHIT, B., ALY, R., and BOURGEOIS, L.D. 1975. A simple single-step immunoimmobilisation method for the detection of *Salmonella* in the presence of large numbers of other bacteria. J. Med. Microbiol. *8,* 173-176.

MORSE, E.V., and DUNCAN, M.A. 1974. Salmonellosis—an environmental health problem. J. Amer. Vet. Med. Assoc. *165,* 1015-1019.

MULDER, R.W.A.W., NOTERMANS, S., and KAMPELMACHER, E.H. 1977. Inactivation of salmonellae on chilled and deep frozen broiler carcasses by irradiation. J. Appl. Bacteriol. *42,* 179-185.

MURTHY, T.R.K., JEMMALI, M., HENRY, Y., and FRAYSSINET, C. 1975. Aflatoxin residues in tissues of growing swine: effect of separate and mixed feeding of protein and protein-free portions of the diet. J. Anim. Sci. *41,* 1339-1347.

NAIK, H.S., and DUNCAN, C.L. 1977. Rapid detection and quantitation of *Clostridium perfringens* enterotoxin by counterimmunoelectrophoresis. Appl. Environ. Microbiol. *34,* 125-128.

NAKAGAWA, M., and NISHIDA, S. 1969. Heat resistance and α-toxigenicity of *Clostridium perfringens* strains in normal intestines of Japanese. Jap. J. Microbiol. *13,* 133-137.

NELSON, J.D., and HALTALIN, K.C. 1972. Effect of neonated folic acid deprivation on later growth and susceptibility to *Shigella* infection in the guinea pig. Amer. J. Clin. Nutr. *25,* 992-996.

NEWBERNE, P.M. 1974. The new world of mycotoxins—animal and human health. Clin. Toxicol. *7,* 161-177.

NICKELSON, R., WYATT, L.E., and VANDERZANT, C. 1975. Reduction of *Salmonella* contamination in commercially processed frog legs. J. Food Sci. *40,* 1239-1241.

NIILO, L. 1971. Mechanism of action of the enteropathogenic factor of *Clostridium perfringens* type A. Infec. Immunity *3,* 100-106.

NIILO, L. 1974. Response of ligated intestinal loops in chickens to the enterotoxin of *Clostridium perfringens.* Appl. Microbiol. *28,* 889-891.

NIILO, L. 1975. Measurement of biological activities of purified and crude enterotoxin of *Clostridium perfringens.* Infec. Immunity *12,* 440-442.

NIILO, L. 1977. Enterotoxin formation by *Clostridium perfringens* type A studied by the use of fluorescent antibody. Can. J. Microbiol. *23,* 908-915.

NISKANEN, A., 1977. Staphylococcal Enterotoxins and Food Poisoning. Production and Detection of Enterotoxins. Publication 19. Technical Research Centre of Finland, Helsinki.

NISKANEN, A., and KOIRANEN, L. 1977. Correlation of enterotoxin and thermonuclease production with some physiological and biochemical properties of staphylococcal strains isolated from different sources. J. Food Protection *40,* 543-548.

NORTHOLT, M.D., VERHULSDONK, C.A.H., SOENTORO, P.S.S., and PAULSCH, W.E. 1976. Effect of water activity and temperature on aflatoxin production by *Aspergillus parasiticus.* J. Milk Food Technol. *39,* 170-174.

O'BRIEN, A.D. *et al.* 1977. Biological properties of *Shigella flexneri* 2A toxin and its serological relationship to *Shigella dysenteriae* 1 toxin. Infec. Immunity *15,* 796-798.

ODLAUG, T.E., and PFLUG, I.J. 1977. Recovery of spores of *Clostridium botulinum* in yeast extract agar and pork infusion agar after heat treatment. Appl. Environ. Microbiol *34,* 377-381.

OEDING, P. 1974. Cellular antigens of staphylococci. Ann. N. Y. Acad. Sci. *236,* 15-21.

OGUMA, K. 1976. The stability of toxigenicity in *Clostridium botulinum* types C and D. J. Gen. Microbiol. *92,* 67-75.

OGUMA, K., IIDA, H., and INOUE, K. 1975. Observations on nonconverting phage, C-n71, obtained from a nontoxigenic strain of *Clostridium botulinum* type C. Jap. J. Microbiol. *19,* 167-172.

OLD, D.C., and PAYNE, S.B. 1971. Antigens of the type-2 fimbriae of salmonellae: "cross-reacting material" (CRM) of type-1 fimbriae. J. Med. Microbiol. *4,* 215-225.

ONG, T.M. 1975. Aflatoxin mutagenesis. Mutat. Res. *32,* 35-53.

ORSKOV, I., ORSKOV, F., JANN, B., and JANN, K. 1977. Serology, chemistry, and genetics of O and K antigens of *Escherichia coli.* Bacteriol. Rev. *41,* 667-710.

OSTOVAR, K. 1973. A study on survival of *Staphylococcus aureus* in dark and milk chocolate. J. Food Sci. *38,* 663-664.

OSTOVAR, K., PEREIRA, R.R., and GALLOP, R.A. 1971. Effects of gamma irradiation on *Salmonella* spp. in smoked lake whitefish (*Coregonus clupeaformis*). J. Fish. Res. Board Can. *28,* 643-646.

OUCHTERLONY, O. 1968. Handbook of Immunodiffusion and Immunoelectrophoresis. Ann Arbor Science Publishers, Ann Arbor, Mich.

PACE, P.J., and KRUMBIEGEL, E.R. 1973. *Clostridium botulinum* and smoked fish production: 1963-1972. J. Milk Food Technol. *36,* 42-49.

PAI, C.H., and MORS, V. 1978. Production of enterotoxin by *Yersinia enterocolitica.* Infec. Immunity *19,* 908-911.

PALMER, E.L., MARTIN, M.L., and MURPHY, F.A. 1977. Morphology and stability of infantile gastroenteritis virus: comparison with reovirus and bluetongue virus. J. Gen. Virol. *35,* 403-414.

PALUMBO, S.A., HUHTANEN, C.N., and SMITH, J.L. 1974. Microbiology of the frankfurter process: *Salmonella* and natural aerobic flora. Appl. Microbiol. *27,* 724-732.

PANALAKS, T., and SCOTT, P.M. 1977. Sensitive silica gel-packed flowcell for fluorometic detection of aflatoxins by high pressure liquid chromatography. J. Assoc. Offic. Anal. Chem. *60,* 583-589.

PARK, H.S., MARTH, E.H., and OLSON, N.F. 1970. Survival of *Salmonella typhimurium* in cold-pack cheese food during refrigerated storage. J. Milk Food Technol. *33,* 383-388.

PARK, H.S., MARTH, E.H., and OLSON, N.F. 1973. Fate of enteropathogenic strains 9f *Escherichia coli* during the manufacture and ripening of Camembert cheese. J. Milk Food Technol. *36,* 543-546.

PARKER, M.T. 1972. Phage-typing of *Staphylococcus aureus. In* Methods in Microbiology, Vol. 7B. J. R. Norris and D. W. Ribbons (Editors). Academic Press, London and New York.

PAYNE, D.N., and WOOD, J.M. 1974. The incidence of enterotoxin production in strains of *Staphylococcus aureus* isolated from foods. J. Appl. Bacteriol. *37,* 319-325.

PAYNE, W.W., and KOTIN, P. 1969. Protecting our foods from environmental intrusion. 1. Microbial and chemical hazards in foods. Food Technol. *23,* 528-530.

PEERS, F.G., and LINSELL, C.A. 1973. Dietary aflatoxins and liver cancer . . . a population based study in Kenya. Brit. J. Cancer *27,* 473-484.

PETHER, J.V.S., and GILBERT, R.J. 1971. The survival of salmonellas on finger-tips and transfer of the organisms to foods. J. Hyg. Camb. *69,* 673-681.

PHILLIPS, S.F. 1972. Diarrhea: a current view of pathophysiology. Gastroenterology *63,* 495-518.

PHILLIPS, S.F. 1975. Diarrhea. Pathogenesis and diagnostic techniques. Postgrad. Med. *57,* 65-71.

PLITMAN, M., PARK, Y., GOMEZ, R., and SINSKEY, A.J. 1973. Viability of *Staphylococcus aureus* in intermediate moisture meats. J. Food Sci. *38,* 1004-1008.

POBER, Z., and SILVERMAN, G.J. 1977. Modified radioimmunoassay determination for staphylococcal enterotoxin B in foods. Appl. Environ. Microbiol. *33,* 620-625.

PORTNOY, B.L., GOEPFERT, J.M. and HARMON, S. M. 1976. An outbreak of *Bacillus cereus* food poisoning resulting from contaminated vegetable sprouts. Amer. J. Epidemiol. *103,* 589-594.

RAEVUORI, M., KIUTAMO, T., NISKANEN A., and SALMINEN, K. 1976. An outbreak of *Bacillus cereus* food-poisoning in Finland associated with boiled rice. J. Hyg. Camb. *76,* 319-327.

RAYMAN, M.K., PARK, C.E., PHILPOTT, J., and TODD, E.C.D. 1975. Reassessment of the coagulase and thermostable nuclease tests as means of identifying *Staphylococcus aureus.* Appl. Microbiol. *29,* 451-454.

READ, R.B., JR., BRADSHAW, J.G., DICKERSON, R.W., JR., and PEELER, J.T. 1968. Thermal resistance of salmonellae isolated from dry milk. Appl. Microbiol. *16,* 998-1001.

READ, R.B., JR., and PRITCHARD, W.L. 1963. Lysogeny among the enterotoxigenic staphylococci. Can. J. Microbiol. *9,* 879-889.

REFAI, M., and ROHDE, R. 1975. New serological investigations into O-antigenic relations between *Escherichia coli, Salmonella, Arizona* and *Citrobacter.* Zbl. Bakt. Hyg. I. Abt. Orig. *A233,* 171-179.

REICHERT, C.A., and FUNG, D.Y.C. 1976. Thermal inactivation and subsequent reactivation of staphylococcal enterotoxin B in selected liquid foods. J. Milk Food Technol. *39,* 516-520.

ROBERN, H., DIGHTON, M., YANO, Y., and DICKIE, N. 1975. Double-antibody radioimmunoassay for staphylococcal enterotoxin C_2. Appl. Microbiol. *30,* 525-529.

ROBERTS, T.A., and SMART, J.L. 1976. The occurrence and growth of *Clostridium* spp. in vacuum-packed bacon with particular reference to *Cl. perfringens* (*welchii*) and *Cl. botulinum.* J. Food Technol. *11,* 229-244.

ROMER, T.R. 1973. Determination of aflatoxins in mixed feeds. J. Assoc. Offic. Anal. Chem. *56,* 1111-1114.

ROMER, T.R. 1975. Screening method for the detection of aflatoxins in mixed feeds and other agricultural commodities with subsequent confirmation and quantitative measurement of aflatoxins in positive samples. J. Assoc. Offic. Anal. Chem. *58,* 500-506.

ROSE, M.J. 1972. Use of gelatinase in the routine sampling of gelatin for *Salmonella* contamination. Appl. Microbiol. *24,* 153-154.

ROSENWALD, A.J., and LINCOLN, R.E. 1966. Streptomycin inhibition of elaboration of staphylococcal enterotoxic protein. J. Bacteriol. *92,* 279-280.

ROUT, W.R., FORMAL, S.B., GIANNELLA, R.A., and DAMMIN, G.J. 1975. Pathophysiology of shigella diarrhea in the Rhesus monkey: intestinal transport, morphological, and bacteriological studies. Gastroenterology *68,* 270-278.

RUDOY, R.C., and NELSON, J.D. 1975. Enteroinvasive and enterotoxigenic *Escherichia coli.* Amer. J. Dis. Child. *129,* 668-672.

SACK, R.B. *et al.* 1977. Enterotoxigenic *Escherichia coli* isolated from food. J. Infec. Dis. *135,* 313-317.

SAITANU, K., and LUND, E. 1975. Inactivation of enterovirus by glutaraldehyde. Appl. Microbiol. *29,* 571-574.

SAKAGUCHI, G., UEMURA, T., and RIEMANN, H.P. 1973. Simplified method for purification of *Clostridium perfringens* type A enterotoxin. Appl. Microbiol. *26,* 762-767.

SAKAGUCHI, G. *et al.* 1974. Cross reaction in reversed passive hemagglutination between *Clostridium botulinum* type A and B toxins and its avoidance by the use of antitoxic component immunoglobulin isolated by affinity chromatography. Jap. J. Med. Sci. Biol. *27,* 161-172.

SAKAZAKI, R., TAMURA, K., and NAKAMURA, A. 1974A. Further studies on enteropathogenic *Escherichia coli* associated with diarrheal diseases in children and adults. Jap. J. Med. Sci. Biol. *27,* 17-18.

SAKAZAKI, R., TAMURA, K., NAKAMURA, A., and KURATA, T. 1974B. Enteropathogenic and enterotoxigenic activities on ligated gut loops in rabbits of *Salmonella* and some other enterobacteria isolated from human patients with diarrhea. Jap. J. Med. Sci. Biol. *27,* 45-48.

SAKAZAKI, R. *et al.* 1974C. Studies on enteropathogenic activity of *Vibrio parahaemolyticus* using ligated gut loop model in rabbits. Jap. J. Med. Sci. Biol. *27,* 35-43.

SAKAZAKI, R. *et al.* 1974D. Enteropathogenicity and enterotoxigenicity of human enteropathogenic *Escherichia coli.* Jap. J. Med. Sci. Biol. *27,* 19-33.

SAKURAI, J., MATSUZAKI, A., TAKEDA, Y., and MIWATANI, T. 1974. Existence of two distinct hemolysins in *Vibrio parahaemolyticus.* Infec. Immunity *9,* 777-780.

SANDEFUR, P.D., and PETERSON, J.W. 1977. Neutralization of *Salmonella* toxin-induced elongation of Chinese hamster ovary cells by cholera antitoxin. Infec. Immunity *15,* 988-992.

SANDERS, A.C. 1975. Report on microbiological methods. J. Assoc. Offic. Anal. Chem. *58,* 246-248.

SASLAW, M.S. *et al.* 1974. Outbreak of foodborne streptococcal disease—Florida. Morbidity Mortality *23,* 365-366.

SATTERLEE, L.D., and KRAFT, A.A. 1969. Effect of meat and isolated meat proteins on the thermal inactivation of staphylococcal enterotoxin B. Appl. Microbiol. *17,* 906-909.

SAUNDERS, G.C., and BARLETT, M.L. 1977. Double-antibody solid-phase enzyme immunoassay for the detection of staphylococcal enterotoxin A. Appl. Environ. Microbiol. *34,* 518-522.

SCHAFFNER, C.P., MOSBACH, K., BIBIT, V.C., and WATSON, C.H. 1967. Coconut and *Salmonella* infection. Appl. Microbiol. *15,* 471-475.

SCHENKEIN, I., GREEN, R.F., SANTOS, D.S., and MAAS, W.K. 1976. Partial purification and characterization of a heat-labile enterotoxin of *Escherichia coli.* Infec. Immunity *13,* 1710-1720.

SCHIFF, L. J. *et al.* 1972. Enteropathogenic *Escherichia coli* infections: increasing awarenesss of a problem in laboratory animals. Lab. Anim. Sci. *22,* 705-708.

SCHOUB, B.D. *et al.* 1977. Rotavirus and winter gastro-enteritis in white South African infants. S. A. Med. J. *52,* 998-999.

SCHROEDER, H.W., and CARLTON, W.W. 1973. Accumulation of only aflatoxin B_2 by a strain of *Aspergillus flavus.* Appl. Microbiol. *25,* 146-148.

SCHROEDER, S.A. 1967. What the sanitarian should know about salmonellae and staphylococci in milk and milk products. J. Milk Food Technol. *30,* 376-380.

SCHULLER, P.L., HORWITZ, W., and STOLOFF, L. 1976. A review of sampling plans and collaboratively studied methods for analysis of aflatoxins. J. Assoc. Offic. Anal. Chem. *59,* 1315-1343.

SEDLOCK, D.M., and DEIBEL, R.H. 1978. Detection of *Salmonella* enterotoxin using rabbit ileal loops. Can. J. Microbiol. *24,* 268-273.

SERAICHEKAS, H.R. *et al.* 1968. Viral depuration by assaying individual shellfish. Appl. Microbiol. *16,* 1865-1871.

SHALITA, Z., HERTMAN, I., and SARID, S. 1977. Isolation and characterization of a plasmid involved with enterotoxin B production in *Staphylococcus aureus.* J. Bacteriol. *129, 317-325.*

SHANK, R.C., BOURGEOIS, C.H., KESCHAMRAS, N, and CHANDAVIMOL, P. 1971. Aflatoxins in autopsy specimens from Thai children with an acute disease of unknown aetiology. Food Cosmet. Toxicol. *9,* 501-507.

SHIH, C.N., and MARTH, E.H. 1972. Production of aflatoxin in a medium fortified with sodium cloride. J. Dairy Sci. *55,* 1415-1419.

SHIH, C.N., and MARTH, E.H. 1974. Some cultural conditions that control biosynthesis of lipid and aflatoxin by *Aspergillus parasiticus.* Appl. Microbiol. *27,* 452-456.

SHOTWELL, O.L., GOULDEN, M.L. and KWOLEK, W.F. 1977. Aflatoxin in corn: evaluation of filter fluorometer reading of minicolumns. J. Assoc. Offic. Chem. *60,* 1220-1222.

SIDOROV, J.J. 1976. Intestinal absorption of water and electrolytes. Clin. Biochem. *9,* No. 3, 117-120.

SKJELKVALE, R., and DUNCAN, C.L. 1975. Enterotoxin formation by different toxigenic types of *Clostridium perfringens.* Infec. Immunity *11,* 563-575.

SKJEKLVALE, R., and UEMURA, T. 1977. Detection of enterotoxin in faeces and anti-enterotoxin in serum after *Clostridium perfringens* food-poisoning. J. Appl. Bacteriol. *42,* 355-363.

SMITH, E.R., and BADLEY, B.W.D. 1971. Treatment of *Salmonella* enteritis and its effect on the carrier state. Can. Med. Assoc. J. *104,* 1004-1006.

SMITH, H. 1977. Microbial surfaces in relation to pathogenicity. Bacteriol. Rev. *41,* 475-500.

SMITH, H.W., and GYLES, C. L. 1970. The relationship between two apparently different enterotoxins produced by enteropathogenic strains of *Escherichia coli* of porcine origin. J. Med. Microbiol. *3,* 387-401.

SMITH, J.L., HUHTANEN, C.N., KISSINGER, J.C., and PALUMBO, S.A. 1975A. Survival of salmonellae during pepperoni manufacture. Appl. Microbiol. *30,* 759-763.

SMITH, J.L., PALUMBO, S.A., KISSINGER, J.C., and HUHTANEN, C.N. 1975B. Survivial of *Salmonella dublin* and *Salmonella typhimurium* in Lebanon bologna. J. Milk Food Technol. *38,* 150-154.

SMITH, L.D.S., DAVIS, J.W., and LIBKE, K.G. 1971. Experimentally induced botulism in weanling pigs. Amer. J. Vet. Res. *32,* 1327-1330.

SMITH, M.R. 1971. *Vibrio parahemolyticus.* Clin. Med. *78,* 22-25.

SMITH, M.V., and MULDOON P.J. 1974. *Campylobacter fetus* subspecies *jejuni (Vibrio fetus)* from commercially processed poultry. Appl. Microbiol. *27,* 995-996.

SMITH, P.B. 1972. Bacteriophage typing of *Staphylococcus aureus. In* The Staphylococci. J. O. Cohen (Editor). Wiley-Interscience, N.Y.

SMITH, R.B., JR., GRIFFIN, J.M., and HAMILTON, P.B. 1976. Survey of aflatoxicosis in farm animals. Appl. Environ. Microbiol. *31,* 385-388.

SOLOMON, H.M., LYNT, R.K., JR., KAUTTER, D.A., and LILLY, T., JR. 1969. Serological studies of *Clostridium botulinum* type E and related organisms. II. Serology of spores. J. Bacteriol. *98,* 407-414.

SPEIRS, J.I., STAVRIC, S. and KONOWALCHUCK, J. 1977. Assay of *Escherichia coli* heat-labile enterotoxin with Vero cells. Infec. Immunity *16,* 617-622.

STAMPER, W.J., and BANWART, G.J. 1974. Effect of various peptones in the growth medium on the agglutination of ten *Salmonella* species with pooled Spicer-Edwards antisera. J. Food Sci. *39,* 80-82.

STEYN, P.S. 1977. The metabolic transformation of the mycotoxin aflatoxin B_1. S. Afr. Med. J. *73,* 200-201.

STOLOFF, L. 1976. Report on mycotoxins. J. Assoc. Offic. Anal Chem. *59,* 317-323.

STRONG, D.H., DUNCAN, C.L., and PERNA, G. 1971. *Clostridium perfringens* type A food poisoning. II. Response of the rabbit ileum as an indication of enteropathogenicity of strains of *Clostridium perfringens* in human beings. Infec. Immunity *3,* 171-178.

STUBBLEFIELD, R.D., and SHOTWELL, O.L. 1977. Reverse phase analytical and preparative high pressure liquid chromatography of aflatoxins. J. Assoc. Offic. Anal. Chem. *60,* 784-790.

SUGII, S., OHISHI, I., and SAKAGUCHI, G. 1977. Intestinal absorption of botulinum toxins of different molecular size in rats. Infec. Immunity *17,* 491-496.

SUGIYAMA, H., and YANG, K.H. 1975. Growth potential of *Clostridium botulinum* in fresh mushrooms packaged in semipermeable plastic film. Appl. Microbiol. *30,* 964-969.

SULLIVAN, R., FASSOLITIS, A.C., and READ, R.B., JR. 1970. Method for isolating viruses from ground beef. J. Food Sci. *35,* 624-626.

SULLIVAN, R., MARNELL, R.M., LARKIN, E.P., and READ, R.B., JR. 1975. Inactivation of poliovirus 1 and coxsackievirus B-2 in broiled hamburgers. J. Milk Food Technol. *38,* 473-475.

SWAMINATHAN, B., DENNER, J.M., and AYRES, J.C. 1978. Rapid detection of salmonellae in foods by membrane filter-disc immunoimmobilization technique. J. Food Sci. *43,* 1444-1447.

TAGER, M. 1974. Current views on the mechanisms of coagulase action in blood clotting. Ann. N. Y. Acad. Sci. *236,* 277-291.

TAKEUCHI, A. 1975. Electron microscope observations on penetration of the gut epithelial barrier by *Salmonella typhimurium. In* Microbiology 1975. D. Schlessinger (Editor). American Society for Microbiology, Washington, D.C.

TAMMINGA, S.K., BEUMER, R.R., and KAMPELMACHER, E.H. 1978. The hygienic quality of vegetables grown in or imported into the Netherlands: a tentative survey. J. Hyg. Camb. *80,* 143-154.

TAMMINGA, S.K., BEUMER, R.R., KAMPELMACHER, E.H., and VAN-LEUSDEN, F.M. 1976. Survival of *Salmonella eastbourne* and *Salmonella typhimurium* in chocolate. J. Hyg. Camb. *76,* 41-47.

TAMMINGA, S.K., BEUMER, R.R., KAMPELMACHER, E.H., and VAN-LEUSDEN, F.M. 1977. Survival of *Salmonella eastbourne* and *Salmonella typhimurium* in milk chocolate prepared with artificially contaminated milk powder. J. Hyg. Camb. *79,* 333-337.

TATINI, S.R. 1976. Thermal stability of enterotoxins in food. J. Milk Food Technol. *39,* 432-438.

TATINI, S.R., CORDS, B.R., and GRAMOLI, J. 1976. Screening for staphylococcal enterotoxins in food. Food Technol. *30,* No. 4, 64, 66, 70, 72-74.

TATINI, S.R., SOO, H.M., CORDS, B.R., and BENNETT, R.W. 1975. Heat-stable nuclease for assessment of staphylococcal growth and likely presence of enterotoxins in foods. J. Food Sci. *40,* 352-356.

TAYLOR, R.J., and BURROWS, M.R. 1971. The survival of *Escherichia coli* and *Salmonella dublin* in slurry on pasture and the infectivity of *S. dublin* for grazing calves. Brit. Vet. J. *127,* 536-543.

THARRINGTON, G., JR., ASHTON, D.H., HATFIELD, J.R., and FRY, F.H. 1978. Nonspecific staining of a *Lactobacillus* by *Salmonella* fluorescent antibodies. J. Food Sci. *43,* 548-552.

THOMAS, K.L., and JONES, A.M. 1971. Comparison of methods of estimating the number of *Escherichia coli* in edible mussels and the relationship between the presence of salmonellae and *E. coli.* J. Appl. Bacteriol. *34,* 717-735.

THOMAS, M.E.M., and MOGFORD, H.E. 1970. Salmonellosis in general practice. Observations of cases and their households in Enfield. J. Hyg. Camb. *68,* 663-671.

THOMASON, B.M., and HEBERT, G.A. 1974. Evaluation of commercial conjugates for fluorescent antibody detection of salmonellae. Appl. Microbiol. *27,* 862-869.

THORNLEY, M.J. 1963. Microbiological aspects of the use of radiation for the elimination of salmonellae from foods and feeding stuffs. *In* Radiation Control of Salmonellae in Food and Feed Products. Tech. Report Series No. 22. International Atomic Energy Agency, Vienna, Austria.

TIERNEY, J.T., SULLIVAN, R., LARKIN, E.P., and PEELER, J.T. 1973. Comparison of methods for the recovery of virus inoculated into ground beef. Appl. Microbiol. *26,* 497-501.

TOMPKIN, R.B., CHRISTIANSEN, L.N., SHAPARIS, A.B., and BOLIN, H. 1974. Effect of potassium sorbate on salmonellae, *Staphylococcus aureus, Clostridium perfringens,* and *Clostridium botulinum* in cooked, uncured sausage. Appl. Microbiol. *28, 262-264.*

TOMPKIN, R.B., and STOZEK, S.K. 1974. Direct addition of liquid propylene oxide to dried materials for destruction of salmonellae. Appl. Microbiol. *27,* 276-277.

TRACI, P.A. and DUNCAN. C.L. 1974. Cold shock lethality and injury in *Clostridium perfringens.* Appl. Microbiol. *28,* 815-821.

TROLLER, J.A. 1976. Staphylococcal growth and enterotoxin production-factors for control. J. Milk Food Technol. *39,* 499-503.

TROLLER, J.A., and STINSON. J.V. 1975. Influence of water activity on growth and enterotoxin formation by *Staphylococcus aureus* in foods. J. Food Sci. *40,* 802-804.

TULLOCH, E.F., JR., RYAN, K.J., FORMAL, S.B., and FRANKLIN, F.A. 1973. Invasive enteropathic *Escherichia coli* dysentery. An outbreak in 28 adults. Ann. Int. Med. *79,* 13-17.

TURNBULL, P.C.B. 1976. Studies on the production of enterotoxins by *Bacillus cereus.* J. Clin. Pathol. *29,* 941-948.

TWEDT, R.M., SPAULDING, P.L., and HALL. H.E. 1969. Morphological, cultural, biochemical, and serological comparison of Japanese strains of *Vibrio parahaemolyticus* with related cultures isolated in the United States. J. Bacteriol. *98,* 511-518.

UEMURA, T., GENIGEORGIS, C., RIEMANN, H.P., and FRANTI, C.E. 1974. Antibody against *Clostridium perfringens* type A enterotoxin in human sera. Infec. Immunity *9,* 470-471.

UEMURA, T., and SKJELKVALE, R. 1976. An enterotoxin produced by *Clostridium perfringens* type D. Purification by affinity chromatography. Acta Path. Microbiol. Scand. Sect. B, *84,* 414-420.

UNDERDAL, B., and ROSSEBO, L. 1972. Inactivation of strains of *Salmonella senftenberg* by gamma irradiation. J. Appl. Bacteriol. *35,* 371-377.

UNE, T. 1977. Studies on the pathogenicity of *Yersinia enterocolitica* I. Experimental infection in rabbits. Microbiol. Immunol. *21,* 349-363.

USDA. 1971. Recommended procedure for the isolation of salmonella organisms from animal feeds and feed ingredients. ARS 91-68-1. U.S. Department of Agriculture, Washington, D. C.

VAN DEN BROEK, M.J.M., and MOSSEL, D.A.A. 1977. Sublethal cold shock in *Vibrio parahaemolyticus.* Appl. Environ. Microbiol. *34,* 97-98.

VANDERZANT, C., and NICKELSON, R. 1969. A microbiological examination of muscle tissue of beef, pork, and lamb carcasses. J. Milk Food Technol. *32,* 357-361.

VANDERZANT, C., and NICKELSON, R. 1972. Survival of *Vibrio parahaemolyticus* in shrimp tissue under various environmental conditions. Appl. Microbiol. *23,* 34-37.

VAN RENSBURG, S.J., KIRSIPUU, A., COUTINHO, L.P., and VAN DER WATT, J.J. 1975. Circumstances associated with the contamination of food by aflatoxin in a high primary liver cancer area. S. Afr. Med. J. *49,* 877-883.

VELASCO, J. 1972. Detection of aflatoxin using small columns of florisil. J. Amer. Oil Chem. Soc. *49,* 141-142.

VELAUDAPILLAI, T., NILES, G.R., and NAGARATNAM, W. 1969. Salmonellas, shigellas, and enteropathogenic *Escherichia coli* in uncooked foods. J. Hyg. Camb. *67,* 187-191.

VENN, S.Z., WOODBURN, M., and MORITA, T. 1973. *Staphylococcus aureus* S-6: growth and enterotoxin production in papain-treated beef and ham and beef gravy. Home Econ. Res. J. *1,* 161-172.

WADSTROM, T., LJUNGH, A., and WRETLIND, B. 1976A. Enterotoxin, haemolysin and cytotoxic protein in *Aeromonas hydrophila* from human infections. Acta Pathol. Microbiol. Scand. *B84,* 112-114.

WADSTROM, T. *et al.* 1976B. Enterotoxin-producing bacteria and parasites in stools of Ethiopian children with diarrhoel disease. Arch. Dis. Child. *51,* 865-870.

WALKER, G.C., and HARMON, L.G. 1966. Thermal resistance of *Staphylococcus aureus* in milk, whey, and phosphate buffer. Appl. Microbiol. *14,* 584-590.

WALTKING, A.E. 1971. Fate of aflatoxin during roasting and storage of contaminated peanut products. J. Assoc. Offic. Anal. Chem. *54,* 533-539.

WELKOS, S., SCHREIBER, M., and BAER, H. 1974. Identification of *Salmonella* with the O-1 bacteriophage. Appl. Microbiol. *28,* 618-622.

WESSEL, J.R., and STOLOFF, L. 1973. Regulatory surveillance for aflatoxin and other mycotoxins in feeds, meat, and milk. J. Amer. Vet. Med. Assoc. *163,* 1284-1287.

WEST, S., WYATT, R.D., and HAMILTON, P.B. 1973. Improved yields of aflatoxin by incremental increases of temperature. Appl. Microbiol. *25,* 1018-1019.

WIENEKE, A.A. 1974. Enterotoxin production by strains of *Staphylococcus aureus* isolated from foods and human beings. J. Hyg. Camb. *73,* 255-262.

WILSON. B.J. 1968. Mycotoxins. *In* The Safety of Foods. J. C. Ayres *et al.* (Editors). AVI Publishing Co., Westport, CT.

WILSON, B.J., CAMPBELL, T.C., HAYES, A.W., and HANLIN, R.T. 1968. Investigation of reported aflatoxin production by fungi outside the *Aspergillus flavus* group. Appl. Microbiol. *16,* 819-821.

WILSON, D.M., and JAY, E. 1975. Influence of modified atmosphere storage on aflatoxin production in high moisture corn. Appl. Microbiol. *29,* 224-228.

WILSON, M.R., and HOHMANN, A.W. 1974. Immunity to *Escherichia coli* in pigs: adhesion of enteropathogenic *Escherichia coli* to isolated intestinal epithelial cells. Infec. Immunity *10,* 776-782.

WOOLFORD, A.L., SCHANTZ, E.J., and WOODBURN, M.J. 1978. Heat inactivation of botulinum toxin type A in some convenience foods after frozen storage. J. Food Sci. *43,* 622-624.

YAMADA, S., IGARASHI, H., and TERAYAMA, T. 1977. Heterogeneity of staphylococcal enterotoxin A on isoelectric focusing and disc electrophoresis. Microbiol. Immunol. *21,* 119-126.

YAMAGISHI, T. *et al.* 1976. Persistent high numbers of *Clostridium perfringens* in the intestines of Japanese aged adults. Jap. J. Microbiol. *20,* 397-403.

YANG, H-Y., and JONES, G.A. 1969. Physiological characteristics of enteropathogenic and non-pathogenic coliform bacteria isolated from Canadian pasteurized dairy products. J. Milk Food Technol. *32,* 102-109.

YAO, M.G., DENNY, C.B., and BOHRER, C.W. 1973. Effect of frozen storage time on heat inactivation of *Clostridium botulinum* type E toxin. Appl. Microbiol. *25,* 503-505.

YOLKEN, R.H., *et al.* 1977. Enzyme-linked immunosorbent assay for detection of *Escherichia coli* heat-labile enterotoxin. J. Clin. Microbiol. *6,* 439-444.

ZABRISKIE, J.B. 1970. Relationship of lysogeny to bacterial toxin production. *In* Microbial Toxins. Vol. 1. Bacterial Protein Toxins. S.J. Ajl, S. Kadis, and T.C. Montie (Editors). Academic Press, New York.

7

Indicator Organisms

Indicator organisms have been used since 1892 when Schardinger tested water for what is now called *Escherichia coli*, instead of testing for *Salmonella typhi*. It is difficult to detect *S. typhi* in a water supply. The presence of *E. coli* indicates that there might have been contamination from sewage, and that *Salmonella* or other intestinal pathogens might be present.

Since the original studies, indicator organisms have been used to determine an objectionable microbial condition of food, such as fecal contamination, the presence of potential pathogens or potential spoilage of foods, as well as the sanitary conditions of food processing, production or storage.

There are certain criteria which determine the value of an indicator organism.

If the presence of an organism is to indicate possible fecal contamination or potential pathogens, it must be associated with feces and/or pathogens.

The indicator should not be present as a natural contaminant of the material being analyzed. If it is usually present on a food product, its detection would not necessarily indicate the presence of either fecal material or pathogens. Hence, for these two criteria, knowing the sources of the indicator organisms is important.

The organism should be easy to grow and differentiate. There should be a relatively simple, accurate, rapid and standard test to enumerate the indicator. If the procedure to determine the indicator is more difficult or less accurate than that for pathogens, then we might as well analyze for the pathogens.

If the indicator multiplies very rapidly at storage temperatures of the food, the actual degree of contamination may be obscured. If it dies very rapidly, its value as an index of contamination is lowered. The indicator organisms should withstand processing treatments of the food in a manner similar to pathogens. If the indicator is less stable, it may disappear before the pathogen. If it is more stable, it may persist long after the pathogen is destroyed.

There are many organisms or groups of organisms that have been suggested as indicator organisms. Fecal microorganisms include the enteric bacteria, viruses and protozoa. In general, viruses and protozoa are more difficult to enumerate than are bacteria. Bacteria, such as coliforms, *E. coli,* Enterobacteriaceae, enterococci, pseudomonads, clostridia, staphylococci and the aerobic plate count have been suggested as indicator organisms. How well do these bacteria or groups of bacteria meet the general criteria needed for a good indicator? We need to know the usual source, the association with enteric pathogens, methods for enumeration, growth conditions, survival during storage and processing and the significance of these organisms in foods or a particular food.

COLIFORMS

As a result of the successful use of coliforms as indicators in water, they have been employed as indicators of possible fecal contamination of foods.

Whenever organisms are given a group terminology, there is a question as to which organisms are included, or at least a definition is needed. Coliforms include all aerobic and facultatively anaerobic Gram negative nonsporeforming bacilli which ferment lactose with gas formation within 48 hr at 35°C. For techniques that are used in the membrane filter method for water analysis, an alternate definition is that coliforms include all organisms which produce a colony on eosin methylene blue agar with a golden green metallic sheen within 24 hr of incubation at 35°-37°C.

With the procedures of the U.S. FDA (1976), the definition of a presumptive coliform is an organism that produces gas in lauryl sulfate tryptose (LST) broth in 48 hr at 35°C. A confirmed coliform is a presumptive coliform that produces gas in brilliant green lactose bile (BGLB) 2% broth in 48 hr at 35°C.

The coliform group includes *Escherichia coli, Citrobacter freundii, Enterobacter aerogenes, Enterobacter cloacae* and *Klebsiella pneumoniae.* There are a few strains of other species that ferment lactose and might be included in a coliform determination.

Sources

Coliforms are common inhabitants of the intestinal tract of humans and animals. However, bacteria in this group are of both fecal and nonfecal origin.

The organisms are found in undisturbed soil, and can be at high levels in fecal contaminated soil. With certain conditions, they tend to die in the soil, but with proper nutrients and moisture, they can increase in number. Soil dust can disseminate coliforms into the atmosphere. Rain carries the surface contamination from soil to streams, rivers or lakes.

Coliforms are found on all types of plant material (foliage, roots, flowers). *Klebsiella* predominated in samples obtained from forest environments and from fresh farm produce (Duncan and Razzell 1972). Most of the fresh vegetables they examined had coliform counts of 10^6 to 10^7 per gram. All of the 30 samples of lettuce examined by Kaferstein (1977) contained coliforms (from less than 10 to over 10^5/g).

Coliforms are found on the shells of freshly laid eggs, and can penetrate through the pores if the egg surface is damp. The organisms can be found in raw milk through contamination of the teat canals by feed or manure. Coliforms cause infection of the mast cells (mastitis) of cattle from which they can be shed into the milk. Milk can be contaminated after milking by air, man or equipment.

Coliforms are present on the feathers of live poultry and the hide, hoofs and hair of other animals. Upon slaughter, the organisms can contaminate the meat from these sources, or from intestinal leakage during evisceration.

Shellfish grown in polluted areas concentrate the organisms so that they are contaminated at higher levels than those present in the water.

With the widespread distribution of coliforms, they can be detected on many types of food products, especially those of animal origin.

Methods for Detection

Many methods have been used for the detection of coliforms. The fermentation of lactose on or in a medium is the primary requirement for an organism to be considered a coliform.

The procedure suggested by FDA (1976) consists of using the MPN method by inoculating tubes of LST broth. Growth from those tubes showing gas production (due to fermentation of lactose) after incubation at 35°C for 48 hr is inoculated into tubes of BGLB (2%) broth. These inoculated tubes are incubated at 35°C for 48 hr and observed for gas production (due to fermentation of lactose). The production of gas is considered to be confirmed positive for coliforms.

Many of the reported coliform investigations are performed by plating samples with violet red bile (VRB) agar. Often, higher counts are obtained using an agar since anaerogenic strains will be counted. These strains are not positive in the MPN tube test. Not all of the organisms that ferment lactose on VRB agar prove to be coliforms. This may be due to lactose fermenting organisms other than coliforms that are not inhibited by the selective agents (bile salts and dyes), or the carry-over of food material may reduce the inhibition of the selective agents, or the sampled food can furnish other carbohydrates besides lactose for the organisms to ferment. These factors also can affect the results of the MPN test. Processed food material often contains stressed coliforms which require a non-inhibition medium for repair. Hartman *et al.* (1975) suggested plating the food sample on VRB without bile salts and dyes, and then overlaying the poured plates with VRB with double strength bile salts and dyes. They reported increased counts with this technique. A similar procedure enumerating injured coliforms was suggested by Ray and Speck (1978). They plated the sample with trypticase soy agar and overlayed the surface with VRB after a one hour incubation at room temperature.

Rosen and Levin (1970) described members of the genus *Vibrio* that appeared as coliforms on VRB agar. Other organisms, such as *Proteus* and *Alcaligenes*, can be confused as coliforms on VRB agar.

The presence of nitrates can inhibit gas production by coliforms. Therefore, care should be used in interpreting the analysis of meat curing brines or of foods containing nitrates when gas production is the criterion used for the assessment of the presence of coliforms.

The membrane filter (MF) technique is used to enumerate coliforms in water. After filtering a standard volume of water, the filter pad is placed directly on a differential Endo-type medium for incubation, or a preliminary enrichment in LST broth can be used. One advantage of the MF system is that organisms can be concentrated from low density material, grown on the filter pads and then isolated and studied.

Besides water analysis, the MF technique can be used for certain food products. This would be especially advantageous if a component of the food is a carbohydrate that could be fermented by other than coliform organisms.

A radiometric method employing ^{14}C labeled lactose to determine coliforms was discussed by Bachrach and Bachrach (1974).

Growth and Survival

Coliforms include psychrotrophic types capable of multiplying at 3°-10° C. Thus, they can multiply in foods even when refrigerated. At 10°C, coliforms may grow faster than other species and become one of the dominant groups of organisms. They grow well on most carbohydrates and proteins.

Members of the coliform group may persist in water, soil or foods for long time periods. Usually, the organisms are not able to withstand the pasteurization of milk. Therefore, the presence of coliforms in pasteurized milk indicates either improper pasteurization or recontamination after the heat treatment.

The organisms are rather hardy under natural conditions and resist drying. Coliforms do not withstand the rigors of freezing or frozen storage very well.

Significance of Coliforms

There are good and bad aspects in the use of coliforms as indicator organisms. Since members of this group are found in the intestines and discharges of man and animals, their presence could be due to fecal contamination and there also could be enteric pathogens present. Unfortunately, there are some members with non-fecal origins. The presence of these members in a food would not indicate fecal contamination or potential pathogens. One can rationalize this predicament by saying that this gives an extra margin of safety. Finstein (1973) called the coliforms the most conservative indicator.

Coliforms in milk products generally are considered to be of animal origin, and tend to indicate the hygienic standard of the product and its keeping quality, rather than the presence of human pathogens. The more carefully the milk is produced and handled, the lower the coliform count. Hartley *et al.* (1968) stated that the coliform count is of only limited value with fresh, high quality milk, and of no value for older or low quality milk.

Since coliforms can multiply outside the animal body, their presence in high numbers in a food product may not be indicative of original contamination, but of improper handling, which allowed multiplication of the organisms.

A good indicator should be present if an enteric pathogen is present. The data of Loken *et al.* (1968) showed that 11.4% of protein feed supplement samples with a coliform count of less than one per gram contained salmonellae. Of 607 samples with less than 10 coliforms per gram, 92 (15.2%) contained salmonellae. Only 42% of the samples with over 100 coliforms per gram revealed salmonellae. Hence, there is some question as to the correlation of the presence or absence of coliforms to salmonellae. After examining their data from raw and ready-to-eat foods, Solberg *et al.* (1976) stated that the coliforms cannot serve as an indicator of potential pathogens in these foods.

Smith (1971) stated that persons with severe diarrhea often do not excrete coliforms, but primarily excrete the enteric pathogen. He questioned the use of coliforms as an index of contamination.

The use of coliforms as an index of contamination of animal products has limitations, since these organisms are part of the natural flora indigenous to animals. However, if coliforms occur in unusually large numbers, then contamination from filth and unsanitary practices would be indicated. For pasteurized or cooked animal products, it is possible that coliforms can be used to indicate contamination after heating.

Since coliforms generally fail to survive freezing and frozen storage, they have little or no value as indicators in frozen foods.

The value of the coliform group as an indicator is both questionable and acceptable. In some products, the coliforms may indicate fecal contamination from either human or animal sources or from soil, poor sanitation of equipment, faulty pasteurization techniques or recontamination after pasteurization or cooking. However, because of their ubiquity in nature, the ability to multiply outside the animal body and the non-fecal habitat of especially the *Enterobacter* species, it can be said that the presence of the coliform group is not a reliable indicator of the presence of potential pathogens in a food.

FECAL COLIFORMS

To overcome the criticism of non-fecal coliforms as indicator organisms, there have been suggestions to determine only the fecal coliforms. In 1904, Eijkman discovered that intestinal coliforms produce gas in glucose broth incubated at 46°C, and non-fecal coliforms fail to grow.

When glucose is fermented, the ratio of hydrogen to carbon dioxide formed is 1:1 for the fecal coliforms and 1:2 for non-fecal strains.

The fecal coliforms can be defined as Gram negative facultative rods that ferment lactose at 44.5°C. The fecal coliforms consist primarily of *E. coli*, but a few *Enterobacter* and *Klebsiella* strains can produce gas in lactose broth at 44.5°C (Duncan and Razzell 1972).

Sources

The fecal coliforms are relatively specific for fecal material of warm-blooded animals. However, fecal coliforms were reported on fresh produce (Duncan and Razzell 1972). There were 5.4 million organisms per 100 g of green onions. Insects, plants, animals, soil and polluted irrigation water are sources of fecal coliforms (Geldreich and Bordner 1971).

Methods

The usual procedure is to inoculate growth from positive presumptive coliform tubes into tubes of EC broth, and incubate them at 44.5°C for

24 hr. The incubated broth is examined for gas production, and the number of fecal coliforms estimated with an MPN table.

The direct inoculation of a sample into EC broth does not reveal all of the fecal coliforms. An enrichment in a presumptive broth is needed to attain a level of cells that will produce gas when inoculated into EC broth and incubated at 44.5°C.

With this test, a small portion of fecal coliforms is excluded and a few non-fecal coliforms are included in positive tubes. Although several media might be acceptable, EC broth gives the most rapid results, requiring only 24 hr incubation, as compared to 48 to 72 hr with most media.

A membrane filter (MF) method can be used. After filtration, the MF is saturated with M-FC broth and incubated at 44.5°C for 24 hr. Blue colonies, observed with low-power magnification, are counted as fecal coliforms. Hufham (1974) stated that he could not recommend the MF system when incubated at 44.5°C because the errors in the method varied from 32% to as high as 92%, and the recovery of fecal coliforms was only 9 to 60%. This may be due to the organisms, collected on the surface of the filter pad, being subjected to a local hypertonic environment. To solve this problem, a special membrane filter was developed by Millipore (1975).

Besides the elevated temperature test, the IMViC tests can be used to differentiate coliforms, according to the reactions as listed in Table 7.1. There are some problems associated with the IMViC system. Both the *Klebsiella* and *Enterobacter* are IMViC − − ++. *Klebsiella* can be found in feces, industrial wastes or nature, while *Enterobacter* usually is found in non-fecal sources. Further, this biochemical testing requires the isolation of a sufficient number of colonies from an agar surface, so that the tests are meaningful. These time-consuming and laborious procedures are not adaptable on a routine basis in many laboratories.

To overcome the questionable aspects of the IMViC differentiation, Wolfe and Amsterdam (1968) proposed the HOC scheme (hydrogen sulfide production, ornithine decarboxylation and citrate utilization). According to them, the reactions for these three tests for the coliforms are: *E. coli* − − − or −+−, *Citrobacter* +−+ or +++, *Enterobacter* −++, and *Klebsiella* − −+. They believed these tests give a better separation of the coliform types than the IMViC reactions. Closs (1971) expanded the HOC scheme to the HOMoC, adding motility. The four genera are variable, +, − and +, respectively, for motility. The tests for hydrogen sulfide, ornithine decarboxylase and motility were combined into one tube. Using Simmons citrate medium for the citrate utilization test, only two tubes are needed, as compared to three tubes for the IMViC tests.

Although only *Citrobacter* is listed as producing H_2S, and *E. coli* is not

TABLE 7.1

USUAL IMViC REACTIONS OF THE COLIFORMS

	Reaction			
Type	I	M	V	C
Escherichia (Type I)	+	+	–	–
Escherichia (Type II)	–	+	–	–
Enterobacter	–	–	+	+
Klebsiella	–	–	+	+
Citrobacter	–	+	–	+

I = indole, M = methyl red, V = Voges-Proskauer, C = citrate utilization

supposed to produce H_2S, there are H_2S positive variants of *E. coli* (Braunstein and Mladineo 1974; Darland and Davis 1974).

A fluorescent antibody method for detection of fecal coliforms was reported by Abshire and Guthrie (1973), a coliphage test was described by Kenard and Valentine (1974) and a gas chromatographic system was suggested by Newman and O'Brien (1975).

Growth and Survival

Although the fecal coliforms are associated primarily with the intestinal contents and discharges of man and warm-blooded animals, they can multiply in an acceptable environment outside of the intestines. Bagley and Seidler (1977) reported that 49% of the environmental *K. pneumoniae* grew in EC broth at 44.5°C and 16% were fecal coliform positive.

The fecal coliforms are destroyed by pasteurization or normal cooking temperatures. In frozen foods, the organisms die rather rapidly, and are in very low numbers or absent from stored frozen foods.

Significance of Fecal Coliforms

The fecal coliforms are more closely related to fecal contamination than are the total coliforms. This should be considered an advantage; however, critics state that testing for only fecal coliforms does not give a margin of safety which is important when using indicator organisms.

The organisms can originate from improperly sanitized working surfaces in a processing plant. In these cases, their presence would reflect the quality of sanitation and not the direct pollution of the product. If proper sanitation of the processing plant is lacking, this is important information.

The report of Geldreich and Bordner (1971) showed that irrigation water with counts of over 1,000 fecal coliforms per 100 ml was quite likely to contain salmonellae (96.4% of the time). Water with fecal coliforms of 1 to 1,000/100 ml contained salmonellae 53.5% of the time. Fecal coliforms and salmonellae have similar death rates in natural waters. Silliker and Gabis (1976) suggested that the use of fecal coliforms as indicators of salmonellae contamination would result in the rejection of salmonellae-negative foods and the acceptance of salmonellae-positive foods.

ESCHERICHIA COLI

E. coli is the most prominent fecal coliform. Hence, we might expect that there is little difference between *E. coli* and fecal coliforms. Since the main habitat of *E. coli* is the intestinal tract of man and warm-blooded animals, it is believed by some microbiologists that only the presence of *E. coli* is indicative of fecal contamination of food.

Sources

E. coli is a common inhabitant of the intestinal tract of man and animals. The number of this organism varies in different animals, but an animal will excrete from 130 million to over 18 billion *E. coli* each day. Hence, there is a close association between *E. coli*, fecal material and, possibly, enteric pathogens.

Due to the close association with animals, it is difficult to keep *E. coli* from contaminating the animal carcasses during slaughter.

Besides being in fecal material, *E. coli* can be found on plant material, although not in great abundance. *E. coli* can be present in undisturbed soil, but if the soil is polluted by fecal material, the prevalence is increased. The presence of *E. coli* is unavoidable in or on products that have been in contact with the soil.

Various nuts can be contaminated with *E. coli.* Marcus and Amling (1973) isolated *E. coli* from 23% of the samples of pecans from orchards where animals grazed, compared to only 4% from ungrazed orchards. However, even in undisturbed soil in almond orchards, Kokal and Thorpe (1969) reported an average *E. coli* incidence of 25%. Although immature almonds contained no *E. coli*, as the almonds matured, *E. coli* appeared and, during harvesting and processing, reached an incidence of about 40% before hulling.

Hall *et al.* (1967) analyzed various foods for coliforms and *E. coli. E. coli* was found in cheese and cheese products (75% of the samples), fish and seafood (30%), raw and frozen vegetables (15.8%), prepared and convenience foods (50%), raw meats (76.5%) and sandwiches (16.3%). Only 30 of 74 raw beef patties contained 10 or more *E. coli* per gram (Surkiewicz *et al.* 1975).

Bettelheim *et al.* (1974) suggested there is a difference in the serological groups of *E. coli* isolated from humans and from animals. The strains of *E. coli* isolated from meat generally resembled animal strains. Typically, human serotypes found on meat are believed due to contamination during handling.

Methods

The Bacteriological Analytical Manual (FDA 1976) test for *E. coli* consists of inoculating growth from presumptive positive coliform LST broth into EC broth and incubating at 44.5°C. Growth from those EC broth tubes showing gas production is streaked onto EMB agar, and incubated (35°C for 18-24 hr). Typical *E. coli* colonies (nucleated with or without a sheen) are tested using the IMViC reactions and Gram stain. The designation in Table 7.1 is used to classify the *E. coli.*

A rapid method requires inoculating and incubating LST broth at 44.5° C, and then streaking on EMB agar. Typical *E. coli* colonies on EMB are confirmed with the IMViC tests.

Although incubation of the EC broth for 48 hr will yield slightly more *E. coli*, the number of false positive cultures increases. Therefore, to obtain optimum specificity, the incubation period should be limited to 24 hr.

Other general procedures for the enumeration of *E. coli* have been described by Anderson and Baird-Parker (1975) and Moran and Witter (1976). Fluorescent antibody techniques were described by Pugsley and Evison (1974).

Survival

Generally, *E. coli* dies in frozen storage. The type of food influences the rate of death. In frozen orange concentrate, the organism may survive one day or as long as one week. In a product such as gravy, the death rate is lower, and some *E. coli* may survive frozen storage for several months. The addition of glycerine to water extends the survival period in the frozen state. In frozen products, the decline in numbers does not proceed at a constant rate, but at a continuously decreasing rate.

E. coli and salmonellae decline in numbers at a similar rate in many refrigerated foods. The reduction in *E. coli* and salmonellae is similar during spray drying. The thermal resistances of *E. coli* and salmonellae are similar. In filter sterilized water, the survival of *E. coli* and *S. typhimurium* are essentially the same over normal seasonal temperatures (Mitchell and Starzyk 1975).

Significance

The source of *E. coli* is generally the intestinal tract and feces of warm-blooded animals and man. Hence, it should be an excellent indicator organism. On the other hand, if fecal coliforms lack a desired margin of safety, it is an even greater problem with the use of *E. coli* alone as an indicator.

The determination of *E. coli* includes any problems associated with the enumeration of coliforms plus the added necessity to do additional procedures to confirm and complete the tests for these specific organisms. Thus, the tests for *E. coli* are not as acceptable as those for coliforms. With the procedures involved, the enumeration of enteric pathogens might be considered, rather than *E. coli*.

According to Evison and James (1973), *E. coli* is a reliable index of fecal pollution of water sources in temperate climates. Gartner *et al.* (1975) reported that with increasing contamination of coastal water with *E. coli*, the proportion of positive salmonellae samples also increases.

In many food products, the survival of *E. coli* is similar to that of salmonellae. However, many pathogens may persist after *E. coli* is destroyed. On the other hand, the presence of *E. coli* does not mean that enteric pathogens also are present. Lovell and Barkate (1969) found over 92% of the samples of crayfish contained *E. coli*, but only 3% contained either coagulase positive staphylococci or salmonellae. Conversely, salmonellae have been found in food samples with no *E. coli*. The presence or absence of *E. coli* in oysters was not indicative of the presence of enteric viruses (Fugate *et al.* 1975).

The finding of high coliform or *E. coli* counts in a food has been the basis for seizure by regulatory agencies on the basis that the food "consists in whole or in part of a filthy, decomposed or putrid animal or vegetable substance."

ENTEROBACTERIACEAE

Since the enteric bacteria that fail to ferment lactose are of more importance in public health than those that do ferment lactose, it has been suggested that the entire family Enterobacteriaceae be used as indicator organisms.

If the use of coliforms as indicators can be criticized because of non-fecal origins, some members of this family deserve to be criticized. Some

Enterobacteriaceae occur as saprophytes in nature, some are parasites for plants, causing blights and soft rots, many are found in the intestinal tract of animals and others are potential pathogens of animals and man. Thus, the direct relationship of the organisms of this entire family to fecal contamination or potential pathogens is questionable.

Cox *et al.* (1975) found Enterobacteriaceae on poultry carcasses at various stages of chilling. *Salmonella* and *Shigella* were not found. As they pointed out, this does not mean that these organisms were not present. The most prevalent isolate was *Escherichia*, accounting for 87-96% of the detected Enterobacteriaceae. Thus, the use of *E. coli* or fecal coliforms as indicators would reveal essentially the same information as the Enterobacteriaceae. There are many foods in which the coliform count and Enterobacteriaceae count are almost the same.

Notermans *et al.* (1977) found the Enterobacteriaceae count reflected the level of fecal contamination on dressed poultry only when scalding of the birds was accomplished at 60°C or higher.

The methods suggested to enumerate the organisms were evaluated by Drion and Mossel (1977). They suggested that testing 2 samples of 1 g of food for Enterobacteriaceae and allowing no positives gives the same degree of safety as testing 60-25 g samples for salmonellae and allowing no positives.

Mossel (1967) described the significance of Enterobacteriaceae in foods. The exact limits of these organisms that are allowed in a food depend upon what is attainable in the best sanitized food plant. A "sliding scale" is needed for each type of food product.

FECAL STREPTOCOCCI ENTEROCOCCI

Since the coliforms, fecal coliforms or *E. coli* are not very resistant to heating or freezing, other organisms have been suggested as indicators. The fecal streptococci and enterococci are potential indicator organisms.

The enterococci are in Lancefield's group D of the genus *Streptococcus*. Since references are made to the group D streptococci, the enterococcus group and the fecal streptococci, some distinction should be made between these categories.

S. faecalis and *S. faecium*, along with two subspecies of *S. faecalis* (*liquefaciens* and *zymogenes*) comprise the enterococcus group. These organisms, plus *S. bovis* and *S. equinus* are the group D streptococci. Buchanan and Gibbons (1974) included *S. suis* and *S. avium* as members of group D. The fecal streptococci include all of these group D streptococci, and *S. mitis* and *S. salivarius*.

S. avium has been listed in group Q, but it also possesses the group D antigen, is prevalent in chicken feces, as well as being present in animal and human feces. In most cases, the growth of *S. faecalis*, *S. faecium* and the group Q streptococci are not distinguishable. Thus, the group Q streptococci (*S. avium*) may be enumerated and reported as enterococci. *S. avium* fits most of the characteristics of the enterococcus group, except it neither tolerates 0.1% methylene blue, nor hydrolyzes arginine.

The definitions and the organisms that are included are not necessarily those listed in some of the literature. The streptococci have been studied for many years, but the status and interrelationships of the species have been, and still are, controversial.

Although these terms sometimes are used interchangeably, in this text, both fecal streptococci and group D streptococci include the enterococcus group. Conversely, the enterococcus group does not include all of the organisms in either fecal streptococci or group D streptococci.

The use of enterococci or fecal streptococci as indices of fecal contamination, or as a means of determining the sanitary history of food products, their alleged association with foodborne illness and their occurrence in human pathological conditions attest to their importance in public health.

Sources

Since enterococci are part of the larger group, fecal streptococci, the sources of enterococci are also sources of fecal streptococci. However, if fecal streptococci are present, there could be organisms not included in the enterococcus group.

The primary source of the enterococci and fecal streptococci is the intestines of man and warm-blooded animals. Levels in human feces may range from 10^4 to 10^9 per gram. The organisms are prevalent on the hide, hair, feathers and hoofs of animals and, upon slaughter, the processing equipment, hands of the workers and the food products become contaminated.

Animals excreting these organisms can cause widespread distribution of the enterococci. They have been reported in water, soil, vegetation and insects. The presence of these organisms on vegetation and insects may cloud their value as indices of fecal contamination.

Fecal streptococci were isolated from over 94% of the samples of crayfish examined by Lovell and Barkate (1969). Insalata *et al.* (1969) reported fecal streptococci in freshly prepared, dried, frozen, baked and canned foods, as well as pet foods, sugar, syrup, flour, sauces and beverages.

Hence, although the primary source of these organisms may be the intestinal tract and fecal material, they are widespread in nature and on and in food products.

Methods

In order to be an acceptable indicator, relatively simple and standard methods should be available. The Bacteriological Analytical Manual (FDA 1976) describes methods for determining fecal streptococci, but no method for the specific enterococcus group.

For samples with large numbers of fecal streptococci, pour plates using KF streptococcal agar are prepared, incubated and observed for dark red colonies or colonies having a red or pink center. These colonies are assumed to be fecal streptococci.

For samples with low numbers of fecal streptococci, the MPN technique using KF streptococcal (KFS) broth or azide dextrose (AD) broth is suggested. Positive KFS broth is turbid and yellow and positive AD broth is turbid. No further testing of the culture is designated.

Abshire (1977) tested a medium he designated as D streptococcus-enterococcus broth. His data showed higher recoveries of group D streptococci with this broth than with either AD or KF broths. Donnelly and Hartman (1978) prepared a gentamicin-thallous-carbonate medium. They reported higher recovery of group D streptococci and less false positive organisms with this medium than with other media they tested.

Since few, if any, media are specific for a particular organism or group of organisms, and furthermore, the species of *Streptococcus* may indicate the source of the organism, further biochemical testing of the organisms is warranted.

The tests include the finding of Gram positive cocci that are catalase negative and reduce 0.1% methylene blue. Also, lactose, sorbitol, glycerol and arabinose fermentation, starch hydrolysis, gelatin hydrolysis, hemolysis and reduction of potassium tellurite and TTC can be determined for differentiation of the species. Facklam and Moody (1970) suggested using the bile-esculin test to segregate group D from other streptococci. Lee (1972) reported the use of a tyrosine decarboxylase medium. The enterococci are tyrosine decarboxylase positive.

Mundt (1973) suggested that the reaction in litmus milk and fermentation of melezitose and melibiose might be employed to distinguish between contamination representing recent pollution of human origin and the presence of *S. faecalis* as a member of plants with no sanitary significance.

Besides these biochemical tests, immunofluorescent tests (Pugsley and Evison 1975), phage typing (Caprioli *et al.* 1975), enterocin typing (Sharma *et al.* 1976) and counterimmunoelectrophoresis (Portas *et al.* 1976) have been suggested for enumerating and identifying the streptococci in these groups.

Growth

The enterococci are able to grow at both 10° and 45°C (Table 3.1). They can grow in broth with 6.5% NaCl and in media at pH 9.6. They can tolerate 40% bile.

The nutritional requirements for enterococci and some other fecal streptococci is such that they may fail to grow on simple culture media. However, most food products contain the nutrients needed for growth of these organisms.

Survival

The fecal streptococci survive outside of the intestinal tract better than do the coliforms. *S. faecalis* remains viable longer than *E. coli* in acid foods. In frozen foods, the fecal streptococci remain viable for long periods. The enterococci survive spray drying and storage in egg powder better than coliforms and *E. coli*. Cool, moist conditions prolong the survival of *S. faecalis* in soil (Kibbey *et al.* 1978). Freeze-thaw cycles are lethal.

The enterococci are thermoduric, withstanding heat that would destroy most nonsporeforming bacteria. The organisms can withstand 60°-65°C for 30 min. *S. faecium* is the most heat resistant species.

Although *S. bovis* survived only three days in storm water stored at 20°C, two strains of enterococci survived for over 2 weeks (Geldreich 1973B). In comparison, coliforms, fecal coliforms and salmonellae failed to survive for 2 weeks.

Enterococci are sensitive to chlorination. They are destroyed in 15 sec with 100 ppm chlorine at pH 8.4 and in 2 min with 10 ppm chlorine at pH 5.8 (Shannon *et al.* 1965).

Significance

The chief source of fecal streptococci is the intestinal tract of man and animals. The occurrence of these bacteria in water or food infers either direct or indirect fecal contamination.

The occurrence of fecal streptococci in water generally suggests fecal pollution. Their absence indicates little or no warm-blooded animal contamination. Fecal streptococci rarely multiply in polluted water; however, this may not be true for food products.

The presence of *S. bovis* or *S. equinus* indicates animal pollution rather than human contamination. Therefore, increased information is obtained if the species of fecal streptococci are identified.

Enterococci survive environmental conditions that destroy other microorganisms of sanitary significance. Due to their resistance to freezing, low pH and moderate heat treatment, the enterococci have been suggested as an indicator in some types of food products.

The finding of enterococci or fecal streptococci in a food is not proof that enteric pathogens are present. Once the enterococci or fecal streptococci contaminate a food processing plant, they can become established and multiply on surfaces contaminated with organic material. From these sources, they can contaminate food products. Thus, the organisms are able to multiply in environments far removed from the original source of fecal contamination.

In this case, the presence of the organisms could be an indication of poor handling or sanitary practices. However, complete elimination of these organisms is extremely difficult, even under good sanitary conditions.

Because of their universal presence, they might not be good indicators. There is no concrete evidence of a health hazard due to the presence of enterococci in a food product.

It has been stated that the enterococci cannot replace *E. coli* as an index for pollution. Also, the presence of the fecal streptococci, per se, gives little information, and either total counts or coliform counts are needed in conjunction with the fecal streptococcus or enterococcus count.

The use of these organisms in food products should be related to specific product practices, handling procedures and the microenvironments of the food. In some products, their presence has some meaning as indicator organisms; in other products, their presence is meaningless.

OTHER INDICATORS

The staphylococci, or *S. aureus*, have been proposed as indicators. The presence of large numbers of *S. aureus* is an indication of a potential health hazard due to staphylococcal enterotoxin, as well as questionable sanitation.

Clostridia have been suggested, but they are not very specific for human feces. Since they are sporeformers, they may persist in foods when most enteric organisms are gone. However, two clostridia, *C. perfringens* and *C. botulinum*, are important in foodborne illnesses.

Spore counts of thermophiles have been used as an index of washing efficiency and cleanliness of certain vegetables.

Machinery mold (*Geotrichum candidum*) has been used as an indicator of sanitation of food processing operations. It readily grows on food adhering to equipment surfaces and contaminates food that moves through the contaminated equipment.

The use of *Pseudomonas aeruginosa* as an indicator has been suggested. The presence of the organism in the intestinal tract of people, as well as its physiological characteristics, makes it a possible indicator of fecal contamination.

The presence of certain phages can be used as indicator systems. Kenard and Valentine (1974) found a high degree of correlation between fecal coliforms and coliphage counts. The presence of phage can be demonstrated in 6-8 hr. In this case, the coliphage indicate the presence of an indicator group, the fecal coliforms.

For a discussion of *Bifidobacterium*, reductase tests and others, the reader is referred to Mossel (1967).

SUMMARY

Indicator organisms are used to avoid the laborious process and expensive equipment needed to isolate pathogenic bacteria and viruses from food.

The significance attached to the presence of intestinal bacteria in water does not necessarily apply to food. The presence of *E. coli* in water indicates recent pollution by human waste. However, *E. coli* in food, whether in raw or heat treated food, is not evidence of direct pollution by either human or animal excreta.

The absence of an indicator in a food sample cannot be regarded as proof that intestinal pathogens also are absent.

Mossel (1967) and Drion and Mossel (1972) listed some reasons why indicators have a place in food microbiology. Bacteria are heterogeneously distributed. Hence, although present in a food, the enteric pathogens might not be present in a particular sample of the food. The detection of some of the enteric pathogens is difficult. The use of indicator groups gives a margin of safety and, with the continual absence of indicator organisms in a series of samples, we could assume that enteric pathogens are absent. In order to determine the presence or absence of enteric pathogens, several organisms need to be considered. To analyze all food samples for the presence of potential pathogens would be a tremendous task. Therefore, the testing for indicator organisms can be of value in assessing the microbial condition of a food. Caution should be used in the evaluation of information regarding the presence of indicators. According to Corlett (1974), the use of indicator organisms is of limited value for evaluating the presence of pathogens, and indicators now are used as evidence of inadequate heat treatment, contaminated equipment and/or recontamination after processing.

BIBLIOGRAPHY

ABSHIRE, R.L. 1977. Evaluation of a new presumptive medium for group D streptococci. Appl. Environ. Microbiol. *33*, 1149-1155.

ABSHIRE, R.L., and GUTHRIE, R.K. 1973. Fluorescent antibody as a method for the detection of fecal pollution: *Escherichia coli* as indicator organisms. Can. J. Microbiol. *19*, 201-206.

ANDERSON, J.M., and BAIRD-PARKER, A.C. 1975. A rapid and direct plate method for enumerating *Escherichia coli* biotype 1 in food. J. Appl. Bacteriol. *39*, 111-117.

BACHRACH, U., and BACHRACH, Z. 1974. Radiometric method for the detection of coliform organisms in water. Appl. Microbiol. *28*, 169-171.

BAGLEY, S.T., and SEIDLER, R.J. 1977. Significance of fecal coliform-positive *Klebsiella*. Appl. Environ. Microbiol. *33*, 1141-1148.

BETTELHEIM, K.A. *et al.* 1974. *Escherichia coli* serotype distribution in man and animals. J. Hyg. Camb. *73*, 467-471.

BRAUNSTEIN, H., and MLADINEO, M.A. 1974. *Escherichia coli* strains producing hydrogen sulfide in iron-agar medium. Amer. J. Clin. Pathol. *62*, 420-424.

BUCHANAN, R.E., and GIBBONS, N.E. 1974. Bergey's Manual of Determinative Bacteriology, 8th Edition. The William and Wilkins Co., Baltimore, Md.

CAPRIOLI, T., ZACCOUR, F., and KASATIYA, S.S. 1975. Phage typing scheme for group D streptococci isolated from human urogenital tract. J. Clin. Microbiol. *2*, 311-317.

CLOSS, O. 1971. Two-tube test for the rapid identification of prompt lactose-fermenting genera within the family *Enterobacteriaceae*. Appl. Microbiol. *22*, 325-328.

CORLETT, D.A., JR. 1974. Setting microbiological limits in the food industry. Food. Technol. *28*, No. 10, 34-40.

COX, N.A., MERCURI, A.J., JUVEN, B.J., and THOMSON, J.E. 1975. *Enterobacteriaceae* at various stages of poultry chilling. J. Food Sci. *40*, 44-46.

DARLAND, G., and DAVIS, B.R. 1974. Biochemical and serological characterization of hydrogen sulfide-positive variants of *Escherichia coli*. Appl. Microbiol. *27*, 54-58.

DONNELLY, L.S., and HARTMAN, P.A. 1978. Gentamicin-based medium for the isolation of group D streptococci and application of the medium to water analysis. Appl. Environ. Microbiol. *35*, 576-581.

DRION, E.F., and MOSSEL, D.A.A. 1972. Mathematical-ecological aspects of the examination for *Enterobacteriaceae* of foods processed for safety. J. Appl. Bacteriol. *35*, 233-239.

DRION, E.F., and MOSSEL, D.A.A. 1977. The reliability of the examination of foods, processed for safety, for enteric pathogens and *Enterobacteriaceae*: a mathematical and ecological study. J. Hyg. Camb. *78*, 301-324.

DUNCAN, D.W., and RAZZELL, W.E. 1972. *Klebsiella* biotypes among coliforms isolated from forest environments and farm produce. Appl. Microbiol. *24*, 933-938.

EVISON, L.M., and JAMES, A. 1973. A comparison of the distribution of intestinal bacteria in British and East African water sources. J. Appl. Bacteriol. *36*, 109-118.

FACKLAM, R.R. 1973. Comparison of several laboratory media for presumptive identification of enterococci and group D streptococci. Appl. Microbiol. *26*, 138-145.

FACKLAM, R.R., and MOODY, M.D. 1970. Presumptive identification of group D streptococci: the bile-esculin test. Appl. Microbiol. *20*, 245-250.

FDA. 1976. Bacteriological Analytical Manual for Foods, 4th Edition. Food and Drug Administration, Washington, D.C.

FINSTEIN, M.S. 1973. Sanitary bacteriology. *In* CRC Handbook of Microbiology. Vol. I. Organismic Microbiology. A.I. Laskin and H.A. Lechevalier (Editors). CRC Press, Cleveland, Ohio.

FUGATE, K.J., CLIVER, D.O., and HATCH, M.T. 1975. Enteroviruses and potential bacterial indicators in Gulf Coast oysters. J. Milk Food Technol. *38*, 100-104.

GARTNER, H., HAVEMEISTER, G., WALDVOGEL, B., and WUTHE, H.H. 1975. Qualitative and quantitative *Salmonella* investigations and their hygienic valuation in connection with the *E. coli* titre, demonstrated with examples from the coastal waters of Kiel Bight (Western Baltic Sea). Zbl. Bakt. Hyg., I. Abt. Orig. *B160*, 246-267.

GELDREICH, E.E., and BORDNER, R.H. 1971. Fecal contamination of fruits and vegetables during cultivation and processing for market. A review. J. Milk Food Technol. *34*, 184-195.

HALL, H.E., BROWN, D.F., and LEWIS, K.H. 1967. Examination of market foods for coliform organisms. Appl. Microbiol. *15*, 1062-1069.

HARTLEY, J.C., REINBOLD, G.W., and VEDAMUTHU, E.R. 1968. Bacteriological methods for evaluation of milk quality. A. review. I. Use of bacterial tests to evaluate production conditions. J. Milk Food Technol. *31*, 315-322.

HARTMAN, P.A., HARTMAN, P.S., and LANZ, W.W. 1975. Violet red bile 2 agar for stressed coliforms. Appl. Microbiol. *29*, 537-539.

HUFHAM, J.B. 1974. Evaluating the membrane fecal coliform test by using *Escherichia coli* as the indicator organism. Appl. Microbiol. *27*, 771-776.

INSALATA, N.F., WITZEMAN, J.S., and SUNGA, F.C.A. 1969. Fecal streptococci in industrially processed foods—an incidence study. Food Technol. *23*, No. 10, 86-88.

KÄFERSTEIN, F.K. 1977. The occurrence of antibiotic-resistant microorganisms (*Escherichia coli* type I and coliforms) in some foods. Zbl. Bakt. Hyg. I. Abt. Orig. *B164*, 111-118.

KENARD, R.P., and VALENTINE, R.S. 1974. Rapid determination of the presence of enteric bacteria in water. Appl. Microbiol. *27*, 484-487.

KIBBEY, H. J., HAGEDORN, C., and MCCOY, E.L. 1978. Use of fecal streptococci as indicators of pollution in soil. Appl. Environ. Microbiol. *35*, 711-717.

KOKAL, D., and THORPE, D.W. 1969. Occurrence of *Escherichia coli* in almonds of nonpareil variety. Food Technol. *23*, 227-232.

LEE, W.S. 1972. Improved procedure for the identification of group D enterococci with two new media. Appl. Microbiol. *24*, 1-3.

LOKEN, K.I., CULBERT, K.H., SOLEE, R.C., and POMEROY, B.S. 1968. Microbiological quality of protein feed supplements produced by rendering plants. Appl. Microbiol. *16*, 1002-1005.

LOVELL, R.T., and BARKATE, J.A. 1969. Incidence and growth of some health-related bacteria in commercial freshwater crayfish (genus *Procambarus*). J. Food Sci. *34*, 268-271.

MARCUS, K.A., and AMLING, H.J. 1973. *Escherichia coli* field contamination of pecan nuts. Appl. Microbiol. *26*, 279-281.

MILLIPORE. 1975. A new membrane filter for fecal coliform analysis. Publication *PB806*. Millipore Filter Corporation, Bedford, Mass.

MITCHELL, D.O., and STARZYK, M.J. 1975. Survival of *Salmonella* and other indicator microorganisms. Can. J. Microbiol. *21*, 1420-1421.

MORAN, J.W., and WITTER, L.D. 1976. An automated rapid test for *Escherichia coli* in milk. J. Food Sci. *41*, 165-167.

MOSSEL, D.A.A. 1967. Ecological principles and methodological aspects of the examination of foods and feeds for indicator microorganisms. J. Assoc. Offic. Anal. Chem. *50*, 91-104.

MUNDT, J.O. 1973. Litmus milk reaction as a distinguishing feature between *Streptococcus faecalis* of human and non-human origins. J. Milk Food Technol. *36*, 364-367.

NEWMAN, J.S., and O'BRIEN, R.T. 1975. Gas chromatographic presumptive test for coliform bacteria in water. Appl. Microbiol. *30*, 584-588.

NOTERMANS, S., VAN LEUSDEN, F.M., and VAN SCHOTHORST, M. 1977. Suitability of different bacterial groups for determining faecal contamination during post scalding stages in the processing of broiler chickens. J. Appl. Bacteriol. *43*, 383-389.

PORTAS, M.R., HOGAN, N.A., and HILL, H.R. 1976. Rapid specific identification of group D streptococci by counterimmunoelectrophoresis. J. Lab. Clin. Med. *88*, 339-344.

PUGSLEY, A.P., and EVISON, L.M. 1974. Immunofluorescence as a method for the detection of *Escherichia coli* in water. Can. J. Microbiol. *20*, 1457-1463.

PUGSLEY, A.P., and EVISON, L.M. 1975. A fluorescent antibody technique for the enumeration of faecal streptococci in water. J. Appl. Bacteriol. *38*, 63-65.

RAY, B., and SPECK, M.L. 1978. Plating procedure for the enumeration of coliforms from dairy products. Appl. Environ. Microbiol. *35*, 820-822.

ROSEN, A., and LEVIN, R.E. 1970. Vibrios from fish pen slime which mimic *Escherichia coli* on violet red bile agar. Appl. Microbiol. *20*, 107-112.

SHANNON, E.L., CLARK, W.S., JR., and REINBOLD, G.W. 1965. Chlorine resistance of enterococci. J. Milk Food Technol. *28*, 120-123.

SHARMA, D.P., SHRINIWAS, and BHUJWALA, R.A. 1976. Enterocin typing of group 'D' streptococci. Jap. J. Microbiol. *20*, 559-560.

SILLIKER, J.H., and GABIS, D.A. 1976. ICMSF methods studies. VII. Indicator tests as substitutes for direct testing of dried foods and feeds for *Salmonella*. Can. J. Microbiol. *22*, 971-974.

SMITH, M.R. 1971. *Vibrio parahemolyticus*. Clin. Med. *78*, 22-25.

SOLBERG, M. *et al.* 1976. What do microbiological indicator tests tell us about the safety of foods? Food Product Development *10*, No. 9, 72-80.

SURKIEWICZ, B.F. *et al.* 1975. Bacteriological survey of raw beef patties produced at establishments under federal inspection. Appl. Microbiol. *29*, 331-334.

WOLFE, M.W., and AMSTERDAM, D. 1968. New diagnostic system for the identification of lactose-fermenting Gram-negative rods. Appl. Microbiol. *16*, 1528-1531.

8

Food Spoilage

As consumers, we depend upon our senses of sight, smell, taste and touch to evaluate food quality. When based upon past experiences, these organoleptic evaluations are used to determine if food is spoiled. Since people have different past experiences and abilities, there are conflicting opinions concerning the point at which a food is no longer acceptable for consumption.

The deterioration of food can be caused by rodents, insects, tissue enzymes, nonenzymatic chemical changes, physical effects and the action of microorganisms. We are concerned primarily with the deterioration of food by microorganisms and microbial enzymes. Tissue enzymes may aid in microbial deterioration of food by altering certain food components so they are more readily available for microbial use. The effect of tissue enzymes on deterioration is quite evident in over-ripe fruits.

The microbial deterioration of a food usually is manifested by alterations in the appearance, texture, color, odor, flavor or by slime formation. The appearance includes color changes, formation of pockets of gas or swelling and microbial growth, especially that of molds. As some food products deteriorate, they tend to become soft or mushy. Degradation of food results in the formation of compounds which have odors and flavors different from those of the fresh food.

MICROBIAL FUNCTIONS

Like humans, the microorganisms need food for energy, cell growth, maintenance and reproduction. In the scheme of nature, one of the important roles of microorganisms is to degrade organic material that is no longer needed, so the elements can be reused. Unfortunately, the microorganisms attack our food resources with equal vigor.

The biochemical changes that occur in a food may be desirable or undesirable. Microorganisms change a food into another product, such as cabbage to sauerkraut, grape juice to wine or alcohol to vinegar. Under controlled conditions, these reactions are desirable, but if we want grape juice, then the formation of alcohol and the production of wine are undesirable. The useful aspects of microorganisms in foods are discussed in the next chapter. The basic reactions, whether desirable or undesirable, are essentially the same.

The system by which undesirable changes occur in a food is determined by the composition of the food (carbohydrate, fat, protein), the types and numbers of microorganisms associated with the food, the factors (intrinsic or extrinsic) that influence the growth of microorganisms and changes in these factors that occur during processing or during deterioration.

The microorganisms present on food include those associated with the raw material, those acquired during harvesting, handling, and processing, or those surviving a preservation treatment and storage. The microorganisms may act singly or in groups to break down complex organic compounds. The active enzyme systems of the dominant microorganisms, as well as the natural enzymes of the food, will determine the course of degradation of a food.

FOOD COMPOSITION

The approximate amounts of carbohydrate, fat and protein of various foods are shown in Tables 4.3 and 4.4. Plant products are primarily carbohydrate. Animal products are protein and fat. Milk is composed of a mixture of carbohydrate, fat and protein. Processed foods, such as candy and baked goods, are carbohydrate.

The data of Watt and Merrill (1963) indicate no carbohydrate for most muscle tissues of animals. However, other publications indicate there might be less than 1% carbohydrate in some muscle tissue. Animal products such as seafoods, liver, milk and eggs do contain carbohydrates. Certain sausage products contain added sugar, which accounts for the carbohydrate in this food.

Carbohydrates are utilized more readily than are proteins or fats. The fermentation of carbohydrates produces acids and lowers the pH. The acidic condition tends to inhibit proteolytic organisms. The term "protein sparing" has been used to describe the metabolism of organisms in a food containing carbohydrates and proteins. Since very few foods are all carbohydrate, all fat, or all protein, more than one type of degradation may occur.

DEGRADATION OF COMPONENTS

Carbohydrates

The general formula of carbohydrates is $C_x(H_2O)_y$. The carbohydrates are divided into monosaccharides, disaccharides and polysaccharides. The monosaccharides are polyhydroxy aldehydes (aldoses) or polyhydroxy ketones (ketoses). Some carbohydrates are listed in Table 8.1.

TABLE 8.1

COMPOUNDS DEGRADED AS CARBOHYDRATES

Monosaccharides	Disaccharides	Polysaccharides
Hexoses	Maltose (glucose + glucose)	Dextrins
Aldoses	Lactose (glucose + galactose)	Starch
Glucose	Sucrose (glucose + fructose)	Glycogen
Mannose	Cellobiose (glucose + glucose)	Inulin
Galactose		Cellulose
Ketoses		Hemicellulose
Fructose		Galactan
Sorbose		Xylan
Polyhydric Alcohols		Complexed Polysaccharides
Mannitol		Glucosides
Glycerol		Salicin
Adonitol		Amygdalin
Dulcitol		Esculin
Sorbitol		Tannins
		Pectins
		Gums
		Mucilages

There are some free monosaccharides, but most naturally occurring carbohydrates are disaccharides and polysaccharides. For utilization, bacteria first break down these complex carbohydrates to their constituent monosaccharides.

Polysaccharides.—Plant cells and tissues have a fibrous substance embedded in an amorphous support matrix. The fibrous material resists tension and the matrix resists compression. In plants, the primary fibrous substance is cellulose.

Plants have a rather wide range of matrix compounds. In higher plants, the matrix polysaccharides are classified into a group of acidic polysaccharides called pectins, and a heterogeneous group of neutral polysaccharides called hemicelluloses. When stiffness is needed, a nonpolysaccharide, lignin, is incorporated. In algae (seaweeds), such things as carrageenans and agar are the matrix materials.

Pectin.—Pectolytic enzymes are produced by a wide range of microorganisms, especially those causing soft rot. The pectic enzymes include pectin esterase (EC3.1.1.11), polygalacturonases (EC3.2.1.15) and pectate transeliminases (EC4.2.99.3). The random splitting of glycosidic bonds results in softening and liquefaction.

The pectic enzymes are found in plant tissues, as well as in many types of microorganisms. The plant enzymes act during the ripening of fruit to cause a softening effect. This is desirable to an acceptable level but continued action by these enzymes causes excessive softness and a mushy product. When a microorganism is isolated from a softened vegetable or fruit, one cannot be certain that the softening effect was due to enzymes from the organism or the plant tissue. The degradation of pectin is perhaps the primary spoilage defect of plant tissue.

Starch.—This polysaccharide is an important component of many plant products. Several bacteria and molds possess an extracellular enzyme, diastase or amylase, which hydrolyzes starch. Starch is hydrolyzed in two stages: first, the disaccharide maltose is produced, and then maltose is hydrolyzed to form two molecules of glucose.

Monosaccharides.—Glucose is the main carbohydrate used as a carbon and energy source. The breakdown of these sugars can proceed by several pathways. Some of the reactions and intermediates of glycolysis are summarized in Figure 8.1.

Pyruvic acid ($CH_3COCOOH$) is a key compound in the metabolism of glucose. Depending upon the metabolic pathways available to the organism, and the environmental conditions, the pyruvate can be involved in many reactions, either fermentative or oxidative. Some of the metabolic products resulting from carbohydrates are listed in Table 8.2. The products include organic acids, alcohols, CO_2, H_2 and H_2O.

In aerobic respiration, pyruvate is converted into CO_2 and H_2O by means of the tricarboxylic acid (TCA) cycle, Krebs cycle or citric acid cycle. To enter this system, the pyruvate is converted to acetate activated with coenzyme A. Only the aerobic and some facultatively anaerobic microorganisms possess an intact TCA cycle.

The pyruvic acid can be decarboxylated to form acetaldehyde (CH_3-CHO) and CO_2. The acetaldehyde can remain or be reduced to ethyl alcohol (CH_3CH_2OH), oxidized to acetic acid (CH_3COOH) or condensed to form acetoin ($CH_3COCHOHCH_3$) or acetylmethylcarbinol (AMC). The AMC can be oxidized to diacetyl ($CH_3COCOCH_3$), which has a butter flavor, or reduced to 2,3 butanediol ($CH_3CHOHCHOHCH_3$). Pyruvate can be aminated to form alanine.

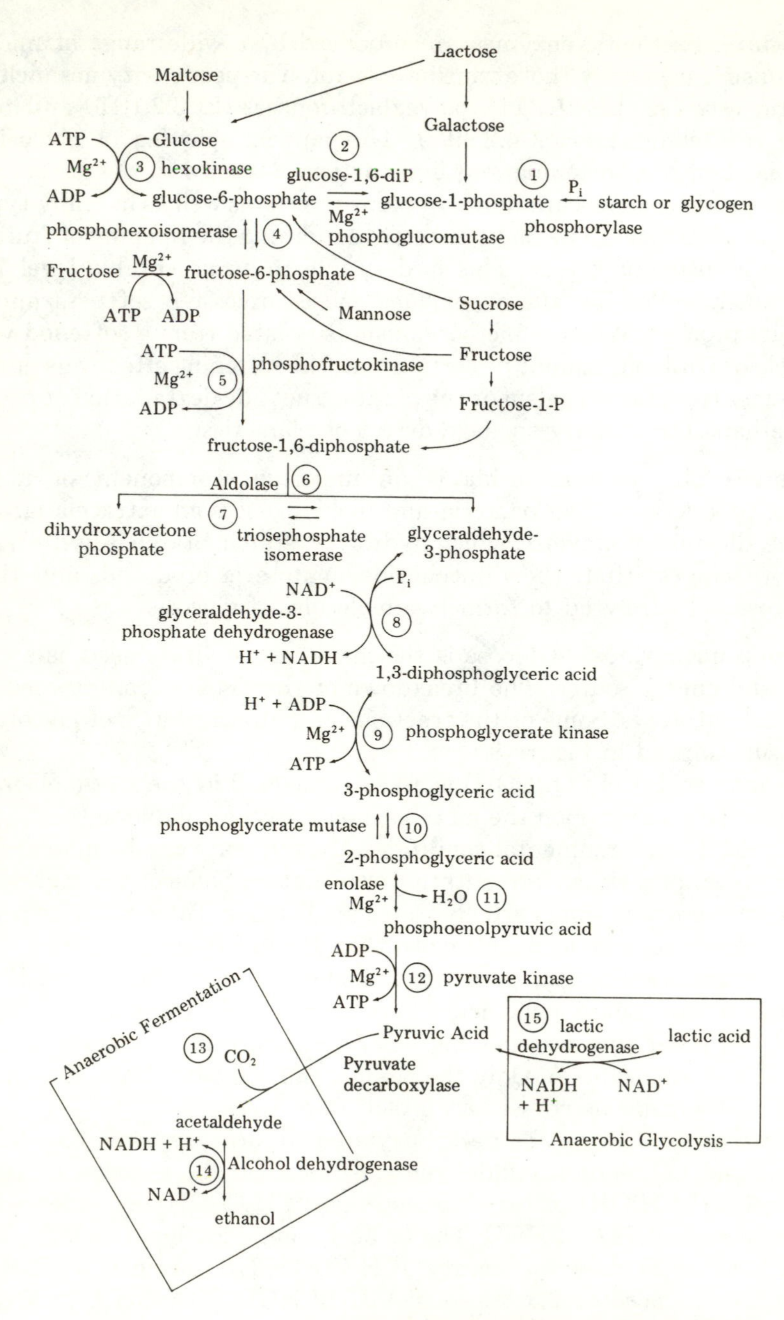

Courtesy of Aurand and Woods (1973)

FIG. 8.1. A GLYCOLYTIC PATHWAY FOR THE DEGRADATION OF VARIOUS SUGARS

Cellulose.—Only a few species of bacteria are able to decompose cellulose. However, these bacteria are widely distributed in soil and in ruminants. Cellulose is hydrolyzed to glucose either directly, or with the formation of an intermediate, cellobiose.

Microbial cellulase production seems to be regulated by the presence of

TABLE 8.2

SOME PRODUCTS OF CARBOHYDRATE METABOLISM

Organisms	Products
Leuconostoc mesenteroides	Lactic acid Ethyl alcohol CO_2
Leuconostoc cremoris	Acetic acid Acetoin Diacetyl CO_2
Saccharomyces (yeast)	Ethyl alcohol CO_2
Clostridium botulinum	Acetic acid Butyric acid Propionic acid Isobutyric acid Isovaleric acid Propyl alcohol Isobutyl alcohol Butyl alcohol Isoamyl alcohol
Propionibacterium	Propionic acid Acetic acid Isovaleric acid Formic acid Succinic acid Lactic acid CO_2
Escherichia coli	Lactic acid Acetic acid Formic acid CO_2 H_2
Bacillus cereus	Acetoin Glycerol 2,3-Butanediol Lactic acid Succinic acid Formic acid Acetic acid CO_2

glucose. Hence, cellulases become important only in the later stages of decay when glucose and the readily available carbohydrates become exhausted. The breakdown of cellulose is of secondary importance to pectin in the deterioration of plant tissue.

Lipids

The principal lipids in foods are fats. Fats are esters of glycerol and fatty acids and are called glycerides. The generalized structure of triglycerides is:

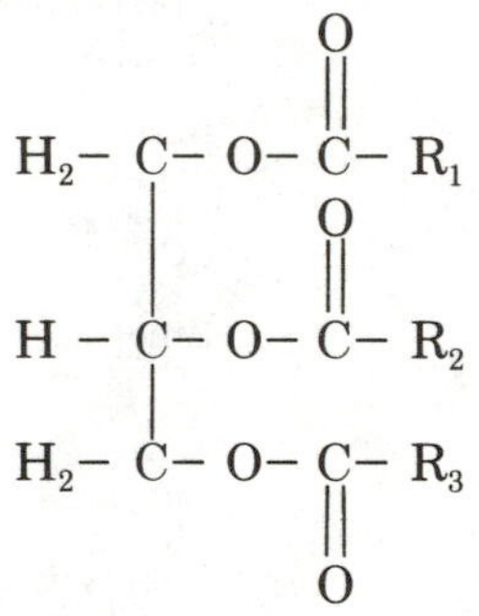

$$\begin{array}{l} H_2-C-O-\overset{\displaystyle O}{\overset{\|}{C}}-R_1 \\ \quad\;\; | \qquad\quad \overset{\displaystyle O}{\|} \\ H\;-C-O-C-R_2 \\ \quad\;\; | \\ H_2-C-O-\underset{\displaystyle O}{\underset{\|}{C}}-R_3 \end{array}$$

The R-groups denote the carbon chain of the fatty acids. Some common fatty acids are listed in Table 4.13. The triglycerides may have the same fatty acid in all three positions, or two or three different fatty acids may be present.

Many foods contain fat that is susceptible to hydrolysis, oxidation and other chemical processes that result in the formation of various compounds. Both desirable and undesirable flavor changes in foods are associated with these compounds.

The enzymes that hydrolyze triglycerides are called lipases. The designation of true lipase is given to enzymes that attack only water insoluble substrates.

The oxidative deterioration of fats involves the reactions of unsaturated fatty acids with oxygen to give hydroperoxides. The hydroperoxides are not flavor compounds, but readily decompose to carbonyl compounds resulting in off-flavor or odor. The carbonyl compounds are mixtures of saturated and unsaturated aldehydes and ketones.

Deteriorated fat is called rancid. The simple release of free fatty acids is called hydrolytic rancidity and oxidative deterioration is oxidative rancidity. The formation of ketones is referred to as ketonic rancidity.

Hydrolytic rancidity is important in fats. Short chain water soluble fatty acids (butyric, caproic and caprylic) cause obnoxious rancid flavors in milk.

Except for the release of volatile fatty acids, the main flavor changes are not due to hydrolysis, but to oxidation. The oxidation of the fat is more often due to tissue enzymes or autoxidation than to microbial activity. Autoxidation is accelerated by heat and light, and is catalyzed by heavy metals and their salts, organometalic compounds, hemoglobin, hematin compounds and photochemical pigments. Certain species of *Aspergillus* and *Penicillium* produce oxidative enzymes that catalyze the oxidation of free fatty acids to methyl ketones.

A pure fat is not attacked by microorganisms, since there must be a nutrient containing aqueous phase in which the organism can grow. However, most of our fatty foods (butter, cream, margarine), contain an aqueous phase associated with the fat.

There are many lipolytic microorganisms. Some of the genera that contain lipolytic species or strains are listed in Table 8.3.

TABLE 8.3

SOME GENERA CONTAINING LIPOLYTIC SPECIES OR STRAINS

Bacteria	Fungi
Acinetobacter	*Absidia*
Aeromonas	*Alternaria*
Alcaligenes	*Aspergillus*
Bacillus	*Candida*
Chromobacterium	*Cladosporium*
Corynebacterium	*Endomyces*
Enterobacter	*Fusarium*
Flavobacterium	*Geotrichum*
Lactobacillus	*Mucor*
Micrococcus	*Neurospora*
Pseudomonas	*Penicillium*
Serratia	*Rhizopus*
Staphylococcus	*Torulopsis*
Streptomyces	

The lipolytic characteristics of microorganisms depend upon the fat in the substrate. Tom and Crisan (1975) showed that although many organisms decomposed tributyrin or tween 80, only a few were able to decompose fish oil added to a medium.

With the breakdown of fats, various products are formed, including glycerol, free fatty acids, aldehydes, ketones and alcohols.

Proteins

Proteins are composed of amino acids combined by peptide linkages. These linkages are broken by the addition of water, catalyzed by enzymes. The proteins are degraded through proteoses, peptones, polypeptides and dipeptides to amino acids. The enzymes that hydrolyze the peptide linkages of proteins are called proteases. These proteases convert the proteins into diffusible polypeptides and dipeptides, which can enter the bacterial cell. The polypeptides and dipeptides are without offensive odor.

The native proteins are resistant to attack by microorganisms. There are low molecular weight compounds, such as dipeptides and free amino acids in fresh meat, fish and poultry tissue. These substances are readily used by the microbial flora. Spoilage of the protein-type foods may be evident before any significant amount of protein is degraded. In advanced spoilage, some protein is hydrolyzed by proteolytic enzymes, whether produced by the tissues or the microorganisms.

Amino Acids.—The degradation of amino acids is of primary importance in the spoilage of protein foods. The products that are formed depend upon: 1) the type of microorganism; 2) the types of amino acids; 3) temperature; 4) the amount of available oxygen; and 5) the types of inhibitors that might be present.

There are many amino acids and hence, many products of amino acid dissimilation. Some products that are formed anaerobically are foul smelling. This form of degradation is called putrefaction. Aerobic degradation, with oxidation of the metabolic products, is called decay.

The two main reactions of microorganisms on amino acids are decarboxylation or deamination. The reactions on amino acids include:

1. *Oxidative deamination.*—

$$2R\text{-}CHNH_2\text{-}COOH + O_2 \rightarrow 2\ R\text{-}CO\text{-}COOH + 2NH_3$$

In this case, an α-keto acid and ammonia are formed. The keto acid formed depends upon the R group. If R is CH_3, then pyruvic acid (CH_3-CO-COOH) is formed. Secondary reactions of the highly reactive keto acids occur, so that they seldom appear as a product.

2. *Reductive deamination.*—

$$R\text{-}CHNH_2\text{-}COOH + 2H \rightarrow R\text{-}CH_2\text{-}COOH + NH_3$$

In this case, an organic acid and ammonia are formed. This reaction is generally used by the strict anaerobes or facultative anaerobes grown under anaerobic conditions.

3. *Hydrolytic deamination.*—

$$R\text{-}CHNH_2\text{-}COOH + H_2O \rightarrow R\text{-}CHOH\text{-}COOH + NH_3$$

In this case a hydroxyacid and ammonia are produced. The reaction occurs under neutral or slightly alkaline conditions.

4. *Hydrolytic deamination and decarboxylation.*—

$$R\text{-}CHNH_2\text{-}COOH + H_2O \rightarrow R\text{-}CH_2OH + CO_2 + NH_3$$

In this case, an alcohol, carbon dioxide, and ammonia are formed.

5. *Deamination and desaturation.*—

$$R\text{-}CH_2\text{-}CHNH_2\text{-}COOH \rightarrow R\text{-}CH{=}CH\text{-}COOH + NH_3$$

In this case, an unsaturated organic acid and ammonia are produced.

6. *Decarboxylation.*—

$$R\text{-}CHNH_2\text{-}COOH \rightarrow R\text{-}CH_2NH_2 + CO_2$$

In this reaction, an amine and carbon dioxide are produced.

7. *Oxidation-reduction.*—

$$3CH_3CHNH_2\text{-}COOH + 2H_2O \rightarrow 2\ CH_3CH_2COOH + CH_3COOH + 3NH_3 + CO_2$$

The overall reaction results in two organic acids (one with one less carbon atom), ammonia, and carbon dioxide.

8. *Anaerobic degradation—release of hydrogen.*—

$$5COOH\text{-}CH_2\text{-}CH_2\text{-}CHNH_2\text{-}COOH + 6H_2O \rightarrow 6CH_3COOH + 2CH_3CH_2CH_2COOH + 5CO_2 + 5NH_3 + H_2$$

The overall reaction involving, in this case, glutamic acid, yields acetic acid, butyric acid, carbon dioxide, ammonia, and hydrogen.

9. *Transamination.*—

Although the amino acid is not degraded in this reaction, transamination does alter the composition of the substrate which will determine the final degradation product.

$$COOH\text{-}CH_2\text{-}CH_2\text{-}CHNH_2\text{-}COOH + COOH\text{-}CH_2\text{-}CO\text{-}COOH \rightarrow COOH\text{-}CH_2\text{-}CH_2\text{-}CO\text{-}COOH + COOH\text{-}CH_2\text{-}CHNH_2\text{-}COOH$$

In this example, glutamic acid reacts with oxaloacetic acid to form α-keto glutaric acid and aspartic acid.

10. *Mutual oxidation and reduction.*—

Some amino acids (alanine, leucine, phenylalanine, and valine) can serve as hydrogen donators and are oxidized, while others (arginine, glycine, hydroxyproline, ornithine, and proline) can serve as hydrogen acceptors and are reduced. This reaction occurs with anaerobic bacteria grown in anaerobic conditions, and is commonly called the Stickland reaction.

These reactions (1-10) account for most of the decay and putrefaction of amino acids. However, special mention should be given to reactions of certain amino acids.

Tryptophan can enter reactions resulting in many products. One product, indole, is used in the identification of certain microorganisms. Another product is skatole. Small amounts of skatole are present in the perfume of the jasmine and orange blossoms. It is one of the two most important nitrogenous substances found in natural perfumes. In larger concentrations, skatole is a foul smelling compound. It is found prominently in fecal material.

When sulfur containing amino acids, such as cystine or methionine are degraded, hydrogen sulfide may be one of the resultant products.

The decarboxylation of the diamino acids, lysine and ornithine, yield cadaverine and putrescine accordingly:

$$NH_2\text{-}(CH_2)_4\text{-}CHNH_2\text{-}COOH \rightarrow NH_2\text{-}(CH_2)_5\text{-}NH_2 + CO_2$$

$$NH_2\text{-}(CH_2)_3\text{-}CHNH_2\text{-}COOH \rightarrow NH_2\text{-}(CH_2)_4\text{-}NH_2 + CO_2$$

Arginine is degraded through ornithine to putrescine.

By the dissimilation of amino acids, many diverse products, such as carbon dioxide, hydrogen, ammonia, hydrogen sulfide, organic acids, alcohols, amines, diamines, mercaptans and organic disulfides may be formed. The production of ammonia and amines tends to cause an increase in the pH. Trimethylamine is more basic than is ammonia. An increase in the pH of a protein food indicates protein degradation, just as a decrease in pH results from the fermentation of carbohydrates.

OTHER DETERIORATION

The metabolism of microorganisms can cause softening of fruits and vegetables by degradation of pectin and cellulose. They can cause changes in the flavor or odor of foods by degradation of carbohydrates, fats and proteins. Besides these aspects, microorganisms create unacceptable foods by changing the appearance of the food, either by mold growth, or by altering the color of a pigment. Microorganisms also can metabolize sugars and produce dextrans or levans which feel like slime on the surface of foods such as meat, poultry or fish, or a condition called ropiness, such as in milk or bread.

Appearance

When microbial growth can be observed on a food, it usually is considered to be unacceptable. Pigmented bacteria can, at times, be observed on a food. Mold mycelia indicate potential spoilage; however, mold is grown on certain types of cheeses, such as Roquefort and Camembert, to develop the distinctive flavors.

Various pigments (chlorophyll, carotene, myoglobin) are present in foods. The degradation of chlorophyll and carotene is related to the lipolytic breakdown of fats. A lipase in peas effectively degrades chlorophyll to pheophytin, and the degradation of both chlorophyll and pheophytin is related to an increase in fat peroxide. Carotene is fat soluble and its destruction is related to fat decomposition.

The myoglobin pigment of meat muscle can be altered by oxidation-reduction reactions. Myoglobin is dark red to purple and is present in the interior part of the meat. When the muscle tissue is cut and exposed to oxygen, the myoglobin is oxygenated to oxymyoglobin, which is bright red and is the usual pigment associated with meat. In myoglobin and oxymyoglobin, the iron is in the ferrous form. When it becomes oxidized to the ferric form, metmyoglobin is formed. This pigment is dark brown.

Although these pigment changes do not necessarily make the food inedible, a consumer may consider the product unacceptable.

Slime Formation

Several bacteria produce microbial polysaccharides (dextrans, levans, or amyloses) from various disaccharides present in food. These polysaccharides form unpleasant slime in and on food, causing the food to be both unpalatable and unacceptable to the consumer. An example of this is the slimy or ropy texture of fruit concentrate or milk when infected with *Leuconostoc mesenteroides, Bacillus subtilis* or *E. coli.* Sucrose and maltose are readily used by *E. coli, L. mesenteroides* or *B. subtilis* to produce amyloses and dextrans. *B. megaterium* and *B. subtilis* produce levans, which are long chain polysaccharides.

Dextran is a glucose polymer. Sometimes it is encountered as a slime in large globular masses during the processing of cane or beet sugar. It increases the viscosity of the sugar solution and retards filtration and crystallization. These globular masses primarily are due to activities of *Leuconostoc mesenteroides* and *L. dextranicum*, although *Streptococcus salivarius* and *S. bovis* are able to synthesize from sucrose and raffinose a dextran-type insoluble carbohydrate.

MICROBIAL DEFECTS IN SPECIFIED FOODS

Presumably the enzymes produced by microorganisms, even if the microorganisms are not multiplying, could, over an extended storage time, decompose the food and cause spoilage. However, spoilage is associated with large numbers of microorganisms. Therefore, the organisms that cause spoilage are those that can multiply and become dominant. The factors affecting growth and dominance are discussed in Chapter 4. These factors are considered in regard to the spoilage of specific types of foods.

Fruits

A large and varied population of microorganisms, including the spores of many types of fungi, contaminate the surface of fruits during the growing season. Relatively few of the fungi are capable of attacking the fruit before harvest. The ripening process increases the susceptibility of the fruit to invasion. Fruits harvested from the ground have a more varied microbial flora than those picked from the tree. Falling from the tree causes bruising, and contact with the grass and soil provides an added source of inoculum.

Fruits have a low pH that inhibits most bacteria. The acid tolerant bacteria are mainly Gram positive lactobacilli and leuconostocs. Fruits are usually spoiled by yeasts and molds which are acid tolerant.

The microbial defects of fruit and fruit products are listed in Table 8.4. *Penicillium* species are important spoilage organisms of most fruits. *P. digitatum* and *P. italicum* cause soft rot in citrus fruits, and *P. expansum* causes blue mold rot of deciduous fruits. Blue mold rot consists of soft, watery, tan to light brown areas that are readily gouged out of the fruit flesh. The tissue has a moldy or musty odor and flavor. There are typical bluish-green spores on the fruit. It is the most important storage decay of apples.

Black rots in apples, citrus fruits, bananas and pineapples are due to species of *Alternaria*, as well as other molds. The tissue becomes soft and watery. The brown rot of fruits has unsunken, decayed areas, turning dark brown to black in the center. The mold spore masses are yellowish-gray. The skin clings tightly to the center of old lesion.

Anthracnose is a defect with scattered black or dark brown sunken spots covering firm decayed tissue. With moist conditions, there are pink spore masses on the spots.

Gray mold rot is evidenced by light brown, fairly firm, watery decay, covered extensively with delicate dirty-white mycelium. Grayish-brown velvety spots may be seen in advanced spoilage.

The loss of berries results from dehydration, discoloration and overripe, mushy, damaged, moldy or spoiled fruit. Berries are spoiled primarily by *Botrytis cinerea* and *Mucor mucedo.* However, certain fungi are prevalent on specific berries at various times during production, harvesting and storage.

The species of several genera of yeasts (*Saccharomyces, Hanseniaspora, Hansenula, Pichia, Torulopsis, Candida, Debaryomyces* and *Kloeckera*) can cause a fermentation defect in fruits.

TABLE 8.4

SOME MICROBIAL DEFECTS OF FRUITS AND THEIR PRODUCTS

Food	Defect	Organisms
Fresh fruit, general	Blue mold rot	*Penicillium expansum*
	Black mold rot	*Aspergillus niger*
	Souring, bitter flavor	*Streptococcus faecalis*
	Soft rot	*Byssochlamys fulva, Penicillium*
Apples	Fermentation	*Torulopsis, Candida, Pichia*
Apricots	Gray mold rot	*Botrytis cinerea*
Bananas	Storage rot	*Colletotrichum, Gloeosporium Botryodiplodia, Erwinia* sp.
	Black rot	*Alternaria*
	Crown rot	*Colletotrichum, Fusarium Verticillium*
Berries, general	Fungal rot	*Botrytis cinerea, Mucor mucedo*
Strawberries	Gray mold rot	*B. cinerea*
	Fermentation	*Kloeckera*
Citrus fruits	Soft rot	*Penicillium*
	Black rot	*Alternaria*
Dates	Fermentation	*Saccharomyces, Candida Hanseniaspora, Torulopsis*
Figs	Souring	*Gluconobacter*
Guava	Anthracnose rot	*Colletotrichum*
Olives	Softening (stem-end shrivel)	*Rhodotorula glutinis, R. minuta, R. rubra, Saccharomyces, Hansenula*
	Sloughing of skin	*Klebsiella, Enterobacter, Escherichia, Aeromonas liquefaciens*
Peaches	Brown rot	*Monilinia fructicola Sclerotinia*
	Decay	*Rhizopus stolonifer*
Canned fruit	Butyric acid	*Clostridium*
	Soft rot	*Byssochlamys fulva, B. nivea*
Apricots	Softening	*Rhizopus stolonifer, R. arrhizus*

TABLE 8.4. (*Continued*)

Food	Defect	Organisms
Banana puree	Gas	*Bacillus licheniformis*
Grapefruit sections	Gas (CO_2)	*Lactobacillus brevis*
Fruit juice	Souring, CO_2	*Lactobacillus*
	Acetification, vinegar	*Acetobacter*
	Moldy surface	*Penicillium*
	Cloudy	Nonfermenting yeasts
	Cloudy, alcohol	Fermenting yeasts
	Buttermilk flavor	*Lactobacillus, Leuconostoc*
Jelly, jam, preserves	Fungal	*Xeromyces bisporus*
	Fermentation	Osmophilic yeasts
Wine	Acetification	*Acetobacter, Gluconobacter*
	Mousy odor, cloudy, slimy	Lactic acid bacteria
	Acetaldehyde	*Saccharomyces oviformis* *S. beticus*
	Flowers	*Candida vini*

The lactobacilli are important in the spoilage of fruit juice. Some strains are quite acid tolerant and can metabolize citric and malic acid. This reduces the acidity, resulting in a bland, rather flat flavor and a loss in astringency. The production of diacetyl by these organisms results in a buttermilk flavor in the fruit juice.

Leuconostoc mesenteroides produces dextrans, resulting in a slimy, unpleasant texture of fruit juices. Yeasts contaminate and ferment fruit juice, especially apple juice. If the sugar concentration is 10-30%, the osmophilic yeasts, *Saccharomyces rouxii* and *S. mellis,* can ferment the sugars to alcohol. *Acetobacter* can convert the alcohol to acetic acid giving the fruit juice a vinegar flavor.

The most common defect of wine is acetification. This is caused by *Gluconobacter* and *Acetobacter.* All alcoholic beverages containing less than 15% w/v ethyl alcohol are subject to this type of spoilage.

Species of *Leuconostoc* and *Streptococcus* (especially *Leuconostoc mesenteroides*), as well as some *Acetobacter*, can produce dextrans in sweet wines. In this case, a ropy texture is developed which then becomes slimy and viscous.

The mature spores of *Byssochlamys fulva* and *B. nivea* are thermoduric and may survive the heat treatment during fruit canning. These fungi can grow at rather low redox potentials, and their pectic enzymes cause spoilage, especially of canned berries.

Vegetables

Since vegetables are harvested from or near the soil, they are subjected to a heterogeneous flora of soil, as well as airborne microorganisms.

In general, the pH of vegetables is near neutrality, so that bacteria, as well as fungi, cause deterioration. Due to the lower pH of tomatoes, the spoilage is similar to that of fruit, although bacterial spoilage also occurs. Microbial defects of vegetables are listed in Table 8.5.

TABLE 8.5

SOME MICROBIAL DEFECTS OF VEGETABLES AND THEIR PRODUCTS

Food	Defect	Organism
Fresh vegetables	Soft rot, mushy	*Erwinia carotovora*, *Pseudomonas fluorescens*
	Soft black rot	*Alternaria*, *Rhizopus nigricans*
	Black mold rot	*Aspergillus niger*
	Blue mold rot	*Penicillium*
Beans, snap	Anthracnose	*Colletotrichum*
	Blight	*Xanthomonas*
Cabbage	Leaf spot	*Alternaria*
	Gray mold	*B. cinerea*
Carrots	Soft rot	*E. carotovora* *Rhizopus stolonifer*
	Fungal rot	*Fusarium*
	Decay, wet rot	*Sclerotinia sclerotiorum* *Rhizoctonia carotae*
Celery	Fungal rot	*Mucor*
	Pink rot	*S. sclerotiorum*
Onions	Neck rot	*Botrytis allii*
	Rot	*Pseudomonas cepacia*
	Brown rot	*P. aeruginosa*
	Black mold	*Aspergillus niger*
Potatoes	Ring rot	*Corynebacterium*
	Dry rot	*Fusarium*
Tomatoes	Ferment	*Candida*, *Pichia*, *Hanseniaspora*, *Kloeckera*
	Fungal rot	*Alternaria*, *Aspergillus*, *Botrytis*, *Colletotrichum*, *Monilia*, *Penicillium*, *Rhizopus*
	Bacterial spot	*Xanthomonas*
	Soft rot	*Byssochlamys fulva*

TABLE 8.5. *(Continued)*

Food	Defect	*Organism*
Canned vegetables		
Corn, green beans, peas	Flat-sour	*Bacillus stearothermophilus*
	Sulfide stinker	*Desulfotomaculum nigrificans*
	Putrid swell	*Clostridium sporogenes*
	Hard swell	*Clostridium thermosaccharolyticum*
Tomato	Flat-sour	*Bacillus coagulans*
	Butyric fermentation	*Clostridium pasteurianum, C. butyricum*
Fermented vegetables		
Brine	Film forming	Yeasts (*Candida, Debaryomyces, Hansenula, Kloeckera, Pichia, Rhodotorula, Saccharomyces*)
Pickles	Soft	*Bacillus*
	Black	*Bacillus*
	Soft, mushy, slimy	*B. subtilis*
	Reduced acidity	Yeasts
Sauerkraut	Pink	*Rhodotorula*
Vegetable juice	Sour	*Lactobacillus, Acetobacter*

Celery plants are infected by strains of *Sclerotinia sclerotiorum*, which produces a pink rot at the base of the plant. Carrots, cabbage and other vegetables are soft rot-spoiled by bacteria, such as *Erwinia carotovora*, as well as the fungi *Rhizopus, Aspergillus* and *Alternaria.*

Canned tomato juice is subject to flat sour spoilage if not properly handled either during preparation or final heat treatment. Flat sour spoilage is attributed to the presence of and growth of *Bacillus coagulans*, which either survives normal heat processes, or recontaminates the product. Detection of spoilage is made initially by flavor, pH and odor, followed by diagnostic tests.

Other Carbohydrate Foods

The microbial defects of these foods are listed in Table 8.6. Cereals, honey, molasses, syrup, biscuits and candy ordinarily have water activities too low to support the growth of bacteria. However, due to storage at high relative humidity or production of water by metabolism, bacteria can be responsible for the spoilage of these products.

TABLE 8.6

SOME MICROBIAL DEFECTS OF MISCELLANEOUS CARBOHYDRATE FOODS

Food	Defect	Organism
Beer	Ropy	*Gluconobacter*
	Gas, slime	*Streptococcus lactis*
	Off-flavor	*Lactobacillus, Pediococcus, Brettanomyces, Candida, Pichia*
	Turbid	*Candida, Brettanomyces*
	Fruity	*Candida*
	Haze, diacetyl	Lactic acid bacteria, pediococci
Bread	Ropy, slime	*Bacillus subtilis*
	Black mold	*Rhizopus nigricans*
	Blue mold	*Penicillium*
	Pink mold	*Neurospora*
	Sour	Lactic acid or coliform bacteria
	Red	*Serratia marcescens*
Candy (chocolate cream)	Fermentation	Yeasts
Cereals, grains	Moldy	*Aspergillus, Penicillium*
	Discoloration	*Rhizopus nigricans*
Wheat	Pink	*Erwinia rhapontici*
Corn	Blue-eye	*Penicillium martensii*
Chocolate sauce	Cloudy	*Xeromyces bisporus*
Coconut	Rancidity	*Micrococcus luteus, Bacillus subtilis*
Dough "refrigerated"	Gas, slime, sour	*Lactobacillus, Leuconostoc, Streptococcus*
Honey	Fermented, yeasty	*Torulopsis*, osmophilic yeasts
Molasses	Gas, frothy	*Clostridium*, osmophilic yeasts
Peanuts	Moldy	*Fusarium, Penicillium, Aspergillus*
Soft drinks	Turbidity	Yeasts
Sugar solutions	Slime	*Leuconostoc mesenteroides*
	Ferment, yeasty	Osmophilic yeasts

Fungi are the primary agents causing deterioration of damp grain. The alterations of grain caused by fungi include discoloration, decreased germination, loss in weight, formation of mycotoxins, mustiness and an increase in temperature.

The examination of fresh and spoiled refrigerated dough products indicates that fungi, although present, are not able to grow due to the low redox potential (Hesseltine *et al.* 1969). The main spoilage organisms are species of *Lactobacillus* and *Leuconostoc.* Gas and swollen containers are the most obvious indications of spoilage. The buildup of gas causes the cardboard containers to split and the dough leaks through the seams. A slightly sour or moldy-sour odor may be noted. In some cases, the spoiled dough appears discolored. *Leuconostoc mesenteroides* and *L. dextranicum* produce dextrans from sugar, and cause slime in these doughs, as well as in sugar solutions.

Most raw sugars have an a_w between 0.65 and 0.75. Only osmophilic yeasts and some xerophilic fungi can grow on these raw sugars.

Soft centered fondants and chocolates can be fermented by yeasts, which produce ethyl alcohol and CO_2. The gas build-up may cause the outer chocolate coating to burst.

Lactobacilli are hops tolerant. These organisms produce lactic acid, dextran hazes and diacetyl in beer. A diacetyl flavor generally is not desirable in the light lager beers produced in the USA. Several yeasts, such as *Brettanomyces, Candida* and *Pichia* can infect finished beer and, with secondary fermentations, cause defects (dextran hazes, films, off-flavors and off-odors) in the beer.

The decarboxylation of hydroxycinnamic acids normally present in barley, by certain enterobacteria, such as *Hafnia*, result in a phenolic off-flavor in beer (Lindsay and Priest 1975).

Bakery products have been spoiled by several types of microorganisms causing various defects. Most of these problems have been controlled except for mold (blue mold or black mold spoilage).

Animal Products

Since fresh animal products are perishable, they are chilled and stored in ice or a refrigerator (0° to 4°C). This means that psychrotrophs become dominant and are the primary cause of spoilage. If these fresh products are mishandled and allowed to remain at room temperature (20°C), a more diverse flora may be present, since not only will many psychrotrophs grow, but also there will be some mesophilic strains.

The organisms most often involved with spoilage of refrigerated fresh meat, poultry, fish and eggs are species of the genus *Pseudomonas*. *Pseudomonas putrefaciens* has been designated as being important in spoilage of meat and poultry. This organism is listed in Addendum III of *Pseudomonas* (Buchanan and Gibbons 1974), with organisms which were assigned to *Pseudomonas*, but which possess characteristics which do not conform to the given definition of *Pseudomonas*. More recently, *P. putrefaciens* has been listed as *Alteromonas putrefaciens* (McMeekin 1977; Peel and Gee 1976).

Red Meat.—In general, the muscle tissue of live, healthy animals is sterile. The organisms that are present in the carcass are concentrated in the lymph nodes. These contaminated nodes are thought to be the source of bacteria for deep spoilage in post rigor meat. After slaughter, the defense mechanisms are slowed or halted. This enables bacteria to multiply and spread throughout the tissues.

As with most foods, spoiled meat lacks a precise definition, since what may be acceptable to one person may be spoiled to another.

Spoilage of meat is due to the growth and metabolism of large numbers of microorganisms on the surface or the interior. Most spoilage is on the surface. Both the number and type of organisms affect the spoilage characteristics of meat. The number of organisms that are present when spoilage is evident varies from 10^6 to 10^8 per cm^2 of meat surface. This variation apparently is due to the activity of the organisms present, as well as the criteria used by the investigators to determine spoilage. Vacuum packaged cured or processed meats may have microbial levels over 10^8/g and be considered satisfactory for consumption.

The most common indications of spoilage are: 1) off-odor and slime, usually due to the growth of aerobic bacteria on the cut surfaces of meat; 2) fungal growth, which is favored at water activities too low for bacterial growth; 3) bone-taint, or deep spoilage, due to anaerobic or facultative microorganisms; and 4) discoloration, primarily due to alterations of myoglobin, the muscle pigment.

There are some differences in the spoilage patterns of fresh and cured meat. There is only limited information on cooked meat spoilage. Generally, cooked meat is spoiled by the few types of organisms that can survive the cooking, or those that gain access to the cooked product. Fresh, chilled meat spoilage is evidenced by off-odor and slime due to *Pseudomonas*, *Acinetobacter* and *Alcaligenes*. Cured meats become sour due to the activity of *Micrococcus*, *Lactobacillus* and *Microbacterium* species. Microbial defects of meat and meat products are listed in Table 8.7.

TABLE 8.7

MICROBIAL DEFECTS OF RED MEATS AND THEIR PRODUCTS

Product	Defect	Organism
Fresh, refrigerated (0°-5°C)	Off-odor, slime, discoloration	*Pseudomonas, Aeromonas, Alcaligenes, Acinetobacter, Microbacterium, Moraxella, Proteus, Flavobacterium, Alteromonas, Saccharomyces*
	Lipolysis, pungent odor	*Pseudomonas*, yeasts
	Moldy	*Penicillium*
	Whiskers	*Thamnidium*
	Black spot	*Cladosporium*
	White spot	*Sporotrichum*
Fresh (15°-40°C)	Bone taint	*Clostridium*
	Gassy	*C. perfringens*
	Foul odor	*C. bifermentans, C. histolyticum, C. sporogenes*
Vacuum packaged	Acid, sweet, rancid	*Lactobacillus, Microbacterium, Enterobacter, Hafnia*
Cured meat		
Bacon	Cheesy, sour, rancid	*Micrococcus*
	Discoloration	Molds
	Slight souring	*Lactobacillus, Micrococcus Vibrio, Alcaligenes, Corynebacterium*
	Putrefaction	*Clostridium sporogenes*
Vacuum packaged	Cabbage odor	*Proteus inconstans*
	Tainted	*Vibrio*
Brines	Turbid	*Debaryomyces, Kloeckera*
Ham	Surface slime	*Micrococcus, Microbacterium*, Yeasts
	Gassy or puffy	*Clostridium*
	Green discoloration	*Lactobacillus, Streptococcus, Leuconostoc*
	Bone and meat "sours"	*Clostridium*
	Surface growth (Dry-cured)	Molds
Sausages	Slime on surface	*Micrococcus*, yeasts
	Gas production (vacuum packed)	*Lactobacillus*
	Greenish discoloration	*Lactobacillus viridescens, Leuconostoc*
Fermented sausage	Slime	Yeasts
	Spots	Molds

TABLE 8.7. (*Continued*)

Product	Defect	Organism
Canned		
Commercially sterile	Gas, putrefaction	Sporeformers (*Bacillus*, *Clostridium*)
Semi-preserved	Souring, discoloration	*Streptococcus*
	Putrefaction, gas	*Bacillus*, *Clostridium*

Fresh Meat.—The rate of spoilage depends upon the numbers and types of organisms initially present, the conditions of storage (temperature) and the characteristics (pH, a_w) of the meat.

In general, the whole carcass presents a surface of fat and connective tissue affording little opportunity for bacterial growth. This is seen in Figure 8.2. The surfaces of meat cuts will support the growth of large numbers of bacteria, and ground meat offers not only ample and desirable surfaces, but a thorough inoculation of the meat during grinding.

The temperature of the meat is perhaps the most important factor that determines the predominant microflora and the resultant spoilage. At temperatures of 50°C or higher, thermophilic bacteria can grow. Usually thermophilic bacteria cause spoilage of heat-processed foods that have been improperly handled during processing. However, these high temperatures may prevail in a closed car parked in the sun on a hot day.

Ingram and Dainty (1971) discussed the spoilage of meat held in warm conditions (25°-40°C). The temperature of the animal is in the upper part of this range at the time of slaughter. After slaughter, the temperature tends to increase and then to change to that of the surroundings. The redox potential also is lowered.

One of the less strict anaerobes, *Clostridium perfringens*, is the first species to produce large numbers. If the temperature remains at 25°-40°C, at cell levels of 10^8/g, gas production is noticeable, and the meat color changes from red through lilac to gray. *C. perfringens* is one of the gas gangrene organisms, and the changes in the carcass resemble the symptoms of the disease. According to Ingram and Dainty (1971), at this stage, there is no particularly obnoxious smell. The more proteolytic clostridia, *C. bifermentans, C. histolyticum* and *C. sporogenes*, can grow when the redox potential is lowered by the growth of *C. perfringens*. These organisms produce the odorous amines and sulfur compounds from amino acids.

Intermediate temperatures (15°-25°C) may occur when the chilling of meat is slow, delayed or the meat is held in this temperature range. In

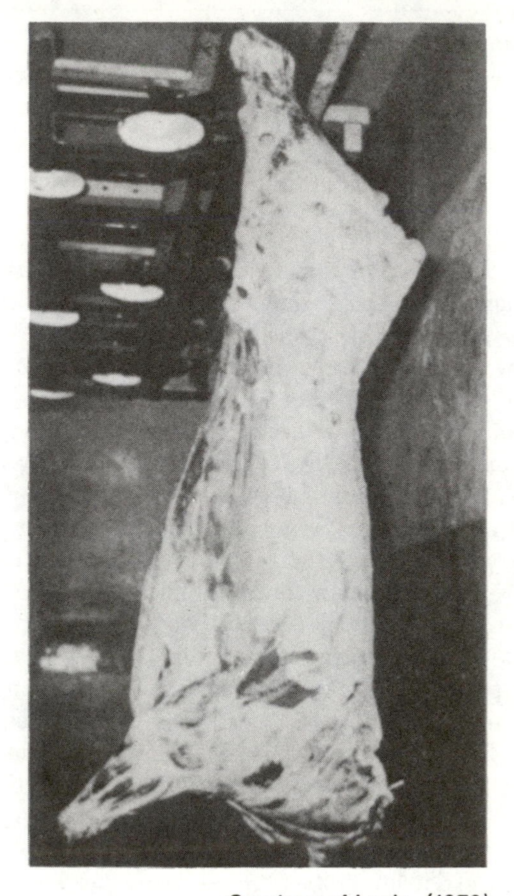

Courtesy of Levie (1979)

FIG. 8.2. A BEEF SIDE

Note the fat covering, protecting the muscle tissue from microbial contamination.

this case, the main spoilage organisms are the mesophilic clostridia. They grow rapidly, within the meat, causing internal spoilage or bone taint in beef.

In order to reduce the biochemical and microbial changes, fresh meat is stored at temperatures near 0°C. At low temperatures, spoilage is evidenced on the surface. In storage conditions maintaining a moist meat surface, spoilage is due to Gram negative bacteria, primarily species of *Pseudomonas.* Species of *Aeromonas, Alcaligenes, Acinetobacter, Moraxella, Flavobacterium, Enterobacter, Microbacterium, Proteus* and *Streptococcus* occasionally are found in the surface flora of spoiled meat.

The defects of the meat are off-odor, slime formation and discoloration. If tasted, the flavor would be unacceptable to most people. Dainty *et al.* (1975) could not detect proteolysis prior to the appearance of spoilage odor and slime.

The important biochemical changes occur in the meat juices which contain free amino acids, nucleotides and peptides. These nutrients are sufficient for microbial growth and the metabolism of these compounds leads to the formation of H_2S, NH_3, indole, cadaverine and other substances which characterize spoilage.

The pigment stability in meat involves factors such as oxygen penetration, bacterial growth, lipolysis and the presence of flavin compounds. Exposure to light can enhance the destruction of oxymyoglobin on the surface of beef. Soft white fluorescent light seems to be the most detrimental (Satterlee and Hansmeyer 1974). Bacteria do not seem to degrade myoglobin. According to Nicol *et al.* (1970), a pseudomonad referred to as *P. mephitica* can produce H_2S, which reacts with myoglobin at low oxygen levels, forming the green colored sulfmyoglobin.

The water activity can play a role in spoilage. Fresh meat will lose moisture if stored below 99% RH. Even at RH of 99%, the growth of *Pseudomonas* is reduced. At an a_w of 0.96 or less, most of the usual microorganisms causing spoilage of fresh meat are inhibited. When the surface of meat is lower than a_w of 0.96, the slower growing fungi become evident. In this situation, *Thamnidium chaetocladioides* and *T. elegans* (cause of "whiskers"), *Cladosporium herbarum* ("black spot") and *Sporotrichum carnis* ("white spot") growth can produce moldy meat. For the most part, the activity of molds is limited to the outer surface of meat where aerobic conditions prevail. Mold growth on meat is not always considered unsatisfactory. The growth of *T. elegans* on beef reportedly tenderizes the meat and imparts a nutty flavor.

The type of atmosphere can influence spoilage of meat. Fresh beef stored in air has the dominant *Pseudomonas* species. However, when vacuum packaged, *Lactobacillus* predominates. Other organisms include *Microbacterium, Enterobacter* and *Hafnia.* At 4°C, any odors produced are acidic, sweet or rancid (Patterson and Gibbs 1977; Seideman *et al.* 1976).

Cured Meat.—Meat that is treated with salts, such as NaCl, $NaNO_2$ and $NaNO_3$, differ from untreated fresh meat. The most apparent differences are in the color and flavor of fresh and cured meat. However, the intrinsic factors of cured meat, such as pH, a_w and the presence of inhibitors, play an important role in the microbial ecology of these products. Besides ham, bacon and corned beef, there are cured comminuted meats (weiners, bolognas).

The a_w of cured meat (0.88 to 0.95) is lower than that of fresh meat (0.98 to 1.00) and the specific inhibitory effects of the added salts may account for the differences in the spoilage patterns of fresh and cured meats.

For bacon, the important types of organisms are *Micrococcus,* yeast, *Vibrio, Acinetobacter, Alcaligenes, Arthrobacter* and *Corynebacterium* (Gardner 1971). It is generally stated that bacon in aerobic conditions is spoiled by micrococci. However, a significant portion of the flora on the lean portion of the bacon is Gram negative rods. As the pH of bacon decreases from 6.0 to 5.5, there is an increase of yeasts in the flora during spoilage. Many molds, such as *Alternaria, Aspergillus, Mucor* and *Rhizopus* can cause surface discolorations.

According to Gardner (1971), chemical changes do not occur in bacon until the total microbial count reaches 10^8/g or higher. The off-odors of spoiled bacon were described as cheesy, putrid or sour by Dempster (1972), or as smoked fish and rancid cheesy by Ingram and Dainty (1971). The rancid cheesy flavor was thought to be due to lipolysis of fat by micrococci.

The predominant microflora of vacuum packaged bacon is determined by the salt content. Species of *Micrococcus* give way to lactic acid bacteria in bacon with low salt (5-7%), while they remain dominant in bacon with higher salt (8-12%).

Yeasts may appear in the flora of packaged bacon. These organisms are believed to assimilate nitrite and may be partially responsible for the loss of nitrite in stored bacon.

For ham, the important type of spoilage is "souring." This includes various spoilage problems from mild degradation to extreme putrefaction. Before the modern short-curing methods, clostridia caused bone sours prior to the meat being chilled or cured. This is the same as the internal spoilage of fresh meat at warm temperatures. *C. putrefaciens* is psychrotrophic and prominent in the spoilage of chilled ham. Proteolytic, gassy or putrefactive souring is still encountered in country-style dry-cured ham.

The short-cure commercial method of curing has greatly reduced the incidence of ham souring. However, gassy or puffy hams can result if the fresh meat is held too long at a moderate temperature prior to cure. This allows clostridia to grow. Proper slaughter and bleeding of hogs, prompt chilling of the meat, prompt handling, good sanitation and use of pump and cover brines that have low microbial counts have aided in improving the microbial quality of ham.

Modern commercial ham has a milder cure (with less salt), milder smoking, a more moist surface, and vacuum packaging, than dry-cured hams. Slicing during packaging creates sources of contamination and more cut surfaces on which microorganisms can grow. Although the dry-salt cured hams suffer from mold growth on the surface, the commercial hams allow the growth of micrococci, microbacteria, various lactic acid bacteria and yeasts.

The microbiological changes in raw hams were reported by Giolitti *et al.* (1971). They found that sour hams had higher amounts of methyl mercaptan, fatty acids, carbonyl compounds, free amino acids, ammonia and H_2S than normal hams. *Arthrobacter* species, detected in spoiled hams, were able to produce methyl mercaptan from sulfur containing amino acids. They stated that *Micrococcus, Arthrobacter, Corynebacterium, Lactobacillus, Pediococcus* and yeasts are able to produce H_2S and might contribute to the deteriorative process.

Fecal streptococci may be found in stored hams and cause a green discoloration and off-flavor.

Comminuted and cured sausage products are subjected to various types of spoilage. These products are generally composed of mixtures of pork, beef, salt, sugar, sodium nitrite and spices. Hence, the flora of these products will be somewhat different than that of fresh meat.

A surface slime of micrococci and yeasts can occur when sufficient moisture is present. As the surface dries, inhibiting the bacteria, mold growth can cause spoilage. When products, such as luncheon meats, are vacuum packaged, aerobic growth is inhibited and lactic acid bacteria become dominant. Facultatively anaerobic yeasts may grow. The lactic acid bacteria produce CO_2 and can cause a swelling of the package, while the yeasts produce a surface slime.

For bologna products, a casing is used. Moisture accumulates at the meat-casing interface, allowing the growth of bacteria. Micrococci can cause a slime layer during growth at this interface.

During storage of these meats, microorganisms of the genera *Micrococcus, Lactobacillus* and *Leuconostoc* can grow and cause souring of the product. The production of organic acids and reducing compounds by the bacteria can cause a fading of the pink color at the outer surface of the product, resulting in a ring of discoloration.

Insufficient heat treatment allows the survival of *Streptococcus faecium* and other lactic acid bacteria that can cause souring of the product.

Green discolorations can occur in cured sausages. The greening may appear as rings, cores or on the surface. *Lactobacillus viridescens* can grow in products with reduced oxygen tension and pH, and is the organism primarily involved in greening of these sausages. Other *Lactobacillus* and *Leuconostoc* species sometimes are present. The bacteria produce peroxides which react with the cured meat pigment causing the green discoloration. The first evidence is the appearance of small greenish spots on the damp surfaces of the cured sausage. These spots tend to spread and cover the sausage surface if favorable conditions exist for a sufficient time. In some cases, a slimy appearance may be noted.

Fermented Sausages.—Some comminuted sausages, such as cervelat, Thuringer and Lebanon bologna, are not only cured, but also are fermented by lactic acid bacteria to produce a tangy flavor. Dependence upon the natural contaminants to ferment the product may cause spoilage, so a starter culture of homofermentative lactic acid bacteria is used. The lower pH of these fermented sausages helps control some types of bacteria, but they are susceptible to spoilage similar to cured sausage products. A surface slime due to yeasts, and discoloration due to mold, may occur.

Canned Meat.—Canned meat products are subject to the same type of spoilage as other low acid foods, if heat resistant spores which survive the process can germinate and grow. In semipreserved or pasteurized cured products, such as canned ham, the curing salts and refrigeration are used to prevent spore germination and outgrowth. If not adequately processed, thermoduric cells, such as *Streptococcus faecium*, may survive and cause souring. This organism may cause rapid discoloration after the product is removed from the can. If the spores of clostridia are able to germinate, and the resultant cells grow, gas may be formed along with extensive putrefaction.

Poultry.—The spoilage of uneviscerated and eviscerated poultry is somewhat different. However, due to the fact that commercial poultry in the United States is eviscerated, only this type of poultry is considered in this section. Uneviscerated poultry is discussed by Barnes and Shrimpton (1957).

Immediately after processing, any of several hundred species of microorganisms might be found. However, as the poultry is chilled and held in cold storage, psychrotrophic microorganisms predominate and cause deterioration.

The main defects are off-odor, which appears at a bacterial load between 10^6 and 10^8 per cm^2, and slime formation, which occurs soon after off-odor is noted. As the number of bacteria increases, the flavor score decreases. Species of *Pseudomonas* are the principal spoilage organisms. Besides *Pseudomonas*, other organisms, similar to those in fresh red meat spoilage, are found on spoiled poultry (Table 8.8). These

TABLE 8.8

MICROBIAL DEFECTS OF POULTRY AND POULTRY PRODUCTS

Product	Defect	Microorganism
Poultry meat	Off-odor, slime	*Pseudomonas, Acinetobacter, Moraxella, Alcaligenes, Aeromonas, Alteromonas*
Shell eggs	Black rot	*Proteus, Aeromonas*
	White rot (colorless)	*Citrobacter, Alcaligenes*
	Sour	*Pseudomonas*
	Green white	*P. fluorescens*
	Musty	*Pseudomonas*
	Moldy	Many types of molds
	Red rot	*Serratia marcescens*
	Custard rot	*Citrobacter, Proteus, Enterobacter*
	Yellow and green rot	*Alcaligenes, Flavobacterium, Cytophaga*
Liquid whole egg	Fishy	*Pseudomonas, Flavobacterium, Chromobacterium*
	Off-odor, sour	*Proteus, Alcaligenes, Escherichia, Flavobacterium, Pseudomonas, Bacillus*
Liquid albumen	Off-odor	*Pseudomonas, Acinetobacter, Enterobacter*

include *Aeromonas, Moraxella, Alcaligenes, Flavobacterium* and *Micrococcus.* When antibiotics were used on poultry, yeasts became the important spoilage microorganisms. However, yeasts are not important on normally processed and eviscerated poultry.

McMeekin (1975) studied the flora of breast meat held at 2°C. He reported the flora consisted of Gram negative, motile rods. The main organisms were *Pseudomonas* (sections I, II and IV), plus an enteric type. During storage, the off-odor producers (principally *Pseudomonas* II) became dominant. By 16 days, 96% of the microbial population was *Pseudomonas* II species. Nonpigmented strains of pseudomonads produce more intense off-odors than do pigmented strains.

The chemical compounds found on spoiled chicken included H_2S, methyl mercaptan, dimethyl sulfide, acetone, toluene, n-heptane, 1-heptene, n-octane, methyl acetate, ethyl acetate, heptadiene, methanol and ethanol (Freeman *et al.* 1976). The temperature of storage influenced the types of compounds present on spoiled poultry.

Eggs.—It is generally accepted that, when laid, the contents of the hen egg are free from bacteria. There are exceptions due to ovarian infections.

The egg contents are protected by the shell and associated membranes and chemical inhibitors in the egg albumen. This means that microorganisms must penetrate these barriers and then be able to grow in the egg contents to cause spoilage. Penetration of eggs is aided by moisture on the shell. If not properly stored or washed, penetration may be quite rapid, and spoilage can occur.

For egg products, the shell and membranes are removed. During processing, the liquid egg is subject to contamination from organisms on the egg shell, equipment, humans, added ingredients and the final container.

Bacterial analysis of shell eggs during storage has revealed molds (*Penicillium, Aspergillus, Cladosporium, Rhizopus* and *Mucor*), yeasts (*Rhodotorula*) and bacteria (*Pseudomonas, Micrococcus, Bacillus, Proteus, Alcaligenes, Flavobacterium, Citrobacter, Escherichia* and *Enterobacter*).

As with other protein foods, species of *Pseudomonas* are the main spoilage organisms of shell eggs and egg products.

When bacteria grow within the egg, they decompose the contents and form by-products. This results in characteristic odors, appearance or colors from which the rots acquire their name (Table 8.8).

The United States Department of Agriculture (USDA 1972) described a loss egg as "an egg that is inedible, smashed or broken so that the contents are leaking, frozen, contaminated, or containing bloody whites, large blood spots, large unsightly meat spots, or other foreign material." Inedible eggs are due to nonmicrobial as well as microbial defects. An egg with a blood ring, embryo or stuck yolk is inedible.

Besides these loss types, shell eggs will absorb odors from the storage atmosphere. Most vegetables and fruits will impart flavors and odors to shell eggs. If stored with apples, the eggs will be bitter and have a cardboard flavor and odor. If stored near gasoline or kerosene, the shell eggs will taste and smell like these compounds.

Eggs that become inedible due to microbial growth are listed by USDA (1972) as black rots, white rots, sour eggs, eggs with green whites, mixed rots, musty eggs and moldy eggs.

Other designations for rotten eggs are fluorescent green, red, custard, colorless, green and yellow, mixed and rusty red rots.

Black Rots.—When viewed with a candling light, black rot eggs are virtually opaque. When broken out, the egg content has a muddy (dark brown) appearance, a repulsive putrid odor and H_2S is evident. In many eggs, an internal gas pressure develops. The bacteria associated with this type of spoilage are species of *Proteus* and *Aeromonas*. Rots of this type are more likely to occur at room temperature (20°C) than in cold storage (4°C or less).

White Rot.—Threadlike shadows may be seen in the thin white, and in later stages, the yolk appears severely blemished when the shell egg is viewed with the candling light. When opened, the egg yolk shows a crusted appearance and frequently has a fruity odor.

This type of inedible egg sometimes is referred to as a colorless rot. Various organisms have been associated with this rot, including *Citrobacter, Salmonella* and *Alcaligenes*.

Sour Eggs.—These eggs are difficult to detect by ordinary candling, but they usually show a weak white and murky shadow around an off-center, swollen yolk. These eggs also are called fluorescent and are quite readily detected by observing with UV light. A green sheen is produced by species of *Pseudomonas*.

Since a green fluorescence is observed, these inedible eggs also have been termed fluorescent green rots.

Green Whites.—This defect is caused mainly by *Pseudomonas fluorescens*. Green whites of broken out eggs fluoresce when observed with UV light. Eggs with green whites may or may not have a sour odor, since the green fluorescence can be observed long before any odor can be detected.

Musty Eggs.—These frequently appear clear and free from foreign material when candled. The musty odor may be caused by odors in the atmosphere being absorbed by the egg contents. Also, some microorganisms occasionally invade shell eggs and produce a musty odor.

Moldy Eggs.—Mold growth is visible as spots on the shell, in checked areas of the shell or inside the egg.

The mold contamination of eggs seems to be due to the re-use of moldy packing materials. Several molds, such as *Penicillium, Alternaria* and *Rhizopus* can grow on eggs. However, molds are not a prominent cause of egg spoilage.

Red Rot.—These eggs are distinguished by a red discoloration of the albumen and the surface of the yolk. An ammoniacal to putrid odor may occur. *Serratia marcescens* has been considered as the cause of red rot.

Custard Rot.—In this rot, the yolk is encrusted with custard-like material and occasionally flecked with olive green pigment. The albumen becomes thin with an orange tint. There may be a slightly putrid to putrid odor. *Citrobacter* and *Proteus vulgaris* have been associated with this type of spoilage.

Mixed Rot.—These addled eggs occur when the vitelline membrane of the yolk breaks and the yolk mixes with the white, resulting in a murkiness throughout the interior of the egg when viewed with a candling light. With no off-odor, these are referred to as odorless mixed rots.

Other Rots.—*Alcaligenes* has been accused of causing both yellow rots and green rots. These rots are similar in odor and in the appearance of albumen. However, the yolk is dark yellow in the former and dark green to black in the latter case.

Rust red rot is associated with growth of *P. vulgaris.*

Egg Products.—The spoilage of egg products is evidenced by off-odors described as sour, fecal or fishy. Organic acids (lactic, succinic) are recognized as indices of decomposition of egg products (Bethea 1970; Staruszkiewicz 1969).

Seafoods.—Fish and other seafoods are subject to contamination of microorganisms in their marine environment, as well as those acquired during catching, handling and processing. However, as with other foods, even though a mixed flora exists, certain types of organisms become dominant during storage of the product. The dominant types are those that can survive and multiply on or in the food, at the temperature that prevails.

The defects of seafoods are listed in Table 8.9.

Fish.—The spoilage pattern depends somewhat on the species and type of fish, the initial microbial flora, the area of catch, the method of catch, processing method and method of storage.

The initial microbial flora is dependent upon the contamination of the water and bottom sediment from the area of catch. Fish caught on a line have lower counts than fish that are trawled by dragging a net along the bottom. The trawl net drags through the bottom sediment which usually has high counts of microorganisms.

The temperature of the water affects the fish microflora. Fish from warm seas tend to have mesophilic strains, whereas from cold seas, the organisms are primarily psychrotrophs.

The methods for handling the fish after catching influence the contamination. Fish do not have a sphincter muscle, so when any pressure is exerted, intestinal discharge occurs. This pressure can happen when the net is hauled aboard, when the fish are piled on the deck or when whole fish are stored in ice.

The caught fish may be left whole, beheaded or gutted. A gutted fish

TABLE 8.9

SOME MICROBIAL DEFECTS OF SEAFOODS

Product	Defect	Microorganism
Fish		
Fresh	Off-odor	*Pseudomonas, Alteromonas, Acinetobacter, Vibrio, Aeromonas, Moraxella, Proteus*
	Fruity	*Pseudomonas*
	Ammoniacal	*Pseudomonas, Alteromonas*
	H_2S odor	*Pseudomonas, Alteromonas*
Salted	Pink	*Halobacterium, Halococcus*
	Dun (red growth)	*Hemispora stellata, Sporendonema epizoum*
	Cheesy, putrefactive	Red halophilic bacteria
Crayfish	Sweet to foul odor	*Pseudomonas, Lactobacillus,* Coryneforms
Oysters	Pink	Yeasts (*Rhodotorula*)
Shrimp	Off-odor	*Pseudomonas*
Squid	Yellow discoloration	*P. putida*
	Red discoloration	*Serratia marcescens*

may spoil somewhat differently than a whole fish, due to the intestinal flora.

The method used to enumerate the microorganisms can influence the flora. Incubation at 37°C reveals fewer bacteria than obtained by incubating at either 0° or 20°C.

The fish can pick up added contamination on shipboard if the holding areas are not maintained in a sanitary manner. The fish must be handled rapidly on the ship, and chilled. The fish are usually mixed with ice for cooling; however, some ships have freezing facilities.

During processing, the organisms in the surface slime layer or the skin can be spread throughout the processing equipment, the workers and onto the flesh of the fillet. Hence, the normal potentially sterile flesh can be inoculated with millions of bacteria. With this contamination, the fish may be spoiled by the time they reach the consumer. The microbial flora on the fish leaving the processing plant may be different from that on the fish entering the plant.

Alur *et al.* (1971) isolated *Pseudomonas, Proteus, Aeromonas* and *Achromobacter* from spoiled fish. *Pseudomonas* and *Proteus* cause putrid and ammoniacal odors, while *Acinetobacter* and *Aeromonas* are associated with unpleasant sweetish or fruity odors. *Micrococcus* inoculated into fish homogenates produces stale odors.

Cytophaga, coryneforms, *Micrococcus, Bacillus, Vibrio, Flavobacterium* and *Clostridium* have been encountered in fresh and spoiling fish. Members of the family Enterobacteriaceae are seldom encountered in fresh or spoiling marine fish, but some may be found in fish caught in polluted areas from both fresh and salt water.

During storage of the fish at 0°C, the number of organisms increases after a 1 to 2 day lag period. During spoilage, the *Pseudomonas* and *Acinetobacter* become increasingly dominant, with *Flavobacterium* showing a transient increase. By 10 to 12 days, *Pseudomonas* may comprise 90% of the population. The species of *Pseudomonas* tend to change as spoilage progresses. Regardless of which species are involved, organisms of the genus *Pseudomonas* are the most active spoilage organisms in fish stored at 0°C. The primary factor influencing the spoilage rate is temperature. As the temperature is increased from 0°C the growth rate of pseudomonads increases. However, as the temperature approaches 20°C, the mesophilic flora becomes evident.

Lipids in fish have a high content of polyunsaturated fatty acids. Hence, oxidative rancidity is more evident in fish than in most other animal tissue.

In marine fish, the trimethylamine oxide (TMAO) is reduced by bacterial and enzymatic action to trimethylamine (TMA), a spoilage product. The odor of trimethylamine at low levels is referred to as a stale fishy odor. Piperidine, δ-amino valeric acid, and δ-amino valeric aldehyde also have a fishy odor.

During spoilage, volatile bases, amines and organic acids are formed by decarboxylation or by deamination of amino acids and organic bases. Hydrogen sulfide, mercaptans and disulfides add to spoilage odors. Spoilage odors have been designated as fishy, stale, musty, rancid, sour, ammoniacal, yeasty, fruity, sweet, acid and putrid.

Vacuum packaging of fish changes the spoilage flora in a manner similar to fresh meat. *Lactobacillus* and *Microbacterium* grow and cause souring, whereas the aerobic pseudomonads are inhibited.

Shewan (1971) stated that bacterial spoilage can be prevented by storing fish below −10°C, so, from a practical viewpoint, the bacteriology of spoilage is somewhat academic. This is also true for spoilage of meat and poultry products. It is unfortunate that so much of our meat, poultry and fish is maintained in the so-called fresh condition, rather than being frozen.

Crayfish.—The spoilage bacteria of freshwater crayfish was studied by Cox and Lovell (1973). They reported that spoilage of the tail occurred when the total aerobic count reached 10^9/g. In the fresh meat, *Micrococcus, Staphylococcus* and *Alcaligenes* were the major genera, while in meat spoilage, at 0°C, *Pseudomonas*, and at 5°C, *Achromobacter* (*Acinetobacter, Moraxella*) predominated. These genera comprised the major bacterial flora of the spoiled meat at either temperature (0° or 5°C).

Clams.—The microbiology of soft shell clams was reported by Cox (1965). The clams were judged to be unacceptable on the basis of black discoloration or a slimy appearance. The predominating spoilage bacteria are Gram negative rods that produce sweet ester-like, musty, oniony or cesspool-like odors on agar media. He suggested that the spoilage flora consisted mainly of *Pseudomonas* types.

Crabs.—The bacteriological spoilage of canned, pasteurized crab cake mix stored at several temperatures was investigated by Loaharanu and Lopez (1970). When stored at 18° or 30°C, *Bacillus* and *Micrococcus* predominated, while at 2°C, *Alcaligenes* became the dominant genus of organisms.

Oysters.—Frozen or fresh oysters spoil by either fermentative, characterized by a sour odor, or putrefactive mechanisms. Canned oyster decomposition is putrefactive. Oysters that are well cooked and then refrigerated spoil due to fat rancidity, rather than microbial decomposition. Red or pink coloration of oysters may be due to *S. marcescens* or to pink yeasts, such as *Rhodotorula*, as well as to phytoplankton or eggs of the oyster crab. Various discolorations can occur in shellfish (Boon 1977).

Shrimp.—These animals are caught at sea by towing a trawl net along the bottom. After one to three hours of trawling, the net is hauled aboard and emptied. Since the bottom sediment is more contaminated with bacteria than the upper water, one would expect harvested shrimp to have rather high bacterial counts. Novak (1973) reported counts of 31,000 to 1,200,000 bacteria per gram of whole shrimp. He further reported that, although the head was 40% of the shrimp weight, it contained 75% of the bacteria. Therefore, heading the shrimp lowered the microbial count.

Bieler *et al.* (1973) reported that the heading of shrimp exposed tissues, allowed the gut to empty and triggered enzymatic changes. Hence, they found that shrimp stored in ice maintained better quality with the head on than when headless.

The predominant organisms on iced raw shrimp are *Moraxella, Pseu-*

domonas, Acinetobacter, Arthrobacter, Flavobacterium and *Cytophaga* (Lee and Pfeifer 1977). They believed the presence of *Arthrobacter* and *Acinetobacter* may indicate inadequate cleaning. The presence of *Moraxella, Flavobacterium* and *Cytophaga* indicates the degree of secondary contamination and *Pseudomonas* indicates the potential shelf-life of processed shrimp.

Dairy Products.—Milk, when obtained from the cow, may not be sterile. During and after milking, the milk is subjected to organisms from various sources, although contaminated equipment used to handle, transport, store and process the milk seems to be the main source of organisms.

With extended storage of milk products at refrigerated temperatures, psychrophilic or psychrotrophic organisms are a cause of spoilage.

The organisms associated with spoilage generally do not survive the heat treatment used to pasteurize milk. Hence, spoilage is usually caused by organisms recontaminating the milk after pasteurization.

The defects that can occur in milk due to microbial growth are: 1) off-flavors; 2) lipolysis with development of rancidity; 3) gas production; 4) fermentation to lactic acid with souring; 5) coagulation of milk proteins; 6) viscous or ropy texture; and 7) discoloration. Some of the defects of milk and dairy products are listed in Table 8.10.

As with other animal products, species of *Pseudomonas* are prominent in producing defects in milk. They are associated with fruity odors and rancidity. The fruity odor has been described as strawberry or the flavor of the May apple. The fruity aroma of milk is apparently a mixture of ethyl butyrate and ethyl hexanoate.

P. fluorescens is associated with rancidity and proteolytic defects, and *A. putrefaciens* with proteolytic defects and surface taint of butter.

Certain lactic acid bacteria are involved with defects in fresh milk. Although the production of lactic acid and the resulting souring are desirable in fermented milk products, they are undesirable in fresh milk. Besides this type of defect, a strain of *Streptococcus lactis* designated as variety *maltigenes* metabolizes leucine and produces 3-methylbutanol, which gives a malty flavor defect in dairy products. At the time the aroma first becomes detectable, counts range from 10^7 to 10^8 per ml. Another variety of *S. lactis* has been associated with a ropy or slimy condition in milk and cream. Ropiness has been ascribed to growth of several microorganisms, such as coliforms and *Alcaligenes.*

Butter.—This is primarily fat with some moisture, carbohydrate and protein. One of the main defects is rancidity. This can be due to oxidation, as well as to microbial growth. Molds are able to grow on the

TABLE 8.10

SOME MICROBIAL DEFECTS IN DAIRY PRODUCTS

Product	Defect	Microorganism
Milk		
Pasteurized, refrigerated	Rancid	*Pseudomonas, Alcaligenes, Staphylococcus*
	Ropy or slimy	Coliforms, *Pseudomonas, Alcaligenes, Micrococcus, Bacillus subtilis*
	Sour (acid, gas)	Lactic acid bacteria
	Discoloration	*Chromobacterium*
	Bitter, fruity	*Pseudomonas, Flavobacterium Alcaligenes, Proteus, Acinetobacter*
Canned	Swelling, gas	*Clostridium sporogenes*
Cream	Foamy	*Candida, Torulopsis*
Butter	Surface taint	*Pseudomonas, Alteromonas*
Cheese	Gassy, butyric acid	*Clostridium tyrobutyricum*
	Gassy, floating or split curd	*Leuconostoc, S. lactis* subsp. *diacetylactis*
	Moldy	*Penicillium, Scopulariopsis Mucor*, other molds
Soft	Black mold (cat's fur)	*Mucor*
	Surface growth	*Torulopsis, Debaryomyces*
Cottage	Slimy curd, putrid odor	*Pseudomonas*
	Discoloration	*Flavobacterium*, yeasts, molds
	Slimy, gelatinous	*Pseudomonas, Alcaligenes, Flavobacterium*, coliforms
	Fruity	Yeasts
Cheddar	Sweet, yeasty, fruity	Yeasts
Swiss	Gassy, sweet	Yeasts (*Torulopsis*)
	Off-odor	*C. sporogenes*
Yogurt	Yeasty	Yeasts (*Torulopsis*)

surface of butter. Putrid, proteolytic and fruity flavors in butters are caused by psychrotrophic bacteria.

Surface taint and yeasty butter are other defects of this product caused by microorganisms. Butter is made from cream. If the cream has a defect, such as rancidity, it can affect the quality of the butter.

Cheese.—Defects that can occur in milk may appear in other dairy products. If a product is made from defective milk, the defect may carry over to the product.

In cottage cheese, the flavor defects of milk may be accompanied by a gelatinous or tapioca curdormation. Lipolytic and proteolytic organisms cause defects in cottage cheese.

P. fragi and *A. putrefaciens* have been associated with a slimy curd defect on the surface of cottage cheese. A fruity, putrid or rancid odor and a fruity or bitter flavor may accompany this defect.

Surface discolorations may occur due to the growth of the pigmented *Flavobacterium. E. coli* can cause barny or unclean flavors and, if the cottage cheese is held at room temperature, the organism can cause a gassy defect.

The yeast, *Rhodotorula*, produces pink spots on the surface of the product. These pink spots eventually become a pink slime. *Torulopsis* also produces a slime, but it is yellow. *Geotrichum* produces off-white, tan or yellow surface discolorations.

Molds multiply rather slowly. Cottage cheese may be spoiled by other organisms before mold growth is evident. However, there are occasions when growth of *Penicillium* and *Mucor* appears on the surface of the product.

A pink discoloration of Romano and other Italian cheese varieties was investigated by Shannon *et al.* (1969). The pink discoloration occurred as a uniform band of color near the cheese surface, or as discoloration throughout the entire cheese. According to them, species of *Lactobacillus* (*L. helveticus* and *L. bulgaricus*) can cause this discoloration. Pink discolorations due to strains of *Propionibacterium*, certain reducing *Bacillus* species and reducing micrococci have been suggested as causing pink to red discolorations in various cheeses.

Fats and Oils

Rancidity, acidity, soapiness and off-flavor defects can occur in fatty foods, such as butter, margarine, lard and vegetable oils. The development of rancidity can be due to autoxidative deterioration, lipolysis by microbial lipases or lipoxidation due to lipoxidase enzymes.

There are several lipolytic microorganisms (Table 8.3). Microorganisms associated with the liberation of free fatty acids in butter and margarine include *Cladosporium, Candida, Pseudomonas, Micrococcus, Penicillium* and *Geotrichum*. The hydrolysis of olive oil, coconut oil and butterfat can be caused by species of *Micrococcus, Pseudomonas, Serratia* and *Staphylococcus*.

Various species of *Aspergillus* and *Penicillium* found on peanuts are lipolytic, and can degrade peanut oil.

The defect characterized by rancidity and acidity is due to the liberation of free fatty acids, especially butyric (C_4), caproic (C_6), caprylic (C_8) and capric (C_{10}), and the production of their corresponding methyl ketones. When lauric and myristic acids are liberated from triglycerides in coconut oil and butterfat, a soapy flavor defect can occur.

Besides the direct liberation of the fatty acids, ketones, volatile acids and secondary alcohols can be produced from free fatty acids by β-oxidation to the corresponding β-keto acid. These can decarboxylate to yield methyl ketones or cleave to form acetyl-coenzyme A and the lower fatty acid, which is two carbon atoms shorter. The reduction of methyl ketones yields secondary alcohols.

The spoilage of mayonnaise and salad dressings was investigated by Kurtzman *et al.* (1971). *Saccharomyces bailii* was the principal cause of mayonnaise spoilage, although *Lactobacillus fructivorans* also was involved. Species of *Bacillus* were isolated from several spoiled samples. The yeast *Zygosaccharomyces* was involved in spoilage of French salad dressing. Rather low numbers of yeast and bacteria were present in the spoiled product. They reported only five yeasts per gram in one spoiled mayonnaise sample. The highest yeast count they reported was 165,-500/g.

Canned Food

Canned foods are discussed in the section on heat preservation, but some recognition should be made here.

Bacterial spoilage of canned food may be due to underprocessing, elevated storage temperature or leakage of the container due to improper closures, rough handling or defective cans.

Underprocessed cans of food are usually spoiled by heat resistant sporeformers (*Bacillus, Desulfotomaculum* and *Clostridium*) which survive the process. Leakage can result in spoilage by nonsporeformers, which could not have survived the heat processing, but recontaminate the food after the heat treatment. Heat resistant thermophilic organisms may survive the processing and, if the food is not cooled after heating, or if the cans are stored at elevated temperatures, these thermophiles can grow. There are four prominent thermophilic organisms causing spoilage of canned food. *B. stearothermophilus* causes flat sour spoilage of canned nonacid foods. It is facultatively anaerobic and can grow at 70°C. *B. coagulans* is less heat resistant, but is more acid tolerant than *B. stearothermophilus.* It causes flat sour spoilage of canned tomatoes.

Clostridium thermosaccharolyticum forms large volumes of gas and resultant hard swells or blown cans. The spores are very heat resistant. Blown cans can be caused by various species of clostridia. *Desulfotomaculum nigricans* produces hydrogen sulfide. This gas reacts with metallic ions from the container or food, forming sulfides, some of which are black.

Molds and yeasts are not significantly important in the spoilage of canned foods, except for those mold spores, such as *B. fulva*, that can survive fruit processing treatments.

The movement of bacteria from contaminated cooling water into the processed canned food may result in various spoilage problems. Bean and Everton (1971) discussed a problem of pigmented bacteria surviving the chlorination treatment given the water used to cool heat processed cans of product. No gas was noted (no swelling of the cans), but a canned pudding developed cheesy to sour flavors and there was separation of the product.

BIBLIOGRAPHY

ALUR, M.D., LEWIS, N.F., and KUMTA, U.S. 1971. Spoilage potential of predominant organisms and radiation survivors in fishery products. Indian J. Exp. Biol. *9*, 48-52.

AURAND, L.W., and WOODS, A.E. 1973. Food Chemistry. AVI Publishing Co., Westport, Conn.

BARNES, E.M., and SHRIMPTON, D.H. 1957. Causes of greening of uneviscerated poultry carcasses during storage. J. Appl. Bacteriol. *20*, 273-285.

BEAN, P.G., and EVERTON, J.R. 1971. Observations on the taxonomy of chromogenic bacteria isolated from cannery environments. J. Appl. Bacteriol. *32*, 51-59.

BETHEA, S. 1970. Note on determination of free succinic acid in eggs by gas-liquid chromatography. J. Assoc. Offic. Anal. Chem. *53*, 468-470.

BIELER, A.C., MATTHEWS, R.F., and KOBURGER, J.A. 1973. Rock shrimp quality as influenced by handling procedures. Proc. Gulf and Caribbean Fisheries Inst. 25th Annu. Session, 56-61.

BOON, D.D. 1977. Coloration in bivalves. A review. J. Food Sci. *42*, 1008-1015.

BUCHANAN, R.E., and GIBBONS, N.E. 1974. Bergey's Manual of Determinative Bacteriology, 8th Edition. Williams and Wilkins Co., Baltimore.

COX, J.R. 1965. Bacteriological studies on the shelf life of soft shell clams (*Mya arenaria*). J. Milk Food Technol. *28*, 32-35.

COX, N.A., and LOVELL, R.T. 1973. Identification and characterization of the microflora and spoilage bacteria in freshwater crayfish *Procambarus clarkii* (Girard). J. Food Sci. *38*, 679-681.

DAINTY, R.H., SHAW, B.G., DEBOER, K.A., and SCHEPS, E.S.J. 1975. Protein changes caused by bacterial growth on beef. J. Appl. Bacteriol. *39*, 73-81.

DEMPSTER, J.F. 1972. Vacuum packaged bacon; the effects of processing and storage temperature on shelf life. J. Food Technol. 7, 271-279.

FREEMAN, L.R., *et al.* 1976. Volatiles produced by microorganisms isolated from refrigerated chicken at spoilage. Appl. Environ. Microbiol. *32*, 222-231.

GARDNER, G.A. 1971. Microbiological and chemical changes in lean Wiltshire bacon during aerobic storage. J. Appl. Bacteriol. *34*, 645-654.

GIOLITTI, G., CANTONI, C.A., BIANCHI, M.A., and RENON, P. 1971. Microbiology and chemical changes in raw hams of Italian type. J. Appl. Bacteriol. *34*, 51-61.

HESSELTINE, C.W., GRAVES, R.R., ROGERS, R., and BURMEISTER, H.R. 1969. Aerobic and facultative microflora of fresh and spoiled refrigerated dough products. Appl. Microbiol. *18*, 848-853.

INGRAM, M., and DAINTY, R.H. 1971. Changes caused by microbes in spoilage of meats. J. Appl. Bacteriol. *34*, 21-39.

KURTZMAN, C.P., ROGERS, R., and HESSELTINE, C.W. 1971. Microbiological spoilage of mayonnaise and salad dressings. Appl. Microbiol. *21*, 870-874.

LEE, J.S., and PFEIFER, D.K. 1977. Microbiological characteristics of Pacific shrimp (*Pandalus jordani*). Appl. Environ. Microbiol. *33*, 853-859.

LEVIE, A. 1979. The Meat Handbook, 4th Edition. AVI Publishing Co., Westport, Conn.

LINDSAY, R.F., and PRIEST, F.G. 1975. Decarboxylation of substituted cinnamic acids by enterobacteria: the influence on beer flavour. J. Appl. Bacteriol. *39*, 181-187.

LOAHARANU, P., and LOPEZ, A. 1970. Bacteriological and shelf-life characteristics of canned, pasteurized crab cake mix. Appl. Microbiol. *19*, 734-741.

MCMEEKIN, T.A. 1975. Spoilage association of chicken breast muscle. Appl. Microbiol. *29*, 44-47.

MCMEEKIN, T.A. 1977. Spoilage association of chicken leg muscle. Appl. Environ. Microbiol. *33*, 1244-1246.

NICOL, D.J., SHAW, M.K., and LEDWARD, D.A. 1970. Hydrogen sulfide production by bacteria and sulfmyoglobin formation in prepacked chilled beef. Appl. Microbiol. *19*, 937-939.

NOVAK, A.F. 1973. Microbiological considerations in the handling and processing of crustacean shellfish. *In* Microbial Safety of Fishery Products. C.O. Chichester and H.D. Graham (Editors). Academic Press, New York.

PATTERSON, J.T., and GIBBS, P.A. 1977. Incidence and spoilage potential of isolates from vacuum-packaged meat of high pH value. J. Appl. Bacteriol. *43*, 25-38.

PEEL, J.L., and GEE, J.M. 1976. The role of micro-organisms in poultry taints. *In* Microbiology in Agriculture, Fisheries and Food. F.A. Skinner and J.G. Carr (Editors). Academic Press, London.

SATTERLEE, L.D., and HANSMEYER, W. 1974. The role of light and surface bacteria in the color stability of prepackaged beef. J. Food Sci. *39*, 305-308.

SEIDEMAN, S.C., *et al.* 1976. Effect of various types of vacuum packages and length of storage on the microbial flora of wholesale and retail cuts of beef. J. Milk Food Technol. *39*, 745-753.

SHANNON, E.L., OLSON, N.F., and VONELBE, J.H. 1969. Effect of lactic starter culture on pink discoloration and oxidation-reduction potential in Italian cheese. J. Dairy Sci. *52*, 1557-1561.

SHEWAN, J.M. 1971. The microbiology of fish and fishery products—a progress report. J. Appl. Bacteriol. *34*, 299-315.

STARUSZKIEWICZ, W.F., JR. 1969. Collaborative study on the quantitative gas chromatographic determination of lactic and succinic acids in eggs. J. Assoc. Offic. Anal. Chem. *52*, 471-476.

TOM, R.A., and CRISAN, E.V. 1975. Assay for lipolytic and proteolytic activity using marine substrates. Appl. Microbiol. *29*, 205-210.

USDA. 1972. Egg Grading Manual. Agricultural Handbook No. 75. U.S. Department of Agriculture, Washington, D.C.

WATT, A.K., and MERRILL, A.L. 1963. Composition of Foods. Agricultural Handbook No. 8. U.S. Department of Agriculture, Washington, D.C.

9

Useful Microorganisms

Microorganisms are used in many facets of the food industry. Desired alterations of food by microorganisms are referred to as fermentations, regardless of the type of metabolism. By definition, fermentation is the anaerobic breakdown of an organic substance by an enzyme system, in which the final hydrogen acceptor is an organic compound. Hence, the aerobic oxidation of alcohol to acetic acid in vinegar production is not a true fermentation. For our purposes, the actions of microorganisms on foods are called food conversions.

Since the enzyme systems of the organisms determine the reactions in the foods, in many cases it is advantageous to use the purified enzymes, rather than the microorganisms.

Microorganisms are used to produce ingredients, such as flavorings for foods. The amino acids, cellular protein and vitamins formed by microorganisms are added as supplements to improve the nutrient value of foods.

It is possible that microorganisms can be used to remove undesirable flavoring compounds or toxic agents from foods. At present, there are more hypothetical possibilities than actual uses of these systems.

To assay for vitamins and other constituents of foods, microorganisms have certain possibilities and advantages over normal chemical techniques.

FOOD CONVERSIONS

The conversion of one food to another may be called controlled degradation. Various types of food conversions are listed in Table 9.1. Pederson (1979), as well as other references in the bibliography, should be consulted for more complete descriptions of food conversions than is possible in this text.

TABLE 9.1

SOME MICROBIAL FOOD CONVERSIONS

Organisms	Substrates	Products
Lactic acid bacteria, species of *Leuconostoc, Lactobacillus, Pedicoccus* and/or *Streptococcus*	Cabbage	Sauerkraut
	Cucumbers	Pickles
	Olives	Olives (green, ripe)
	Vanilla beans	Vanilla
	Red meat	Sausages, (salami, Thuringer, Lebanon bologna, cervelat, summer, pepperoni)
	Milk products	
	Cream	Sour cream, cultured butter, ghee
	Milk	Cultured milk, acidophilus, yogurt
	Milk	Cheese—unripened (cottage, pot, cream)
	Milk	Cheese—ripened (Cheddar, American, Edam, Cheshire)
Propionibacterium	Unripened cheese	Cheese (Swiss, Emmenthaler, Gruyere)
Brevibacterium linens	Unripened cheese	Cheese (limburger, brick, Trappist)
Penicillium roqueforti	Unripened cheese	Cheese (Roquefort, blue, Stilton, Gorgonzola)
P. camemberti	Unripened cheese	Camembert cheese
Lactic acid bacteria with yeasts	Taro	Poi
	Flour (dough)	Sour dough bread, pancakes
	Ginger plant	Ginger beer
	Beans	Vermicelli
	Rice, black gram	Idli
Saccharomyces, Aspergillus oryzae	Soybeans, wheat	Shogu (soy sauce), miso
A. flavus or *Mucor rouxii,* yeasts	Kaffir corn	Kaffir corn beer

TABLE 9.1. (*Continued*)

Organisms	Substrates	Products
Yeasts	Malt	Beer, ale, stout, lager, bock, porter, Pilsner
	Fruit	Wine, vermouth
	Wines	Brandy
	Molasses	Rum
	Grain mash	Whiskey
	Flour (dough)	Bread
	Fruit peel	Citron
Yeasts with *Acetobacter* or *Gluconobacter*	Sugar, fruit, potatoes, honey, malt, grain alcohol	Vinegar
Halophilic bacteria	Fish	Nuoc-mam-ngapi
Bacillus subtilis	Soybeans	Natto
Actinomucor elegans, Mucor	Soybean curd	Sufu (Chinese cheese, bean cake)
Rhizopus oligosporus	Soybeans	Tempeh, tempe
R. oligosporus	Coconut press cake	Tempeh bongkrek
Neurospora sitophila	Peanut press cake	Ontjom

Lactic Acid Bacteria

The organisms referred to as lactic acid bacteria include certain species in the genera *Streptococcus, Pediococcus, Leuconostoc* and *Lactobacillus.* The first two genera are homofermentative, the leuconostocs are heterofermentative and the lactobacilli include both homofermentative and heterofermentative types. The homofermenters convert carbohydrates primarily to lactic acid, while the heterofermenters produce lactic acid and substances such as acetic acid, ethyl alcohol and carbon dioxide. This means that the homofermenters use primarily one metabolic pathway, whereas the heterofermenters use more than one pathway.

The lactic acid bacteria usually grow in a sequence in a food product. Generally, the leuconostocs or streptococci begin the fermentation and are followed by pediococci and lactobacilli.

The production of lactic acid from sugar is used in vegetable, fruit and dairy products fermentations. The organisms normally present in or on the foods can be utilized. However, there are cultures available for use in controlled fermentations. These starter cultures are preferred in the fermentation of milk.

Although there are many similarities in the lactic acid fermentations, due to the various substrates, there are some differences.

Sauerkraut.—This means acid cabbage. According to definition, sauerkraut is the clean, sound product of characteristic flavor, obtained by full fermentation, chiefly lactic, of properly prepared and shredded cabbage in the presence of not less than 2%, nor more than 3% salt. It contains, upon completion of the fermentation, not less than 1½% of acid, expressed as lactic acid. Sauerkraut which has been rebrined in the process of canning or repacking contains not less than 1% of acid, expressed as lactic acid.

Sound heads of cabbage are prepared by washing and removing the outer leaves and any defective leaves. The core is removed and the leaves are shredded. Shredding the leaves gives a larger total surface area and allows the extraction of juice. About 2.25% salt is added uniformly to layers of shredded cabbage. The combination of salt and packing down of the cabbage expels the juice from the cabbage, and a brine results. The juice contains sugars and other nutrients from the cabbage. Depending upon the variety of cabbage and other factors, the sugar content is about 3-6%. The amount of sugar influences the fermentation and the final acidity.

When the vat or tank is essentially full, a plastic sheet is used as a cover to keep out dirt and air. The environmental conditions, numbers and kinds of microorganisms, cleanliness of cabbage and vat, salt concentration and distribution, temperature and covering influence the fermentation (Pederson 1979).

Many types of microorganisms are associated with raw cabbage. Most of these are not involved in the fermentation. Cabbage contains substances which are inhibitory to Gram negative bacteria (Fig. 9.1). With the inhibitors, salt and an anaerobic environment, the lactic acid bacteria tend to dominate. The fermentation is started by *Leuconostoc mesenteroides.* It converts the sugar to lactic acid, acetic acid, alcohol, CO_2 and other products which contribute to the flavor of sauerkraut. The CO_2 helps maintain anaerobic conditions in the fermenting cabbage.

As the acids accumulate, *L. mesenteroides* is inhibited, but the fermentation continues with *Lactobacillus brevis, Pediococcus cerevisiae* and finally, *Lactobacillus plantarum.* The sequence of organisms and acid production during a typical sauerkraut fermentation are listed in Table 9.2.

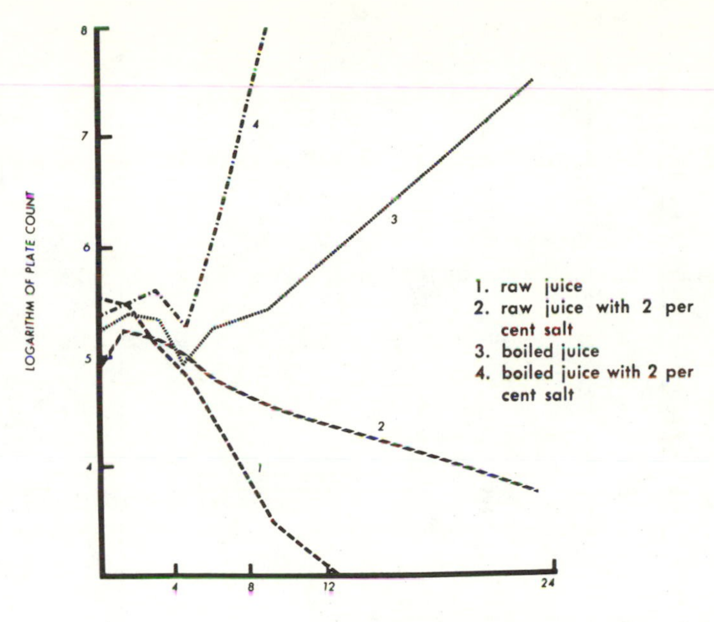

Courtesy of Pederson (1979)

FIG. 9.1. BACTERICIDAL ACTION OF CABBAGE JUICE TOWARDS A GRAM NEGATIVE BACTERIUM OBTAINED FROM CABBAGE SURFACE

Bottom scale represents time in hours.

Temperatures of 18°C or less (Pederson 1979), to 22°C (Weiser *et al.* 1971) are desirable for a good fermentation. At temperatures over 22°C, the growth of lactobacilli is favored, and a rapid fermentation results. The lack of heterofermentation causes the quality of this kraut to be poor. The flavor is inferior, resembling acidified cabbage. This kraut has a relatively short shelf life.

The proper concentration of salt favors the growth of the lactic acid bacteria in the correct sequence. Too little salt results in poor flavor and soft kraut. Too much salt inhibits the lactic acid bacteria and may result in an acid flavor, darkening, and growth of yeasts (Pederson 1979).

Although salt influences certain organisms, the inhibition of undesirable bacteria is due primarily to the acid produced by the fermentation of sugars.

Keeping air from the fermentation is important in controlling molds and yeasts. Air allows the growth of these microorganisms on the cabbage surface and can result in softening, darkening and the development of undesirable flavors.

Pickles.—There are many pickle products produced from cucumbers.

TABLE 9.2

DEVELOPMENT OF ACID AND CHANGE IN BACTERIAL FLORA IN SAUERKRAUT FERMENTATION

				Estimated Number of Each Type × 10^5 per ml					
Time, Days	Total Acid	pH	Total Plate Count × 10^5 per ml	Aerobes	*Leuconostoc mesenteroides*	*Lactobacillus brevis*	*Lacto-bacillus plantarum*	*Pedio-coccus cere-visiae*	Yeasts
5/6	0.15	4.48	1320	1	1319				
1-1/6	0.30	4.23	4400		4400				
1-5/6	0.74	3.93	9650		9150		500		
2-1/6	0.77	4.00	4660		4660				
2-5/6	0.97	3.87	4490		4440	50			
4	1.18	3.67	1250		687	62	313	188	
5	1.19	3.71	970			49	436	145	
6	1.16	3.63	1410			70	1270	70	
8	1.42	3.68	4670			750	3170	250	
10	1.57	3.59	2300			1035	690		
13	1.65	3.59	1200			200	1000		
17	1.61	3.58	546			200	1000		
21	1.78	3.51	251				248		3

Courtesy of Pederson (1979)

Because of their general acceptance, pickles were the first fermented vegetable product made on a large commercial scale. Over 40% of the cucumber crop is not fermented, but is made into fresh-pack products. The fresh cucumbers are packed with acetic or lactic acid (acetic acid is preferred), and various spices are added to give the characteristic flavor of the particular product. These pickles are then pasteurized for preservation. Monroe *et al.* (1969) found an acidity of 0.6% (acetic acid), and heating to an internal temperature of 71°-77°C prevented spoilage of the product. Below 71°C, acidities up to 1.0% did not prevent spoilage, and temperatures above 77°C resulted in damage to the internal structure of the pickles.

Natural Fermentation.—Cucumbers from the field are washed and placed in a salt brine of about 25° salometer and allowed to ferment. The salt inhibits the growth of undesirable microorganisms, and allows the salt-tolerant lactic acid bacteria to ferment the sugars to lactic acid. Sugars and other nutrients diffuse from the cucumbers into the brine and are used by the microorganisms. The salt diffuses into the cucumbers during the holding period in the brine.

The formation of sufficient lactic acid is an important factor in the quality and preservation of the fermented pickle. The rate of acid production and the total acid produced depend on the variety and size of cucumber, initial salt concentration, the temperature and the natural microflora of the cucumbers. During fermentation, the pH is lowered to about 3.5. During storage, salt is gradually added to a salometer level of 45°-60°. This level of salt, with the low pH, halts enzymatic and bacterial activities and preserves the pickles.

The natural microflora of the cucumbers is quite variable, and includes bacteria, yeasts and molds. The salt and a lowered redox potential favor the growth of facultatively anaerobic organisms. The vats are covered with plastic, and UV rays from sunlight or UV lamps are used to prevent surface growth of film-forming yeasts.

The initial microflora may contain molds primarily associated with parts of the flower still attached to the cucumbers. Molds do not survive due to the salt and to the low redox potential. However, the pectinolytic enzymes from molds have been involved with softening of brined cucumbers.

There is a miscellaneous group of yeasts at the beginning of the fermentation. Yeasts can affect the fermentation by utilizing sugars that would otherwise be metabolized to lactic acid by the lactic acid bacteria. Also, the yeasts can utilize the produced lactic acid, raise the pH and allow other potential spoilage types of microorganisms to grow. The yeasts produce large amounts of gas. This is associated with bloater or hollow cucumber formation. However, there are other causes for this defect.

The sugars that diffuse from the cucumbers are fermented sequentially by *Leuconostoc mesenteroides, Pediococcus cerevisiae, Lactobacillus brevis* and *Lactobacillus plantarum.* Depending on the condition of fermentation, about 0.6 to 1.2% lactic acid is formed in about 7-14 days. As the pH is lowered to 3.2, the metabolism of *L. plantarum* is inhibited, and, in the study of Etchells *et al.* (1975), about 0.25% sugar remained after lactic acid formation had ceased. It is desirable to ferment as much of the sugar as possible to retard the growth of yeasts.

Controlled Fermentation.—The cucumbers are washed and sanitized with a chlorine solution which removes most of the undesirable microorganisms. After brining (20°-25° salometer), the cover brine is acidified with acetic acid and buffered with sodium acetate or sodium hydroxide. The brine is purged with N_2 to remove dissolved CO_2. A culture of *L. plantarum* is added for the fermentation. With controlled fermentation, the need to add more salt during storage is reduced. This is important since the Environmental Protection Agency has an objective for a zero salt discharge from pickling operations by 1982.

Defects.—The two main defects of fermented pickles are bloaters and softening.

Bloaters are those fermented and cured pickles that float on the brine or are hollow or have large air spaces in the interior. Bloater formation is due to the accumulation of gas inside the cucumber during fermentation (Fig. 9.2). This can be caused by several factors. In the early stages of the fermentation, the coliforms and certain halophilic bacteria can produce hydrogen and cause bloaters. The fermentation of sugar by yeasts results in formation of much gas. Some bloaters are formed in fermentations without yeasts being present. The respiration of cucumber tissue plus the fermentation by the homofermentative *P. cerevisiae* or *L. plantarum* produces sufficient CO_2 to cause bloater formation (Fleming *et al.* 1973; Etchells *et al.* 1975). Piercing of the fruit prior to brining controls this defect. Purging of CO_2 from cucumber brines with nitrogen can reduce the amount of bloater damage (Fleming *et al.* 1975). Controlled fermentation, increasing the depth of the vat and purging with nitrogen reduce the incidence of bloaters (Fleming *et al.* 1977).

Bloaters are not a complete loss, since they can be used in cut pickle and relish products. However, their value is reduced by about 50%.

Softening of the salt-stock pickles is attributed to pectinolytic enzymes which degrade the cucumber tissue. The main source of these enzymes is molds that enter the vat with the cucumbers, especially with portions of flowers that may remain attached to the cucumbers. Pectin degrading enzymes are naturally present in the cucumber (fruit and seeds). The softening can be essentially controlled by removing any flowers attached to the cucumbers. Pasteurization of the fermented pickles at about 82°C for 25-30 min will inactivate the pectinolytic enzymes.

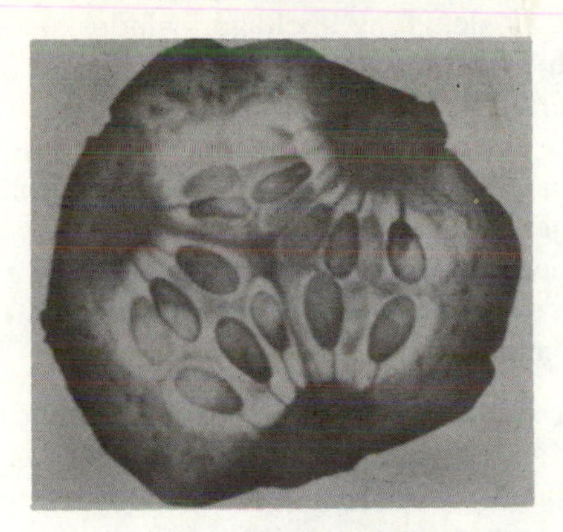

Courtesy of Pederson (1979)

FIG. 9.2. SLICE OF LARGE CUCUMBER SHOWING TYPICAL SEPARATION OF LOCULES, THE FIRST STAGE IN BLOATING

Olives.—Olives are brined and fermented in a manner similar to cucumbers. Before brining, the olives are soaked in a 1.25-2.0% lye solution. This is necessary to hydrolyze oleuropein, a bitter factor in the olive. After treatment, the lye is removed by rinsing the olives in fresh water. Some nutrients are lost by this washing treatment. Therefore, excessive, unneeded washing is not desirable. If too much carbohydrate is lost, there will not be enough to develop sufficient acidity. To overcome this problem, reduced washing, addition of lactic acid to neutralize residual lye or adding sugar to the brined olive for acid production has been suggested. The lye also may affect the microbial flora. If so, it is necessary to add cultures of desirable organisms.

After washing, the olives are brined. According to Pederson (1979), the brine concentration varies from 5 to 15% salt, depending upon the variety and size of the olives. As water diffuses out of the olives, and salt penetrates, the brine concentration is reduced. Additional salt is added to maintain the concentration. The vats, barrels or other containers are covered to maintain a low redox level.

The entire fermentation process may take from two weeks to several months. The same organisms active in the fermentation of cucumbers also are involved in olive fermentations. A level of at least 0.6-0.7% acid is needed for proper preservation and flavor of the product. Depending on the variety and type of treatment given the olives, the acidity varies from 0.18 to 1.27% (Samish *et al.* 1968).

Various factors, such as the origin, maturity and variety of olives, treatment prior to brining, brine strength, sugar content, acidity, available desired microflora and temperature influence the fermentation.

A defect known as sloughing spoilage includes severe softening, skin rupture and flesh sloughing. The defect is caused by Gram negative pectinolytic bacteria (Patel and Vaughn 1973), and occurs during the washing to remove the lye prior to brining. They described the bacteria involved as strains of *Xanthomonas* and various coliforms.

As with cucumber fermentations, yeasts can grow in or on the brine during fermentation of olives. Pink yeasts (*Rhodotorula*) and fermenting, pectinolytic yeasts (*Saccharomyces* and *Hansenula*) can cause softening of olives (Vaughn *et al.* 1969, 1972).

Red Meat.—Fermentation of sugars to lactic acid is utilized in the production of certain semidry and dry sausages (Fig. 9.3). Various formulations of chopped meat are mixed with spices, sugar, salt, sodium nitrate and/or sodium nitrite.

Courtesy of Pederson (1979)

FIG. 9.3. CURED OR FERMENTED SAUSAGES ARE PREPARED IN A VARIETY OF SIZES

Included here are variations from a moist Thuringer to a dry pepperoni; variously called Thuringer, summer sausage, beerwurst, beer summer sausage, Lebanon bologna, Genoa sausage, and many others.

The mixture is held at a temperature that allows the desirable bacteria to produce lactic acid from the sugars. These bacteria are members of *Pediococcus* and *Lactobacillus.* Everson *et al.* (1970) stated that *Pediococcus acidilactici* is the species used in sausage products. Although it is possible to depend on naturally occurring organisms, the use of starter cultures of *Pediococcus* and *Lactobacillus* is preferred. The culture also may contain a *Micrococcus.* Starter cultures result in better color, aroma, flavor and texture of the fermented product. Also, the processing time is reduced with a more rapid drop in pH and the yield is increased. Both frozen and freeze dried cultures are available. The fermented sausages are smoked and dried, during which the desirable characteristic flavor is developed.

Poultry Meat.—Chicken and/or turkey meat, either alone or combined with beef, has been suggested for use in the production of dry fermented sausages. The fermentation is similar to that with red meat. Acton and Dick (1975) reported that fermented and dried turkey sausage, on a fat free basis, had a lactic acid content of 3.1-3.2% and a pH of 4.6.

Dairy Products.—The production of lactic acid from the lactose in milk is important in the manufacture of fermented dairy products. The main lactic acid formers are the homofermenting streptococci, *S. lactis* and *S. cremoris.*

Strains of these organisms vary in the rate of acid production. Also, the rate is influenced by the temperature, pH, antibiotics, bacteriophage, stimulants, inhibitory compounds, milk composition, available nutrients, the condition of the culture, strain compatibility and strain dominance.

Some strains of *S. cremoris* can degrade citric acid with the formation of diacetyl, an important flavor component of fermented milks. *S. lactis* subsp. *diacetylactis* produces not only considerable lactic acid, but also degrades citrate to diacetyl. Other flavor components include volatile acids, dimethyl sulfoxide, methyl ketones and lactones.

The organisms that constitute the aroma and flavor producers are strains of *Leuconostoc* species (*L. cremoris, L. dextranicum* and *L. mesenteroides*). *S. lactis* subsp. *diacetylactis* is included as a flavor producer. Some strains of *L. dextranicum* and *L. mesenteroides* produce lactic acid in milk, but rather slowly and in low amounts. *L. cremoris* produces diacetyl only in acidic substrates and is optimal at pH 4.3.

S. lactis subsp. *diacetylactis* degrades citric acid faster and produces more carbon dioxide than the leuconostocs. The excess gas gives fermented milk drinks a desirable effervescence, but is undesirable in cheese manufacture, since it may crack the cheese.

Besides streptococci and leuconostocs, certain species of lactobacilli are used in milk fermentations. *Lactobacillus lactis* coagulates milk with an acidity of about 1.6% lactic acid. Lactobacilli that have been found in fermented milk products include *L. helveticus, L. bulgaricus, L. acidophilus* and *L. casei.*

The natural bacterial contaminants have been used to ferment raw milk. However, to assure a proper fermentation and an acceptable product, after pasteurization or heat treatment of milk, cultures containing the desirable organisms are used to inoculate the milk. In order to inoculate a large vat of milk, sufficient culture must be produced, first in mother cultures and then in bulk culture. The propagation of cultures must be done under conditions that minimize contamination with undesirable bacteria or with phage that can infect the desirable bacteria.

Both acid formers and flavor producers are important in the fermentation and hence, in the culture. If insufficient acid is produced, a product with poor keeping quality results. Without aroma and flavor producers, the fermented product is flat or acid, and may have a metallic flavor.

The consumption of fermented milk has been suggested for people who have an intolerance for lactose, due to the lack of lactase (β-galactosidase). There is less lactose in fermented milk than in normal fresh milk.

Cultured Buttermilk.—This can be made from whole milk, reconstituted nonfat dry milk, partially skim milk or skim milk. One or two percent fat is desirable in cultured buttermilk, since it improves the consistency and palatability.

The milk is heated to 85°-90°C for 30-60 min. Heating destroys many bacteria, inactivates natural bacterial inhibitors and helps prevent "wheying-off" of the product.

After heating, the milk is cooled to about 21°C and inoculated with a culture. Since good cultured buttermilk contains lactic acid and flavor compounds, the culture must contain a lactic acid producer (*S. lactis* or *S. cremoris*) and a flavor producer (a *Leuconostoc, S. lactis* subsp. *diacetylactis,* or both). The inoculated milk is incubated at 21°-22°C until the titratable acidity (as lactic acid) reaches about 0.85% with a pH of 4.4-4.5. This requires about 14-16 hr. A desirable cultured milk should contain at least 2 mg diacetyl per liter. Hence, it is desirable to add citrate (about 0.25%) to the milk so that an acceptable level of flavor compounds is produced during fermentation. When the cultured buttermilk reaches the desirable stage, the fermentation is halted by cooling the product to 10°C or less.

Acetaldehyde, produced by *S. lactis* subsp. *diacetylactis* and some other lactic acid formers, causes a defect in cultured milk which is called "green" or "yogurt-like" flavor. *L. cremoris* reduces the acetaldehyde to alcohol and prevents this defect.

Rather than fermenting milk to make cultured buttermilk, an acidified buttermilk can be produced. Reportedly, the acidified, flavored product is consistently good, has a long shelf life, and can be made in less time, with fewer people.

Cultured Sour Cream.—This product is pasteurized and homogenized cream fermented in a manner similar to cultured buttermilk. It contains not less than 0.20% lactic acid and 18% butterfat.

Cultures in Butter.—A lactic culture is added to cream and fermentation proceeds at 18°-20°C. An acidity of 0.5 to 0.6% may be attained prior to churning. This butter is sold as the unsalted product. If cultured acid cream is used to manufacture salted butter, the product tends to deteriorate with a fishy flavor. Salted butter has better shelf life if the pH is near neutral.

For salted butter, a distillate from a culture can be added to give the desirable flavor.

Acidophilus Milk.—This is milk fermented with *Lactobacillus acidophilus.* The milk is heated to 120°C for 15 min, cooled to 37°C and inoculated with *L. acidophilus.* Incubation is continued at 37° until coagulation is evident. The fermented milk has a clean acid flavor, but is not especially palatable.

The organism supposedly has therapeutic value for various disorders. The therapeutic capabilities are probably overestimated. Most lactic cultures, including *L. acidophilus,* do not survive in the fermented product.

Yogurt.—In some countries, goat, ewe, mare or cow milk is used to produce yogurt. In the USA, either whole or skim milk from cows is used. The milk is standardized to 10.5 to 11.5% solids, heated to about 90°C (30 to 60 min), and then cooled (40° to 45°C). The inoculum is a mixed culture of *Streptococcus thermophilus* and *Lactobacillus bulgaricus* in a 1:1 ratio. The combined action of these two organisms is needed to obtain the desired flavor and acid in the product.

The flavor depends to some extent on the production of acetaldehyde. Together, the cultures will produce about 25 mg/liter, but either organism alone produces about 8 mg/liter or less.

In a survey of 152 yogurt samples, Arnott *et al.* (1974) found only 15.1% had a desired 1:1 ratio of *S. thermophilus* and *L. bulgaricus.* According to Moon and Reinbold (1976), *S. thermophilus* tends to outgrow *L. bulgaricus.* Symbiotic growth is inferred, but they found commensalism and competition between these organisms.

Dutta *et al.* (1972) suggested an incubation temperature of 42°C for yogurt manufacture, so that symbiotic growth and the proper ratio of the organisms is maintained.

To halt the fermentation, the product is cooled to 5°-10°C.

An acidophilus yogurt is made by substituting *L. acidophilus* for *L. bulgaricus.*

A continuous process for yogurt production using *S. thermophilus* and *L. bulgaricus* was described by Driessen *et al.* (1977).

Cottage Cheese.—This is a soft, unripened cheese. Cottage cheese manufacture consists of coagulating the casein of skim milk by acid, cutting the coagulum into cubes, heating to reduce the moisture in the curd, washing to remove residual whey and cooling of the curd. A cream dressing may be added for texture and flavor. The product should not have more than 80% moisture.

The skim milk is pasteurized (62.8°C for 30 min or 71.7°C for 15 sec). Overheating can result in a soft coagulum and a lower yield. The inoculum should be able to produce lactic acid rapidly. *S. lactis, S. cremoris* or a combination of these organisms is the culture of choice. *S. lactis* subsp. *diacetylactis* is not satisfactory since it produces large amounts of carbon dioxide. This gas can result in a floating curd defect.

The incubation temperature may vary from 20° to 33°C. For an eight hour setting time, a temperature of 30° to 33°C is used, while for a 12 hr incubation, from 20° to 24°C is used.

Rather than using a lactic culture, the skim milk can be acidified by direct addition of lactic acid. The coagulation can be accomplished with acid, rennet, rennet plus pepsin, other commercial preparations or a combination of acid and enzymes. A large-grained, low-acid cheese is made by adding rennet to the milk, cutting the curd into large cubes and washing the curd to reduce the acid flavor.

The curd is ready to cut when it is firm but not hard and brittle. The size of the cubes determines to some extent the size of the particles in the finished cheese. Cutting the curd is done when the pH is between 4.6 and 4.8. In this pH range, the curd will expel moisture most readily when stirred and heated, since this pH range includes the isoelectric point of casein.

Heating of the curd halts the fermentation and aids in expelling water. When the curd has attained the proper firmness, the whey is drained and the curd is washed with cool tap water. A final washing is made with ice cold water. When the curd is firm and dry, it is salted for flavoring.

The curd may be creamed to enhance the flavor. The flavor is obtained by adding starter distillate or cultured skim milk to the creaming mixture. There is a preference for cottage cheese creamed with a mixture prepared with a culture concentrate rather than a normal culture (Gilliland *et al.* 1970).

The cottage cheese can be contaminated by the wash water or by poor sanitation during creaming and packaging. With poor sanitation, the shelf life may be only three to four days, rather than three to four weeks. Surface spoilage (slimy defect) can be caused by *Pseudomonas* and *Alcaligenes.* Yeasts and molds indicate poor sanitation in the processing plant and can cause spoilage of cottage cheese with pH less than 5.0.

Cheddar Cheese.—This cheese was first made in the village of Cheddar in Somersetshire, England. So much of this cheese is made in the USA that it is often called American cheese, or American Cheddar cheese. Cheddar is a firm ripened cheese, ranging in color from nearly white to yellow-orange, depending upon the amount and type of coloring added.

In making Cheddar cheese, the lactic acid culture is added to a vat of pasteurized whole milk. The culture may be a pure culture of *S. lactis* or *S. cremoris*, but preferably, a mixture of these organisms. *L. cremoris* may be added for flavor and its presence increases the rate of acid formation by the lactic acid culture.

After a short fermentation (about one hour), at 30° to 32°C, a coagulant, such as rennet, is added. The acidity developed by the lactic culture plus the coagulant causes the curd to form and to solidify into a custard-like consistency. The whey is separated from the curd, aided by heating (36°-46°C) and cheddaring. The curd is then milled, salted, colored (if desired) and packed into cheese hoops or metal containers for pressing. After pressing overnight, the hoops are removed and the cheese is placed in a curing room for three to four days to dry and cure. Additional curing and aging require two months to two years. The shorter the curing period, the more mild the flavor, with a smooth waxy body. In the longer curing periods, the flavor becomes more sharp and the body of the cheese becomes broken down. The flavor is regarded as a blend of fatty acids, organic acids, amino acids, carbonyl compounds, esters, alcohols and sulfur compounds.

The Cheddar cheese flavor is not developed if *S. lactis* is the only organism present. The number of *S. cremoris* declines during cheddaring and this organism disappears during aging of the cheese. When *Streptococcus faecalis* is added to the culture, there is an increased level of cheddar flavor. Lactobacilli are present in small numbers in raw milk. Some are destroyed during heating, but recontamination occurs from air and equipment. During the curing and aging process, the lactobacilli increase and reach a maximum in three to six months. Some lactobacilli aid in the production of flavor of Cheddar cheese, while other strains and species result in an inferior cheese. Pediococci, micrococci and enterococci may play some part in the development of flavor and aroma of this product.

The defects in Cheddar cheese include bitter, fruity or acid flavors or lack of flavor. Bitter tasting peptides are a result of proteolysis of the casein and the inability of the culture to degrade the peptides to amino acids. Fast acid producers have a tendency to yield bitter flavored cheese. Fruit flavored cheese tends to contain high levels of ethyl butyrate, ethyl hexanoate and ethyl alcohol. Carbon dioxide produced by microorganisms causes a slit-open defect in Cheddar cheese.

Rosenau *et al.* (1978) described systems for a processed cheddar-like cheese using chemical additives, rather than cultures to manufacture the product.

Lactic Acid Bacteria and Other Bacteria

In some fermented products, not only is the production of lactic acid important, but also other compounds involved with flavor or other characteristics. Both dairy and vegetable products are manufactured with a combination of lactic acid bacteria and other microorganisms.

Other Dairy Products.—As stated previously, lactic acid production is part of the process for all fermented dairy products. In some cases, acidulation with lactic acid can be used.

Swiss Cheese.—This cheese is manufactured by using lactic acid bacteria and propionic acid bacteria. Swiss cheese is a hard cheese characterized by a sweet flavor, and by gas holes or eyes distributed throughout the cheese.

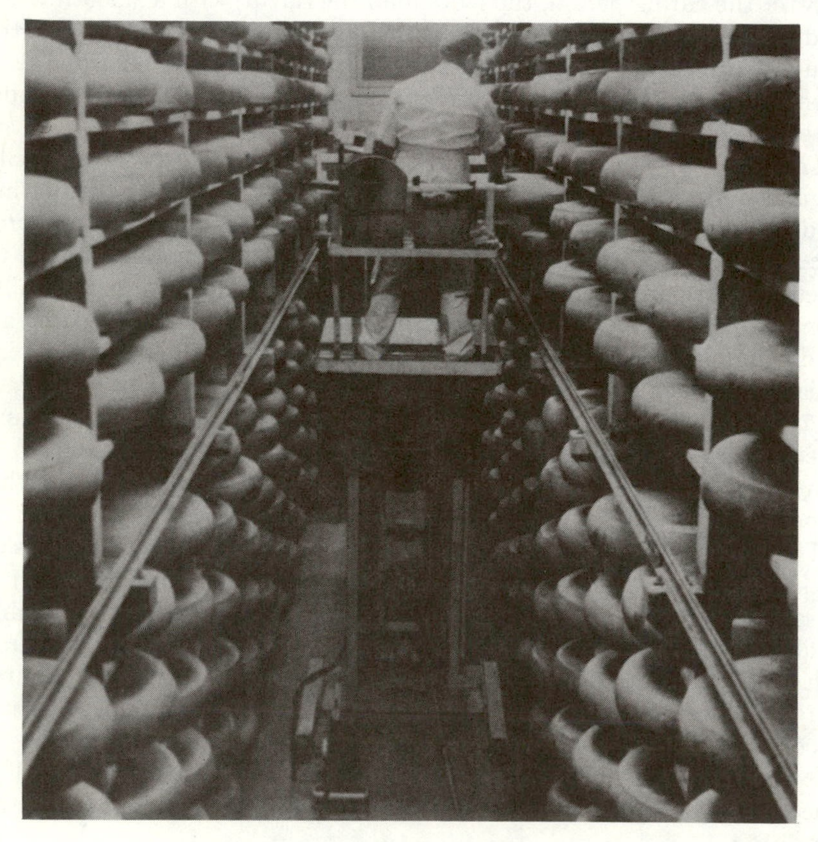

Courtesy of Swiss Cheese Union, Inc., Berne, Switzerland

FIG. 9.4. SWISS CHEESE IN CURING ROOM

The organisms used in Swiss cheese manufacture are the lactic acid producers (*S. thermophilus* and *L. bulgaricus*) and a species of *Propionibacterium* which produces propionic acid. *Propionibacterium freudenreichii* subsp. *shermani* is usually given credit as the species in Swiss cheese. However, *Propionibacterium freudenreichii* subsp. *freudenreichii* and *globosum*, as well as *P. acidi-propionici* and *P. jensenii*, may be found in Swiss cheese. The propionibacter are responsible for the characteristic flavor and eye formation in Swiss cheese.

After coagulation and separation of the whey, the curds are put into a cheese hoop and pressed lightly. The cheese is turned several times, and is pressed more firmly each time. During the first two or three weeks of ripening, the cheese is salted and turned every one to three days.

Next, the cheese is moved to a prewarming cellar (17° to 20°C) for 10 to 14 days, and then to a fermentation room for 6 to 8 weeks. The fermentation room is 22° to 23°C, with a relative humidity of 80 to 90%. During this time, the cheese is turned and washed with salt water two to three times per week (Fig. 9.4).

The final process consists of curing at 7° to 10°C and high relative humidity. After four months, the cheese is mild, and has the typical sweet nutty flavor. Stronger flavors are developed by leaving the cheese in the curing room for 8 to 12 months.

Courtesy of Swiss Cheese Union, Inc., Berne, Switzerland

FIG. 9.5. TYPICAL EYE FORMATION IN QUALITY SWISS CHEESE

The proprionibacter produce CO_2, which is not able to escape, so it concentrates in various places in the cheese. The pressure produced by this gas results in the eyes in Swiss cheese (Fig. 9.5). The turning, salting, washing, temperature and humidity are essential to obtain uniformity of these holes. Undesirable organisms, such as butyric acid bacteria, produce off-flavors, as well as gases such as hydrogen, which results in poor eye formation.

Swiss cheese is one of the more difficult cheeses to manufacture, since defects can occur to the aroma, flavor and eye formation with improper care and handling (Hettinga and Reinbold 1975; Langsrud and Reinbold 1974).

Limburger.—This is a semisoft surface-ripened cheese with a distinct odor and flavor. It is regarded as one of the most delicate and difficult cheeses to make. After the milk is coagulated, the curd is packed into rectangular forms and residual whey is allowed to drain. When the cheese is firm enough to retain its shape, it is removed from the form and salted and turned frequently.

At the beginning of ripening, fungi predominate. They reduce the acidity so that bacteria can grow. After the fungi, micrococci may be present along with *Brevibacterium linens.* This organism (*B. linens*) is common in the slime of surface-ripened cheese. It is thought that proteolytic enzymes secreted by the organism diffuse into the cheese and cause the characteristic softening and flavor.

Although yeasts are listed as the initiators, these have been described as *Geotrichum candidum, Oidium lactis* or *Oospora.* At the present time, *Oidium lactis* is *Geotrichum candidum*, which is a mold. Some *Oospora* species are now included in the yeast genus, *Trichosporon.*

Lactic Acid Bacteria with Yeasts

Some fermented milks, such as kumiss and kefir, utilize a simultaneous production of lactic acid and alcohol. The alcohol is a product of yeast metabolism. The alcoholic content of kefir is about 0.3 to 1.0%, while that of kumiss is from 1 to 2%.

The lactic acid bacteria are *Streptococcus lactis* and *Lactobacillus bulgaricus.* The yeast is a lactose fermenter (perhaps *Kluyveromyces lactis*).

Other products in which lactic acid bacteria and yeasts are involved include sour dough, idli and ogi.

Sour Dough.—This type of dough was so prominent among prospectors of the Old West that they were named sour doughs. San Francisco is still noted for sour dough bread.

The sour dough process uses a lactic acid bacterium for the souring and a yeast for leavening. Sugihara *et al.* (1971) screened 200 yeast isolates from sour dough. They reported two types of yeast, *Saccharomyces exiguus* and *Saccharomyces inusitatus.* The lactic acid bacteria isolated from sour dough were difficult to grow, and their characteristics did not fit any known species (Kline and Sugihara 1971). They suggested the name *Lactobacillus sanfrancisco* for this organism.

Lactic Acid Bacteria with Molds

Molds are involved in the aging and curing of several cheeses (Table 9.1). Two of the more familiar cheeses are Roquefort and Camembert.

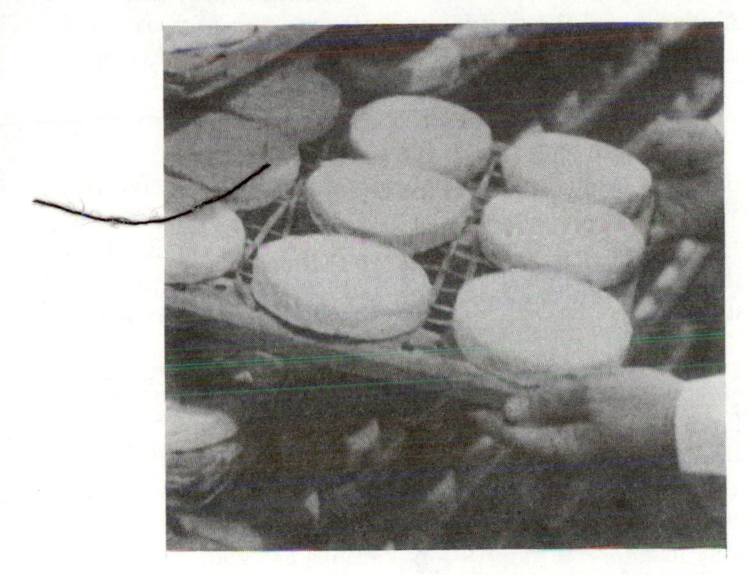

Courtesy of The Borden Co.

FIG. 9.6. CAMEMBERT CHEESE SHOWING WHITE SURFACE MOLD DURING CURING

Roquefort Cheese.—Roquefort cheese is made from sheep's milk and is ripened in the famous caves in Roquefort, France. A similar cheese that is common in the USA is made with cow's milk and is called Blue cheese.

One of the characteristics of this cheese is the greenish blue marbling of its soft, creamy white interior. This marbled appearance is due to the growth of *Penicillium roqueforti*, a blue-green mold, throughout the cheese. So that oxygen is available to favor interior growth, the cheese is pierced with needles. *P. roqueforti* is able to grow at a lower redox potential than many other molds.

The characteristic sharp, biting flavor of blue cheese is generally attributed to hydrolysis of the cheese fat to fatty acids, and then the conversion of these fatty acids to methyl ketones. The development of blue cheese flavor was discussed by Jolly and Kosikowski (1975) and Kinsella and Hwang (1976).

Camembert Cheese.—This is a soft, surface-ripened cheese (Fig. 9.6). The coagulated curd is placed into hoops and allowed to settle for about 2 days. The cheese is then removed from the hoop, salted and inoculated with *Penicillium camemberti.* The cheese is cured at about 12°C. Curing requires at least 60 days. The mold plus bacteria and yeasts ripens the cheese from the outside toward the center. During curing, a grayish-white felt-like growth of mold is followed by a secondary growth that produces a sliminess and the surface becomes reddish to russet colored. The cheese has a mild to pungent flavor.

Yeasts

The metabolism of sugar by yeast yields alcohol and carbon dioxide. The production of alcohol is important in various beverages (beer, wine, liquor), the CO_2 is used in the baking industry and the yeast cells are a source of protein, lipid and various useful chemicals and enzymes. Although yeasts are the main organisms utilized in the production of these foods, other organisms may play a role.

Alcohol.—Alcohol is produced during the metabolism of various sugars by yeasts. There are many types of alcoholic beverages.

Alcoholic beverages can be made from many products. If a starchy product is fermented, the starch is first converted through maltose to glucose. If the sugar content is low, sugar is added to obtain the desired amount for the alcoholic fermentation. Besides alcoholic beverages, industrial alcohol, which is produced by chemical systems, is used for many purposes. With the potential shortage of oil and gasoline, alcohol may be produced from wastes. According to Trevelyan (1975) we could never produce enough alcohol to replace petroleum products. He estimated that alcohol produced by fermentation of agricultural wastes could replace only about six percent of the gasoline now used in the USA.

Various strains of species of *Saccharomyces* are involved in the production of alcoholic beverages. Beverages involving distillation are discussed by Harrison and Graham (1970).

Beer.—Beer is produced in several steps. Barley is cleaned, graded, washed and steeped. It is then allowed to germinate for about five days. After germination, heat is used to stop the sprouting process and to dry the grain. The rootlets are screened out of the barley malt.

An adjunct, such as ground rice or corn, is mixed with the barley malt, wetted with clear, filtered water, and cooked. This mash is then strained into tanks. A clear amber liquid called wort is obtained. Hops, the seasoning of beer, are added to the wort and this mixture is boiled to obtain the correct delicate hop flavor. By this time, amylase has degraded the starch to maltose.

The hopped wort is strained, cooled and then pumped to fermentation tanks. Here, yeast is added. The fermentation changes the sugars in the wort into alcohol and carbon dioxide.

For aging and natural carbonation, the beer is pumped into tanks which may contain substances such as beechwood chips to provide surfaces for the yeast. Some freshly yeasted wort is added, and the beer is allowed to become naturally carbonated by a slow and quiet fermentation.

During this secondary fermentation, chill-proofing of beer can be accomplished by adding a proteolytic enzyme, such as papain. This hydrolyzes residual protein which would otherwise precipitate and cloud the beer when chilled. Chemical systems are available to precipitate proteins from beer (Gorinstein 1978).

The beer is then filtered to remove yeast cells. The beer is either pasteurized (60°C for 15-20 min) after packaging, or it is packaged aseptically after bulk pasteurization or microfiltration. Bulk beer in kegs is not pasteurized, so it must be refrigerated.

Some yeasts settle to the bottom of the fermenting vat and are called bottom yeasts. These are strains of *Saccharomyces uvarum* and are used to produce lager beer. Other yeasts, strains of *S. cerevisiae*, tend to collect at the surface and are called top yeasts. The product of this fermentation is called ale. Each brewery has certain strains of these yeast species for its particular products. Not all strains will produce an acceptable beverage. Hence, when a good strain is obtained, it is handled very carefully. It is a problem to maintain these strains of yeast. With continuous transfer they may become contaminated or they may mutate. Wellman and Stewart (1973) found that cultures frozen and stored in liquid nitrogen remained stable over a three-year test period. However, strains of the organisms did vary in their stability in the frozen state.

Besides alcohol and carbon dioxide, there are various metabolic products that can affect the flavor of the beer. These compounds include ethyl acetate and other esters, fusel alcohols (pentanol, isopentanol and isobutanol), diacetyl, 2,3-pentanedione, sulfur compounds (sulfites, sulfides, mercaptans, mercaptals, thioaldehydes and thioketones) and leakage of amino acids and nucleotides from the yeast cells. Yeasts can produce hydrogen sulfide in beer, wine and other alcoholic beverages (Wainwright 1971).

The amount of these compounds depends upon the conditions of fermentation, wort composition and yeast strain. Top fermented ale usually contains more fusel alcohols than bottom fermented lagers. The amount of H_2S produced is rather small, and most of it is removed by the purging effect of CO_2. Excessive amounts of H_2S or other sulfur compounds in the beer result in an unacceptable flavor. There are factors in beer that limit the types of organisms that can grow. These factors include a low pH and redox potential, the isohumulones of hops that inhibit Gram positive bacteria and the alcohol produced by the yeast. The contaminants are usually acetic acid bacteria, lactic acid bacteria, coliforms and wild yeasts.

The acetic acid bacteria (*Acetobacter* and *Gluconobacter*) can oxidize ethyl alcohol to acetic acid. In the anaerobic environment of active fermentation, this is not possible.

Some lactic acid bacteria (*Lactobacillus* and *Pediococcus*) have a tolerance to the isohumulones of hops. They are microaerophilic and tolerate acids. They produce lactic acid, diacetyl, off-flavors, turbidity and ropiness of beer.

The coliforms (*Klebsiella* and *Escherichia*) grow only when the pH is above 4.3. They impart various odors and flavors to the wort. If the wort is stored for future fermentation, coliforms can become a problem.

Not only the so-called wild yeasts, but also the various strains of brewers' yeasts can cause contamination. Bottom yeasts may become contaminated with top yeasts and vice versa. This will result in an alteration of the desired product.

A bacterium, *Zymomonas anaerobia*, has been found in beer and cider. It produces acetaldehydes and H_2S. The beer has a very unpleasant odor and flavor.

Wine.—Wine is considered to be the oldest fermented alcoholic beverage. The term wine is applied to the product made by alcoholic fermentation by yeasts of grapes or grape juice, with an aging process. However, the products of fermentation of berries, fruit and such things as honey, palm juice, rhubarb and dandelion also are called wines. These are designated by the substance from which they were made (blueberry wine, dandelion wine).

Only about three percent of the world's wine is produced in the USA. Most of this wine is made in California.

The quality of the finished wine depends upon the grapes, fermentation techniques and aging and blending. Anyone can make wine, but only the experts can produce high quality wine.

Some of the factors influencing the quality of the grape are climate, soil conditions (temperature, fertility, type and drainage) and variety of grape. For a discussion of these factors, see Amerine *et al.* (1967). A year with a good climate is known as a vintage year. If the climate is too cold, the grapes do not mature as well, so there is less sugar and more acid.

Courtesy Wine Institute, Amerine et al. (1972)

FIG. 9.7. STAINLESS STEEL FERMENTING TANKS

Note their location outdoors.

The grapes are crushed and pressed to release the juice, which is called the must. The sugar content is determined with a hydrometer. The sugar content can be adjusted by adding sugar, mixing high and low musts together or by adding water. Multiplying the hydrometer reading in degrees Balling by 0.575, gives an estimate of the percentage alcohol of the finished product. The amount of acid can be determined and adjusted to some extent by addition of a sugar solution (amelioration).

The must contains many types of microorganisms. In order to control these contaminants, the must is treated with sulfur dioxide (SO_2), or potassium metabisulfite. The yeast primarily involved in the fermentation has been called *Saccharomyces ellipsoideus, S. vini* or *S. cerevisiae* var *ellipsoideus.* However, Van der Walt (1970) classified the yeast as *Saccharomyces cerevisiae.* There are many strains of this yeast species.

Kunkee and Amerine (1970) also listed other *Saccharomyces* (*S. fermentati, S. bayanas* and *S. oviformis*) as wine yeasts. Van der Walt (1970) included *S. oviformis* as a strain of *S. bayanas.* Besides the various strains of *Saccharomyces*, Kunkee and Amerine (1970) listed about 160 species of yeasts that have been found on grapes, in musts or in wine. These also may influence the overall fermentation of the must, especially if they are resistant to SO_2 treatment.

Courtesy Wine Institute, Amerine et al. (1972)

FIG. 9.8. TABLE WINE BOTTLE AGING CELLAR IN CALIFORNIA WINERY

For red wine, the must is allowed to ferment with the skin and pulp of the red grapes. The fermentation is allowed to proceed until the correct amount of color is extracted from the skin. During this early stage, aeration is used to promote the growth of the yeast.

After this initial fermentation, the must is drawn off, which is called racking. The must is put into closed tanks (Fig. 9.7) to provide anaerobic conditions for the main fermentation. The produced CO_2 tends to purge the must of oxygen and inhibits aerobic organisms, such as *Acetobacter*. The fermentation proceeds at a temperature of 21° to 29°C, and may require from one to four weeks.

The wine is pumped into barrels, vats or tanks for aging. The aging time varies with the type of wine. During aging, the bouquet and aroma of the wine are developed. Various compounds, such as esters, are formed. Most wines are blended to maintain flavor consistency. This is usually done during the aging period.

After aging, the wine may be pasteurized at 60°C for 30 min and bottled or, in some cases, packed in kegs for shipment, and bottled elsewhere. When properly aged wine is bottled, it will continue to improve for several years (Fig. 9.8).

There are special procedures for certain products, such as dessert wine and champagne. Dessert wines are sweet and have an alcohol content of about 20%. To attain these conditions, the fermentation is halted while there is sufficient sugar remaining, by adding brandy. The higher alcoholic content of brandy increases the alcohol in the resultant wine.

For champagne and other sparkling wines, the wine is put through a secondary fermentation. New must, sugar and yeast are added to the stock wine, which is then bottled (Fig. 9.9) or put into a closed tank. The CO_2 that is produced remains in the wine, which causes the effervescence.

Courtesy of Wine Institute, Amerine et al. (1972)

FIG. 9.9. CHAMPAGNE BOTTLES ON RACKS

Although acids in grape juice contribute to the organoleptic quality of wine, retard microbial spoilage and stabilize color, too much acid is undesirable. Amelioration not only dilutes the acid, but dilutes flavor and other components.

A fermentation called the malo-lactic fermentation, can be used to reduce the acidity. Malic and tartaric acids account for about 90% of the acidity in grapes. Many lactic acid bacteria are able to convert malic acid to lactic acid and CO_2. This fermentation usually follows the normal alcoholic fermentation; thus, it is a secondary fermentation. The general equation of the conversion is:

$$\underset{\text{Malic Acid}}{HOOC\text{-}CH_2\text{-}CHOH\text{-}COOH} \rightarrow \underset{\text{Lactic Acid}}{CH_3\text{-}CHOH\text{-}COOH} + CO_2$$

The reaction goes through an intermediate, pyruvic acid, which is then reduced to lactic acid. The monocarboxylic lactic acid is not as acid as the dicarboxylic malic acid. Hence, the acidity of the wine is reduced.

Not all lactic acid bacteria can tolerate the alcohol and low pH in order to convert malic to lactic acid. *Leuconostoc oenos* can grow in 10% ethyl alcohol at pH 4.8. With lower alcohol levels, growth occurs at pH 4.2 or less. The lactobacilli are rather tolerant of acids and, according to Eskin *et al.* (1971), *L. plantarum, L. brevis* and *L. trichodes* can tolerate up to 15 to 20% ethyl alcohol.

The yeast, *Schizosaccharomyces pombe*, utilized all of the malic acid if the must pH was over 3.0 (Yang 1973). Even at pH 2.5, about 70% of the malic acid was metabolized. With *S. pombe*, ethyl alcohol and carbon dioxide are the end products.

There are certain defects that can occur in wine. The lower the sugar content, the less likely that spoilage will occur. Thus, dry wines are more stable than other types of wine.

To prevent microbial spoilage of the finished wine, it is important to deactivate any residual microorganisms before or after bottling. This can be accomplished by pasteurization, addition of inhibitors, such as SO_2, or by filtration. The delicate flavors of some wines are harmed by heating or by adding SO_2. For these wines, filtration is the preferred method of removing microorganisms.

Baked Products.—The leavening or raising of dough or batters is due to the incorporation of air into the product. This may be accomplished by whipping, such as egg whites in angel cakes, by adding chemical agents that react to produce gas, or by fermentation of sugars by microorganisms to produce carbon dioxide.

Leavening gives the finished product a cellular structure which improves the texture and flavor and increases the volume as compared to the unleavened product.

Although many microorganisms ferment sugars with the release of carbon dioxide, *Saccharomyces cerevisiae*, or bakers' yeast, is best adapted for leavening of bakery products. Yeast fermentation has certain advantages over chemicals, since it can contribute a characteristic flavor and aroma to the product and gas evolution can continue over a longer time. The products of yeast fermentation also affect the texture of the dough.

Bakers' yeast is used in the manufacture of bread, rolls, sweet dough, pretzels and crackers (Fig. 9.10). *S. cerevisiae* can ferment many simple monosaccharides and can hydrolyze sucrose and maltose to simple sugars. Yeast cannot hydrolyze the starch present in flour. Enzymes (amylases) naturally present in flour, or added enzymes (diastatic malt or amylase), hydrolyze the starch to maltose and glucose, making it available to the yeast.

Courtesy of Pederson (1979)

FIG. 9.10. SOME COMMON LEAVENED BREAD-LIKE PRODUCTS: PRETZELS, DOUGHNUTS, SALTINE CRACKERS, CINNAMON CRISP, RYE WAFERS, AND GARLIC BREAD STICKS

In continuous bread-making processes, the yeast is grown and fermentation occurs in a preferment or brew, which contains little or no flour. The fermentation products (alcohol, acids, aldehydes, carbon dioxide) in the preferment contribute to the flavor of the finished product. The leavening action is accomplished during proofing (holding the mixed dough for about 45 min in pans before baking).

Bakers' yeast is available as dried or compressed yeast. Commercial dried yeast has about 8% moisture and compressed yeast about 70%

moisture. Since lactic acid bacteria are added to the substrate used for growing yeast, these bacteria are found in commercial yeast.

The higher the moisture, the less stable is the yeast. Temperature and oxygen also affect the stability. A moisture level of 8% in dried yeast seems to be a compromise between stability and activity, since there is evidence that activity is lost at a very low moisture level. Rehydration in water at 35° to 40°C results in a greater activity and loaf volume than if water at 5° to 10°C is used for rehydration.

The heat to which the fermented dough is exposed during baking kills the yeast cells, drives off the alcohol and sets the flour protein surrounding the entrapped carbon dioxide.

Yeasts with Acetic Acid Bacteria

This combination of organisms is used in the production of vinegar and in the fermentation of cacao beans and citron.

Vinegar.—Vinegar is produced by an alcoholic fermentation followed by oxidation of the alcohol to acetic acid. The vinegar must contain at least 4 g of acetic acid per 100 ml. The strength of vinegar is referred to in grains, with 10 grains equal to one percent. Hence, 4 g/100 ml is 4% or 40 grain vinegar.

Vinegar can be produced from any food product that can be fermented by yeast to ethyl alcohol. Among the materials used are fruits, fruit juices, tubers, cereal grains, molasses, honey, coconut, beets, malt and refiners' syrup. Vinegars are classified according to the raw material from which they were made. In the USA, the term vinegar implies that it was made from apples. The names cider vinegar and apple vinegar also are used for this product. Wine vinegar is made from grapes. Distilled, grain or spirit vinegar is made from an alcoholic solution that has been distilled.

It is essential that alcoholic fermentation be complete before vinegar formation, since the acetic acid interferes with further alcohol production. The overall chemical changes can be represented by the reaction:

$$C_2H_5OH + O_2 \rightarrow CH_3COOH + H_2O$$

Acetobacter

Several side reactions occur which alter the final composition of the vinegar.

There are various systems for converting alcohol to vinegar. All of the methods provide a means of bringing together the alcohol, air and *Acetobacter*. The air supplies the oxygen needed by the strictly aerobic *Acetobacter*, and for oxidizing the alcohol through acetaldehyde to acetic acid. The efficiency of aeration determines the rate of conversion of alcohol to acetic acid. There are three basic methods: 1) the slow, open-vat method; 2) the trickle, generator method; and 3) submerged or bubble method.

Open-vat.—In the open-vat method, the substrate is placed into vats or barrels and inoculated with fresh vinegar or with mother of vinegar (a thick gelatinous skin formed during acetification). Since aeration is inefficient, the conversion may require several months. The system developed in Orleans, France was a slow process adapted from the older methods and is known as the Orleans process.

Generator.—In the quick process, the vinegar is manufactured in a generator in which the substrate is trickled over a packing material (beechwood shavings, coke, charcoal, ceramics, corn cobs) on which the *Acetobacter* is inoculated. The packing material allows a large surface area for the components to come together. The heat caused by the reaction results in air moving from the bottom through the generator and out of the top. When excess heat is produced, cooling of the generator system is needed. One pass of the substrate through the generator is not sufficient to convert all the alcohol to acetic acid, so the product is recirculated to complete the conversion. Not only is the generator method quicker than the vat method, but also higher grain vinegars (over 100 grain) may be obtained. The amount of acetic acid in the vinegar depends on the alcohol content of the substrate. Theoretically, 1.0% alcohol should yield 1.3% acetic acid. However, actual yields vary from 0.8% to 1.0%, due to inefficiencies and residual alcohol in the vinegar.

Bubble.—The submerged culture, or bubble method, was developed from the methods used for antibiotic production. In this system, air is pumped into the vat containing the stirred substrate (alcohol) and *Acetobacter.*

Besides acetic acid and water, vinegar contains residual ethanol, traces of ethyl acetate, fusel alcohols and their acids and traces of other compounds.

Biologically produced acetic acid in vinegar can be distinguished from that produced chemically from petroleum or coal by measuring the ^{14}C content with a scintillation counter (Kaneko *et al.* 1973). The basis is that fossil fuels have a lower ^{14}C content than recently produced carbon.

Yeasts with Molds

In the production of saké, molds are used to hydrolyze the starch of rice to carbohydrates fermentable by yeasts to alcohol. Although many microorganisms may be present and affect the reactions and product, the principal organisms are *Aspergillus oryzae* and a strain of *Saccharomyces cerevisiae.* Kodama (1970) discussed the manufacture of saké and the various organisms that might be involved.

Molds

The molds are used in many fermentations, especially in Oriental countries. The most readily recognized mold produced product is soy sauce.

Soy Sauce.—This is a dark brown liquid that is an important flavoring material for the normally bland rice diet of the Far East. Also, it is used on vegetables, meats, poultry and fish. Of all the fermented foods of Asian countries, soy sauce has received the greatest acceptance in the USA.

Soaked, steamed soybeans are mixed with roasted wheat in a ratio of about 2 to 1. This is inoculated with a mold koji. The koji is cooked rice that is inoculated with a species of *Aspergillus* and incubated for three to five days. The mold species that have been used are *A. oryzae, A. soyae* or *A. japonicus.* This mixture is stored until the mold covers the surface of the soybean-wheat mixture. Brine (1 kg salt/4 liters of water), is added at the rate of about 2 liters/kg of soybeans. This blend, or moromi, is fermented for 3-6 months, first at 15°C, then at 25°C. The mold enzymes hydrolyze the carbohydrates and proteins. Bacteria appear in the early stages of the fermentation and lower the pH from 6.7 to 5.0. These bacteria are in the genera *Pediococcus* and/or *Lactobacillus.*

In the later stages, the osmophilic yeast *Saccharomyces rouxii* ferments part of the sugars (from the hydrolysis of carbohydrates) to produce alcohol (about 2.5%). Besides *Saccharomyces*, other yeasts that may be involved include *Hansenula* and *Torulopsis.*

In the final phase of the fermentation, the pH is lowered to about 4.5 to 4.8. At this time, the fermented mash is filtered and the liquid obtained (soy sauce) is pasteurized at 80° to 85°C and bottled for use.

There are several variations in this procedure that are used for the production of soy sauce in Asian countries (Pederson 1979). The flavors in shoyu or soy sauce were discussed by Nunomura *et al.* (1976A,B).

Tempeh.—Tempeh is a fermented food characteristic of Indonesia. The product made from soybeans is called tempeh kedalee, while that made from coconut press cake is called tempeh bongkrek.

There are several methods for making tempeh. Briefly, for the soybean product, the beans are soaked, boiled and the hulls removed. The beans may be cooked, and the bean mash made into cakes, inoculated with tempeh mold, wrapped in banana leaves and fermented. There are several species of *Rhizopus* that can ferment the beans to a desirable product, but *R. oligosporus* is considered to be the tempeh mold. At a temperature of 30° to 40°C, the tempeh is ready to eat in 24-48 hr.

Tempeh has been described as having a bland flavor, is quite nutritious and has simple low cost processing techniques.

Miscellaneous Fermentations

There are certain fermentations in which individual or many microorganisms may be involved. Among these is the fermentation of glucose in egg white.

Egg White.—Egg white contains about 0.5% glucose, a reducing sugar. In commercially dried egg white (3 to 9% moisture), glucose will combine with amino acids to form brown insoluble products, due to the Maillard reaction.

The simple solution to this problem is to remove the glucose. There are several systems that can be used for this purpose. Many microorganisms ferment or oxidize glucose, so that it is not reactive with the amino acids. In a natural fermentation, the liquid albumen is allowed to remain at room temperature and the contaminants acquired during the breaking and separation of the egg white will eliminate the glucose. With this system, there may be undesirable organisms that degrade the protein, or those that are health hazards, such as the salmonellae, which also grow in the fermenting product.

Seeding of the albumen with specific organisms was used, so that there would be some semblance of a controlled fermentation. The organisms included coliforms (mainly *Escherichia coli* or *Aerobacter* [*Klebsiella*]), yeasts (species of *Saccharomyces* and *Candida*) and streptococci. Although these fermentations were acceptable, due to the lack of microbiologists in many egg products operations, control tended to suffer.

An enzyme isolated from *Aspergillus niger* is called glucose oxidase since, in its presence, glucose is oxidized to gluconic acid. The advantages of using this system are that the microbial load is not increased as in fermentation and the glucose is not lost, but remains in the product as gluconic acid. This means the product yield is greater. This enzyme system is discussed further in the section on enzymes.

ENZYME SYSTEMS

Enzymes are organic catalysts which allow reactions to occur under relatively mild conditions. Hence, they are well suited for use in the food industry. There are cases in which enzymes can be used more effectively than live microbial cells. When a series of reactions or several types of reactions are desired, such as in wine production, the intact microorganism is the system of choice, but if only one reaction is needed, separated enzymes may be beneficial. Also, some reactions or new products or processes may be possible only through enzyme activity.

Industrial enzymes are obtained from plant and animal tissues and from microorganisms. Since microorganisms can be grown in large amounts in controlled conditions, they offer an unlimited source of many enzyme systems.

Microorganisms produce intracellular and extracellular enzymes. Extracellular enzymes (carbohydrases and proteases) are easier to obtain than intracellular enzymes (glycolytic and tricarboxylic acid cycle enzymes). While extracellular enzymes can be obtained by filtering to remove the microbial cells, for intracellular enzymes, the cells must be broken up and the enzymes extracted with water or buffer to separate them from the cell debris.

On a comparative basis, enzymes are rather expensive. If they are added directly to a food, they are difficult or impossible to recover, so are used only once. Since enzymes are catalysts, they should be reusable. To accomplish this, processes have been developed for immobilizing enzymes on inert carriers. Immobilization may be accomplished by adsorption, chemical attachment (covalent bonding), microencapsulation (enclosing within a material), covalent crosslinking, entrapment or copolymerization. Various carriers or supports have been suggested for immobilized enzymes (Brotherton *et al.* 1976; Kawashima and Umeda 1976; Monsan *et al.* 1978; Morikawa *et al.* 1976). An enzyme sandwich was described by Vorsilak *et al.* (1975). Since both enzymes and substrates differ in various properties, no one system of immobilization is useful for all enzyme systems. Each of the immobilization methods has certain advantages and disadvantages, which were discussed by Stanley and Olson (1974). The problems concerned with immobilized enzymes include the reduction of enzyme activity upon insolubilization, loss of activity after repeated use, mechanical degradation of insoluble enzymes or its support after use and microbial growth with the enzyme acting as the substrate.

In general, the stability of enzymes is increased by immobilization. Although immobilized enzymes can be reused, the activity tends to decrease. In some systems, the enzyme becomes lost to the substrate solution, and the components of some substrates can block the active sites of the enzymes. Hence, there may be a limit as to how long the enzymes can be used.

Besides stability and reuse, other advantages of immobilization include using a continuous process rather than a batch operation, better control of the process, halting the reaction by removing the enzyme and less product inhibition. One can conceivably use immobilized enzymes in a series so that several reactions can be obtained during the treatment.

Immobilized enzyme systems are useful primarily for liquid foods. The physical and chemical characteristics of some liquid foods pose problems in using the enzymes. Thus, there are research opportunities in this field.

The processes for immobilizing enzymes have been adapted to immobilize microbial cells for fermentations (Ohmiya *et al.* 1977; Schnarr *et al.* 1977).

Some microbial enzymes that are useful in the food industry are listed in Table 9.3.

TABLE 9.3

SOME MICROBIAL ENZYMES USEFUL IN FOODS

Enzyme	Source	Application
Amylases	*Aspergillus, Rhizopus, Bacillus*	Conversion of starch to fermentable components in baking, brewing and syrup manufacture Clarification of fruit juices Scrap candy recovery Vegetable canning
Catalase	*Micrococcus, Aspergillus*	Decomposition of H_2O_2 in dairy and egg products
Cellulase	*Trichoderma, Sporotrichum, Chaetomium*	Conversion of cellulose to fermentable components Clarification of fruit juices and wine
β-Galactosidase (lactase)	*Kluyveromyces fragilis*	Conversion of lactose to glucose and galactose in dairy products
α-Glucosidase (maltase)	*Saccharomyces uvarum*	Conversion of maltose to glucose in brewing
Glucose isomerase		Conversion of glucose to fructose in corn syrup
Glucose oxidase	*Aspergillus, Penicillium*	Conversion of glucose to gluconic acid in liquid eggs Remove oxygen from juices or from head space of containers
Invertase	*Saccharomyces cerevisiae, Candida utilis*	Conversion of sucrose to glucose and fructose Prevents granulation in soft centered confections Used in production of artificial honey
Lipase	*Aspergillus, Rhizopus, Penicillium, Candida*	Conversion of fat to glycerol and fatty acids Flavor production in cheese Removal of egg yolk from egg white

TABLE 9.3. (*Continued*)

Enzyme	Source	Application
Pectic enzymes (pectinases, pectin esterases, polygalacturonase, polymethylgalacturonase)	*Aspergillus, Fusarium oxysporum, Corynebacterium, K. fragilis*	Clarification of wine and fruit juices; release of juice from fruit for increased yield, removal of pectin for concentrated fruit juices
Proteases (peptidases, rennin)	*A. oryzae, A. nidulans, Penicillium chrysogenum, B. subtilis, B. stearothermophilus*	Conversion of proteins to peptides and amino acids (improves dough in baking, meat tenderizing, milk clotting, enzymes for cheese, chillproofing of beer)

Amylases

These enzymes hydrolyze starch. Most starches are composed of amylose and amylopectin. Amylose is an unbranched polysaccharide with glucose units joined by 1,4-α-glucosidic linkages. Amylopectin is a branched polysaccharide with the glucose units as in amylose and 1,6-α-glucosidic linkages at the points of branching. The products of hydrolysis of starch depend upon the specific amylase that is present.

Commercial amylases are not pure and contain small amounts of maltase and other carbohydrases. In general, amylases can hydrolyze gelatinized starch more readily than raw starch.

The dextrins and oligosaccharides released from starch by the amylases are important in the color and texture of crust and the shelf life of bread. These enzymes can replace the diastatic malt in doughs where the flavor of malt is not needed.

Alpha Amylase (3.2.1.1).—These enzymes are obtained from *Aspergillus oryzae, Aspergillus niger, Rhizopus oryzae, Bacillus subtilis* and *Bacillus polymyxa.* They rapidly fragment starch to reducing dextrins. They do not hydrolyze the 1,6-α-glucosidic linkages of amylopectin.

The enzyme facilitates filtration by converting gelatinous starch to soluble dextrins. It clarifies fruit juices which are turbid due to starch. It is useful in the baking, brewing and alcohol industries, converting starch to fermentable components. By its action on cocoa starch, it helps ensure stability to chocolate syrup.

Beta Amylase (3.2.1.2).—These enzymes hydrolyze starch from the non-reducing ends of the glucosidic chains and release maltose. They do not attack the 1,6 linkages. Alone, the beta amylases are of little use. In combination with alpha amylases, they are used in the brewing, fermentation and baking industries.

Glucoamylase (3.2.1.3).—This is a glucose liberating amylase. Glucoamylase is used to convert partially hydrolyzed corn starch (treated with acid and alpha amylase) to corn syrup containing more than 90% glucose on a solids basis (Walton and Eastman 1973). They discussed the use of immobilized amylases for this conversion process.

The enzymes can be obtained from organisms such as *Aspergillus niger, A. oryzae, Rhizopus delemar, R. oryzae* and *R. niveus.*

Catalase (1.11.1.6)

Catalase is produced by many organisms. Commercially, it is obtained from a strain of *Aspergillus niger* or *Micrococcus lysodeikticus.* This enzyme catalyzes the reaction in which hydrogen peroxide is decomposed to water and oxygen. When hydrogen peroxide is added to milk for cold sterilization, catalase is used to remove the excess hydrogen peroxide. Catalase is present in commercial glucose oxidase preparations to release O_2 from H_2O_2 for the oxidation of glucose to gluconic acid, such as in the process for liquid egg white.

Cellulase (3.2.1.4)

Cellulase catalyzes the hydrolysis of the 1,4 linkages of cellulose to form dextrins and more usable sugars. Several fungi, including the thermophilic fungi, are a source of cellulase. The fungi include species or strains of *Myrothecium verrucaria, Stachybotrys atra, Trichoderma viride, Aspergillus niger, Chaetomium thermophile, Sporotrichum, Talaromyces, Thermoascus* and *Humicola. T. viride* is the most important commercial source of the enzyme.

Cellulase is used to remove the cellulose cloud and clarify citrus juices, to improve the body of beer, to aid in extraction of flavoring compounds (such as essential oils), to increase the digestibility and nutritive value of plant products and to treat cellulosic wastes for potential use as feed or food.

Beta Galactosidase (3.2.1.23)

The common name for this enzyme is lactase. It catalyzes the hydrolysis of lactose to glucose and galactose. This enzyme can remove the lactose from milk, making it available to people with lactose intolerance. Lactose is less soluble than the hydrolysis products. Therefore, treating milk with the enzyme improves the solubility of dried milk and the consistency of concentrated milk, ice cream bases and frozen milk. The relative sweetness of the hydrolysis products is higher than lactose, which is beneficial in many products.

Many bacteria, molds and yeasts produce the enzyme. The usual commercial source is the yeast *Kluyveromyces fragilis.* The molds, *Aspergillus niger* and *A. oryzae*, are sources of this enzyme.

Alpha Glucosidase (3.2.1.20)

This enzyme, also called maltase, hydrolyzes maltose to glucose. This is reportedly the rate-limiting step in beer processing. By using the enzyme, it is believed the rate of the brewing process can be increased. The enzyme is obtained from the yeast *Saccharomyces uvarum.*

Glucose Isomerase

Kooi and Smith (1972) discussed the conversion of glucose to fructose by means of glucose isomerase obtained from a strain of *Streptomyces olivochromogenes.* Other organisms producing isomerase include species of *Pseudomonas, Enterobacter, Lactobacillus* and *Bacillus.*

Glucose is obtained by hydrolysis of starch, such as corn starch. The isomerization of glucose to fructose is important since the relative sweetness of fructose is 2½ times that of glucose, and one of the main reasons for adding sugar to foods is to sweeten the product.

Glucose Oxidase (1.1.3.4)

This enzyme catalyzes the oxidation of glucose to gluconic acid. The excess hydrogen peroxide is then catalyzed to water and oxygen by the presence of catalase. Hence, the enzyme system is called the glucose oxidase-catalase system.

Glucose oxidase is obtained from *Aspergillus niger*, although other organisms, such as *Aspergillus oryzae, Penicillium glaucum* and *P. notatum* also produce this enzyme. Nakamatsu *et al.* (1975) reported that the best source of the enzyme was a strain of *Penicillium purpurogenum.*

This enzyme can be used to remove glucose from egg white, whole eggs or pork prior to dehydration. This improves the quality and shelf life of the dried product.

Since it is specific for glucose, the enzyme can be used as an analytical tool to estimate the amount of glucose in foods. This is especially useful in the estimation of glucose formed by the hydrolysis of lactose or starch in the food industry. Also, the assay is used clinically for determining glucose in blood and urine.

Since the enzyme uses oxygen to react with glucose, it can be employed as an oxygen scavenger for food systems that deteriorate in an oxygen atmosphere. The removal of oxygen by this system may increase the shelf life or aid in retaining the quality of foods.

Immobilization of the glucose oxidase enzyme system was discussed by Atallah and Hultin (1977).

Invertase (3.2.1.26)

Invertase catalyzes the hydrolysis of sucrose to fructose and glucose. The resultant invert sugar is sweeter than sucrose.

Although several microorganisms possess the ability to hydrolyze sucrose, the yeasts have been used as a source of invertase. The yeasts include *S. cerevisiae, S. rouxii, K. fragilis* and *Candida utilis.*

Invertase can be used whenever it is desirable to have sucrose hydrolyzed. It is used in confections to convert solid sucrose mixtures to a fluid consistency after covering with a substance such as chocolate, to produce soft-centered confections.

Limonoate Dehydrogenases

Limonin is a bitter component in citrus products. If this bitter principle could be controlled, there might be an increased demand for citrus products. Brewster *et al.* (1976) compared the limonoate dehydrogenases of an *Arthrobacter* and a *Pseudomonas.* The activity of the enzyme from *A. globiformis* was not stable at pH 3.5, while that of the pseudomonad was relatively stable at this pH. The enzyme from the pseudomonad is inactivated at 50°C (Hasegawa 1976).

Lipases (3.1.1.3)

Lipase acts on lipids hydrolyzing them to glycerol and free fatty acids. The application of microbial lipases was reviewed by Seitz (1974). Lipolytic microorganisms produce lipases. The commercial source includes *Aspergillus niger* and *A. oryzae*, as well as *Candida lipolytica, C. cylindracea, Penicillium roqueforti, Rhizopus delemar, Mucor pusillus* and *Torulopsis ernobii.*

Lipases isolated from microorganisms have been used to develop flavor in cheese or cheese-like foods. Lipases have been suggested for treating milk fat for use in ice cream, margarine and butter. These enzymes could be used to improve the flavor of certain foods and to eliminate yolk fat in egg white and improve its whipping properties.

Pectic Enzymes

These enzymes hydrolyze pectin and pectic substances to lower molecular weight compounds. Various names have been used for these enzymes. The usual enzyme system is pectinase, which generally is a mixture of polygalacturonase (3.2.1.15) and pectic methylesterase (3.1.1.11). Pectic lyase (4.2.2.3) and pectin trans-eliminase (4.2.99) were suggested as clarifying agents for fruit juices (Ishii and Yokotsuka 1973).

Pectic methylesterase causes demethylation, while the polygalacturonase catalyzes the hydrolysis of α-1-4-galacturonide bonds of pectin. The release of methyl groups exposes carboxyl groups, which, in the presence of calcium or other multivalent ions, form insoluble salts which can be removed.

The commercial sources are *Aspergillus niger* and *Rhizopus oryzae;* however, several other fungi, such as *Aspergillus soyae, A. japonicus, A. wentii, Penicillium glaucum* and *P. expansum,* may be used.

Pectin remaining in fruit juices tends to hold other substances in suspension as a colloidal system. By hydrolyzing the pectin, the protective colloidal action is destroyed, so that suspended substances will settle out and can then be removed from the juice by filtration.

The addition of pectic enzymes to fruits during crushing or grinding increases the yield of pressed juice. According to Codner (1971), pectic enzymes are needed to express the juice from black currants.

When fruit juices are concentrated, it is necessary to remove the pectin so that a liquid rather than a gel is obtained. For jams and jellies, it is desirable to remove the natural pectin so that a standardized amount of pectin can be added to make uniform products.

Proteases

This is a group of enzymes that catalyze the hydrolysis of peptide bonds in proteins and yield peptides of lower molecular weight. Most proteases split specific peptide linkages. Hence, it is necessary to select the appropriate protease or combination of enzymes so that the desired reaction is obtained.

The usual commercial source of proteases is *Aspergillus oryzae* and *Bacillus subtilis.* However, proteases can be obtained from any proteolytic organism. Some organisms that have been used include *Aspergillus niger, Streptomyces griseus* and species of *Rhizopus.*

The applications of proteases include the chillproofing of beer, meat tenderization, the production of protein hydrolysates, to mellow the dough in bakery products, the production of fish protein solubles and to improve protein recovery from oil seed meals.

Fujimaki *et al.* (1971) used an acid protease from *Aspergillus saitoi* to liberate the odor producing substances from soy protein concentrate. They found the resultant hydrolysates were bitter, but a treatment with carboxypeptidase eliminated the bitterness.

In baked products, the use of proteases in the sponge dough reduces mixing requirements. The treated doughs are more extensible and can be molded more easily than untreated dough. Proteases are of value in controlling the pliability, eliminating buckiness and assuring proper machinability of the dough, as well as increasing loaf volume.

Proteases are added to beer or ale during the finishing operation to prevent an undesirable haze when these beverages are cooled. The hydrolysis of high molecular weight proteins prevents the formation of the haze. Bliss and Hultin (1977) suggested the use of immobilized protease to inactivate enzymes in foods.

Milk-clotting Enzymes

In cheese manufacture, rennet is added to the vat of milk during lactic acid fermentation to aid in the clotting process. The usual source of rennet has been the fourth stomach of suckling calves. Due to the lack of sufficient calves to supply the demand for rennet, other sources have been investigated. Although most proteolytic enzymes can cause clotting, in some cases, the proteolytic activity is such that the cheese curd is digested. Hence, the best enzymes are those that have high clotting activity and low proteolytic activity.

The microorganisms used for commercial production of milk clotting are *Endothia parasitica, Mucor miehei* or *M. pusillus.* Other organisms from which milk-clotting enzymes have been isolated and studied include *Byssochlamys fulva* and various species of *Bacillus, Aspergillus* and *Penicillium.*

The coagulation of milk with immobilized protease enzymes was reviewed by Taylor *et al.* (1976).

CHEMICALS

Some organisms do not participate in food fermentations directly, but are used to produce chemicals that are added to foods. Such chemicals as vitamins and amino acids upgrade the nutrient value of food, and gums and dextrins improve the physical characteristics of foods. Demain (1971) listed alcohols, amino acids, antibiotics, antioxidants, coloring agents, enzymes, nucleotides, organic acids, plant growth regulators, polyols, polysaccharides, protein, sugars and vitamins as fermentation products used in the food or feed industries.

Amino Acids

Amino acids can be produced by chemical synthesis, but both optical isomers are obtained. In contrast, the synthesis by microorganisms results in the L-isomer which is the form used by biological systems. The microbial sources of certain amino acids are shown in Table 9.4.

Lysine and methionine are important as additives to plant proteins to increase the nutrient value. Glutamic acid is used in the production of monosodium glutamate, which is a flavor enhancer in many foods.

Other Organic Acids

Various organic acids are added to foods. The main supply of citric acid was lemons, but now it is obtained by fermentation of sugars with *Aspergillus niger.* Acetic and lactic acids are discussed in regard to

TABLE 9.4

SOME MICROBIAL SOURCES OF AMINO ACIDS

Amino Acids	Microbial Sources
Alanine	*Corynebacterium, Brevibacterium*
Aspartic acid	*Escherichia, Serratia, Bacillus, Pseudomonas*
Glutamic acid	*Corynebacterium, Brevibacterium, Micrococcus, Microbacterium, Bacillus, Pseudomonas, Streptomyces, Aeromonas, Flavobacterium*
Lysine	*Micrococcus, Bacillus, Escherichia, Torulopsis, Saccharomyces, Proteus, Erwinia, Candida, Corynebacterium*
Methionine	*Pseudomonas, Rhodotorula, Micrococcus*
Ornithine	*Micrococcus*
Phenylalanine	*Escherichia, Micrococcus*
Threonine	*Escherichia*
Tyrosine	*Micrococcus*
Tryptophan	*Claviceps, Serratia, Micrococcus, Rhizopus, Hansenula, Candida, Escherichia*
Valine	*Enterobacter, Escherichia, Klebsiella, Micrococcus, Brevibacterium*

vinegar production and vegetable and dairy fermentations. In lactic acid production, *Lactobacillus delbrueckii* is used since the reaction can be accomplished at 50°C. This high temperature suppresses many other organisms (psychrotrophs and mesophiles). Some microbial sources of organic acids are listed in Table 9.5.

Microbial Gums

Gums are used in the food industry in aqueous systems when there are problems involving suspension, thickening, emulsion stabilization and rheological modifications. One of the sources of gums is microorganisms. Several microbial gums have been produced and tested. One that is available commercially is xanthan gum. It is an optional emulsifying and stabilizing ingredient in French dressing and a stabilizing ingredient in certain cheeses and cheese products.

In the production of xanthan gum, *Xanthomonas campestris* is grown in a well aerated medium containing glucose, a nitrogen source, dibasic potassium phosphate and trace elements. The gum is produced as an exocellular polysaccharide coat surrounding the cell wall.

TABLE 9.5

SOME MICROBIAL SOURCES OF ORGANIC ACIDS

Acid	Microorganism
Acetic	*Acetobacter*
Citric	*Aspergillus, Penicillium, Trichoderma*
Gluconic	*Penicillium*
Lactic	*Lactobacillus delbrueckii*
Propionic	*Propionibacterium*
Pyruvic	*Pseudomonas*
Succinic	*Brevibacterium*

Vitamins

Vitamins are added to various foods for enrichment. Only a few vitamins are produced by fermentation. Riboflavin, produced chemically, is used in food, while that produced by fermentation is used in feeds. Some microbial sources of vitamins are listed in Table 9.6.

Flavoring Compounds

Autolyzed yeast extract is used as a flavor substitute for meat extracts. Flavor enhancing nucleotides can be produced by strains of streptomycetes.

MICROBIAL PROTEINS

Due to the potential increase in population and the limited supply of

TABLE 9.6

SOME MICROBIAL SOURCES OF VITAMINS

Vitamin	Microorganism
Ascorbic acid	*Pseudomonas*
β-Carotene	*Rhodotorula*
B_{12}	*Bacillus, Propionibacterium, Streptomyces, Pseudomonas, Escherichia, Nocardia, Rhizobium*
Riboflavin	*Aspergillus niger, Eremothecium ashbyii, Ashbya gossypii, Candida, Streptomyces*

land for agriculture, alternative sources of food have been investigated. One of these sources is microbial cells. Various names, such as novel protein, unconventional protein, minifoods, petroprotein and single cell protein (SCP), have been used for these products. Actually, the products are microbial cells or microbial proteins. Microbial protein (MP) is used in animal feed more than directly as human food.

Either live or dead microorganisms are present in nearly all of the food that we eat. In some products, such as yogurt, the presence of certain microorganisms is used as a sales gimmick. Thus, the use of microorganisms for food should not be objectionable if they are not pathogenic and produce no harmful toxins.

Advantages

Microorganisms have many advantages over other types of food sources. Microorganisms do not require much space; they do not compete for the arable land that is available; they can be genetically manipulated to produce certain types of products; they have a rapid rate of growth and, since growing conditions can be easily controlled, there are no environmental problems as in normal agriculture, such as freezing, floods or drought. Besides these advantages, microorganisms can be used to digest and convert wastes into useful products. This also reduces the problems of waste disposal.

Substrates

Various substrates ranging from CO_2 with sunlight for algae to complex substances for bacteria, yeasts and molds, have been proposed for the production of microbial protein. These substances are derived from petroleum, natural gas, industrial products, agricultural crops and various wastes.

According to Calam and Russell (1973), to be economical, the minimum capacity of a processing plant should be 50,000 tons of SCP/year. This means that an adequate supply of substrate is needed for microbial growth and SCP production. Petroleum and natural gas products, such as kerosene, fuel oil, gas oil, n-alkanes, methanol and ethanol were considered desirable since they were readily available in an apparently unlimited supply. Due to the recent petroleum and natural gas shortages and the increasing prices, some of these substrates now may seem less attractive. There are certain problems in utilizing fuel oil, gas oil and n-alkanes. Since they are not miscible with water, the mixture has to be constantly agitated to keep the cells in contact with the substrate. Also, a high volume of oxygen is needed and ammonium salts or nitrates must be added as a nitrogen source for the microbial cells. About one-half of the hydrocarbon is converted to cells, the rest being oxidized to carbon dioxide and water.

With the known petroleum sources estimated to be exhausted in the foreseeable future, we might question the use of petroleum products to produce SCP.

Other problems with petroleum usage include the need to remove the excess heat due to the oxidation of the hydrocarbons and also the separation of the cells from substrates, such as fuel oil or gas oil. There are potentially carcinogenic polycyclic hydrocarbons in oil-based fermentations.

Methanol is regarded as a very promising substrate. This alcohol can be produced from natural gas that would normally be burned at remote oil wells. Also, it can be produced from coal or wood. Methanol is relatively cheap, there is an ample supply, it is essentially pure (99.5 to 99.8%), it is miscible with water, less oxygen is required, less heat removal is needed, it is easily stored and handled, contamination is minimized due to restrictive use by only a few microorganisms and the cells are readily separated from this substrate (Cooney *et al.* 1975; Cooney and Makiguchi 1977).

Ethanol can be obtained from ethylene, a petroleum product, or by fermentation of carbohydrates. The use of ethanol as a substrate is similar to methanol (Laskin 1977).

Methane, derived from fossil fuels or by degradation of waste products, can be used as a substrate. The process was discussed by Wilkinson (1971).

Some agricultural crops which are primarily carbohydrate could be used as substrates to produce SCP, and this protein could then be used to upgrade the nutritional value of the food crop.

The disposal of our wastes is a problem which is increasing every year.

If our wastes could be used to produce a usable product such as SCP, we might be solving two problems, waste disposal and a food source.

For the USA, organic agricultural wastes have been estimated at over a billion metric tons annually (Bellamy 1974). The annual world production of straw is estimated at a billion tons. Generally only ⅓ to ½ of the agricultural crop is harvested and the remainder is usually left in the field. This residue is often a harborage for pests and plant pathogens. For some vegetables, such as peas, the entire plant is hauled to the processing plant and, after the peas are removed mechanically, the pea vines accumulate and may become a disposal problem.

Crop residues, such as corn stalks and cobs, straw, pea and bean vines and other vegetable processing wastes contain from 30 to 60% cellulose, 10 to 30% hemi-celluloses, 5 to 20% ash, 4 to 18% lignin and relatively little amounts of protein and fat.

Urban and industrial organic wastes accumulate at about 200 million metric tons annually. Some of these wastes are not biodegradable and others may contain toxic chemicals which make them unsuitable as substrates.

The suitability and economics of these wastes depend upon their availability (the amount and location), composition, biodegradability, treatment(s) needed, presence of toxic materials (including pesticides) and various costs (product gathering, transportation and storage).

Some of the wastes that have been considered as potential substrates are animal feed-lot wastes, straw, sugar cane bagasse and sewage. Various wastes from food processing plants have been investigated as potential substrates for SCP production. Some of these food processing wastes are seasonal and diluted with water. Hence, there are costs to concentrate the dilute wastes and accumulate enough to use over the entire yearly operation of a processing plant.

Some wastes, especially cellulosic wastes, need to be treated with acid or alkali so that they can be used by microorganisms. A source of nitrogen (ammonium salts or nitrates) and minerals such as calcium, iron, phosphorus and magnesium, need to be added for acceptable growth of microorganisms. Since these different types of wastes vary in their composition, it is difficult to establish a plant and procedure to utilize all of them. Only a few accumulate in a volume capable of supplying an SCP facility and even then, the gathering and transportation costs must be considered.

Microorganisms

All types of microorganisms (bacteria, yeasts, molds, higher fungi, algae and protozoa) have been suggested as potential sources of protein. For economic, as well as other reasons, the organisms should be able to grow on simple media (preferably waste products) which are inexpensive; it is helpful if they can grow in a continuous culture; they should be easily separated at harvest; pure cultures with known genetic factors and means of altering these factors should be available; there should be no toxic or allergenic factors in the culture; the product should be palatable or readily disguisable, contain high quality protein and be easily packaged and stored; and the remaining effluent should be of a quality that is easily disposed of, with little or no further treatment.

There are certain advantages and disadvantages for using each type of microorganism in this process. Bacteria have the fastest growth rate, but being the smallest, they are more difficult to harvest. Molds are large and are the easiest to harvest, but they have a slower growth rate than either bacteria or yeasts. The yeasts are easier to harvest than are the bacteria.

The type of microorganism, as well as the species of microorganism, used in the process depends upon the substrate, conditions for growth, equipment and other factors.

Yeasts.—As a food for humans, the yeasts have been more apparent than have the other microorganisms. There is a precedent for using yeast. Yeasts have been used as a food source since workers in Germany developed the process during World War I. The first successful commercial process of yeast production was from molasses and ammonia. Since this beginning, several types of substrates have been used to grow yeasts for feed and food. Yeast extracts are used in soups, sauces, relishes and some processed meats.

Although some yeasts can multiply by a sexual process of spore fusion, under most circumstances, they reproduce by budding. The time required for a mother cell to produce a daughter (the doubling time) for most yeasts with favorable growing conditions varies from one to three hours. Cooney *et al.* (1975) cited references showing that for three species of yeast on a methanol substrate, the doubling time varied from three to nine hours.

Various yeasts have been utilized as potential food sources, including species of *Candida, Saccharomyces, Rhodotorula, Torulopsis, Hansenula, Kluyveromyces, Debaryomyces, Pichia* and *Kloeckera.*

Besides the hydrocarbons and alcohols, various wastes, such as paper mill (sulfite waste liquor), dairy, potato starch, sauerkraut, soybean, spent solubles, cannery and citrus have been used to propagate yeast.

Yeasts can be used for producing vitamins, enzymes, high-quality protein and other physiological essentials. Since yeasts can be bred for selective purposes, it is possible to obtain special yeasts which are richer than ordinary yeasts in certain nutritional components, such as high protein yeasts.

In considering the nutritive value of yeast, the proximate analysis of the yeasts is usually published. The protein content can vary from 30 to 65% depending upon the yeast and substrate. The amino acid content of these proteins varies considerably. In general, the amino acid analysis of the yeast protein compares favorably with animal protein, except that the methionine content is low. By adding methionine to yeast protein, the protein efficiency ratio (PER) of 0.9 to 1.4 is almost doubled to 2.0 to 2.5. Although this PER is still lower than whole egg (PER 3.0), it is similar to casein (PER 2.5). According to Shacklady (1972), the biological value (BV) of dried whole egg is 90, while that of yeast protein varied from 46 to 61. When 0.3% methionine was added to the yeast protein, the BV increased to values of 91 to 96, indicating the supplemented yeast compares favorably to plain dried whole egg, or dried skim milk (BV of 87). The relatively high lysine content of yeast would improve the nutrient value if added to grains such as corn, wheat or rice. It would not be as useful to add yeast to soybeans, since they too are low in methionine. Macmillan and Phaff (1973) listed *Rhodotorula gracilis* as a source of methionine.

One problem with SCP in general is the high nucleic acid content (6-16%) of microbial cells. The end product of purine metabolism is uric acid. Although we can tolerate low levels, when the daily nucleic acid intake is over 2 g, the uric acid, being insoluble at body pH levels, can cause conditions such as gout or kidney stones. Methods for reducing or removing the nucleic acids from yeast protein have been reported (Tajima and Yoshikawa 1975; Trevelyan 1976; Zee and Simard 1975) and reviewed by Litchfield (1977).

In a review, Harrison (1970) reported that ingestion of live yeast cells is not desirable, since they can remove vitamins and other nutrients from the digestive tract. Whole dead cells are not easily digested because of the cell wall. With this inability to digest the cells, the protein is not usable. To make the protein available, the cells can be disintegrated mechanically or by enzymes.

DeGroot *et al.* (1975) fed yeasts to rats over a two year period and found no adverse effect on mortality, rate of body weight gain, hematology, kidney function, fertility, histological changes or incidence of tumors. Sinai *et al.* (1974) found an increased resistance to respiratory diseases when monkeys were fed brewers' yeast. They believed the yeast stimulated phagocytosis.

Chickens fed 5-10% yeast diets were significantly lighter than those on a control diet (Shannon *et al.* 1976). This was thought to be due to less feed intake when yeast was incorporated.

Yeast, per se, has not replaced meat proteins in the human diet, primarily on the basis of flavor and other quality attributes. Dried yeast has very few functional properties, but yeast protein has several. The refined proteins can be spun or otherwise processed to function in various ways and can be added to foods in the same manner as vegetable proteins (Hayakawa and Nomura 1978; Huang and Rha 1978A).

Bacteria.—The use of bacterial cells or protein as food lags behind yeasts. However, they have been consumed in fermented foods such as yogurt.

Generally, bacteria multiply faster than do yeasts. Most bacterial SCP has a higher level of protein and a higher BV than yeast SCP. Although bacteria, being smaller, are more difficult to recover than are yeasts, there are methods using flocculation, filtration and centrifugation that allow recovery, but at added expense.

The type of bacteria that is used depends upon the substrate. Species of *Pseudomonas* are grown on substrates of methanol, kerosene, fuel oil or gas oil (Lipinsky and Litchfield 1974). Abbott *et al.* (1973, 1974) described the growth of *Acinetobacter calcoaceticus* on ethanol. They stated that the composition of an organism is strongly dependent upon its growth rate and environment. They obtained a higher protein content of the cells when the growth rate was increased. The fermentation of rice straw with a species of *Cellulomonas* and *Alcaligenes faecalis* was discussed by Han (1975) and Han and Callihan (1974). It was suggested that the microbial protein could be used for nonruminant feeding and the residue could be fed to ruminants.

Rhodopseudomonas gelatinosa is a photosynthetic bacterium utilizing CO_2 and sunlight. *Thermomonospora* and *Thermoactinomyces* are thermophilic bacteria that produce protein on cellulose substrates. *Methylococcus* and *Hyphomicrobium* species utilize methane. *Hydrogenomonas* can assimilate hydrogen and carbon dioxide for SCP production (Bellamy 1974; Litchfield 1977).

The dry weight of bacterial cells is 47-86% protein. The nitrogen digestibility and weight gain of rats were improved when cells of *Pseudomonas* were homogenized before feeding (Yang *et al.* 1977). At levels of 10-20% of the diet, methanol-grown bacterial cells resulted in reduced growth rates and other adverse effects on young chicks (D'Mello and Acamovic 1976).

Besides the problem of harvesting, the nucleic acid content of bacterial cells appears to be higher than that of yeast cells. Nucleic acid content up to 20% was suggested (Shacklady 1972). This can be reduced with known processing systems. The feeding of bacterial cells to humans has not been successful due to the development of gastrointestinal problems.

Molds.—Molds are complex, multicellular aerobic organisms. They grow over a wide range of pH, temperature and substrates. They contain the vitamin B complex and from 13 to 60% crude protein. The methionine content of mold protein is rather low, as is true for yeast protein. Also, the tryptophan and cystine content is low. For other amino acids, it compares favorably with milk and fish meal proteins.

Although molds are the easiest of the microorganisms to separate from the substrate, they have been suggested for use to upgrade various agricultural wastes for feeding to animals. There has been little effort in using mold protein as human food.

The cellulolytic fungus, *Trichoderma viride* and other *Trichoderma* have been grown on feedlot wastes, cereal straws and other cellulosic wastes (Griffin *et al.* 1974; Han and Anderson 1975; Peitersen 1975A). Other molds grown on various substrates include *Sporotrichum pulverulentum, Fusarium moniliforme, Pleurotus, Agrocybe, Aureobasidium, Stropbaria, Aspergillus, Alternaria, Cephalosporium* and *Rhizopus* (Eriksson and Larsson 1975; Gregory *et al.* 1976; Han *et al.* 1976; Macris and Kokke 1977; Willetts 1975; Zadrazil 1977).

Mushrooms.—Higher fungi, or mushrooms, have been known as a source of food since the history of man began. Their efficiency of converting carbohydrates to protein is approximately 65%. As with other fungal protein, there is a low content of methionine and tryptophan in mushrooms.

Mixed Cultures.—In nature, mixed cultures are present during the degradation of cellulosic and other wastes. In these wild, mixed cultures, the organisms consume each other, so that a large amount of protein does not accumulate. However, in a controlled mixed culture, two organisms may grow better and produce more protein.

Mixed cultures of *Trichoderma viride* and a yeast (*Candida utilis* or *Saccharomyces cerevisiae*) were used with alkali treated barley straw (Peitersen 1975B). The overall rate of protein production was increased by use of the mixed culture.

Algae.—The algae are the simplest plants that contain chlorophyll. Their use as a potential source for food should be readily apparent, since they begin the food chain for sea life (Ryther and Goldman 1975).

The algae vary in size from microscopic forms to the giant seaweed or kelp that may be 40 or more feet long. The familiar habitat of algae is any body of sunlit water, ponds, streams, lakes or oceans. There are also terrestrial algae.

Since algae contain chlorophyll, they can utilize solar energy by photosynthesis, like the higher plants, to produce food from CO_2, H_2O and inorganic salts. They have many of the advantages listed for the other microorganisms. Generally, the algae have a slower growth rate than other microorganisms. The algae require light for growth, so either artificial light is needed or growth is limited to areas with warm sunny climates. When grown outdoors in ponds or on sewage, as has been suggested, there is the problem of contamination with undesirable bacteria. *Spirulina maxima* is grown successfully in the naturally alkaline water of Lake Texcoco in Mexico.

Algae have been harvested and used by the people in the Chad Republic of Africa for centuries, with no apparent ill effects. The red algae have been harvested for their agar and carrageenin. The brown algae, *Laminaria* and *Macrocystis,* are sources of algin. Agar is used for solidifying media, as well as in gels for dessert. Carrageenin and algin are added to food products to improve the texture. A red alga, *Porphyra*, has been used as a food in Japan.

The protein concentration of dried algae varies from 5 to 64%. Cells of *Chlorella* contain about 60% protein (Endo *et al.* 1974). The different levels are due to differences in algae and in growth conditions. The PER of algal protein ranges from 1.25 to 2.6 and the BV from 54 to 72. The analysis of the proteins of algae shows that they are low in the sulfur-containing amino acids (cystine and methionine).

There have been toxic effects noted due to certain algae. The toxic effect of "red tide" on the killing of fish is well known. Some species of blue-green algae have caused the death of animals. Shilo (1967) reviewed algal toxins, but no toxins were discussed for the microalgae, *Chlorella, Scenedesmus* or *Spirulina*. Humans can tolerate some algal protein, but if the daily intake of algae exceeds 100 g/day, there may be gastrointestinal distress.

For good digestibility, it is necessary that the cells be broken. For human acceptance, the protein will need to be purified and processed into other foods.

Due to the many difficulties in growing, processing and use in human foods, other microorganisms show more promise as potential sources of protein than do algae.

The Potential of Microbial Protein

Many people are optimistic about the future prospects of microbial protein. With improved processing techniques, the potential will, no doubt, also increase. The main use of SCP is in animal feed, but some has been used in human food and this use should increase in the future.

The two main concerns are the safety and cost. As with any situation or product, there is always a risk factor. Feeding trials with rats and other animals indicate that the product is safe. However, for human consumption, it is necessary to reduce the nucleic acid content of the product. Gastrointestinal disturbances have occurred in humans fed certain SCP. Until problems such as these are solved, only low levels of SCP can be used as human food.

The cost of producing SCP is such that it may not be able to compete with other plant proteins, but it can compete with many animal proteins. The nutritive value of SCP is generally lower than animal proteins, but the SCP is improved by processing the cells, isolating the protein and supplementing it with the limiting amino acids.

If wastes, which cost money for disposal, can be used to produce SCP, we might consider the savings in the disposal cost in the overall economics of the SCP. Since SCP can be spun into fibers like plant proteins, it could be used to make synthetic meats, or added as extenders of meat products.

With our present and expanded future needs for protein, the use of microorganisms as a source of feed or food cannot be overlooked.

TABLE 9.7

ASSAY OF VITAMINS

Vitamin	Microorganism
Biotin	*Lactobacillus plantarum* *Saccharomyces cerevisiae*
Choline	*Neurospora crassa* (mutant)
B_6 (pyridoxine)	*Saccharomyces uvarum* *Neurospora sitophila* (mutant)
Nicotinic acid	*Leuconostoc mesenteroides* *Lactobacillus plantarum*
B_{12}	*Lactobacillus leichmannii* *Escherichia coli* (mutant) *Euglena gracilis* *Ochromonas malhamensis*
Thiamin	*Lactobacillus viridescens*
Pantothenic acid	*Saccharomyces uvarum* *Pediococcus acidilactici* *Lactobacillus plantarum*
Folic acid	*Streptococcus faecalis* *Lactobacillus casei*
Riboflavin	*Lactobacillus casei* *Lactobacillus helveticus*
Inositol	*Neurospora crassa* (mutant) *Saccharomyces uvarum*

ASSAY

Microorganisms can be used to determine or assay the amount of vitamins and amino acids in food products.

The use of living cells for analytical purposes gives a high degree of sensitivity as well as biological specificity due to the particular responses of the metabolic processes of the cell. The reagents are the test organism and the medium used for growth. The reaction is the metabolic response, or lack of response, by the cell. By using serial dilutions of the sample in the growth medium, inoculating with the test organism, incubating and determining the response of the cell, we can determine the amount of substance being assayed. A semiautomatic system for vitamin assay was described by Berg and Behagel (1972), and an automated rapid assay for amino acids was discussed by Itoh *et al.* (1974).

Some of the microorganisms and vitamins assayed are listed in Table 9.7. The test microorganisms and corresponding amino acids are listed in Table 9.8. Special strains, or mutants of the microorganisms, are used for these assays.

TABLE 9.8

ASSAY OF AMINO ACIDS

Amino Acid	Microorganism
Arginine	*Streptococcus faecalis*
Aspartic acid	*Leuconostoc mesenteroides*
Cystine	*Leuconostoc mesenteroides*
Glutamic acid	*Lactobacillus plantarum*
Glycine	*Leuconostoc mesenteroides*
Histidine	*Leuconostoc mesenteroides*
Isoleucine	*Lactobacillus plantarum*
Leucine	*Lactobacillus plantarum*
Lysine	*Leuconostoc mesenteroides*
Methionine	*Leuconostoc mesenteroides*
Phenylalanine	*Leuconostoc mesenteroides*
Proline	*Leuconostoc mesenteroides*
Serine	*Leuconostoc mesenteroides*
Threonine	*Streptococcus faecalis*
Tryptophan	*Lactobacillus plantarum*
Tyrosine	*Leuconostoc mesenteroides*
Valine	*Lactobacillus plantarum*

Besides vitamins and amino acids, microorganisms can be used to assay materials for antibiotics, some types of pesticides and bacteriolytic enzymes, such as lysozyme.

Various microbial enzymes can be used to determine chemicals in clinical specimens and other material (Fishman and Schiff 1976). Butane-2,3-diol dehydrogenase from a strain of *Sarcina* was used to analyze for butanediol and acetoin in wine (Masuda and Muraki 1975; Muraki and Masuda 1976).

BIBLIOGRAPHY

ABBOTT, B.J., LASKIN, A.I., and MCCOY, C.J. 1973. Growth of *Acinetobacter calcoaceticus* on ethanol. Appl. Microbiol. *25*, 787-792.

ABBOTT, B.J., LASKIN, A.I., and MCCOY, C.J. 1974. Effect of growth rate and nutrient limitation on the composition and biomass yield of *Acinetobacter calcoaceticus.* Appl. Microbiol. *28*, 58-63.

ACTON, J.C., and DICK, R.L. 1975. Improved characteristics for dry, fermented turkey sausage. Food Prod. Develop. *9*, No. 8, 91-94.

AMERINE, M.A., BERG, H.W., and CRUESS, W.V. 1972. The Technology of Wine Making, 3rd Edition. AVI Publishing Company, Westport, Conn.

ARNOTT, D.R., DUITSCHAEVER, C.L., and BULLOCK, D.H. 1974. Microbiological evaluation of yogurt produced commercially in Ontario. J. Milk Food Technol. *37*, 11-13.

ATALLAH, M.T., and HULTIN, H.O. 1977. Preparation of soluble conjugates of glucose oxidase and catalase by cross-linking with glutaraldehyde. J. Food Sci. *42*, 7-10.

BELLAMY, W.D. 1974. Single cell proteins from cellulosic wastes. Biotechnol. Bioeng. *16*, 869-880.

BERG, T.M., and BEHAGEL, H.A. 1972. Semiautomated method for microbiological vitamin assays. Appl. Microbiol. *23*, 531-542.

BLISS, F.M., and HULTIN, H.O. 1977. Enzyme inactivation by an immobilized protease in a plug flow reactor. J. Food Sci. *42*, 425-428.

BREWSTER, L.C., HASEGAWA, S., and MAIER, V.P. 1976. Bitterness prevention in citrus juices. Comparative activities and stabilities of the limonoate dehydrogenases from *Pseudomonas* and *Arthrobacter.* J. Agr. Food Chem. *24*, 21-24.

BROTHERTON, J.E., EMERY, A., and RODWELL, V.W. 1976. Characterization of sand as a support for immobilized enzymes. Biotechnol. Bioeng. *18*, 527-543.

BUCHANAN, R.E., and GIBBONS, N.E. 1974. Bergey's Manual of Determinative Bacteriology, 8th Edition. Williams and Wilkins, Baltimore, Md.

CALAM, C.T., and RUSSELL, D.W. 1973. Microbial aspects of fermentation process development. J. Appl. Chem. Biotechnol. *23*, 225-237.

CODNER, R.C. 1971. Pectinolytic and cellulolytic enzymes in the microbial modification of plant tissues. J. Appl. Bacteriol. *34*, 147-160.

COONEY, C.L., LEVINE, D.W., and SNEDECOR, B. 1975. Production of single-cell protein from methanol. Food Technol. *29*, No. 2, 33, 36, 38, 40, 42.

COONEY, C.L., and MAKIGUCHI, N. 1977. An assessment of single cell protein from methanol-grown yeast. Biotechnol. Bioeng. Symp. No. 7, 65-76.

DEGROOT, A.P., DREEF-VAN DER MEULEN, H.C., TIL, H.P., and FERON, V.J. 1975. Safety evaluation of yeast grown on hydrocarbons. IV. Two-year feeding and multigeneration study in rats with yeast grown on pure *n*-paraffins. Food Cosmet. Toxicol. *13*, 619-627.

DEMAIN, A.L. 1971. Microbial production of food additives. Symposia Soc. Gen. Microbiol. *21*, 77-101.

D'MELLO, J.P.F., and ACAMOVIC, T. 1976. Evaluation of methanol-grown bacteria as a source of protein and energy for young chicks. Brit. Poultry Sci. *17*, 393-401.

DRIESSEN, F.M., UBBELS, J., and STADHOUDERS, J. 1977. Continuous manufacture of yogurt. I. Optimal conditions and kinetics of the prefermentation process. Biotechnol. Bioeng. *19*, 821-839.

DUTTA, S.M., KUILA, R.K., ARORA, B.C., and RANGANATHAN, B. 1972. Effect of incubation temperature on acid and flavor production in milk by lactic acid bacteria. J. Milk Food Technol. *35*, 242-244.

ENDO, H., NAKAJIMA, K., CHINO, R., and SHIROTA, M. 1974. Growth characteristics and cellular components of *Chlorella regularis*, heterotrophic fast growing strain. Agr. Biol. Chem. *38*, 9-18.

ERIKSSON, K.E., and LARSSON, K. 1975. Fermentation of waste mechanical fibers from a newsprint mill by the rot fungus *Sporotrichum pulverulentum*. Biotechnol. Bioeng. *17*, 327-348.

ESKIN, N.A.M., HENDERSON, H.M., and TOWNSEND, R.J. 1971. Biochemistry of Foods. Academic Press, New York.

ETCHELLS, J.L., *et al.* 1975. Factors influencing bloater formation in brined cucumbers during controlled fermentation. J. Food Sci. *40*, 569-575.

EVERSON, C.W., DANNER, W.E., and HAMMES, P.A. 1970. Bacterial starter cultures in sausage products. J. Agr. Food Chem. *18*, 570-571.

FISHMAN, M.M., and SCHIFF, H.F. 1976. Enzymes in analytical chemistry. Anal. Chem. *48*, 322R-332R.

FLEMING, H.P., *et al.* 1973. Bloater formation in brined cucumbers fermented by *Lactobacillus plantarum*. J. Food Sci. *38*, 499-503.

FLEMING, H.P., ETCHELLS, J.L., THOMPSON, R.L., and BELL, T.A. 1975. Purging of CO_2 from cucumber brines to reduce bloater damage. J. Food Sci. *40*, 1304-1310.

FLEMING, H.P., THOMPSON, R.L., BELL, T.A., and MONROE, R.J. 1977. Effect of brine depth on physical properties of brine-stock cucumbers. J. Food Sci. *42*, 1464-1470.

FUJIMAKI, M., KATO, H., ARAI, S., and YAMASHITA, M. 1971. Application of microbial proteases to soybean and other materials to improve acceptability, especially through the formation of plastein. J. Appl. Bacteriol. *34*, 119-131.

GILLILAND, S.E., ANNA, E.D., and SPECK, M.L. 1970. Concentrated cultures of *Leuconostoc citrovorum*. Appl. Microbiol. *19*, 890-893.

GORINSTEIN, S. 1978. Different forms of nitrogen and the stability of beer. J. Agr. Food Chem. *26*, 204-207.

GREGORY, K.F., *et al.* 1976. Conversion of carbohydrates to protein by high temperature fungi. Food Technol. *30*, No. 3, 30-32, 35.

GRIFFIN, H.L., SLONEKER, J.H., and INGLETT, G.E. 1974. Cellulase production by *Trichoderma viride* on feedlot waste. Appl. Microbiol. *27*, 1061-1066.

HAN, Y.W. 1975. Microbial fermentation of rice straw: nutritive composition and in vitro digestibility of the fermentation products. Appl. Microbiol. *29*, 510-514.

HAN, Y.W., and ANDERSON, A.W. 1975. Semisolid fermentation of ryegrass straw. Appl. Microbiol. *30*, 930-934.

HAN, Y.W., and CALLIHAN, C.D. 1974. Cellulose fermentation: effect of substrate pretreatment on microbial growth. Appl. Microbiol. *27*, 159-165.

HAN, Y.W., CHEEKE, P.R., ANDERSON, A.W., and LEKPRAYOON, C. 1976. Growth of *Aureobasidium pullulans* on straw hydrolysate. Appl. Environ. Microbiol. *32*, 799-802.

HARRISON, J.S. 1970. Miscellaneous products from yeast. *In* The Yeasts. Vol. 3. Yeast Technology. A.H. Rose and J.S. Harrison (Editors). Academic Press, London and New York.

HARRISON, J.S., and GRAHAM, J.C.J. 1970. Yeasts in distillery practice. *In* The Yeasts. Vol. 3. Yeast Technology. A.H. Rose and J.S. Harrison (Editors). Academic Press, London and New York.

HASEGAWA, S. 1976. Metabolism of limonoids. Limonin D-ring lactone hydrolase activity in *Pseudomonas*. J. Agr. Food Chem. *24*, 24-26.

HAYAKAWA, I., and NOMURA, D. 1978. Effect of surfactants on spinnability and rheological properties of single cell protein (SCP). Agr. Biol. Chem. *42*, 17-23.

HETTINGA, D.H., and REINBOLD, G.W. 1975. Split defect of Swiss cheese. II. Effect of low temperatures on the metabolic activity of *Propionibacterium*. J. Milk Food Technol. *38*, 31-35.

HUANG, F., and RHA, C. 1978. Formation of single-cell protein filament with hydrocolloids. J. Food Sci. *43*, 780-782, 786.

ISHII, S., and YOKOTSUKA, T. 1973. Susceptibility of fruit juice to enzymatic clarification by pectin lyase and its relation to pectin in fruit juice. J. Agr. Food Chem. *21*, 269-272.

ITOH, H., MORIMOTO, T., and CHIBATA, I. 1974. Automated rapid method for microbioassay of amino acids. Anal. Biochem. *60*, 573-580.

JOLLY, R., and KOSIKOWSKI, F.V. 1975. Blue cheese flavor by microbial lipases and mold spores utilizing whey powder, butter, and coconut fats. J. Food Sci. *40*, 285-287.

KANEKO, T., OHMORI, S., and MASAI, H. 1973. An improved method for the discrimination between biogenic and synthetic acetic acid with a liquid scintillation counter. J. Food Sci. *38*, 350-353.

KAWASHIMA, K., and UMEDA, K. 1976. A method for preparing bead shaped immobilized enzyme. Agr. Biol. Chem. *40*, 1143-1149.

KINSELLA, J.E., and HWANG, D. 1976. Biosynthesis of flavors by *Penicillium roqueforti*. Biotechnol. Bioeng. *18*, 927-938.

KLINE, L., and SUGIHARA, T.F. 1971. Microorganism of the San Francisco sour dough bread process. II. Isolation and characterization of undescribed bacterial species responsible for the souring activity. Appl. Microbiol. *21*, 459-465.

KODAMA, K. 1970. Sake yeast. *In* The Yeasts. Vol. 3. Yeast Technology. A.H. Rose and J.S. Harrison (Editors). Academic Press, London and New York.

KOOI, E.R., and SMITH, R.J. 1972. Dextrose-levulose syrup from dextrose. Food Technol. *26*, No. 9, 57-59.

KUNKEE, R.E., and AMERINE, M.A. 1970. Yeasts in wine-making. *In* The Yeasts. Vol. 3. Yeast Technology. A.H. Rose and J.S. Harrison (Editors). Academic Press, London and New York.

LANGSRUD, T., and REINBOLD, G.W. 1974. Flavor development and microbiology of Swiss cheese—A review. IV. Defects. J. Milk Food Technol. *37*, 26-41.

LASKIN, A.I. 1977. Ethanol as a substrate for single cell protein production. Biotechnol. Bioeng. *7*, 91-103.

LIPINSKY, E.S., and LITCHFIELD, J.H. 1974. Single-cell protein in perspective. Food Technol. *28*, No. 5, 16, 18, 20, 22, 24, 40.

LITCHFIELD, J.H. 1977. Single-cell proteins. Food Technol. *31*, No. 5, 175-179.

MACMILLAN, J.D., and PHAFF, H.J. 1973. Yeasts. General survey. *In* Handbook of Microbiology. Vol. I. Organismic Microbiology. A.I. Laskin and H.A. Lechevalier (Editors). CRC Press, Cleveland, OH.

MACRIS, B.J., and KOKKE, R. 1977. Kinetics of growth and chemical composition of *Fusarium moniliforme* cultivated on carob aqueous extract for microbial protein production. Eur. J. Appl. Microbiol. *4*, 93-99.

MASUDA, H., and MURAKI, H. 1975. Enzymatic determination of acetoin in wines. J. Sci. Food Agr. *26*, 1027-1036.

MONROE, R.J., *et al.* 1969. Influence of various acidities and pasteurizing temperatures on the keeping quality of fresh-pack dill pickles. Food Technol. *23*, 71-78.

MONSAN, P., DUTEURTRE, B., MOLL, M., and DURAND, G. 1978. Use of papain immobilized on spherosil for beer chillproofing. J. Food Sci. *43*, 424-427.

MOON, N.J., and REINBOLD, G.W. 1976. Commensalism and competition in mixed cultures of *Lactobacillus bulgaricus* and *Streptococcus thermophilus.* J. Milk Food Technol. *39*, 337-341.

MORIKAWA, Y., *et al.* 1976. Dichloro-s-triazinyl resin as a carrier of immobilized enzyme. Agr. Biol. Chem. *40*, 1137-1142.

MURAKI, H., and MASUDA, H. 1976. Enzymatic determination of butane-2,3-diol in wines. J. Sci. Food Agr. *27*, 345-350.

NAKAMATSU, T., AKAMATSU, T., MIYAJIMA, R., and SHIIO, I. 1975. Microbial production of glucose oxidase. Agr. Biol. Chem. *39*, 1803-1811.

NUNOMURA, N., SUSAKI, M., ASAO, Y., and YOKOTSUKA, T. 1976A. Identification of volatile components in shoyu (soy sauce) by gas chromatography—mass spectrometry. Agr. Biol. Chem. *40*, 485-490.

NUNOMURA, N., SASAKI, M., ASAO, Y., and YOKOTSUKA, T. 1976B. Isolation and identification of 4-hydroxy-2(or 5)-ethyl-5(or 2)-methyl-3(2H)-furanone, as a flavor component in shoyu (soy sauce). Agr. Biol. Chem. *40*, 491-495.

OHMIYA, K., OHASHI, H., KOBAYASHI, T., and SHIMIZU, S. 1977. Hydrolysis of lactose by immobilized microorganisms. Appl. Environ. Microbiol. *33*, 137-146.

PATEL, I.B., and VAUGHN, R.H. 1973. Cellulolytic bacteria associated with sloughing spoilage of California ripe olives. Appl. Microbiol. *25*, 62-69.

PEDERSON, C.S. 1979. Microbiology of Food Fermentations, 2nd Edition. AVI Publishing Company, Westport, Conn.

PEITERSEN, N. 1975A. Production of cellulase and protein from barley straw by *Trichoderma viride.* Biotechnol. Bioeng. *17*, 361-374.

PEITERSEN, N. 1975B. Cellulase and protein production from mixed cultures of *Trichoderma viride* and a yeast. Biotechnol. Bioeng. *17*, 1291-1299.

ROSENAU, J.R., CALZADA, J.F., and PELEG, M. 1978. Some rheological properties of a cheese-like product prepared by direct acidification. J. Food Sci. *43*, 948-953.

RYTHER, J.H., and GOLDMAN, J.C. 1975. Microbes as food in mariculture. Annu. Rev. Microbiol. *29*, 429-443.

SAMISH, Z., COHEN, S., and LUDIN, A. 1968. Progress of lactic acid fermentation of green olives as affected by peel. Food Technol. *22*, 1009-1012.

SCHNARR, G.W., SZAREK, W.A., and JONES, J.K.N. 1977. Preparation and activity of immobilized *Acetobacter suboxydans* cells. Appl. Environ. Microbiol. *33*, 732-734.

SEITZ, E.W. 1974. Industrial application of microbial lipases: a review. J. Amer. Oil Chem. Soc. *51*, 12-16.

SELLARS, R.L., and BABEL, F.J. 1970. Cultures for the Manufacture of Dairy Products. CHR. Hansen's Laboratory, Milwaukee, Wisconsin.

SHACKLADY, C.A. 1972. Nutritional qualities of single-cell proteins. *In* Health and Food. G.G. Birch, L.F. Green, and L.G. Plaskett (Editors). Applied Science Publishers, London.

SHANNON, D.W.F., MCNAB, J.M., and ANDERSON, G.B. 1976. Use of an n-paraffin-grown yeast in diets for replacement pullets and laying hens. J. Sci. Food Agr. *27*, 471-476.

SHILO, M. 1967. Formation and mode of action of algal toxins. Bacteriol. Rev. *31*, 180-193.

SINAI, Y., KAPLUN, A., HAI, Y., and HALPERIN, B. 1974. Enhancement of resistance to infectious diseases by oral administration of brewer's yeast. Infec. Immunity *9*, 781-787.

STANLEY, W.L., and OLSON, A.C. 1974. Symposium: Immobilized enzymes in food systems. The chemistry of immobilized enzymes. J. Food Sci. 660-666.

SUGIHARA, T.F., KLINE, L., and MILLER, M.W. 1971. Microorganisms of the San Francisco sour dough bread process. I. Yeasts responsible for the leavening action. Appl. Microbiol. *21*, 456-458.

TAJIMA, M., and YOSHIKAWA, S. 1975. Changes of nucleic acids in the preparation process of cell protein isolates from yeast (*Saccharomyces cerevisiae*). Agr. Biol. Chem. *39*, 611-616.

TAYLOR, M.J., RICHARDSON, T., and OLSON, N.F. 1976. Coagulation of milk with immobilized proteases: a review. J. Milk Food Technol. *39*, 864-871.

TREVELYAN, W.E. 1975. Renewable fuels: ethanol produced by fermentation. Trop. Sci. *17*, 1-13.

TREVELYAN, W.E. 1976. Autolytic methods for the reduction of the purine content of baker's yeast, a form of single-cell protein. J. Sci. Food Agr. *27*, 753-762.

VAN DER WALT, J.P. 1970. Genus 16. *Saccharomyces* Meyen emend. Reess. *In* The Yeasts. A Taxonomic Study. J. Lodder (Editor). North-Holland Publishing Company, Amsterdam-London.

VAUGHN, R.H., *et al.* 1969. Some pink yeasts associated with softening of olives. Appl. Microbiol. *18*, 771-775.

VAUGHN, R.H., STEVENSON, K.E., DAVE, B.A., and PARK, H.C. 1972. Fermenting yeasts associated with softening and gas-pocket formation in olives. Appl. Microbiol. *23*, 316-320.

VORSILAK, P., MCCOY, B.J., and MERSON, R.L. 1975. Enzyme immobilized in a membrane sandwich reactor. J. Food Sci. *40*, 431-432.

WAINWRIGHT, T. 1971. Production of H_2S by yeasts: role of nutrients. J. Appl. Bacteriol. *34*, 161-171.

WALTON, H.M., and EASTMAN, J.E. 1973. Insolubilized amylases. Biotechnol. Bioeng. *15*, 951-962.

WEISER, H.H., MOUNTNEY, G.J., and GOULD, W.A. 1971. Practical Food Microbiology and Technology, 2nd Edition. AVI Publishing Co., Westport, Conn.

WELLMAN, A.M., and STEWART, G.G. 1973. Storage of brewing yeasts by liquid nitrogen refrigeration. Appl. Microbiol. *26*, 577-583.

WILKINSON, J.F. 1971. Hydrocarbons as a source of single-cell protein. *In* Microbes and Biological Productivity. D.E. Hughes and A.H. Rose (Editors). Twenty-first Symposium of the Soc. Gen. Microbiol. The University Press, Cambridge, England.

WILLETTS, A. 1975. Fungal growth on C_1 compounds: a study of the amino acid composition of a methanol-utilizing strain of *Trichoderma lignorum.* Appl. Microbiol. *30*, 343-345.

YANG, H.H., YANG, S.P., and THAYER, D.W. 1977. Evaluation of the protein quality of single-cell protein produced from mesquite. J. Food Sci. *42*, 1247-1250.

YANG, H.Y. 1973. Effect of pH on the activity of *Schizosaccharomyces pombe.* J. Food Sci. *38*, 1156-1157.

ZADRAŽIL, F. 1977. The conversion of straw into feed by *Basidiomycetes.* Eur. J. Appl. Microbiol. *4*, 273-281.

ZEE, J.A., and SIMARD, R.E. 1975. Simple process for the reduction in the nucleic acid content in yeast. Appl. Microbiol. *29*, 59-62.

10

The Control of Microorganisms

The control of microorganisms is one of the major concerns of food microbiologists. This control is needed to retard or prevent spoilage and to reduce or eliminate health hazards associated with foods. The control of contaminants also aids in obtaining better results when specific microorganisms or enzymes are used in food processing.

Four basic systems are used to aid in the control of microorganisms in foods. These are: 1) prevent contamination (asepsis); 2) remove contaminants; 3) inhibit growth; and 4) destroy contaminants.

In most food products, two or more of these systems are used to control the microbial level. Preventing contamination is practiced for all foods, but since contamination will still occur, other safeguards are needed.

Besides controlling the microorganisms, foods must be protected from reactions that are catalyzed by inherent enzymes, from chemical degradation, such as fat oxidation, loss of nutrients and from the destruction by pests, such as insects and rodents. There are various chemicals and procedures that will control microorganisms and pests, but cannot be used for foods. This is because the food must be safe for consumption, be of acceptable organoleptic quality and the nutritional value of the food must be maintained.

Although the control of microorganisms in food is usually relegated to the food processor, everyone involved in the production, processing, handling, warehousing, retailing, preparation and serving should be involved in the control process as well as in maintaining a safe and nutritious food supply.

The need for this overall effort is evident from the fact that only a few outbreaks of foodborne illness are caused by problems at the processing level. Most of the outbreaks are caused by mishandling and contamination of foods at food service establishments or in the home.

In some cases, we assume that a food will be handled and used in a particular manner; the assumption may not always be correct. For example, we assume that people will cook meat prior to consumption. However, a series of outbreaks of salmonellosis revealed that some people thought that eating raw hamburger would make them healthy, strong and vigorous. Due to the misuse of foods by some people, it is necessary to take precautions beyond normal requirements. A food microbiologist must be aware of all aspects of a food, from production to consumption.

One of the primary functions of food processing is the preservation of foods. This preservation ranges from short periods of a few days or weeks to long terms of a year or more. Preservation by the removal, inhibition or destruction of microorganisms is easier and the results are more satisfactory, when the original microbial index of the food is low. Hence, keeping the contamination low by sanitary means is very important.

CONTROL BY ASEPSIS

If all of the microorganisms in a food were either useful or inert, we would have no concern. Unfortunately, the spoilage and health hazard types are quite prevalent. Generally, the higher the microbial load, the greater is the possibility that these undesirable types are included. The exceptions are foods in which specific microorganisms are grown to produce desirable products, and a high level of the useful microorganism is needed.

It is much easier to inhibit or destroy low numbers than high numbers of microorganisms. A food with a low microbial load will generally have a longer shelf life than a food with a high microbial level. The shelf life is the time after packaging in which a food maintains its best quality if it is stored under proper conditions of humidity and temperature.

Due to the ubiquitous nature of microorganisms, it is impossible to keep them from contaminating food. However, we can reduce contamination by controlling the potential sources of microorganisms.

The control of contamination is referred to as sanitation. Sanitation may be defined as a modification of the environment in such a way that a maximum of health, comfort, safety and well-being occurs to man. The sanitation involved in the food industry is only a small part of overall sanitation.

The control of the microbial quality of food must begin with the production and harvesting of food. Then it must carry over to the processor and on to the ultimate consumer.

Production and Harvesting

The raw materials used by the food processor affect the quality of the finished product. In order to have adequate supplies of high quality raw materials with acceptable microbial loads, the food processor must work closely with the producer, or become the producer. In some cases, being the producer seems to be the best method. As a result, we now have companies that own or control food producing and processing systems. By owning a system, the company can integrate all of the processes to end up with a uniform, high quality product.

With many foods, the processor must rely on the producer to deliver an acceptable raw material. Some ingredients are obtained from foreign countries where the level of sanitation is different from ours. Hence, the processor sometimes has to work with raw materials that are produced in areas beyond his control. Sometimes a visit to the foreign supply areas will help in obtaining more acceptable products.

When the raw materials are obtained from local areas, the processor can usually visit the sources of supply, make evaluations and suggest procedures that improve the quality. Sometimes the processor obtains help from local, state and federal regulatory agencies that inspect the producing areas.

Milk.—Many years ago, the occurrence of epidemics of various diseases due to milk caused the health authorities to decide that milk production should be regulated. To help control the spread of tuberculosis, cows were tested and reactors eliminated. From this beginning, many regulations and codes of operations have been devised to control the production and handling of milk at the farm. No other commodity is produced with the amount of regulation as grade A milk. As a result, tuberculosis and brucellosis in milk are controlled. At present, the main infectious problem at the farm is mastitis.

Besides the health of the dairy cow, standards have been developed for the housing and milking areas, so that they can be maintained in a satisfactory manner. There are requirements for the equipment used to obtain, handle and store the milk, and for the equipment to be properly cleaned, sanitized and maintained in a satisfactory manner. Grade A milk must be cooled to 10°C or less within two hours after milking to maintain the quality. The requirements are more stringent for the production of grade A milk than for manufacturing grade milk. This latter milk is used in various dairy products other than liquid milk.

Due to the requirements imposed on the production of milk, the small and marginal producers have gone out of business. They simply could not afford the facilities and equipment needed to maintain proper sanitation. It is rare when small producers or processors and government regulations are compatible.

Insect and rodent control are needed at the farm level, as well as at the processing level. Only approved insecticides can be used in dairy barns or around other food products. The removal of potential breeding places helps control insects and rodents.

Animal Production.—Healthy, disease-free animals should be maintained in a relatively clean, disease-free environment. Modern husbandry practices have tended to increase rather than decrease infections in domestic animals. The practice of crowding animals into feed lots, broiler houses, laying houses or holding pens, increases the potential for spreading diseases, including salmonellosis. Sanitation of all animal quarters on the farm may help, but will not prevent, infection.

With meat-type chickens (broilers), flocks that are *Salmonella*-free tend to result in carcasses that are *Salmonella*-free. When salmonellae are isolated from birds prior to slaughter, the carcasses tend to contain these organisms.

The transportation of animals to market or from one farm to another causes stress, and animals held in dirty pens until slaughter shed salmonellae at a higher rate than on the farm. This indicates that stress induces shedding of organisms already present in the intestinal tract, or it makes the animals very susceptible to *Salmonella* infection, or both. One solution may be to hold animals at the farm and then move them in clean, uncrowded conveyances to the slaughter house for immediate slaughter.

One of the sources of contamination of carcasses during slaughter is of fecal origin. Withdrawal of feed and water 8-10 hr before slaughter reduces the fecal content of chickens and might reduce the potential contamination of the carcasses.

Shell Eggs.—Regardless of whether laying hens are housed on the floor or in cages, it is desirable to keep the egg shells clean. The incidence of spoilage is much greater in dirty eggs than in clean eggs, whether the shell is washed or not. When eggs are washed, the temperature of the wash water should be 20°C higher than the temperature of the egg and an approved germicide should be added to the cleaning solution. The addition of a germicide reduces spoilage by about 50% of that when eggs are washed in water alone. Only odorless, nontoxic cleaner-sanitizers should be used for washing and sanitizing egg shells.

Fruits and Vegetables.—The production of quality plant products starts with quality plant varieties and seeds. The disease resistance and the amount and quality of food produced are important considerations for the selection of plant varieties.

The World Health Organization (WHO 1969) recommended several hygienic practices for fruits and vegetables. The environmental sanitation during growing included sanitary disposal of human and animal wastes, sanitary quality of irrigation water and control of animals, pests and diseases. The sanitary harvesting aspects include the equipment and containers which should be cleaned and maintained, and not be a source of microorganisms. The unfit produce should be segregated and disposed of in a satisfactory manner. The product should be protected from contamination by animals, insects, birds, chemicals or microorganisms.

The transportation of produce should be done in conveyances that can be cleaned and not be a source of contamination. All handling should be with care and, if needed, refrigeration or ice of sanitary quality should be used.

The contamination of plant products by soils is important to the food processor, since soils contain bacterial spores, including those of *Clostridium botulinum.* Bacterial spores are resistant to various sanitizing agents, as well as food preservation treatments. Hence, it would be a definite advantage to the processor if spores could be kept out of the raw materials.

Seafoods.—Seafoods are contaminated by the environment in which they grow, as well as during harvesting and transportation to the processing plant.

The main problem of contamination during growth involves the bivalve mollusks, such as oysters, clams and mussels. They grow in estuaries or protected coastal waters. These are also the most convenient places for the disposal of sewage and other wastes of coastal communities. The mollusks obtain nourishment by pumping water through a complex system of gills which filter and concentrate food, such as bacteria and marine organisms. With the dumping of sewage, the bacteria may include potential pathogens. With oysters and clams, the entire animal, including the intestinal tract and its contents, is consumed, frequently raw, or after superficial heating. If the bivalves have consumed enteric pathogens, they can serve as vehicles for organisms that cause foodborne illness.

Due to a series of illnesses, culminating in a major typhoid fever epidemic in 1924 from eating contaminated oysters, a voluntary cooperative program for the certification of interstate shellfish shippers was established in 1925 by the U.S. Public Health Service (USDHEW 1965A,B). The coastal states are charged with the duty of adopting laws to regulate the shellfish industry. The states make sanitary and bacteriological surveys of growing areas, delineate and patrol restricted areas, prevent illegal harvesting, inspect shellfish plants and conduct other activities that are needed to insure that the shellfish reaching the consumer have been grown, harvested and processed in a sanitary manner.

The survey of the growing areas has three parts: 1) the area is sufficiently removed from major sources of contamination so that shellfish are not subject to fecal contamination; 2) the area is free from pollution by potentially harmful industrial wastes; and 3) the median coliform level does not exceed 70/100 ml of water and not more than 10% of the samples exceed 230/100 ml for a 5-tube MPN, or 333/100 ml for a 3-tube MPN. With these safeguards, no outbreaks of foodborne illness have occurred from eating shellfish from approved waters. There have been outbreaks of infectious hepatitis from eating raw oysters and clams that were obtained from areas not approved for harvesting.

A standard for water samples not to exceed 14 fecal coliforms per 200 ml has been proposed (Andrews *et al.* 1975).

If harvested from water containing the organisms, shellfish may contain *Vibrio parahaemolyticus* and fish may contain *Clostridium botulinum* type E. Fish obtained from polluted water may contain various enteric organisms, including *Salmonella, Shigella, E. coli* and *Clostridium perfringens.*

Spoilage of seafood begins after removal from the water. Hence, the treatment that these foods get aboard the catching ship will determine the quality that is obtained by the consumer.

Huss *et al.* (1974) found that fish, as caught, have low bacterial loads. The hygienic standards on board the ship determined the level of contamination of the fish. In order to reduce the growth rate of the bacteria, proper refrigeration, icing or freezing must be used. Processing of the fish should be done as soon as possible after catching. Some ships have processing systems on board. In these cases, seawater is used to wash and process the fish. If not treated with chlorine, this water may serve as a source of microorganisms. The sanitary conditions of the working areas and equipment, and the health and hygiene of workers on the ship should be the same as in processing plants on land.

Processing

If care is taken to produce and deliver a high quality product to the processing plant, it is the duty of the processor to maintain the quality at a high level.

The U.S. Food, Drug and Cosmetic Act of 1968 states "A food shall be deemed to be adulterated if it contains any poisonous or deleterious substance which may render it injurious to health; or if it consists in whole or in part of any filthy, putrid, or decomposed substance; or it has been prepared, packed, or held under insanitary conditions whereby it *may* have become contaminated with filth or rendered injurious to health." With these stipulations, the presence of any microorganisms or chemical that may cause illness or that would indicate the food may contain harmful agents makes the food adulterated and subject to seizure. The presence of filth, such as insects, rodent hairs or droppings, is evidence of an adulterated product. However, there are tolerances for these agents. Even if the food does not contain filth or potentially harmful agents, if conditions are such that these substances may be present, the food can be called adulterated. Although this act applies only to foods in interstate commerce, many states have copied the federal regulations.

The buildings, environment and processing facilities, as well as raw materials, personnel, equipment (including cleaning and sanitizing) and refuse disposal facilities, are important to the overall plant operations.

The level of sanitation maintained in the plant has a direct bearing on the quality of the final product, and should be a basic part of the quality control program of the processor.

One author stated that sanitation is an added cost which is included in the price of the product to the consumer. A more realistic statement would be that unsanitary conditions result in added costs that are passed on to the consumer. Added costs, such as those caused by loss of food due to a short shelf life and spoilage of perishable products, medical bills of consumers ingesting pathogenic organisms or poisons, legal fees and settlement of court cases due to illness or death of former consumers, are passed on to present consumers. Regulatory agents may seize a product that violates the regulations. As a result, the least that can happen is to have inventory tied up. The seizure may result in destruction of the product, corporation or personal fines, prison terms and the closing of the processing plant. Hence, although sanitation may be a cost, the lack of sanitation can create a greater cost. Sanitation is a necessity in order to stay in business and has resulted in advantages to both the processor and consumer.

Proper sanitation reduces or eliminates potential spoilage hazards and creates a longer shelf life of perishable products. This reduces losses due to spoilage, increases the efficiency of plant operations, results in easier maintenance of equipment, develops better employee relationships, workmanship and safety, increases consumer acceptance and public relations, improves the quality and safety of foods for the consumer and, with better consumer acceptability, there is a potential increase in sales. This results in faster turnover and less spoilage of perishable foods. As a result of these benefits, the actual cost of the product may decrease.

There are some people who believe that we are too sanitation conscious and have established ridiculous requirements. There are statements that regulations against rodent leavings and insects in foods are only aesthetic, since there is no safety factor involved. Undoubtedly, there is a point beyond which the costs outweigh any benefits of the program.

The *may have* section (Section 402 (a) (4), of the U.S.FD and C Act) describes adulterated food. To interpret and to establish criteria to aid industry in complying with this section of the Act, the FDA has published regulations called good manufacturing practices (GMPs) (GSA 1977). Amendments including GMPs for the processing of specific foods have been adopted. These GMP regulations cover essentially all operations of a food processing plant from the outside appearance, equipment and processes to personnel.

Other government agencies also have regulations which spell out the sanitary requirements in food processing operations, such as meat (USDA 1976), poultry (USDA 1970) and eggs and egg products (USDA 1971).

Each food processing operation has certain points at which failure to prevent contamination can be detected by laboratory tests with maximum assurance and efficiency. These are known as critical control points. Bauman (1974) defined hazard analysis as the identification of sensitive ingredients, critical process points and relevant human factors, as they affect product safety. Together these form the concept called hazard analysis critical control points (HACCP).

Foods have been placed into five categories in terms of health hazards, which are based on three hazard characteristics (NAS 1969). These are described in Chapter 6 in regard to controlling *Salmonella* in foods. These hazard characteristics are used as the basis for the HACCP inspections.

According to Kauffman (1974), the HACCP inspection approach is used to determine the points in the process which are critical to the safety of the product, and then these can be used by the processor to identify critical points. HACCP inspections, guided by the GMPs, will tell the FDA and industry what needs to be done to assure safe, high quality food production and distribution.

The critical control points are determined for each type of food process and for each food processing plant. The critical control points have been discussed for frozen foods (Peterson and Gunnerson 1974) and for canned foods (Ito 1974).

With this concept of inspection, not only is visible evidence of filth (insects, rodents) or unclean equipment and facilities important, but also the microbial analysis for potential hazard or spoilage types is used at critical control points to determine the problems involved in the process. This concept increases the importance of the microbiologist.

Although the GMP and HACCP concepts provide a basis for satisfactory operation for food processors, they do not state how to meet the requirements in all cases, or why some of the GMPs are needed. Therefore, it seems worthwhile to discuss some of these factors.

Plant and Grounds.—The plant should be designed so that materials can be received, stored, processed, warehoused and shippped in an efficient and satisfactory manner.

Some operations, such as washing dirt off vegetables or slaughtering animals, are not considered to be clean, and must be separated from areas where processed food is exposed. The segregation of facilities should be such that no cross contamination can occur between raw products and processed products.

Air.—Fresh air is needed in food processing areas. Since air is a potential source of microorganisms, a system or systems are needed for microbial control. The sources of microorganisms and the means of dispersal throughout the plant must be considered in designing a method for control.

The ventilating system is the main place to control microorganisms in the air. Filters have been developed to remove microorganisms, as well as other materials from air. Viruses were effectively removed from air by filters with an efficiency rating of 93% or over (Roelants *et al.* 1968). There are filters available with ratings of 99.9% and an ultra filter of 99.999%. The filter should be rated by the dioctyl-phthalate filter test using 0.3 μm particles to challenge the filter.

By pumping in more filtered air than is removed by the outlets, a positive air pressure is established in the room. With positive air pressure, when a door is opened for personnel or product to move in or out of the room, unfiltered air will not come in, since air is moving out of the room.

The clean, fresh air should enter the cleanest areas of the processing plant, move to the less clean areas and leave the plant from the areas that are "dirty." This procedure reduces the potential airborne contamination of processed product from these less clean areas.

To eliminate turbulence factors, laminar air flow (unidirectional) has been developed. Laminar air flow devices are effective in reducing and controlling airborne contamination. Due to the relatively higher cost, they have not been readily adopted for use in the food industry. They should find usage in cabinets or special rooms where contamination of products must be tightly controlled, such as in aseptic packaging.

Besides mechanical filters, the air may be treated by chemical germicides, UV radiation, electronic air cleaners, thermal sterilization, centrifuging or "scrubbing." When meticulous air cleaning is essential, electrostatic precipitation might be employed. The larger the particles, the easier it is to remove them. Thus, molds and yeasts can be removed by electrostatic precipitation more easily than bacteria or viruses. Due to the high cost of installation, this system has been used little, if any, in the food processing industry. Electrostatic precipitation is used to remove dust and other pollutants from effluent air.

Water.—The water may be from a municipal supply, in which case it should be potable. That is, it is free from potential pathogens. However, municipal water is not tested for potential spoilage organisms, and reservoirs and pipelines at the plant can serve as a place for multiplication. Some pseudomonads grow in distilled water. Hence, it may be necessary to treat the water by chlorination at the processing plant.

The chemicals in the water influence the type of cleaning agents that are used in the processing plant. When water is an ingredient of the food, chemicals causing odors and flavors may affect the characteristics of the food product. In these cases, the water should be monitored and given the necessary treatments that are indicated by the analysis and by the food that is being processed. The treatment of water ranges from filtration, flocculation, softening, demineralization, reverse osmosis, distillation, to various combined processes to obtain pure water. A water softener is not a means of removing organisms and, in some cases, the resins in softeners may act as a source of inoculum of the soft water.

Equipment and Utensils.—The equipment should be designed so that food does not become lodged on or in it to form pockets of microbial growth for continual contamination of foods moving over or through it. The sanitary design of equipment results in easier cleaning and sanitizing, thereby saving labor and materials.

The equipment surfaces in contact with food must be nontoxic, nonabsorbent, nonporous and noncorrosive. Although equipment is made of several types of material, stainless steel is preferred for food contact surfaces (Fig. 10.1).

Stainless steel is not just one metal, but is a group of alloys of iron with chromium, nickel, carbon, molybdenum, titanium, silicon, phosphorus, manganese and sulfur.

Although stainless steel is corrosion resistant, it is not corrosion proof. Stainless steel is protected from corrosion by its self-repairing surface film of chromium oxide. When this film is broken down by cleaning, it will reform when exposed to the air. If abrasive materials are used in cleaning, the metal surface will be scratched, which allows corrosion to occur. If harsh chemicals are used in cleaning, they can cause pitting of the metal. Scratches and pitting make it difficult to clean, sanitize and maintain the equipment in a satisfactory manner.

Certain materials should not be used in equipment that contacts food. Due to its porous nature and difficulty in maintenance, wood should not be used. Copper and its alloys accelerate rancidity of fats, discolor food products and contaminate foods with copper salts. Cadmium is toxic. Lead should not be used, except in small amounts where soldering is needed. Aluminum is attacked by acids and alkalis and it has a tendency to rub off and cause black marks. Iron can cause off-flavor in products,

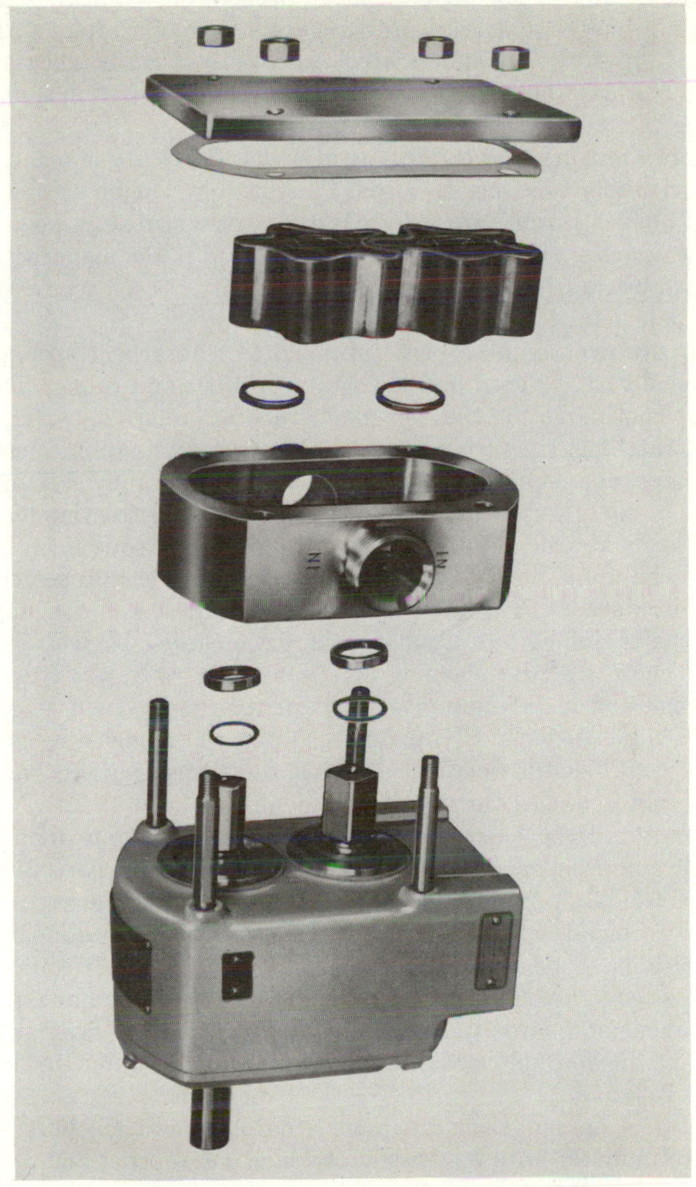

Courtesy of CREPACO, Inc.

FIG. 10.1. SANITARY POSITIVE ROTARY PUMP

From the bottom the parts are 1) Gear Case; 2) Shaft Sleeve Seals; 3) Shaft Sleeves; 4) Pump Body; 5) Shaft Seals; 6) Rotors—Rubber Covered; 7) Cover Gasket; 8) Front Cover; and 9) Hex Nuts.

such as milk, and it is impossible to clean because of rusting and pitting. Zinc and antimony are not acceptable metals for food contact surfaces.

As an alternative to stainless steel, glass (usually borosilicate glass pipelines) or glass fiber reinforced polyester resin are used in some food processing plants.

There are some areas or conditions in which flexibility is needed and plastics or rubber products are used. Plastics may not be acceptable if they contain free phenol, formaldehyde or any other substance which can transfer to or alter the food. Since additives used in the manufacture of plastics can be transferred to foods, the suitability of each type of plastic must be determined for each particular food.

Belting with exposed absorbent fabric core or absorbent sponge type rubber should not be used in food plants. Rubber and rubber-like materials are considered to be acceptable if they are relatively inert, resistant to scratching, fracturing, peeling, chipping, scoring and distortion by the temperature, chemicals and methods to which they are normally subjected in operation or the cleaning and sanitizing treatment.

Ease of assembly and disassembly is important if the equipment is to be cleaned by hand methods. When possible, equipment should be designed for clean-in-place (CIP) or automatic methods. These systems may cost more at the beginning but usually pay for themselves by requiring less labor and more effective cleaning and sanitizing.

The design and installation of equipment for proper drainage are important. This is needed in CIP systems, so that soil removed by cleaning solutions, as well as the cleaning solutions, sanitizing solutions and rinse waters can be removed from the equipment.

In order to aid the dairy industry in obtaining equipment made of acceptable material, many years ago the International Association of Milk and Food Sanitarians established 3A-Sanitary Standards for Dairy Equipment. Since then, this organization (now the International Association of Milk, Food and Environmental Sanitarians) in conjunction with various government agencies and trade organizations, has expanded the equipment standards to other food products. These standards can be found in the Journal of Milk and Food Technology and the Journal of Food Protection.

Processing equipment does not have a normal microbial flora. It becomes contaminated with food residues which can support high levels of microorganisms. As the buildup of the microbial load progresses, it is a source for contaminating food that contacts it. Hence, it is necessary that the equipment be cleaned and sanitized at appropriate intervals to keep the microbial load at a reasonable level.

Cleaning.—Equipment that is less than 100% clean is still dirty. It is difficult, if not impossible, to sanitize equipment that is not completely clean. Therefore, cleaning is a very important part of the operation.

Before cleaning, the food must be removed from the line and protected from the possible contamination by the cleaning and sanitizing agents, as well as the water and soil removed from the equipment.

In some processing plants, all of the equipment is cleaned and sanitized manually. In other plants, some or all of the cleaning and sanitizing may be semi-automated or fully automated. In any of these systems, the human factor is the most important ingredient, since even with a fully mechanized system, it is necessary to set up the operation and supply the system with the cleaning and sanitizing agents.

The first part of any cleaning operation is to remove waste material and to rinse loose soil from the surfaces with clean water. For fatty materials, hot water may aid in this initial rinse, but cool water is desirable for protein soils.

Even if most of the soil can be removed by hand and by water rinsing, the equipment is not clean, since there is a residue that is difficult to remove. An input of energy is needed to remove this residual soil. The energy forms are thermal (hot water, steam), chemical (cleaning agents) and mechanical (high pressure spray, manual scrubbing).

The cleaning process consists of bringing the cleaning solution into intimate contact with the soil on the equipment surface, removal of the soil from the equipment, dispersion of the soil into the cleaning solution and prevention of redisposition of the soil, as well as scale due to hard water, onto the cleaned surface. Various cleaning agents are available that perform these activities. Cleaning agents are designed to reduce the amount of work needed to remove the adhering soil.

Cleaning Agents.—A cleaner may consist of a single compound or be a complex mixture of various chemicals.

Only approved cleaners and sanitizers should be used in food processing plants. Various government agencies compile lists of compounds considered to be acceptable and safe for washing food processing equipment, as well as the floors and walls of plants.

A cleaning agent should have the following characteristics:

1. Be quickly and completely soluble.
2. Be noncorrosive to metal surfaces.
3. Be nontoxic.
4. Be easily rinsed.
5. Be economical.
6. Be stable during storage.
7. Not be harsh on the hands if used for manual cleaning.
8. Possess germicidal action, if required.

The types of cleaners and their properties are listed in Table 10.1.

TABLE 10.1

RELATIVE CLEANING VALUES OF VARIOUS DETERGENT INGREDIENTS

KEY TO CHART

- A High Value
- B Medium Value
- C Low Value
- D Negative Value
- 1 Via Precipitation
- 2 Via Sequestration
- 3 Also Stable to Heat

	INGREDIENTS	COMPARATIVE ABILITY											
		Emulsification	Saponification	Wetting	Dispersion	Suspension	Peptizing	Water Softening	Mineral Deposit Control	Rinsability	Suds Formation	Non-corrosive	Non-irritating
Basic Alkalies	Caustic Soda	C	A	C	C	C	C	C	D	D	C	D	D
	Sodium Metasilicate	B	B	C	B	C	C	C	C	B	C	B	D
	Soda Ash	C	B	C	C	C	C	C	D	C	C	C	D
	Trisodium Phosphate	B	B	C	B	B	B	A[1]	D	B	C	C+	C–
Complex Phosphate	Sodium Tetraphosphate	A	C	C	A	A	A	B[2]	B	A	C	AA	A
	Sodium Tripolyphosphate	A	C	C	A	A	A	A[2]	B	A	C	AA	B
	Sodium Hexametaphosphate	A	C	C	A	A	A	B[2]	B	A	C	AA	A
	Tetrasodium Pyrophosphate	B	B	C	B	B	B	A[2]	B	A	C	AA	B
Organic Compounds	Chelating Agents	C	C	C	C	C	A	AA[3]	A	A	C	AA	A
	Wetting Agents	AA	C	AA	A	B	B	C	C	AA	AAA	A	A
	Organic Acids	C	C	C	C	C	B	A[3]	AA	B	C	A	A
	Mineral Acids	C	C	C	C	C	C	A[3]	AA	C	C	D	D

Source: Klenzade Dairy Sanitation Handbook

Substances that lower the surface tension of a liquid, or the interfacial tension between two liquids, are known as surface active agents, or surfactants. Chelating agents are chemicals that can form a soluble ring-like complex with a metallic cation. In this form, the metallic ions do not form insoluble films on the equipment, but are removed from the equipment by rinsing. The chief organic chelating agents, or sequestrants, are sodium salts of EDTA and related compounds, the hydroxy carboxylic compounds (citric, gluconic, tartaric and hydroxyacetic acids) and the aminohydroxy compounds (such as tetraethanolamine). The important inorganic sequestrants are the sodium polyphosphates (sodium pyrophosphate, sodium tripolyphosphate, sodium hexametaphosphate). They have a softening effect on hard water. The chelating agents can increase the effectiveness of sanitizing agents.

Sanitizing.—If the cleaning procedure removed all of the organic material, the equipment would not need to be sanitized. The sterilization of all equipment would be ideal, but neither practical nor economical. Thus, we are satisfied if the equipment is sanitized.

In some cases steam, hot water or hot air is used for sanitizing. In most processing plants, these agents are not practical, so chemical sanitizers are used. Sanitizing does not mean that all living bacterial cells are destroyed or removed. A sanitized surface does mean, however, that all pathogenic or disease producing bacteria, as well as a large percentage of nonpathogenic ones, are destroyed.

Some sanitizing agents are added to detergents (detergent-sanitizers) so that the bacteria are killed during cleaning. This single operation of cleaning and sanitizing saves time and labor. When chlorinated cleaners are used, the chlorine not only kills bacteria, but also aids in the removal of soil and reduces redeposition.

The effectiveness of a sanitizer depends upon the concentration and the time of exposure. The pH modifies the sanitizing action by affecting both the bacteria and the chemical. An increase in temperature usually increases the rate of sanitizing. The effectiveness of the sanitizer depends upon the types and numbers of microorganisms, as well as their previous history. The use of a high concentration of a sanitizing agent, as a substitute for thorough cleaning is inefficient, expensive and, in some cases, may cause corrosion of the metal. Chemicals that leave bacteriostatic or bactericidal residues on surfaces may help with sanitation programs in food processing plants.

There are thousands of chemicals that have germicidal powers, but only those compounds that are approved by a regulatory agent can be used as sanitizers. The three most popular types of sanitizers are the chlorine compounds, iodine or iodophors and quaternary ammonium compounds. Acid-anionic surfactants also have a sanitizing action.

Chlorine.—The most widely used sanitizers are the chlorine containing compounds. Six factors determine the effectiveness of chlorine as a sanitizing agent. They are: 1) type and concentration of chlorine compound used; 2) pH; 3) temperature; 4) contact period; 5) types of organisms; and 6) presence of organic material.

The types of chlorine compounds include the sodium and calcium hypochlorites, the salts of isocyanuric acid, the hydantoin derivatives and chlorinated trisodium phosphates, as well as gaseous chlorine. In powdered chlorinated cleaners the source of chlorine is usually chlorinated trisodium phosphate, potassium dichlorocyanurate, dichlorodimethylhydantoin or chloramine T. The first two give the most active chlorine in the cleaning solution. Besides these compounds, mixtures of chlorine and bromine, as well as chlorine dioxide, have been used as sanitizing agents. The most widely used agents in the food processing area are the hypochlorites.

The advantages and disadvantages of the use of hypochlorites as sanitizing agents are listed in Table 10.2. The hypochlorites are effective at relatively high dilutions and they are active against a wide range of bacteria and bacterial spores, as well as molds, yeasts, bacteriophages and some viruses. The hypochlorites are considered to be more effective against Gram negative than against Gram positive bacteria. Viruses are more resistant than bacteria to the action of chlorine.

TABLE 10.2

ADVANTAGES AND DISADVANTAGES OF HYPOCHLORITES AS SANITIZERS

Advantages	Disadvantages
Relatively inexpensive	Unstable during storage
Quick acting	Inactivated by organic compounds
Not affected by hard water salts	Corrosive if misused
Harmless residue, does not form a film	Irritates skin
Effective at high dilution	Odor may be undesirable
Active against a wide variety of microorganisms, including spores and phage	Precipitate in iron waters
Relatively non-toxic at use dilutions	Effectiveness decreases with increasing pH of solution
Non-staining	May remove carbon from rubber parts of equipment
Colorless	
Easy to prepare and apply	
Concentration easily determined	
Can be used for water treatment	

In moderately strong concentrations, the hypochlorites can cause irritation of the skin and may cause an objectionable odor. In some cases, unacceptable flavors may develop in foods exposed to hypochlorites.

There is a reduction of efficiency of hypochlorites by organic compounds and oxidants. Due to their high corrosion potential, they should not be used at high temperatures or be allowed to contact the surfaces of equipment for excessively long periods.

Chlorine compounds are used in various aspects of food processing. They are added to water used for washing and conveying raw food products, for cleaning and sanitizing food-handling equipment and for cooling heat-sterilized cans of food.

In-plant chlorination is accomplished by injecting gaseous chlorine or liquid sodium hypochlorite into the water supply of a processing plant by an automatic proportioner. Sufficient chlorine is added to satisfy the demand of substances present in the water (the breakpoint), as shown in Fig. 10.2, and to give a free available residual chlorine of from 2 to 20 mg/liter. The free available chlorine is in the form of hypochlorous acid (HOCl) and hypochlorous ions (OCl^-). The combined chlorine residual is that chlorine reacted with amino nitrogen (chloramines). The total available chlorine includes the combined and free chlorine.

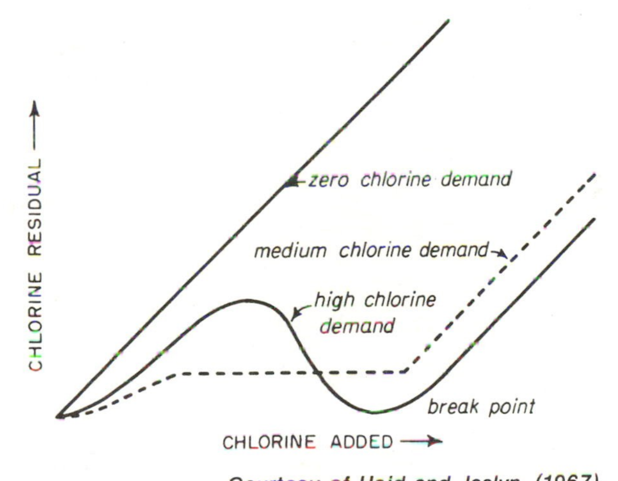

Courtesy of Heid and Joslyn (1967)

FIG. 10.2. CHLORINE DEMAND CHARACTERISTICS OF WATER

In-plant chlorination provides a continuous bactericidal action of chlorine on food preparation equipment during its operation. The use of chlorinated water sprays at selected points reduces or prevents accumulation of microbial slimes and off-odors in food processing plants. The total microbial load of the finished product tends to be lower in food plants with in-plant chlorination. Chlorination cannot substitute for good sanitary practices and good plant operations.

Sodium hypochlorite reacts with water to form hypochlorous acid and sodium hydroxide:

$$NaOCl + H_2O \rightleftarrows HOCl + NaOH$$

Thus, the solution is alkaline, and the hypochlorous acid ionizes to form hydrogen ions and hypochlorous ions:

$$HOCl \rightleftarrows H^+ + OCl^-$$

The hypochlorous acid can dissociate to form hydrochloric acid and nascent oxygen:

$$HOCl \rightarrow H^+ + Cl^- + O$$

The HOCl, OCl^- and O have germicidal properties. In an acid solution, the hypochlorous acid tends to remain as HOCl which is more bactericidal than the hypochlorous ion. Hence, the bactericidal efficiency of chlorine solutions is higher in acid than in alkaline solutions (Fig. 10.3) and is increased with temperature increases.

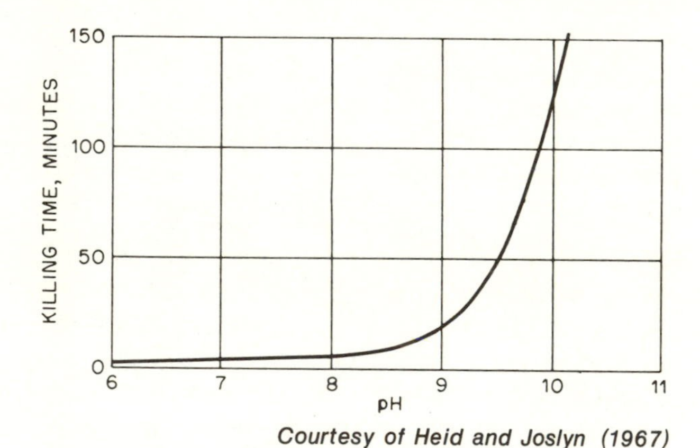

Courtesy of Heid and Joslyn (1967)

FIG. 10.3. EFFECT OF pH ON GERMICIDAL EFFICIENCY OF HYPOCHLORITE SOLUTION

It is thought that the HOCl can penetrate the bacterial cell wall and interfere with the sulfhydryl containing enzymes involved in glucose metabolism. The inhibition of essential enzyme systems results in the death of the cell. Chlorine may damage the cell membrane causing leakage of components. It also forms substitution products with proteins and amino acids.

Rosenkrantz (1973) suggested that NaOCl is capable of reacting with the deoxyribonucleic acid of living cells. This chemical caused mutations of *Salmonella typhimurium* by oxidation of purine and pyrimidine moieties (Wlodkowski and Rosenkrantz 1975).

The effect of chlorine on spores of *Bacillus* and *Clostridium* was investigated by Wyatt and Waites (1975). They found that chlorine disrupts the spore coat by combining with and removing protein. This inactivated the germination mechanism, but if spores were able to germinate, most were damaged to the extent that no outgrowth occurred. The heat resistance of the spores was reduced by a chlorine treatment. Treatment of *B. cereus* spores with a solution of 0.25% NaOCl at 20°C for 19.5 min resulted in a 99% loss in viability (Kulikovsky *et al.* 1975). They found the spores lost calcium ions, dipicolinic acid, ribonucleic acid and deoxyribonucleic acid. The primary effect was the degradation of the outer spore coats which led to a disruption of the normal permeability barriers. In general, *Bacillus* spores are more resistant than *Clostridium* spores to the action of chlorine.

For sanitizing equipment, regulatory agencies recommend a chlorine concentration of 100 to 200 mg/liter with a holding time of 2 min. A 200 mg/liter hypochlorite solution circulated for 2-5 min is usually satisfactory for sanitizing pipes cleaned in place. Enclosed vats or tanks can be sanitized by fogging with a chlorinated (200 mg/liter) solution. With open vessels or vats, the sanitizer can be pressure sprayed onto the surfaces.

The use of chlorine has received a great deal of bad publicity due to its reactions with organic compounds and the resultant mutagenic and carcinogenic potentials. Due to these aspects, it has been suggested that, in some cases, the use of chlorine may create a health hazard.

Iodine.—Iodine is well recognized for its germicidal properties against a wide variety of microorganisms. However, its corrosiveness, toxicity, instability, low solubility in water and other unacceptable characteristics have limited the usefulness of iodine in food plants.

Loose combinations formed between iodine and surface active agents are called iodophors. The surface active agents act as carriers and aid in solubilization of the iodine. The surface active agent can be nonionic, cationic, or anionic, but certain nonionic types are preferred. The advantages and disadvantages of using iodophors as sanitizers are listed in Table 10.3.

As with hypochlorites, the iodophors show greater bactericidal activity at an acidic pH. Hence, phosphoric acid buffers are added to the sanitizer to maintain the pH at 4.0 to 5.0.

In the iodophor combination, the germicidal properties of iodine are maintained or enhanced but the characteristic odor of iodine is eliminated. At the recommended dilution, the iodophor solution has a rich amber color and, as it is used, the color fades. Hence, the iodophors have a built-in indicator of effectiveness.

TABLE 10.3

ADVANTAGES AND DISADVANTAGES OF IODOPHORS AS SANITIZERS

Advantages	Disadvantages
Good stability	Effectiveness decreases with increase in pH, such as carryover of alkaline detergent solutions
Long shelf life	Less effective than chlorine against bacterial spores and bacteriophages
Surface activity	May cause off-flavors in dairy products
Generally active against all microorganisms except bacterial spores and bacteriophages	Should not be used at temperatures above 49°C
Destroy yeast cells at a greater rate than do hypochlorites	May cause discoloration
Not affected by hard water salts	More expensive than chlorine
Relatively non-toxic	Causes staining of some material such as plastic hoses
Not corrosive	
Do not penetrate the skin	
Good penetrating and spreading properties	
Acid nature prevents mineral film formation	
Built-in color indicator	
Concentration easily determined	
Less sensitive than hypochlorites to organic matter	
Easily dispensed and controlled	

The iodophors have good stability with a long shelf life. A high degree of surface activity gives them good penetrating and spreading power and aids in quick draining. Residues or films of minerals are prevented since they are solubilized by the acid nature of the iodophors.

Iodophors cannot be used at temperatures above 49°C due to sublimation of the iodine, which results in loss of effectiveness.

Although the iodophors have good germicidal properties, they are less effective than hypochlorites against bacterial spores and bacteriophages.

There is less reaction of iodophors than hypochlorites with organic matter. However, the germicidal power is reduced by starch.

Quaternary Ammonium Compounds.—The quaternary ammonium salts (quats) have the general formula:

$$
\begin{array}{ccccc}
 & & R_2 & & \\
 & & | & & \\
R_1 & - & N & - & R_3 \quad X \\
 & & | & & \\
 & & R_4 & &
\end{array}
$$

The R stands for organic groups ranging from methyl (CH_3) to long chain aliphatic groups (C_8H_{17} to $C_{18}H_{37}$) and phenyl groups. For bactericidal properties, at least one R is a plain or substituted long chain aliphatic group. The X is an anion, such as a halide (Cl, Br), sulfate or acetate.

The quats may be bacteriostatic in low concentration and bactericidal in high concentration. Their lethal effect has been attributed to various activities, including reactions with cell membranes, denaturation of essential cell proteins or enzyme inactivation.

The effects on the cell membrane may be due to membrane lysis, membrane enzyme inactivation or interference with the transport system.

TABLE 10.4

ADVANTAGES AND DISADVANTAGES OF QUATERNARY AMMONIUM COMPOUNDS AS SANITIZERS

Advantages	Disadvantages
Stable	Incompatible with anionic agents in detergents
Long shelf life	Expensive
Stable to temperature changes	Low activity in hard water
Effective in alkaline conditions	Less effective in activity against spores and bacteriophage, as well as coliforms and psychrotrophs
Noncorrosive	Need to rinse residual film from equipment
Odorless	Problem with foam during mechanical application
Less affected by organic matter	
Residual bacteriostatic effect	
Nonirritating to the skin	
Easily dispensed and controlled	
Control off-odors	
Nontoxic	
Active against many microorganisms, especially thermoduric types	
Good penetration qualities	
Combined with nonionics for detergent sanitizing agents	

Naturally occurring microorganisms may be more resistant than laboratory cultures to quats or other sanitizers. Some strains of *Pseudomonas* and *Xanthomonas* grow with certain quats as the sole source of carbon (Dean-Raymond and Alexander 1977).

The quaternary ammonium compounds have certain advantages and disadvantages when used as sanitizing agents. These are listed in Table 10.4.

The quats are not as effective as the hypochlorites in destroying bacterial spores or Gram negative bacteria, including coliforms and psychrotrophs. They are definitely inferior in activity against bacteriophages. However, the addition of certain sequestering agents enhances the viricidal activity. In general, the quats are very effective against Gram positive bacteria.

Generally, residues of quats are not allowed on equipment when food is being processed. Therefore, it is necessary to rinse the equipment, preferably with chlorinated water, prior to processing food.

Raw Materials.—In order to control the raw material, it may be necessary for the processor to become the producer. For ingredients not directly controlled, specifications are needed, if for no other reason than that rejects of another company are not acquired. The specifications must be written so that they are flexible, but rigid enough so that they are useful. Some conditions for microbiological specifications of raw materials were described by Davies (1968) and Mossel (1969). The allowable number of the significant types of microorganisms, as well as the processing treatment that these ingredients are given, should be considered. The presence of microorganisms that may be a public health hazard should be of concern to the processor, especially if the processing treatments do not destroy these potential pathogens. Spoilage organisms cannot be overlooked. The presence of a significant number of resistant bacterial spores in ingredients such as sugar, flour or spices may cause spoilage of canned foods.

Before worrying about microbial analysis, the raw materials should be given a preliminary examination with normal human senses for evidence of filth or decomposition. The presence of rodent pellets or hairs, insect infestation, field rot or decay of fruit and vegetables, a diseased condition of animals prior to slaughter, parasitic infestation of fish or decomposition of the material are causes for rejection. Since the quality generally does not improve with processing, only clean and sound materials should be brought into the processing plant.

After the raw material is inspected and passed, it should be processed as soon as possible. When delays are unavoidable, facilities for storage should be maintained to protect the raw material from contamination and infestation, as well as deterioration. Some fresh raw materials can be modified by storage to obtain a more uniform product. An example is the storage of potatoes for making chips. At low temperatures, reducing sugars will accumulate and result in a dark potato chip. By conditioning these potatoes for a short time at high temperatures, the reducing sugars are returned to a low level.

Many foods have an outside peel, skin or shell to protect the interior portion of the food from microorganisms. However, the organisms on the outside surface serve as a source of contamination when the protective covering is removed during processing. Hence, the raw material should be washed, when needed, to remove field dirt, other material and many of the microorganisms from the outside surface.

Fish that are delivered to the cutting table may be covered with millions of spoilage bacteria. Washing the fish will remove most of these surface organisms and reduce contamination of the equipment and finished product.

The washing of the raw food material should be done in an area separated from the rest of the processing facilities. The water should be free of microorganisms that may cause foodborne illness or spoilage.

Operations.—It is important that the equipment is maintained, kept in good repair and is properly cleaned and sanitized. When fruits and vegetables are cut or peeled, or an animal is killed and the hide removed, the natural defenses are broken down and the food is subject to contamination and the beginning of a chain of events that lead to quality loss and finally to destruction of the food.

Bacteria multiply rapidly. Therefore, the speed of handling becomes important. The operations should be timed so there is no backup of product, since long delays in processing can cause microbial problems. Bacteria develop more rapidly as the temperature rises; hence, the temperature should be kept as low as possible in keeping with processing requirements.

When organisms attain a high level in a food, even if they are inhibited or killed, their enzymes may remain active and cause changes in texture, flavor and color.

Equipment, such as trays or vats, must not be used interchangeably for raw and cooked products unless this equipment is completely cleaned and sanitized before being used for cooked products.

Equipment is designed to reduce the handling of foods by humans. This equipment causes less contamination than humans if it is properly cleaned and sanitized.

To reduce the temperature of poultry carcasses, continuous chillers are used. With proper sanitation during processing, washing of the carcasses before chilling, maintaining low water temperatures, chlorination of the chill water with an acceptable input of fresh water and ice with resultant overflow of excess water, there is a reduction of microbial contamination of the carcasses.

Highly significant reductions in bacterial numbers on beef and sheep carcasses are obtained by washing with high pressure sprays (Kotula *et al.* 1974; Patterson 1972). The microbial load of beef carcasses was lowered when washed with chlorinated (100 to 400 mg/liter) water (Emswiler *et al.* 1976). However, Stevenson *et al.* (1978) found the bacterial counts were not lowered significantly when beef carcasses were sprayed with a solution with 200 mg/liter chlorine. Pedraja (1973) recommended that fish or seafoods be passed under a spray of chlorinated (10 mg/liter) water before and after processing to control bacteria.

Packaging.—Food packages are made from paper, plastic, glass and metal, as well as combinations of these materials. The package acts as a container for the food. The package should identify and advertise the contained product, as well as present an attractive appearance to give it sales appeal. The package usually gives the product a standardized shape. The main function of packaging is to separate the food from the surrounding environment so that the food is protected from recontamination of microorganisms, from pests and from physical and chemical damage due to dirt, atmospheric oxygen, light, storage odors and moisture.

There are certain requirements of packaging materials. The package should be easily filled and emptied. It must be strong and durable since damage to the package will result in contamination. The material used for packaging must not contaminate the food with foreign odors, flavors or toxic substances. Further, the food must not corrode or weaken the package. The package must be able to withstand the processing treatment of the food (heating, freezing, cold storage). The package must be obtainable at a reasonable cost.

If not properly handled and protected, the package may be a source of microorganisms. The food may acquire chemicals from the packaging materials, such as plastics or waxes either alone or coated onto paper. Makinde *et al.* (1976) listed chemicals such as monomers (ethylene, vinyl, propylene, styrene) and catalysts used in preparing plastics or various stabilizers, pigments, antistatic, antifogging, bactericidal and antifungal agents incorporated into plastics. Polycyclic aromatic hydrocarbons or carcinogens such as 3,4-benzopyrene may be associated with waxes. Polyvinyl chloride plastics are made from vinyl chloride, a human carcinogen.

With packaging, the greatest number of people can be supplied with a clean, wholesome supply of food. Some environmentalists consider packages to be an unnecessary waste and a danger to the environment. The questionable sanitary habits of store personnel and customers in handling nonpackaged food should not be tolerated. The so-called health food store personnel are the main offenders.

There is a cry for returnable containers. However, in the past, returnable bottles have served around the home as handy containers for things such as shellac, paint, kerosene and gasoline. Other debris such as cigarette butts, paper and various filth is found in returned bottles. The ecology of packages and packaging materials was discussed by Baribo (1972) and Thomka (1971).

Different packaging materials vary in their physical and chemical properties. The permeability to moisture, oxygen, gases, light and even microorganisms varies with different materials. With the various properties of packaging materials, the characteristics of a package can be engineered to fit almost any food product or processing and storage requirement.

Personnel.—Humans are an important cause of microbial contamination of foods. The GMPs stress: 1) disease control; 2) cleanliness; 3) education and training; and 4) supervision. The concern of the plant personnel for sanitation and personal hygiene must begin with plant management. Management must set a good example, furnishing the proper environment and establishing within the personnel a desire to comply with the requirements necessary to maintain sanitation and personal hygiene.

All personnel should be involved in keeping the processing plant in a neat and orderly manner. However, of special concern are those people that handle ingredients, foods or food contact surfaces. In the handling of cream-filled pastries, cooked meat and poultry products, milk products and salads, the hygiene of workers is a critical control point (Bryan 1974).

Disease Control.—Even with the development of equipment for various purposes, there is still direct handling or close contact of food by humans. Although equipment is a major source of microorganisms, these are usually spoilage types normally associated with foods. Humans are a potential source of pathogens. In this respect, humans are more important than equipment as a source of food contaminants.

Anyone who is sick, has a communicable disease, is a carrier of such disease, or has boils, sores or infected wounds, must not be allowed to work in any area of a food plant where there is a possibility of contaminating the food or a food-contact surface.

Cleanliness.—Employees should maintain a clean appearance. The GMPs suggest the wearing of clean outer garments. Even though the employee dresses in clean clothing at home, it is not sterile. Further, by the time of arrival at the plant, the clothing has been subjected to various sources of contamination, which can include both food spoilage and disease-producing microorganisms. Hence, in some cases, it is required that employees wear special protective outer clothing which is not worn outside the processing plant. In the general guidelines for handling ready-to-eat products, employees that work on raw and the ready-to-eat products must change any garment that has contacted raw product. Due to the shedding of microorganisms by humans, special clean-room clothing may be needed in selected areas. This clothing is essentially that worn in operating rooms, including coveralls or smocks with hand, foot and head coverings. In many food plants, the employees wear company furnished uniforms. Different colored uniforms should be provided for different departments. In this way, there will be less chance for cross contamination of food by personnel.

The hands are used for many things. Therefore, it is essential that before working with food, the hands are washed and, if necessary, sanitized. Simply rinsing the hands is not sufficient. Washing with soap and water does not remove all of the microorganisms; however, a thorough washing with soap and water will remove most of the transient organisms. Antimicrobial soaps will aid in removal of the resident flora. Sanitizing solutions for hand dips may contain chlorine, iodine or any other approved bactericidal agent.

Not only are hand washing and sanitizing needed before beginning work, but also after each break, after leaving and returning to work or when the hands may be contaminated by handling contaminated equipment or picking up debris from the floor.

The handwashing facilities should be provided with foot operated controls to obtain warm water. The tap handles have been found to be highly contaminated. It does little good to wash one's hands and then promptly contaminate them with enteric-type organisms from the tap handles.

In some cases, gloves are worn to control potential contamination. These should be disposable, rubber gloves and should be maintained in a sound, clean and sanitary condition.

Hair is a source of microorganisms, as well as loose hair that contaminates food. Hence, hair coverings, such as hair nets, beard nets, headbands, caps or other effective hair restraints must be worn by employees.

Personnel should not have extraneous materials in the food processing area. Smoking, chewing of gum or tobacco, spitting, eating or drinking cannot be tolerated in areas that will affect the food. Also, the food must be protected from contaminants, such as perspiration, cosmetics and medicants.

With the regulations of various government agencies relative to such things as health and safety, no one except authorized plant employees or government inspectors should be allowed in the food handling or processing areas.

People tend to have bad habits, forget what they are supposed to be doing and get careless. People forget to wash their hands, or merely rinse them. Food is left at room temperature rather than being put into the refrigerator or freezer. Some people believe there is nothing wrong with salvaging unfit food. It is relatively easy to observe examples of poor sanitation, as well as questionable practices when eating in a restaurant or other food service unit.

In a processing plant, the cleanup crew must do a complete and thorough job of cleaning and sanitizing the plant and equipment. If they are not careful, they can contaminate the equipment more than they clean it. The food processing crew must follow rules for acceptable operations.

Government inspectors should be clean and set an example for the plant personnel. It is illogical for a slovenly unkempt person to try to explain sanitation and personal hygiene to other people.

Training.—Many food processing jobs are seasonal and often at relatively low salary. Hence, the food industry tends to obtain young people, part-time housewives or various workers who did not make it elsewhere. Some of them are poorly trained people in the lower socio-economic group. With this combination of jobs and types of people, the turnover rate is rather high and sanitary training is difficult.

The employees must be properly trained and retrained in the correct use of equipment, sanitation and personal hygiene. They must be informed of the perishable character of the food product, sources of contamination and their role in the production of a top quality product.

Since a majority of outbreaks of foodborne illness are due to contamination or mishandling of food at food service establishments and in the home, it seems logical that the general population needs training in the care and handling of food products and in personal hygiene. A course concerning these aspects given in high school might be beneficial.

Supervision.—The supervision of the plant sanitation and the sanitary training of personnel should be under the direction of a well-trained sanitarian.

The sanitarian and other plant supervisors must be able to recognize unsatisfactory employee habits or health problems. If employees cannot or will not follow the necessary rules, the supervisor must remove them from critical areas, since the supervisor is ultimately responsible for compliance with the GMPs or regulations that are applicable to the product being processed.

Microbial Evaluation

A visual inspection of raw material, equipment, packages and finished product can detect gross filth, but microbial analyses are needed to determine the levels of microbial contamination.

Samples for analysis should be obtained from the equipment surfaces, hands, raw material, the material during processing (on-line samples), finished processed product and other things such as air, water or packaging materials that might be involved in product contamination. Routine tests will show day to day variation and help to find weaknesses in raw materials, ingredients, equipment sanitation or personal hygiene.

Not only the total number of organisms but also pathogenic types, such as *Salmonella* or *Staphylococcus* or indicators such as coliforms, *Escherichia coli* or enterococci, should be determined. In some processing plants the spore forming bacteria (*Bacillus, Clostridium* and *Desulfotomaculum*), yeasts, molds, psychrophiles (*Pseudomonas*) or other spoilage types might be evaluated. In the citrus industry, a colorimetric test for the presence of diacetyl and acetylmethylcarbinol is effective in quality control. These are metabolic products of the degradation of citric acid by species of *Lactobacillus* and *Leuconostoc.*

The total aerobic plate count of samples obtained from various stages of production may be used as a general index of plant sanitation. However, processing treatments such as heating will decrease the count and the addition of certain ingredients may increase the number of microorganisms. If sharp increases in numbers occur in a particular area it is evidence that something may be amiss and needs investigation and corrective actions.

A significant increase in the day to day microbial level of equipment surfaces indicates faulty cleaning and/or sanitation. Close observation of the cleaning and sanitizing may reveal the problem so it can be resolved.

Microbial surveys of various foods reveal products handled with good sanitary conditions usually have significantly lower counts than products produced by plants operating with poor sanitary conditions. There are times when poor sanitary practices are not revealed by microbial numbers of the final product, but with sufficient sampling over several sampling periods and analysis of on-line samples, a correlation can be obtained between plant sanitation and bacterial loads. When corrections were made in frozen meat and gravy plants, follow-up samples had significantly lower bacterial levels (Surkiewicz *et al.* 1973). When analyzed, the coliforms, *E. coli* and coagulase positive staphylococci counts generally reveal the sanitation of the plant and/or personal hygiene of the workers.

Food Service Establishments

Of the reported outbreaks of foodborne disease, in which the place of mishandling is known, most are due to food service establishments. It is evident that if the health authorities are really interested in reducing the number of foodborne outbreaks, then the food service establishments need more regulation and observation than has been evident. It is unfortunate that people must be forced to do an acceptable job. Too many owners of food service businesses have no background in sanitation and the proper handling of foods.

The same general requirement for housekeeping, equipment, cleaning, sanitizing and personnel problems and regulations that apply to food processing plants also are required for food service establishments. However, the problems are compounded since many different types of food are utilized at one location and more handling is needed.

In many food processing plants, one or only a few foods are handled and processed, whereas in a restaurant all types of foods in various stages of processing, are handled and utilized. Some raw foods are served with only a little preparation, such as in salads. Other foods such as meats are cooked. The common organisms (salmonellae and staphylococci) that cause foodborne illness are destroyed by cooking meat to an internal temperature of 74°C (Bryan and McKinley 1974). Hamburgers or other meats may not receive sufficient heat treatment during the cooking process to destroy pathogens. Cooked meats for sandwich use may be left at room temperature rather than being refrigerated. The same knives or other equipment used to handle raw meat may be used for cooked meat. Hence, there is cross contamination between raw and cooked products.

Foods that are prepared just prior to serving normally do not cause health problems. It is food that is prepared a day or two before serving and then not properly refrigerated, or food that, after preparation, is held for serving during a two or three hour lunch or dinner period that can cause problems of foodborne illness.

To prevent growth of microorganisms, warming tables should keep hot foods at or above 60°C and cold tables should keep cold foods at or below 7°C. If foods are to be held for any period over two or three hours, they should be refrigerated.

Various accessories (salt, pepper, catsup, sugar, steak sauce, syrup) may be found on the table or served to several customers from one table to another. The customers can be a source of microbial contamination, including potential pathogens, of these supplies.

Certainly things have improved since the days when only a superficial rinse was given to many utensils and glasses. However, further improvements are needed in the food service industry in order to eliminate contamination and foodborne illness.

Due to several outbreaks of gastroenteritis on cruise ships during 1973, the U.S. Public Health Service developed a set of recommendations which include: 1) improving availability and use of handwashing facilities by food service personnel; 2) improving refrigeration facilities and practices; 3) holding all perishable cooked food at 7°C at all times; and

4) upgrading facilities and practices for washing and sanitizing dishes, glasses and utensils. These recommendations are valid for other food service units.

Retail Stores

Mishandling of foods can occur at the retailer with subsequent problems of foodborne illness.

Both raw and processed meats are handled in the meat department. There can be cross contamination so that pathogens associated with raw meats may contaminate processed meats or other foods that will receive no further treatment. Butchers handling raw meats should wash their hands before handling cooked or processed meats.

Sanitation at Home

The practice of sanitation and personal hygiene in the home will go far toward preventing problems in other food handling establishments. People seem to be unaware of their responsibility for proper home storage, hygienic food preparation and sanitary practices in the kitchen. Efforts to improve food protection in the home must overcome family customs and personal habits.

For a foodborne illness to occur, a potentially pathogenic organism must be in an acceptable food environment for growth or production of toxin. A sufficient amount of this contaminated food must be eaten by a susceptible person for illness to occur. Thus, not all unsanitary acts will result in illness. The person who prepares the food becomes complacent and careless. Besides carelessness, other factors that may be involved in foodborne outbreaks are apathy, ignorance and poor judgment.

Food protection in the home encompasses essentially the same systems as in the food processing plant. It begins with the selection of food and continues with transportation, storage, preparation, serving and handling leftovers and wastes.

At the store, we can make a judgment about certain foods by their appearance. Damaged or opened packages should not be selected. If produce has a shriveled appearance, discoloration, evidence of mold or other microbial growth, it should be avoided. In the frozen food counter, do not select packages that are piled above the freezer. Canned foods that show bulging or the leaking of contents should be pointed out to the manager. Although, in general, the food in dented cans may be satisfactory, this type of can does show evidence of rough handling. This could affect the can seams and result in spoilage or a health hazard.

Perishable food should be protected while being transported home. Allowing food to remain in a hot trunk of a car parked in a sunny parking lot for several hours will cause problems. Many people own cooling chests which will give some protection to perishable foods during the period of transportation.

The home should have adequate space for dry storage as well as refrigerated storage. These areas should be kept neat and clean. The foods that are brought home must be promptly stored in the proper area, especially those that require refrigeration or freezing.

The food preparer should practice good personal hygiene. Thoroughly washing the hands is essential, and rewashing is needed when food preparation activities are interrupted for any reason. It is expected that animals should be kept out of processing plants, but some people allow their house pets to run around the kitchen and think little about this. Rodent and insect control are as important at home as in the processing plant.

The kitchen should be kept clean. This includes the floors, walls, cupboard, stove, refrigerator and other accessories. Pots and pans should be cleaned thoroughly. Have you ever seen anyone sanitize the utensils at home? This may account for some of the problems. Although wiping cloths in processing plants and food service establishments are considered to be unsatisfactory, it is unusual to find a home in which dish towels are not used to wipe at least some of the utensils or dishes.

The fresh meats, poultry and fish should be cooked sufficiently to destroy pathogenic types of microorganisms. Eating raw animal products is not recommended. There should be no cross contamination between raw meats and prepared foods.

With most meals, more food is prepared than is eaten. Hence, there are leftovers. If these are to be used again, they should be placed promptly in the correct storage conditions, and perishable foods should be refrigerated. It is not considered acceptable to prepare food ahead of time and allow it to remain at room temperature before serving. Sooner or later, this practice will result in illness.

Summary

The care in producing quality food and handling it to prevent contamination and growth of microorganisms will extend the storage life. Some basic considerations of sanitation are presented in the text, but for further information the reader is referred to other publications, such as FAO (1974), GSA (1977), Thorner (1973), USDA (1970, 1971, 1976) and USDHEW (1965B).

CONTROL OF MICROORGANISMS BY REMOVAL

One means of controlling microbial levels in food is the removal of microorganisms. The systems used commercially for this procedure include washing, centrifugation and filtration.

Washing

Washing of fruits and many vegetables is used in the preparation of these foods for canning, freezing or fresh consumption. Washing removes soil, microorganisms and some residual pesticides. The removal of soil and associated heat resistant spores is important when vegetables are to be heat processed.

After evisceration of broilers, the carcasses go through a spray washer. This removes many of the microorganisms from the carcass surface. Due to the benefits found by spray washing poultry carcasses, this process has been extended to large animal carcasses.

Whether fruits, vegetables or animal carcasses, washing by high pressure water spraying is more effective than merely rinsing.

Washing may not be beneficial in all cases. Fellers and Pflug (1967) reported that washing of pickling cucumbers "reduced the storage life to half that of unwashed fruit." The polishing of fruit with saran brushes can cause invisible injuries to the peel. This can result in increased decay during storage.

If the washing of shell eggs is done improperly, there is increased spoilage during storage.

Centrifugation

This operation involves the separation of a solid from a liquid or a liquid from a liquid by means of centrifugal force. This process has been used in the preparation of sugar, clarification of fruit juice and the separation of cream from milk.

By increasing the gravitational force of the milk separator, it is possible to separate bacteria from milk. This apparatus has been called a bactofuge and the process as bactofugation.

In the bactofuge, the denser bacterial cells, as well as sediment, are thrown toward the vertical wall of the bowl. The outer edge of the bowl has two small holes (0.3 mm) through which about 1.5% of the milk containing from 90 to 97% of the bacteria is removed. By putting two bactofuges in series, a further reduction in bacteria is accomplished. As a result, 99% or more of the bacteria in the original milk is removed.

Although originally described for milk, other applications have been developed, including the separation of microorganisms from substrate in the production of single cell protein, the purification of molasses for use in the production of bakers' yeast (*Saccharomyces cerevisiae*) and the separation of distillers' yeast from the substrate prior to distillation of the alcohol.

When bacteria are removed by centrifugation, there is no damage to the flavor or other quality factors such as occurs with heat. Also, the cells are removed, not just destroyed.

Spores are more dense than bacterial cells and are readily removed from liquid foods. The removal of over 99% of the heat resistant spores and vegetative cells makes it easier to destroy the remaining spores by heat treatment.

There are problems with some applications. The centrifugation also may remove some protein, calcium and phosphorus from milk.

In order to run viscous raw molasses through the bactofuge, it is diluted and heated. Also, heat is applied to other foods to aid in separation.

The use of centrifugation is limited to liquid foods and then only when the density of the bacteria is greater than the food components. Since not all of the microorganisms are removed, these liquid foods cannot be considered as sterile. However, by reducing the microbial load, it is easier to achieve sterilization by heat treating the food.

Filtration

Filtration is the separation of solids from a liquid or gas by passing the substance through a porous material, the filter. The size of the filter pores determines the particles that are trapped on the filter and those that go through.

The materials used for filtration of aqueous solutions, oils or organic solutions include asbestos pad filters, diatomaceous earth discs and cellulose membranes (acetyl-cellulose or nitrocellulose).

Bacterial filters have pores of 0.45 μm. However, if all but the smallest viruses are to be eliminated, a pore size of 0.1 to 0.2 μm is needed.

Membrane filters show a typical sponge structure. Although they are often described as a two-dimensional sieve with uniform sized pores, this is not the case (Wilke 1974). According to him, the pore size of a membrane filter, as stated by the manufacturer, represents the mean of a more or less ideal pore size distribution curve. A 0.45 μm filter was found to have pores ranging from 0.3 μm to over 0.7 μm. On the other hand, data listed by a manufacturer indicate the pore size varies from about 0.43 to 0.47 μm.

A major problem with membrane filters is they become blocked very readily. According to Wilke (1974), even filtering demineralized water causes rapid blockage unless the membrane filter is protected by prefilters. When he used a glass fiber mat as a prefilter, an eight-fold increase in output was obtained before clogging. By replacing the glass fiber mat with an asbestos depth filter, he obtained an additional 10-fold output.

Wilcke (1974) stated that a load of 1.5×10^8 organisms/cm^2 completely blocked the membrane. Besides microorganisms, proteins can hamper the filtration.

There are some advantages in using membrane filters. The inert behavior, high initial flow rate, particle retention due to the "sieve" effect, no shedding of fibers, negligible loss of solution and no effect on the concentration, pH, color or flavor of the filtrate are good features of a membrane filter (Grubert 1974).

Asbestos pad filters are made in various mixtures of washed asbestos fiber with cotton and other filler materials. The efficiency of the filter is determined by the amount of asbestos. This type of filter is highly absorbent due to its negative charge and will remove substances such as viruses, enzymes, proteins and various complexes. This may be an asset or a disadvantage. The major disadvantage is that asbestos fibers are liable to shed from these filters. Consumer advocates have complained about the possible presence of asbestos fibers in beverages filtered with these filters. The concern is due to a chronic lung infection acquired by asbestos workers. Also, these workers tend to have a high incidence of stomach and intestinal cancer. Since the asbestos filters have been used as prefilters for membrane filters for over 20 years, if any health hazards exist, they should be evident by now. However, other fiber-type filters with no asbestos have been developed for prefiltering beverages.

Before the membrane or other filter is used to remove bacteria from substances, it must be sterilized. This usually is accomplished by means of steam heat or with ethylene oxide gas. After filtration, any pipes, filling apparatus and packaging material must be sterile or the product will be recontaminated.

Filtration is used in the sterilization of media or media components that are adversely affected by heat treatment. It is used in the microbial analysis of liquids, such as potable water with low microbial counts (membrane filtration method). The filters concentrate the organisms to a countable level.

Millipore (1970) listed several uses for membrane filters. These include: 1) the filtration of beer or wine for cold pasteurization or for analysis of yeasts and bacteria; 2) filtering air to determine microbial contamination or to remove microorganisms and other particles from air used in contact with food products, in fermentation vats and in packaging food products; 3) in the analysis of additives, pipelines, packaging equipment and foods, such as soft drinks, syrups and liquid sugar for microbial contamination; 4) remove yeasts and bacteria for the biological sterilization of champagne; and 5) remove microorganisms or particulate material from water, such as that used in blending of alcoholic beverages, to achieve clarity.

Cold pasteurization or sterilization by filtration can be used for clear fruit juices, vinegar, soft drinks, syrups and vegetable oils.

Although some enzymes might be removed by asbestos prefilters, other types may have little effect on enzymes. Since filtration does not destroy enzymes, if these cause undesirable reactions in the food, heating must be used to inactivate these organic catalysts.

Filtration has some advantages, as well as disadvantages, as a means to control microorganisms in foods. Among the advantages are that there is less energy needed to filter than to heat liquid foods. Not only are the microorganisms removed, but also particulate matter, so that clarity is improved. Problems include the prefiltering or clogging of the filters, the residual enzymes, the necessity for sterilizing the filter system prior to use, and the necessity for observing aseptic handling and packaging.

BIBLIOGRAPHY

ANDREWS, W.H. *et al.* 1975. Comparative validity of members of the total coliform and fecal coliform groups for indicating the presence of *Salmonella* in the Eastern oyster, *Crassostrea virginica.* J. Milk Food Technol. *38*, 453-456.

BARIBO, L.E. 1972. The ecology of milk packaging. J. Milk Food Technol. *35*, 121-125.

BAUMAN, H.E. 1974. The HACCP concept and microbiological hazard categories. Food Technol. *28*, No. 9, 30, 32, 34, 74.

BRYAN, F.L. 1974. Microbiological food hazards today—based on epidemiological information. Food Technol. *28*, No. 9, 52-66.

BRYAN, F.L. and MCKINLEY, T.W. 1974. Prevention of foodborne illness by time-temperature control of thawing, cooking, chilling and reheating turkeys in school lunch kitchens. J. Milk Food Technol. *37*, 420-429.

DAVIES, A. 1968. Control of raw materials. J. Food Technol. *3*, 431-436.

DEAN-RAYMOND, D. and ALEXANDER, M. 1977. Bacterial metabolism of quaternary ammonium compounds. Appl. Environ. Microbiol. *33*, 1037-1041.

EMSWILER, B.S., KOTULA, A.W. and ROUGH, D.K. 1976. Bactericidal effectiveness of three chlorine sources used in beef carcass washing. J. Anim. Sci. *42*, 1445-1450.

FAO. 1974. Fish and Shellfish Hygiene. Report of a WHO Expert Committee Convened in Cooperation with FAO. Food and Agricultural Organization of the United Nations, Rome.

FELLERS, P.J. and PFLUG, I.J. 1967. Storage of pickling cucumbers. Food Technol. *21*, 74-78.

GRUBERT, G. 1974. The membrane filter technique for sterilizing filtration and for particle removal, with particular reference to the removal of particles from parenteral solutions. *In* International Symposium on Sterilization and Sterility Testing of Biological Substances. R.H. Regamey, F.P. Gallardo and W. Hennessen (Editors). S. Karger, Basel.

GSA. 1977. Code of Federal Regulations. Title 21. Food and Drugs. Office of the Federal Register, General Services Administration. Washington, D.C.

HEID, J.L. and JOSLYN, M.A. 1967. Fundamentals of Food Processing Operations: Ingredients, Methods, and Packaging. AVI Publishing Co., Westport, CT.

HUSS, H.H. *et al.* 1974. The influence of hygiene in catch handling on the storage life of iced cod and plaice. J. Food Technol. *9*, 213-221.

ITO, K. 1974. Microbiological critical control points in canned foods. Food Technol. *28*, No. 9, 46-48.

KAUFFMAN, F.L. 1974. How FDA uses HACCP. Food Technol. *28*, No. 9, 51, 84.

KOTULA, A.W., LUSBY, W.R., CROUSE, J.A. and DEVRIES, B. 1974. Beef carcass washing to reduce bacterial contamination. J. Anim. Sci. *39*, 674-679.

KULIKOVSKY, A., PANKRATZ, H.S. and SADOFF, H.L. 1975. Ultrastructural and chemical changes in spores of *Bacillus cereus* after action of disinfectants. J. Appl. Bacteriol. *38*, 39-46.

MAKINDE, M.A., GILBERT, S.G. and LACHANCE, P.A. 1976. Nutritional implications of packaging systems. Food Product Develop. *10*, No. 7, 112-119.

MILLIPORE. 1970. The Scope of Millipore Technology. Catalog No. LTMT 003 BB. The Millipore Corporation, Bedford, Mass.

MOSSEL, D.A.A. 1969. Microbiological quality control in the food industry. J. Milk Food Technol. *32*, 155-171.

NAS. 1969. An Evaluation of the *Salmonella* Problem. Pub. No. *1683*. National Academy of Sciences, Washington, D.C.

PATTERSON, J.T. 1972. Hygiene in meat processing plants. 5. The effect of different methods of washing, drying and cooling on subsequent microbial changes in sheep carcasses. Record Agr. Res. *20*, 7-11.

PEDRAJA, R.R. 1973. Current status of the sanitary quality of fishery products in the Western Hemisphere. *In* Microbial Safety of Fishery Products. C.O. Chichester and H.D. Graham (Editors). Academic Press, New York and London.

PETERSON, A.C. and GUNNERSON, R.E. 1974. Microbiological critical control points in frozen foods. Food Technol. *28*, No. 9, 37-44.

ROELANTS, P., BOON, B. and LHOEST, W. 1968. Evaluation of a commercial air filter for removal of viruses from the air. Appl. Microbiol. *16*, 1465-1467.

ROSENKRANZ, H.S. 1973. Sodium hypochlorite and sodium perborate; preferential inhibitors of DNA polymerase-deficient bacteria. Mutat. Res. *21*, 171-174.

STEVENSON, K.E., MERKEL, R.A. and LEE, H.C. 1978. Effects of chilling rate, carcass fatness and chlorine spray on microbiological quality and case-life of beef. J. Food Sci. *43*, 849-852.

SURKIEWICZ, B.F., HARRIS, M.E. and JOHNSTON, R.W. 1973. Bacteriological survey of frozen meat and gravy produced at establishments under federal inspection. Appl. Microbiol. *26*, 574-576.

THOMKA, L.M. 1971. Plastic packages and the environment. J. Milk Food Technol. *34*, 485-491.

THORNER, M.E. 1973. Convenience and Fast Food Handbook. AVI Publishing Co., Westport, CT.

USDA. 1970. Federal Facilities Requirements for Existing Poultry Plants. U.S. Department of Agriculture, Washington, D.C.

USDA. 1971. Eggs and egg products inspection. Federal Register *36*, No. 104, 9814-9834.

USDA. 1976. U.S. Inspected Meatpacking Plants. A Guide to Construction, Equipment, Layout. Agr. Handbook No. 191, U.S. Department of Agriculture, Washington, D.C.

USDHEW. 1965A. Sanitation of Shellfish Growing Areas. National Shellfish Sanitation Program Manual of Operations. Part I. U.S. Department of Health, Education, and Welfare. Washington, D.C.

USDHEW. 1965B. Sanitation of the Harvesting and Processing of Shellfish. National Shellfish Sanitation Program Manual of Operations. Part II. U.S. Department of Health, Education and Welfare, Washington, D.C.

WHO. 1969. Recommended International Code of Hygienic Practice for Canned Fruit and Vegetable Products. Secretariat of the Joint FAO/WHO Food Standards Programme, FAO, Rome.

WILKE, H. 1974. Properties of filter sheets made of cellulose-asbestos mixtures and filter membranes. Combination of both filter media in practice. *In* International Symposium on Sterilization and Sterility Testing of Biological Substances. R.H. Regamey, F.P. Gallardo, and W. Hennessen (Editors). S. Karger, Basel.

WLODKOWSKI, T.J. and ROSENKRANZ, H.S. 1975. Mutagenicity of sodium hypochlorite for *Salmonella typhimurium*. Mutat. Res. *31*, 39-42.

WYATT, L.R. and WAITES, W.M. 1975. The effect of chlorine on spores of *Clostridium bifermentans, Bacillus subtilis* and *Bacillus cereus*. J. Gen. Microbiol. *89*, 337-344.

11

Control of Microorganisms by Retarding Growth

By altering the factors discussed in Chapter 4, to retard or inhibit the growth of microorganisms, the storage life of foods can be extended. These factors included temperature, water activity, acidity, oxidation-reduction potential and chemical inhibitors.

There are cases in which there is a rather fine line separating inhibition and death. Some chemicals are bacteriostatic at low concentrations and bactericidal at higher levels. Low temperatures will reduce or inhibit microbial growth while high temperatures will kill the cells.

The retardation of growth means that the food product is not sterile and, if mishandled or abused, the microorganisms may grow and cause spoilage, or be a health hazard.

LOW TEMPERATURE STORAGE

Temperature is one of the most important environmental factors influencing the growth and activity of microorganisms. Both microbial and biochemical activity can be reduced by lowering the temperature of a food and holding it in a refrigerator or freezer. The lower the temperature, the lower the rate of biochemical reactions or microbial activity. Therefore, it might be assumed that freezing and freezer storage would be best for all foods. However, freezing damages food tissues, and, for some types of foods, this damage is unacceptable. For some foods, low temperature refrigeration can cause injury. Thus, the temperature at which a food can be stored depends upon the type of food as well as economic factors. Removing heat and maintaining a controlled cold environment are expensive. Although many types of foods or ingredients might benefit from low temperature storage, we are concerned mainly with perishable foods.

Refrigeration

The minimum growth temperature for most mesophilic organisms is about 10°C. Since refrigeration generally refers to temperatures below 10°C, these mesophiles do not grow and are not a problem with refrigerated food. Some mesophilic microorganisms are psychrotrophic and can grow on refrigerated foods. However, the optimum growth temperature of these organisms is 25° to 30°C and reduced growth occurs at temperatures of 10°C or less.

The main organisms of concern on refrigerated foods are the psychrophiles. These organisms can grow at temperatures as low as −15°C and some reportedly have an optimum growth temperature as low as 10°C.

Although psychrotrophic and psychrophilic microorganisms can grow at low temperatures, they grow slower as the temperature is reduced. The lower growth rate is very evident as the minimum growth temperature of an organism is approached. Also, as the refrigerator temperature is reduced from 10°C, fewer strains can grow and cause spoilage. Hence, food will spoil about four times as fast at 10°C and twice as fast at 5°C as at 0°C.

Some microorganisms causing foodborne illness are psychrotrophic, but most will not grow or produce toxins below 4.4°C. None of these organisms grow below 1.7°C. For safety, perishable foods should be held below these temperatures. Under no circumstances should perishable foods be left at room temperature longer than necessary, as dictated by good handling procedures.

Quite often, other systems for preserving food are used in conjunction with refrigeration. Since psychrophiles are primarily aerobic, vacuum or carbon dioxide packaging with the exclusion of oxygen will delay microbial spoilage. Salting, curing, smoking or other chemical systems, as well as mild heat treatments, may be used to inhibit or reduce the microorganisms on refrigerated food.

Cold Shock.—When microorganisms are transferred rapidly from one temperature to another, they are subjected to a thermal stress. The cells are the most sensitive to cold shock when they are in the early exponential growth phase. The DNA of cold shocked cells contains more single-strand breaks or nicks than either unshocked or recovered cells (Sato and Takahashi 1970). There is a leakage of amino acids and nucleotides from the cells. However, the type of cellular response depends upon the rate of cooling and warming of the cells.

Various substances, such as amino acids, sucrose, magnesium, calcium and manganese, have a protective effect on the microorganisms.

Rapid Chilling.—It is necessary to remove the field heat from fruits and vegetables, and the animal heat from milk, eggs or meat, to prevent spoilage of these foods. There are many ways in which the temperature of a food can be lowered rapidly. Liquid foods can be put through a plate or other type of heat exchanger. Other foods are rapidly cooled by forced air cooling, spray chilling, hydrocooling, icing or in continuous chillers in slush ice. Fluidized beds are used for cooling vegetables after blanching and prior to freezing.

It is recommended that foods be refrigerated in small containers, so that thorough cooling is possible within a short time. Bulk refrigeration can be a contributing factor in foodborne illness.

When food is placed in storerooms, it should not be stacked tightly, since allowance must be made for air circulation. This will aid in keeping a uniform temperature and humidity in the warehouse, storeroom, refrigerator or freezer.

TABLE 11.1

APPROXIMATE FREEZING POINTS OF SELECTED FOODS

Food	Temperature °C	Food	Temperature °C
Beef	−1.6 to −2.2	Oranges	−2.2
Pork	−1.6 to −2.2	Peaches	−1.4
Poultry	−2.8	Pears	−1.9
Fish	−0.6 to −3.3	Pineapples	−1.2 to −1.6
Shellfish	−2.2 to −2.7	Asparagus	−1.2
Shell eggs	−2.2 to −2.8	Beans	−1.1 to −1.3
Egg yolk	−0.6 to −0.7	Beets	−2.8
Egg white	−0.4 to −0.5	Carrots	−1.3
Milk	−0.5 to −0.6	Celery	−1.1
Cheeses	−2.2	Corn, sweet	−1.7
Brick	−8.7	Cucumbers	−0.8
Cottage	−12.7	Lettuce	−0.4
Ice cream	−2.8	Melons	−1.5 to −1.7
Apples	−2.0	Peas, green	−1.1
Apricots	−2.2 to −2.3	Peppers, green	−1.1
Avocados	−2.7	Potatoes	−1.7
Bananas	−1.0 to −3.3	Pumpkins	−1.1
Berries	−0.9 to −1.7	Rhubarb	−2.0
Cherries	−2.2 to −4.2	Spinach	−0.9
Grapes	−2.5 to −3.8	Sweet potatoes	−1.9
Lemons	−2.2	Tomatoes	−0.9
Limes	−1.5	Turnips	−0.8
Nuts	−4.5 to −10.5	Fruit pies	−3.0 to −4.0
Olives	−1.9	Meat pies	−1.4 to −1.9

Latent-zone Chilling.—Refrigeration usually is considered to be storage at temperatures from 0° to 5°C. The freezing point of most foods is below 0°C (Table 11.1). Hence, storage below 0°C, but above the freezing point has several advantages (Smith 1976). At these colder temperatures there is a longer shelf life than obtained with normal refrigeration. The energy requirement is lower and there is less product alteration or damage than when the food is frozen.

Types of Food.—Due to the different characteristics of foods, the cooling and refrigeration requirements vary. The potential refrigerated shelf life is influenced by the chemicals in a food.

Meat.—It is desirable to cool and hold meat below 3°C and as close to −2°C as possible to prevent growth of potential pathogens and to retard growth of spoilage organisms and maintain the meat in a fresh condition (Table 11.2). Below −2°C, meat will freeze.

From a microbiological viewpoint, it is desirable to remove the body heat from carcasses as soon as possible after slaughter. The body temperature varies from 38° to 40.5°C, with an average of 38.9°C, depending on the animal.

The rate at which a carcass is cooled depends not only on the chilling room temperature, but also on the size, heat capacity and amount of fat covering of the carcass and the velocity at which the air is circulated in the room (Fig. 11.1). Light carcasses (lamb, pork, veal and small beef) will cool in ⅓ to ½ the time required for heavy beef carcasses. Increasing the air velocity will reduce the cooling time, but also may increase shrinkage due to moisture loss. A high humidity will help reduce shrinkage, but may stimulate microbial growth on the carcass surface. Although microbiologically the carcasses should be chilled as rapidly as possible, this may result in less tender meat.

Beef carcasses are generally aged to increase the tenderness of the meat. Smith *et al.* (1971) suggested that chilling the carcasses for the first 16-20 hr at 16°C appears to be the most practical system to use to develop tenderness.

In beef aging rooms, the use of UV light has made possible higher room temperatures to obtain a more rapid tenderization of the meat. The UV light, as well as the ozone produced by these lamps, helps retard microbial growth.

The use of vacuum packaging for aging beef at 0° to 4.4°C was investigated by Minks and Stringer (1972). The bacterial counts and the shrinkage losses are significantly lower, with beef packed in vacuum as compared to beef that is not packaged.

TABLE 11.2

STORAGE CONDITIONS OF SELECTED ANIMAL PRODUCTS

Product	Storage Temp °C	Conditions Humidity %	Approximate Storage Life
Meat			
Beef			
Fresh	−2.0 to 1.11	88-92	1-6 weeks
Pork			
Fresh	−2.0 to 1.11	85-90	3-12 days
Bacon	1.11 to 4.44	85	2-4 weeks
Sausage			
Smoked	4.44 to 7.22	85-90	
Lamb			
Fresh	−2.0 to 1.11	85-90	5-12 days
Poultry			
Fresh	−2.0 to 0		1 week
Eggs			
Shell	−1.7 to 0.6	85-90	8-9 months
Dried	1.67	Low	6-12 months
Cheese	1.67	65-70	
Brick	−1.11 to 1.11	50	
Cottage	0 to 1.11	45	
Fish			
Fresh	0.56 to 4.44	90-95	5-20 days
Smoked	4.44 to 10.0	50-60	6-8 months

Source of Data: Farrall (1976)

Lower grade carcasses are boned and the meat used in sausage and other processed products. The USDA requires that all fabrication areas (beef breaking, pork cutting or boning rooms) be maintained at 10°C or lower to retard bacterial growth.

Fresh meat is packaged for retail sale and put into a refrigerated display case. The salable life for this meat is about three days. In the home, refrigeration (2.2° to 4.4°C) of meat should be limited for maximum quality. Ground meats, liver and poultry should not be stored more than two days, roasts and steaks for five days and ham or bacon for seven days (Table 11.3). For longer storage, the meat should be frozen.

Cured Meats.—The psychrophilic flora is inhibited by the curing salts so these products are more stable than fresh meats during refrigerated storage. They often receive a pasteurization treatment that also increases the stability. Vacuum packaging aids in preventing discoloration and the anaerobic environment inhibits aerobic bacteria and molds.

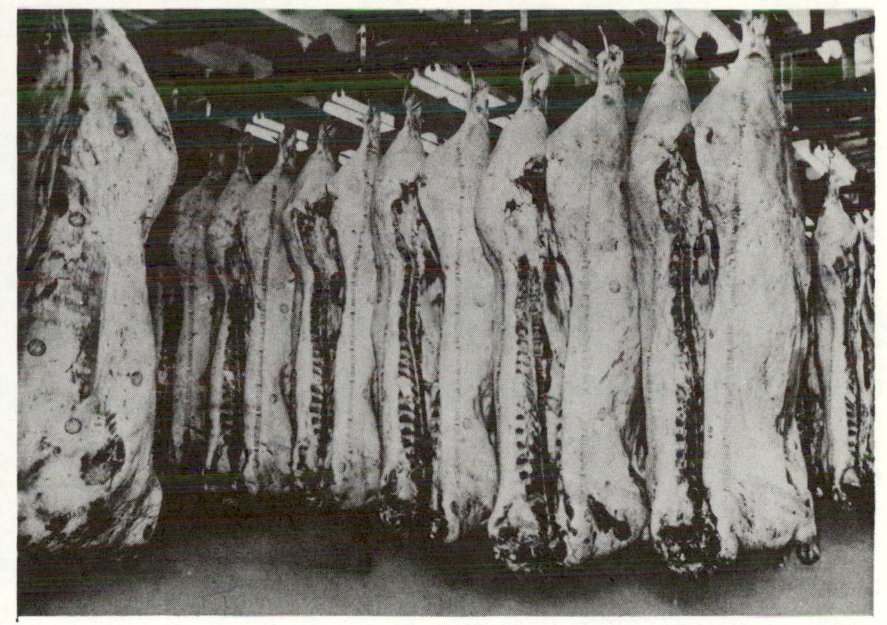

Courtesy of USDA

FIG. 11.1. CATTLE CARCASS COOLER

Note the fat covering on the carcasses.

Poultry.—The storage life of poultry carcasses is limited by microbial activity on the surface. The initial bacterial load and temperature of storage influence the storage life of poultry carcasses.

After processing, the rapid removal of body heat is important. This may be accomplished by continuous immersion chilling, spray chilling or air chilling. Immersion chilling is the usual practice in the USA. It is more efficient, effective and economical than other systems.

For retail sale, the carcasses are packed on a fiberboard base and overwrapped or are packed in a plastic bag. They may be whole or cut-up. Freshly chilled poultry carcasses packaged and held at −0.6°C spoil in 17 to 19 days and at 3.3°C, the storage life is about 10 to 11 days. Poultry carcasses are more subject to spoilage than beef, pork or lamb carcasses. The surface to volume ratio is greater for poultry carcasses which may account for some of this difference. Also, the method of slaughter and processing may result in a greater contamination of poultry carcasses by psychrophilic spoilage types of microorganisms.

TABLE 11.3

HOME STORAGE CHART FOR MEAT AND POULTRY

Product	Storage Period (to Maintain Quality) Refrigerator 1.7°–4.4°C Days	Freezer −17.8°C Months
Meat, fresh		
Roasts (beef and lamb)	3 to 5	8 to 12
Roasts (pork and veal)	3 to 5	4 to 8
Steaks (beef)	3 to 5	8 to 12
Chops (lamb and pork)	3 to 5	3 to 4
Ground and stew meats	1 to 2	2 to 3
Variety meats	1 to 2	3 to 4
Sausage (pork)	1 to 2	1 to 2
Meat, processed		
Bacon	7	1
Frankfurters	7	½
Ham (whole)	7	1 to 2
Ham (half)	3 to 5	1 to 2
Ham (slices)	3	1 to 2
Luncheon meats	3 to 5	Freezing not recommended
Sausage (smoked)	7	Freezing not recommended
Sausage (dry and semi-dry)	14 to 21	Freezing not recommended
Meat, cooked		
Cooked meats and meat dishes	1 to 2	2 to 3
Gravy and meat broth	1 to 2	2 to 3
Poultry, fresh		
Chicken and turkey	1 to 2	12
Duck and goose	1 to 2	6
Giblets	1 to 2	3
Poultry, cooked		
Pieces (covered with broth)	1 to 2	6
Pieces (not covered)	1 to 2	1
Cooked poultry dishes	1 to 2	6
Fried chicken	1 to 2	4

Source: USDA (1970)

Shell Eggs.—As soon as an egg is laid, its quality begins to decline. With our present system of producing, distributing and marketing shell eggs, the main concern is with maintaining quality rather than with microbial spoilage. Low temperature, shell treatments, thermostabilization, carton overwraps, controlled humidity and various other innovations have been suggested or used to help maintain egg quality.

When shell eggs are stored at −1.1°C, the rate of moisture loss is decreased, the amount of carbon dioxide lost is reduced, the rate of chemical change is slowed and microbial growth is retarded.

Egg Products.—These are obtained by carefully breaking the egg shell and removing the liquid. Most of the egg products are obtained by using egg breaking machines. The highly perishable nature of egg yolk and whole eggs makes it necessary that this liquid be cooled to 4.4°C or less within 1½ hr after mixing and straining (to remove egg shell particles). Egg whites can be held at a higher temperature than yolk or whole egg due to the microbial inhibitors and the less available nutrients in egg white.

Liquid eggs can be held at 0° to 5°C with relatively little change in the microbial load up to 24 hr. However, at a temperature of 10°C or more, significant increases occur in the number of microorganisms.

Milk and Milk Products.—When obtained, milk is at the body temperature of the cow (about 39°C).

Raw milk should be held as close to 0°C as possible. The coliform count tends to decrease at 3.3°C or less, while the psychrophiles can show a 10-fold increase in 3 days at 7.2°C. Pasteurized milk should be held at 5°C or less to obtain the maximum storage life.

Buttermilk, yogurt, cottage cheese and other soft cheeses are chilled promptly after manufacture to near 0°C. With holding at 0°C, they have a shelf life of a week or more. Unfortunately, due to the practice of stacking of cartons in the display case, those packages at the top may reach 5° or even 10°C. This shortens the useful life of the product.

The refrigerator in the home may not be as cold as desired and the food is removed from the refrigerator during preparation of the meal. Allowing these products to remain at room temperature for one hour each day can reduce the storage life.

Some dairy products such as hard cheeses can have a refrigerated storage life of several months or more. Mold growth can occur on the surface of hard cheese. This growth is not desirable due to the formation of off-flavors in the immediate area of mold growth. Also, the potential production of mycotoxins by the molds must be considered.

Fishery Products.—These products are very perishable. Fishery products must be cooled promptly and held in this condition until consumed or used in further processed items. The methods for cooling include ice, refrigerated sea water and cold-air refrigeration.

There is a marked extension in the storage life of fish by small reductions in the temperature in the range of 0°C. A reduction in temperature from 3° to 0°C doubles the storage life and a reduction from

10° to 0°C increases the storage life by a factor of 5 or 6. Fish spoils rapidly at temperatures of 15° or 20°C.

Ocean whiting can be stored in ice or refrigerated sea water (−1.1°C) and maintain good quality for 7 days. Unfrozen sardines can be ice-packed and held at −1.1° to −0.5°C for only three days. The storage life of trout and herring at 0°C can be extended from one to two weeks by vacuum packaging (Hansen 1972). This retards oxidative rancidity of the fat, as well as microbial growth.

According to Cobb *et al.* (1974), shrimp boats remain at sea for 4 to 20 days. To maintain the quality, the shrimps are packed in ice. As with other fishery products, icing can cause a certain amount of leaching of nutrients from the shrimp. With ice storage, the shrimp in the lower layers tend to have a higher bacterial load due to the melting ice carrying bacteria from the upper layers. The weight of the shrimp and ice will tend to crush the shrimp at the bottom of the pile. This crushing, plus the bacteria washing down from the upper layers, increases the rate of deterioration.

Fruits and Vegetables.—Substantial losses occur in the shipping, distribution and retailing of fresh fruits and vegetables. These losses are due to biochemical, enzymatic and bacteriological activities that cause over-ripeness and decay. Prolonging the storage life of fresh produce is advantageous, not only to the food industry, but also to the consumer, since it is the consumer who pays for the losses in the form of higher prices.

To help control these losses, it is essential that the field heat be removed from the fresh produce as soon as possible after harvesting. Then the fresh fruit produce should be maintained in a cool condition.

Two systems for removing the field heat are hydrocooling and vacuum cooling. Hydrocooling is applicable to both fruits and vegetables. In this system, cold water which is sprayed over the fresh product reduces the temperature of the produce at a very high rate. The warmed water is recirculated through ice or a refrigeration system to cool it before re-spraying. In some cases, hydrocooled products have developed more decay than untreated produce. The addition of chlorine helps control the microorganisms in the water, as well as those on the surface of the produce.

Vacuum cooling is adaptable to products that have a large surface to volume ratio. This system works in a manner similar to a human perspiring. The evaporation of water from the surface produces a cooling effect. To prevent dehydration of the fresh produce, water is sprayed onto it. Then the vacuum treatment evaporates this sprayed-on water and cools the produce. The amount of cooling is roughly proportional to the water loss. Products in which vacuum cooling is beneficial include lettuce, spinach and cabbage.

Besides cold storage, controlling the atmospheric gases (oxygen and carbon dioxide) and the relative humidity are beneficial in maintaining quality during storage of fruits and vegetables (Tables 11.4 and 11.5).

TABLE 11.4

RECOMMENDED STORAGE TEMPERATURES, RELATIVE HUMIDITIES AND APPROXIMATE STORAGE LIFE OF FRUITS AND NUTS

Commodity	Storage Temperature, °C	Relative Humidity, %	Approximate Storage Life
Apples	−1.1 to −0.6	85 to 90	—
Apricots	−0.6 to 0	85 to 90	1 to 2 weeks
Avocados	2.8 to 8.9	85 to 90	—
Bananas	11.7 to 15.6	85 to 90	1 to 3 weeks
Berries			
Blackberries	−0.6 to 0	85 to 90	7 to 10 days
Dewberries	−0.6 to 0	85 to 90	7 to 10 days
Gooseberries	−0.6 to 0	85 to 90	3 to 4 weeks
Loganberries	−0.6 to 0	85 to 90	7 to 10 days
Raspberries			
Black	−0.6 to 0	85 to 90	7 to 10 days
Red	−0.6 to 0	85 to 90	7 to 10 days
Strawberries	−0.6 to 0	85 to 90	7 to 10 days
Cherries	−0.6 to 0	85 to 90	10 to 14 days
Coconuts	0 to 1.7	80 to 85	1 to 2 weeks
Cranberries	2.2 to 4.4	85 to 90	1 to 3 months
Figs	−0.6 to 0	85 to 90	10 days
Grapefruit	0 to 10	85 to 90	—
Grapes			
Vinifera	−1.1 to −0.6	88 to 92	3 to 6 months
American	−0.6 to 0	80 to 85	3 to 8 weeks
Lemons	12.8 to 14.4	85 to 90	1 to 4 months
Limes	7.2 to 8.9	85 to 90	6 to 8 weeks
Mangoes	10.0	85 to 90	15 to 20 days
Nuts	0 to 2.2	65 to 70	8 to 12 months
Olives	7.2 to 10	85 to 90	4 to 6 weeks
Oranges	−1.1 to 1.1	85 to 90	8 to 10 weeks
Papayas	7.2	85 to 90	15 to 20 days
Peaches	−0.6 to 0	85 to 90	2 to 4 weeks
Pears	−1.7 to −0.6	88 to 92	2 to 7 months
Persimmons	−1.1	85 to 90	2 months
Pineapples			
Mature green	10 to 15.6	85 to 90	3 to 4 weeks
Ripe	4.4 to 7.2	85 to 90	2 to 4 weeks
Plums, prune	−0.6 to 0	85 to 90	3 to 8 weeks
Pomegranates	−0.6 to 0	85 to 90	2 to 4 months
Quinces	−0.6 to 0	85 to 90	2 to 3 months

Adapted from Desrosier and Desrosier (1963).

TABLE 11.5

RECOMMENDED STORAGE TEMPERATURES, RELATIVE HUMIDITIES AND APPROXIMATE STORAGE LIFE OF VEGETABLES

Commodity	Storage Temperature °C	Relative Humidity, %	Approximate Storage Life
Artichokes, Jerusalem	0	90 to 95	2 to 5 months
Asparagus	0	90 to 95	3 to 4 weeks
Beans			
Green	7.2	85 to 90	8 to 10 days
Lima—unshelled	0	90 to 95	2 to 4 weeks
shelled	0	90 to 95	15 days
Beets			
Topped	0	90 to 95	1 to 3 months
Not topped	0	90 to 95	10 to 14 days
Broccoli, Italian	0	90 to 95	7 to 10 days
Brussels sprouts	0	90 to 95	3 to 4 weeks
Cabbage	0	90 to 95	3 to 4 months
Carrots			
Topped	0	90 to 95	4 to 5 months
Not topped	0	90 to 95	10 to 14 days
Cauliflower	0	90 to 95	2 to 3 weeks
Celery	−0.6 to 0	90 to 95	2 to 4 months
Corn, green	−0.6 to 0	90 to 95	4 to 8 days
Cucumbers	7.2 to 10	90 to 95	10 to 14 days
Eggplants	7.2 to 10	90 to 95	10 days
Endive	0	90 to 95	2 to 3 weeks
Garlic	0	70 to 75	6 to 8 months
Horseradish	0	90 to 95	10 to 12 months
Leeks, green	0	85 to 90	1 to 3 months
Lettuce	0	90 to 95	2 to 3 weeks
Melons (ripe)			
Watermelon	2.2 to 40	80 to 85	1 to 2 weeks
Muskmelon	4.4 to 10	80 to 85	10 to 14 days
Honeydew	4.4 to 10	80 to 85	2 to 4 weeks
Casaba, Persian	4.4 to 10	80 to 85	4 to 6 weeks
Mushrooms	0	85 to 90	5 days
Okra	10	85 to 95	2 weeks
Onions	0	70 to 75	6 to 8 months
Parsnips	0	90 to 95	2 to 4 months
Peas, green	0	85 to 90	1 to 2 weeks
Peppers, green	7.2	85 to 90	8 to 10 days
Potatoes	3.3 to 4.4	85 to 90	6 to 9 months
Pumpkins	10 to 12.8	70 to 75	2 to 6 months
Rhubarb	0	90 to 95	2 to 3 weeks
Rutabagas	0	90 to 95	2 to 4 months
Spinach	0	90 to 95	10 to 14 days
Squashes			
Summer	4.4 to 10	85 to 95	2 to 3 weeks
Winter	10 to 12.8	70 to 75	4 to 6 months
Sweet potatoes	12.8 to 15.6	80 to 85	4 to 6 months
Tomatoes			
Ripe	4.4 to 10	85 to 90	7 to 10 days
Mature green	12.8 to 21.1	80 to 85	3 to 5 weeks
Turnips—topped	0	90 to 95	4 to 5 months

Adapted from Desrosier and Desrosier (1977).

As is generally true of all foods, most fruits and vegetables keep their fresh quality longer if stored at slightly above their temperature of freezing. The freezing temperature of fruits varies about −3.9° to −1.0°C and vegetables about −3.7° to −0.5°C (Table 11.1).

Some fruits and vegetables are injured by low temperature storage (Table 11.6). When injured produce is returned to room temperature, the respiration rate rises, then declines rapidly, due to the death of cells in the tissues.

TABLE 11.6

COMMODITIES SUSCEPTIBLE TO COLD INJURY WHEN STORED AT ONLY MODERATELY LOW TEMPERATURES

Commodity	Approximate Lowest Safe Temperature, °C	Character of Injury When Stored Between 0°C and Safe Temperature
Apples		
certain varieties	1.1 to 2.2	Internal browning, soggy breakdown.
Avocados	7.2	Internal browning.
Bananas		
green or ripe	13.3	Dull color when ripened.
Beans (snap)	7.2 to 10	Pitting increasing on removal, russeting on removal.
Cranberries	1.1	Low-temperature breakdown.
Cucumbers	7.2	Pitting, water-soaked spots, decay.
Eggplants	7.2	Pitting or bronzing, increasing on removal.
Grapefruit	7.2	Scald, pitting, watery breakdown, internal browning.
Lemons	12.8 to 14.4	Internal discoloration, pitting.
Limes	7.2	Pitting.
Mangoes	10	Internal discoloration.
Melons		
Cantaloupe	7.2	Pitting, surface decay.
Honey dew	4.4 to 10	Pitting, surface decay.
Casaba	4.4 to 10	Pitting, surface decay.
Crenshaw and Persian	4.4 to 10	Pitting, surface decay.
Watermelons	2.2	Pitting, objectionable flavor.
Okra	4.4	Discoloration, water-soaked areas, pitting decay.
Olives, fresh	7.2	Internal browning.
Oranges, California	1.7 to 2.8	Rind disorders.
Papayas	7.2	Breakdown.
Peppers, sweet	7.2	Pitting, discoloration near calyx.
Pineapples		
mature–green	7.2	Dull green when ripened.
Potatoes		
Chippewa and Sebago	4.4	Mahogany browning.
Squash, winter	10 to 12.8	Decay.
Sweet potatoes	12.8	Decay, pitting, internal discoloration.
Tomatoes		
mature–green	12.8	Poor color when ripe; tendency to decay rapidly.
ripe	10	Breakdown.

Adapted from Desrosier and Desrosier (1977).

Tropical and subtropical fruits are especially subject to low temperature injury. Even some fruits that are grown in temperate climates can be affected.

If grapefruit or lemons are held at 10° to 12.5°C for several weeks, they develop abnormalities of the skin and flesh. Limes can be stored for about eight weeks at 7.5° to 10°C. Oranges are less susceptible than other citrus fruits to chilling injury.

Mature green tomatoes and peppers are injured by long exposure to low temperatures. If ripe, tomatoes may be held a few days near 0°C, but for storage of a week or longer, 10°C is preferred. Green tomatoes should not be stored below 12.5°C. Injured green tomatoes do not ripen satisfactorily and they are subject to infection by microorganisms that cause rot and decay. On the other hand, storage of green tomatoes above 21°C accelerates ripening and the development of soft rot decay.

When cucumbers are stored at 7°C or lower for 10 days or more, they develop pits, dark colored watery patches and tissue collapse. These damaged tissues can readily become infected with microorganisms that cause rotting. A temperature of 10°C and 95% RH are the best conditions for storing and transporting cucumbers (Etchells *et al.* 1973).

Potatoes can heal injured areas of the skin if held at normal room temperatures. This wound healing is very slow at 10°C. Sweet potatoes need a curing treatment at about 29.5°C and a humidity of 85% to heal cuts and abrasions. After this treatment, 12.5°C is recommended, since chilling injury occurs at lower temperatures. The injury causes darkening of the tissues and makes the potatoes susceptible to rotting.

Some vegetables deteriorate rapidly if not placed promptly into low temperature storage. Quick cooling and storage at 0.5° to 1.5°C are needed for corn, peas and green lima beans. Leafy vegetables, beets and carrots need quick cooling and storage at 0.5° to 4.5°C (the temperature depending on the vegetable) to maintain their quality.

Cereals.—The cold storage of grain is possible but Hyde and Burrell (1969) concluded that using refrigeration in temperate climates was only marginally better than ventilation with selected ambient air.

Canned Food.—Various biochemical reactions can occur in canned foods during storage. These changes include loss in organoleptic quality, alterations in chemical constituents and losses in nutrients. These reactions are generally reduced when the canned food is held in cold storage (10°C or less), than at ambient-temperature storage. The extent of any benefit depends upon the type of food.

Freezing

Foods are frozen in order to extend their storage life over that obtained in a refrigerator. The food is preserved by using temperatures low enough to stop or greatly reduce the deterioration caused by microorganisms, enzymes or chemicals such as oxygen. It is generally recognized that freezing offers one of the best means of maintaining the fresh color, flavor and appearance of many foods.

General Aspects.—Pure water at atmospheric pressure freezes at 0°C. As the temperature is lowered below 0°C, the vapor pressures of water and ice are reduced so that the water activity below −10°C is lower than that necessary for the growth of most bacteria (Table 11.7).

TABLE 11.7

VAPOR PRESSURES AND WATER ACTIVITY OF WATER AND ICE

Temperature °C	Vapor Pressure Liquid Water mm Hg	Vapor Pressure Ice mm Hg	Water Activity $\frac{VP_{ice}}{VP_{water}}$
0	4.579	4.579	1.00
−5	3.163	3.013	0.95
−10	2.149	1.950	0.91
−15	1.436	1.241	0.86
−20	0.943	0.776	0.82
−25	0.607	0.476	0.78
−30	0.383	0.286	0.75
−40	0.142	0.097	0.68
−50	0.048	0.030	0.62

The fluids in plant or animal tissues are aqueous solutions. The freezing point of these solutions may range from 0° to −10°C (Table 11.1). When food is subjected to a temperature below freezing and the extracellular water begins to freeze, the solutes in this water tend to migrate to the remaining liquid. This increases the concentration of solutes and lowers the freezing point of this liquid. The extent of solute concentration is influenced by the product characteristics and the rate of freezing. Some solid material is occluded within or among the ice crystals, the amount depending upon the freezing rate.

As the solute concentration of the extracellular liquid increases, an osmotic differential occurs and water tends to leave the plant, animal or microbial cells. Slow freezing causes a greater degree of solute concentration and dehydration of the cells than does rapid freezing.

When a liquid is cooled slowly, solidification commences at a relatively small number of places, and the resultant crystals grow to large sizes. Rapid cooling, on the other hand, results in the initiation of a large number of smaller crystals. The rate of freezing, therefore, may alter the physical structure of the tissue. Rapid freezing of food is desired.

The rupture of the cellular structure may result in a poor texture of the thawed food. The exudation of juices, resulting in "drip," may occur in frozen and thawed plant and animal tissue. It contains dissolved protein, other nitrogenous substances, vitamins and minerals.

Although most of the water in a food is frozen at −20°C, there is usually a certain amount of bound water that does not freeze. This unfrozen water is considered to be not available for biological activity, so it should not affect microbial deterioration of the frozen food.

Systems for Freezing.—Heat can be removed from foods by convection, conduction, evaporation or radiation. Commercially, freezing systems use convection, conduction or combinations of these. The product may be frozen before packaging, after packaging or even after casing. Packaged foods freeze more slowly than unpackaged foods, given the same conditions of freezing, due to the insulating effect of the packaging material.

Many systems have been developed for freezing foods (Table 11.8). The system selected for use depends on the type of food, speed of freezing and cost factors. Although relatively slow, tunnel air blast freezing can be used for most foods.

TABLE 11.8

SYSTEMS FOR FREEZING FOOD

Still air (sharp freezer)
Air blast (room)
Tunnel (air blast)
Fluidized bed
Plate freezers (single, double or pressure)
Slush freezing
Liquid immersion (various liquids—brine, glycol)
Freezant spray (dichlorodifluoromethane)
Liquified gases (CO_2, N_2)

Fluidized bed freezing is a means of rapidly freezing small particulate material. The food to be frozen is placed on a mesh belt and very cold air is blown upward through the mesh. The velocity of the air is such that the food particles seem to float. The moving and turning of the food allow all of the food surfaces to be exposed so that the food is frozen in a few minutes. One advantage of this system is that the food particles are frozen individually. The entire carton of food does not need to be thawed when only a small portion is needed.

Liquid immersion freezing is used to rapidly freeze poultry and other similar products. Liquid is a better conductor of heat than is air, so that more rapid freezing is accomplished than in air blast freezing. Various substances have been used as liquid freezants.

Direct contact freezing, using a fluorocarbon, dichlorodifluoromethane (Freon), combines the speed of cryogenic freezing with the low cost of air blast freezing. The freezant at −22°C is used for immersion or a spray to freeze food. The system is enclosed so that the freezant can be recycled, but losses can occur.

For rapid freezing, liquified gases (carbon dioxide or nitrogen) are used. There are advantages for liquid nitrogen freezing. The system is shown in Fig. 11.2. Since the food is frozen in an inert atmosphere of nitrogen, surface oxidation is prevented. The rapid freezing and low temperature prevent microbial growth. There are problems associated with this cryogenic freezing. The sudden reduction in the temperature of food by direct contact with liquid nitrogen can cause shattering of the food. Other materials such as plastic, rubber or glass become brittle and can crack or shatter due to the thermal shock caused by liquid nitrogen.

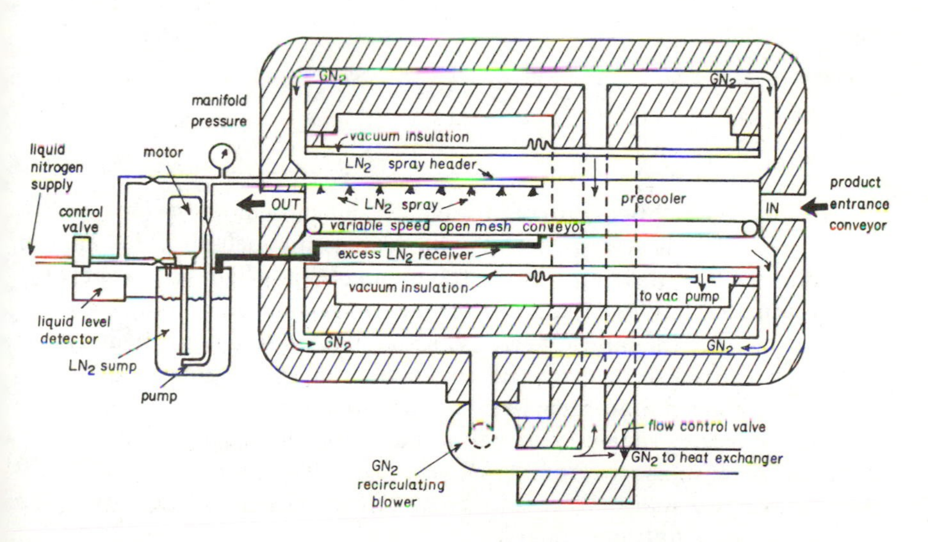

Courtesy of Farrall (1976)

FIG. 11.2. LIQUID NITROGEN FLASH FREEZER

Stability of Frozen Foods.—Some aspects that affect the storage life of frozen foods are the treatments prior to freezing, the type of packaging, the temperature, fluctuations in the temperature and conditions for thawing of the frozen food. The frozen storage life of selected foods is listed in Table 11.9.

TABLE 11.9

APPROXIMATE STORAGE LIFE (MONTHS) FOR QUALITY MAINTENANCE OF SELECTED FOODS

Product	Temperature °C −17.8	−12.2	−6.7
Orange juice (blanched)	27	10	4
Peaches	12	<2	0.2
Strawberries	12	2.4	0.3
Cauliflower	12	2.4	0.3
Green beans	11-12	3	1
Green peas	11-12	3	1
Spinach	6-7	<3	0.75
Raw chicken	27	15.5	<8
Fried chicken	<3	<1	<0.6
Turkey pies or dinners	<30	9.5	2.25
Beef (raw)	13-14	5	<2
Pork (raw)	10	<4	<1.5
Lean fish (raw)	3	<2.25	<1.5
Fat fish (raw)	2	1.5	0.8

The lower the temperature of storage, the longer will be the storage life of the product. The storage life may be doubled for each 10°C decrease in temperature. The deterioration of foods can occur at high freezer storage temperatures. Many enzymes are not inactivated by these temperatures. To help control enzymatic activity, some foods are given a heat treatment (blanching) to destroy the enzymes prior to freezing.

Good packaging materials prevent the diffusion of oxygen into the package to retard rancidity or oxidation. Also, the package will prevent evaporation of moisture from the frozen food. The addition of ascorbate or various antioxidants to certain frozen foods enhances the storage life by retarding oxidative changes.

The freezing rate affects the storage life of frozen foods. Slow freezing conditions may allow a food to remain at temperatures above 0°C for several hours. The quality lost due to slow freezing may be equivalent to that lost during several months of storage at −18°C.

Nutritive Losses.—The retention varies with the nutrient, the nature of the food product, the prefrozen processing, the method of freezing and the temperature and time of storage.

Kramer (1974) found that over 90% of the vitamin C is retained in beans, peas and asparagus stored for 12 months at −20°C, while peaches, spinach, broccoli and cauliflower may lose 20 to 50% of their vitamin C content during this storage.

Frozen foods retain a higher proportion of their original nutritive value than foods preserved by almost any other method.

Defrosting.—If a malfunction occurs during storage, transportation, distribution or retailing, the frozen food can be subjected to defrosting temperatures. In many cases, it is necessary to defrost the frozen food prior to use. Thus, defrosting is of concern whether it is accidental or deliberate.

If a frozen food has been held at defrost temperatures, the growth and activities of enzymes of psychrophilic bacteria may be evident by the resultant offensive odors or sour flavor. When subjected to defrost temperatures, vegetables may become shriveled, bleached and softened.

Besides these natural indicators, there are several man-made indicators to detect temperature abuse of frozen food. Perhaps the simplest indicator is to include an ice cube in the package. If the ice cube is melted, it indicates that the food was subjected to defrost temperatures, since the ice will thaw at a temperature higher than the thaw temperature of the food. For a discussion of defrost indicators, see Kramer and Farquhar (1976). These indicators tell only if the product was subjected to a condition greater than some predetermined temperature at some point in time. They give no information about the condition or quality of the food.

If a food is only partially defrosted, there is usually no harmful effect, at least from a public health viewpoint. This is also true concerning a product which is thawed, but still cold. These foods usually can be refrozen with no potential problems. If a food becomes thawed and warmed to room temperature, it may or may not be safe to refreeze, depending upon the type of food and the time that it has been at room temperature.

For usage, fast thawing of frozen foods helps retard bacterial growth, as well as loss of quality attributes. The rate of thawing usually is lower than the rate of freezing. This is due to the lower temperature differential during thawing than during freezing, and the rate of heat transfer is lower through the defrosted surface than through the frozen surface during freezing.

Some foods can be cooked from the frozen state without intermediate thawing. With other foods, thawing in the refrigerator is usually suggested since the temperature of the defrosted food will remain low enough to prevent the growth of potential pathogens. Thawing in warm water will hasten the process, but if the warm water is in direct contact with the food, nutrients from the food will be dissolved into the water and lost. With this method of thawing a packaged frozen turkey, the outer part of the turkey will be warm enough to support the growth of spoilage and potentially pathogenic organisms while the inside is still frozen. A microwave oven is useful for tempering frozen foods used in further processed items.

Advantages and Disadvantages of Freezing.—The advantages of freezing as a method of preserving food are that it: 1) neither adds nor removes any components; 2) imparts no new flavor or alters the natural flavor; and 3) does not diminish the digestibility or cause a significant loss of nutrient value.

Some disadvantages of this process are: 1) microorganisms may be reduced in number, but not destroyed; 2) spores are very resistant and toxins are not destroyed; and 3) frozen foods not properly wrapped dehydrate very rapidly which causes marked deterioration in the flavor and general appearance of the food.

Specific Foods.—Besides the general aspects of freezing foods, there are some special problems, conditions or characteristics for specific types of foods.

Red Meats.—The storage life of frozen meat depends upon the initial quality and condition, pH of the meat, the interval between slaughter and freezing, the method of preparation, the temperature of holding before freezing, package method and materials, method and temperature of freezing and the temperature and other conditions of frozen storage.

The quality of red meats can be maintained if the meat is packaged in a wrapper impermeable to oxygen and moisture, rapidly frozen and then stored below −18°C. Beef and lamb can be stored for one or more years if properly protected from dehydration (Table 11.9). Pork, especially smoked and/or cured pork, has poorer storage life. However, it can be held for six months or more if oxygen is excluded from the product and dehydration is prevented. The fatty components of meats are susceptible to changes in odor and flavor with pork fat being more susceptible than beef or lamb.

Fresh meats are seldom frozen for retail sale to consumers. Surveys indicate consumers prefer to buy fresh meat. However, about 80% of this fresh meat is then frozen in home freezers. It is obvious that freezing in commercial freezers would produce a more satisfactory product.

Meats preserved by freezing are stored at temperatures low enough to prevent microbial growth. Freezing with subsequent thawing will kill some microbial cells. Microorganisms tend to die slowly during storage, but those that survive will grow in the thawed meat at a rate similar to that exhibited by the same strain in raw meat (at the same temperature) which had never been frozen. The rate of microbial growth in thawed meat is dependent upon the temperature of the meat surface.

A condition called freezer burn may occur on the surface of frozen meat. It appears as patches of light colored tissue and gives the meat a bleached, unattractive appearance. The cause is similar to a freeze-drying system in which ice crystals are sublimated. The control of freezer burn is aided by packaging in a moisture impermeable film.

Poultry.—Poultry meat that is frozen should be fresh, well finished and properly packaged. Although the quality is not improved, freezing should maintain the quality as closely as possible to the value at the time of freezing. Freezing can affect the appearance, the exudation of fluid (drip) on thawing and a progressive loss of flavor, tenderness and juiciness.

Freezer burn can occur on frozen poultry in a manner similar to red meat. With poor packaging, this dehydration can occur in one month or less of storage.

The long bones in young frozen poultry (under 15 weeks of age) usually are discolored. The freezing and thawing processes hemolyze the red blood cells, releasing hemoglobin. The hemoglobin is able to pass out of the inner porous bone. This darkening is a normal condition found in frozen poultry and in no way affects the food value or flavor of the meat. It does affect the attitude of the consumer concerning frozen poultry. Bone darkening was nearly absent when poultry was frozen in liquid nitrogen at −196°C (Cunningham 1974).

There was little change in the quality of broilers through five thawings and refreezings (Baker *et al.* 1976). There was a slight increase in the microbial load after the five thawings and refreezings. The bacterial load on the chickens before freezing was 3.8 (10^7) CFU/chicken, and after five thawings the loads were 2.1 (10^8) and 3.0 (10^8) CFU/bird for fast and slow freezings, respectively. Thawed poultry does not spoil more rapidly than poultry which has not been frozen.

Chickens are generally retailed in the refrigerated fresh form, but turkeys are usually packaged in an evacuated, tight fitting plastic film in the frozen condition. Turkey in evacuated packages freezes at a faster rate than that in unevacuated packages.

Seafoods.—The storage life of frozen fish is significantly longer than that of iced or refrigerated fish. It is desirable to lower the temperature rapidly. The frozen fish should be stored at or below −20°C and preferably at −30°C.

Fish and other seafoods are more delicate than red meat or poultry. Some of the deteriorative alterations of seafoods are manifested as dehydration, color changes, rancidity, toughening or loss of tenderness, formation of drip, flavor changes and gaping. Some of these changes are due to protein denaturation and breakdown of fat.

The storage life is generally shorter for seafoods than for meat or poultry (Table 11.9). The storage life of frozen silver salmon steaks was extended substantially by vacuum packaging in low oxygen permeable film with freezing and storage at −18°C (Yu *et al.* 1973).

Egg Products.—When egg yolk is frozen and thawed, it becomes a viscous, rubbery mass. This irreversible increase in apparent viscosity is known as gelation. Certain additives, such as sugar (sucrose), fructose, galactose or salt, will reduce gelation of the frozen and thawed yolk. Proteolytic enzymes added to the liquid will partially inhibit gelation. If the egg yolk is frozen rapidly, such as with liquid nitrogen, there will be less gelation than when it is frozen by slower freezing systems. Rapid thawing also reduces gelation. Rapid freezing, combined with rapid thawing, is more effective in preventing gelation of frozen and thawed egg yolk than either one alone.

When sugar or salt is added to yolk, the freezing point is lowered. The temperature of the liquid must be reduced to −12°C within 60 hr after pasteurization. The eggs freeze from the outside of the carton toward the middle, so it is a race between ice formation and microbial growth. If the bacteria win, the product becomes sour. With pasteurization to destroy *Salmonella*, many of the spoilage bacteria are destroyed. Hence, there should be less spoilage today than in the pre-pasteurization era.

Dairy Products.—These are frozen as a means of preservation, or in the manufacture of ice cream or ice milk.

The freezing point of milk varies within relatively narrow limits. The average freezing point is considered to be −0.55°C. The determination of the freezing point provides a method of detecting added water. If water is added to milk, the freezing point will be closer to 0°C.

The deterioration in physical stability is one of the principal defects encountered in frozen milk. During frozen storage, the proteins become unstable and when the milk is thawed, the proteins settle out. Besides the appearance, the milk has an undesirable chalky texture.

Fruits and Vegetables.—Most of these products can be frozen and stored at −17.8°C for several months. The quality losses include color, texture and flavor. The factors that influence the quality of the frozen product include the initial quality, the type and variety of the fruit or vegetable, processing operations, freezing temperature, packaging and storage temperature.

Some varieties of certain fruits and vegetables withstand freezing and frozen storage much better than other varieties. There is no improvement in quality. Therefore, only high quality products should be frozen.

The texture of tomatoes is ruined with ordinary freezing and thawing. It has been suggested that thawed tomato slices frozen with liquid nitrogen have acceptable color, flavor and texture. Levine and Potter (1974) studied various methods for freezing tomatoes. Of all the treatments they studied, tomato slices dipped into 0.1% calcium chloride and then quick frozen with liquid nitrogen were the most acceptable after thawing. Even then the flavor and texture were poorer when compared to fresh, unfrozen tomato slices.

Most vegetables must be heat treated (blanched) before freezing to inactivate the enzymes, or during storage these catalysts may cause flavor or color deterioration.

Baked Goods.—Bakery products are especially adaptable to frozen storage. Although products such as crackers and cookies generally are not frozen commercially, home freezing of cookies can be used to extend the storage life.

The freezing point of most bakery products is between −7° and −2°C. The richer doughs freeze at the lower temperatures. Some adjuncts, such as jams or frostings, require even lower temperatures for freezing.

Of all the frozen baked goods, pies have had the greatest success. Frozen fruit pies can be stored at −17.8°C for 6 to 18 months. The time depends upon the type of fruit and the method of processing, packaging and freezing. Above −17.8°C, the quality of frozen pies will suffer due to the crust becoming soggy.

Miscellaneous Foods.—The freezer cases in the retail grocery stores contain frozen dinners, pot pies, pizzas and various mixtures of many types of foods. Although there has been much concern about the possible health hazards of certain precooked foods, they have had a good record for noninvolvement in foodborne disease outbreaks.

Microbiological Aspects.—Although freezing is not a method for sterilizing food, many of the associated microbial cells are killed or damaged by freezing, frozen storage and thawing.

When frozen food is thawed, it should be handled at least as carefully as fresh food, so that the associated microorganisms do not grow and cause spoilage or become a health hazard.

The microorganisms associated with frozen foods are influenced by pretreatments before freezing. Heat treatments, such as pasteurization, blanching or cooking, result in the destruction or damage to many microbial cells.

Calcott and MacLeod (1974) listed reasons for the cryoinjury of cells. These are: 1) thermal shock; 2) effect of concentration of extracellular solutes; 3) toxic action of concentrated intracellular solutes; 4) dehydration; 5) internal ice formation; and 6) attainment of a minimum cell volume.

In frozen foods, we are concerned with the lethal effects since we would like to destroy as many cells as possible without destroying the organoleptic or physical quality of the food. On the other hand, freezing is used to preserve cultures for research as well as those organisms used in food fermentations, as a source of enzymes, or as a source of single cell protein.

There are many conflicting reports in the literature concerning the effects of freezing on microbial cells. Part of the problem may be due to the use of different types of microbial cells in various suspending media which are cooled at various or unknown rates. The response of microorganisms is dependent upon many factors. Those that have been considered include: 1) the type, species and strain of the microorganism; 2) the age, population density and nutritional status of the cells; 3) the cooling rate; 4) the minimum temperature to which the cells are cooled; 5) composition, pH and type of suspending medium, including the presence of protective agents; 6) the temperature and time of storage; 7) the rate of warming or thawing; and 8) the methods used to determine cell viability.

Bacterial spores are the most resistant microbial entity to freeze damage. Gram positive cells are more resistant to freezing and thawing than are Gram negative cells. Survival of microorganisms is greater in a supercooled than in a frozen environment.

The effect of the cooling rate on the survival of microorganisms is depicted in Fig. 11.3. As the cooling rate increases, the survival increases; but, as the cooling rate increases further, the survival decreases. After a low point, the survival again increases at the ultrafast cooling rates. This curve indicates at least two aspects of the effect of freezing on microbial cells. At low cooling rates, the extracellular water begins to freeze and form ice crystals. The solutes are concentrated into the remaining water. This causes the emigration of water from the cells due to the osmotic pressure gradient between the intracellular and extracellular solutions or the vapor pressure difference between the supercooled intracellular water and the extracellular ice. The slower the process of cooling and the greater the permeability of the cells, the greater will be the loss of water from the cells. The movement of water out of the cells concentrates the solutes of the cells, lowers the intracellular vapor pressure and dehydrates the cells, the cell volume decreases and solutes precipitate.

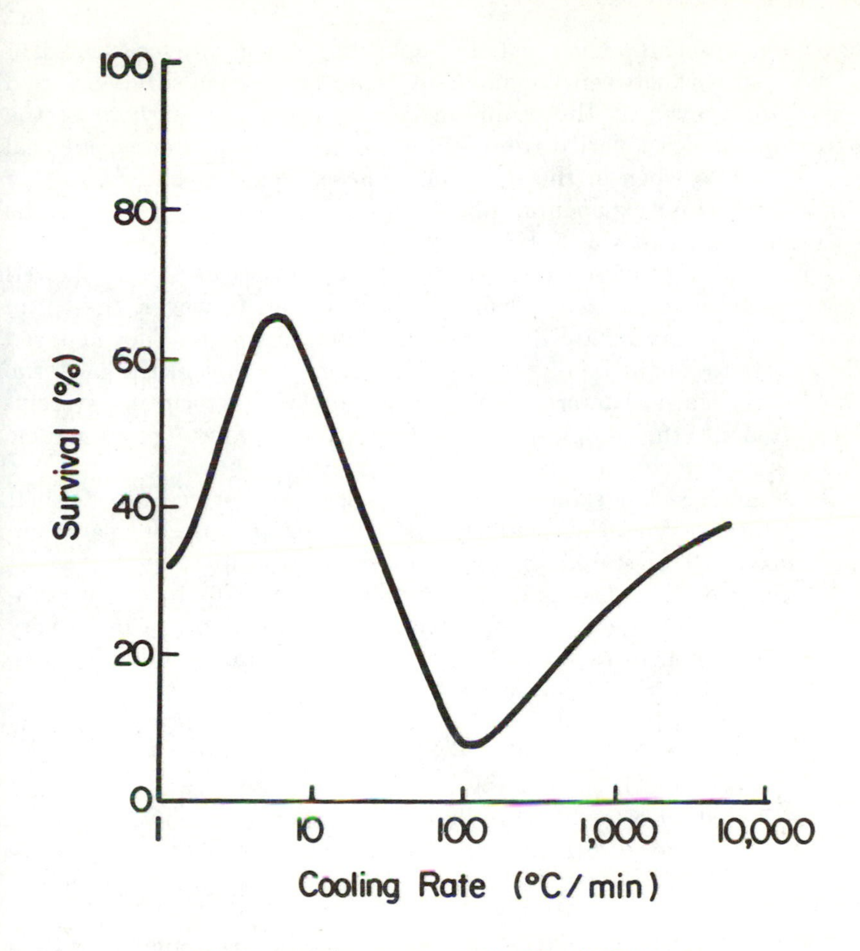

FIG. 11.3. COOLING RATE AND CELL SURVIVAL

The withdrawal of water from the cell contents can result in irreversible changes in the protoplasm which may be fatal. Intracellular water becomes supercooled, but does not freeze at low cooling rates.

The maximum survival of *Escherichia coli* is at a cooling rate of about 6°C/min, and the minimum at about 100°C/min (Calcott and MacLeod 1974). Calcott *et al.* (1976) determined the survival of *Streptococcus faecalis, Salmonella typhimurium, Klebsiella aerogenes, Pseudomonas aeruginosa* and *Azotobacter chroococcum* at various cooling rates. The cooling rate for optimum survival varied from 7°C/min for *A. chroococcum* to 11°C/min for *P. aeruginosa.* The minimum survival for these five organisms was found to be at about 100°C/min. At higher cooling rates, the survival increased.

Meyer *et al.* (1975) studied the effect of cooling rate on a mesophilic yeast and a psychrophilic yeast. The optimum cooling rate for survival of both yeasts was between 4.5° and 6.5°C/min. The level of survival varied considerably even at the optimum cooling rate. The survival of the psychrophilic yeast varied from 5%, when cooled from the exponential phase, to 71% when in the stationary phase. The mesophilic *Candida utilis* survival for exponential phase cells was 47% and for those in the stationary phase, it was 71%.

Four Gram negative bacteria studied by Calcott *et al.* (1976) showed less survival when frozen in saline than in water. However, the Gram positive *S. faecalis* did not show this reduction in survival. They believed the difference might be due to the respiratory systems of the bacteria. The Gram negative bacteria respire aerobically with cytochrome systems integrated into the cytoplasmic membranes. The *S. faecalis* is anaerobic and has no cytochrome system.

At very high cooling rates, small crystals of ice are formed. Due to their high surface energies, the small ice crystals tend to enlarge during warming. Mazur (1970) stated that a 40-fold decrease in survival of yeast cells occurred when the time for warming from −70° to 0°C was increased from 0.06 sec to 3.6 sec. The damage apparently can be very rapid, and is thought to be due to the growth of the intracellular ice crystals, rather than to the initial ice formation.

The enlarging intracellular ice crystals exert sufficient force to rupture plasma membranes. Cells killed by intracellular freezing suffer membrane damage and become leaky.

Injury by either effect, extracellular freezing or intracellular freezing, can cause damage to surface or internal membranes. The damage from solution effects can be reduced by the presence of protective agents that seem to act at the cell surface.

High warming rates result in greater survival of ultracooled cells than do low rates of thawing (Calcott and MacLeod 1974). The enhanced recovery of slowly frozen cells by slow thawing is thought to be due to the ability of the cells to reconstitute more readily during slow thawing. Rapid thawing of the dehydrated cells causes a rapid entrance of water which may cause excessive internal pressure and resultant rupture of the membranes.

Damage to the cell membranes and cell wall is indicated by the leakage of components out of or into the cell. The leakage of large molecular weight material indicates that not only is the plasma membrane damaged, but also the cell wall is altered.

Cryoprotective Agents.—There are many substances that protect the microbial cell during freezing and thawing. Some, such as glycerol and dimethylsulfoxide, penetrate the cells while others do not. Besides glycerol and dimethylsulfoxide, protective agents include egg white, carbohydrates, peptides, serum albumin, malic acid, milk, glutamic acid, yeast extract, diethylene glycol, Tween 80, dextran, sodium glutamate, glucose, polyethylene glycol and erythritol.

Meryman *et al.* (1977) suggested that cryoprotectants act by reducing the amount of ice formed in the cell or by increasing the time required for the water to leave the cell, by increasing the viscosity of the extracellular solution.

Glycerol reduces damage to the cell wall and membrane, while Tween 80 prevents only membrane damage (Calcott and MacLeod 1975). The mechanism of action of glycerol is not fully understood. The presence of glycerol decreases the mole fraction of solutes in the suspending fluid at freezing temperatures and prevents the high concentration of solutes which might affect the cell wall and membranes. Glycerol may protect by preventing large extracellular ice crystals.

Goldberg and Eschar (1977) concluded that freezing damage to cells of lactic acid bacteria was correlated with the cellular fatty acids.

Microorganisms in Frozen Foods.—Frozen foods receive no terminal heat treatment for sterilization either by the processor or by the consumer.

It is evident that many factors exert an influence on microorganisms in foods that are frozen. During freezing, the solutes and some microorganisms become concentrated in the unfrozen portion. Until the temperature is reduced below the minimum temperature for growth, presumably some microorganisms can multiply. However, the solutes can become concentrated to the level that microbial growth is inhibited. Microbial growth can occur in frozen foods stored at −7°C and possibly even at −12°C.

The greatest reduction in the microbial load occurs during or shortly after the freezing of foods. There is a gradual reduction in numbers during frozen storage.

Although food that is being thawed is susceptible to microbial growth, there is generally a long lag phase before growth commences if the temperature is kept near 0°C.

Although salmonellae, staphylococci and other potential pathogens can survive freezing and frozen storage, the saprophytic flora tends to inhibit their growth. During freezing and thawing of food, the temperature favors the growth of the psychrophilic organisms. Hence, in nearly all cases, if frozen product is mishandled, spoilage is apparent before the food becomes a health hazard.

There have been suggestions that freezing of foods disrupts the cellular structure of flesh foods and that bacteria can grow more readily on thawed meat that was frozen than on unfrozen meat. Sulzbacher (1952) found no indication that frozen meat was more perishable after thawing than was fresh meat. Elliot and Straka (1964) reported that frozen and thawed chicken decomposes at about the same rate as unfrozen chicken. Thawed meats, like fresh meats, are perishable and should not be held very long before cooking.

In some studies of the microorganisms associated with frozen foods, the total number of aerobic bacteria was determined by incubation at 37°C. At this temperature, mesophilic types are enumerated, but the psychrophiles are the important potential spoilage organisms in frozen foods. Incubation should be conducted at 20°C or less to obtain an estimate of psychrophilic types.

Although there is much evidence that freezing reduces the bacterial load on foods, Gunaratne and Spencer (1974) reported that freezing does not destroy a significant number of microorganisms on chicken. They used four freezing methods and found no significant difference in the reduction of bacteria on chicken thighs.

The microbial characteristics of frozen seafoods are similar to those of red meat and poultry. During freezing of seafoods, from 60 to 90% of the bacterial population is inactivated. During frozen storage, there is a gradual decline in the number of cells. Gram positive cells survive frozen storage better than Gram negative cells.

As with red meat and poultry, defrosted fish spoils at about the same rate as unfrozen fish.

Poliovirus inoculated into oysters showed a gradual decline in plaque forming units (PFU) during frozen storage at −36°C (DiGirolamo *et al.* 1970). About 10% of the original PFU survived 12 weeks of storage.

Vibrio parahaemolyticus is a potential pathogen associated with fishery products. It is generally stated that this organism does not survive freezing and frozen storage. *V. parahaemolyticus* cells were inoculated into oysters, fillets of sole and crabmeat (Johnson and Liston 1973). Although there was a sharp reduction in vibrios during freezing, these organisms persisted in these seafoods stored at either −15° or −30°C, with better survival at −30°C. They were found in oysters after 130 days, the end of the experiment.

The microbiology of frozen egg products is similar to other foods. During freezing, there may be microbial growth and, in poorly frozen product, spoilage may occur. With proper freezing, there is usually a reduction in the microbial load. During frozen storage at −15° to −18°C, the microbial load may show a slight decline. There is a 50 to 85% reduction in bacteria within 60 days of frozen storage.

The microbial load on raw vegetables may reach 10^6/ g. Washing and blanching (95°C for one to three minutes) will reduce the number of microorganisms. In commercial practice, the vegetables are recontaminated by equipment, so that the microbial load at the time of freezing may be higher than that of the raw vegetables.

When slow cooling rates are used for freezing vegetables, such as corn or peas, microbial growth may occur before the product is frozen. However, the inactivation of cells during the freezing process helps compensate for any increase in numbers. Overall, there may be a slight decrease or increase in the microbial load. Fast cooling rates and freezing result in essentially no change in the microbial count of vegetables.

There is a greater reduction in the bacterial load if the product is stored at −12°C rather than at either −18° or −23°C. However, quality may suffer at the warmer temperature. A temperature of −4°C is not sufficiently low to inhibit microbial growth in vegetables, such as peas.

Frozen vegetables held at a thawing temperature of 2°C do not show an increase in microorganisms for about two days.

Some of the dominant organisms in frozen vegetables are species of *Micrococcus, Flavobacterium, Streptococcus, Pseudomonas, Lactobacillus* and *Enterobacter.*

When frozen fruits are held at −4°C or higher, yeasts and molds have a tendency to grow. Growth usually does not occur if the fruit is packed in an airtight container. At these temperatures, the respiration of the fruit replaces the oxygen with carbon dioxide. This, along with the low pH, inhibits the growth of most types of microorganisms.

DRYING

Drying is thought to be the oldest method used to preserve food. Early in time, man observed that naturally dried seeds could be stored from one season to the next. The sun drying of meat, fish and fruit has been practiced for many centuries. These dried foods tended to have their own characteristic flavor, aroma and texture. The drying process caused irreversible changes, for example, grapes became raisins, and the rehydrated food was different from its fresh counterpart.

In order to compete with fresh foods or with other preservation methods, it is necessary to keep irreversible reactions of dried foods to a minimum. At the present time, most commercially dried foods are produced with controlled conditions of temperature, humidity and air flow. Since removal of moisture by heat is relatively expensive and uses much energy, systems such as screening, pressing, centrifuging, settling, filtration, reverse osmosis, osmotic solutions or freeze concentration may be used to eliminate water prior to a final drying of the food.

Reasons for Drying

The obvious reason for drying food is to prevent microbial growth. However, there are other reasons including: 1) to preserve the product from physical or chemical changes induced or supported by excess moisture; 2) to reduce the packaging, storage and transportation costs; 3) to prepare a material for a process when only dry materials can be used; 4) to remove moisture added in previous operations; 5) to bring the product to a moisture level at which it is normally graded, bought and sold; 6) to recover waste products; and 7) to provide convenience and a saving of time to users.

Dried foods weigh less and there is a reduction in bulk. A 30-dozen case of shell eggs weighs about 26 kg, while the dried product weighs about 5.1 kg. The dried eggs occupy about 16% of the space of the fresh product. Thus, there is a lower cost of transportation and storage.

If properly packaged, most dried foods can be stored at ambient conditions, with no special freezing or refrigeration requirements. One advantage is that the package of dried food can be opened, part of it removed for use and the remainder can be expected to have an acceptable storage life. Due to their concentrated character, dried foods can be used as ingredients to create new types of foods.

The reduction in weight and bulk, the good storage life and versatility are important, especially during wartime. Hence, in the past, dried foods were exploited during wartime and then more or less forgotten. However, after World War II, although the drying of food decreased, the technology that had been developed was utilized to produce consumer goods, such as instant dried milk and dried potato products. Several dried mixes, such as cake mixes, also have been developed.

Pretreatments

Only food of good quality should be used for drying. Generally, quality is not improved by the process.

The usual and necessary washing, cleaning, breaking, peeling, slicing, and other operations are carried out prior to drying. The operations needed depend upon the type of food to be dried.

Some foods require a heat treatment prior to drying. Quite often, foods such as meats are cooked before dehydration. Cooking meat can reduce the moisture content by 20% or more. Cooking before drying adds a useful convenience factor to dried foods. Also, cooking reduces the number of viable microorganisms.

Potatoes are cooked prior to drying so that the dried product is ready for use merely by adding water and mixing. Instant rice is prepared by a cooking process and redrying of the product.

Blanching is an important pretreatment for dehydrated vegetables. This process inactivates enzymes, expels air from tissues, yields a more tender product on reconstitution, minimizes loss of palatability, delays the development of odors, reduces color loss, helps retain carotene and ascorbic acid during dried storage and lengthens the storage life.

The peel, as well as the waxy coating on fruits, inhibits the removal of water from fruits. Dipping of fruits and berries into a weak sodium carbonate or lye solution produces fine cracks in the skin, removes the outside waxy covering and facilitates drying.

Fruits are given a sulfite treatment to prevent browning during drying and storage. Antioxidants may be added to foods to help prevent oxidative changes in the dried product.

For eggs and egg white, the reducing sugars are removed by fermentation or treatment with glucose oxidase. This is to prevent the Maillard reaction in the dried products.

Other pretreatments may consist of concentration by reverse osmosis or other methods. Milk may be concentrated to 20 to 40% solids by evaporation in vacuum pans before spray drying.

Drying Aspects

Water is associated with foods as: 1) water containing solids in solution (dissolved salts or other solutes); 2) water containing solids in suspension; 3) water occupying the pores or interstices of a solid; 4) water adhering to the surface of a solid; 5) chemically combined water (water of hydration); and 6) hydroscopic water which is naturally absorbed by a substance from an atmosphere containing water vapor.

The removal of water during drying requires the addition of heat for evaporation. The water migrates within the material to the surface. Here it is evaporated into the atmosphere surrounding the product and then removed from the vicinity of the food. The rate of removal of moisture from the surface is dependent upon the temperature, humidity and movement of the surrounding air. The size, shape and other characteristics of the food particle will influence moisture migration from the center to the surface.

The usual drying curve shows a constant rate period followed by a falling rate period (Fig. 11.4). During the constant rate period, moisture travels to the surface quite rapidly so that the surface is kept moist. When the rate of migration of water from the interior cannot maintain a sufficient flow to the surface to sustain the initial rate of evaporation, the constant rate period is terminated. There is a decline in the drying rate and it becomes very small as the water content of the food approaches equilibrium.

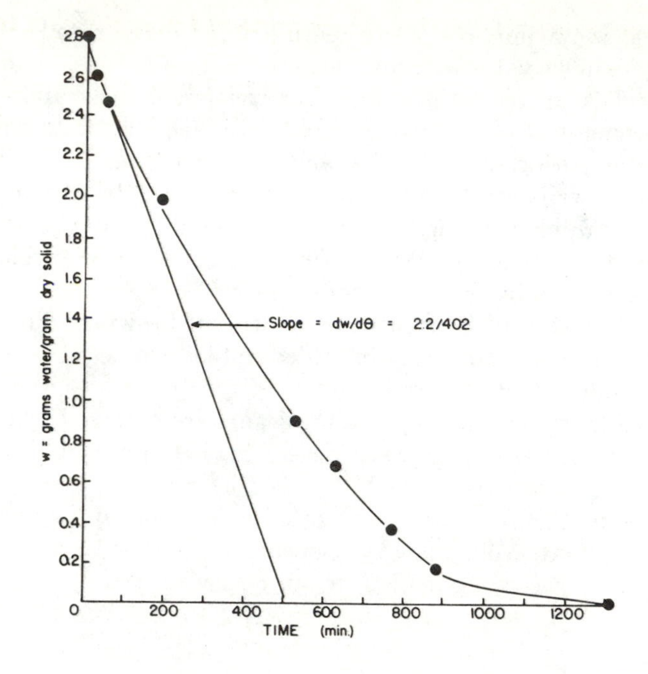

Courtesy of Charm (1978)

FIG. 11.4. DRYING CURVE SHOWING SLOPE FOR CONSTANT RATE PERIOD

During the drying process, some portions of the food will dry more quickly than others. The wet areas will be at lower temperatures than the dry portion, due to evaporative cooling.

In dehydrated foods, the sugars, acids and inorganic salts are concentrated, which also lowers the a_w of the foods and increases the preservative action.

Chemical Stability

Dried foods may have adverse quality characteristics when they are compared to fresh foods. The alterations in quality may be due to enzymatic activities, protein denaturation, oxidation of lipids or other reactions during drying, storage or reconstitution. The three factors that generally determine the type and extent of deterioration of dried food are residual moisture, oxygen content and storage conditions (time and temperature).

There is an optimum moisture content for the best storage stability of each food. Complete dehydration may result in destruction of the food. The monomolecular layer of moisture, referred to as bound water, may be regarded as a protective film, which protects the particles of food from attack by oxygen.

Dried foods must be packaged in moisture impermeable materials or stored in an atmosphere with a relative humidity similar to that of the food. In mixtures of various dehydrated foods, there is a transfer of moisture from items of high moisture vapor to those of lower moisture vapor, until equilibrium is attained. Quite often, dried mixes will include smaller packages of ingredients within the main package in order to maintain the components at different moisture levels.

Drying preserves food as long as the material is kept dry. Reconstituted dry products should be used as soon as possible since, after reconstitution, these foods are perishable, being subject to microbial spoilage.

Courtesy of USDA

FIG. 11.5. GRAPES ON TRAYS DRYING IN THE SUN

Types of Drying Systems

There are several systems that can be used to remove water from foods. The types of drying systems are listed in Table 11.10 and several are depicted in Fig. 11.5–11.10. The choice of the method depends upon the qualities that must be maintained, the sensitivity of the food to heat damage, the rehydration characteristics and the cost of the process. In some cases, only one of the processes will yield a commercially acceptable or marketable product.

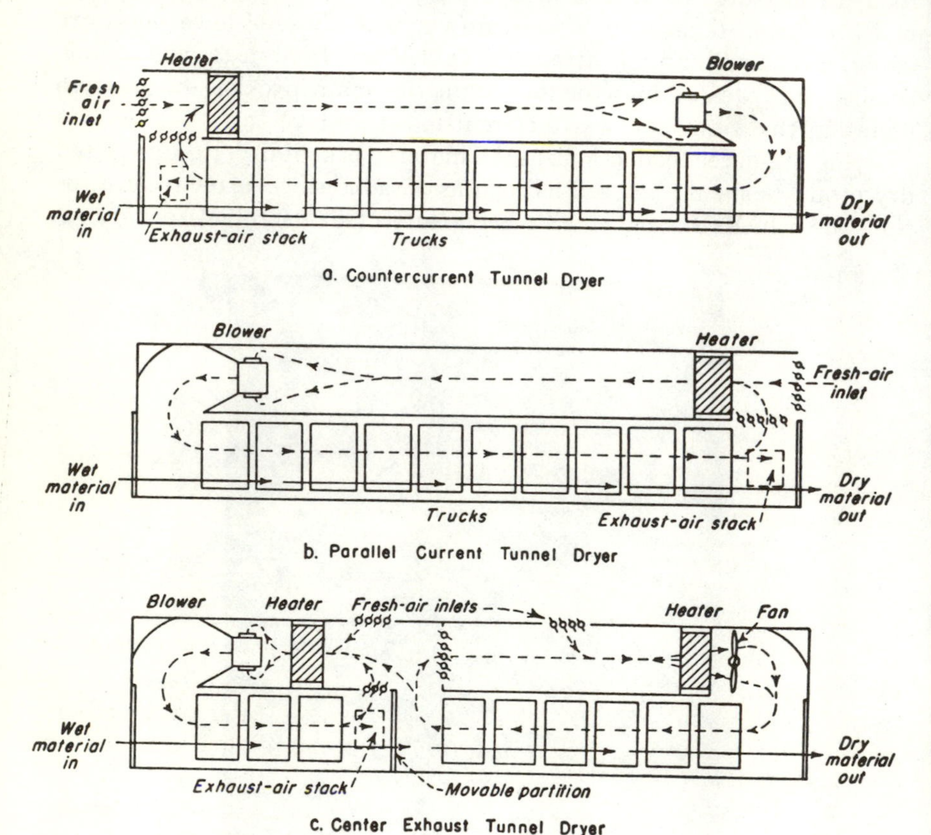

Courtesy of Charm (1978)

FIG. 11.6. THREE TYPES OF TUNNEL DRIERS

The four principal methods of drying are: 1) hot air drying, for foods such as vegetables; 2) spray drying for liquids and semiliquids; 3) vacuum drying for juices; and 4) freeze drying. There are various types of driers that use these principal methods. For some products, combinations of these methods are used.

Natural.—Natural air drying is simply the exposure of the product to the sun and the wind.

Fish and meats are dried by natural methods in some areas of the world. Since natural drying of fish is a slow process, if the temperature is too high, microbial growth can cause spoilage before the moisture content is reduced to a level for preservation. It is difficult to preserve fatty fish by this system of drying. Raisins can be produced by sun drying grapes.

TABLE 11.10

SOME SYSTEMS FOR THE DRYING OF FOODS

Types of Drier	Usual Food Types
Natural (solar)	Grapes (raisins), meats, fish
Bin	Grain, peanuts
Kiln	Apples, some vegetables
Cabinet or compartment	Fruits and vegetables
Tray or pan	Egg white
Tunnel	Fruits and vegetables
Continuous belt (atmospheric or vacuum)	Vegetables
Fluidized bed	Small pieces or granular
Roller or drum (atmospheric or vacuum)	Milk, fruit and vegetable purees
Spray	Milk, eggs, pureed foods
Foam-mat	Fruit and vegetable juices
Freeze-vacuum	Chicken, shrimp, meat, coffee
Microwaves	Pasta products, finish drying of foods

Controlled.—Controlled air drying is used to reduce the moisture content of various crops to a level acceptable for storage. Warm or hot air is blown through bins of the product. High temperatures affect the enzymes, germination and nutritive value of grains. Barley may become unacceptable for malting when used in beer-making if it is subjected to high air temperatures.

Air suspension drying is used as a finishing stage of some spray drying operations to obtain very low moisture levels of food. This is similar in principle to the fluidized bed dryer.

The vibrating conveyor dryer is used to redry powdered products that are moistened in the instantizing process.

Vacuum Drying.—In this process, the food is placed in a chamber that can be evacuated. Water is evaporated at a lower temperature than at atmospheric pressure. At the lower temperatures, there is less quality damage to the food.

Tunnel Drying.—This system uses drying tunnels through which carts or trucks containing trays of product to be dried are moved. Hot air is blown parallel or countercurrent to the movement of the food.

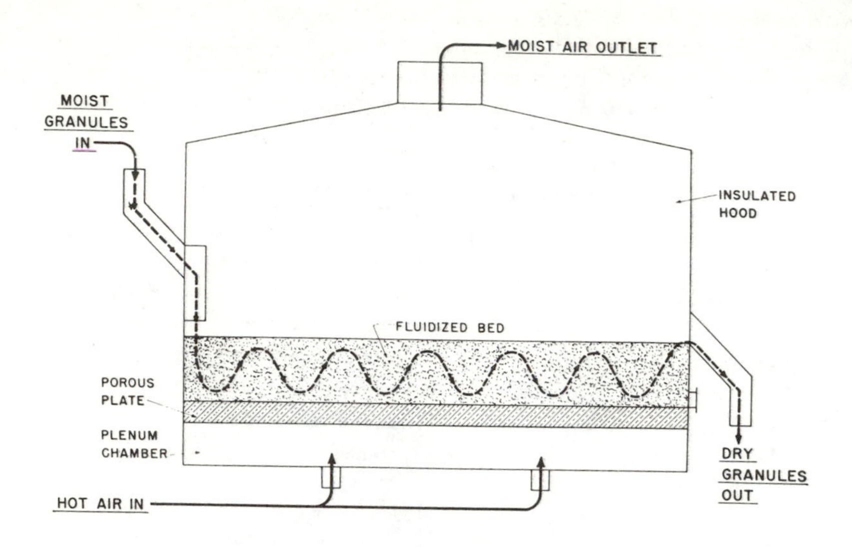

Courtesy of USDA

FIG. 11.7. FLUIDIZED BED DRIER

Fluidized Bed Drying.—Fluidized beds are used for freezing, blanching, transporting and other procedures, as well as for drying.

Fluidized bed dryers hold granular material in a turbulent suspension by aeration. Hot air supports and drys the food particles. Due to the intimate mixing, the process has a high thermal efficiency. It is particularly suited to final drying of partially dried products.

To overcome problems of the fluidized bed, centrifugal fluidized beds have been developed (Hanni *et al.* 1976).

In this bed, during fluidization, the food particles are subjected to a centrifugal force greater than gravitational force. This increases the apparent density of the particles and, by varying the centrifugal force, achieves smooth fluidization at any air velocity. With the increased air velocity, heat transfer is improved even at moderate temperatures. This helps prevent scorching or surface heat damage of the food.

Drum Drying.—In this system, the liquid or liquified food is run as a thin layer or film onto a revolving heated drum. The moisture is removed almost instantaneously. The dried product is scraped off in sheets or broken into flakes as the roller is turned slowly.

The drum drier is used in the production of dried foods, such as sweet potato flakes, tomato flakes, instantized breakfast cereals, starch and soup mixes.

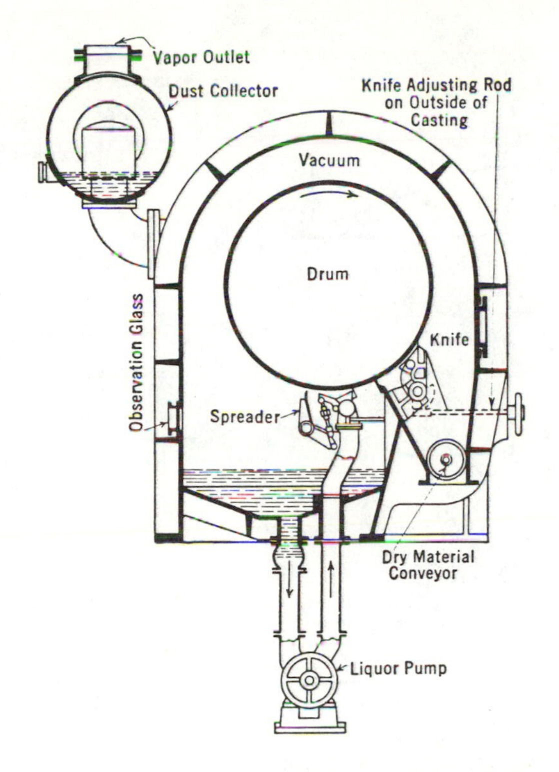

Courtesy of Farrall (1976)

FIG. 11.8. VACUUM ROLLER DRYER

Spray Drying.—In this process, a liquid is sprayed with heated air into a drying chamber. The atomized substance gives up its water very rapidly. The dry powder falls to the floor or is separated by a bag or cyclone collector, and the air stream goes out of the drying chamber carrying the water vapor.

Since the drying occurs almost instantaneously (5 to 30 sec), spray drying is useful for heat-sensitive as well as other liquids. The very rapid evaporation of water from the atomized particles produces a cooling effect which helps counterbalance the heating effect of the hot air.

Spray drying can be used to dry any liquid that can be atomized. The main foods are milk and eggs, but coffee, tea, fruit juices, and other foods have been spray dried.

Foam Spray Drying.—This process is applicable to liquids and slurries. The injection of an inert gas into the feed line of a conventional spray drier produces a foam that dries rapidly.

Foam spray drying is said to increase the efficiency and capacity of a

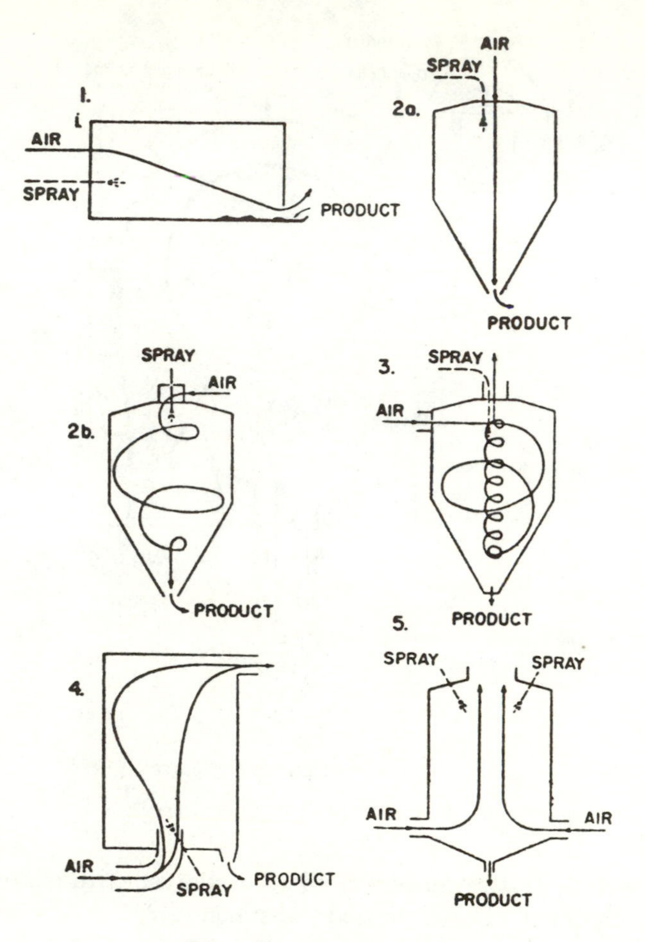

Courtesy of Van Arsdel and Copley (1973)

FIG. 11.9. TYPES OF SPRAY DRIERS

spray drier. The product has enhanced flowability, is more bulky and reconstitutes readily even in cold water. Milk and eggs have been dried by this system.

Foam-mat Drying.—In this process, the liquid food is concentrated, converted to a stable foam, spread in a thin layer on a supporting surface, cratered by blowing air or an inert gas through the foam, and dried with warm air.

The process has been used for drying a wide range of foods, including fruit juices, egg white, whole egg, whole milk, fruit purées, potatoes and tomato juice. Most fruit juices should be concentrated to 40 to 45% solids prior to foaming.

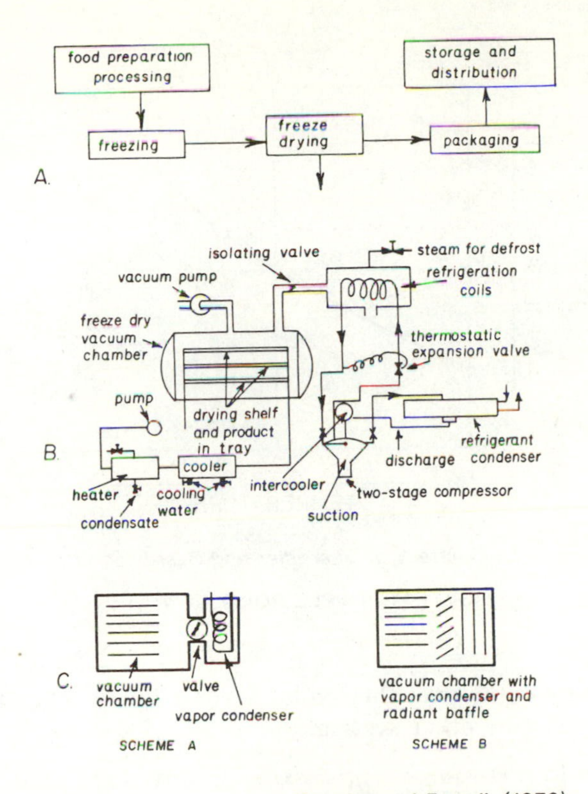

Courtesy of Farrall (1976)

FIG. 11.10. FREEZE DRYING SYSTEMS

Freeze drying relationships: A—the freeze drying cycle; B—process and cycle plant flow diagram; C—arrangement of vacuum freeze drying chamber and condenser.

Freeze Drying.—In this process, frozen food is dried under a high vacuum. This also is called sublimation drying, lyophilization or vacuum contact plate drying.

Sublimation is the conversion of ice to water vapor without passing through the liquid phase (Fig. 11.11). At normal atmospheric pressure, sublimation is a very slow process, but it can be speeded up by reducing the pressure. As ice is sublimed, energy is used. Therefore, it is necessary to supply heat to the food to compensate for this latent heat of sublimation. The heating must be carefully controlled so that the food remains frozen.

Due to the low temperatures involved, freeze drying is especially suited for the dehydration of heat sensitive foods. Freeze drying occurs with a minimum of discoloration and loss of flavor and nutrients, such as vita-

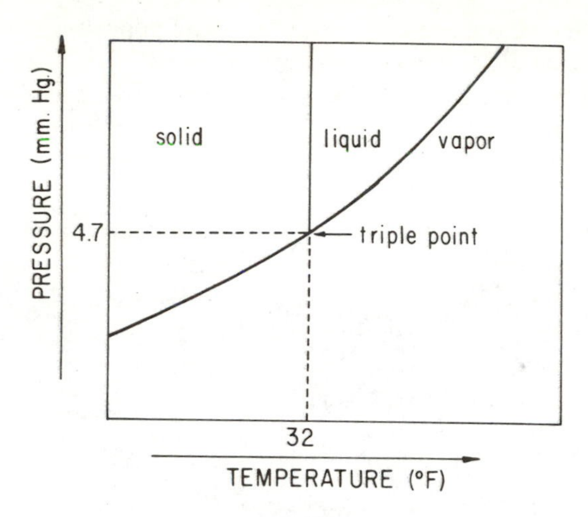

Courtesy of Desrosier and Desrosier (1977)

FIG. 11.11. TRIPLE POINT OF WATER

Considerations in freeze drying.

mins. Foods retain more of their volatile constituents when freeze dried than when dried by other systems.

Microwave Drying.—In this process, microwaves are used to heat the product being dried. With microwave heating, the material being treated is exposed to a high frequency field. This causes polar molecules of the material to oscillate rapidly. It is this molecular motion that produces heat within the material rather than in the air surrounding the material. The internal heating permits drying from the inside out, and avoids extreme surface hardness (case hardening).

Liquid water selectively absorbs the radiation. Hence, as the water is removed from the food, there are fewer water molecules to absorb the microwave energy and the process becomes self regulating. In conventional drying with hot air, the entire food mass, even when dry, absorbs the heat energy.

Heat loss to the surrounding air is minimal, since the heat generated is inside the food product. Maurer *et al.* (1971) indicated the cost of microwave drying was only ¾ of the cost of conventional drying and Anon. (1974) determined that the energy used was only ½ of that in conventional drying. Due to the potential for case hardening, conventional air drying of noodles, macaroni or other pasta products requires from 5 to 10 hr. With microwave heating and drying, this time is reduced to between 15 and 36 min (Anon. 1974; Maurer *et al.* 1971).

Reverse Osmosis.—The removal of water by heating or freeze drying requires a phase change of the water. To change phases, an expenditure of energy is needed. In reverse osmosis, with sufficient pressure to overcome osmotic effects, water is forced through a membrane which limits the passage of other molecules. Hence, it is a form of molecular filtration. Some small molecules, such as salt and acids, can pass through the membrane, although with difficulty.

The cost of reverse osmosis equipment is rather high. This may be offset by the savings in energy to remove water from the food. However, the system only concentrates the material and does not dry it. Therefore, in most cases, the food is still perishable and must be dried, frozen or given some other treatment for preservation.

Food Products

Dried milk products and potato products account for most of the commercially dried foods produced in the USA. Milk represents the main dehydrated food in countries in which dairies are prominent. There are many other types of dehydrated foods including both animal and plant products. Condiments, such as spices, garlic and onions, along with modified starches, coffee and beans are prominent dried foods.

Animal Products.—Dried meat, poultry and fishery products are microbiologically stable, but are subject to oxidative and browning reactions. The oxidation of the lipid portion of meats can result in rancidity. The removal of all visible fat and the packaging of the dried product in inert gas or vacuum help retard oxidative rancidity.

Moorhouse and Salwin (1970) reported that freeze drying of shrimp removes not only the moisture, but also volatiles that are associated with deterioration. When partially decomposed shrimp were freeze dried, the rehydrated product had a better odor than control samples. When shrimp in the first stage of decomposition were freeze dried, they became passable. Those in advanced stages of decomposition were improved by freeze drying, but not to a passable stage. They concluded that chemical tests for decomposition have limited value in assessing the degree of decomposition of freeze dried shrimp.

Agglomeration is used to change dry powders into those that reconstitute readily, such as instant dry skim milk. In this process, the spray dried product is moistened by steam or hot humid air or a water spray. The milk powder is partially dissolved to cause a sticky condition, so there is adhesion between contacting particles. A stream of warm air immediately dries these agglomerates.

Fruit.—Dried fruit is an important commodity. For the USA, most of the dried fruit is produced in California. Fruits are sun dried, air dried, vacuum dried and freeze dried.

Raisins are produced by drying grapes with various systems. The clusters of grapes may be cut from the vines and placed on paper directly in the rows between the vines, and sun dried. A newer system is to spray the grape clusters on the vine with an ethyl ester of a fatty acid. An emulsion of ethyl oleate is easy to handle and is quite effective. This treatment modifies the waxy covering on the grapes so that the grapes dry on the vine and then are harvested by hand or by mechanical methods.

Also, the grape clusters can be harvested, dipped into the emulsion of ethyl oleate and dried with hot air in a tunnel drier (Bolin *et al.* 1975; Ponting and McBean 1970).

Raisins are sticky when they exude syrup. As they lose moisture, they become hard. Moisture loss can be slowed by coating the raisin with a thin layer of beeswax or other such substance. Beeswax can be made more plastic for coatings by adding peanut oil, cottonseed oil or acetylated monoglycerides. Protein coatings reduce moisture loss from raisins (Bolin 1976).

The main deterioration of dried fruit is due to browning. This may be accompanied by impairment of flavor, texture and nutrient value. The rate of browning of dried fruits is dependent upon the moisture content, availability of oxygen, concentration of added sulfur dioxide and storage temperature.

If the fruit is treated with SO_2, enzymatic browning is inhibited and nonenzymatic browning is retarded. Browning may be greatly retarded by refrigerating the dried fruit or by drying to 1% moisture or less.

Vegetables.—There are many systems that have been used for drying vegetables. Good quality dried vegetables can be obtained by air drying at atmospheric pressure (cabinet or continuous belt systems) or by the vacuum contact plate method. Potato products are usually drum dried.

Most freshly dried vegetables are of high quality, but if the moisture content is not reduced to low levels, the flavor, color and nutrients deteriorate during storage. With low moisture and refrigeration, some vegetables can be stored for three years or longer.

Cereals.—Moisture is the most important of the various factors influencing the rate of deterioration of grain. If the moisture content is maintained at a sufficiently low level (below 13%), grain can be stored for many years with little deterioration, even under otherwise unfavorable storage conditions. However, grain, as it is harvested, or the milled products, may have moisture levels near or above the amounts for safe storage. Cereals are dried by forcing warm air through them. The relative humidity of the air in contact with the grain may be more important than the moisture content of the grain. The equilibrium relative humidity should be no greater than 65% if grain is to be stored for two or three years. Cold temperatures and reduced oxygen levels improve storage life, since molds will be inhibited.

Microbial Aspects

In considering the microbial aspects of dried foods, we should examine the growth or death of cells before drying, during drying, during storage, during rehydration and in the reconstituted product.

Before Drying.—Foods that are to be dried should be considered as perishable foods. They should be handled so that the microbial load is kept as low as possible prior to drying.

Heat treatments, such as the blanching of fruits and vegetables, the cooking of meat, poultry or fish, or the pasteurization of liquid eggs or milk reduces the total microbial count. Freezing in the freeze drying system and atomization during spray drying may lower the number of viable microorganisms in foods.

Drying.—Drying removes the moisture from a food and lowers its water activity (a_w). If no death occurred during drying, there would be a higher microbial load per gram of dried product than in the original food, due to the loss of water. Usually, the dried product has a lower microbial level than the original food due to the death of microorganisms. Of those organisms that survive drying, some have been stressed and have sublethal lesions.

Although the viable microbial load is decreased during drying, the dried product is not sterile, regardless of the system used for drying. In fact, drying is used to preserve cultures of microorganisms.

The lethal action during drying is influenced by the organism (species, strain, physiological age, state of the cells, cell concentration), the condition of drying (freeze, spray, roller, air, moisture, temperature, time, the rate and extent of water loss, residual water content) and the type of food or suspending medium (protective factors, pH, inhibitors).

If drying proceeds at a slow rate and at acceptable temperatures, the microorganisms may multiply during the process. In general, foods being dried should be either at a high or low temperature, so that bacterial growth cannot occur until the water content is reduced to a level that will inhibit microbial action. This is not always possible since some products become overheated and lose their physical qualities if temperatures are too high.

Freeze drying is the most gentle of the drying methods, both with respect to foods and to the microbial population. A high microbial count in the frozen product will be reflected in the dried food. During freezing, drying, storage and rehydration, there can be a shift in the microbial population. Gram negative bacteria such as *Pseudomonas, Escherichia* and *Vibrio* do not survive freeze drying as well as do Gram positive cocci and the spore forming bacteria.

Freeze drying reduced the total viable count of 10^4/g in acceptable shrimp to 10^3/g (Moorhouse and Salwin 1970). Shrimp of questionable quality or spoiled shrimp with total counts of 10^6/g had viable counts of 10^3 to 10^4/g after freeze drying. Although no specific microorganisms were determined, it can be assumed the spoiled shrimp contained pseudomonads.

Freeze drying includes a number of stresses that may cause injury to microorganisms. Freeze drying can injure the cell membrane. Cell membrane damage may result in death if the organism cannot repair the injury. The formation of mutants by freeze drying indicates that cellular DNA is altered or damaged. In a series of experiments it was observed that cells not capable of DNA repair had a much higher death rate than those cells that could repair DNA (Tanaka *et al.* 1975).

During spray drying, aerosols of the liquid food are dried in hot air. Microorganisms are affected by the aerosolization, heating and drying. When cells are aerosolized there may be changes in the outer layer of the cell envelope, cytoplasmic constituents may leak out and they may lack control of ion transport. The exact nature of the lesions resulting in permeability changes is unclear.

The lower the temperature of spray drying liquid egg, the better are the physical qualities of the powder, but the greater is the survival of bacteria, including the salmonellae. Over 99% of the salmonellae in liquid egg are killed during spray drying if the inlet air temperature is 121°C and the outlet temperature is about 60°C. There is a difference in the survival rate of various strains or serotypes of *Salmonella.* Rapid cooling of the powder is essential to maintain the quality of the powder. However, this allows greater survival of the microorganisms.

The plate count of dried vegetables is usually 10^3/g or less. The flora varies with the vegetables. The organisms include molds, actinomyces, micrococci, staphylococci, bacilli, enterococci and coliforms. Clostridia may be found in some samples.

The microorganisms associated with various dehydrated foods were determined by Powers *et al.* (1971). Of 129 samples, 93% had less than 10,000/g (total aerobes), 98% had less than one coliform in 10 g and 99% revealed no fecal coliforms. Of 104 samples, all of the 5 g subsamples were negative for coagulase positive *S. aureus* and 98% of the 10 g subsamples yielded no *Salmonella.* Only 102 samples were tested for fecal streptocoocci and 93% had less than 100/g.

Graikoski (1973) reviewed the microbiology of dried fish. The dried product could be produced with less than 100 microorganisms per gram. The majority of surviving bacteria were Gram positive heat resistant micrococci with Gram negative cocci comprising the remainder.

The presence of coliforms in dried cooked meat indicates contamination and poor sanitary practices during preparation and handling.

The contamination of dried fruit with mold spores is difficult to prevent due to the prevalence of molds in air around fruit processing plants.

Storage.—The removal of moisture lowers the a_w of a food. The a_w is related to the spoilage rate and type of spoilage that may occur in a food during storage. Lowering of the a_w can result in the selective growth of microorganisms. Thus, it is common to see mold growth on dried products stored in humid conditions.

At 0.70 a_w and below, most microbial growth is inhibited. However, some xerophilic fungi can grow at a_w of 0.65 and osmophilic yeasts at an a_w of 0.60. In order to recommend safe a_w limits for foods, consideration must be given to the nature of the food, its history, processing, packaging and conditions of storage and handling.

Foods are usually dried below the moisture levels that will inhibit microbial growth to retard chemical deterioration which occurs more readily at 0.65 to 0.70 a_w than at lower a_w levels.

The unequal distribution of moisture can occur in various types of dried foods. Thus, there may be areas or pockets that contain more moisture than is indicated by the analysis of the entire sample. Molds, or even bacteria, might grow in these pockets of moisture.

Factors affecting survival include the species and strain of the organism, morphology (spores), age and mass of the culture, treatments before storage, moisture content of dried material, composition of the food or substrate and the conditions of storage (temperature, time and atmosphere).

The driest environment does not always give the greatest survival of microorganisms. Both *S. newport* and *P. fluorescens* dried in papain digest broth, die rapidly at either a_w 0.00 or 0.40 when stored in air (Marshall *et al.* 1973). *S. newport* survives drying and storage better than does *P. fluorescens.*

Over 99% of the *P. fluorescens* cells surviving drying of whole egg lose their viability in two weeks storage at 35°C. However, components of the cells, such as enzymes, remain active after death of the organisms and contribute to the undesirable changes in egg products. Keeping the initial bacterial count of the liquid egg low is important in the maintenance of the quality during storage of the dried product.

Species of *Salmonella* that survive drying of eggs die during storage of the dried product, even at low temperatures. The storage of dried albumen at elevated temperatures (50° to 60°C) will inactivate any salmonellae in a matter of a few days to a week. The time needed for this inactivation is influenced by number of survivors, the temperature and the moisture content or a_w of the dried powder. Inactivation is greater at 10% moisture than at 2% moisture.

The initial death rates of salmonellae in dried fishmeal were not influenced appreciably by a_w from 0.54 to 0.71 (Doesburg *et al.* 1970). Lowering the a_w to 0.34 reduced the death rates, especially in meal stored at 15° or 20°C.

During storage of dried milk, the viable bacteria decrease in number. *Streptococcus durans* dies quite rapidly in spray dried milk powder, while *S. thermophilus* remains viable for longer periods. Micrococci survive better than streptococci and, as the storage time is lengthened, aerobic sporeformers (bacilli) tend to dominate the viable population. Storage of dried milk for three months will inactivate over 70% of the bacteria.

Survival is highest for organisms stored in vacuum and lowest for those stored in air or oxygen (Cox and Heckly 1973). Following dehydration, oxygen can kill *Serratia marcescens* (Cox *et al.* 1974A). The site for the toxic action of oxygen lies in the interspace between the cytoplasmic membrane and the cell wall. The presence of reducing substances helps prevent the lethal action of oxygen.

Due to the death of microorganisms during the storage of dried foods, it is difficult to correctly assess the microbial quality of these products. This is especially true if there is an unknown interval between drying and examination for viable microorganisms. A total microscopic count may be useful in determining the microbial character of dried foods.

Rehydration.—Normally, no bacterial growth occurs in dried food. Reconstitution or rehydration of such food results in a perishable product. This rehydrated food should be used immediately or refrigerated just as any other perishable food.

The growth of microorganisms in rehydrated food is influenced by the initial number and types that are present, as well as the temperature and time for rehydration.

The microbial population of dried animal products becomes essentially mesophilic types. When rehydrated and held at 4°C, there is a shift in types so that psychrotrophs dominate. If the rehydrated food is held at 20° to 30°C mesophilic types, including potential pathogens, will grow and become prominent.

With the total viable population reduced in dried food, any surviving potential pathogen will have less competition and microbial interactions in the rehydrated food. If not properly handled before use, the food may become a vehicle for foodborne illness.

Due to osmotic forces and plasmolysis, rehydration can cause the apparent death of some cells. Plasmolysis is considered to be the result of the retraction of the plasma membrane away from the cell wall due to osmotic forces.

Due to sublethal injury, some cells will not be able to grow as readily as others in the rehydrated food. Since rehydration is necessary for microbial analysis, the effect of rehydration on these injured cells or other cells is important.

The microbial analysis of dried food should be interpreted with certain facts in mind. If inhibitory media are used, injured cells will not grow and be detected. Bacteria play no significant part in the microbiological deterioration of dried foods under normal conditions. This is because there is not sufficient water for their growth. The bacteria that dominate the microbial flora are important only when the food is rehydrated and stored at temperatures acceptable for their growth.

INTERMEDIATE MOISTURE FOODS

An intermediate moisture (IM) food is one that is preserved by lowering the water activity by drying or adding solutes, often in combination with chemical preservatives and/or anaerobic packaging. IM foods have been defined as having moisture levels ranging from 15 to 50% and a_w values of 0.6 to 0.9. Regardless of the exact moisture or a_w value, IM foods can be eaten without rehydration, but they are microbiologically stable without refrigeration.

IM foods have certain advantages:

1. They are relatively low in moisture. Hence, they can be considered as concentrated in regard to bulk, weight and caloric content.
2. Their plastic characteristic allows them to be molded into uniformly shaped cohesive blocks to facilitate packaging and storage.
3. They can be consumed with no preparation by the user.
4. They are similar to normal foods in texture.

Foods with a_w 0.60 or less are essentially free from microbial growth. If solutes are used to obtain the low a_w, the food may have unacceptable flavors due to the high concentration of solutes needed.

Foods with a_w over 0.85 readily support the growth of microorganisms. Between a_w 0.60 and 0.85, most of the microbial growth is due to molds, with some yeasts growing in this range. By adding antimycotic agents to the food, the microbial growth can be controlled. Since molds are primarily aerobic, reducing the availability of oxygen will limit their growth. Thus, there can be good storage stability at room temperature.

The types of IM foods include high moisture dried fruits (figs, dates, raisins, prunes, apricots, apples), fig newtons, marshmallows, soft candies, jams, jellies, molasses, honey, syrups, fruit cake, brownies, fruit filled toaster products, diet bars, pepperoni and other dry sausages, pemmican, jerky and the semi-moist pet foods.

Most of these foods are well known. With the creation of semi-moist pet foods, there has been an effort to develop more human foods that have IM characteristics.

There are many ways in which an IM food can be developed. So far, systems that have been used are: 1) Mix ingredients so that an acceptable moisture level and a_w are obtained. Usually antimycotic agents and solutes to keep the a_w low are included in the formula. 2) Dry a food below a_w 0.85 and, if needed, add antimycotic agents for preservation. 3) Mix ingredients, such as sugar added to fruit, and then evaporate water to a desired consistency and sugar concentration, as in the manufacture of jam. 4) Dry a food and then rehydrate it in a solution containing solutes and antimycotic agents so that the final food has a normal moisture content, but a lowered a_w and preservatives. 5) Soak small pieces of food in a low a_w solution of various chemicals and antimycotic agents. Equilibration will produce a food with lowered a_w and preservatives. 6) Add water binding agents to maintain a higher water content but a lower a_w. Or 7) use mild drying conditions to lower water content in conjunction with the water binding capacity of natural plant colloids (Gee *et al.* 1977).

By incorporating an effective antimycotic agent, heating to destroy vegetative microorganisms and adjusting the a_w to 0.85, pet foods sealed in plastic pouches have an excellent record for stability under market conditions.

Equilibration of foods by soaking in solutions of low a_w was discussed by Brockman (1970) for tuna, pork, beef, carrots, celery, pineapple and macaroni. Water activities of 0.81 to 0.86 were obtained for these foods.

The usual antimycotic agent is potassium sorbate. However, other substances also have antimycotic activities. Propylene glycol acts as a plasticizing humectant for texture of the food as well as contributing to the soluble solids of the aqueous phase. It is also an antimycotic agent. Together, potassium sorbate and propylene glycol act as an antimycotic system to protect IM foods from mold and yeast deterioration. Calcium propionate also may be used to aid in antimycotic activity.

With IM foods, small temperature changes may cause moisture condensation inside the package. Microbial growth can occur in these condensates which have high a_w and usually sufficient nutrients.

One problem with the bacteriological picture of IM foods is the possibility of growth of *S. aureus.* This organism can begin to grow at a_w 0.83

to 0.84, and to produce enterotoxin at 0.86 (Tatini 1973; Troller 1973). The growth and toxin production depend upon other growth factors being at acceptable levels.

A reduction in the bacterial load has been reported for IM foods stored at 38°C (Brockmann 1970; Kaplow 1970). *S. aureus* at a level of 10^4/g was inoculated into chicken a la king or ham sauce at a_w 0.85. The number of *S. aureus* was reduced to less than 2,000/g in one month and to about 0.4/g after four months. Inoculation with *E. coli*, *C. perfringens* or *Salmonella* gave comparable results.

Labuza *et al.* (1972) challenged IM foods with various microorganisms. The IM foods were prepared by direct mixing or by freeze drying, and then absorption of moisture to the desired level. The microorganisms showed better growth in the direct mix than in the freeze dried food system even though both foods were at the same a_w. *S. aureus* grew at 25°C in the direct mix food at a_w 0.84. In the freeze dried system at 0.84, there was death, although there were survivors evident for one month. Pseudomonads showed growth at a_w 0.84 in direct mixed food, but no growth in food adjusted after freeze drying, even at a_w 0.90. These data indicate that microbial growth is affected by more than the a_w of a food. Presumably the moisture content must be considered.

CHEMICALS

The chemical preservation of food has a long history. The preservative action of acids, salt and sugar was recognized by primitive people. As discussed in Chapter 4, there are chemical inhibitors naturally present in some foods.

Public Concern

Some consumer groups have argued against the presence of chemicals in foods. There will always be chemicals in foods, simply because foods are composed of chemicals. Some chemicals naturally present in foods are toxic to humans (Chapter 1).

Food Additives

There are many definitions for food additives. According to the U.S. Food, Drug and Cosmetic Act, a food additive is a substance the intended use of which results or may be reasonably expected to result directly, or indirectly, in its becoming a component or otherwise affecting the characteristics of any food. In 1958, Congress amended the Act to restrict food additives to substances specifically approved after that date by the

FDA. Because many good substances had been in use for years without FDA approval, Congress exempted those which, through long use, were considered to be safe by experts (GRAS).

Food additives have been defined by the FAO/WHO as substances which are added intentionally to food, generally in small quantities, to improve the appearance, flavor, texture or other storage properties. This definition does not include substances which are not deliberately added, but find their way into food.

Additives may be intentional or incidental. Intentional additives are added on purpose to perform specific functions. Incidental additives have no function in the finished food, but become part of it through some phase of production, processing, storage or packaging. Normally, they are present only in trace amounts.

Chemicals Added to Food

There are several major classes of chemicals that are added to foods. These have been listed variously, but usually the function in food is considered. Examples are acidulants, antioxidants, colors, enzymes, flavors or flavor enhancers, functional protein additives, leavening agents, nutritive agents or supplements, preservatives, stabilizers or thickeners, surfactants and miscellaneous (including anticaking agents, dough conditioners and sequestrants). According to an FDA press conference in in 1973, the FDA uses 31 categories for food ingredients or additives. Of the total food additives, chemical preservatives comprise only 3% or less on either a weight or dollar basis.

According to Middlekauff (1974) sugar is the main substance added to foods, being used at an annual rate of about 46 kg/person. Of the other major substances per person, we use salt (6.8 kg/year), corn syrup (3.8 kg/year) and dextrose (1.9 kg/year). All of the other substances total about 4.1 kg/year for each person. Of this amount, the next 32 additives, including nutrient supplements (vitamins and amino acids), account for 3.5 kg/year. Although there are thousands of additives, most of them account for a very small portion of the 910 kg of food that each of us consumes each year.

Reasons for Using Additives

There are certain reasons for adding and for not adding chemicals to foods. The reasons for adding chemicals are to:

1. Maintain or improve the nutritional quality of food.
2. Reduce waste by enhancing the keeping quality.
3. Make foods more attractive to the consumer.

4. Provide essential aids in food processing.

With our present technology of food preservation, the use of chemicals is essential. Without food additives, the cost of food would increase. Besides the increased cost, most foods would be less wholesome and lack the present keeping quality.

Reasons to Not Allow Additives

Chemicals cannot be added to foods if they:

1. Disguise faulty processing or handling methods.
2. Deceive the consumer.
3. Reduce the nutritive value.
4. Replace good manufacturing practices which could accomplish the same effect.
5. Are unsafe for human consumption.

Strong flavors cannot be added to foods to disguise partially spoiled or spoiled food.

Safety of Chemicals

The safety of chemicals in foods is very important, regardless of whether the chemical is added or is naturally present in the food. The absolute proof of safety of a chemical for human use is difficult to demonstrate. It is much easier to show that a chemical may be unsafe than to prove absolute safety. Due to this, consumer action groups have a definite advantage over the food industry in the battle of food additives.

To petition for the use of a chemical additive for foods, certain information must be submitted to the regulatory agency. This information includes the chemical identity of the substance, the purity, a quantitative method of assay for the amounts that would be expected in a food, its reactivity and stability, solubility, manufacturing process, its intended purpose and, if it accomplishes this purpose, an environmental impact analysis concerning its manufacture, use and consumption, suggested tolerances for use, toxicology, biochemistry and allergic responses.

A typical evaluation of safety includes several studies to determine acute, subacute and chronic toxicity factors. In the acute studies, at least two species of animals are fed the material to determine the LD_{50}. This is a lethal dose for 50% of the test animals. The safety of the chemical if it is inhaled or contaminates the skin is determined.

The subacute tests consist of 90-day feeding tests using two species, one of which is not a rodent. Dogs are usually the nonrodent species. The chemical is fed at three or more levels. The animals are examined to determine complete biochemical and histopathological effects. The no-effect level and the high level that cause effects are determined.

The chronic feeding tests require a minimum of two years, but may go on forever. Two species, such as the rat and dog, are used. A reproduction study through three generations may be needed. As in the subacute studies, there is a complete work-up of biochemical and histopathological effects. The metabolic fate of the chemical should be determined, such as the absorption, distribution, concentration in organs or tissues and metabolic products that result. The mutagenic, carcinogenic and teratogenic properties of the chemical are determined. The possible interactions with other chemicals in foods or drugs, as well as dietary restrictions, must be known.

It should be evident that to acquire all of this information requires considerable time and money. Even if the tests indicate the chemical is safe, there is no assurance that it will receive favorable consideration.

When a chemical is deemed to be safe, a tolerance is established. The tolerance or limitation does not exceed the smallest amount needed for the intended purpose, even though a higher amount would be safe. After determining the amount of the chemical that produces any undesirable effect on the test animals, only 1/100th of that amount is normally the maximum allowed for humans. It is likely that humans might react differently than the test animals to the chemical.

Further information about food safety and the systems used to evaluate the safety of chemicals in foods was presented by Anon. (1978), Brown *et al.* (1978) and Visek *et al.* (1978).

Chemical Preservatives

A chemical preservative can be defined as a substance that is capable of inhibiting, retarding or arresting the decomposition of food. This definition includes microbial inhibitors, as well as antioxidants, acidulants and sequestrants. In this text, we are concerned mainly with controlling the deterioration of foods by microorganisms.

Besides the requirements described for all chemical additives, there are requirements for food preservatives as listed in Table 11.11.

Chemical preservatives should be used only when other methods to control microorganisms are either lacking, damaging to the product or are very expensive. It has been suggested that, except in a limited number of cases, the use of chemical preservatives would be unnecessary if strict attention were paid to sanitation and hygiene during processing, and adequate refrigeration were used during the various stages after processing. Chemical preservatives do add a margin of safety from possible abuses by people involved at any of the post processing stages. This is applicable especially to the actions of the consumer.

TABLE 11.11

REQUIREMENTS FOR USE OF CHEMICAL PRESERVATIVES

1. Provide an economical means of preservation.
2. Be used only when other preservation methods are inadequate or not available.
3. Extend the storage life of food.
4. Not lower the quality (color, flavor, odor) of the food.
5. Be readily soluble.
6. Exhibit antimicrobial properties over the pH range of the particular food.
7. Be safe at levels that are needed.
8. Be readily identified by chemical analysis.
9. Not retard the action of digestive enzymes.
10. Not decompose or react to form compounds of greater toxicity.
11. Be easily controlled and uniformly distributed in the food.
12. Have a wide antimicrobial spectrum that includes the spoilage types of organisms associated with the food to be preserved.

In considering the possible hazards associated with the use of chemical preservatives, it is important to consider the foods in which they are used. The possible adverse effects to humans are likely to be greater if used in foods such as bread, meat, milk or fresh fruits and vegetables, which may be consumed in substantial amounts daily, than in foods that are eaten only occasionally in small amounts.

Ideally, the chemical will inhibit or kill the important microorganisms and then break down to harmless, non-toxic substances. The chemical should not decompose so fast that it is ineffective. Slow inactivation of microorganisms may lead to unsuccessful preservation. Nonsterilizing amounts of antimicrobial agents are selective and influence the type of microbial flora that can survive in a food system. The degree of inhibition varies with the preservative and the amount of inhibition is determined by the concentration of the chemical.

Activity of Preservatives.—The factors that affect the microbial activity of chemical preservatives include such things as the type of chemical and its concentration, the type of organism and its physiological state, the numbers of organisms, the composition of the food and its pH and the temperature of storage.

Bacterial spores are the most resistant type of microorganism. Fungal spores are more resistant than vegetative cells to the action of preservatives. In many cases, molds are more susceptible than yeasts to inhibitors. There is a variation in resistance to chemicals between species and between strains of the same species. The higher the microbial load, the greater the amount of chemical preservative that is needed to accomplish inhibition or death of the cells.

Actively growing microbial cultures are susceptible to killing agents. As the culture ages and becomes inactive, the cells tend to become more resistant to antimicrobial conditions. Sporicidal agents can be considered as chemical sterilizers, since they have the potential to destroy all microbiological forms of life. The apparent lethal or inhibitory action on microbial spores may be due to either inhibition of germination or of outgrowth. Only a few agents are capable of killing all types of microorganisms, including spores.

Many preservatives have an increased activity in acid foods. Organic material of the food may react with the preservative, rendering it less toxic or inert to microorganisms. Liquid foods allow better contact between the inhibitor and the microorganisms than do solid foods. In cases in which the microorganisms are inside food particles, they are protected from the effects of the chemical preservatives.

In general, increasing the temperature increases the effect of preservatives on microorganisms. If, however, a low temperature is increased toward the optimum of the organism, the stimulatory effect on growth may outweigh the increased action of the preservative. If the temperature is above the optimum for growth, the increased preservative effect is more pronounced.

Time is an important factor in the use of chemical agents. Chemicals may react with the organisms quite rapidly or they might require several hours to achieve the desired action on the microorganisms. Hence, the longer the contact time, the more effective will be the action of the chemical preservative.

Chemical preservatives may inhibit the growth (bacteriostat, fungistat), or kill (bactericide, fungicide, sporicide or virucide) microorganisms. Whether a chemical is bacteriostatic or bactericidal may be relative. In very dilute amounts some chemicals may act as food sources for the microorganisms. Increasing the amount may be inhibitory, and still higher levels may kill some or all of the microbial cells. In general, the more concentrated the chemical agent, the more effective will be its action. Usually, very high levels are not desired due to the potential adverse effect on food quality and toxicity to humans consuming the food.

We determine the viability of a microbial cell by its ability to multiply. The apparently lethal action on cells may be reversible, so that there is only a bacteriostatic effect. In some cases, the number of killed cells is balanced by the number of cells that are able to multiply, so that, although there is a lethal effect, the action appears to be only bacteriostatic.

The bacteriostatic agents most frequently used are those that lower the a_w below the level needed for bacterial growth. Sugar and salt are two of these important compounds. Besides retarding growth by affecting the chemical or physical nature of foods, preservatives affect microorganisms by several mechanisms, including the denaturation of protein, inhibition of enzymes, alteration or destruction of DNA, the cell wall or the cytoplasmic membrane, suppression of cell wall synthesis or competition with essential metabolites.

Interference with Genetic System.—In this case, the chemical enters the cell. Some chemicals can combine with, or attach to, the ribosome and inhibit protein synthesis. If genes are affected by the chemical, enzyme synthesis, which the genes control, will be inhibited.

If the logarithm of surviving bacteria is plotted against time, at least a portion of the curve is a straight line. This indicates that the lethal action is due to the reaction of a single molecule of the cell. The alteration or destruction of the DNA fits the monomolecular reaction requirement. Hence, the effect of inhibitors on the genetic mechanism was thought to be the primary action on the cell. The fact that the curve is not exactly a straight line function indicates that other actions occur.

Cell Walls or Membranes.—The chemical does not need to enter the cell to inhibit growth. A reaction on the cell wall or membranes can alter the permeability of the cell. This can impair the passage of nutrients into the cell, as well as allow leakage of cellular constituents from the cell.

Damage to the cell wall alone does not kill the microbial cell. Due to the increased permeability, the chemical is able to enter the cell. Once inside, the chemical may lyse the cell membrane or coagulate the cytoplasm of the cell. Damage to the cell membrane can occur by preservatives reacting with chemically active sites or by dissolving lipid constituents.

Since cell walls are complex polymers, some chemicals can interfere with cell wall production, by affecting synthesis of simple components, inhibiting the polymerization of the components or, the cell wall, when developed, may not be able to satisfy the requirements of the cell.

Inhibition of Enzymes.—Since enzymes are proteins, they can be denatured by heavy metals, alcohols, phenols or surface active substances. Generally, these chemicals are too toxic for food preservatives.

Wetting agents alter the colloidal properties of enzymes and interfere with enzyme action. High concentrations of salts will curtail the biological activity of enzymes. Sufficient alteration of the pH, either up or down, will inhibit the action of enzymes and prevent the multiplication of microorganisms. Oxidizing agents can inactivate certain enzyme systems that contain active sulfhydryl (-SH) groups. These -SH groups, when oxidized, form disulfide (S-S) bonds which cause the enzymes to become inactive. Reducing agents split the S-S bonds and reproduce the -SH form. The halogens, hydrogen peroxide and ozone, can oxidize the -SH group. Other mild oxidizing substances also may inhibit microbial cells in this manner.

On the other hand, certain active systems have S-S bonds. These enzymes can be inactivated by reducing agents.

The hydroxyl (OH) group is essential for the activity of certain enzymes. This group can react with chemicals and cause inactivation of the enzyme. Other groups of enzymes that can react with chemical inhibitors include the amine, amide and carboxyl groups, as well as metals in respiratory enzyme systems.

Another action on enzyme systems is due to antienzymes, such as the trypsin inhibitor of soybeans. These antienzymes are found in a number of natural products, especially in cereals and legumes.

Some chemicals selectively inhibit certain enzyme systems. Although microorganisms and humans can perform a given enzymatic reaction, the actual enzymes involved, and even the associated coenzymes, can be quite different in many features. Thus, it is possible to selectively inhibit specific enzyme reactions of microorganisms without affecting those of humans.

Binding of Essential Nutrients.—Since microorganisms differ in their nutrient requirements, the binding of essential nutrients affects different organisms in different ways. If an organism has few requirements, it will be less affected than an organism that needs many preformed nutrients.

Although hundreds of antimetabolites (analogues of vitamins, amino acids, purines and pyrmidines) have been synthesized, only a few have been useful to inhibit microbial growth. The sulfonamides that prevent the conversion of para-aminobenzoic acid to folic acid are the classic example of these antimetabolites.

Action on Humans.—The toxicity of chemicals to humans may be either due to the same mechanism(s) that causes the effect on microorganisms or the person can develop a hypersensitivity to the compound. This results in an allergenic reaction. There will always be an occasional person who develops a hypersensitivity to any type of substance that is introduced into the body.

When a foreign chemical enters the body, it has three possible fates.

The chemical may undergo spontaneous reactions to other products without the intervention of enzymes.

The chemical may be metabolized and transformed into other compounds by means of enzymes of the body. This is the fate of most compounds. The enzymes that metabolize foreign chemicals are located in the liver, intestines, kidneys and lungs. Bacteria in the gut play a role in this metabolism.

The inhibitor may be excreted with no alteration. These compounds are usually highly polar types. Many other foreign compounds are metabolized to highly polar types which then are eliminated from the body.

Activity of Combination Treatments.—In certain circumstances, combinations of two or more chemical preservatives are more effective than indicated by testing each one individually. For the treatment of diseases, combinations of antibiotics may be used, since, if there is resistance to one antibiotic, there may not be resistance to the second or third antibiotic.

Combinations of chemicals and other treatments may increase the potential for control of microbial spoilage of food. Using low water activity to control bacteria and antimycotic agents to control fungi is discussed for IM foods.

Types of Chemical Preservatives.—Some chemicals that help prevent bacterial and fungal spoilage have been used for centuries. Sodium chloride was used by primitive man to prevent spoilage of meat and fish. Sugar has been used for many years in the making of jams and jellies.

Before the use of various chemical preservatives, mold was common on bread. Mold inhibitors, such as the sodium or calcium salts of propionic acid or sodium diacetate, are helpful in controlling mold in bakery products. Mold control in bread was a serious problem for the baker, both in the bakery and after the bread left the bakery. Chemical agents give a margin of safety in cases such as this.

Besides adding chemicals directly to food, some are applied to the wrapping material to control contamination on the surface of the food. Antimycotic agents that have been used for food wraps include sodium benzoate, sodium and calcium propionates, methyl and propyl parahydroxybenzoates, sorbic acid and diphenyl.

Some chemicals are added to water that is used to wash food. In these cases, some chemical residual may remain on the food. Fumigants, such as SO_2, used for grapes and in the drying of fruit, leave a residual material on the food.

Some chemicals that are used as preservatives, such as salt and sugar, from a legal viewpoint, are not classed as preservatives. Chemicals that are antioxidants and may or may not have antimicrobial properties, are listed as chemical preservatives. There are other chemicals that are added to foods for various reasons that have some antimicrobial properties.

Acids.—Acids have many functions in foods. An acid condition is unfavorable for the growth of many microorganisms. The preservative effect of acids may be due to the pH, the undissociated acid molecule or the anion. With short chain fatty acids, a decrease in pH enhances the activity to a greater degree than is the case with the acids with longer chains. At low pH levels, the undissociated molecules of the weak, short chain organic acids enter the cell and interfere with intracellular enzymes. The ionic form does not pass through the cell membrane as well as does the undissociated form. These acids are more active in substrates that are low in vitamins and amino acids.

Branched chain fatty acids are less active than the corresponding straight chain acids. Unsaturated acids have been reported to have the same activity or to be more active than the corresponding saturated acids. The substitution of a hydroxyl (OH) group for a hydrogen has been reported to both increase and decrease the activity of the fatty acids. Kabara *et al.* (1972) reported that the OH group was particularly active, but a -SH group caused loss of activity. They found that amine derivatives of fatty acids showed activity against both Gram positive and Gram negative microorganisms.

It has been suggested that short chain fatty acids can inhibit or kill both Gram positive and Gram negative bacteria. The long chain fatty acids affect primarily Gram positive bacteria since they cannot penetrate the lipopolysaccharide layer of Gram negative bacteria.

The growth inhibition of bacteria by these lipophilic acids has been attributed to the effect on membrane transfer systems of amino and keto acids (Freese *et al.* 1973). The long chain lipophilic acids are more readily partitioned into the cell membrane. This attachment is reversible.

It is well established that an increase in acidity or in hydrogen ion activity inhibits the growth or even destroys certain microorganisms, reduces the germination and outgrowth of spores and reduces the heat resistance. The thermal death point of yeasts and molds is less affected by low pH than is that of bacteria.

The pH and the type of acid are important in the inhibitory or lethal action of these chemicals. Acetic acid has greater inhibitory and lethal effects on the basis of pH than either hydrochloric or lactic acids.

The exact order of effectiveness depends upon various factors such as the type of microorganism, whether inhibition or death is determined, the pH, temperature and other environmental conditions of the substrate, as well as the concentration of acid that is used. Comparisons have been made on molarity, normality, percentage, as well as at specified pH levels. These various systems influence the relative amounts of the acids that are present to inhibit the microorganisms. By selecting the correct microorganism, conditions of study and the basis for comparison, any acid may be or may not be as effective as any other acid as an antimicrobial agent.

Respiratory viruses are generally acid labile. Acids may be quite important in preventing the transmission of viruses by foods.

Doty (1968) reported that a mixture of 10% propionic, 10% lactic, 5% citric acids in 75% water killed salmonellae. When this mixture was added to meat meal at the rate of 5%, all salmonellae were destroyed.

Acetic Acid (CH_3COOH).—This acid is soluble in water and has a pK value of 4.76. The main reasons for using acetic acid as a food preservative are its low cost, availability and its low toxicity. Acetic acid is listed as GRAS and can be added to food in accordance with good manufacturing practice. The inhibitory effect on microorganisms is in excess of that due to pH alone. The undissociated molecule can penetrate the cell and exert its toxic action, such as by lowering the internal pH of the cell.

Acetic acid is effective against a wide spectrum of bacteria. It is bactericidal to coliforms and salmonellae.

Acetic acid is the principal organic component of vinegar. Vinegar is used in many foods, including mayonnaise, salad dressing, prepared mustard, pickles, marinades, catsup and pickled beets.

Although it is an effective inhibitor, acetic acid cannot be used on some foods because of its pungent odor. Treatment of beef carcasses with a 4% solution of acetic acid significantly reduces the microbial load (Bala *et al.* 1977). Acetic acid can cause discoloration and textural changes of poultry skin and animal carcasses.

Dehydroacetic acid inhibits bacteria, yeasts and molds at concentrations of 0.005 to 0.4%. It appears to be a better fungistatic agent than bacteriostatic agent. Autoclaving of the chemical causes only a slight decrease in biological activity.

The sodium salt of dehydroacetic acid at pH 5.0 is twice as active as sodium benzoate against *Saccharomyces cerevisiae* and 25 times more effective against *Penicillium glaucum* or *Aspergillus niger*.

The pharmacological data on this chemical tend to be contradictory. There are some who believe that the margin between the possible toxic level and the amounts likely to be used is too small to justify the use of this chemical or its salts as food preservatives. On the other hand, there is evidence that the chemical is safe.

Dipping certain fruits and vegetables in a solution of 0.1 to 0.2% dehydroacetic acid extends the storage life of the produce. At 0.1% it inhibits undesirable secondary fermentations in alcoholic beverages, such as beer and wine. In bread and baked pastries, a level of 0.2% inhibits microbial growth.

Sodium diacetate is a loose, equimolecular compound of acetic acid and sodium acetate. It tends to enhance the antimicrobial activity of acetic acid. The action of sodium diacetate is not as pH dependent as is that of acetic acid. This compound is listed as GRAS and can be used at 0.4 parts per 100 parts of flour for use in bread.

Ascorbic Acid.—This is vitamin C which is present in many foods, especially citrus fruits, and is safe to use as a food additive.

Although ascorbic acid inhibits pseudomonads in a liquid substrate, there is no measurable effect when sprayed onto a meat surface. In this case, the ascorbic acid may be reduced to below the level needed for inhibition, the meat acts as a buffer to maintain a higher pH level, the ascorbic acid may not reduce the redox potential of the meat surface to the same extent as in a liquid or the ascorbate may be oxidized by substances in the meat.

The antibacterial action may be due to breakdown products of oxidized ascorbic acid, or ascorbic acid may cause inhibition in a manner similar to other organic acids.

Benzoic Acid.—Besides benzoic acid, sodium benzoate and the parahydroxy esters (parabens) are used as chemical preservatives. Due to its better solubility, sodium benzoate is preferred to benzoic acid for commercial applications as a preservative.

The FAO/WHO unconditional acceptable daily intake of benzoic acid for humans is 5 mg/kg body weight. Cats and rats are particularly susceptible to high levels of benzoic acid. This chemical is excreted by many mammalian species either as hippuric acid or as benzoyl glucuronide. Benzoic acid is present naturally in many plant products, sometimes exceeding 0.1%, the level established for benzoate in foods by the USFDA.

The antimicrobial activity of benzoates is dependent upon the pH of the substrate, since it is the molecular form that is effective in preservation. Using the formula $\text{pH} = \text{pK} + \log\frac{(\text{ionized})}{(\text{molecular})}$, and a pK of 4.19, the pH at which various amounts of undissociated acid are present can be calculated, and is listed in Table 11.12. The undissociated or molecular form passes through the cell membrane more readily than the dissociated form of benzoic acid.

TABLE 11.12

THE pH VALUES NEEDED FOR VARIOUS LEVELS OF UNDISSOCIATED ORGANIC ACIDS

	Acids		
Undissociated Acid (%)	Benzoic	Propionic	Sorbic
99	2.19	2.87	2.75
95	2.91	3.59	3.47
90	3.24	3.92	3.80
80	3.59	4.27	4.15
70	3.82	4.50	4.38
60	4.01	4.69	4.57
50 (pK)	4.19	4.87	4.75
40	4.37	5.05	4.93
30	4.56	5.24	5.12
20	4.79	5.47	5.35
10	5.14	5.82	5.70
1	6.19	6.87	6.75
0.5	6.49	7.17	7.05

The antimicrobial effect of sodium benzoate is due either to alteration of the permeability of the cell membrane, or to competition with coenzymes.

Benzoic acid and sodium benzoate are used primarily to inhibit yeasts and molds. Most of the important bacteria are inhibited by the low pH levels needed for the benzoate to be effective. The yeast *Saccharomyces bailii* is resistant to benzoic and sorbic acids (Warth 1977).

Benzoates have been used in fruit juice, jellies, jams, soft drinks, pureés and concentrates. A level of 0.05% sodium benzoate helps preserve carbonated beverages. Carbonation increases the activity of benzoate against certain microorganisms.

Combinations of benzoate with other preservatives may be beneficial. Sulfur dioxide helps prevent discoloration and flavors attributed to benzoate. Combinations of benzoate and sorbate are believed to be more effective than either chemical used alone.

The esters of parahydroxybenzoic acid have been suggested for preservation of foods. It is only the esters and their salts that are of value as preservatives. The free parahydroxybenzoic acid is of little use in practical concentrations. The antimicrobial activity of the parahydroxybenzoates increases with the length of the alkyl chain of the alcohol moiety, up to the amyl ester. Some mixtures of the esters show greater activity than would be expected by their individual action.

These esters have been called PHB esters, parabens or parasepts. The parabens have certain advantages over benzoic acid. They are considered to be safer for human consumption and are reported to be more effective as preservatives. They are more effective at higher pH levels than benzoic acid and even have antimicrobial activity in neutral substrates. The esters from methyl to butyl are active against both Gram negative and Gram positive bacteria, as well as fungi. Esters with five or more carbons in the alkyl chain are not as active toward Gram negative bacteria. The fungal activity is maximum with the hexyl, heptyl and octyl esters.

The addition of heptyl paraben to beer in place of pasteurization gives a microbiologically stable product. It is approved for use in fermented malt beverages in a concentration not to exceed 12 $\mu g/g$. Combinations of propyl and heptyl esters control microbial growth in wine at a normal level of contamination. Butyl paraben (50 $\mu g/g$) and heptyl (5 $\mu g/g$) effectively inhibit the growth of microorganisms in new wine.

Parker (1969) suggested the parabens inhibit germination rather than the outgrowth of bacterial spores. The methyl and propyl parabens allow a limited degree of germination of *B. subtilis* spores followed by some outgrowth of vegetative cells.

Citric Acid.—Citric acid ($COOH\text{-}CH_2\text{-}C(OH)\text{-}(COOH)\text{-}CH_2\text{-}COOH$) is a common constituent of citrus fruits. It is an acidulant with unique flavoring characteristics and is water soluble.

Although citric acid has been tested as an antimicrobial agent, it is considered to be not as effective as other acids or preservatives. However, at 0.3%, citric acid significantly lowered the level of inoculated salmonellae remaining on poultry carcasses (Thomson *et al.* 1967). The microbial load of fish flesh was lowered to a greater extent by citric acid than potassium sorbate (Debevere and Voets 1972). Citric acid (0.1 or 0.3%) added to fish flesh inhibits the formation of total volatile nitrogen and trimethylamine.

Lactic Acid.—This acid ($CH_3\text{-}CHOH\text{-}COOH$) is a natural constituent of some foods and is formed by lactic acid bacteria in several fermented foods.

Depending upon the substrate and microorganisms, lactic acid has been reported as having very good, just average or rather poor antimicrobial properties.

Malic Acid.—This is hydroxysuccinic acid ($COOH\text{-}CHOH\text{-}CH_2\text{-}COOH$), a natural component of apples. It is permitted by the FDA as an optional acidifying ingredient in mayonnaise and other salad dressings. In these products, it has a microstatic action against yeasts and certain bacteria.

Peracetic Acid.—This is known also as peroxyacetic acid (CH_3-COOOH) or acetyl hydroperoxide. It is a strong oxidizing agent and will explode if heated to 110°C.

It is freely soluble in water and the aqueous solutions are quite stable. It reacts slowly with water to form acetic acid and hydrogen peroxide, but shows greater bactericidal action than either of these. A 1% solution is active against vegetative bacteria, bacterial spores, fungi and viruses.

The main problem with the use of this chemical is the pungent odor of its vapors and the off-flavor associated with acetic acid that occurs due to residues.

Propionic Acid.—This acid (CH_3-CH_2-COOH) is soluble in water and has a pK of 4.87. It is produced by propionibacters during the aging of Swiss cheese. Propionic acid is a constituent of human body fluids as the free acid or its salts. It is listed as GRAS to be used in accordance with good manufacturing practice. The calcium and sodium salts are GRAS and may be used in bread at the rate of 0.32% of the weight of the flour.

The bacteriostatic and fungistatic activity is greater in acid than in neutral or alkaline substrates. indicating the antimicrobial action is due to undissociated acid.

The propionates are effective against molds, have a limited antibacterial activity and relatively no effect on yeasts. Since propionates do not affect yeast fermentation, they are ideal in the preservation of bread.

Propionates are used to control mold in various types of cheese, malt extracts, chocolate, fruits, vegetables, bread, cakes, pies, jellies and preserves.

Sorbic Acid.—This unsaturated carboxylic acid (CH_3-CH=CH-CH=CH-COOH) is water soluble (0.25% at 30°C). The sodium and potassium salts are more soluble than the acid. Sorbic acid is more soluble in fat or oil than in water.

The sorbates are listed as GRAS. These chemicals may be added to foods in accordance with good manufacturing practice. They are used in foods at levels from 0.02 to 1.6% (Table 11.13). In humans, sorbic acid is metabolized in a manner similar to other fatty acids. All fatty acids of even numbered carbon atoms are converted to ketone bodies which accumulate and are then excreted as such in the urine. These ketone bodies represent intermediates in fatty acid oxidation. Under the influence of carbohydrates, they are oxidized completely to carbon dioxide and water.

TABLE 11.13

PROPOSED USE LEVELS OF SORBIC ACID AND POTASSIUM SORBATE IN FOODS

Food	Use Level (%)
	Sorbic Acid
Baked goods, baking mixes	0.23
Soft candies	1.4
Cheese	0.3
Other foods	0.2
	Potassium Sorbate
Baked goods	0.1
Non-alcoholic beverages	0.1
Cheese	0.3
Confections and frosting	0.2
Fats and oils	1.6
Gelatin, puddings, fillings	0.8
Sweet sauces, toppings, syrups	0.7
Other foods	0.12

Sorbic acid may be applied to foods in a variety of ways, including direct incorporation into the food, dipping the food in a solution or slurry of the acid, spraying a solution onto the food, application by means of impregnated packaging materials or addition as a solution in ethanol, propylene glycol or vegetable oil.

Treatment of salad vegetables by immersion in potassium sorbate (0.2%) solution is not effective and, in some cases, is detrimental in maintaining these foods during refrigerated storage (Priepke *et al.* 1976).

The dipping of dates in a 2% solution of potassium sorbate does not prevent yeast growth during storage (Bolin *et al.* 1972). This is due to the pH of dates being near 5.9, so that the sorbates are less effective than in other fruits. The addition of methyl bromide in conjunction with the sorbate dip is effective in preserving dates.

The growth of yeasts in apple juice is retarded by addition of 0.025 to 0.035% sorbic acid. However, this level does not control certain species of *Acetobacter*. For protection of apple juice, levels of 0.05 to 0.10% potassium sorbate in conjunction with refrigeration have been recommended.

Hayatsu *et al.* (1975) found that the sodium nitrite added to an ice-chilled, stirred suspension of sorbic acid in an acidic medium caused a rapid reaction. The products of the reaction contained no detectable sorbic acid, but were separated into seven fractions by thin layer chromatography. The product mixture and one fraction exhibited a significant mutagenic activity on *E. coli.* Three fractions showed strong bactericidal action at levels of 50 $\mu g/g$. It might be these fractions that inhibit *C. botulinum* in cured meat with added sorbate (Anon. 1976B). Due to the mutagenic effect, it is evident that further studies are needed on these products to determine the safety of using a combination of sorbic acid and nitrite in cured meats.

Acids of the tricarboxylic acid cycle, in equimolar concentrations, can neutralize the inhibitory action of sorbic acid. Certain molds can metabolize sorbic acid, especially if it is present in suboptimal quantities or the mold contamination is excessively high. Hence, the use of sorbates is not a substitute for good plant sanitation.

Succinic Acid.—This acid ($HOOC-CH_2-CH_2-COOH$) was as effective as acetic acid in its ability to inhibit microorganisms on poultry carcasses (Mountney and O'Malley 1965). Although treatment with a 3 or 5% succinic acid solution at 60°C effectively reduced the microbial load of poultry, the appearance was impaired (Cox *et al.* 1974B).

Acetaldehyde.—This chemical (CH_3-CHO) has a boiling point of 21°C and is naturally present in some fruit tissues. The vapors are irritating to mucous membranes of humans. The chemical has a general narcotic action. Large amounts may cause respiratory paralysis and result in death.

Acetaldehyde vapor has bactericidal and fungicidal properties. It is effective in the control of various microorganisms causing post harvest decay of fruits and vegetables, as well as yeasts associated with spoilage of fruit juice (Barkai-Golan and Aharoni 1976). The extent of fungicidal properties depends upon the concentration of the vapor and the length of exposure.

Ethyl Alcohol.—This is one of the most widely used substances to control microorganisms. It acts by denaturing protein.

Alcohol is bactericidal, but not sporicidal. It has poor penetrating power and is readily inactivated by organic matter. It is not considered to be a food preservative, although it can increase the antimicrobial effectiveness of other substances, such as sorbic acid in wine.

Benomyl.—Benomyl is the name given to methyl-1-(butylcarbamoyl)-2-benzimidazolecarbamate. Benomyl dips reduce the rotting of apples by *Gloeosporium* species. Applied to tomatoes as a dipping treatment, benomyl controlled *Botrytis, Penicillium* and *Cladosporium* (Domenico *et al.* 1972). Benomyl did not control *Alternaria* which seemed to grow better with benomyl treatment. This was believed due to reduced competition from the inhibited fungi.

Butylated Hydroxyanisole (BHA).—This chemical is added to foods as an antioxidant. Its presence can affect the microbiological population.

At 0.1%, BHA prevented the growth and aflatoxin production by *Aspergillus parasiticus* spores (Chang and Branen 1975). They also reported that BHA has an antimicrobial effect on *S. aureus, E. coli* and *S. typhimurium*. Dawson *et al.* (1975) found increased microbial counts in ground turkey containing BHA.

Chlorine.—The use of chlorine in the sanitation of processing plants and equipment is discussed in Chapter 10.

There have been many hypotheses regarding the site of action of hypochlorite on the bacterial cell. It is generally agreed that the action of chlorine involves the penetration of the cell. There are suggestions that once inside, it reacts with cellular protoplasm which causes inactivation. There are other suggestions that the chlorine reacts with key enzyme systems and prevents normal respiration. These reactions may involve oxidation or denaturation of the protein. The -SH groups of essential enzymes are possible sites of action. The oxidation theory has been questioned, since oxygen from other sources does not kill bacteria as readily as does chlorine.

There have been suggestions that cell membranes are altered or disrupted by chlorine. This allows diffusion of cell contents out of the cell or toxic materials into the cell.

Sodium hypochlorite preferentially inhibits polymerase deficient bacteria. Polymerase I is active in repair of damaged DNA. Rosenkranz (1973) suggested that the hypochlorite enters the cell and reacts with cellular DNA. This action is shared with known mutagens and carcinogens. The mechanism of the reaction with DNA remains to be elucidated.

Diphenyl.—This is also biphenyl or phenylbenzene ($C_{12}H_{10}$). It is colorless with a pleasant but peculiar odor and is insoluble in water.

In experimental animals, the diphenyl is excreted mainly as 4-hydroxydiphenyl. High levels cause an intoxication with increased respiration rate, loss of appetite and body weight, muscular weakness, paralysis and convulsion. Respiration difficulties can terminate in coma or death. The LD_{50} for rats is 2.2 g/kg.

Diphenyl is effective against molds but is not very active against yeasts or bacteria. It is used on the inside of shipping containers of citrus fruits, especially oranges. The diphenyl gradually volatilizes and inhibits the growth of molds on the fruit surface.

Although diphenyl may migrate into the peel, very little is found in the juice (usually less than 1 μg/ml). The amount allowed in the whole fruit (including the peel) is 110 μg/g (0.011%).

Ethylenediaminetetraacetic Acid (*EDTA*).—This chemical is a chelating agent. The chemical formula is $(HOOCCH_2)_2\text{-}NCH_2CH_2N\text{-}(CH_2COOH)_2$. The principal function of EDTA is to chelate trace metals that have an adverse effect on the quality of foods. EDTA is colorless, odorless and tasteless.

The FDA approved the use of EDTA and its sodium and calcium salts as additives in various foods.

The research concerning potential toxic effects of EDTA was reviewed by Anon. (1972). There are conflicting reports concerning the effect of EDTA on animals and humans. In mammals, 90% or more of the oral dose passes through the intestinal tract unabsorbed. In the chicken, high concentrations of EDTA have been found in the liver. Liver damage and nephrosis have been reported in rats injected with $CaNa_2EDTA$. On the other hand, a diet with 5,000 μg/g EDTA revealed no toxic effects in dogs or rats. According to this review (Anon. 1972), when a diet of 2% EDTA was fed to pregnant rats, the litter size was normal, although the babies were small and 7% were malformed. With 3% EDTA, no fetuses survived and they were malformed.

Although these chelating agents are not classed as germicides, they act as synergists for germicidal action. By chelating the trace metals, they deprive the microorganisms of these necessary cellular components, as well as affecting metal containing enzymes.

Although EDTA had only a slight effect on the total bacterial count of fresh fish (Baltic herring and rainbow trout), there was a favorable effect on the keeping quality (Kuusi and Loytomaki 1972). The volatile basic nitrogen, trimethylamine and hypoxanthine, increased at a slower rate in EDTA treated fish than in the untreated controls.

Winarno *et al.* (1971) found that a concentration of EDTA of 5mM inhibited germination and outgrowth of spores of *C. botulinum* type A and also toxin production in fish homogenate. The inhibitory action increased with increasing pH. When $CaCl_2$ or $MgCl_2$ at equimolar concentrations was added, the inhibitory action of EDTA was eliminated.

Hydrogen Peroxide (H_2O_2).—Hydrogen peroxide is available as stabilized solutions at various concentrations. Aqueous solutions of H_2O_2 gradually deteriorate to water and oxygen. Excess H_2O_2 can be removed from food by adding catalase to produce water and oxygen. The addition of H_2O_2 to certain foods is a simple and efficient method to inactivate certain undesirable microorganisms.

Hydrogen peroxide has been approved by the FDA for use as an antimicrobial agent in raw milk to be used to make certain types of cheese. This milk can be treated with no more than 0.05% H_2O_2. Added to liquid egg white, H_2O_2 is an aid in pasteurization to destroy salmonellae. Milk or liquid egg treated with H_2O_2 can be pasteurized at a lower temperature than that normally used for this purpose. The use of H_2O_2 has been suggested for improving the quality of milk in developing countries when refrigeration is not as available as in the USA. It may be useful in treating containers for use in aseptic packaging of foods.

The effectiveness of H_2O_2 is affected by the concentration, the bacterial population, the temperature, the time of contact and prior treatment given the microorganisms, such as a heat treatment.

The actual effect of H_2O_2 on cells is not completely known, other than its oxidative action. According to Rosenkranz (1973), H_2O_2 is a mutagen and an inducer of chromosomal aberrations. It is capable of reacting with DNA and its bases. Hydrogen peroxide affects certain enzyme systems.

Lysozyme.—This is an enzyme that is naturally present in various biological systems. Egg white lysozyme was suggested as a preservative for saké (Yajima *et al.* 1968). The minimum inhibitory concentration against lactobacilli in saké varied from 1.25 $\mu g/ml$ for *L. fermentum* to 100 $\mu g/ml$ for *L. acidophilus.* In saké, lysozyme was very stable. Lysozyme has its optimal activity near pH 7.0.

Spores treated with reagents which rupture the S-S bonds in the spore coat become sensitive to lysozyme.

Systems have been suggested that allow lysozyme to attack the cell walls of Gram negative bacteria. The use of EDTA in combination with lysozyme was suggested as a means of controlling salmonellae (Anon. 1976A). Miller (1969) suggested that treating Gram negative bacteria with a mixture of H_2O_2 and ascorbic acid renders the organisms sensitive to lysozyme.

The relatively high cost of lysozyme limits the use of this enzyme as an antimicrobial agent for food preservation.

Methyl Bromide.—This chemical is used for fumigation of dried fruits to prevent insect infestation. Methyl bromide has weak bactericidal properties. It is active only on fungi and vegetative bacterial cells.

At a level at which methyl bromide protected dates against microorganisms, the dates developed an off-odor (Bolin *et al.* 1972). A combination of sorbate and methyl bromide gave good protection of dates with no detectable odor or flavor changes. Methyl bromide is allowed in dates if the residual inorganic bromide is below 100 $\mu g/g$.

Beta-propiolactone (BPL).—This chemical can be considered as an internal ester of β-hydroxypropionic acid. The structure is:

$$\underbrace{CH_2—CH_2—C}_{O}=O$$

Compounds with a four membered ring generally have high chemical activity due to the tendency of the ring to open. Because of its high reactivity, it must be handled with caution. BPL is toxic, but is hydrolyzed rapidly in water to the nontoxic β-hydroxypropionic acid.

At sufficient concentration (0.5 to 1%), it is active against all types of microorganisms, including bacterial spores and viruses.

BPL was used to inactivate porcine parvovirus associated with commercial trypsin (Croghan and Matchett 1973). Treatment with BPL did not affect the enzymatic activity of the trypsin.

Murata and Kanegawa (1973) suggested that the primary and major action of BPL seemed to be the prevention of DNA replication.

Frequent topical or subcutaneous applications of dilute BPL produce tumors in mice and rats. It was suggested that BPL is a weak carcinogen.

Propylene Glycol.—This chemical (CH_3-CHOH-CH_2-OH) has many functions in the food industry. It has some antimicrobial properties and has been used as a food preservative. The chemical is relatively nontoxic.

Propylene glycol exhibits activity toward Gram positive and Gram negative bacteria. Acott *et al.* (1976) found the only approved chemical effective in controlling aspergilli and staphylococci in an IM food was propylene glycol. Usually a mixture of 2% propylene glycol and 0.3% potassium sorbate is used as the antimycotic system in IM meat products.

Sodium Chloride (NaCl).—Treating food with salt was one of the earliest methods of food preservation. Salt preservation was important in the development of the fishing industry. In some areas of the world, salted fish remains an important food product.

The concentration of salt needed to prevent microbial growth is related to pH, water content, type of microorganism, temperature, chemical composition of substrate and the presence of other inhibitory substances. The concentration of salt in the water phase is the most important aspect (Table 11.14).

The reason for inhibition of bacteria by salt is not readily apparent. It is thought to be primarily a plasmolytic effect, although other suggestions have been made. These include dehydration, removal of oxygen, interference with enzymes, altering of pH or, at high concentration, that sodium or chloride ions become toxic.

Sato *et al.* (1972) suggested that the lethal effect of NaCl is related to the loss of magnesium ions from the cells. The basis for this was finding Mg ions released from *E. coli* cells when incubated with NaCl and the recovery of salt-injured cells by incubation with Mg ions. The addition of Mg ions prevented the degradation of RNA in *E. coli* cells caused by 0.15M NaCl solution (Ito *et al.* 1977).

TABLE 11.14

SALT CONCENTRATION IN THE WATER PHASE (OR BRINE) IN FOODS WITH VARIOUS WATER CONTENT

Salt Added	Brine Concentration with Water Content of Food (%)				
%	40	50	60	70	80
2.5	5.88	4.76	4.00	3.45	3.03
3.0	6.98	5.66	4.76	4.11	3.61
3.5	8.05	6.54	5.51	4.76	4.19
4.0	9.09	7.40	6.25	5.41	4.76
5.0	11.11	9.09	7.69	6.67	5.88
6.0	13.04	10.71	9.09	7.89	6.98
7.0	14.89	12.28	10.45	9.09	8.04
8.0	16.67	13.79	11.76	10.26	9.09
9.0	18.37	15.25	13.04	11.39	10.11

Microorganisms vary in their tolerance to salt. The least tolerant are organisms such as *Spirillum*, which can tolerate only about 1% NaCl. Most normal organisms have a slightly greater salt tolerance. Generally these are inhibited by 3 to 10% salt. The mesophilic Gram negative rods and psychrophiles are inhibited by 4 to 10% salt. The lactic acid producing bacteria vary in tolerance from 4 to 15% salt. Sporeforming bacteria can tolerate from 5 to 16% salt. The organisms in these upper ranges are called halotolerant. Although they grow well in low levels of salt, they can tolerate these high levels of salt and even display some growth. These organisms include the micrococci, staphylococci and sporeforming bacteria.

The halophilic (salt loving) bacteria need relatively high concentrations of salt for growth. Some may grow at levels approaching saturation (26.4%). Most of the halophiles contain red, pink or other pigments and are classified in the genera *Halobacterium* or *Halococcus*.

Yeast species vary greatly in their tolerance to NaCl. The two most important factors are thought to be a mechanism to prevent entrance of NaCl to retain a low concentration of salt within the cells and the ability of enzyme systems to react in the presence of high levels of salt. A number of yeast types are able to propagate in pickle brine with 19-20% NaCl. They are more resistant to salt at pH 3.0 to 5.0 than at higher or lower values.

Salt has been reported to both increase lethality due to heat and to have a protective effect for the cells. Generally, low levels (1-2%) of salt may protect, while higher levels (6-10%) may increase lethality of heat.

Reports indicate that a higher concentration of salt is needed to inhibit the outgrowth of type A or B *C. botulinum* spores than those of type E.

The growth and toxin production of various strains of *C. botulinum* type A and B spores were determined in an experimental pickle fermentation. The organisms produced toxin with 4%, but not with 8%, salt in the brine (Ito *et al.* 1976).

There are three main types of microbial spoilage of salted fish. These are sliming, pink (red) and dun. Sliming is characterized by a semigreasy, sticky, glistening layer which is yellow-gray or beige and results in a pungent odor. The dominant organisms involved are Gram negative rods, although Gram positive and Gram negative cocci also are present. These organisms grow optimally with 6 to 8% salt.

The pink or red defect is due to growth of red halophilic bacteria, which are species of *Halobacterium* or *Halococcus*. Besides the red color, an off-odor is evident due to the production of indole.

The defect called dun consists of peppered spots ranging in color from chocolate to fawn. It is caused by a halophilic mold, *Sporendonema epizoum*, which has optimum growth at 10-15% salt.

Sodium Phosphates.—Sodium and phosphate combine to form a number of compounds such as sodium tetraphosphate, sodium hexametaphosphate, sodium tripolyphosphate and tetrasodium pyrophosphate. These are referred to as polyphosphates.

The polyphosphates are added to foods for various reasons. The antimicrobial effects are mainly due to the chelation of metallic ions.

Soaking chicken carcasses in solutions of 8% polyphosphates retards the production of off-odor, slime and discoloration during refrigerated storage. Lipolytic types of bacteria on the poultry seem to be inhibited to a greater extent than proteolytic types.

Dipping of cherries in 10% polyphosphate solution inhibited fungal growth up to 30 days at 1.1°C (Post *et al.* 1968). Untreated fruit showed fungal growth at 14 days. Sodium tetraphosphate appeared to be the most effective antimycotic of the polyphosphates that they tested.

Sucrose.—Although there are many sugars, the common use of the name sugar denotes sucrose. Low levels of sucrose are utilized by many microorganisms as a source of energy, but high levels of this chemical can have inhibitory properties. The preservative effect of sucrose is due to the development of osmotic forces and a lowered water activity.

Sugar is used in the preservation of several foods, including sweetened condensed milk, jams and jellies. Sweetened condensed milk is made by adding up to 40% sucrose to fresh milk. The milk is heated in evaporating pans to remove water and then canned as a product with 70 to 75% solids. Yeasts and molds are killed by the heat treatment during evaporation, the can keeps out further contamination and the high sugar content prevents the growth of surviving microorganisms.

The main preservative factor in normal jams is the combination of sucrose concentration and low pH. In an acid medium, sucrose is inverted to glucose and fructose. This increases the number of solute molecules which assist in lowering the a_w of the product. The low a_w, along with a pH of about 3.0, prevents the growth of most microorganisms.

The control of microbial growth is helped by filling these products while still hot, by heating the surface by steam injection, spraying the surface with a chemical fungicide or using a cover paper impregnated with a fungicide.

Sulfites.—The sulfites include sulfur dioxide (SO_2), sulfurous acid (H_2SO_3) and the sulfite, hydrogen sulfite or metabisulfite salts of sodium (Na_2SO_3, $NaHSO_3$ or $Na_2S_2O_5$), potassium and calcium.

The sulfites have been on the GRAS list, with levels set for specific foods. Dried fruits contain the highest levels while dried vegetables range from a few hundred to 2,000 $\mu g/g$. There are lesser amounts in wine and beer (0 to 400 $\mu g/g$). Thus, the amount of sulfite that can be used depends upon the type of food and, in some cases, the further processing that may occur. Besides any legal limits of sulfites, there are technological limits, since, at high levels, they may produce off-odors or off-flavors in certain foods.

The antimicrobial action is pH dependent. Sulfur dioxide combines with water to form sulfurous acid ($SO_2 + H_2O \rightarrow H_2SO_3$). This acid has two dissociation constants. These are near pH 2 and pH 7. At pH 7, the HSO_3^- and $SO_3^=$ are in about equal proportions. At pH 5, most of the compound is in the form of HSO_3^-. At lower pH values, protonation of the bisulfite ion results in molecular sulfur dioxide. The lower the pH, the higher is the proportion of SO_2. The free, molecular SO_2 is about 60 times more inhibitory than is the bound form. The molecular SO_2 form has the greatest and the $SO_3^=$ the least antimicrobial effect. The undissociated H_2SO_3 is more active than HSO_3^-.

Sulfites inactivate certain enzyme systems, such as cytochrome oxidase. The cytoplasmic membrane of bacterial cells is attacked by SO_2 which alters the permeability.

Sulfites combine with acetaldehyde, the normal hydrogen acceptor required for glycolysis, and inhibit fermentation. There is evidence that bisulfite reacts with pyrimidine bases, so the antimicrobial activity may involve the RNA and DNA of microbial cells.

The sulfites show antimicrobial activity against both fungi and bacteria. Gram negative bacteria are more susceptible than Gram positive types. Some lactic acid bacteria are fairly resistant to the action of SO_2, even at high concentrations. Yeasts are more resistant to SO_2 than either bacteria or molds.

The use of SO_2 on stored grain inhibits mold proliferation and subsequent production of mycotoxins.

Fresh ground beef treated with sodium metabisulfite, equivalent to 400 to 450 μg/g SO_2, and stored at 5°C, showed retarded proteolytic breakdown, but an increased rate of fat hydrolysis (Pearson 1970). The sulfite treatment of meat increased the storage life by several days. The SO_2 treatment of fresh meat is not allowed in the U.S., since it destroys thiamin and is said to maintain the red color of meat, which could be deceptive. In England, the use of SO_2 is restricted to fresh ground meat that contains cereal. Ground meat has a relatively short shelf life and SO_2 is thought to extend this period. When SO_2 treated sausage becomes spoiled, the dominant microorganisms are bacilli. According to the review of Kidney (1974), sulfiting of meat helps retard or destroy organisms that cause foodborne illness (salmonellae, *S. aureus* and *C. perfringens*).

There is some loss of sulfite through oxidation to sulfate and through volatilization and evaporation of SO_2 during heat treatment or during storage.

There have been some objections to the use of sulfites in foods on the basis of their potential toxicity to humans. Estimates of the daily per capita consumption of SO_2 have ranged from less than 10 mg to up to 600 mg. Beer and wine drinkers consume more SO_2 than nondrinkers. The United Nations FAO/WHO has established the acceptable daily level of intake as 0.70 mg SO_2/kg body weight.

The sulfites that are ingested are oxidized to sulfates which are excreted in the urine. Most of the sulfite intake is from natural sources of sulfite or sulfur-containing amino acids.

Various studies indicate that dietary sulfites are not very toxic. A diet with rather high sulfite and low thiamin can cause undesirable effects, due to a lack of thiamin.

Thiabendazole (TBZ).—TBZ has antifungal properties. According to Edney (1973), the use of TBZ has been widely adopted by apple growers to reduce the rotting of apples by *Gloeosporium* species.

Antibiotics.—The therapeutic value of antibiotics in medicine stimulated research to determine their effectiveness as food preservatives. This revealed that broad spectrum antibiotics, such as chlortetracycline (CTC), or oxytetracycline (OTC) had greater potential for use in foods than antibiotics, such as penicillin. In Nov. 1955, a tolerance of 7 μg/g for CTC in or on uncooked poultry was established. The following year, the same tolerance was established for OTC. The petition requesting the establishment of the tolerance stated that the cooking of poultry destroyed residues of CTC when used at the accepted level. In 1959, the authorization to use CTC or OTC was extended to raw, whole or gutted fish, shucked scallops and unpeeled shrimp. A maximum tolerance of 5 $\mu g/g$ was established for these products.

The original tolerances were established due to the antibiotics having a useful function in the food and their use in food was safe. However, due to the hazard of the development of resistant strains of pathogens, the increasing resistance of these organisms, the potential for hypersensitivity of humans to the antibiotics, the finding of residual antibiotics even after cooking (although these levels were very low and probably insignificant), the lack of significant benefits when used in commercial operations and the costs and difficulties involved in monitoring all of these aspects, the tolerance of CTC, OTC or similar antibiotics on or in food was rescinded and they are no longer permitted in foods in the U.S. Antibiotics were never approved for use in fresh meat.

Although antibiotics are not permitted as direct additives, this does not mean that there are no antibiotics in our food, nor does it mean that all antibiotics will be banned in the future.

Several antifungal antibiotics have been suggested to control mold or yeast damage on crops in the field or on fruits and vegetables immediately postharvest. Since the antibiotics are on the surface, they can be washed off quite readily, so that there should be little, if any, residual in the food.

As long as microorganisms are associated with food, there is a possibility that antibiotics are present since, after all, microorganisms produce antibiotics.

An antibiotic that occurs naturally in food products such as cheese, and has attracted much attention, is nisin. This antibiotic is a polypeptide produced by certain strains of *Streptococcus lactis.* It has inhibitory properties against Gram positive microorganisms.

Some advantages concerning the use of nisin in foods are that it is not toxic to humans, it has no use in medicine and it does not form cross resistant mutants. Its medical value is of no significance since it is practically insoluble in blood at physiological pH. It is digested by intestinal enzymes and bacteria and the large size of the molecule precludes absorption if intra-muscular injection is used. It is unlikely that nisin has any effect on the intestinal flora. Nisin has no effect on Gram negative bacteria and does not inhibit all of the Gram positive bacteria. Some bacteria produce the enzyme, nisinase, which inactivates nisin.

The primary application of nisin is in heat processed foods. This is due to the action of nisin on spores. Heat fixes nisin on spores and thereby reduces their heat resistance. This makes it possible to heat sterilize foods with reduced heat treatments.

Nisin does not influence the heat resistance of some spores, but its continued presence in the recovery medium reduces the count of survivors. In sufficient concentration, nisin either inhibits the germination of spores or the outgrowth of the cells. According to the review of Hurst (1972), nisin prevents the outgrowth rather than the germination of spores. However, the germination of spores of three species of *Bacillus* was significantly lower in the presence of 100 μg/ml of nisin (Gupta *et al.* 1972). There is evidence that nisin is effective mainly against heat damaged spores.

Clostridium botulinum is one of the more nisin-resistant species. Therefore, with canned foods, sufficient time and temperature of process are needed to assure that *C. botulinum* spores are controlled. Any reduction in the heat process could affect the safety of canned food. Hence, the use of nisin would need to be evaluated for each type of heat processed food.

Nisin is not allowed as a food additive in the U.S.; however, some countries allow its use in one or more foods.

Since antibiotics can become associated with foods in many ways, some countries, and the Food and Agricultural Organization of the World Health Organization of the United Nations, have set tolerances for some of these agents. For the FAO, these include ampicillin (0.06 μg/g), penicillin (0.06 μg/g), oxytetracycline (0.25 μg/g) and neomycin (0.50 μg/g).

Sodium Nitrite (*$NaNO_2$*).—The history of meat curing by nitrites and nitrates is discussed by Binkerd and Kolari (1975). Besides these chemicals, various other ingredients including NaCl are used in this process. The reduction of nitrates produces nitrites.

Any additive should have some useful function in the food and be safe for the consumer. Sodium nitrite has certain useful functions in cured meat, but the safety of this chemical has been questioned.

Nitrite is added to the curing mixture to develop the typical color of cured meat. In acid conditions, nitrite forms nitrous acid, which is reduced to nitric oxide. By a series of reactions, the nitric oxide combines with the muscle pigment myoglobin to form nitric oxide myoglobin. Upon heating, this pigment forms the more stable red pigment, nitrosyl hemochrome, associated with cooked ham.

Hams were cured with NaCl long before any benefits were credited to $NaNO_2$. According to Kemp *et al.* (1974), good quality hams can be produced without either nitrate or nitrite. Hams cured with 182 μg/g nitrite had a more typical ham color than those cured with 91 μg/g nitrite (Brown *et al.* 1974). Eakes and Blumer (1975) obtained hams with acceptable color when 90 μg/g nitrite was used in the cure. Higher levels of nitrite produced higher concentrations of the cured pigment, but these higher values were not considered to be necessary for color acceptability.

Hence, relatively low levels of nitrite are sufficient for color development. Nitrite levels of 10 μg/g and 50 μg/g of meat have been suggested as sufficient for complete color development of cured meat. Reducing agents such as ascorbic acid accelerate color formation and increase color uniformity and stability.

The main ingredient and preservative in the curing mixture is NaCl. At the level at which salt can be used without developing an over-salty product, it cannot be relied upon to inhibit all microorganisms and preserve the meat. Although the nitrates have no apparent inhibitory properties, the nitrites do have antimicrobial activity. Even then, cured meats are susceptible to spoilage.

The action of nitrite on microorganisms is due to the presence of undissociated nitrous acid (HNO_2). Nitrous acid can react with a wide variety of substances, including myoglobin, ascorbic acid, phenols, secondary amines, amino groups and thiol groups. The antimicrobial action of nitrous acid is pH dependent.

Above pH 7.5, nitrite may aid bacterial growth. Between pH 6.0 and 7.0, there is only a small portion of undissociated HNO_2 present, so there is only a slight antibacterial effect. Between pH 4.5 and 5.5, the presence of nitrous acid shows a definite antimicrobial effect. Below pH 4.5, the nitrous acid appears to decompose to nitric acid, nitric oxide and water. Neither nitric oxide nor nitric acid shows the same activity against microorganisms as nitrous acid. In the presence of oxygen, the nitric oxide is oxidized to nitrogen dioxide. This combines with water to form nitric acid and nitrous acid. In canned or vacuum-packaged meat, as well as in the interior of meat, there would not be sufficient oxygen for these reactions to occur. Below pH 4.5, nitrite is less effective, but pH effects will help control bacterial growth.

The main benefit of nitrite is the inhibition of *C. botulinum* in canned cured meat. If sufficient heat is used to destroy *C. botulinum* spores, the quality of ham may be affected. The nitrite concentration in vacuum-packed hams delays the onset of toxin formation and spoilage. Despite the fact that both aerobic and anaerobic spores can be grown from canned cured meats, according to a USFDA Consumer Memo, there have been no known outbreaks of botulism caused by processed food containing nitrites.

The inhibition of *C. botulinum* involves the interaction of the concentration of salt and sodium nitrite, the pH, the amount of heating, the number of spores that are present and the temperature of storage. Apparently, heat injury is not necessary for $NaNO_2$ to inhibit toxigenesis of *C. botulinum* (Pivnick *et al.* 1967). Curing salts do not alter the heat resistance of spores, but they tend to inhibit the subsequent outgrowth of surviving spores.

Christiansen *et al.* (1974) stored cured bacon inoculated with *C. botulinum* at 7° and 27°C. No toxin was found in samples stored at 7°C for 84 days. However, samples formulated with 120 μg/g nitrite or less became toxic when held at 27°C. At low levels of inoculation (210 spores before processing and 52 spores/g after processing), levels of 170 μg/g nitrite or more prevented toxin formation. At higher levels of inoculation (4,300 spores/g after processing), toxin was detected even with 340 μg/g nitrite added to the meat. Increased levels of nitrite decreased the rate of toxin production. The level of nitrite needed to prevent toxin production depended upon the number of *C. botulinum* spores.

The importance of a low pH on nitrite inhibition of toxin production was shown by Christiansen *et al.* (1975). At normal pH levels (pH 5.8-5.9) in a summer sausage, nitrite levels of 150 μg/g did not prevent toxin production. At lower pH (4.2-4.7), no toxin was evident with nitrite at 50 μg/g.

A combination of 200 μg/g $NaNO_2$ and 500 μg/g sodium erythorbate prevented the growth of two strains of *Bacillus cereus* (Raevuori 1975). McLean *et al.* (1968) reported that $NaNO_2$ up to 200 μg/ml did not affect growth or enterotoxin B production by *Staphylococcus aureus* in brain heart infusion broth at pH 7.0. However, 120 μg/ml nitrite, 2% salt and 200 μg/ml nitrate reduced the amount of enterotoxin produced.

Many organisms, such as some strains of lactobacilli, pediococci, enterococci and various Gram negative species, including salmonellae, are relatively resistant to nitrite.

Media containing nitrite become more inhibitory to bacterial growth when heated (Perigo and Roberts 1968). With heating at pH 6.0, levels of 80 μg/ml or less of nitrite inhibited the growth of several strains of *C. botulinum* and *C. perfringens*, as well as five other species of clostridia. Most of these organisms were inhibited by 20 μg/ml or less of residual nitrite. Although there are doubts that this effect occurs in heat processed cured meats, Chang *et al.* (1974), reported an inhibitory effect to *C. botulinum* remained in canned pork luncheon meat with only 2 μg/g of residual nitrite.

The concentration of nitrite decreases during storage of cured meat and may disappear after several months. Sodium nitrite is considered to be heat labile and some of it is destroyed during the heat processing used in cooked or canned cured meats. Growth of organisms in stored cured meat may be due to a tolerance for nitrite, or it can be due to the loss of nitrite to such a low level that there is essentially no antibacterial activity.

The loss of nitrite is important, not only because there is a reduction in the amount of inhibitor, but also the type of compounds that are formed may be desirable or undesirable. In a meat model system, Miwa *et al.* (1978) reported about 74% of the nitrogen from added nitrite was found as nitrite, nitrate, nitrosothiol, denatured nitrosomyoglobin and gaseous nitrogen. The other 26% was found in unidentified compounds. Woolford *et al.* (1976) suggested that about 30% of the added nitrite is bound to protein and Woolford and Cassens (1977) reported between 10 and 15% of the nitrite was in the adipose tissue of bacon. They found a difference in distribution of nitrite nitrogen in bacon with or without added ascorbate. Gilbert *et al.* (1975) suggested several compounds that may be formed by reactions of food constituents with nitrite.

Sodium nitrite is readily absorbed from the intestinal tract and rapidly disappears from the bloodstream. Some 30 to 40% of this nitrite is excreted unchanged in the urine. Some $NaNO_2$ undoubtedly reacts with hemoglobin in the blood, oxidizing it to methemoglobin (CDC 1975). The normal reducing systems of the blood can handle some increase in methemoglobin, but when the system becomes overtaxed, cyanosis occurs, followed by methemoglobinemia which can be fatal. Numerous cases of infant methemoglobinemia have been related to excessive nitrite in the diet.

In recent years, the main concern has been the reaction of nitrite with secondary and tertiary amines to form nitrosamines. Some, but not all, nitrosamines are mutagenic, teratogenic and carcinogenic. The mutagenic and carcinogenic aspects of nitrosamines were reviewed by Olajos (1977). Various carcinogenic and noncarcinogenic nitrosamines were listed by Nagao *et al.* (1977).

The widespread use of nitrites, coupled with the natural occurrence of amines in foods, has resulted in the possibility that these carcinogens are present in our food supply.

There are many nitrosamines, but the common ones are N-nitropyrrolidine, N-nitrosodimethylamine and N-nitrosopiperidine (Fig. 11.12). Nitrosamines were reported in salted fish (Fong and Chan 1976), various Chinese foods (Fong and Chan 1977; Gough *et al.* 1977) and fermented beverages (Bassir and Maduagwu 1978).

When nitrosamines are found in cured meats, they are usually at levels below 10 ng/g (Fiddler *et al.* 1975; Panalaks *et al.* 1974). However, some products had levels of nitrosodimethylamine from 90 to 105 ng/g. Nitrosamines rarely have been found in frankfurters, but dimethylnitrosamine was reported by Wasserman *et al.* (1972), in one sample, at 84 ng/g.

The analysis of spice-cure mixtures yielded nitrosamines at levels of 2 μg/g (Havery *et al.* 1976) to 25 μg/g (Sen *et al.* 1974). Due to findings such as these, the USFDA and USDA ordered meat packers to quit mixing sodium nitrite with spices in advance of placing the curing mixture into meat products. Federal regulations prohibit the marketing of spice-curing mixtures unless the nitrites and spices are separated by either chemical buffer or packaging.

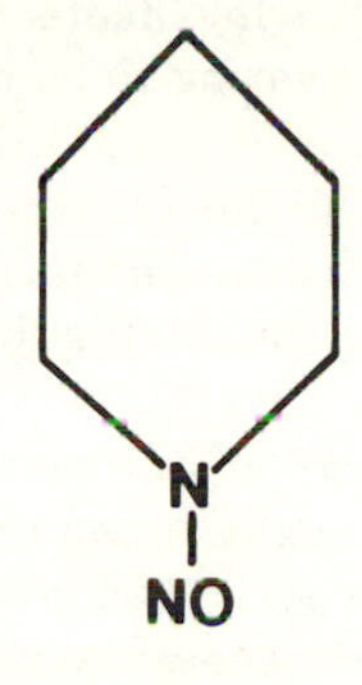

Nitrosopiperidine

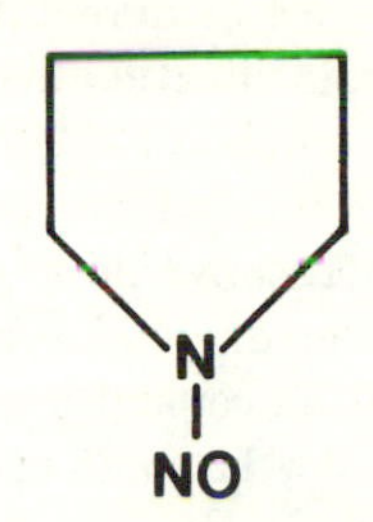

Nitrosopyrrolidine

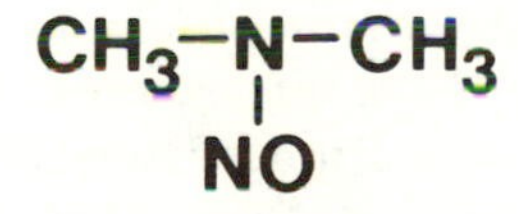

Dimethylnitrosamine

FIG. 11.12. NITROSAMINES

Ordinarily, nitrosamines are not found in raw bacon, but they are detected in cooked bacon and the cooked out fat (Fazio *et al.* 1973; Havery *et al.* 1976). Levels as high as 139 ng/g of N-nitrosopyrrolidine in the cooked bacon and 920 ng/g in the cooked out fat have been reported. The formation of nitrosamines is favored at low pH and by heating. The cooking of bacon induces the formation of nitrosamines, but the method by which it is formed has not been definitely established. The addition of either putrescine or proline to ground pork containing sodium nitrite increased the amount of N-nitrosopyrrolidine that was formed (Warthesen *et al.* 1976). They suggested that a significant portion of the nitrosamine may be volatilized and not detected in either cooked bacon or rendered fat. Nakamura *et al.* (1976) found that bacon heated to 200°C contained a higher level of nitrosopyrrolidine than when heated to either 175° or 225°C. At 200°C or over, the longer the heating time, the lower was the amount of detectable nitrosamine. This was apparently due to the rate of volatilization becoming greater than the rate of formation. At temperatures below 125°C, the amount of the nitrosamine increased as the time of heating increased. They suggested pathways for the formation of N-nitrosopyrrolidine in bacon.

Pensabene *et al.* (1974) reported different levels of nitrosamine formation in bacon cooked by various methods. Baking of bacon resulted in the detection of 12 to 35 ng/g, while microwave-cooked bacon contained from 0 to 3 ng/g. Broiling, pan frying and the baconer system resulted in levels of nitrosamine between 7 and 20 ng/g.

There has been concern that the mere presence of nitrites and amines in food can result in formation of nitrosamines by reactions in the food, by microorganisms or in the acidic environment of the stomach. It has been suggested that these precursors should be at a low level in the diet. Nitrates are converted to nitrites by bacteria in the mouth. It has been estimated that almost 80% of the nitrite that reaches the stomach is from saliva. There is a direct relationship between salivary nitrite and ingested nitrate (Spiegelhalder *et al.* 1976). Many vegetables contain nitrate and nitrite.

It has been reported that microorganisms in food or in the gastrointestinal tract can produce nitrosamines (Coloe and Hayward 1976; Hashimoto *et al.* 1976). Besides forming nitrosamines, intestinal bacteria have been reported to degrade nitrosamines (Rowland and Grasso 1975).

Some work indicates that feeding of secondary amines and nitrites fails to produce tumors in experimental animals. Sen *et al.* (1975) reported that simultaneous ingestion through the drinking water of diethylamine hydrochloride and sodium nitrite for up to 30 months, failed to produce tumors in guinea pigs. When diethylnitrosamine was in the drinking water, 18 of 20 guinea pigs developed liver tumors.

Sen *et al.* (1975) believed that when nitrite and secondary or tertiary amines are mixed in a diet, some nitrosamines may be formed and animals fed this diet are more likely to develop tumors than when the compounds are fed separately. They suggested that it is desirable to keep the precursors of the nitrosamines at a low level in the diet.

There are statements in the literature that if nitrosamines are proved to be carcinogenic, nitrites will not be allowed as additives in cured meats. There is documentation that nitrosamines are carcinogenic, but nitrites have, so far, not been banned as additives. One argument is that the use of nitrites has never been shown to cause cancer in humans. Several additives have been banned from foods with less data than accumulated regarding the carcinogenic character of nitrosamines. There was a statement that a person would need to eat 20 tons of bacon to be equivalent to the amount of nitrosamine that caused cancer in rats. This has not been a deterrent in banning other potentially carcinogenic chemicals from foods or food use.

Various ingredients used in cured meats may either stimulate or inhibit nitrosamine formation. When glucono-delta-lactone (GDL) was added to a model system, there was an increase in dimethylnitrosamine (Fiddler *et al.* 1973A). They found that sodium ascorbate, ascorbic acid or sodium erythorbate (isoascorbate) markedly inhibited nitrosamine formation in the model system as well as in frankfurters (Fiddler *et al.* 1973B). The increased nitrosamine formation due to GDL was nullified by ascorbate or erythorbate. They found that sodium acid pyrophosphate or sodium tripolyphosphate resulted in a lower level of nitrosamine formation.

Compounds such as sodium bisulfite, tannic acid, hydroquinone, ammonium sulfamate, methionine, cysteine, glutathione and propyl gallate can block the nitrosation reaction (Gray and Dugan 1975). Sen *et al.* (1976) reported that treatment of bacon with 1,000 μg/g of propyl gallate, piperazine, ascorbyl palmitate or sodium ascorbate just prior to frying, reduces the formation of nitrosamines during heat treatment.

Besides inhibiting nitrosamine formation, according to Raevuori (1975), erythorbate increases the antimicrobial effect of nitrite. Tompkin *et al.* (1978) reported 50 μg/g nitrite with added erythorbate was as effective as 156 μg/g of $NaNO_2$ alone to inhibit *C. botulinum.*

Incze *et al.* (1974) stated that the danger of botulism is the only acceptable argument by health authorities in favor of the continued use of nitrite or nitrate.

Incze *et al.* (1974) found cysteine-nitrosothiol to be a more effective inhibitor than nitrite when *C. sporogenes*, salmonellae or *Streptococcus faecium* was tested. They did not report the use of *C. botulinum* spores in their tests.

Potassium sorbate has been proposed as an adjunct to help control the growth of *C. botulinum* (Anon. 1976B). Bacon samples were inoculated with 1,100 spores of *C. botulinum* per gram and treated with 0.2% potassium sorbate and 40 μg/g sodium nitrite. When stored at 27°C, *C. botulinum* was inhibited at least as well with this mixture as when 156 μg/g nitrite was used alone.

Wierbicki and Heligman (1973) suggested using lower levels of nitrite in conjunction with radiation pasteurization (radappertization). Radiation treatment of foods for microbial destruction has not been approved in the USA.

The legal limits of nitrite have been lowered in recent years. Originially, the limit was 200 μg/g residual nitrite. With this limit, higher levels of nitrite could be added, as long as the residual was 200 μg/g or less. After research showed that nitrosamines were formed in cooked bacon and may be present in other cured meats, an expert panel was established to advise the Secretary of Agriculture.

With the panel's recommendation, the U.S. Department of Agriculture originally proposed:

1. A limit of 2,183 μg/g of sodium nitrate or 2,597 μg/g of potassium nitrate to be added to dry cured products; and 1,716 μg/g of sodium nitrate or 2,042 μg/g of potassium nitrate to be added to fermented sausages.
2. A limit of 624 μg/g of sodium nitrite or 768 μg/g of potassium nitrite to be added to dry cured products and 156 μg/g sodium nitrite or 192 μg/g potassium nitrite to be added to fermented sausages.
3. Whether nitrate, nitrite or a combination is used in these products, the maximum residual nitrite level as sodium nitrite would be 200 μg/g of meat.
4. In canned cured products (perishable, shelf stable or sterile), in cooked sausages and other perishable products (other than bacon) a limit of 156 μg/g of sodium nitrite or 192 μg/g of potassium nitrite introduced by pumping into solid pieces of meat or otherwise incorporated into comminuted products.
5. Canned cured sterile products would be limited to a residual 50 μg/g sodium nitrite. All other canned cured products and products prepared with curing solutions were limited to 125 μg/g and cooked sausage to 100 μg/g residual sodium nitrite.
6. The use of sodium or potassium nitrate or nitrite would not be permitted in meats for commercial preparation of infant (strained) or junior (chopped) foods.
7. The maximum amount of sodium nitrite permitted in bacon to be limited to 125 μg/g. A requirement would be made that ascorbate or erythorbate be added at the maximum rate permitted by regulation.
8. Salt alone would be permitted as a preservative when added at a sufficient amount so that the finished product has a minimum brine concentration of 10% or a water activity no greater than 0.92.

It is to be expected that these recommended levels will be reduced as more information is obtained. In 1977, the expert panel suggested ingoing nitrite levels of 100 μg/g (with 300 μg/g of nitrate if needed) for dry cured products; 156 μg/g for canned, cured, perishable and shelf-stable products; 100 μg/g in cooked sausages; 100 μg/g in fermented sausage (with nitrate to produce an organoleptically acceptable product); and 120 μg/g for bacon. The allowable residual nitrite levels would vary from 50 to 125 μg/g as determined before storage.

Thus, at least for now, nitrites may remain in most cured meats, but generally at reduced levels from those used previously. The analysis of cured meat before these proposals showed that most had residual nitrite levels of less than 100 μg/g of meat. With nitrite added at only 120 μg/g for bacon, it will be at a low residual level and perhaps cause no problem with nitrosamine formation. To help reduce the possibility of toxin production by *C. botulinum*, the expert panel suggested adding ascorbate or erythorbate at 550 μg/g of bacon.

Smoking.—In ancient times it was noticed that meat exposed to smoke from a fire had improved keeping qualities. Due to refrigeration, the preservation of meat by smoking is no longer as important. However, smoking is used to develop distinctive and desirable flavors in certain foods, such as meat, fish, poultry and cheese. The chemical composition of wood smoke depends upon factors such as the type of wood, the amount of air, the temperature and time. Hickory or other hardwoods are preferred. Soft woods cause unpleasant flavors in the food.

The preservative effect is a result of the combination of drying and the deposition of the chemicals resulting from the thermal decomposition of wood. The amount of smoke deposited on a food is a complex function of the composition and concentration of the smoke, the temperature and humidity and the nature of the food surface. To enhance the deposition of particles of wood smoke onto the food surface, electrostatic charges can be used.

Several hundred chemicals have been identified in the smoke condensate. These chemicals include acids, alcohols, aldehydes, ketones, phenols, carbonyl compounds, waxes, resins and tars (Table 11.15). The acids, phenols and carbonyls are recognized as the components which contribute most significantly to the odor and flavor of smoke. Volatile phenolic compounds in the vapor phase of the smoke are primarily responsible for the smoke flavor.

Several of the chemicals found in smoke or on smoked foods normally would not be allowed as food additives. Hence, smoking is one means of using these chemicals in foods.

A number of investigators have detected 3,4-benzopyrene and 1,2,5,6-dibenzanthracene in wood smoke. These polycyclic hydrocarbons are carcinogenic to laboratory animals. Although it is thought that smoked foods are unlikely to be a health hazard, according to Moodie (1970), there is a relatively high incidence of carcinoma of the alimentary tract in populations in which smoked foods are consumed in large quantities.

In a survey, Malanoski *et al.* (1968) detected polycyclic aromatic hydrocarbons in many smoked foods at low levels (0.5 to 7.0 ng/g). Fretheim (1976) reported benzopyrene at levels of 0.15 ng/g, as well as several other polycyclic hydrocarbons in smoked sausages. The carcinogenic polycyclic hydrocarbons are derived from the lignin portion of wood. Their production is minimized when the temperature of combustion is maintained below 350°C.

TABLE 11.15

APPROXIMATE COMPOSITION OF SMOKE FROM VARIOUS WOODS

	Acids	Aldehydes and Ketones	Bases	Phenol
Beech	4.0	2.7	4.0	2.5
Oak	5.2	1.8	1.6	2.7
Plane tree	5.1	5.2	1.8	4.0
Birch	4.0	2.3	3.9	3.4
Pine	12.3	0.1	4.1	3.3
Alder	4.3	3.4	2.8	3.1
Lime	6.0	4.8	3.1	3.8
Aspen	9.1	5.2	4.1	5.2
Fir	6.8	0.2	3.8	4.6

Source: Heid and Joslyn (1967)

Using liquid smoke is one means of developing a smoked flavor in meat. In liquid smoke production, the smoke is fractionally distilled and the selected fraction diluted with water. This results in a product reportedly essentially free of the water insoluble 3,4-benzopyrene. However, White *et al.* (1971) found concentrations of benzopyrene ranging from 25 to 3,800 ng/g in condensates which settled out of liquid smoke flavors during storage.

Formaldehyde, acids and phenols are antimicrobial agents. They are regarded as being responsible for much of the antibacterial action in smoked meats. As would be expected, bacterial spores are quite resistant to smoking. Many vegetative bacteria are destroyed by smoking for two or three hours, but most of the spores survive smoking for seven or more hours. The destruction of bacteria is correlated to some extent with smoke density, duration of smoking, temperature of the meat, fish or poultry and the humidity. Spoilage of smoked meat is often due to contamination that occurs after processing, with subsequent exposure to conditions which permit microbial growth.

Since smoking does not control spores, including those of *C. botulinum*, the good manufacturing practices for smoked fish (GSA 1977) require heating at 82.2°C for 30 min.

Liquid smoke at 0.08% retards but does not stop growth of bacteria. At a level of 0.1%, liquid smoke has a preservative effect and gives a mild smoke flavor. At levels above 0.4%, there are excellent preservative effects but there is a strong smoke flavor, and this level of usage is not economical.

Other Chemicals.—There are many other antimicrobial agents. Some of these are agricultural fungicides that are used to treat fruits in the field or postharvested fruits. These include Maneb, Captan, Dichloran, Nabam (Disulfiram), Dithane, Ziram, Zineb, Thiram and *o*-phenylphenol.

CONTROLLED ATMOSPHERES

Fruits and vegetables respire during storage and this respiration contributes to the loss of quality. As the temperature of storage is decreased toward 0°C, the rate of respiration is reduced, and storage life is increased. During respiration, O_2 is used and CO_2 is given off. The regulation of the concentration of O_2 and CO_2 for storage of fresh food is called controlled atmosphere (CA) storage. Since CA storage slows respiration, less heat is produced by the stored product and less energy is needed for refrigeration.

Developing the desired atmosphere by natural respiration is satisfactory for fruits, such as apples. However, this is not satisfactory for perishable vegetables. Deterioration occurs before the optimum atmosphere is obtained. Generators are used to rapidly produce the desired CA for storage of vegetables.

The low O_2 level in CA affects the growth of some molds and the CO_2 level limits the growth of some microorganisms. Also, microorganisms are less able to attack ripening fruit as readily as overripened fruit. Hence, the microbial spoilage is retarded by the slowing of the ripening process. The direct inhibition of microorganisms due to gas concentrations seems to be secondary.

Different types of fruits and vegetables, and even the varieties of the same type, have different optimum concentrations of O_2 and CO_2 for storage. When the O_2 concentration is reduced low enough to appreciably reduce mold growth, many fruits are affected adversely.

High CO_2 and low O_2 concentrations become a safety hazard for workers. Therefore, controlled atmospheres are not practical in warehouses when produce is constantly being moved in and taken out.

Edney (1973) reviewed the storage of apples in CA. Refrigeration and CO_2 will retard the rate of apple rotting. However, if the CO_2 concentration is increased beyond the optimum, increased rotting may result. There were fewer rotten apples at 8% CO_2 than at either 4 or 12% CO_2. The lowering of the O_2 level from the 21% in air to 10% reduced the rotting of apples, and a further reduction in rots was noted as the O_2 concentration was reduced from 10 to 3%.

Wilson *et al.* (1975) stored high-moisture corn (19.6 or 29.4% moisture) inoculated with *Aspergillus flavus* in various mixtures of O_2 and CO_2. They found that reduced O_2 and/or increased CO_2 delayed deterioration but growth was not completely stopped for *A. flavus*. In their modified atmospheres, the level of aflatoxin in the corn was always less than 20 μg/kg. They suggested that, in naturally infested corn, the *A. flavus* might do better than these inoculated cultures in the modified atmospheres.

Carbon Dioxide

When the concentration of CO_2 is increased slightly above normal, there is generally a stimulation of microbial growth. At high concentrations, CO_2 has been recognized as a means to control the growth of microorganisms in carbonated beverages, beef, poultry, fruits, vegetables and shell eggs.

Lillehoj *et al.* (1972) studied the effect of CO_2 on germination and growth of *Penicillium martensii* on corn. At 30°C, the germination fell from 36% in air to 2% in 60% CO_2, and no germination was found with the combination of 10°C and 60% CO_2. With either 40% or 60% CO_2, and storage at 5° or 10°C, there was essentially no penicillic acid formed by this mold. Penicillic acid is a carcinogenic mycotoxin.

King and Nagel (1967) found a linear correlation between CO_2 concentration and the amount of inhibition on the rate of growth of *Pseudomonas aeruginosa*. The inhibition by CO_2 was observed only when other environmental factors (O_2 tension, pH, nutrients and temperature) were not optimum.

The susceptibility to CO_2 varies with the organism. The psychrotrophic organisms that cause spoilage of refrigerated flesh foods are particularly sensitive to CO_2, while mesophiles are less sensitive.

Vacuum Packaging

This type of package is evacuated before sealing, either by inserting a vacuum probe into the neck of the package, or by placing the package into a chamber and evacuating. With flexible films, this results in a package that is drawn tightly around the contents.

Since the development of oxygen impermeable films, vacuum packaging has become the method for packing table-ready meat items. Even if a complete vacuum is not attained, the small amount of residual oxygen will soon be used by the microorganisms and replaced with CO_2.

Sliced processed meats that are vacuum packed have a longer shelf life than when packaged at atmospheric pressure. Various microorganisms have been reported to be inhibited on vacuum packed cured meat. The microorganisms on vacuum packaged meats are primarily micrococci, as well as sporeformers and lactobacilli.

Vacuum packaging retards the growth of common aerobic spoilage bacteria, such as *Pseudomonas* species, on refrigerated fresh meat, poultry and fish. This reduces putrefaction and slime formation.

The differences reported for growth of aerobic organisms in vacuum packed meat are due to the levels of vacuum attained in the packages or the types of film used for packaging. As absolute vacuum is approached, the amount of oxygen remaining in a package reaches a very low value. Commercial systems do not reach absolute vacuum, so there is some residual oxygen present. At the high degrees of vacuum, there is better maintenance of quality and a greater inhibition of microorganisms.

Ozone

Ozone contains three atoms of oxygen. The third atom is loosely bound to the other two. This third atom easily unites with cellular material resulting in inhibition or death.

Ozone is formed by electric discharge, such as an electric arc through air or by UV radiation at wavelengths of 200 to 210 nm. This production occurs especially at higher altitudes.

Since ozone is a powerful oxidizing agent, it has been postulated that part of the cellular damage is due to the oxidation of -SH bonds. It is likely that the ozone reacts with the double bonds of cell wall lipids, since ozone causes leakage of cellular components and lysis of some microbial cells. The leakage or lysis of the cells depends on the extent of the reaction with cell wall lipids.

The lethal threshold concentration of ozone for washed cells of *Bacillus cereus* was 0.12 mg/liter and for *E. coli* and *B. megaterium*, it was 0.19 mg/liter. For spores of the bacilli, the lethal threshold concentration was 2.29 mg/liter (Broadwater *et al.* 1973). Unwashed cells were not killed during five minutes exposure up to 0.71 mg/liter of ozone. Apparently, organic material associated with unwashed cells affected the activity of the ozone.

A part of the microbial effect of UV lamps in meat storage rooms is due to the production of ozone and the resultant inhibition or death of bacteria. An ozone level of 0.1 mg/liter has little effect on microorganisms in air, but on moist meat surfaces, it exerts a bacteriostatic effect. The continual exposure of the organism to sublethal doses of ozone may eventually cause the death of the cell, particularly at low temperatures.

The ozone level that is toxic to small animals is about 3 to 12 mg/liter. At 0.02 to 0.04 mg/liter, most people can detect the odor of ozone, and at

0.1 mg/liter, it may cause headache, dryness and irritation of the throat, respiratory passages and eyes. Ozone is a powerful mutagenic and oxidizing agent.

MICROBIAL INTERACTIONS

The use of microbial fermentations as a means of preservation is well known and discussed in Chapter 9. Besides the production of acids or alcohols, some microorganisms form bacteriocins and antibiotics that inhibit other bacteria (Chapter 4).

The inoculation of foods with certain microorganisms to control other microorganisms might be unique, although this has been used in fermentation such as in dairy products. The addition of a lactic culture to ground meat significantly retarded the growth of Gram negative spoilage bacteria (Reddy *et al.* 1975). There was a less dramatic effect when the lactic culture was used on beef steaks. Besides a lower count of spoilage bacteria, the pH, volatile nitrogen and free fatty acids were lower in inoculated than in control samples of meat.

This type of inhibition is limited to certain types of foods. In most instances, it would probably be more beneficial to add the chemicals directly than to depend on inoculated microorganisms to produce them in the food.

BIBLIOGRAPHY

ACOTT, K., SLOAN, A.E., and LABUZA, T.P. 1976. Evaluation of antimicrobial agents in a microbial challenge study for an intermediate moisture dog food. J. Food Sci. *41*, 541-546.

ANON. 1972. The complexities of EDTA. Food Cosmet. Toxicol. *10*, 697-700.

ANON. 1974. Compact 3-stage microwave/conventional dryer. Drying time cut up to 90%, energy use in half. Food Proc. *35*, No. 9, 25-26.

ANON. 1976A. Enzyme destroys salmonellas in poultry meat. Broiler Business *27*, No. 1, 38.

ANON. 1976B. Potassium sorbate proposed as answer to nitrite reduction in cured meats. Food Proc. *37*, No. 6, 8.

ANON. 1978. The risk/benefit concept as applied to food. Food Technol. *32*, No. 3, 51-56.

BAKER, R.C., DARFLER, J.M., MULNIX, E.J., and NATH, K.R. 1976. Palatability and other characteristics of repeatedly refrozen chicken broilers. J. Food Sci. *41*, 443-445.

BALA, K., STRINGER, W.C., and NAUMANN, H.D. 1977. Effect of spray sanitation treatment and gaseous atmospheres on the stability of prepackaged fresh beef. J. Food Sci. *42*, 743-746.

BARKAI-GOLAN, R., and AHARONI, Y. 1976. The sensitivity of food spoilage yeasts to acetaldehyde vapors. J. Food Sci. *41*, 717-718.

BASSIR, O., and MADUAGWU, E.N. 1978. Occurrence of nitrate, nitrite, dimethylamine, and dimethylnitrosamine in some fermented Nigerian beverages. J. Agr. Food Chem. *26*, 200-203.

BINKERD, E.F., and KOLARI, O.E. 1975. The history and use of nitrate and nitrite in the curing of meat. Food Cosmet. Toxicol. *13*, 655-661.

BOLIN, H.R. 1976. Texture and crystallization control in raisins. J. Food Sci. *41*, 1316-1319.

BOLIN, H.R., KING, A.D., JR., STANLEY, W.L., and JURD, L. 1972. Antimicrobial protection of moisturized Deglet Noor dates. Appl. Microbiol. *23*, 799-802.

BOLIN, H.R., PETRUCCI, V., and FULLER, G. 1975. Characteristics of mechanically harvested raisins produced by dehydration and by field drying. J. Food Sci. *40*, 1036-1038.

BROADWATER, W.T., HOEHN, R.C., and KING, P.H. 1973. Sensitivity of three selected bacterial species to ozone. Appl. Microbiol. *26*, 391-393.

BROCKMANN, M.C. 1970. Development of intermediate moisture foods for military use. Food Technol. *24*, 896-900.

BROWN, C.L., HEDRICK, H.B., and BAILEY, M.E. 1974. Characteristics of cured ham as influenced by levels of sodium nitrite and ascorbate. J. Food Sci. *39*, 977-979.

BROWN, J.P., ROEHM, G.W., and BROWN, R.J. 1978. Mutagenicity testing of certified food colors and related azo, xanthene and triphenylmethane dyes with the salmonella/microsome system. Mutat. Res. *56*, 249-271.

CALCOTT, P.H., LEE, S.K., and MACLEOD, R.A. 1976. The effect of cooling and warming rates on the survival of a variety of bacteria. Can. J. Microbiol. *22*, 106-109.

CALCOTT, P.H., and MACLEOD, R.A. 1974. Survival of *Escherichia coli* from freeze-thaw damage: a theoretical and practical study. Can. J. Microbiol. *20*, 671-681.

CALCOTT, P.H., and MACLEOD, R.A. 1975. The survival of *Escherichia coli* from freeze-thaw damage: the relative importance of wall and membrane damage. Can. J. Microbiol. *21*, 1960-1968.

CDC. 1975. Acute nitrite poisoning. Morbidity Mortality *24*, 195.

CHANG, H.C., and BRANEN, A.L. 1975. Antimicrobial effects of butylated hydroxyanisole (BHA). J. Food Sci. *40*, 349-351.

CHANG, P., AKHTAR, S.M., BURKE, T., and PIVNICK, H. 1974. Effect of sodium nitrite on *Clostridium botulinum* in canned luncheon meat: evidence for a Perigo-type factor in the absence of nitrite. Can. Inst. Food Sci. Technol. J. *7*, 209-212.

CHARM, S.E. 1978. The Fundamentals of Food Engineering, 3rd Edition. AVI Publishing Co., Westport, CT.

CHRISTIANSEN, L.N. *et al.* 1974. Effect of sodium nitrite on toxin production by *Clostridium botulinum* in bacon. Appl. Microbiol. *27*, 733-737.

CHRISTIANSEN, L.N. *et al.* 1975. Effect of sodium nitrite and nitrate on *Clostridium botulinum* growth and toxin production in a summer style sausage. J. Food Sci. *40*, 488-490.

COBB, B.F., VANDERZANT, C., and HYDER, K. 1974. Effect of ice storage upon the free amino acid contents of tails of white shrimp (*Penaeus setiferus*). J. Agr. Food Chem. *22*, 1052-1055.

COLOE, P.J., and HAYWOOD, N.J. 1976. The importance of prolonged incubation for the synthesis of dimethylnitrosamine by enterobacteria. J. Med. Microbiol. *9*, 211-223.

COX, C.S., GAGEN, S.J., and BAXTER, J. 1974A. Aerosol survival of *Serratia marcescens* as a function of oxygen concentration, relative humidity, and time. Can. J. Microbiol. *20*, 1529-1534.

COX, C.S., and HECKLY, R.J. 1973. Effects of oxygen upon freeze-dried and freeze-thawed bacteria: viability and free radical studies. Can. J. Microbiol. *19*, 189-194.

COX, N.A. *et al.* 1974B. Evaluation of succinic acid and heat to improve the microbiological quality of poultry meat. J. Food Sci. *39*, 985-987.

CROGHAN, D.L., and MATCHETT, A. 1973. β-propiolactone sterilization of commercial trypsin. Appl. Microbiol. *26*, 832.

CUNNINGHAM, F.E. 1974. Effect of freezing temperature on bone darkening in cooked broilers. Poultry Sci. *53*, 425-427.

DAWSON, L.E., STEVENSON, K.E., and GERTONSON, E. 1975. Flavor, bacterial and TBA changes in ground turkey patties treated with antioxidants. Poultry Sci. *54*, 1134-1139.

DEBEVERE, J.M., and VOETS, J.P. 1972. Influence of some preservatives on the quality of prepackaged cod fillets in relation to the oxygen permeability of the film. J. Appl. Bacteriol. *35*, 351-356.

DESROSIER, N.W. and DESROSIER, J.N. 1977. The Technology of Food Preservation, 4th Edition. AVI Publishing Co., Westport, CT.

DIGIROLAMO, R., LISTON, J., and MATCHES, J.R. 1975. Survival of virus in chilled, frozen, and processed oysters. Appl. Microbiol. *20*, 58-63.

DOESBURG, J.J., LAMPRECHT, E.C., and ELLIOT, M. 1970. Death rates of salmonellae in fishmeals with different water activities. I. During storage. J. Sci. Food Agr. *21*, 632-635.

DOMENICO, J.A., RAHMAN, A.R., and WESTCOTT, D.E. 1972. Effects of fungicides in combination with hot water and wax on the shelf life of tomato fruit. J. Food Sci. *37*, 957-960.

DOTY, D.M. 1968. Inhibiting growth of *Salmonella* with chemicals in animal by-product meals. Nat. Prov. *158*, No. 22, 8-10.

EAKES, B.D., and BLUMER, T.N. 1975. Effect of various levels of potassium nitrate and sodium nitrite on color and flavor of cured loins and country-style hams. J. Food Sci. *40*, 977-980.

EDNEY, K.L. 1973. Post harvest deterioration of fruit. Chem. Ind. *1973*, 1054-1056.

ELLIOT, R.P., and STRAKA, R.P. 1964. Rate of microbial deterioration of chicken meat at 2°C after freezing and thawing. Poultry. Sci. *43*, 81-86.

ETCHELLS, J.L. *et al.* 1973. Influence of temperature and humidity on microbial, enzymatic, and physical changes of stored, pickling cucumbers. Appl. Microbiol. *26*, 943-950.

FARRALL, A.W. 1976. Food Engineering Systems, Vol. I—Operations. AVI Publishing Co., Westport, CT.

FAZIO, T., WHITE, R.H., DUSOLD, L.R., and HOWARD, J.W. 1973. Nitrosopyrrolidine in cooked bacon. J. Assoc. Offic. Anal. Chem. *56*, 919-921.

FIDDLER, W., PENSABENE, J.W., KUSHNIR, I., and PIOTROWSKI, E.G. 1973A. Effect of frankfurter cure ingredients on N-nitrosodimethylamine formation in a model system. J. Food Sci. *38*, 714-715.

FIDDLER, W. *et al.* 1973B. Use of sodium ascorbate or erythorbate to inhibit formation of N-nitrosodimethylamine in frankfurters. J. Food Sci. *38*, 1084.

FIDDLER, W. *et al.* 1975. Dimethylnitrosamine in souse and similar jellied cured-meat products. Food Cosmet. Toxicol. *13*, 653-654.

FONG, Y.Y., and CHAN, W.C. 1976. Methods for limiting the content of dimethylnitrosamine in Chinese marine salt fish. Food Cosmet. Toxicol. *14*, 95-98.

FONG, Y.Y., and CHAN, W.C. 1977. Nitrate, nitrite, dimethylnitrosamine and N-nitrosopyrrolidine in some Chinese food products. Food Cosmet. Toxicol. *15*, 143-145.

FREESE, E., SHEU, C.W., and GALLIERS, E. 1973. Functions of lipophilic acids as antimicrobial food additives. Nature (London) *241*, 321-325.

FRETHEIM, K. 1976. Carcinogenic polycyclic aromatic hydrocarbons in Norwegian smoked meat sausages. J. Agr. Food Chem. *24*, 976-979.

GEE, M., FARKAS, D., and RAHMAN, A.R. 1977. Some concepts for the development of intermediate moisture foods. Food Technol. *31*, No. 4, 58-64.

GILBERT, J., KNOWLES, M.E., and MCWEENY, D.J. 1975. Formation of *C*- and *S*-nitroso compounds and their further reactions. J. Sci. Food Agr. *26*, 1785-1791.

GOLDBERG, I., and ESCHAR, L. 1977. Stability of lactic acid bacteria to freezing as related to their fatty acid composition. Appl. Environ. Microbiol. *33*, 489-496.

GOUGH, T.A., WEBB, K.S. PRINGUER, M.A., and WOOD, B.J. 1977. A comparison of various mass spectrometric and a chemiluminescent method for the estimation of volatile nitrosamines. J. Agr. Food Chem. *25*, 663-667.

GRAIKOSKI, J.T. 1973. Microbiology of cured and fermented fish. *In* Microbial Safety of Fishery Products. C.O. Chichester and H.D. Graham (Editors), Academic Press, New York.

GRAY, J.I., and DUGAN, L.R., JR. 1975. Inhibition of N-nitrosamine formation in model food systems. J. Food Sci. *40*, 981-984.

GSA. 1977. Code of Federal Regulations. Title 21. Food and Drugs. General Services Administration, Washington, D.C.

GUNARATNE, K.W.B., and SPENCER, J.V. 1974. Effect of certain freezing methods upon microbes associated with chicken meat. Poultry Sci. *53*, 215-220.

GUPTA, K.G., SIDHU, R., and YADAV, N.K. 1972. Effect of various sugars and their derivatives upon the germination of *Bacillus* spores in the presence of nisin. J. Food Sci. *37*, 971-972.

HANNI, P.F., FARKAS, D.F., and BROWN, G.E. 1976. Design and operating parameters for a continuous centrifugal fluidized bed drier (CFB). J. Food Sci. *41*, 1172-1176.

HANSEN, P. 1972. Storage life of prepacked wet fish at 0°C. II. Trout and herring. J. Food Technol. *7*, 21-26.

HASHIMOTO, S., YOKOKURA, T., KAWAI, Y., and MUTAI, M. 1976. Dimethylnitrosamine formation in the gastro-intestintal tract of rats. Food Cosmet. Toxicol. *14*, 553-556.

HAVERY, D.C. *et al.* 1976. Survey of food products for volatile N-nitrosamines. J. Assoc. Offic. Anal. Chem. *59*, 540-546.

HAYATSU, H., CHUNG, K.C., KADA, T., and NAKAJIMA, T. 1975. Generation of mutagenic compound(s) by a reaction between sorbic acid and nitrite. Mutat. Res. *30*, 417-419.

HEID, J.L., and JOSLYN, M.A. 1967. Fundamentals of Food Processing Operations: Ingredients, Methods, Packaging. AVI Publishing Co., Westport, CT.

HURST, A. 1972. Interactions of food starter cultures and food-borne pathogens: the antagonism between *Streptococcus lactis* and sporeforming microbes. J. Milk Food Technol. *35*, 418-423.

HYDE, M.B., and BURRELL, N.J. 1969. Control of infestation in stored grain by airtight storage or by cooling. Proc. 5th British Insecticide and Fungicide Conf., 412-419.

INCZE, K., FARKAS, J., MIHALYI, V., and ZUKAL, E. 1974. Antibacterial effect of cysteine-nitrosothiol and possible precursors thereof. Appl. Microbiol. *27*, 202-205.

ITO, K., NAKAMURA, K., IZAKI, K., and TAKAHASHI, H. 1977. Degradation of RNA in *Escherichia coli* induced by sodium chloride. Agr. Biol. Chem. *41*, 257-263.

ITO, K.A. *et al.* 1976. Effect of acid and salt concentration in fresh-pack pickles on the growth of *Clostridium botulinum* spores. Appl. Environ. Microbiol. *32*, 121-124.

JOHNSON, H.C., and LISTON, J. 1973. Sensitivity of *Vibrio parahaemolyticus* to cold in oysters, fish fillets and crabmeat. J. Food Sci. *38*, 437-441.

KABARA, J.J., SWIECZKOWSKI, D.M., CONLEY, A.J., and TRUANT, J.P. 1972. Fatty acids and derivatives as antimicrobial agents. Antimicrob. Agents Chemother. *2*, 23-28.

KAPLOW, M. 1970. Commercial development of intermediate moisture foods. Food Technol. *24*, 889-893.

KEMP, J.D., FOX, J.D., and MOODY, W.G. 1974. Cured ham properties as affected by nitrate and nitrite and fresh pork quality. J. Food Sci. *39*, 972-976.

KIDNEY, A.J. 1974. The use of sulphite in meat processing. Chem. Ind. *1974*, 717-718.

KING, A.D., JR., and NAGEL, C.W. 1967. Growth inhibition of a *Pseudomonas* by carbon dioxide. J. Food Sci. *32*, 575-579.

KRAMER, A. 1974. Storage retention of nutrients. Food Technol. *28*, No. 1, 50-60.

KRAMER, A., and FARQUHAR, J.W. 1976. Testing of time-temperature indicating and defrost devices. Food Technol. *30*, No. 2, 50-53, 56.

KUUSI, T., and LÖYTÖMAKI, M. 1972. On the effectiveness of EDTA in prolonging the shelf life of fresh fish. Z. Lebensmittel-Untersuch. Forsch. *149*, 196-204.

LABUZA, T.P., CASSIL, S., and SINSKEY, A.J. 1972. Stability of intermediate moisture foods. 2. Microbiology. J. Food Sci. *37*, 160-162.

LEVINE, M.B., and POTTER, N.N. 1974. Freeze-thaw stability of tomato slices: effects of additives, freezing, and thawing rates. Food Prod. Develop. *8*, No. 9, 76, 78, 80, 82, 90.

LILLEHOJ, E.B., MILBURN, M.S., and CIEGLER, A. 1972. Control of *Penicillium martensii* development and penicillic acid production by atmospheric gases and temperatures. Appl. Microbiol. *24*, 198-201.

MALANOSKI, A.J. *et al.* 1968. Survey of polycyclic aromatic hydrocarbons in smoked foods. J. Assoc. Offic. Anal. Chem. *51*, 114-121.

MARSHALL, B.J., COOTE, G.G., and SCOTT, W.J. 1973. Effects of various gases on the survival of dried bacteria during storage. Appl. Microbiol. *26*, 206-210.

MAURER, R.L., TREMBLAY, M.R., and CHADWICK, E.A. 1971. Microwave processing of pasta. Improves product, reduces cost and production time. Food Technol. *25*, 1244-1249.

MAZUR, P. 1970. Cryobiology: the freezing of biological systems. Science *168*, 939-949.

MCLEAN, R.A., LILLY, H.D., and ALFORD, J.A. 1968. Effects of meat-curing salts and temperature on production of staphylococcal enterotoxin B. J. Bacteriol. *95*, 1207-1211.

MERYMAN, H.T., WILLIAMS, R.J., and DOUGLAS, M.S.J. 1977. Freezing injury from "solution effects" and its prevention by natural or artificial cryoprotection. Cryobiology *14*, 287-302.

MEYER, E.D., SINCLAIR, N.A., and NAGY, B. 1975. Comparison of the survival and metabolic activity of psychrophilic and mesophilic yeasts subjected to freeze-thaw stress. Appl. Microbiol. *29*, 739-744.

MIDDLEKAUFF, R.D. 1974. Legalities concerning food additives. Food Technol. *28*, No. 5, 42-44, 46, 48.

MILLER, T.E. 1969. Killing and lysis of Gram-negative bacteria through the synergistic effect of hydrogen peroxide, ascorbic acid, and lysozyme. J. Bacteriol. *98*, 949-955.

MINKS, D., and STRINGER, W.C. 1972. The influence of aging beef in vacuum. J. Food Sci. *37*, 736-738.

MIWA, M., OKITANI, A., KURATA, T., and FUJIMAKI, M. 1978. Reaction between nitrite and low salt-soluble, diffusible fraction of meat. Partial purification and some properties of unknown reaction products. Agr. Biol. Chem. *42*, 101-106.

MOODIE, I.M. 1970. Reduction of 3,4-benzopyrene content in curing smoke by scrubbing. J. Sci. Food Agr. *21*, 485-488.

MOORHOUSE, B.R., and SALWIN, H. 1970. Effect of freeze-drying and cooking on shrimp quality. J. Assoc. Offic. Anal. Chem. *53*, 899-902.

MOUNTNEY, G.J., and O'MALLEY, J. 1965. Acids as poultry meat preservatives. Poultry Sci. *44*, 582-586.

MURATA, A., and KANEGAWA, T. 1973. Effect of β-propiolactone on *Lactobacillus casei*-phage J1 system. J. Gen. Appl. Microbiol. *19*, 467-480.

NAGAO, M. *et al.* 1977. Mutagenicity of N-butyl-N-(4-hydroxybutyl) nitrosamine, a bladder carcinogen, and related compounds. Cancer Res. *37*, 399-407.

NAKAMURA, M. *et al.* 1976. Pathways of formation of N-nitrosopyrrolidine in fried bacon. J. Food Sci. *41*, 874-878.

OLAJOS, E.J. 1977. Biological interactions of N-nitroso compounds: a review. Ecotoxicol. Environ. Safety *1*, 175-196.

PANALAKS, T. *et al.* 1974. Further survey of cured meat products for volatile N-nitrosamines. J. Assoc. Offic. Anal. Chem. *57*, 806-812.

PARKER, M.S. 1969. Some effects of preservatives on the development of bacterial spores. J. Appl. Bacteriol. *32*, 322-328.

PEARSON, D. 1970. Effect on various spoilage values of the addition of sulphite and chlortetracycline to beef stored at 5°C. J. Food Technol. *5*, 141-147.

PENSABENE, J.W. *et al.* 1974. Effect of frying and other cooking conditions on nitrosopyrrolidine formation in bacon. J. Food Sci. *39*, 314-316.

PERIGO, J.A., and ROBERTS, T.A. 1968. Inhibition of clostridia by nitrite. J. Food Technol. *3*, 91-94.

PIVNICK, H. *et al.* 1967. Effect of sodium nitrite and temperature on toxinogenesis by *Clostridium botulinum* in perishable cooked meats vacuum-packed in air-impermeable plastic pouches. Food Technol. *21*, No. 2, 100-102.

PONTING, J.D., and MCBEAN, D.M. 1970. Temperature and dipping treatment effects on drying rates and drying times of grapes, prunes and other waxy fruits. Food Technol. *24*, No. 12, 85-88.

POST, F.J., COBLENTZ, W.S., CHOU, T.W., and SALUNKHE, D.K. 1968. Influence of phosphate compounds on certain fungi and their preservative effects on fresh cherry fruit (*Prunus cerasus*, L.). Appl. Microbiol. *16*, 138-142.

POWERS, E.M., AY, C., EL-BISI, H.M., and ROWLEY, D.B. 1971. Bacteriology of dehydrated space foods. Appl. Microbiol. *22*, 441-445.

PRIEPKE, P.E., WEI, L.S., and NELSON, A.I. 1976. Refrigerated storage of prepackaged salad vegetables. J. Food Sci. *41*, 379-382.

RAEVUORI, M. 1975. Effect of nitrite and erythorbate on growth of *Bacillus cereus* in cooked sausage and in laboratory media. Zbl. Bakt. Hyg. I. Abt. Orig. *B 161*, 280-287.

REDDY, S.G., CHEN, M.L., and PATEL, P.J. 1975. Influence of lactic cultures on the biochemical, bacterial and organoleptic changes in beef. J. Food Sci. *40*, 314-318.

ROSENKRANZ, H.S. 1973. Sodium hypochlorite and sodium perborate: preferential inhibitors of DNA polymerase-deficient bacteria. Mutat. Res. *21*, 171-174.

ROWLAND, I.R., and GRASSO, P. 1975. Degradation of N-nitrosamines by intestinal bacteria. Appl. Microbiol. *29*, 7-12.

SATO, M., and TAKAHASHI, H. 1970. Cold shock of bacteria. IV. Involvement of DNA ligase reaction in recovery of *Escherichia coli* from cold shock. J. Gen. Appl. Microbiol. 16, 279-290.

SATO, T., IZAKI, K., and TAKAHASHI, H. 1972. Recovery of cells of *Escherichia coli* from injury induced by sodium chloride. J. Gen. Appl. Microbiol. *18*, 307-317.

SEN, N.P., DONALDSON, B., CHARBONNEAU, C., and MILES, W.F. 1974. Effect of additives on the formation of nitrosamines in meat curing mixtures containing spices and nitrite. J. Agr. Food Chem. *22*, 1125-1130.

SEN, N.P., SMITH, D.C., MOODIE, C.A., and GRICE, H.C. 1975. Failure to induce tumours in guinea-pigs after concurrent administration of nitrite and diethylamine. Food Cosmet. Toxicol. *13*, 423-425.

SEN, N.P. *et al.* 1976. Inhibition of nitrosamine formation in fried bacon by propyl gallate and L-ascorbyl palmitate. J. Agr. Food Chem. *24*, 397-401.

SMITH, D.P. 1976. Chilling. Food Technol. *30*, No. 12, 28, 30, 32.

SMITH, G.C., ARANGO, T.C., and CARPENTER, Z.L. 1971. Effects of physical and mechanical treatments on the tenderness of the beef longissimus. J. Food Sci. *36*, 445-449.

SPIEGELHALDER, B., EISENBRAND, G., and PREUSSMANN, R. 1976. Influence of dietary nitrate on nitrite content of human saliva: possible relevance to *in vivo* formation of N-nitroso compounds. Food Cosmet. Toxicol. *14*, 545-548.

SULZBACHER, W.L. 1952. Effect of freezing and thawing on the growth rate of bacteria in ground meat. Food Technol. *6*, 341-343.

TANAKA, Y., OHNISHI, T., TAKEDA, Y., and MIWATANI, T. 1975. Lethal effect of freeze-drying on radiation-sensitive mutants of *Escherichia coli*. Biken J. *18*, 267-269.

TATINI, S.R. 1973. Influence of food environments on growth of *Staphylococcus aureus* and production of various enterotoxins. J. Milk Food Technol. *36*, 559-563.

THOMSON, J.E., BANWART, G.J., SANDERS, D.H., and MERCURI, A.J. 1967. Effect of chlorine, antibiotics, β-propiolactone, acids, and washing on *Salmonella typhimurium* on eviscerated fryer chickens. Poultry Sci. *46*, 146-151.

TOMPKIN, R.B., CHRISTIANSEN, L.N., and SHAPARIS, A.B. 1978. Enhancing nitrite inhibition of *Clostridium botulinum* with isoascorbate in perishable canned cured meat. Appl. Environ. Microbiol. *35*, 59-61.

TROLLER, J.A. 1973. The water relations of food-borne bacterial pathogens. A Review. J. Milk Food Technol. *36*, 276-288.

USDA. 1970. Meat and Poultry. Care Tips for You. Home and Garden Bull. *174*, U.S. Dep. Agric., Washington, D.C.

VAN ARSDEL, W.B., and COPLEY, M.J. 1973. Food Dehydration, Vol. 2. Practices and Applications. AVI Publishing Co., Westport, CT.

VISEK, W.J., CLINTON, S.K., and TRUEX, C.R. 1978. Nutrition and experimental carcinogenesis. Cornell Vet. *68*, 3-39.

WARTH, A.D. 1977. Mechanism of resistance of *Saccharomyces bailii* to benzoic, sorbic and other weak acids used as food preservatives. J. Appl. Bacteriol. *43*, 215-230.

WARTHESEN, J.J., BILLS, D.D., SCANLAN, R.A., and LIBBEY, L.M. 1976. N-nitrosopyrrolidine collected as a volatile during heat-induced formation in nitrite containing pork. J. Agr. Food Chem. *24*, 892-894.

WASSERMAN, A.E. *et al.* 1972. Dimethylnitrosamine in frankfurters. Food Cosmet. Toxicol. *10*, 681-684.

WHITE, R.H., HOWARD, J.W., and BARNES, C.J. 1971. Determination of polycyclic aromatic hydrocarbons in liquid smoke flavors. J. Agr. Food Chem. *19*, 143-146.

WIERBICKI, E., and HELIGMAN, F. 1973. Shelf stable cured ham with low nitrite-nitrate additions preserved by radappertization. Proc. Int. Symp. Nitrite in Meat Products. Zeist, The Netherlands.

WILSON, D.M., HUANG, L.H., and JAY, E. 1975. Survival of *Aspergillus flavus* and *Fusarium moniliforme* in high-moisture corn stored under modified atmospheres. Appl. Microbiol. *30*, 592-595.

WINARNO, F.G., STUMBO, C.R., and HAYES, K.M. 1971. Effect of EDTA on the germination and outgrowth from spores of *Clostridium botulinum* 62-A. J. Food Sci. *36*, 781-785.

WOOLFORD, G., and CASSENS, R.G. 1977. The fate of sodium nitrite in bacon. J. Food Sci. *42*, 586-589, 596.

WOOLFORD, G., CASSENS, R.G., GREASER, M.L., and SEBRANEK, J.G. 1976. The fate of nitrite: reaction with protein. J. Food Sci. *41*, 585-588.

YAJIMA, M., HIDAKA, Y., and MATSUOKA, Y. 1968. Studies on egg white lysozyme as a preservative of sake. J. Ferment. Technol. *46*, 782-788.

YU, T.C., SINNHUBER, R.O., and CRAWFORD, D.L. 1973. Effect of packaging on shelf life of frozen silver salmon steaks. J. Food Sci. *38*, 1197-1199.

12

Control of Microorganisms by Destruction

The death of some microbial cells occurs during refrigeration, freezing, drying or chemical treatment. However, these systems are not expected to produce a sterile food. Sterilization is considered to be a process by which all forms of life are destroyed. When subjected to a lethal process, a bacterial culture is reduced at a rate which is approximately logarithmic. This indicates a first order reaction. The inactivation or death of cells can be determined by the formula:

$$K = \frac{\log a - \log b}{t}$$

where K is the death rate, a is the initial number of organisms and b is the number of cells remaining at time (t).

Considering the microorganisms of public health significance, *C. botulinum* is the most difficult to destroy. The main objective is to eliminate this organism from processed and packaged foods.

The inactivation of microorganisms has been attributed to gases, thermal processing, ionizing radiation and light waves.

GAS TREATMENTS

Gases can be used to sterilize materials which cannot withstand the high temperatures of heat sterilization. Many organic compounds begin to degrade at 71°C and this degradation accelerates as the temperature is increased to 121°C. Volatile food flavors can be altered by heat. Some plastic materials are softened by heat.

Gaseous sterilization offers a means for packaging heat sensitive products in an essentially microbe-free condition. A gas not only can affect airborne and exposed surface bacteria, but also it can penetrate porous materials to attack the microbial cells. Each gaseous substance has certain advantages and disadvantages.

It is necessary to evaluate the physiological effectiveness and the toxic hazards, as well as the retained residuals of any particular gas. Important aspects in the use of gas sterilization are: comparative efficiency of different methods of application, effect of humidity and atmospheric gases, resistance of microorganisms, synergism or potentiation of gas mixtures, the mode of action, reaction kinetics, physical and chemical sorption, effects on food quality, photodecomposition, air pollution, worker safety, permeability of materials such as packages and the nature and physiochemical properties of the material being treated (Berck 1965).

Besides the gases used to inactivate microorganisms, fumigants such as phosphine, ethylene dibromide and methyl bromide are employed to destroy insects in stored grain.

Ethylene Oxide

Ethylene oxide is a cyclic ether with the structural formula:

$$\underset{\diagdown \ O \ \diagup}{H_2C - CH_2}$$

Ethylene oxide is a gas at ambient temperature and pressure. It boils at 10.7°C and freezes at −111°C. It dissolves in and reacts with water to form ethylene glycol. Mixtures of 3 to 100% ethylene oxide in air are highly flammable and explosive. To eliminate this hazard, ethylene oxide is mixed with dichlorodifluoromethane, carbon dioxide or formic acid methyl ester.

For an effective sterilization process, the gas concentration, time, temperature and humidity are parameters that must be considered and controlled. Factors such as stratification effects of various gases, diffusion barriers, moisture reducing effects, temperature gradients and ancillary gaseous reactions can enhance or limit sterilization. The sterilizing action of ethylene oxide requires three hours or more, but is effective against spores as well as vegetative bacteria, yeasts, molds and viruses. Some advantages and disadvantages of ethylene oxide treatment are listed in Table 12.1.

The ability of ethylene oxide to penetrate certain packaging materials (paper and some flexible plastics) allows foods to be sterilized in sealed containers. The packaging material must be gas permeable, but impermeable to bacteria.

The mechanism by which ethylene oxide inactivates microorganisms is not completely known. A commonly accepted hypothesis is that the microbicidal action is directly related to alkylating activity. The enzymes and other proteins essential for cellular metabolism contain carboxyl, amino, sulfhydryl, hydroxyl and phenolic groups. Ethylene oxide reacts with these groups by replacing an available hydrogen atom with the alkyl hydroxyethyl group. The bacterial cell cannot utilize the new chemical and therefore it dies. According to Ernst (1974), the biologically critical reaction occurs with RNA and DNA within the cells. Winarno and Stumbo (1971) observed an impairment of RNA and protein synthesis. They considered this to be an indirect effect resulting from alkylation of DNA components. This alkylation reaction is thought to be important in the inactivation of spores. Spores are usually several times more resistant than vegetative cells to chemical agents. With ethylene oxide, the resistance of some spores is comparable to that of some vegetative cells, while the more resistant spores are only about five times as difficult to inactivate as the less resistant vegetative cells. The ability of the gas to penetrate may play some role in the effective action against spores.

TABLE 12.1

ETHYLENE OXIDE STERILIZATION

Advantages	Disadvantages
Good penetration of materials	Flammable
Attacks and kills all microorganisms	Process is slow
Little, if any, damage to materials (compared to heat)	Residues (analysis may be needed)
Effective at room temperature	Expensive
Effective with dry products	Each process must be closely monitored
Sterilize packaged items	Toxic
	May alter nutrients and other quality factors of foods

Bacillus subtilis var *niger* spores are more resistant than other spores to ethylene oxide. The decimal reduction times (D) of *Micrococcus radiodurans* and *Streptococcus faecalis* were comparable to spores, except those of *B. subtilis* var *niger* (Kereluk *et al.* 1970A). Increasing the temperature decreases the D. In commercial practice, temperatures of 30°-65°C are used.

Kereluk *et al.* (1970B) reported that spore destruction rates varied little between 15 and 90% RH. Kuzminski *et al.* (1969) found a lower D for *C. botulinum* at 3% RH than at levels between 23 and 73%. The optimum RH depends upon the material and the procedures used in the treatment.

Ethylene oxide and propylene oxide have been used to sterilize spices, cereals, colloid gums, fruits, dried yeasts and other items.

Spices, especially imported spices, have high bacterial counts. When used as ingredients, they may contribute to spoilage of the food. Ethylene oxide treatment can reduce the microbial load by 90 to 100%. This treatment is rather costly and time consuming. Also, ethylene oxide may result in unfavorable effects on some spices.

The use of ethylene oxide to sterilize food has been questioned. Ethylene oxide is chronically toxic at levels not detected by smell. It is a skin irritant, producing erythema and edema with a potential for sensitization. Various toxic effects of ethylene oxide, including death, were discussed by White (1977). Embree *et al.* (1977) reported mutagenic effects of ethylene oxide on various cells, including mammalian cells.

Besides formation of the toxic ethylene glycol when hydrated, Wesley *et al.* (1965) reported that ethylene oxide reacts with naturally occurring chlorides in food to form chlorohydrins. These are toxic, nonvolatile and persist in the food during processing. They reported ethylene chlorohydrin at levels of over 1 mg/g in spices sterilized with ethylene oxide.

Ethylene oxide can be used to control microorganisms and insects in ground spices or other natural seasoning materials, except mixtures to which salt has been added. The residue cannot exceed 50 μg/g of spice. Other food industry uses include dried coconut, black walnuts, starch and packaging materials (FDA 1977).

Ethylene oxide sterilization should be used only when other sterilization techniques are not practical.

Propylene Oxide

Propylene oxide has the structural formula:

$$CH_2 - CH\,CH_3 \quad (\text{ring closed through } O)$$

It is a colorless liquid with a boiling point of 35°C. It is used like ethylene oxide since it must be diluted with inert gases for safety. It has weak penetrating ability, but it is microbicidal. It requires a higher temperature or longer treatment time to accomplish the same microbial kill as ethylene oxide.

Propylene oxide has been used in various foods to kill 90% or more of the microorganisms. Propylene oxide treatments of pecans reduced the surface flora by 80 to 92% and the internal microorganisms by 64% (Blanchard and Hanlin 1973). Even with high concentrations of propylene oxide, neither bacteria nor fungi could be eliminated completely.

The hydration of propylene oxide yields propylene glycol. This should be acceptable since propylene glycol is used to lower the a_w in IM foods.

However, as with ethylene oxide, propylene oxide can react with chlorides to form chlorohydrins in foods (Wesley *et al.* 1965).

Propylene oxide can be used as a package fumigant on or in dried prunes or glacé fruit. It can be used on bulk quantities of cocoa, gums, processed spices, starch and processed nutmeats (except peanuts) when these foods are to be further processed. There are limitations on temperature, time and residues (FDA 1977).

Methyl Bromide

Methyl bromide (CH_3Br) is a colorless gas. The boiling point at atmospheric pressure is 4.5°C. It is not flammable in air but burns in oxygen. Methyl bromide is used for insect control in stored grain and other commodities.

As a gas, methyl bromide is very toxic. It is absorbed readily through the skin and the lungs. The skin may become red and itch and blisters may appear. Inhalation of methyl bromide causes fatigue and blurred vision. High concentrations can cause respiratory paralysis and heart failure.

Salmonella species and *E. coli* are quite susceptible to methyl bromide. Micrococci are more resistant than the Gram negative rods. Bacterial spores are resistant to the action of this gas. This fumigant is effective against some, but not all, fungi and viruses.

Bolin *et al.* (1972) reported that the best overall treatment, of those they tested for controlling yeast in dates, was to dip the dates in potassium sorbate and inject methyl bromide into the package. Schade and King (1977) suggested this gas can eliminate *E. coli* and salmonellae from whole nut kernels. Methyl bromide is listed as an acceptable fumigant for grain and methyl formate can be used on raisins and dried currants (FDA 1977).

Formaldehyde

Formaldehyde (HCHO) gas is present in wood smoke, especially in wood used for smoking ham or fish. It is flammable in air at concentrations of 7 to 73% by volume. Like ethylene oxide, it is an alkylating agent. As such, bacterial spores are only slightly more (2-15 times) resistant than are vegetative cells. Optimum activity occurs at about 80 to 90% RH. Below 50% or above 90% RH, there is a rapid reduction in activity.

Formaldehyde vapor is not very penetrating when used under the normal conditions of fumigation. Its action is confined primarily to airborne bacteria and those on exposed surfaces. Even a thin film of organic matter will protect microorganisms from formaldehyde.

HEAT

The most common method of killing microorganisms is to subject them to a heat treatment. When properly done, this is an effective means of improving the microbiological quality of foods. Sometimes, relatively mild heat treatments are used in conjunction with other processes, such as refrigeration, freezing, drying or acidification. Since the destruction of microorganisms is incomplete, poor sanitation and mishandling of these mildly heat treated products may result in a higher total load of microorganisms than was present before the heat treatment.

The destruction of spoilage microorganisms increases the storage life of the product. However, if this food is then recontaminated with pathogenic types of microorganisms, the natural competing flora is not present, and the food may become a health hazard.

Compared to other means of sterilization, heat may be considered as more efficient and convenient. Besides killing microorganisms, heating will inactivate enzymes (tissue or microbial) that can cause deterioration of food during storage. The one condition needed in using heat is that the material being sterilized must be resistant to heat damage.

Food microbiologists must be concerned not only with the preservation of foods, but also with the sterilization of laboratory materials. The heat treatment of instruments used for sampling and analysis is discussed in Chapter 2. Moist heat (steam or water) is much more effective in destroying microorganisms than is dry heat. The effect of heat on microorganisms is generally believed to be due to enzyme inactivation, protein denaturation, or both.

There are many processes in which foods are subjected to heat. Each of these systems may influence the microbial population. The extent of this influence depends on the type of heat treatment (heat source, temperature, time), the type of food and the types of microorganisms that are present.

Heat may be transferred to and through foods by conduction, convection or radiation. The type of heat transfer depends upon the source of heat and the type of food being heated.

Heating Systems

The processes in which heat is applied to foods include cooking, scalding, pasteurizing, blanching and canning, as well as drying, distillation, evaporation and concentration.

In most of the processes employing heat, complete destruction of all microbial forms is neither attempted nor attained. Only in thermal processing of foods with steam under pressure, such as in canning, is an attempt made to approach a sterilized product. When sufficient heat is applied to assure a sterile food, the organoleptic and nutritional properties of the food suffer. In many cases, the food would no longer be acceptable to the consumer.

Cooking.—Although some fresh fruits and vegetables are eaten raw, most of our foods are given a heat treatment prior to consumption. Cooking is an example of temporary preservation.

Normal cooking may destroy most of the vegetative bacteria, except for thermophiles. Most of the spores will survive normal cooking procedures. The guidelines established by the USDA for cooked meats suggest a minimum end-point temperature of 63°C. This temperature will destroy some pathogenic organisms but will not affect bacterial spores, the enterotoxin of *S. aureus* or the neurotoxin of *C. botulinum.* The data of Blankenship (1978) indicate that some salmonellae may survive in beef roasts heated to a temperature of 64°C.

The effect of cooking oysters in destroying inoculated polioviruses was reported by DiGirolamo *et al.* (1970). Stewing (placing oysters in boiling milk and stewing for eight minutes), frying (placing oysters in oil at 177°C and frying for 10 min), baking (20 min in an oven at 121.5°C) and steaming (held in flowing steam for 30 min) did not eliminate the inoculated viruses. From 7 to 10% of the viruses survived these treatments. Coxsackievirus B2 and poliovirus 1 were inactivated by broiling hamburger patties to 60°C and holding at room temperature for 3 min (Sullivan *et al.* 1975). At high inoculation levels (10^7 pfu/g), Sabin type 1 poliovirus survived in ground meat held at 80°C for 5 min (Filppi and Banwart 1974).

Since cooking does not eliminate all microorganisms, it is essential that the cooked food be held above 50°C or promptly cooled below 4°C to prevent the growth of potential pathogens.

Microwave Heating.—Microwave ovens are becoming prominent in home kitchens throughout the USA. There are many industrial uses for microwaves. Most microwave ovens use either 915 or 2,450 megacycles. At 915 megacycles, the current is reversed 915 million times each second.

Polar molecules, such as water, have positive and negative charges concentrated at opposite ends of the molecule. As the microwaves pass through food, these molecules attempt to become aligned with the alternating positive and negative field of the microwaves. This rapid movement back and forth creates molecular friction which appears as heat.

Microwave heating differs from the more conventional methods of heating. Heating occurs internally rather than from the outside. It is an extremely rapid heating system. Heating can be controlled since it commences when the power is turned on and stops the instant the power is turned off.

Other heating sources must be at a higher temperature than the food so that the heat will move from the source to the food. Microwaves, per se, have no temperature, so there is no over-heating or burning of the food surface. Only the polar molecules become hot directly due to the microwaves. The nonpolar molecules are heated indirectly by conduction or convection from the heated polar molecules.

There have been several reports inferring that microwave heating is more bactericidal than conventional cooking methods. In a review, Rosen (1972) very convincingly reported that the killing of microorganisms is due only to thermal effects. An explanation for the diverse results is that hot spots occur from the lack of uniform distribution of microwaves. These hot areas will kill more microorganisms than would be expected by determining the average temperature.

Janky and Oblinger (1976) found no apparent differences between heating turkey rolls in hot water or a microwave oven in reducing the microbial populations. The treatment of dry materials, such as spices, with microwaves has no significant effect on the bacterial population.

Pasteurization.—The heat treatment of foods below temperatures needed for sterilization may be referred to as pasteurization. Quite often, a temperature below 100°C is called pasteurization, while above 100°C, the process is sterilization. In most of the pasteurization processes, the food is heated to between 60° and 85°C for a few seconds up to an hour. Generally, the heat treatment given to a food is designed to inactivate specific types or groups of microorganisms. With pasteurization, some organisms are killed, some are attenuated (sublethal injury) while the spores may be stimulated to germinate. The lethal effect depends upon the heat resistance of the organisms. Pasteurization can be used for foods in which quality is affected adversely by a more severe heat treatment.

The process is named for Pasteur who found that heating wine to 50° or 60°C for a short time inactivated spoilage microorganisms without seriously affecting the quality of the wine. Since this work, the main objective of pasteurization has evolved to destroy certain pathogenic microorganisms in specific foods. When spoilage organisms are not very heat resistant, pasteurization can extend the storage life of products, especially if the treated food is refrigerated, frozen or otherwise treated to control surviving microorganisms.

Pasteurization is used to control microorganisms in foods such as liquid egg products, dairy products, alcoholic beverages (beer, wine), crab, smoked fish and certain high-acid products (fruit juice, pickles, sauerkraut, vinegar).

Egg Products.—The USDA regulations stipulate that all liquid, frozen and dried whole egg, yolk and white be pasteurized or otherwise treated to destroy all viable salmonellae. Depending on the nature of the liquid egg product, various times and temperatures are used, as listed in Table 12.2. Salmonellae are less heat resistant in egg white than in whole egg or egg yolk. Higher temperatures are needed to pasteurize egg yolk products than liquid whole egg. A variety of egg products are produced and they have different degrees of heat sensitivity. Salted eggs used in salad dressings do not have to be pasteurized. They must be properly labeled and the salad dressing must contain not less than 1.4% acetic acid and have a pH of 4.1 or lower. The final product must be held for 72 hr. Salmonellae will die at room temperature under these conditions.

TABLE 12.2

PASTEURIZING CONDITIONS FOR EGG PRODUCTS

Product	Temperature °C	Average Holding Time (min)
Whole egg, plain	60	3.5
Yolk		
Plain	60 or 62.2	7.0 3.5
Sugared or salted	62.2 or 64.4	7.0 3.5
Egg white		
Plain, pH 9.0	56.7	3.5
Plain, pH 9.0, treated with H_2O_2	51.7	3.5
Stabilized with $Al_2(SO_4)_3$ at pH 7.0	60	3.5

These pasteurization conditions ordinarily will reduce the standard plate count by about 99.9% and the number of salmonellae essentially to zero. The heat sensitivity of salmonellae is affected by the pH. When 60°C is used, the D value at pH 9.0 is about 0.1 min and at pH 5.5, it is about 1.0 min. Hence, pasteurization of egg white at pH 9 requires less heat treatment than at lower pH levels. The addition of 10% salt or sugar to egg yolk increases the heat stability of salmonellae by 5 to 10 times. Fortunately, the stability of the egg proteins to heat also is increased by sugar or salt.

Successful pasteurization is based on a critical time-temperature relationship. If the temperature is lowered by 1°C, the efficiency of pasteurization is decreased. If the temperature is allowed to increase, there is danger of coagulating the egg, forming a film on the heat exchanger surfaces (Fig. 12.1) and/or damaging the functional properties of the treated egg.

Courtesy of CREPACO, Inc.

FIG. 12.1. PLATE HEAT EXCHANGER OPENED TO SHOW ARRANGEMENT OF PLATES AND GASKETS

As liquids flow through the holding tubes, the flow may be laminar, turbulent or transitional, which has some characteristics of both laminar and turbulent flow. In any flow, the material in the center of the tube is flowing at a faster rate than that near the side of the tube. Hence, not all of the liquid egg passes through the pasteurizer at the same rate. Scalzo *et al.* (1969) suggested that the holding tube requirements should be based on the fastest, rather than the average, particle to traverse the tube. The pasteurization specifications (USDA 1969) require that every particle be held for at least a specified time and temperature. The normal control of plant operations uses the average holding times. These times are considered to be twice that of the fastest particle time. Hence, the minimum time for whole egg at 60°C would be 1.75 min, and the average time 3.5 min.

Liquid egg white is more heat sensitive than whole egg or yolk. If egg white is heated above 56° to 57°C, there are problems of coagulation. The addition of hydrogen peroxide to egg white reduces the heat resistance of the salmonellae so that lower pasteurization temperatures can be used.

Except for conalbumin, the proteins in egg white are sufficiently heat stable at pH 7.0. Adding an aluminum salt such as aluminum sodium sulfate stabilizes the conalbumin so that it can be heated to 60°C without coagulation (Cunningham and Lineweaver 1965).

Shafi *et al.* (1970) found *Micrococcus, Alcaligenes, Pseudomonas, Bacillus* and unidentified bacteria in pasteurized egg white. These genera, as well as species of *Staphylococcus, Streptococcus, Arizona* (*Salmonella*) and unidentified yeasts and molds, were reported in pasteurized whole egg (Shafi *et al.* 1970). They reported similar types in pasteurized egg yolk.

Milk.—Before milk pasteurization was a requirement, diseases due to drinking contaminated milk were a common occurrence. There were epidemics of typhoid fever, scarlet fever, diphtheria and infant diarrhea, as well as outbreaks of tuberculosis, brucellosis and other illnesses. Today, pasteurized milk is one of our safest foods. Although the elimination of diseased animals, better sanitation and refrigeration have helped improve our milk supply, much of the credit should go to pasteurization.

Milk is pasteurized to destroy pathogenic organisms, to reduce total bacterial numbers, to extend the storage life and to inactivate enzymes that can affect milk flavors adversely.

When properly done, pasteurization renders harmless or destroys all disease producing microorganisms known to be transmitted through milk, and does not significantly impair or alter the flavor or food value of the milk. Milk proteins are much more heat stable than egg proteins. Therefore, a higher temperature can be used.

In the vat process, the temperature is increased to 62.7°C and maintained for 30 min. In the high temperature short time (HTST) process, the milk is heated rapidly to 71.7°C and held for 15 sec. Equivalent temperatures and times for HTST approved by the FDA (1977) are:

1) 89°C for 1 sec; 2) 90°C for 0.5 sec; 3) 94°C for 0.1 sec; 4) 96°C for 0.05 sec; or 5) 100°C for 0.01 sec.

Ultra-pasteurized (UHT) dairy products are heated at or above 138°C for at least 2 sec (FDA 1977). This product has an extended refrigerated shelf life. The time-temperature relationships of various pasteurization systems are shown in Fig. 12.2.

Some particularly heat resistant micrococci and microbacteria as well as spores of *Bacillus* and *Clostridium* may survive UHT. Heating milk to 149°C is expected to kill even some of the most heat resistant spores.

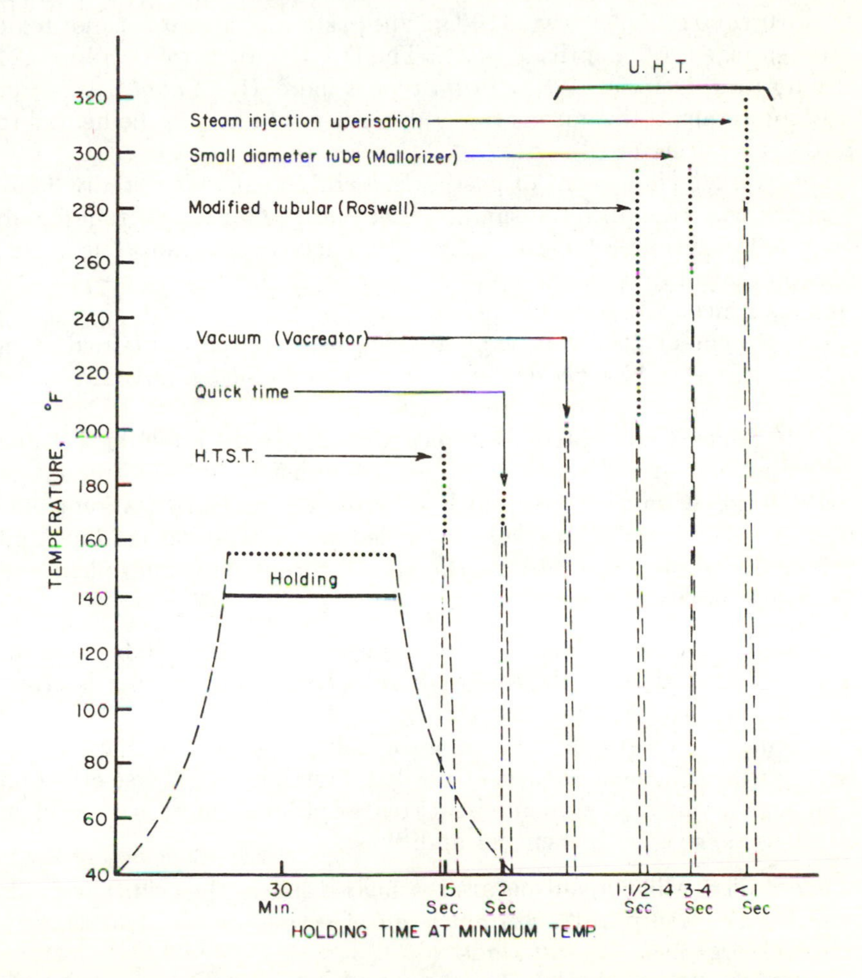

FIG. 12.2. PASTEURIZATION TIME AND TEMPERATURE

Pasteurization of milk destroys several enzyme systems. Milk is analyzed for residual alkaline phosphatase to determine the adequacy of heat treatment. One problem with using this enzyme assay is the reactivation of the enzyme following heating by the HTST system. The reactivation is small as compared to the activity in raw milk.

Animal Tissue.—Crab meat may be given heat treatments somewhat below those used for cooking in order to eliminate a sizeable proportion of the microbial population.

Cockey and Tatro (1974) found pasteurization at 85°C for 1 min was sufficient to reduce the level of inoculated *C. botulinum* type E spores from 10^8/100 g to 6 or fewer/100 g. The pasteurized meat was not toxic after storage for 6 months at 4.4°C. The $D_{82.2°C}$ varied from 0.49 to 0.74 min for four strains of *C. botulinum* type E spores (Lynt *et al.* 1977). For a 12D treatment, this means the crab meat would need to be heated to 82.2°C for 8.9 min.

The average plate count of pasteurized crab meat is reportedly 3,000 bacteria per gram. In retail samples, Lee and Pfeifer (1975) found crab meat to range from 2.1×10^4 to 8.2×10^7 bacteria per gram. The anaerobes plus facultative anaerobes were reported at about 10^5/g for pasteurized crab meat (Ward *et al.* 1977).

The prominent genera in canned crab meat samples obtained from retail stores were *Moraxella, Pseudomonas, Acinetobacter, Arthrobacter* and *Micrococcus* (Lee and Pfeifer 1975).

The problems associated with pasteurized fishery products became very apparent in 1963 with 17 cases of botulism and 5 deaths due to consumption of smoked fish that had been given a pasteurization treatment. *C. botulinum* spores were not killed and, due to mishandling and holding at warm temperatures, the spores germinated, multiplied and produced the lethal toxin.

Alcoholic Beverages.—Pasteurization is used to destroy spoilage organisms in beer and wine. Heating of beer in the final container is effective, but is costly and inefficient.

Pasteurization just before bottling controls the growth of microorganisms in wine. The heat treatment needed often has an adverse effect on high quality wines. As with any heat treated product, there is no residual effect as there is with chemical additives.

Acid Foods.—Microorganisms are less heat stable as the acidity of foods is increased. Hence, psteurization is an effective means of destroying the microorganisms in acid foods.

For fruit juices, the heat treatment selected for pasteurization often is based on inactivation of enzymes rather than destruction of microorganisms. The microbial population usually consists of yeasts, molds, lactic and acetic acid bacteria which are generally heat sensitive. Other organisms usually are inhibited by the acid in fruit juices. Even intermediate acidities are suitable for the growth of only a few sporeformers, such as *B. thermoacidurans* (*B. coagulans*) or *C. pasteurianum.*

In grape juice, yeasts are destroyed in a few minutes at 55° to 65°C. Mold spores are more resistant with some withstanding temperatures of 74° to 80°C for up to 15 min. Commercially, grape juice is flash heated to 75° to 85°C.

Fresh pickles packed in solutions of acetic or lactic acids are called fresh-pack pickles. These pickles are given a heat treatment to aid in preservation. Underpasteurization results in spoilage, while overpasteurization can cause a softening of the pickles. Heating to a temperature of 74°C for 15 min was recommended for pasteurization of acidified fresh-pack pickles. Pickles covered with acetic or lactic acids at pH 3.7 remained firm when heated to 74°C for 30 min (Bell *et al.* 1972).

Dry Heat.—Moist heat is more effective than dry heat for destroying microorganisms. Dry heat is believed to cause death of the cells by the destructive oxidation of cell components.

There are significant reductions in the total count, and salmonellae can be eliminated when dried egg white is stored at temperatures of 50° to 70°C. Commercially, the dried albumen is stored at 51.6° to 54.4°C for 5 to 7 days, which not only eliminates salmonellae, but also improves the functional performance of the product.

Blanching.—This is a heat treatment given to many foods prior to drying, freezing or canning. Blanching inactivates enzymes, aids in cleaning, expels internal gases, removes mucilage-like substances which contribute to off-flavors, helps preserve the organoleptic quality during storage, aids in filling of containers by wilting, shrinking or softening the food, helps assure adequate can vacuum, is essential when the specific gravity of certain products is used in grading and destroys many vegetative microbial cells.

Blanching can reduce the total microbial load of raw vegetables by 99.9%. The product generally has a lower microbial count when steam blanched than when hot water blanched. A steam blancher is shown in Fig. 12.3.

Canning.—This is the main system used for the long term preservation of food. The food is packaged in hermetically sealed containers of metal, glass or thermostable plastic, either before or after heat treatment.

Flowing steam is at a temperature of 100°C or less. The temperature of the steam increases as the pressure increases. The usual temperature for sterilizing media in a bacteriology laboratory is 121.1°C. This corresponds to a gauge pressure of 103.5 kPa or absolute pressure of 207 kPa. At this

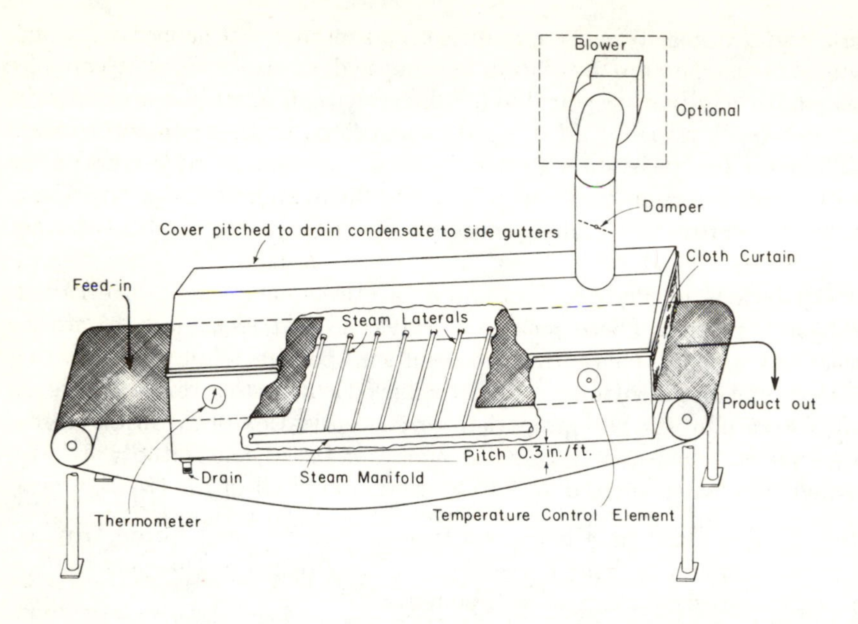

Courtesy of Tressler et al. (1968)

FIG. 12.3. CONTROLLED-TEMPERATURE STEAM BLANCHER

temperature (121.1°C) steam will kill bacterial spores in 15 min. Saturated steam is more penetrating than hot air. Because of its dryness, superheated steam is about as effective as hot air for sterilizing.

It is important to remove all of the air from the sterilizing chamber. If 50% of the air remains, a pressure of 207 kPa will be only 112°C instead of 121°C.

The temperatures used for heat processing or canning vary from 100°C for high acid foods to 121°C for low acid foods. For HTST processes, temperatures in the range of 120° to 150°C, or higher, may be used. A short time at a high temperature generally results in a product of better quality than equivalent processes at lower temperatures.

The use of heat for food preservation and the history of canning was reviewed by Goldblith (1971, 1972).

Fully sterilized canned products generally are less palatable and less nutritious than less severely heat treated products. Complete sterilization of food may not be attained and generally, it is not needed. The heat processed products are considered to be commercially sterile. This means that all organisms that might be a health hazard (*C. botulinum*), as well as spores that might cause spoilage under normal nonrefrigerated storage conditions, are destroyed. Some extremely heat resistant thermophilic spores might remain in the product. With normal storage temperatures, these will not germinate and grow. However, there may be problems in tropical areas, in vending machines that dispense hot foods, or when heat processed food is not cooled properly before warehousing.

The time necessary to obtain commercial sterility at a specified temperature depends upon the number and type of sporeforming organisms that are present. By practicing good sanitation and keeping the heat resistant spores at a low level, less time is needed to obtain a commercially sterile product.

When the food is processed in a jar or a can, a good vacuum inside the container is desired. A vacuum reduces the strain on the container during heat processing, holds the ends in a collapsed concave position during subsequent storage and reduces the amount of headspace oxygen. The vacuum is obtained by exhausting with steam or by vacuum sealing. The loss of vacuum or springer formation is one of the principal types of pack failure. The evolution of gas within the container causes the can ends to become distorted beyond the normal concave position. The formation of gas may be either from hydrogen produced by corrosion on the inside of the can by the food product or by microbial action on the food product. Microbial degradation of the food is usually recognized by the odor or appearance of the product when the container is opened. If gas formation due to microorganisms occurs in canned products stored below 38°C, the food is not commercially sterile, since the spores that germinate and grow at these temperatures should have been destroyed.

It is not the intent of this text to suggest times and temperatures for processing foods. Such things as the type of retort or process used for heating, the size of the container, the composition of the product, thermal conductivity of the food (rate of heat penetration), the number and heat resistance of bacterial spores, the initial food temperature and the temperature of the retort must be considered when developing a thermal process to obtain a commercially sterile product.

There are various types of retorts for heat processing canned foods. These include still retorts (vertical or horizontal) (Fig. 12.4) agitating retorts, hydrostatic retorts (Fig. 12.5), rotary cookers, helical pump can sterilizers, as well as aseptic canning systems and flame sterilizers.

The food may be heated directly by steam either before filling into the container or by injection of steam through the open top of the container prior to closing and sealing. When heated prior to filling, the food product is flash sterilized by direct steam injection or in a high pressure continuous heating system at 120° to 150°C or higher for brief periods. This process is particularly valuable for food to be packed in large cans which, with still retorting, require a long process time. Heat sterilization before filling is limited to foods which can be pumped. The heated foods are usually cooled slightly and filled into sterile containers in an aseptic manner. This process is called aseptic canning.

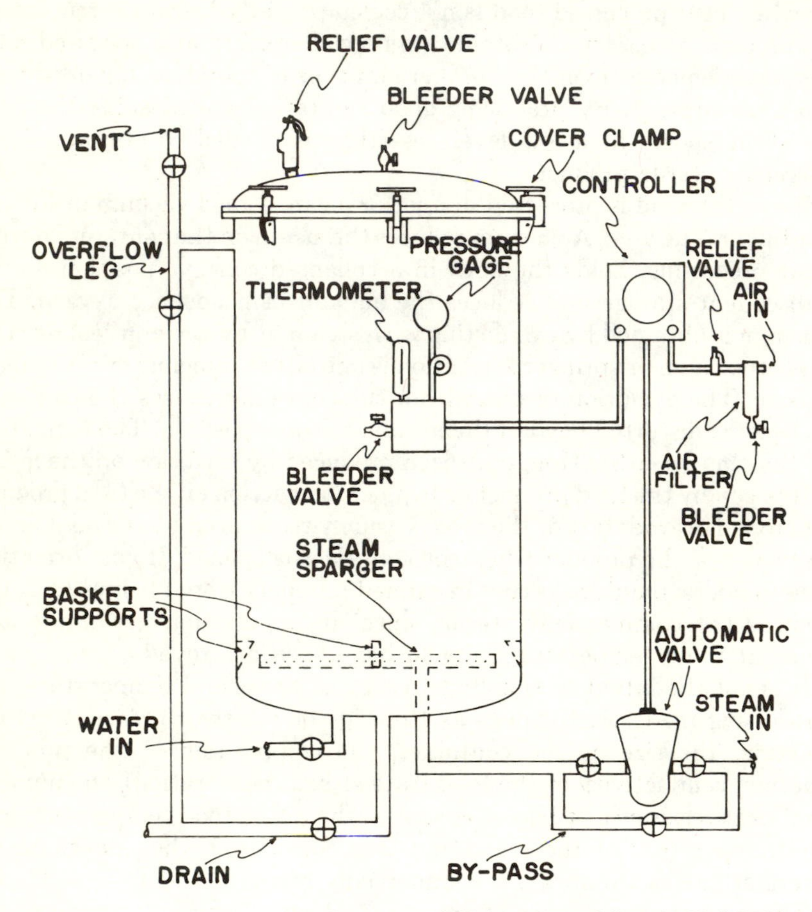

Courtesy of Nickerson and Ronsivalli (1979)

FIG. 12.4. CONVENTIONAL RETORT

Certain canned foods can be heated directly with a flame. In this process, the cans are rotated rapidly to prevent burning of the food and to induce convection heating. The direct flame process reportedly yields processed products with improved quality as compared to those conventionally processed.

The high temperatures of flame sterilization destroy the microorganisms before there is extensive heat damage to the product. The direct flame process is suitable for products that heat by convection, or by combined convection-conduction.

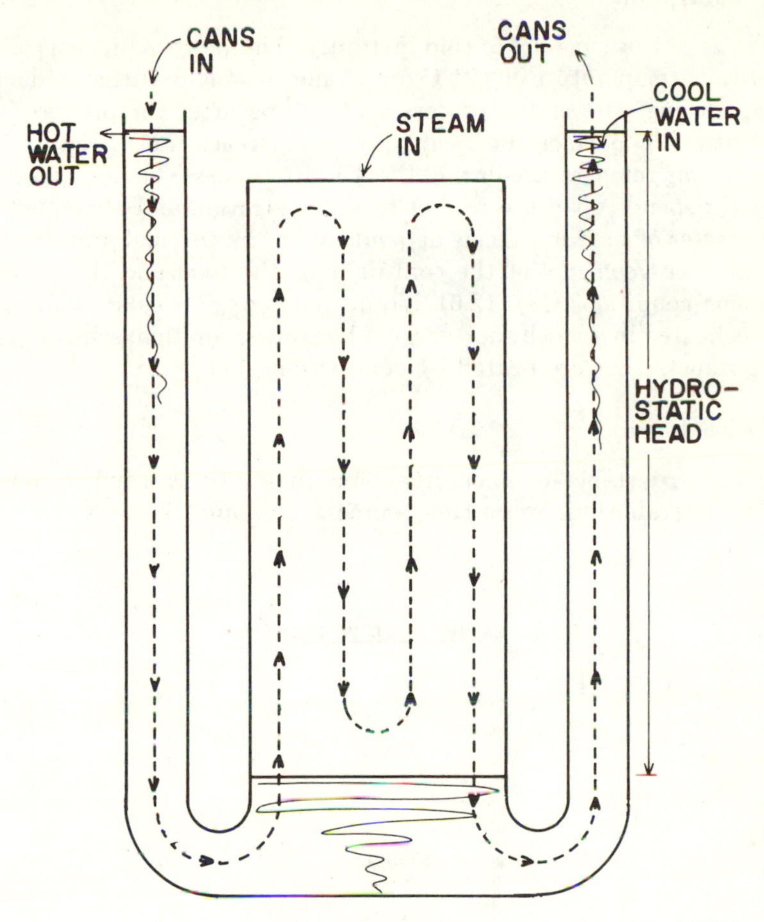

Courtesy of Nickerson and Ronsivalli (1979)

FIG. 12.5. HYDROSTATIC RETORT

Mathematical evaluations can be used to determine either the sterilizing effect of a heat process or the necessary heat process that will produce a specific sterilizing effect. Many suggestions have been presented for estimating either of these two considerations, as well as heating and other aspects of thermal processing of foods (Clark 1978; Hayakawa 1978; Lund 1978; Merson *et al.* 1978; Shiga 1976; Stumbo *et al.* 1975).

After processing, the cans are cooled in water to a temperature of 36° to 42°C. If cans are cooled much below 36°C, they may not dry thoroughly and rusting will result. If the cans are cased at temperatures much over 42°C, thermophilic spoilage may occur.

Heat Penetration

Foods do not become hot or cold instantly. Therefore, a process calling for a retort temperature of 121°C for 15 min means that the product is heating during part of the process and cooling after the process. The temperature in part of the food may never reach 121°C, but it approaches this temperature during the 15 min process. The temperature attained depends upon the rate of heat penetration and other factors. The transfer of heat into foods depends upon the thermal properties of the food, the geometry of the container of the food and the thermal processing conditions (Fig. 12.6). Liquid foods or particulate matter in fluids is heated in a much shorter time by convection than solid or semisolid products that are heated by conduction.

Heat Resistance

The heat resistance of microorganisms in a food is a major consideration in establishing processing temperatures and times.

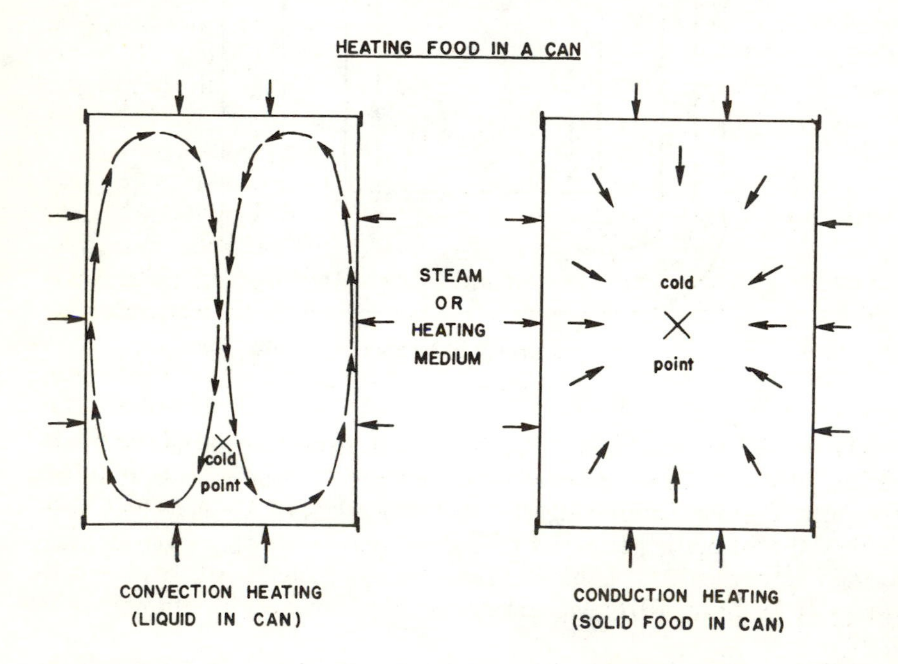

Courtesy of Desrosier and Desrosier (1977)

FIG. 12.6. COLD POINT OF HEATING FOR CONVECTION AND CONDUCTION TYPE PRODUCTS

Several methods have been described and used to determine the heat resistance of microorganisms. These include the thermal death time tube or can methods, the rate of destruction method, capillary tube, special heat exchangers and the miniature retort system. These various methods attempt to minimize the come-up time and cooling effects on destruction of the microorganisms.

Vegetative cells of bacteria, molds and yeasts are destroyed by temperatures 10° to 15°C above the optimum temperatures for growth. For appropriate times at temperatures of 60° to 80°C, most vegetative cells, as well as viruses, are destroyed. Somewhat higher temperatures may be needed for thermophilic or thermoduric microorganisms.

All vegetative cells are killed in 10 min at 100°C, and many spores are destroyed in 30 min at 100°C. However, some spores will resist heating at 100°C for several hours.

Spores are as much as 100,000-fold more heat resistant than the corresponding vegetative cells (Gould and Dring 1975A). Some mold spores are only slightly more resistant, while others are much more resistant than the mycelium cells. Reportedly, spores of *Byssochlamys fulva* can survive 5 hr at 88°C. Bacterial spores are the most resistant entities. Some of these spores, such as those of *Bacillus stearothermophilus*, may survive at 100°C for up to 20 hr with an initial population of 10^5 to 10^6 spores/g. The spores of *B. stearothermophilus* are of particular interest to the canning industry because of their high heat resistance and the flat-sour spoilage of low acid canned foods caused by this organism.

It has been suggested that the heat resistance of bacterial spores may result from dehydration of the central protoplast (Gould and Dring 1975A,B). This dehydration is brought about and maintained by osmotic activity and the cortex may act as an osmoregulatory organelle. Acids may interfere with this osmoregulation and render spores heat sensitive (Gould 1977).

Not only do different microorganisms differ in their heat resistance, but also different strains of the same species show diffferences. The conditions under which the organisms are grown, the nature of the product in which the microorganisms are suspended during the test, and the methods used to determine the survivors can influence the apparent heat resistance of a microbial culture.

Effect of Heat on Microbial Cells

Although moist heat has been used for many years as a method of sterilization, the primary lethal event in thermal inactivation of microorganisms has not been fully determined. The denaturation or coagulation of proteins, breaks in deoxyribonucleic acid (DNA), lesions in ribonucleic acid (RNA) and damage to the cytoplasmic membrane have been suggested as possibilities.

The denaturation or coagulation of proteins involved in cell respiration or cell multiplication usually is suggested as a cause of death.

In the range of 50° to 60°C, the leakage of cellular components into the suspending medium indicates there is damage to the permeability barrier of the cell. However, at higher temperatures, death precedes leakage. Scheie and Ehrenspeck (1973) suggested that heat caused denaturation of protein(s) in the cell envelope of *E. coli*. This resulted in a weakened peptidoglycan layer which would be sufficient to prevent multiplication with the conditions they provided. Further injury results from internal osmotic pressure rupturing the cell membrane at the weakened areas. As long as heat is applied, the rupture allows cellular material to escape. Hence, it might be assumed that destruction of cells at temperatures that cause sublethal injury may involve the cell membrane. If death precedes leakage at higher temperatures, apparently other mechanisms are involved.

Russell and Harries (1968) reported that for nonsporeforming cells such as *E. coli*, RNA degradation is closely related to heat induced death. Both RNA and membrane lesions were reported in *Pseudomonas fluorescens* (Gray *et al.* 1973). For commencement of growth, the repair of both the cell membrane and RNA damage was required.

The heating of spores at 70° to 100°C results in a loss of dipicolinic acid (DPA), proteins and other cellular constituents. This indicates damage to the spore coat. However, reportedly, the death of spores proceeds faster than the release of DPA.

The apparent death of spores may be due to the inability of the spore to germinate or, after germination, the inability for outgrowth to occur with the production of more cells. Hashimoto *et al.* (1972) suggested the primary cause of thermal inactivation of spores is due to physical and chemical alterations which interfere with the imbibition of water into the core during germination. The data of Flowers and Adams (1976) suggested a spore component destined to become cell membrane or wall was the site of injury.

DiGioia *et al.* (1970) concluded that thermal inactivation of phage or viral infectivity was due to protein coagulation.

Protein coagulation, if not the primary lethal event, is important in the thermal destruction of microorganisms. The denaturation of protein can occur in several areas of the cell or spore.

Factors Affecting Heat Resistance

The heat resistance of microorganisms is influenced by the type of heat treatment, the type of cell, as well as the conditions of growth and age, the suspending medium and the method for enumerating the survivors.

Type of Cell.—Spores are more heat resistant than vegetative cells usually by a factor of 10^4 to 10^5. Although there are many theories for the high resistance of spores, the exact mechanism is still conjectured. Even with spores, the method of production can affect the apparent heat resistance.

Growth or Sporulation Medium.—According to a review by Gould and Dring (1974), calcium and manganese ions in sporulating media increase the heat resistance of the spores. They reported no direct evidence implicating monovalent ions in increasing the thermal properties of intact spores. No increase in heat resistance was observed with manganese ions in the sporulation medium of *Bacillus stearothermophilus* (Cook and Gilbert 1968).

The pH of the sporulation medium affects the heat resistance of resultant spores.

The type of peptone, carbon source and fatty acids in the medium influence the apparent heat resistance of microorganisms. Spores are more resistant when produced in cooked meat medium than in raw meat. Doyle and Marth (1975) found *Aspergillus* conidia produced on media low in protein and high in glucose were more heat resistant than those produced in high protein-low glucose media.

Temperature of Growth.—Spores produced at high temperatures are generally more heat resistant than those produced at low temperatures. Hence, spores produced by thermophiles tend to be more heat resistant than spores produced by either mesophiles or psychrophiles. Although heat resistance is directly related to the growth and sporulation temperature, there are some limitations and exceptions. Phages react similarly to spores. A higher propagation temperature enhances the thermostability of viruses.

Age of Cells.—Young, actively growing cells are more sensitive to almost any stress than older, more mature cells. Young spores tend to be less heat resistant than older spores.

Suspending Medium.—The chemical and physical characteristics of the suspending medium in which microorganisms are heated influence the heat resistance. It is well established that spores are significantly more resistant when heated in distilled water than when heated in various buffers.

Perhaps the most important aspect that influences heat resistance is the pH of the suspension. Besides this, the a_w, the presence of various solutes, fatty materials and proteins affect the heat resistance of microorganisms.

Since foods vary in their chemical and physical characteristics, it would be expected that the heat resistance of a microorganism varies when suspended in different foods. Sullivan *et al.* (1971B) found different resistances for reovirus I when suspended and heated in raw milk, sterile milk or raw chocolate milk. Fewer differences in resistance were detected when herpes simplex virus was the test organism. The most protective substrate for these viruses was ice cream mix.

Effects of pH.—Generally, microorganisms have a greater heat resistance near pH 7.0 than above pH 8.0 or below pH 6.0 (Fig. 12.7).

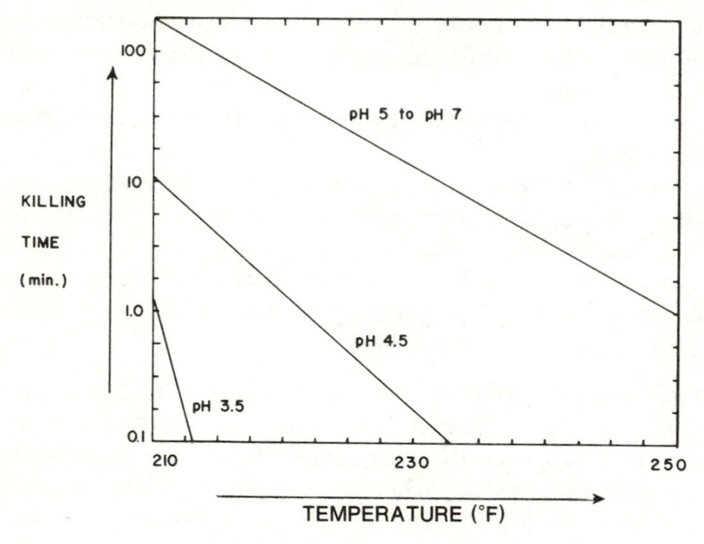

Courtesy of Desrosier and Desrosier (1977)

FIG. 12.7. INFLUENCE OF pH OF HEATING MEDIUM ON HEAT RESISTANCE OF SPORES

The more acid the substrate, the less the heat resistance of spore suspensions.

To take advantage of the lower heat resistance at low pH values, acids are added to certain food products prior to processing. Due to the variable pH of foods such as tomatoes, acids are added to adjust the pH to acceptable levels. At pH 3.9, tomatoes can be processed at 100°C for 34 min to kill a normal spore load, but at pH 4.8, a process of 110 min at 100°C is required.

Part of the preservative effect of low pH may be due to damage to the cells or spores with sublethal treatments and part due to the inability for growth of these damaged cells at low pH values.

Water Activity.—The water activity of the suspending medium in which the organisms are heated influences the heat resistance. In general, the microorganisms are more susceptible to heat in foods with high a_w. In foods with very low a_w, the heat effect is similar to that of dry heat.

One of the effects of heat on microbial cells is the coagulation of protein. Coagulation occurs more readily at high a_w and is reduced as the a_w is lowered.

With moist conditions, spores of *B. stearothermophilus* are about 50,000 times more heat resistant than spores of *C. botulinum* type E. However, at a_w values less than 0.5, the ratio falls to about 10 times (Murrell and Scott 1966). The heat resistance of spores of *B. stearothermophilus* in egg powder, fish protein concentrate and wheat flour was greater at a_w 0.33 than at 0, 0.68 or 0.99 (Harnulv *et al.* 1977). In an egg macaroni product, Hsieh *et al.* (1976) found that both *Salmonella anatum* and *Staphylococcus aureus* had a maximum heat resistance in the a_w range of 0.75 to 0.80.

The water activity of the suspending medium is often adjusted by the addition of salt, carbohydrates or other solutes. These substances can influence the heat resistance of microorganisms. Hence, the effect of these adjusted suspensions on heat resistance may be due to the solutes, to lowered a_w or to both. When salmonellae were heated in solutions of sugars and polyols, there was no direct relationship between heat resistance and a_w (Corry 1974).

Carbohydrates.—These substances can influence the heat resistance of microorganisms (Fig. 12.8). The thermostability of *C. botulinum* spores is directly related to the concentration of sucrose in the suspending medium. As the concentration of sucrose increases, the heat resistance increases.

The protection afforded the cells by sucrose may be due to the reduction of the a_w, the medium surrounding the cells, as well as extraction of water from the cell. Corry (1976B) found the degree of protection of the solutes was correlated with plasmolysis or cell shrinkage.

Corry (1976A) determined the heat resistance of *Saccharomyces rouxii* and *Schizosaccharomyces pombe* in solutions of sugars and polyols. The resistance was maximum with sucrose, less in sorbitol and least in solutions of glucose, fructose or glycerol.

The presence of higher carbohydrates, such as starch and pectin, reportedly gives some protection to spores and cells and increases the heat resistance.

Proteins.—There is a tendency for proteins to protect microorganisms from the lethal effect of heat. Peptones, yeast extract and albumin can provide protection for certain organisms.

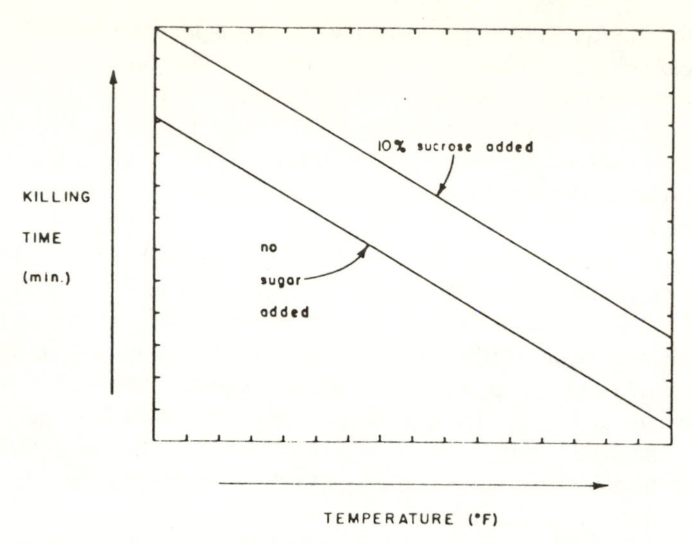

Courtesy of Desrosier and Desrosier (1977)

FIG. 12.8. INFLUENCE OF SUGAR ON THE HEAT RESISTANCE OF BACTERIAL SPORES

High concentrations of sugar protect spores.

Antimicrobial Substances.—Various antibacterial substances tend to increase the destruction of microorganisms by heat. An example is hydrogen peroxide used in conjunction with heat pasteurization of liquid egg white.

The addition of sulfur dioxide to fruit and fruit products tends to lower the heat resistance of the microorganisms. The effect is less apparent with yeast than with bacteria. Yeasts are relatively tolerant of sulfur dioxide.

Lipids.—Microorganisms suspended in fat or oil are more difficult to destroy than when they are in an aqueous medium (Fig. 12.9). Because of the poor heat conductivity of oil and the reduction in moisture, microorganisms in a fatty medium have a much greater resistance to heat than those heated in an aqueous medium without fat. The heat destruction in fats resembles dry heat sterilization.

Increasing the fat content of hamburger increased the D value of poliovirus (Filppi 1973). This protection of the virus was more evident at 80°C than at 50°C.

Enumeration.—When spores or vegetative cells receive a sublethal heat treatment, they may be injured in some manner. The spores may remain dormant for extended periods before germination. Dormancy can

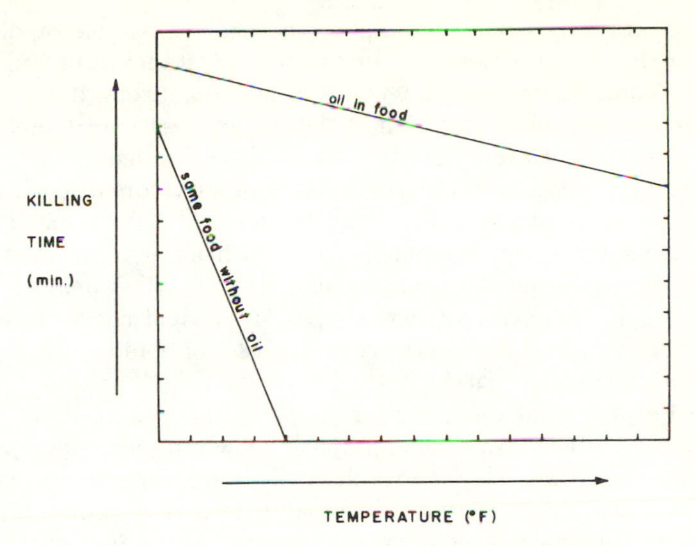

Courtesy of Desrosier and Desrosier (1977)

FIG. 12.9. INFLUENCE OF OIL ON HEAT RESISTANCE OF YEASTS

Organisms trapped in oil phase are killed by dry heat, and have much heat resistance in comparison with organisms in water phase.

be reduced or eliminated by using a suitable recovery medium. The addition of starch is thought to absorb fatty acids that inhibit outgrowth of germinated spores. Vegetative cells that are sublethally damaged require special environmental conditions for repair, growth and multiplication.

Since the recovery medium can affect the number of survivors of a heat treatment, the apparent heat resistance of an organism may vary to a significant extent.

Spores that survive sublethal heat treatments generally are recovered better at a lower temperature than the optimum for unheated spores. The addition of lysozyme to recovery media aids the germination and colony formation of spores surviving extended heat treatments. The exposure of heated spores to EDTA sensitized the spores to lysozyme and increased the number of recovered survivors (Adams 1974). Calcium dipicolinate also can initiate germination of heat treated spores.

Survival Curve

Since microorganisms differ in their resistance to heat and this resistance can vary according to age of cells, growth medium and suspending medium, we need a system to mathematically express survival. Such a system makes it possible to compare the heat resistance of different species at the same temperature, or the heat resistance of one species at different temperatures, and in various suspending media. It also enables the establishment and evaluation of thermal processes for foods in relation to the types of organisms that must be controlled.

By plotting the logarithm of the number of survivors at each time of sampling, a survival curve is obtained. Generally, bacteria exhibit a logarithmic survival curve when exposed to heat or other destructive agents. If a microorganism is either alive or dead, then death is a first order reaction. However, various shapes of survival curves have been reported. Various explanations have been given for nonlogarithmic death (Han *et al.* 1976; Cerf 1977).

Another aspect is the tendency for the survival curves to show a tailing effect (Fig. 12.10). As survivors approach a low number, they appear to be very heat resistant. This can result in microorganisms surviving a heat process that is in excess of that calculated to destroy the entire microbial population. The exact reason for the tailing effect is not known.

Although deviations from the logarithmic order of death do occur, for the purposes of further discussion, we will assume that a straight line for the survival curve is obtained (Fig. 12.11).

The number of survivors is directly proportional to the initial number present for any given treatment. The higher the initial number of cells, the longer the time necessary to cause their complete destruction. Since the logarithm of survivors never reaches zero, theoretically complete destruction of a microbial suspension can never be achieved.

Decimal Reduction Time (D)

The decimal reduction time is the time required to reduce the microbial population by 90% at a specified temperature. The decimal reduction time is designated by D. This is also the time required for the survival or death rate curve to traverse one logarithmic cycle (Fig. 12.11). In this case, D is 9 min. D can be determined mathematically by the formula:

$$D = \frac{t}{\log a - \log b}$$

a is the initial number of cells and b is the number remaining after time, t. When the difference between a and b is one log cycle, then (log a − log b) is equal to 1, and D = t.

The D value is not constant, but varies with species and strains of microorganisms, temperatures, and other factors that affect the heat

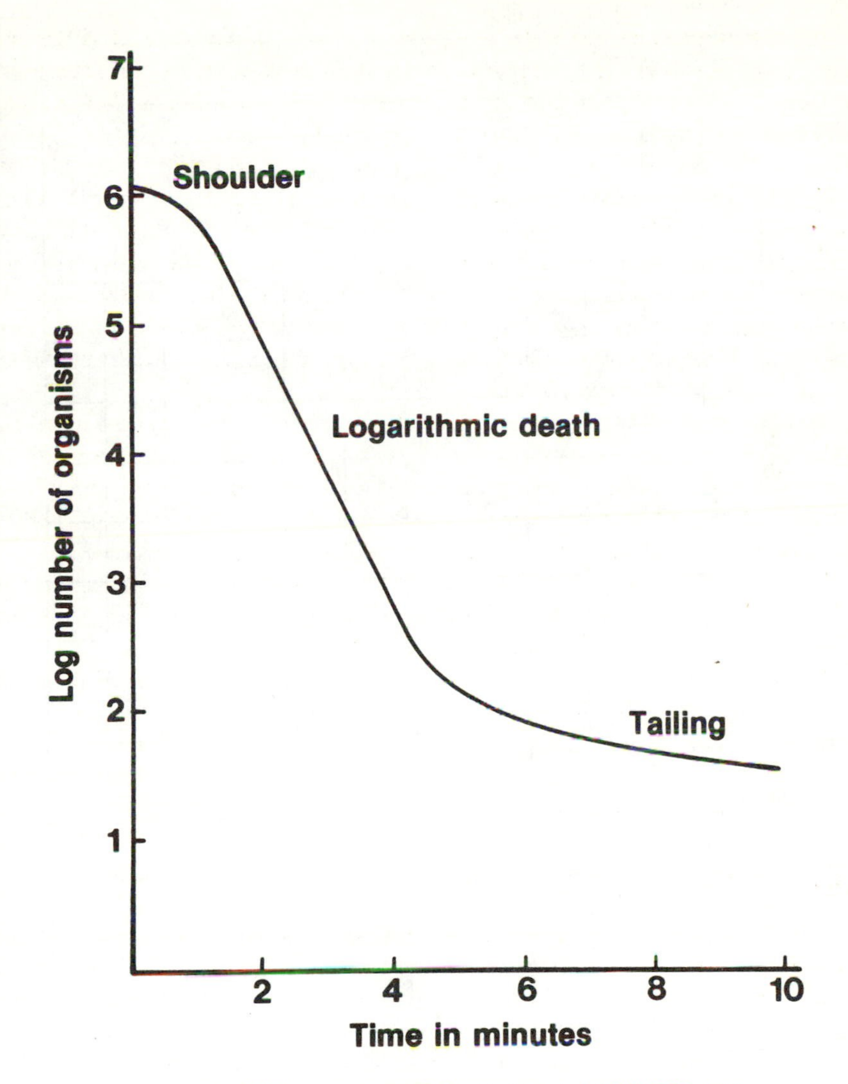

FIG. 12.10. SURVIVAL CURVE SHOWING A SHOULDER AND TAILING

resistance. The D for several microorganisms determined in various suspending media at certain temperatures is listed in Table 12.3. The temperature at which a D is determined may be designated by a subscript. For example, $D_{121°C}$ = 1.0 min means that 90% of the spores are destroyed in 1.0 min at 121°C, and 99% are destroyed in 2 min.

In thermally processed canned foods, the destruction of spores of *C. botulinum* is a major concern. To have reasonable assurance that the spores of *C. botulinum* are destroyed, a 12 D concept was established.

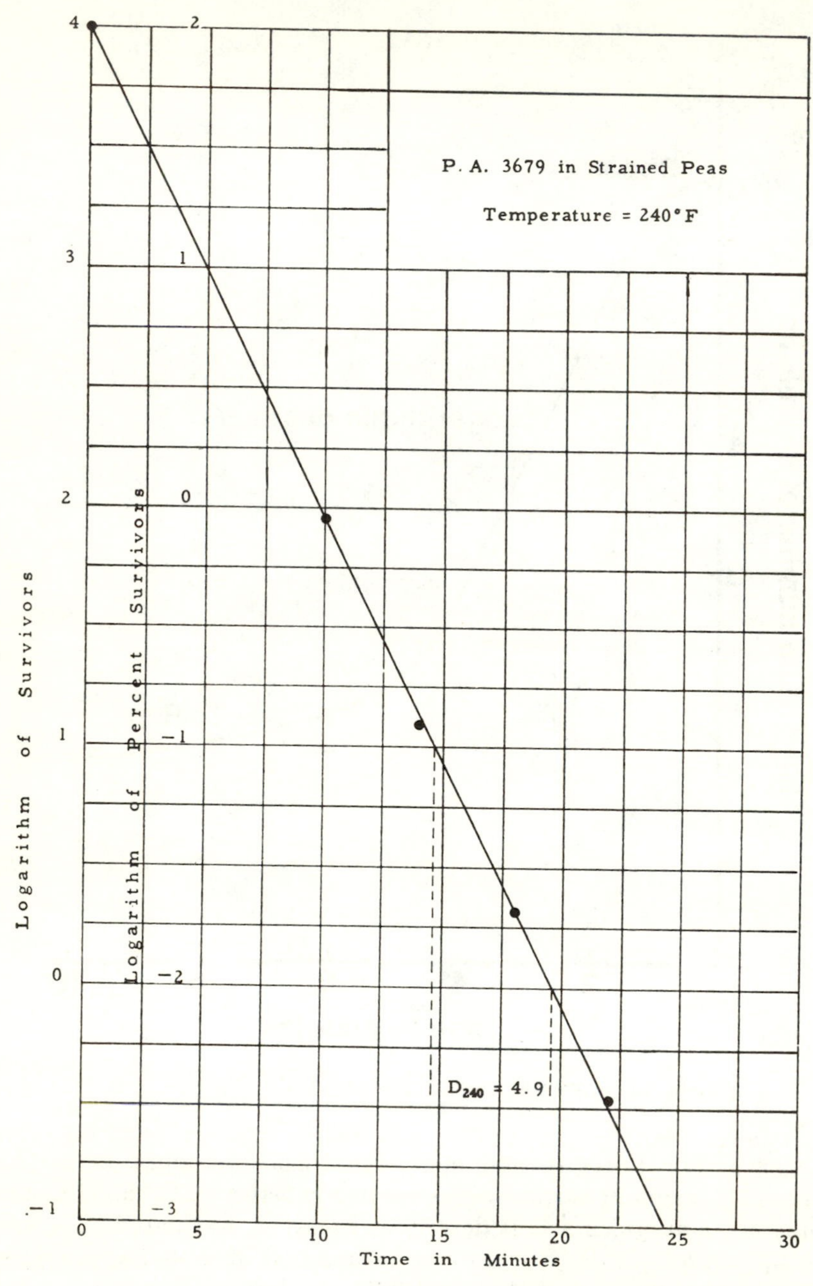

Courtesy of National Canners Association Research Laboratories (1968)

FIG. 12.11. DEATH RATE CURVE SHOWING THE DETERMINATION OF D

TABLE 12.3

DECIMAL REDUCTION TIMES (D) FOR HEATING VARIOUS MICROORGANISMS

Organisms	Suspending Medium	Temp. °C	D Value min.
Vegetative Cells			
Salmonella senftenberg 775W	Custard	60	11.3
	Chicken a la king	60	9.6
Salmonella manhattan	Custard	60	2.4
	Chicken a la king	60	0.40
Staphylococcus aureus	Custard	60	7.7-7.8
	Chicken a la king	60	5.2-5.4
Escherichia coli	Raw milk	57.3	1.3
	Ice cream mix	57.3	5.1
Spores			
Aspergillus flavus (conidia)	Phosphate buffer pH 7.0	50	16.2
		55	3.1
Clostridium botulinum type A	Phosphate buffer pH 7.0	104.4	17.6
		110.0	4.4
		115.6	1.3
		121	0.2-0.4
	Tomato juice pH 4.2	104.4	6.0
		110.0	1.6
		115.6	0.4
type B		110.0	0.7
type E	Phosphate buffer pH 7.0	70.0	29.3-37.5
		80.0	0.4-3.3
Bacillus stearothermophilus	Water	115.0	17.5-18.3
	4% NaCl	115.0	11.3-12.7
	Buffer—pH 7.0	115.0	9.2-11.3
	Buffer—pH 5.0	115.0	4.2

Sources: Cook and Gilbert (1968, 1969); Doyle and Marth (1975); Odlaug and Pflug (1977); Read *et al.* (1961).

This means that the number of spores will be reduced by 12 log cycles. Hence, with an original level of 10^6 *C. botulinum* spores, the 12 D thermal process would reduce this to 10^{-5} spores. It is evident that the lower the original contamination, the lower the number of spores after the process and the less the chance that a can of food will contain a potentially viable spore. Even if one spore does survive the process, there is a chance that it is injured or damaged and is not able to germinate, outgrow and produce the neurotoxin in the environment of the canned food. The spores of *C. sporogenes* and *B. stearothermophilus* are more heat resistant than those of *C. botulinum.*

The thermal inactivation of viruses has been difficult to characterize. It has been suggested that, like bacteria, viruses are inactivated by denaturation of protein. If first order kinetics prevail, one would expect the survival curve to be a straight line. The tailing effect, or decrease in inactivation rate as the time is extended, is a common occurrence in viral survival curves. The formation of aggregates or clumps of viruses is a prominent suggestion for the occurrence. Clumping tends to protect the viruses located in the middle of the clump from the heating effect. During analysis, the clumps are disintegrated to release the surviving viruses (Cliver 1971; Sullivan *et al.* 1971B). It is not likely that viruses can withstand the heat treatment given to thermally processed canned foods.

Thermal Death Time

The thermal death time (TDT) is the time required to achieve sterility of a suspension containing a known number of cells or spores at any predetermined temperature. At any particular temperature, the TDT depends upon the resistance of the cells, as well as the number of cells in the suspension.

Thermal death times are determined for suspensions of a particular organism at several temperatures. When the times necessary for killing are plotted on a logarithmic scale and the corresponding temperatures plotted on a linear scale, a straight line is obtained. This is the TDT curve for that particular organism.

When the D values are plotted on a logarithmic scale, with the corresponding temperatures on a linear scale, a phantom TDT curve is obtained. This should be a straight line (Fig. 12.12). These lines can be described by a D or TDT at some reference temperature (usually 60° or 121°C) and the slope (z). The symbol z is defined as the temperature necessary to bring about a ten-fold change in the TDT or D value. The value of z corresponds to the number of degrees of temperature passed over by the curve in traversing one logarithmic cycle. In Fig. 12.12, the z is listed for each curve. The symbol F is used to designate the time necessary to destroy a given number of microorganisms at a reference temperature, usually 121°C for spores, or 60°C for vegetative cells. To avoid confusion, the temperature can be added as a subscript to F. Hence, F_{60} is the TDT at 60°C and F_{121} is the TDT at 121°C. In Fig. 12.12, °F are used, and the F_{250} for curve A_1 is 5.8 min and for curve A_2 is 7.4 min.

For spores of *C. botulinum* suspended in phosphate buffer, the value of z is about 10°C. Crisley *et al.* (1968) reported the z values ranged from

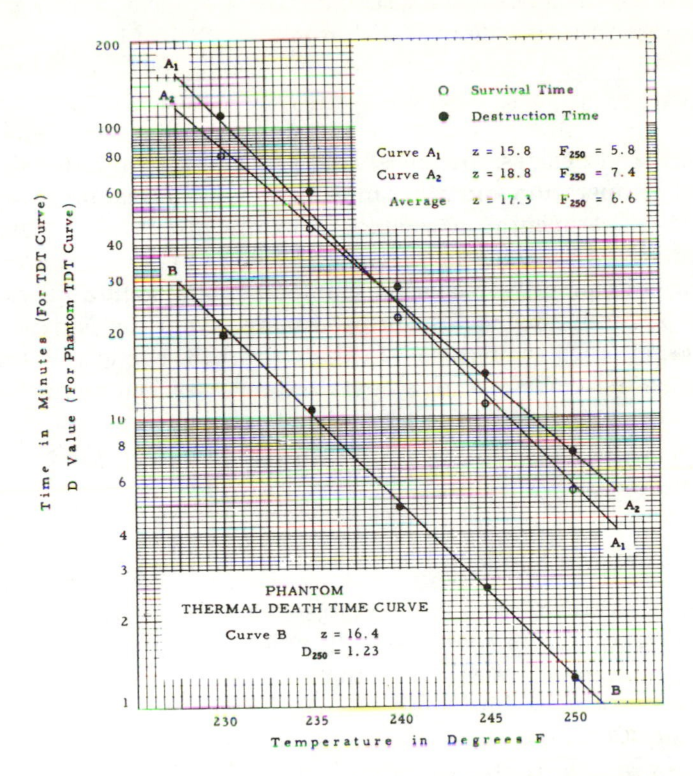

Courtesy of National Canners Association Research Laboratories (1968)

FIG. 12.12. THERMAL DEATH TIME CURVE

5.7° to 7.6°C for five strains of *C. botulinum* type E in fish paste.

For the microorganisms of most importance in canned foods, the z values are usually between 6° and 14°C. The z value varies for different strains and types of microorganisms, as well as the chemical and physical characteristics of the suspending medium. A strain of *Clostridium thermosaccharolyticum* had a z of 6.94°C in distilled water and 10°C in molasses (Xezones *et al.* 1965).

The work by Esty and Meyer in 1922 showed the F was 2.78 min for spores of *C. botulinum*. Due to this, the generally accepted minimum heat treatment for a 12 D destruction of spores of *C. botulinum* in low acid foods is considered to be 3 min at 121°C, or its equivalent. This is termed the minimum safe process or botulinum cook. Commercially canned cured meat products are given thermal processes well below this treatment. In these products, the heat treatment is supplemented by the effects of curing salts on the outgrowth of the spores, so that the treated products are not likely to be a health hazard. Low acid noncured meats are subjected to a treatment of F_{121} 6 min.

Thermal Processes

With D, F and z values, the rate of heat penetration, the pH of the food, the retort temperature and other information, a thermal process can be calculated. Experimental inoculated packs are processed to determine the validity of the calculations. It is evident that the determination of acceptable, safe processing schedules for foods can become quite complex. Thermal processing schedules can be obtained from the National Canners Association, Washington, D.C. Also, the U.S. Food and Drug Administration has guidelines for acceptable processes. Due to the complex interactions of factors concerning the determinations of processing schedules, computers have become an important tool for the calculation of needed processes.

Contaminants and Spoilage

Heat processed canned foods are subject to various types of spoilage. The appearance of the container will indicate probable spoilage. The internal formation of gas will cause conditions of the can termed flipper, springer, soft swell or hard swell.

For a flipper, only one end of the container is slightly bulged, or the can end bulges when the container is struck on a hard surface. In either case, by applying finger pressure, the end may be pressed back to remain flat.

For cans that are swelled, both ends bulge and cannot be pressed flat. Bulged ends of soft swells yield slightly to finger pressure, but those of hard swells do not.

Even if there is no bulging of the can, the contents may be spoiled due to flat-sour, sulfide or mold spoilage.

Spoilage may occur prior to processing or it may be due to inadequate heat treatment, improper holding temperature or to leakage through defective closures, other can defects or rough handling of the cans.

The organisms causing spoilage of underprocessed products are usually heat resistant spores of bacteria that survive the process, germinate and grow during storage of the canned food. Inadequate cooling or high storage temperatures can result in spoilage by the growth of thermophilic bacteria. The normal commercial heat treatment is not designed to destroy the spores of all obligate thermophilic organisms.

Spoilage due to leakage involves nonsporeforming bacteria, yeasts and molds that could not have survived the heat process. They enter the can after processing. Also, organisms involved in foodborne illness may enter cans after processing. Leakage into the container occurs when the hot cans are cooled in contaminated water. When hot, the sealant for the lids and seams is soft and not set so that organisms can enter. This entrance of organisms is aided by the vacuum created in the can during cooling.

Sterility Testing

Canned foods should be commercially sterile. That is, there should be no microorganisms present that have public health significance or that are capable of reproducing in the food under normal nonrefrigerated conditions of storage and distribution.

Contamination of the canned product during sampling and analysis has caused many erroneous results. One of the main sources of contamination is the investigator (Denny 1972; Evancho *et al.* 1973). These workers discussed the necessary precautions that must be taken for sterility testing of canned foods, as well as the media, equipment and procedures for this test.

IONIZATION RADIATION

Shortly after Roentgen discovered X-rays in 1895, it became known that ionizing radiations kill microorganisms. However, it was not until the early 1940's that research employed these radiations as a means of food preservation.

Ionizing radiation consists of corpuscular and electromagnetic radiations (Fig. 12.13) with sufficient energy to strip or cause ejection of electrons from atoms or molecules. Corpuscular radiation is due to subatomic particles of various types which can transfer their kinetic energy to anything they strike. The energy of these subatomic particles is based upon their velocity. Beams of electrons that have been accelerated to speeds approaching that of light are called cathode rays. These are related to the beta particles emitted from atomic nuclei. The other subatomic particles are not usable in food preservation. Alpha particles do not have sufficient penetrating ability (Fig. 12.14). Neutrons cannot be used, since they can induce radioactivity into the food.

Electromagnetic waves produced by high voltage equipment are called X-rays. Gamma radiation is an electromagnetic wave from atomic nuclei and is essentially the same as X-rays. The electromagnetic waves have an energy that is inversely related to the wavelength of the radiation. Thus, the shorter the wavelength, the higher the energy content. The gamma rays, X-rays and electron beams (beta and cathode rays) are the ionizing radiations useful in food preservation. The longer wavelength

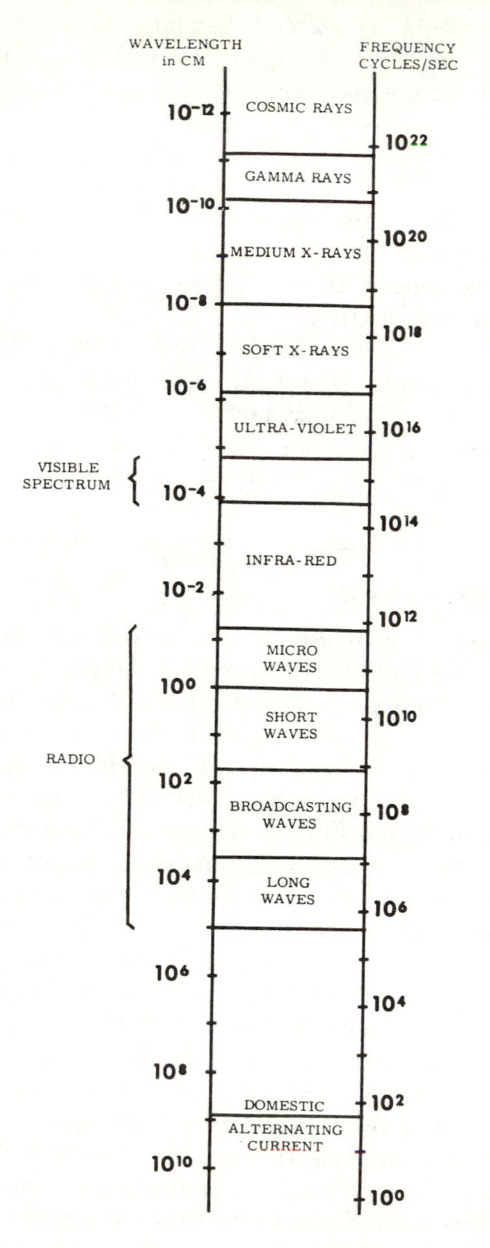

Courtesy of
Robert Taft Sanitary Engineering Center

FIG. 12.13. THE ELECTROMAGNETIC SPECTRUM

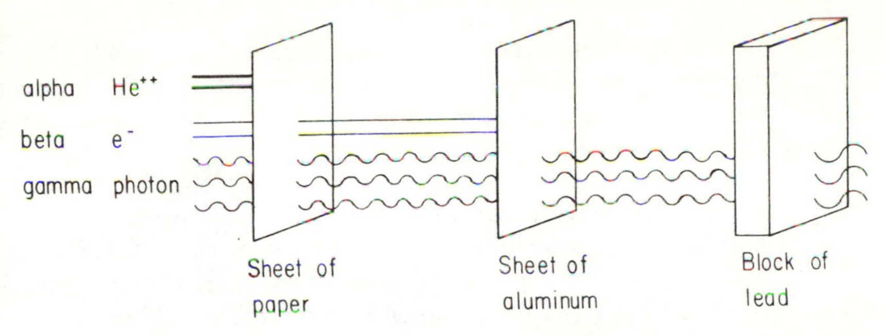

Courtesy of Desrosier and Desrosier (1977)

FIG. 12.14. RELATIVE PENETRATING POWER OF ALPHA, BETA, AND GAMMA RADIATIONS

Usefulness of various radiations is dependent upon requirements. Where depth of penetration is a factor, gamma radiations and high energy electrons must be used.

electromagnetic radiations (infrared, microwave, radio waves) are nonionizing.

The process of food preservation uses gamma rays from radioactive isotopes, such as cobalt-60 or cesium-137, X-rays, or electrons from linear accelerators. The radionuclides emit both gamma and beta rays. However, the energy level of the beta rays is relatively low for food sterilization, so that gamma rays are usually the only ones considered. The electrons are limited in their ability to penetrate food materials. The depth they penetrate depends upon their energy level. Electron beams over 10 million electron volts may induce radioactivity into some of the food constituents or the container, so are not considered to be practical for food preservation. Cobalt-60 has been the most acceptable radiation source because of its availability, price and properties.

The energy of radiation is expressed in rad units. A dose of one rad is obtained when 0.01 Joule (J) of radiation energy is absorbed per kilogram of material, or one rad is equal to 100 ergs of absolute energy absorbed per gram of material. For convenience, the kilorad (Krad) for 1,000 rads and megarad (Mrad) for 1,000,000 rads are used. A typical sterilizing dose is 4.5 Mrad, which is equal to 45 kJ/kg.

For the successful application of any method of food preservation, it must be proved that it is technically advantageous, it is economically competitive (the benefits are greater than the costs) and the treated food is wholesome and safe for human consumption. For this purpose, the biological effects, the stability of the treated foods in storage, the organoleptic and nutrient effects, the safety of the treated food and the relative costs of the process are discussed.

Biological Effects

The lethal action of ionizing radiations on living cells is due to either a direct action on genetic material or to an indirect effect. With indirect action, there is an initial change in the suspending medium or some nongenetic molecule.

Ionizing radiation causes breaks in the DNA. Restitution can occur in most of the single-strand breaks, as long as the repair process is operative. The presence of water increases the extent of the DNA damage (Auda and Emborg 1973). The mutations produced by X-rays are directly proportional to the dose and independent of the rate of dosage.

Generally, the more complex the organism, the more sensitive it is to ionizing radiations (Table 12.4). Hence, humans are very sensitive to ionizing radiations, and microorganisms are the most resistant of all living cells.

TABLE 12.4

APPROXIMATE LETHAL DOSES OF IONIZING RADIATION

Organism	Krad Level
Man	0.6-1.0
Insects	25-100
Bacteria	100-1,000
Fungi	100-1,000
Bacterial spores	1,000-5,000
Viruses	3,000-5,000

Uses in Foods

Foods can be irradiated to inhibit sprouting (potatoes, onions), delay ripening (fruits), destroy insects and their larvae and eggs (grain, papayas), destroy parasites (pork, beef, fish) or destroy microorganisms to extend shelf life, to preserve or to control foodborne illness. Some of these processes have been approved for use in certain countries, as listed in Table 12.5. Although much of the research on irradiation of foods has been done in the U.S., approval for use of this process has been rather slow. The U.S. FDA considers ionizing radiation treatment of foods a food additive, rather than a process.

In 1963, the irradiation of bacon was approved for public consumption. This approval was withdrawn in 1968.

Food irradiation has many potential uses. The realization of these uses could help assure a plentiful and healthful supply of a wide range of foods. With ionizing radiations, it is possible to treat a solid food product after being packed and sealed in the final container, so that recontamination does not occur. It is expected that ionizing radiation will not supplant our other methods of food preservation, but rather it can be used in conjunction with these systems.

Inhibition of Sprouting.—One of the major problems of storage of foods such as tubers or bulbs is sprouting.

When properly applied, certain chemicals can reduce the tendency of potatoes to sprout during storage. The addition of chemicals to foods is being questioned on public health aspects. Hence, irradiation might be another solution. It has been suggested that inhibition of potato sprouting by gamma rays is due to the effect of these rays on the synthesis of nucleic acid in the meristematic tissue (eyes) of the potato.

The irradiation level needed to inhibit sprouting ranges from 5 to 15 Krad, depending on the cultivar. Limited sprouting has been observed on some types of potatoes receiving 5 to 10 Krad of irradiation.

High level irradiation tends to increase the tendency of potatoes to rot. The rotting progressively increases as the irradiation dose increases. There is a tendency for gamma irradiated potatoes to darken after cooking.

When performed within one or two weeks after harvest, the irradiation of onions with as little as 2 Krad inhibits sprouting. However, if stored for several weeks before treatment, even irradiation at high levels will not prevent sprouting, but will soften the onions. The usual gamma ray treatment is with doses of 7 to 15 Krad. Since chemical treatment is not practical for prevention of yam sprouting, irradiation may be beneficial, especially in developing countries in tropical areas.

Delay Ripening.—Irradiation can be used to extend the storage life of many fresh fruits by controlling the rate of maturation. In order to transport fruit to the market, it is necessary to pick it while in the green stage and sell it before becoming overripe. Often, the ripening of the fruit is so rapid that the area of marketing is limited.

With irradiation, an extension of 10 to 20 days of market life has been obtained.

Destruction of Insects.—Insects can consume or ruin 25% or more of the grain stored in some countries. Chemicals can be used to destroy the insects, but there has been a concern by the public that residues of insecticides may be a health hazard. The irradiation of grain not only will kill the insects, but also will destroy the eggs and larvae. For most storage insects, irradiation with 25 Krad is lethal to all pre-adult stages and will sterilize the adults. A level of 20 to 50 Krad was approved for wheat and wheat flour in the U.S. Canada allows a maximum of 75 Krad. According to Tilton *et al.* (1974), the irradiation of heavily infested flour with 40

TABLE 12.5

USES[1] OF IRRADIATION APPROVED[2] FOR VARIOUS COUNTRIES

Country	Product													
	Potatoes	Onions	Garlic	Dried Fruits	Mushrooms	Wheat	Dry Food	Concentrates	Strawberries	Shallots	Animal Feed	Spices	Beef, Pork, Rabbit	Milk Cartons
Bulgaria	+	+	–	–	–	–	–	–	+	–	–	–	–	–
Canada	+	+	–	–	–	+	–	–	–	–	–	–	–	–
Czechoslovakia	+	+	–	–	+	–	–	–	–	–	–	–	–	–
Denmark	+	–	–	–	–	–	–	–	–	–	–	–	–	–
France	+	+	+	–	–	–	–	–	–	+	+	–	–	–
Hungary	+	+	–	–	–	–	–	–	+	–	–	–	–	–
Israel	+	+	–	–	–	–	–	–	–	–	–	–	–	–
Italy	+	+	+	–	–	–	–	–	–	–	–	–	–	–
Japan	+	–	–	–	–	–	–	–	–	–	–	–	–	–
Netherlands	+	+	–	–	+	–	–	–	+	–	–	+	–	+
Philippines	+	–	–	–	–	–	–	–	–	–	–	–	–	–
Spain	+	+	–	–	–	–	–	–	–	–	–	–	–	–
Thailand	–	+	–	–	–	–	–	–	–	–	–	–	–	–
USA	+	+	–	–	–	+	–	–	–	–	–	–	–	–
USSR	+	+	–	+	–	+	–	+	–	–	–	–	+	–

[1]Any food used for consumption by patients who require a sterile diet as an essential factor in their treatment—Approved in Great Britain.

[2]Approval may be unlimited with specified doses, or temporary, conditional or experimental to determine feasibility.

Krad can control the species of insects likely to occur in this product. Once treated, the food must be protected from reinfestation.

Destruction of Parasites.—The most common parasitic diseases transferable from meat to man are cysticerocosis of cattle and trichinosis of pigs. Cysticerci are devitalized by 400-500 Krad. A dose of 300 Krad followed by storage at 2°C is fatal for the cysts in beef. A dose of about 10 Krad will destroy trichinae in pork.

Various nematode larvae are found in fish. When salted fish are eaten in the raw state, these larvae can infect the intestinal tract of humans. In fish with 6% salt, irradiation with only 1 Krad can result in a 100% kill of these larvae.

Destruction of Microorganisms.—Ionizing radiations, especially from linear accelerators, are not necessarily uniform. Therefore, all treated microbial cells might not be exposed to equal levels of these random radiations. The death or injury of microorganisms is due to either direct or indirect effects. The direct effect is due to a direct hit with the radiation causing ionization within the microorganism while the indirect effect results from ionization and diffusion of free radicals and peroxides produced in the vicinity of the organism.

The ionizing radiation doses for destruction of microorganisms vary with the type and concentration of cells, the type of substrate or suspending medium, oxygen tension, temperature or the presence of protective compounds. Certain pre- and post-irradiation treatments can influence the inactivation of microorganisms.

Cellular death from irradiation is related directly to lesions in the nuclear DNA. The damage caused to DNA is in the form of single or double-strand breaks, cross-linking of strands, base transfer or removal and pyrimidine dimer formation. The resistance and survival of microorganisms depend upon the capacity to repair damage to the genetic material.

Death due to radiation is considered to be a first order reaction, so that if the logarithm of survivors is plotted with the corresponding radiation dose on a linear scale, a straight line should be obtained. As with destruction by heat, various types of survivor curves are obtained, including shoulders at the beginning of irradiation and a tailing effect.

The D values should be relatively constant, regardless of dose. However, Anellis *et al.* (1969) found considerable variation with different levels of irradiation. Various methods have been suggested for calculating D values, none of which seems to be entirely satisfactory.

The D values can be used to show the relative resistances of microorganisms. Some estimations that have been reported are listed in Table 12.6.

TABLE 12.6

REPORTED D VALUES FOR IRRADIATION OF SOME MICROORGANISMS

Microorganism	Suspending Medium	D Value (Krad)
Spores		
Clostridium botulinum		
Type A	Cured ham	218-235
	Bacon	214
Type B	Cured ham	164-175
	Pork loin	286-336
	Chicken	370
Type E	Water	80-160
Type F	Water	250
C. sporogenes	Water	160-220
C. perfringens	Water	120-340
Bacillus cereus	Broth	250
B. stearothermophilus	Buffer	252
	Buffer (anoxic)	425
Vegetative cells		
Escherichia coli	Buffer	9
Salmonella typhimurium	Buffer (air)	20
	Buffer (anoxic)	60
	Dried bone meal	91-100
	Dried fish meal	170
Staphylococcus aureus	Buffer	20-25
	Crab meat	40
Micrococcus radiodurans	Buffer	194-308
	Raw beef	350-600
Vibrio parahaemolyticus	Buffer, pH 7.0	5-14
Molds		
A. flavus (conidia)	Water	38
	Dry	50-55
Penicillium citrinum		
(conidia)	Wet	18
(mycelium)	Dry	55
Yeasts		
Saccharomyces cerevisiae	Buffer	36
Trichosporon oryzae	Buffer	120-160
Candida sp.	Buffer	32

TABLE 12.6. (*Continued*)

Microorganism	Suspending Medium	D Value (Krad)
Viruses		
Coxsackievirus B-2	Eagles medium	440-690
	Water (−90°C)	530
	Cooked beef (16°C)	700
	Cooked beef (−90°C)	810
Adenovirus 2	Eagles medium	410
Adenovirus 3	Eagles medium	490
Echovirus 18	Eagles medium	440
Toxins		
Botulinum		
Type E toxin		
washed cells	Buffer	1,700-2,100
cell extract	Buffer	210
purified	Buffer	40
Type A toxin		
purified	Buffer	50-200
purified	Broth	4,000
purified	Cheese	6,000
Staphylococcal		
Enterotoxin B purified	Buffer	2,700
	Milk	9,700

Sources: Anellis *et al.* (1965, 1967, 1969); Ito *et al.* (1974); Kopelman *et al.* (1968); Miura *et al.* (1970); Read and Bradshaw (1967); Roberts (1968); Sullivan *et al.* (1971A, 1973).

Relative Resistances.—The radiation resistance of strains and species will vary widely. Generally, bacterial spores exhibit the highest resistance to irradiation with the exception of some Gram positive cocci. The Gram positive bacteria are more resistant than are Gram negative types. Yeasts and molds tend to be somewhat resistant to irradiation, while viruses, in general, are equivalent to or more resistant than bacterial spores. Young cells are more sensitive than old cells to ionizing radiation. In the absence of oxygen, the cells have an increased resistance to radiation. In this condition there is a restriction of oxidized radical formation.

Certain chemicals, such as reducing compounds, protect the cells. Higher doses are needed to destroy microorganisms in foods than in buffer, with the exception of cured meat. Different strains of the same species reveal considerable differences in their D values.

The removal of water generally increases the resistance of microorganisms to irradiation. This is thought to be due to reduced formation of H* and *OH radicals, the reduction in the mobility of free radicals that are formed and to the protection given the cells by the dried substances.

Some of the least resistant bacteria are species of *Pseudomonas, Proteus, Escherichia* and *Vibrio.* Ito and Iizuka (1971) isolated from rice an organism which they named *Pseudomonas radiora.* In a buffer, strains of the organism had D_{10} values of 60 to 140 Krad. These are 10 to 40 times higher than found for other pseudomonads.

Escherichia coli is quite sensitive to ionizing radiation, although there are resistant strains. Due to the rapid demise of *E. coli,* this organism is a poor indicator of sanitation for irradiated food. Nearly all of the organisms involved in foodborne illness have a higher resistance to irradiation.

Serotypes of *Salmonella* vary in their resistance to irradiation. The relative resistance of serotypes shows that the mechanisms are different for the destructive action by heat and radiation. *S. senftenberg* 775W is the most heat resistant strain, while the radiation resistance of this organism is more similar to other strains and serotypes of salmonellae.

A radiation resistant *Micrococcus* was isolated from meat by Anderson *et al.* (1956). This organism is similar to *M. roseus,* but there are some differences. The organism was named *Micrococcus radiodurans.* This is not recognized as a species in Bergey's Manual (Buchanan and Gibbons 1974). At a level of 10^7 cells/ml, a gamma radiation dose of 6 Mrad was required to destroy the microorganism (Anderson *et al.* 1956). Since then, other radiation resistant micrococci have been isolated. The *Micrococcus* isolated by Lewis (1971) was more resistant to gamma irradiation than *M. radiodurans* by several orders of magnitude.

Reportedly, the high radiation resistance of *M. radiodurans* is due to the efficient capacity of the cells to repair DNA lesions, including double stranded scissions.

Moraxella and *Acinetobacter* are more resistant than pseudomonads to radiation. Ito *et al.* (1976) reported the D values for *Moraxella* species to vary from 44 to 54 Krad and *Acinetobacter* to be 12 Krad. These values are in the range expected rather than the 477 to 1,000 Krad for *Moraxella* and 405-814 Krad for *Acinetobacter* as reported by Welch and Maxcy (1975).

C. botulinum spores are generally more resistant than other clostridial spores. This is very important if ionizing radiations are used for sterilization of food. Spoilage types of clostridia could be eliminated from the food while *C. botulinum* spores could remain viable and produce toxins. Hence, the consumer might not be forewarned of possible danger by gas formation or putrid odors normally developed in contaminated food by spoilage types.

The required minimum radiation dose to achieve commercial sterility of food is based on the resistance of *C. botulinum* spores. In general, Type A spores are more resistant than type B spores. Type A and B spores are more resistant to radiation than are type C, D, E or F spores.

The temperature at which irradiation occurs influences the resistance of *C. botulinum* spores. According to Grecz *et al.* (1971), from 65°C down to −196°C, the spore resistance increases with decreasing temperature.

The presence of curing salts (sodium nitrite, sodium chloride and sodium nitrate) reduces the irradiation dose needed to inactivate spores of *C. botulinum*. This is evidenced in the lower D values needed for ham and bacon as compared to pork loin or chicken.

Spores of *C. botulinum* that have been killed by irradiation still contain active toxin. This indicates that spores are more sensitive to irradiation than is spore toxin. The implication of this is that food that is highly contaminated with *C. botulinum* spores may be toxic even after receiving a dose of irradiation that will destroy the spores.

Enzymes are resistant to ionizing radiation. Fernandez *et al.* (1969) suggested that sufficient enzymes may remain active to synthesize botulinum toxin. They found an increase in toxin level after irradiation which suggested that damaged cells may not be able to show growth and multiplication, but may produce toxin.

The presence of molds that produce mycotoxins in food or feed is undesirable. Strains of *Aspergillus flavus* and *A. parasiticus* that produce aflatoxins are especially important. Low levels of 100-300 Krad may cause a reduction or an increase in the production of mycotoxins. At doses needed to destroy molds, the germinating ability of grain also is destroyed. Further, above 200 Krad, there is an alteration of the grain with production of off-odors, flavors and colors.

The irradiation of some plant products makes them more susceptible to attack by molds. It is thought that this is due to tissue breakdown as a result of pectin depolymerization, and a weakening of plant tissue.

Yeasts show a large variation between species to resistance to ionizing radiation. In general, they are more resistant than filamentous fungi. Some species display a resistance that is comparable to some bacterial spores.

Viruses tend to be more resistant than other organisms to gamma irradiation. Viral inactiviation demonstrates a one-hit or first order reaction curve, when the suspending medium contains free radical scavengers or is in a frozen state. Freezing increased the resistance in water and cooked beef, but this increase was not observed in the case of raw beef (Sullivan *et al.* 1973). Fortunately, viruses are not found in most foods. If present, they tend to be at low levels.

DiGioia *et al.* (1970) suggested that the irradiation inactivation of Newcastle disease virus at low temperature (2.2°C) was due to nucleic acid degradation, while at higher temperatures (60°C), it was due to protein denaturation. Survival is enhanced in the presence of impurities. These impurities apparently bind free radicals and peroxides giving protection to the viruses.

Radappertization.—Radiation sterilization or radappertization is the destruction of all or practically all of the organisms. In order to accomplish commercial sterility, relatively high levels of irradiation are required (see preceding). Of all the microorganisms of concern that are associated with food, the spores of *C. botulinum* are the most resistant to irradiation. The spores of *C. botulinum* are more resistant to ionizing radiation than spores of organisms that have a high heat resistance and cause spoilage.

For ionizing radiation as with heat treatment, a 12 D concept is used for producing commercially sterile foods that are reasonably safe from *C. botulinum* spores or toxin. If the D used is 0.375 Mrad, for commercial sterility, a 12 D treatment is 4.5 Mrad. According to Josephson (1969), the minimum radiation dose (MRD) needed for sterilization varies from 2.3 Mrad for bacon irradiated at ambient temperature to 5.70 Mrad for beef irradiated at −80°C. The petition for approval of irradiated bacon included a 4.5-5.6 Mrad treatment. The determination of 12 D doses needed for radappertization is not a simple calculation (Ross 1976).

Radiation doses for radappertization can cause losses in the organoleptic quality of food. The higher the dose, the greater is the potential for development of undesirable flavors, odors and colors, as well as losses of nutrients.

Of the common meats, beef is the most and pork is the least sensitive to the production of off-flavor. Although these changes in foods due to irradiation have been of concern, often they are less than the alterations due to heat processing.

To minimize alterations of foods due to irradiation, various combination processes have been suggested. These processes include addition of chemicals that are free radical acceptors, use of anaerobic conditions, elimination of moisture or irradiation at freezing temperatures. Each of these systems reduces the objectionable activity of the free radicals.

The amount of nitrite needed in cured ham can be reduced when radappertization is used to control *C. botulinum* (Wierbicki and Heiligman 1973). For this treatment, the ham is frozen and treated at −30°C with 3.7 Mrad. The nitrite was reduced to 25 μg/g, but to prevent fading of the cured color, 100 μg/g of nitrate was added.

Even though nearly all of the bacteria are killed by radappertization, enzymes are not destroyed as readily. Hence, enzyme activity continues, especially when the food is held at ambient temperature.

Radappertization of meat involves a preirradiation treatment with heat (65° to 80°C) to inactivate proteolytic enzymes, sealing in a package under a partial vacuum, freezing to −30°C or lower, and exposing the packaged food to ionizing radiation until the desired dose is absorbed.

Since it is necessary to heat the food to inactivate the enzymes, the technology of radappertization is primarily on partially cooked or cooked foods that include meat, poultry, codfish cakes and shrimp products.

None of the other major classes of foods can withstand the required high doses of irradiation.

Freezing prior to irradiation reduces the rate at which free radicals diffuse through the tissue. This minimizes the side reactions of irradiation and reduces the alteration of odor and flavor. Generally, the lower the temperature of irradiation, the more the quality factors are protected. However, as the temperature is lowered, the microorganisms, including *C. botulinum* spores, have increased resistance to irradiation, so that higher doses are needed to sterilize frozen meats than meats at ambient temperatures.

Radappertization processes and problems have been described for meat, poultry and seafoods (Anellis *et al.* 1972, 1975; Rowley *et al.* 1974).

Radappertization appears to be beneficial for sterilizing and storage of foods when refrigeration and frozen storage are not available. There are disadvantages and problems associated with this treatment. These include the need for heating and freezing during the process, the alterations in organoleptic and nutrient quality and the cost of the operation. Thus, it will be difficult for radappertization to compete with the present preservation procedures used in the United States.

Radurization.—The use of low doses of radiation to destroy a sufficient number of microorganisms and enhance the storage life of foods is called radiation pasteurization or radurization.

The doses of radiation used for radurization (100 to 1,000 Krad) will destroy from 90 to 99% of the microorganisms. Some resistant vegetative cells and spores will survive. Hence, refrigeration or some other combination treatment is needed to retard growth and possible toxin production of the survivors.

The dose used for radurization is a compromise. High doses prolong storage life, but deleterious effects and costs increase as the dose increases. Quite often, the dose used is the maximum amount of irradiation which produces no detectable change in the product. The other possibility is to determine the storage life needed to market a product and fit the dose of irradiation to this requirement.

The psychrophilic pseudomonads are the main spoilage organisms of protein foods (meat, poultry, fish). These bacteria are very sensitive to irradiation, being essentially eliminated by doses used for radurization. Although the elimination of the pseudomonads will extend the refrigerated storage life of the food, spoilage will occur eventually. Some of the surviving organisms can grow at refrigerated temperatures, but they have a slower growth rate than do the pseudomonads.

The treatment of fishery products with 100 to 600 Krad can increase the refrigerated storage life. The effectiveness of radurization varies with the type and initial quality of the product and the associated microbial flora. One problem is the presence of *C. botulinum* type E in fishery products. The spores may survive the low irradiation treatment and growth can occur at temperatures near 4°C with extended storage. The extent of this potential hazard with irradiated fish is not yet known. An unusual effect of irradiation was reported by Segner and Schmidt (1970). They inoculated haddock with spores of type E *C. botulinum,* irradiated it with 100 or 200 Krad and stored it at 4.4° to 10°C. Toxin was detected sooner in the haddock treated with 200 Krad than in that treated with 100 Krad. Some nonproteolytic type B and F cells of *C. botulinum* are present in fishery environments and are capable of growth at low temperatures. These types, as well as type E, must be considered regarding the potential hazard of radurized foods.

With carcasses, a dose of 400 Krad extends the storage life. If held at −2°C, the carcass can be stored for 17 weeks. This allows the shipment of carcasses over long distances, even overseas, and still leaves time for distribution, sales and consumption before spoilage. With storage above 10°C, there is little, if any, benefit from radurization treatment of red meat.

The foods that rank highest in merit for radurization are shrimp, fin fish (haddock, flounder, cod), blue crab, strawberries and mushrooms. These foods are perishable and have high spoilage rates during marketing. They are of relatively high value and, at some marketing stage, they have large volume concentrations.

Radicidation.—This is a low level irradiation treatment to destroy viable nonsporeforming pathogenic types of organisms to reduce or eliminate the problem of foodborne illness. Spores of *C. botulinum* or *C. perfringens* are not destroyed at levels normally used for radicidation. This term was suggested to describe the ionizing radiation treatment needed for elimination of salmonellae from food and feed.

The elimination of salmonellae from poultry and red meat reportedly can be accomplished with doses of 400 to 600 Krad. Vacuum packaged or frozen poultry and meat products require a higher dosage than the fresh product packed in air. For frozen poultry carcasses, doses of 700 Krad eliminate salmonellae.

Heat treatment can be used to eliminate salmonellae from dried feed, such as meat, bone, fish and blood meal, but there is a potential for recontamination prior to packaging and heating decreases the available protein. A dose of 800-1,000 Krad can reduce salmonellae in packaged dry feeds to below detectable levels. Irradiation of animal feeds to eliminate salmonellae would be an important step in breaking the salmonellae cycle, and might reduce the involvement of animal foods in human salmonellosis.

The treatment of animal feed with 2,000 Krad can elminate *Bacillus anthracis*, the cause of anthrax in animals and man.

Dose levels of 200 Krad are sufficient to destroy all expected levels of *Staphylococcus aureus* in most food products. The problem with *S. aureus* is the recontamination by people handling the food during preparation for serving. Radicidation cannot prevent this from occurring.

Thermoradiation.—This is a combination treatment of heat and radiation. Enzymes and spores of *C. botulinum* are relatively resistant to radiation, but are more sensitive to heat. Conversely, thermophilic spores are very heat resistant, while relatively more sensitive to radiation. A combination of heat and radiation, either in sequence or simultaneously, may be of value in food preservation.

Preheating to sublethal temperatures can sensitize vegetative cells but not spores to irradiation. Preirradiation sensitizes spores to subsequent heat treatments. In some cases, simultaneous heating and irradiation are more effective than applying the treatments in sequence. Heat and irradiation are synergistic when used together for killing spores of anaerobes. Thus, it is possible that this synergism can be used to sterilize food, so that the quality of the food is less affected than with the level needed for either treatment applied alone.

Shrimps subjected to heat at 121°C for 8 min and irradiated with 100 Krad can be stored up to two months at ambient temperatures (Savagaon *et al.* 1972). Canned products with excellent quality can be obtained by thermoradiation.

Synergistic effects of temperature and radiation have been determined for several organisms, including *B. stearothermophilus*. For this organism, the D of 210 Krad at 27°C was reduced to 157 Krad at 60°C (Tulis *et al.* 1973).

Inactivation of Toxins.—Microbial toxins reportedly are inactivated at radiation levels comparable to bacterial spores. However, the amount of radiation needed to inactivate botulinum toxin is greater than that needed to destroy the organisms (Table 12.6).

The purer the toxin the more sensitive it is to irradiation (Miura *et al.* 1970). This indicates that impurities protect the toxin in much the same way as cells are protected. They found serum albumin, casein, DNA and RNA protected purified and activated toxin against radiation inactivation. Sugars or ascorbic acid showed little or no protection.

Staphylococcal enterotoxin is more heat stable than botulinum toxin. This higher stability tends to be true for inactivation by ionizing radiation. With the D values for inactivation of enterotoxin determined by Read and Bradshaw (1967), it would be impossible to rid a food of significant amounts of this toxin without destroying the quality of the food. Low doses of irradiation would have little or no effect on either botulinum or staphylococcal toxins.

Inactivation of Enzymes.—The complete, or at least partial, inactivation of enzymes has been considered essential for preservation of foods. The inactivation of enzymes requires higher doses of radiation than that for microbial destruction.

To determine if milk has been heat pasteurized, the presence of phosphatase enzyme is determined. During irradiation of milk, phosphatase is not affected.

Heat treatment of meats to inactivate the enzymes prior to irradiation improves the storage stability of these products. This preheating treatment applies not only to red meats, but also to poultry and seafoods. Even 4.5 to 5.2 Mrad of radiation will inactivate only about 75% of the proteolytic activity of beef (Losty *et al.* 1973).

The heat treatment needed to inactivate enzymes in fruits is sufficient for preservation, so the use of ionizing radiation is unnecessary. There is little advantage in using radiation for the preservation of vegetables.

Since enzymes are more resistant than microorganisms to irradiation, there is a possibility that this process can be used to sterilize or at least significantly decrease the microbial level in commercial enzyme preparations. Ionizing radiations can yield enzymes with low microbial levels without affecting enzyme activity.

Quality of Irradiated Foods

At some level of ionizing radiation, microorganisms, toxins and even some enzymes can be destroyed or inactivated. However, at the doses necessary to destroy viruses, spores or even some vegetative cells, the quality of the food may suffer. At doses as low as 50 Krad, some flavor changes can be detected by trained personnel. As the dose is increased to 200 Krad, the off-flavors become intensified so that they can be detected by untrained people.

The effect of ionizing radiation varies with the type of food. Some foods, such as milk, are very susceptible to even low doses of radiation. Other foods, such as cured meats, can be treated with relatively high doses of radiation without seriously affecting the quality.

Changes in the protein and fat in the food cause undesirable flavor and odor development. The loss of texture of fruits is associated with changes in the pectin.

Ionizing radiation can cause undesirable effects in food; however, other systems of preservation, especially heating, can change the organoleptic and nutrient factors in food. In many cases, the effect of radiation is less drastic than is that of heating.

Undesirable effects resulting from radiation are generally due to free radicals produced when ionizing radiations react with water in the food. Thus, the effects on quality can be reduced by irradiation in the frozen state or the dried state, by the use of free radical acceptors or by irradiation in inert atmosphere or in vacuum. Although these procedures will help reduce, they do not necessarily prevent changes in food.

Irradiation of white pepper, nutmeg and ginger at a level of 0.5 Mrad decreased the number of microorganisms and 1.58 Mrad sterilized these spices (Tjaberg *et al.* 1972). They found a slight increase in the amount of volatile constituents with increasing doses of radiation.

A reduction in the microbial contamination of spices and other seasoning ingredients to the levels required by the canning and meat industries can be achieved with doses of 400 to 500 Krad.

Wholesomeness

In the United States, as well as many other countries, statutes require that proof of safety for consumption (wholesomeness) must be adequate to convince the appropriate health-regulating officials before irradiation of foods will be approved. The factors considered are: nutritional adequacy, microbiological safety, absence of induced radioactivity and absence of carcinogenic or other toxic substances which may be formed by exposure of the food to ionizing radiations.

Ionizing radiation is legally defined as a food additive in the U.S. Under the provision of the Food, Drug and Cosmetic Act, the FDA must approve the use of ionizing radiation for each food subjected to this treatment. For meats and poultry, the USDA has legal responsibility.

The wholesomeness of irradiated food is determined by laboratory analysis of constituents in the food, as well as by using animal feeding studies. Among the considerations of these studies are the effects on growth, food consumption and efficiency, reproduction performance and lactation, life span (longevity), hematology, urology, organ weight, lesions or tumors, and histopathology. Also considered are mutagenesis, teratogenesis and toxicity of volatile radiolysis products.

The many wholesomeness studies have shown that irradiated foods are as wholesome and nutritious as thermally processed foods.

Nutritional Adequacy.—There is concern about the effect of radiation on the proteins, energy sources and vitamins. In general, the loss of any nutrient is dependent on the dose of radiation, as well as the environment during irradiation. The low doses of radiation used for sprout and insect control would be expected to cause insignificant nutritional damage. At high doses (4.5-5.5 Mrad) needed for sterilization, the nutritional adequacy of irradiated foods is comparable to conventional heat-processed foods.

Microbiological Safety.—With radappertized foods that are shelf stable at ambient temperatures, it is essential that all cells and spores of *C. botulinum* be destroyed. This is also true of thermally processed canned foods. In either case, residual, viable spores can result in a toxic food.

There is a possibility that food sterilized by radiation may contain or develop toxin after processing. The amount of radiation required to inactivate the toxin is greater than that needed to destroy the organisms.

The main concern is the potential hazard resulting from the use of low radiation doses to extend the refrigerated storage life of meat and fish. These low doses do not destroy *C. botulinum* spores. The type E strains, as well as some type A and B strains, can grow at low temperatures. The prolongation of storage life of the food increases the potential for toxin production. Due to the selective antimicrobial action by radiation, the typical spoilage pattern of the food is changed and might not be recognized by the consumer. This may lead to mishandling and a potential hazard.

Inasmuch as radiation sterilization does not have the many years of widespread distribution and consumer use as does thermally processed food, it is difficult to compare the potential hazards of the two treatments. In general, it is agreed that with proper controls throughout processing and distribution, irradiated foods will present no microbiological hazards to the consumer.

By virtue of eliminating organisms such as salmonellae, *S. aureus, Vibrio parahaemolyticus* and shigellae, irradiation can be used to reduce the number of foodborne illnesses.

Induced Radioactivity.—No radioactivity is induced when suitable radiation sources are used. Electron accelerators with energy levels above 10 million electron volts (Mev) may induce small, but significant, amounts of radioactivity. Therefore, this treatment is considered unsuitable for irradiation of food. Below 10 Mev no induced radioactivity can be detected. Great Britain limits the accelerator level to 5 Mev.

Irradiation of food with gamma rays from cobalt^{60}, even at high dose rates, induces less radioactivity than can be detected with the most sensitive instruments available. The subject of induced radioactivity is amply reviewed by Meyer (1965).

Carcinogens and Other Toxic Factors.—The effect of ionizing radiation on the chemical components in food and an estimation of the concentration of radiolytic products was discussed by Diehl and Scherz (1975). Numerous animal feeding studies have been used to determine the wholesomeness of irradiated foods. As they pointed out, these tests have revealed no toxic effects of foods irradiated below a dose of 5 Mrad.

Most of the radiolytic products that have been identified also occur naturally in a wide variety of foods. The concentration of radiolytic products is such that they are generally not considered to be of toxicological significance. Usually, exaggerated levels of chemicals are fed to animals. With irradiated foods, this is a difficult task. One approach is to use food irradiated at a higher dose than normal. However, the yield of radiolytic products is not always related to dosage.

An entire diet can be irradiated and tested for toxicity, but if an individual food is to be tested, there is a limit as to how much an animal will eat. Also, we must consider nutritional imbalance if one food is the main component of the diet. Regardless, these practices fall short of the 100-fold factor usually considered as necessary in the testing of chemical additives. Since radiation is considered as an additive by the USFDA, it must pass the tests considered necessary for chemical additives.

Animals fed irradiated foods have developed tumors. However, the incidence of tumors has been no higher with a diet of irradiated than unirradiated foods. Proof of the absence of carcinogens or other toxic substances in food due to irradiation is extremely difficult. No demonstrable carcinogenicity has been shown in the extensive tests that have been done.

Reports from India have questioned the safety of wheat irradiated for insect control. Very few of the early animal feeding studies considered mutagenic or teratogenic effects of irradiated food.

Rats fed for 12 weeks on diets containing wheat, irradiated 20 days or less before feeding, developed polyploid cells in their bone marrow (Vijayalaxmi and Sadasivan 1975). Polyploidy is a condition characterized by more than the normal number of chromosomes in cells. Rats maintained on low levels of dietary protein were more subject to chromosomal abnormalities than were well-fed rats. Vijayalaxmi (1975) found that polyploidy did not occur in rats fed irradiated wheat stored for 12 weeks after irradiation prior to the feeding trial.

Irradiated wheat was fed to 10 children and unirradiated wheat to 5 children suffering from protein-calorie malnutition (Bhaskaram and Sadasivan 1975). Those children consuming unirradiated wheat showed no polyploid cells while those on the irradiated diet did show polyploid cells. A lower incidence was obtained with stored wheat than with freshly irradiated wheat. When the irradiated wheat was withdrawn from the diet, there was a gradual reversal in the incidence of polyploid cells to the original basal level. The precise significance of polyploidy is not clear.

A mutagenic effect of dried food pellets irradiated with 3 Mrad was reported by Johnston-Arthur *et al.* (1975). They used the host mediated assay method with *Salmonella typhimurium* TA 1530 as the cell culture and Swiss albino mice as the treated mammal. There was no difference in mutation frequency between the control group and that fed 0.75 Mrad and irradiated food. When they used *S. typhimurium* G46, negative results were obtained.

Economic Aspects

Compared to the other methods of sterilization (heat or gas fumigation), the use of ionizing radiation is relatively expensive. The relative costs for processing meat were estimated by Wierbicki *et al.* (1965). With the continually increasing costs for natural gas, petroleum products and electricity, the relative cost for radiation might compare more favorably today than in 1965. With radappertization, the heating needed to inactivate enzymes adds to the expense of radiation treatment. To prevent unsatisfactory organoleptic characteristics, freezing of the food prior to treatment is recommended. The cost of treatment increases as the temperature is lowered. One must decide if increased quality is worth the added expense. The most favorable balance between quality loss and cost is to irradiate food at −30°C.

The irradiation cost per unit depends to a large extent upon the annual volume of an irradiation plant, the type of irradiation source (including its efficiency and price), the capital investment and the proximity of the processing plant to the food source.

The centralization of a food industry is necessary before irradiation can be accomplished at a reasonable cost per unit. On the other hand, the equipment needed for thermal processing or freezing is available for various sized operations. Only certain foods are accumulated in large volume in one location.

With most plant crops, irradiation would be difficult due to the widespread distribution and rather short period that the facility would be used each year. Although mobile units are available, the cost for these is higher than for larger stationary units.

There are some benefits of radiation treatments. No other processing system can eliminate salmonellae from frozen meat or poultry products. No one system will eliminate salmonellosis completely, but if 50% of the cases of foodborne illness could be eliminated by irradiation of certain foods, a considerable savings could be envisioned.

Some benefits that have been suggested for irradiation of food include the money saved by reducing food spoilage, the extension of storage life, a saving in transportation and storage costs and a reduction in the number of cases of foodborne illness. With irradiation, it has been suggested there would be a larger variety of foods essentially free of pathogens and parasites. The losses due to insects and sprouting would be reduced. There would be an increased availability of foods to undernourished peoples, expanded export potential and market stabilization. These advantages would be for both the processor and the consumer.

The energy requirements for radappertized and radurized foods should be compared with thermal processing, freezing, drying or other preservation methods. If radiation treatments save energy, they may be necessary in the future if we hope to maintain some semblance of our present standard of living. When all of the benefits are considered, for some foods they outweight radiation costs by a wide margin. However, for other foods, irradiation is not practical due to quality alterations or the cost involved.

Summary

No method of preservation has been subjected to the critical examination prior to use that has been given to ionization radiation. Even with all the research and testing, the irradiation of foods for destruction of microorganisms has not been approved in the U.S.

Preservation methods, including irradiation, will not improve the quality of a food, but they will tend to retain the attributes of high quality food. Irradiation has the potential to reduce the enormous loss of food so that more food will be available to the ever increasing population.

Radiation is not expected to have any substantial effect on the current systems of processing and distribution. However, radiation has the potential for reducing food loss due to sprouting, insects and microorganisms, of decreasing hazards, of increasing food quality and convenience, and of decreasing costs of storage, distribution and marketing.

Radiation will be of considerable importance in eliminating salmonellae from frozen meats and poultry products, as well as from bulk animal feeds. No other process we now have can accomplish this task. In 1976, the World Health Organization approved the radiation of poultry with doses up to 700 Krad. In Canada, levels up to 750 Krad are approved for controlling salmonellae on frozen poultry. The Netherlands has approved irradiated poultry with doses up to 300 Krad. It is expected that the USFDA and the USDA will approve the use of irradiation to eliminate salmonellae from poultry and red meats. At the present time, these foods are contaminating kitchens throughout the USA with salmonellae.

In the U.S., we have highly acceptable frozen, dried and thermally processed foods. However, in less developed areas in which these or other systems are not as adequate, radappertized foods have a greater importance. With the continued problem of energy sources, it may be that irradiation of food will be an important preservation method in the future of the human race, regardless of the country.

ULTRAVIOLET RADIATION

The antimicrobial effect of ultraviolet (UV) light has been known for about 100 years. The lethal action of UV light varies with the intensity of the light and the time of exposure. The temperature, pH, relative humidity and amount of microbial contamination influence the effectiveness of UV light.

Useful Aspects

The energy of UV light is much lower than that of ionizing radiations. This means that UV light has low penetrating power. Ultraviolet light is effective against microorganisms in air, liquids which are clear or in thin films and on surfaces that do not produce shadows. Microorganisms are protected by dust in the air, as well as by dust on the UV light bulbs. Food particles or other substances that UV cannot penetrate protect the microorganisms. Clumping of microbial cells allows those in the central part of the clump to escape the biocidal effect of UV light. Over 50% of the radiation energy is lost at a depth of 5 cm in clear water. Penetration into milk is only about one or two millimeters.

The suggested or actual uses of UV light include the control of microbial growth in bulk-stored maple sap, tenderizing, aging and retailing of meat, the removal of ethylene from banana storage areas, sterilization of packaging materials used in aseptic packaging, control of surface growth on bakery products, the treatment of seawater used for depuration of shellfish, preservation of reverse osmosis membranes, destruction of thermophiles in refined sugars intended for use in canned food, sanitation of equipment and air purification. The production of ozone by certain wavelengths of UV light gives an added germicidal effect.

A 99% reduction in microbial level was obtained in a thin film of apple cider exposed for 40 sec to UV irradiation (Harrington and Hills 1968). The treatment did not affect the flavor. A short exposure of a thin layer of sugar crystals to UV light causes the destruction of large numbers of spores of organisms that can cause spoilage of canned foods.

Relative Microbial Resistance

With some exceptions, UV irradiation is about equally effective against yeasts and either Gram positive or Gram negative bacteria. Certain species of *Micrococcus*, such as *M. radiodurans*, have exceptionally high resistance to UV as well as ionizing radiations (Lewis and Kumta 1972). This is believed to be due to a very efficient repair system. *Streptococcus lactis* appears to be more resistant than other streptococci, as well as the staphylococci. *Staphylococcus aureus*, salmonellae and shigellae have similar low resistances to UV light.

Bacterial spores are more resistant than the corresponding vegetative cells to UV radiation. Most mold spores have as high a resistance as bacterial spores, while some mold spores are ten times as resistant. Viruses vary in their resistance. Some are about as sensitive as vegetative bacterial cells, while others such as tobacco mosaic virus are 100 to 200 times as resistant.

It has been suggested that some molds are protected by fatty or waxy secretions that inhibit the penetration of UV radiations. The pigmentation of certain microorganisms reportedly has a protective effect.

Light rays can cause chemical changes and biological damage only if they are absorbed. Light that passes through a cell has no effect. Due to the high absorption of light with wavelengths between 210 and 300 nm, there is a strong biocidal effect. Several "most effective" wavelengths have been listed in the literature. The wavelength most often suggested is 253.7 nm; however, the wavelength with optimum biocidal effect varies with different microorganisms.

Ultraviolet radiations between 210 and 300 nm are absorbed by proteins and nucleic acids. As a result of chemical reactions, there is chromosome breakage, genetic mutation and enzyme inactivation which can result in cellular death. There is a variety of genetic alterations including base-pair substitution and frameshift mutations, deletions, mitotic crossing over and mitotic gene conversion induced by UV light (Hoffman and Morgan 1976).

Ultraviolet light produces potentially lethal photoproducts in the cellular DNA. A large portion of the bactericidal activity of UV results from nucleotide dimer formation. These dimers are the most abundant and stable of the photoproducts. They inhibit DNA synthesis and, to a lesser extent, RNA and protein synthesis.

Light in the near UV and visible regions (366 to 578 nm) has a variety of effects on cells. In general, little is known of the photochemistry that results in the destructive effects. Usually, this light has a negligible effect on survival. It seems to cause transitory inhibition of respiration, growth, DNA, RNA and protein synthesis and cell division.

With far UV light, the respiration, protein synthesis and RNA synthesis of cells proceed for about 60 min after which they cease for several hours (Swenson *et al.* 1975). Survival is reduced to about 0.5%.

Repair of Lesions

The irradiation of bacteria with ultraviolet light produces a variety of pyrimidine dimers in the DNA. The repair of these lesions can be accomplished by photoreactivation, excision (dark) repair or postreplication repair. Both photoreactivation and dark repair mechanisms have been described in a variety of microorganisms.

Photoreactivation is the reversal of short wavelength UV damage by postirradiation exposure to long wavelength light. Photoreactivation is enzymatic and functions by splitting the dimers in situ. The photoreactivation of damage caused by UV light clearly establishes a major role of pyrimidine dimers in the inactivation of various cells, especially radiation-sensitized mutants. The enzymes involved in this repair are called the photoreactivating enzymes.

Excision-repair of pyrimidine dimers is considered to be a sequence of enzymatic steps. This involves incision breaks, removal of the dimers, resynthesis of the excised regions and bonding between the newly synthesized and preexisting DNA to restore the continuity of the repaired DNA strand.

Strains deficient in excision repair are more sensitive to UV radiation than are repair-proficient strains.

At high fluences, lethality may be enhanced by damage to the excision and recombination repair systems (Webb and Brown 1976). Certain enzymes can be inactivated by light. The inactivation of photoreactivating enzymes by UV light has biological significance. One of the causes of death of cells is the disappearance of the enzyme responsible for the final step in the repair of DNA.

The rate of inhibition and death depends, at least partially, on the rate of repair. With large doses of radiant energy, the repair cannot maintain a balance and photoinhibition occurs. The occurrence of sigmoidal inactivation curves could be related to such effects.

The repair of DNA was reviewed by Friedberg (1975) and Prakash 1976).

Protective Agents

Plasmids may confer partial protection against the bactericidal effects of UV irradiation. Siccardi (1969) reported that of 31 plasmids, 15 protected, 11 had little or no effect and 5 increased the susceptibility of *E. coli* K-12 to UV irradiation. How these plasmids afford protection is not known. However, the authors offered suggestions as to the possible mechanisms for this occurrence.

Interactions

The lethal interaction of UV radiation and mild heat on *E. coli* was studied by Tyrrell (1976). Irradiation with most wavelengths sensitized the cells to temperatures of 45° or 48°C. He proposed that, in addition to DNA damage, both mild heating and near UV treatment interfere with DNA repair mechanisms. Hence, the combination of the two treatments leads to a strong positive interaction.

The simultaneous use of UV and ozone produced synergistic effects on molds but not on bacteria (Kaess and Weidemann 1973).

The interaction of UV light and ionizing radiation was studied by Schneider and Kiefer (1976). Submitting *Saccharomyces cerevisiae* to UV light followed by X-rays and also, in the reverse order, revealed differing reactions. With UV, then X-ray treatment, synergism was observed with all strains tested. With the reverse treatment, differences in synergism were obtained with different strains.

Other studies have revealed an X-ray-ultraviolet synergism for diploid yeast, certain strains of *E. coli, Aspergillus nidulans, Clostridium botulinum* and *Serratia marcescens.*

OTHER SYSTEMS

There are other systems that can inactivate microorganisms, but they have not been developed for commercial usage. This is probably because, at present, there are few, if any, practical applications for these processes.

Visible Light

The effects of visible light on microorganisms were reviewed by Harrison (1967). He listed the types of damage induced by light as loss of colony-forming ability, photolysis, lengthening of lag period, loss of fermentative ability and reduction of ability to support phage growth. The presence of carotenoid pigments in the bacteria sometimes enhances the survival of bacteria when exposed to light.

Light above 400 nm selectively inhibited certain enzymes and cellular processes of *E. coli* (D'Aoust *et al.* 1974). They suggested that visible light can be inhibitory or lethal to a wide variety of procaryotic or eucaryotic microorganisms.

The photodynamic alterations of bacteria are due to simultaneously exposing the organisms to light, a photosensitizing compound (usually a dye) and to air (oxygen).

Laser Energy

Pulsed laser energy effectively inhibited the growth of *E. coli* (Takahashi *et al.* 1975). Other research has indicated the requirement of a photosensitizing dye for inhibition of microorganisms to occur. However, Takahashi *et al.* (1975) found no difference in inhibition with or without added toluidine blue, and less inhibition when acridine orange was present. They stated that technology was just beginning to produce tunable laser irradiation powerful enough for microbiological research.

Ultrasonic Energy

Ultrasonics is the transmission of vibrational energy above the level of audible sound. When alternating current is applied to a crystal, the shape of the crystal changes with the electrical field. These continuous changes in shape cause pulsations or waves which travel through a liquid. These waves have alternate compressions and rarefactions. If the amplitude is high enough, a phenomenon known as cavitation is produced. This is the making and breaking of microscopic bubbles. These bubbles tend to become larger and when they grow to what is known as resonant size, they collapse instantly and violently, producing high local pressures. This pressure is the force that alters the microbial cells.

When microorganisms are subjected to ultrasonics, the overall result may be molecular bond breakage, physical damage by shock waves, denaturation, reactions with free radicals or a combination of these effects.

Microbial cells vary in their ease of ultrasonic disruption. Cells that are relatively easy to disrupt include *Pseudomonas, E. coli,* salmonellae and vegetative cells of *B. cereus.* Cells that are considered as difficult to disrupt include *Staphylococcus aureus,* streptococci, lactobacilli, *Serratia marcescens, B. cereus* spores, *Candida albicans* spores and poliovirus. Generally, the smaller and rounder the cell, the more difficult it is to disrupt. Spores are not greatly affected by ultrasonic treatment.

In a laboratory, suspensions of cells can be inactivated after long exposure to ultrasonics. However, commercially, this process seldom provides a 100% kill of microorganisms. Reportedly, another potential use of ultrasonics is the inactivation of toxins.

The combination of ultrasonics and UV radiation acts synergistically to destroy microbial cells. Synergism has been noted with chemicals, such as chlorine or hydrogen peroxide and ultrasonic treatments (Ahmed and Russell 1975). The time required for the fumigant gases ethylene oxide or propylene oxide to kill microorganisms is reduced by simultaneous treatment with ultrasonic waves.

Ultrasonic treatment lowers the resistance of *M. radiodurans* and *S. faecalis* to ionizing radiation. According to Burgos *et el.* (1972), the heat resistance of *B. cereus* and *B. licheniformis* spores decreased markedly after ultrasound treatment. Ordonez and Burgos (1976) reported that ultrasonic waves do not affect the heat resistance of some strains of *Bacillus* species.

Electricity

Stersky *et al.* (1970) used various voltages to inactivate microorganisms aerosolized into a chamber. As the voltage increased from 6,000 volts to 20,000 volts, the reduction of the bacterial population increased. The highest mean reduction they reported was 59.2%.

Edebo (1969) found the microbicidal effect of transient electric arcs on *E. coli* in tap water was similar to that of UV radiation, both biologically and chemically.

BIBLIOGRAPHY

ADAMS, D.M. 1974. Requirement for and sensitivity to lysozyme by *Clostridium perfringens* spores heated at ultrahigh temperatures. Appl. Microbiol. *27*, 797-801.

AHMED, F.I.K., and RUSSELL, C. 1975. Synergism between ultrasonic waves and hydrogen peroxide in the killing of microorganisms. J. Appl. Bacteriol. *39*, 31-40.

ANDERSON, A.W. *et al.* 1956. Studies on a radio-resistant *Micrococcus*. 1. Isolation, morphology, cultural characteristics, and resistance to gamma radiation. Food Technol. *10*, 575-578.

ANELLIS, A., BERKOWITZ, D., JARBOE, C., and EL-BISI, H.M. 1967. Radiation sterilization of prototype military foods. II. Cured ham. Appl. Microbiol. *15*, 166-177.

ANELLIS A., BERKOWITZ, D., JARBOE, C., and EL-BISI, H.M. 1969. Radiation sterilization of prototype military foods. III. Pork loins. Appl. Microbiol. *18*, 604-611.

ANELLIS, A., BERKOWITZ, D., SWANTAK, W., and STROJAN, C. 1972. Radiation sterilization of prototype military foods: low-temperature irradiation of codfish cake, corned beef, and pork sausage. Appl. Microbiol. *24*, 453-462.

ANELLIS, A. *et al.* 1965. Radiation sterilization of bacon for military feeding. Appl. Microbiol. *13*, 37-42.

ANELLIS, A. *et al.* 1975. Low-temperature irradiation of beef and methods for evaluation of a radappertization process. Appl. Microbiol. *30*, 811-820.

AUDA, H., and EMBORG, C. 1973. Studies on post-irradiation DNA degradation in *Micrococcus radiodurans*, strain $R_{11}5$. Radiat. Res. *53*, 273-280.

BELL, T.A., TURNEY, L.J., and ETCHELLS, J.L. 1972. Influence of different organic acids on the firmness of fresh-pack pickles. J. Food Sci. *37*, 446-449.

BERCK, B. 1965. Fumigation of cereals and cereal products-research and practice. Cereal Sci. Today *10*, No. 4, 112-117.

BHASKARAM, C., and SADASIVAN, G. 1975. Effects of feeding irradiated wheat to malnourished children. Amer. J. Clin. Nutr. *28*, 130-135.

BLANCHARD, R.O., and HANLIN, R.T. 1973. Effect of propylene oxide treatment on the microflora of pecans. Appl. Microbiol. *26*, 768-772.

BLANKENSHIP, L.C. 1978. Survival of a *Salmonella typhimurium* experimental contaminant during cooking of beef roasts. Appl. Environ. Microbiol. *35*, 1160-1165.

BOLIN, H.R., KING, A.D., JR., STANLEY, W.L., and JURD, L. 1972. Antimicrobial protection of moisturized Deglet Noor dates. Appl. Microbiol. *23*, 799-802.

BUCHANAN, R.E., and GIBBONS, N.E. 1974. Bergey's Manual of Determinative Bacteriology, 8th Edition. Williams and Wilkins Co., Baltimore, Md.

BURGOS, J., ORDÔÑEZ, J.A., and SALA, F. 1972. Effect of ultrasonic waves on the heat resistance of *Bacillus cereus* and *Bacillus licheniformis* spores. Appl. Microbiol. *24*, 497-498.

CERF, O. 1977. Tailing of survival curves of bacterial spores. J. Appl. Bacteriol. *42*, 1-19.

CLARK, J.P. 1978. Mathematical modeling in sterilization processes. Food Technol. *31*, No. 3, 73-75.

CLIVER, D.A. 1971. Transmission of viruses through foods. Crit. Rev. Environ. Control *1*, 551-579.

COCKEY, R.R., and TATRO, M.C. 1974. Survival studies with spores of *Clostridium botulinum* type E in pasteurized meat of the blue crab *Callinectes sapidus*. Appl. Microbiol. *27*, 629-633.

COOK, A.M., and GILBERT, R.J. 1968. Factors affecting the heat resistance of *Bacillus stearothermophilus* spores. II. The effect of sporulating conditions and nature of the heating medium. J. Food Technol. *3*, 295-302.

COOK, A.M., and GILBERT, R.J. 1969. The effect of sodium chloride on heat resistance and recovery of heated spores of *Bacillus stearothermophilus*. J. Appl. Bacteriol. *32*, 96-102.

CORRY, J.E.L. 1974. The effect of sugars and polyols on the heat resistance of salmonellae. J. Appl. Bacteriol. *37*, 31-43.

CORRY, J.E.L. 1976A. The effect of sugars and polyols on the heat resistance and morphology of osmophilic yeasts. J. Appl. Bacteriol. *40*, 269-276.

CORRY, J.E.L. 1976B. Sugar and polyol permeability of *Salmonella* and osmophilic yeast cell membranes measured by turbidimetry, and its relation to heat resistance. J. Appl. Bacteriol. *40*, 277-284.

CRISLEY, F.D., PEELER, J.T., ANGELOTTI, R., and HALL, H.E. 1968. Thermal resistance of spores of five strains of *Clostridium botulinum* type E in ground whitefish chubs. J. Food Sci. *33*, 411-416.

CUNNINHGAM, F.E., and LINEWEAVER, H. 1965. Stabilization of egg-white proteins to pasteurization temperatures above 60°C. Food Technol. *19*, No. 9, 136-141.

D'AOUST, J.Y. *et al.* 1974. Some effects of visible light on *Escherichia coli.* J. Bacteriol. *120*, 799-804.

DENNY, C.B. 1972. Collaborative study of a method for the determination of commercial sterility of low-acid canned foods. J. Assoc. Offic. Anal. Chem. *55*, 613-616.

DESROSIER, N.W., and DESROSIER, J.N. 1977. The Technology of Food Preservation, 4th Edition. AVI Publishing Co., Westport, CT.

DIEHL, J.F., and SCHERZ, H. 1975. Estimation of radiolytic products as a basis for evaluating the wholesomeness of irradiated foods. Int. J. Appl. Radiat. Isotop. *26*, 499-507.

DIGIOIA, G.A., LICCIARDELLO, J.J., NICKERSON, J.T.R., and GOLDBLITH, S.A. 1970. Thermal inactivation of Newcastle disease virus. Appl. Microbiol. *19*, 451-454.

DIGIROLAMO, R., LISTON, J., and MATCHES, J.R. 1970. Survival of virus in chilled, frozen and processed oysters. Appl. Microbiol. *20*, 58-63.

DOYLE, M.P., and MARTH, E.H. 1975. Thermal inactivation of conidia from *Aspergillus flavus* and *Aspergillus parasiticus.* 1. Effects of moist heat, age of conidia, and sporulation medium. J. Milk Food Technol. *38*, 678-682.

EDEBO, L. 1969. Production of photons in the bactericidal effect of transient electric arcs in aqueous systems. Appl. Microbiol. *17*, 48-53.

EMBREE, J.W., LYON, J.P., and HINE, C.H. 1977. The mutagenic potential of ethylene oxide using the dominant-lethal assay in rats. Toxicol. Appl. Pharmacol. *40*, 261-267.

ERNST, R.B. 1974. Ethylene oxide gaseous sterilization. *In* International Symposium on Sterilization and Sterility Testing of Biological Substances. R.H. Regamey, F.P. Gallardo and W. Hennessen (Editors). S. Karger, Basel.

EVANCHO, G.M., ASHTON, D.H., and BRISKEY, E.J. 1973. Conditions necessary for sterility testing of heat processed canned foods. J. Food Sci. *38*, 185-188.

FARRALL, A.W. 1976. Food Engineering Systems, Vol. 1. Operations. AVI Publishing Co., Westport, CT.

FDA. 1977. Code of Federal Regulations. Title 21. Food and Drugs. Food and Drug Administration, Government Services Administration, Washington, D.C.

FERNANDEZ, E., TANG, T., and GRECZ, N. 1969. Toxicity of spores of *Clostridium botulinum* 33A in irradiated ground beef. J. Gen. Microbiol. *56*, 15-21.

FILPPI, J.R. 1973. Thermal characterization of poliovirus type 1 in ground beef containing three levels of fat. Ph.D. Thesis. The Ohio State University, Columbus.

FILPPI, J.A., and BANWART, G.J. 1974. Effect of fat content of ground beef on the heat inactivation of poliovirus. J. Food Sci. *39*, 865-868.

FLOWERS, R.S., and ADAMS, D.M. 1976. Spore membrane(s) as the site of damage within heated *Clostridium perfringens* spores. J. Bacteriol. *125*, 429-434.

FRIEDBERG, E.C. 1975. DNA repair of ultraviolet-irradiated bacteriophage T4. Photochem. Photobiol. *21*, 277-289.

GOLDBLITH, S.A. 1971. A condensed history of the science and technology of thermal processing—part 1. Food Technol. *25*, 1256-1262.

GOLDBLITH, S.A. 1972. A condensed history of the science and technology of thermal processing—part 2. Food Technol. *26*, No. 1, 64-69.

GOULD, G.W. 1977. Recent advances in the understanding of resistance and dormancy in bacterial spores. J. Appl. Bacteriol. *42*, 297-309.

GOULD, G.W., and DRING, G.J. 1974. Mechanisms of spore heat resistance. Adv. Microbial Physiol. *11*, 137-164.

GOULD, G.W., and DRING, G.J. 1975A. Heat resistance of bacterial endospores and concept of an expanded osmoregulatory cortex. Nature *258*, 402-405.

GOULD, G.W., and DRING, G.J. 1975B. Role of an expanded cortex in resistance of bacterial endospores. Spores *6*, 541-546.

GRAY, R.J.H., WITTER, L.D., and ORDAL, Z.J. 1973. Characterization of mild thermal stress in *Pseudomonas fluorescens* and its repair. Appl. Microbiol. *26*, 78-85.

GRECZ, N., WALKER, A.A., ANELLIS, A., and BERKOWITZ, D. 1971. Effect of irradiation temperature in the range −196° to 95°C on the resistance of spores of *Clostridium botulinum* 33A in cooked beef. Can. J. Microbiol. *17*, 135-142.

HAN, Y.W., ZHANG, H.I., and KROCHTA, J.M. 1976. Death rates of bacterial spores: mathematical models. Can. J. Microbiol. *22*, 295-300.

HÄRNULV, B.G., JOHANSSON, M., and SNYGG, B.G. 1977. Heat resistance of *Bacillus stearothermophilus* spores at different water activities. J. Food Sci. *42*, 91-93.

HARRINGTON, W.O., and HILLS, C.H. 1968. Reduction of the microbial population of apple cider by ultraviolet irradiation. Food Technol. *22*, 1451-1454.

HARRISON, A.P., JR. 1967. Survival of bacteria. Harmful effects of light, with some comparisons with other adverse physical agents. Annu. Rev. Microbiol. *21*, 143-156.

HASHIMOTO, T., FRIEBEN, W.R., and CONTI, S.F. 1972. Kinetics of germination of heat-injured *Bacillus cereus* spores. Spores 5, 409-415.

HAYAKAWA, K. 1978. A critical review of mathematical procedures for determining proper heat sterilization processes. Food Technol. *32*, No. 3, 59-65.

HOFFMANN, G.R., and MORGAN, R.W. 1976. The effect of ultraviolet light on the frequency of a genetic duplication in *Salmonella typhimurium*. Radiat. Res. *67*, 114-119.

HSIEH, F., ACOTT, K., and LABUZA, T.P. 1976. Death kinetics of pathogens in a pasta product. J. Food Sci. *41*, 516-519.

ITO, H., and IIZUKA, H. 1971. Taxonomic studies on a radio-resistant *Pseudomonas*. Agr. Biol. Chem. *35*, 1566-1571.

ITO, H., IIZUKA, H., and SATO, T. 1974. A new radio-resistant yeast of *Trichosporon oryzae* nov. sp. isolated from rice. Agr. Biol. Chem. *38*, 1597-1602.

ITO, H., SATO, H., and IIZUKA, H. 1976. Study of the intermediate type of *Moraxella* and *Acinetobacter* occurring in radurized Vienna sausages. Agr. Biol. Chem. *40*, 867-873.

JANKY, D.M., and OBLINGER, J.L. 1976. Microwave *versus* water-bath precooking of turkey rolls. Poultry Sci. *55*, 1549-1553.

JOHNSTON-ARTHUR, T. *et al.* 1975. Mutagenicity of irradiated food in the host mediated assay system. Stud. Biophys. *50*, 137-141.

JOSEPHSON, E. 1969. Nuclear applications in the food industry. Proc. 44th Annu. Conf. Am. Industrial Develop. Counc., 134-143.

KAESS, G., and WEIDEMANN, J.F. 1973. Effects of ultraviolet irradiation on the growth of micro-organisms on chilled beef sides. J. Food Technol. *8*, 59-69.

KERELUK, K., GAMMON, R.A., and LLOYD, R.S. 1970A. Microbiological aspects of ethylene oxide sterilization. II. Microbial resistance to ethylene oxide. Appl. Microbiol. *19*, 152-156.

KERELUK, K., GAMMON, R.A., and LLOYD, R.S. 1970B. Microbiological aspects of ethylene oxide sterilization. III. Effects of humidity and water activity on the sporicidal activity of ethylene oxide. Appl. Microbiol. *19*, 157-162.

KOPELMAN, M., MARKAKIS, P., SCHWEIGERT, B.S. 1968. Effect of ionizing radiations on resting conidia of *Aspergillus flavus*. J. Food Sci. *32*, 694-696.

KUZMINSKI, L.N., HOWARD, G.L., and STUMBO, C.R. 1969. Thermochemical effects influencing the death kinetics of spores of *Clostridium botulinum* 62A. J. Food Sci. *34*, 561-567.

LEE, J.S., and PFEIFER, D.K. 1975. Microbiological characteristics of Dungeness crab (*Cancer magister*). Appl. Microbiol. *30*, 72-78.

LEWIS, N.F. 1971. Studies on a radio-resistant coccus isolated from Bombay duck (*Harpodon nehereus*). J. Gen. Microbiol. *66*, 29-35.

LEWIS, N.F., and KUMTA, U.S. 1972. Evidence for extreme UV resistance of *Micrococcus* sp. NCTC 10785. Biochem. Biophys. Res. Commun. *47*, 1100-1105.

LOSTY, T., ROTH, J.S., and SHULTS, G. 1973. Effect of γ-irradiation and heating on proteolytic activity of meat samples. J. Agr. Food Chem. *21*, 275-277.

LUND, D.B. 1978. Statistical analysis of thermal process calculations. Food Technol. *32*, No. 3, 76-78, 83.

LYNT, R.K., SOLOMON, H.M., LILLY, T., JR., and KAUTTER, D.A. 1977. Thermal death time of *Clostridium botulinum* type E in meat of the blue crab. J. Food Sci. *42*, 1022-1025, 1037.

MERSON, R.L., SINGH, R.P., and CARROAD, P.A. 1978. An evaluation of Ball's formula method of thermal process calculations. Food Technol. *32*, No. 3, 66-72, 75.

MEYER, R.A. 1965. Induced Radioactivity in Food and Electron Sterilization. Technical Report *FD-6*. U.S. Army Natick Laboratories, Natick, Mass.

MIURA, T., SAKAGUCHI, S., SAKAGUCHI, G., and MIYAKI, K. 1970. Radiosensitivity of type E botulinum toxin and its protection by proteins, nucleic acids, and some related substances. *In* Proceedings of the First U.S.-Japan Conference on Toxic Micro-organisms. M. Herzberg (Editor). U.S. Department of the Interior, Washington, D.C.

MURRELL, W.G., and SCOTT, W.J. 1966. The heat resistance of bacterial spores at various water activities. J. Gen. Microbiol. *43*, 411-425.

NATIONAL CANNERS ASSOCIATION RESEARCH LABORATORIES. 1968. Laboratory Manual for Food Canners and Processors, Vol. 1. Microbiology and Processing. AVI Publishing Co., Westport, CT.

NICKERSON, J.T.R., and RONSIVALLI, L.J. 1979. Elementary Food Science, 2nd Edition. AVI Publishing Co., Westport, CT.

ODLAUG, T.E., and PFLUG, I.J. 1977. Thermal destruction of *Clostridium botulinum* spores suspended in tomato juice in aluminum thermal death time tubes. Appl. Environ. Microbiol. *34*, 23-29.

ORDONEZ, J.A., and BURGOS, J. 1976. Effect of ultrasonic waves on the heat resistance of *Bacillus* spores. Appl. Environ. Microbiol. *32*, 183-184.

PRAKASH, L. 1976. The relation between repair of DNA and radiation and chemical mutagenesis in *Saccharomyces cerevisiae*. Mutat. Res. *41*, 241-248.

READ, R.B., JR., and BRADSHAW, J.G. 1967. γ-Irradiation of staphylococcal enterotoxin B. Appl. Microbiol. *15*, 603-605.

READ, R.B., JR., SCWHARTZ, C., and LITSKY, W. 1961. Studies on thermal destruction of *Escherichia coli* in milk and milk products. Appl. Microbiol. *9*, 415-418.

ROBERTS, T.A. 1968. Resistance of spores of *Clostridium welchii*. *In* Elimination of Harmful Organisms from Food and Feed by Irradiation. International Atomic Energy Agency, Vienna.

ROSEN, C. 1972. Effects of microwaves on food and related materials. Food Technol. *26*, No. 7, 36-40, 55.

ROSS, E.W., JR. 1976. On the statistical analysis of inoculated packs. Food Sci. *41*, 578-584.

ROWLEY, D.B., ANELLIS, A., WIERBICKI, E., and BAKER, A.W. 1974. Status of the radappertization of meats. J. Milk Food Technol. *37*, 86-93.

RUSSELL, A.D., and HARRIES, D. 1968. Damage to *Escherichia coli* on exposure to moist heat. Appl. Microbiol. *16*, 1394-1399.

SAVAGAON, K.A. *et al.* 1972. Radiation preservation of tropical shrimp for ambient temperature storage. 1. Development of a heat-radiation combination process. J. Food Sci. *37*, 148-150.

SCALZO, A.M., DICKERSON, R.W., JR., READ, R.B., JR., and PARKER, R.W. 1969. Residence times of egg products in holding tubes of egg pasteurizers. Food Technol. *23*, 678-681.

SCHADE, J.E., and KING, A.D., JR. 1977. Methyl bromide as a microbicidal fumigant for tree nuts. Appl. Environ. Microbiol. *33*, 1184-1191.

SCHEIE, P., and EHRENSPECK, S. 1973. Large surface blebs on *Escherichia coli* heated to inactivating temperatures. J. Bacteriol. *114*, 814-818.

SCHNEIDER, E., and KIEFER, J. 1976. Interaction of ionizing radiation and ultraviolet-light in diploid yeast strains of different sensitivity. Photochem. Photobiol. *24*, 573-578.

SEGNER, W.P. and SCHMIDT, C.F. 1970. Inoculated pack studies on *Clostridium botulinum* type E in unirradiated and irradiated haddock. *In* Proc. First U.S.-Japan Conference on Toxic Micro-organisms. M. Herzberg (Editor). U.S. Department of the Interior, Washington, D.C.

SHAFI, R., COTTERILL, O.J., and NICHOLS, M.L. 1970. Microbial flora of commercially pasteurized egg products. Poultry Sci. *49*, 578-585.

SHIGA, I. 1976. A new method of estimating thermal process time for a given F value. J. Food Sci. *41*, 461-462.

SICCARDI, A.G. 1969. Effect of R factors and other plasmids on ultraviolet susceptibility and host cell reactivation property of *Escherichia coli*. J. Bacteriol. *100*, 337-346.

STERSKY, A., HELDMAN, D.R., and HEDRICK, T.I. 1970. The effect of a bipolar-oriented electrical field on microorganisms in air. J. Milk Food Technol. *33*, 545-549.

STUMBO, C.R., PUROHIT, K.S., and RAMAKRISHNAN, T.V. 1975. Thermal process lethality guide for low-acid foods in metal containers. J. Food Sci. *40*, 1316-1323.

SULLIVAN, R., MARNELL, R.M., LARKIN, E.P., and READ, R.B., JR. 1975. Inactivation of poliovirus 1 and coxsackievirus B-2 in broiled hamburgers. J. Milk Food Technol. *38*, 473-475.

SULLIVAN, R. *et. al.* 1971A. Inactivation of thirty viruses by gamma radiation. Appl. Microbiol. *22*, 61-65.

SULLIVAN, R. *et al.* 1971B. Thermal resistance of certain oncogenic viruses suspended in milk and milk products. Appl. Microbiol. *22*, 315-320.

SULLIVAN, R. *et al.* 1973. Gamma radiation inactivation of coxsackievirus B-2. Appl. Microbiol. *26*, 14-17.

SWENSON, P.A., IVES, J.E., and SCHENLEY, R.L. 1975. Photoprotection of *E. coli* B/γ: respiration, growth, macromolecular synthesis and repair of DNA. Photchem. Photobiol. *21*, 235-241.

TAKAHASHI, P.K. *et al.* 1975. Irradiation of *Escherichia coli* in the visible spectrum with a tunable orange-dye laser energy source. Appl. Microbiol. *29*, 63-67.

TILTON, E.W., BROWER, J.H., and COGBURN, R.R. 1974. Insect control in wheat flour with gamma irradiation. Int. J. Appl. Radiat. Isotop. *25*, 301-305.

TJABERG, T.B., UNDERDAL, B., and LUNDE, G. 1972. The effect of ionizing radiation on the microbiological content and volatile constituents of spices. J. Appl. Bacteriol. *35*, 473-478.

TRESSLER, D.K., VAN ARSDEL, W.B., and COPLEY, M.J. 1968. The Freezing Preservation of Foods, Vol. 3. Commercial Food Freezing Operations—Fresh Foods. AVI Publishing Co., Westport, CT.

TULIS, J.J., FOGARTY, M.G., and SLIGER, J.L. 1973. Thermoradiation as a sterilization method. Develop. Ind. Microbiol. *14*, 49-56.

TYRRELL, R.M. 1976. Synergistic lethal action of ultraviolet-violet radiations and mild heat in *Escherichia coli.* Photochem. Photobiol. *24*, 345-351.

USDA. 1969. Egg Pasteurization Manual. Publication No. *ARS 74-48.* U.S. Department of Agriculture, Albany, California.

VIJAYALAXMI. 1975. Cytogenic studies in rats fed irradiated wheat. Int. J. Radiat. Biol. *27*, 283-285.

VIJAYALAXMI, and SADASIVAN, G. 1975. Chromosomal aberrations in rats fed irradiated wheat. Int. J. Radiat. Biol. *27*, 135-142.

WARD, D.R., PIERSON, M.D., and VAN TASSELL, K.R. 1977. The microflora of unpasteurized and pasteurized crabmeat. J. Food Sci. *42*, 597-600, 614.

WEBB, R.B., and BROWN, M.S. 1976. Sensitivity of strains of *Escherichia coli* differing in repair capability to far UV, near UV and visible radiations. Photochem. Photobiol. *24*, 425-432.

WELCH, A.B., and MAXCY, R.B. 1975. Characterization of radiation-resistant vegetative bacteria in beef. Appl. Microbiol. *30*, 242-250.

WESLEY, F., ROURKE, B., and DARBISHIRE, O. 1965. The formation of persistent toxic chlorohydrins in foodstuffs by fumigation with ethylene oxide and propylene oxide. J. Food Sci. *30*, 1037-1042.

WHITE, J.D. 1977. Standard aeration for gas-sterilized plastics. J. Hyg. Camb. *79*, 225-232.

WIERBICKI, E., and HEILIGMAN, F. 1973. Shelf stable cured ham with low nitrite-nitrate additions preserved by radappertization. International Symposium "Nitrite in Meat Products." Zeist, The Netherlands, Sept. 11-14.

WIERBICKI, E., SIMON, M., and JOSEPHSON, E.S. 1965. Preservation of meats by sterilizing doses of ionizing radiation. *In* Radiation Preservation of Foods. Publication *1273*. National Academy of Sciences. National Research Council, Washington, D.C.

WINARNO, F.G., and STUMBO, C.R. 1971. Mode of action of ethylene oxide on spores of *Clostridium botulinum* 62A. J. Food Sci. *36*, 892-895.

XEZONES, H., SEGMILLER, J.L., and HUTCHINGS, I.J. 1965. Processing requirements for a heat-tolerant anaerobe. Food Technol. *19*, 1001-1002.

13

Regulations and Standards

NEED FOR REGULATIONS

When the United States began to change from a rural to an urban society, food science was nonexistent, microbiology was an infant and little was known of the chemistry of foods. The people in the food business were not trained in the need for sanitation. The adulteration of food was common.

As people became aware of the problems in their food supply, they demanded some action. In 1906, Congress passed the Pure Food and Drugs Act, as well as the Meat Inspection Act.

Some regulations are needed to eliminate unfair competition. For example, a process such as pasteurization will help control a health hazard, but increases the cost of production. If every processor is not required to pasteurize a particular product, it is difficult for those willing to do it to compete. The inspection of meat and poultry, with elimination of diseased animals, is a cost that an unscrupulous person would not be willing to pay without government regulations.

PURPOSE OF REGULATIONS

The basic purpose of food laws and food regulatory agencies is to ensure that all foods reaching the consumer are safe, wholesome and are truthfully labeled. The production and marketing of wholesome foods should be important to everyone affiliated with the food industry.

There is general agreement that as consumers, we want clean, wholesome food that is produced and processed in a sound, sanitary manner and is free of microbial pathogens, toxins or harmful chemicals. Some consumer advocates demand a 100% assurance of safety of all consumer goods. There is no way that absolute safety of food or any other commodity can be proven. A product should be made as carefully as possible and then be considered to be safe if the risk is at an acceptable level.

Morrison (1976) defined risk as a probability of occurrence of a deleterious effect in man. Risk can be estimated by scientific experimentation. Morrison (1976) defined acceptability as the benefit/cost ratio. Benefits include convenience, economic gain, increased nutrition, enhancing the quality of life or relief of suffering. The benefits must be important to the consumer, not merely to the manufacturer. The cost is estimated by multiplying the risk times loss. Losses include economic loss and human suffering. These are related to any illness that might occur, the type of illness, duration, possible treatments and other considerations. As Morrison pointed out, peoples' knowledge about risk is too often based on obviously biased and prejudiced sources, seeking to make a case rather than to present the facts.

Since we have an abundant supply of food in the United States, the present demands for food safety overshadow the hunger in many other countries. A hungry person is more interested in satisfying his appetite than in the esthetic aspects of the food he eats. If the predicted population and resultant famines occur, certain regulations and requirements may need to be relaxed to increase the food supply. However, with the resultant high prices associated with scarce commodities, the unscrupulous characters may take over, making regulations even more important.

FOOD LAWS

The enforcement of the Pure Food and Drugs Act and Meat Inspection Act of 1906 was delegated to the Secretary of Agriculture in the United States Department of Agriculture.

The Pure Food and Drugs Act made illegal the adulteration of food entering into interstate commerce. The individual states maintained the sole authority to regulate intrastate production and marketing of food.

The enforcement of the law improved the quality of the food supply. However, as advances were made in technology, it became apparent that a major revision was needed. In 1933, a bill was introduced in Congress and, after many revisions, was passed in 1938. It became the Food, Drug and Cosmetic Act. This act is still the basic legislation, although it has been amended several times. The law applies to exports, imports, commerce between states, in the District of Columbia and the territories. The act prohibits adulteration of food and requires certain information on the food label. It authorizes definitions or standards for food. The specific standards are the standards of identity, standards of quality and standards of fill of container.

Among the amendments to the act, the Miller Pesticide Amendment of 1954 allows the establishment of tolerances of residues of pesticides on raw agricultural commodities shipped in interstate commerce.

The Hale Amendment of 1954 simplifies procedures for establishing food standards. The Factory-Inspection Amendment of 1954 provides that inspections of processing plants and other establishments are authorized upon presentation of credentials and a written notice to the owner, operator or agent in charge.

The Food Additives Amendment of 1958 requires proof of safety before a chemical additive can be used in food. It allows the use of substances which are safe at the levels of intended use. This amendment put the burden of proof of the safety of additives on the chemical and food industries rather than the regulatory agencies. The Delaney Clause in this amendment forbids the use of any substance in any amount if it is found to induce cancer in man or animals. This clause has resulted in much controversy.

The Fair Packaging and Labeling Act of 1966 requires the use of retail packages that are not deceptive. The label must contain certain meaningful information about the value of the food and the ingredients. The name and address of the manufacturer or distributor are required.

The Meat Inspection Act of 1906 requires the inspection of meat processing plants that ship products in interstate commerce. The Wholesome Meat Act of 1967 requires that state inspection services for intrastate meat processors have regulations and requirements comparable to those of the federal regulatory agency.

Prior to 1959, poultry was inspected on a voluntary basis with the poultry industry paying for the service. The Poultry Products Inspection Act of 1959 made poultry inspection mandatory for any processor that ships products in interstate commerce. The Wholesome Poultry Act of 1968 requires comparable regulations for intrastate processors.

In the past, foods not in interstate commerce were not subject to federal laws and regulations. However, this was changed with the Wholesome Meat Act and the Wholesome Poultry Act. The food services on ships, trains, airplanes, buses and on interstate highways are subject to federal inspection. Many food service facilities are national chains, some using foods prepared in central kitchens and shipped across state lines. The next step is the inspection of local cafes and restaurants.

Besides the laws directly aimed at the food industry, there are many other federal laws and agencies that regulate and affect our food supply. There is no limit to the number of laws and agencies that can be created to regulate every segment of our life, including foods.

Besides all of the laws and regulations imposed at the national level, the food processor must comply with city, county and state laws and regulations. For those engaged in exporting and importing, there are international considerations. Through organizations such as the Codex Alimentarius Commission and the International Commission of Microbiological

Specifications for Foods, foods are being standardized at the international level. There are some 50 international organizations that are directly involved with food.

The purpose of the Codex Alimentarius Commission is to protect the health of consumers and ensure fair practices. The primary means of accomplishing this is through establishment of food standards and codes of practice which, hopefully, will be adopted by cooperating countries.

THE ENFORCERS

There are a number of government agencies that are involved in the enforcement of laws and regulations that affect the food industry either directly or indirectly. Every year, these agencies, as well as their rules and regulations, expand. There is no evidence that this trend will be halted in the near future. Due to expansions, reorganizations and empire building, the available information is subject to change. According to Sloan (1976), there are 50 regulatory agencies in the federal government, many of which exert influence or some control on the food industry.

Food and Drug Administration (FDA)

This is the agency generally recognized as the main enforcer of the regulations concerning food. The FDA is part of the Consumer Protection and Environmental Health Service (CPEHS), which is located in the Department of Health, Education and Welfare (HEW).

The FDA enforces the Food, Drug and Cosmetic (FD&C) Act, as amended. It is responsible for regulating all foods except red meats, poultry and eggs. These foods are regulated by the USDA.

The FDA also administers the Fair Packaging and Labeling Act as related to foods and household products.

The FDA devotes part of its effort to the sanitation of milk processing, shellfish, restaurants and interstate travel facilities included in the Public Health Service Act.

FDA Activities.—The food processor has the prime responsibility for the safety, wholesomeness and nutritional quality of his food. The role of the FDA is to see that industry is meeting its responsibility.

About 40% of the FDA budget is allocated to food programs. In these food activities, most of the effort is devoted to food safety and the remainder to economic aspects. The food safety programs include food sanitation control, chemical contaminants, mycotoxins, food and color additives, and nutrition. The economic aspects include food standards and controlling unfair competition due to economic cheating.

In food sanitation control, there are efforts to reduce and control the incidence of foodborne illness due to salmonellae, the toxins of *S. aureus* and *C. botulinum*, as well as other pathogenic and toxic agents. The food sanitation and control program is responsible for the development and enforcement of the good manufacturing practice regulations.

The FDA periodically inspects processing and storage plants to determine if they are sanitary. The inspectors check the wholesomeness of ingredients and finished food products and the legality of packages and labels.

It is quite readily admitted that it is impossible for the FDA to regulate the entire food industry. Hence, the agency must depend upon self-regulation by a responsible and responsive industry. To facilitate this, the FDA has established two projects. These are the quality assurance self-certification project and the quality assurance systems development project. Each processor is required to have an acceptable quality assurance program, and keep records to show that the program is followed and that it is effective. All records are subject to observation by representatives of the FDA.

The Consumer Food Act requires each food manufacturing establishment to develop and provide to the FDA a safety assurance program.

Another program of the FDA is to maintain surveillance on commodities which have demonstrated a potential for mycotoxin contamination. Products that are contaminated above acceptable levels are removed from the market.

The FDA enforces safe levels of use for chemical additives. Standards are established for classes of foods and all such products must then meet the minimum requirements.

Products Not in Compliance.—When a product that is not in compliance with the various laws and regulations is discovered, the FDA has certain courses of action. Among the possible actions it might take are recall, seizure, injunction and prosecution.

Recall.—Some recalls begin when the firm finds a problem, others are conducted at FDA's request. Recalls may involve the physical removal of products from the market or correction of the problem where the product is located.

The recall of a product from the market has become the most effective method of removing all units found to be adulterated, misbranded, dangerous to health, grossly fraudulent or deceptive or materially misleading to the detriment of the consumers' health and welfare. The food industry can recall a product more efficiently and effectively than the FDA. If the firm refuses to recall the product, or the recall is not effective, seizure actions and FDA publicity are alternative considerations.

According to the FDA Weekly Report of Seizures, Prosecution, Injunctions, Field Corrections and Recalls, as well as the Weekly FDA Enforcement Report, during the 23-month period from July 23, 1975 to June 29, 1977, there were some 659 recalls involving foods. Almost 50% of the recalls involved filth (rodent, insect or bird contamination or insanitary conditions), and about 13% involved labeling or short weight. Recalls directly involving microorganisms constituted about 20% of the recalls, while about 10% were due to inadequate processing or failure to control pH. In this latter 10% of the recalls, the products had a potential microbial problem.

Aspects of the recall procedure can be found in the Regulatory Procedures Manual of the FDA (1976B).

Seizure.—This is an action taken to remove a product from commerce because it is in violation of the law. FDA initiates a seizure by filing a complaint with the U.S. District Court where the goods are located. A U.S. marshal is then directed by the court to take possession of the goods until the matter is resolved.

Injunction.—This is a civil action filed by FDA against an individual or company seeking, in most cases, to stop a company from continuing to manufacture or distribute products that are in violation of the law.

Prosecution.—This is a criminal action filed by FDA against a company or individual charging violation of the law.

Natural or Unavoidable Defects.—Growing crops in an open field can result in natural or unavoidable defects. It has never been possible, nor is it now possible, to prevent certain levels of contamination. With this realization, the FDA has developed food defect action levels. The alternative to allowing certain defects would be to use increased levels of pesticides. This could result in higher residues of pesticides in our foods. Growing one's own food organically does not ensure the absence of insect or other contamination.

In Table 13.1 are selected values from the list of defect action levels published in October 1973 (FDA 1973). These levels represent the limit at or above which FDA will take legal action against the product to remove it from the market. The defect action levels are under constant review and are periodically lowered as technology improves.

United States Department of Agriculture (USDA)

The USDA has many functions in food production and processing. The agency most directly associated with the food industry is Food Safety and Quality Service (FSQS). The FSQS oversees the federal meat and poultry inspection program and the standardization and voluntary grading services for meat, poultry, fruit, vegetable and dairy products, as well as egg products inspection and the various food purchasing operations of the Department of Agriculture's school lunch and other family food service programs.

TABLE 13.1

LEVELS FOR NATURAL OR UNAVOIDABLE DEFECTS IN FOOD FOR HUMAN USE THAT PRESENT NO HEALTH HAZARD

Product	Defect
Chocolate and chocolate liquor	Average of 60 microscopic insect fragments per 100 grams when 6/100 gram subsamples are examined or if any one subsample contains 100 insect fragments.
	Average of 1.5 rodent hairs per 100 grams when 6/100 gram subsamples are examined or if any one subsample contains 4 rodent hairs.
Corn meal	Average of one whole insect (or equivalent) per 50 grams. Average of 25 insect fragments per 25 grams. Average of one rodent hair per 25 grams. Average of one rodent excreta fragment per 50 grams.
Raisins	Average of 40 milligrams of sand and grit per 100 grams of natural or Golden Bleached raisins.
	10 insects or equivalent and 35 drosophila eggs per 8 ounces of Golden Bleached raisins.
Apple butter	Average of 4 rodent hairs per 100 grams. Average of 5 whole insects or equivalent (not counting mites, aphids, thrips, scales) per 100 grams.
Peanut butter	Average of 50 insect fragments per 100 grams. Average of 2 rodent hairs per 100 grams.
Pepper	Average of 1% insect infested and/or moldy pieces by weight. Average of 1 milligram of excreta per pound.
Broccoli (frozen)	Average of 60 aphids, thrips, and/or mites per 100 grams.
Brussels sprouts (frozen)	Average of 30 aphids and/or thrips per 100 grams.
Mushrooms	Average of 20 larvae per 100 grams of drained mushrooms and proportionate liquid. Average of five larvae, 2 mm or longer, per 100 grams of drained mushrooms and proportionate liquid.

National Marine Fisheries Service (NMFS)

This organization is under the National Oceanic and Atmospheric Administration (NOAA) in the Department of Commerce. Through its Fishery Product Inspection and Safety Program (FPISP), it offers a voluntary fee for service inspection to processors of fishery products. This service includes inspection for plant sanitation, inspection of the product for safety and wholesomeness, laboratory analyses, product grading, lot inspection and consulting services. Although inspectors of the NMFS are in a processing plant, the FDA regulates fishery products. If seafoods are mixed with meat or poultry products, they become subject to the USDA regulations for meat and poultry.

Occupational Safety and Health Administration (OSHA)

This organization was created in the Department of Labor to enforce the Occupational Safety and Health Act. The OSHA develops and promulgates safety and health standards and regulations, investigates to determine compliance with these standards and regulations and issues citations with proposed penalties for noncompliance. Every employer who is engaged in interstate commerce is subject to the act and to the regulations of OSHA. The law requires detailed reports from the employers concerning every category of safety, noise and other working conditions.

Research on safety and health is conducted by the National Institute for Occupational Safety and Health. This is located in HEW.

Environmental Protection Agency (EPA)

This agency was established in 1970 to coordinate programs of the federal government concerning the environment. This includes air, water, solid wastes, pesticides, radiation and noise. As a regulatory agency, it establishes and enforces environmental standards. Two of the acts it enforces are the Insecticide, Pesticide and Rodenticide Act and the Water Pollution Control Act. These are important to food processors. The food industry uses large quantities of water. After use, the water contains organic matter from the food materials, as well as detergents used during cleaning operations. The food industry pays for the reconditioning of this effluent either with its own facilities or by paying municipal sewage plants where applicable. Due to the level of organic material, the fees for this service can be rather high for the food processor.

Federal Trade Commission (FTC)

The FTC is an independent agency that enforces various laws. It includes the Bureau of Competition, Bureau of Consumer Protection and the Bureau of Economics. In the consumer protection area, the food industry is affected by rules concerning marketing practices and national advertising.

A major activity of the FTC is concerned with antitrust. The food industry, from the producer to the consumer, has been under investigation.

It seems rather incongruous that, due to the high cost of various government agency regulations, small businesses cannot succeed. Then the government has another agency to investigate why larger corporations are able to continue operations.

PROBLEMS OF REGULATIONS

No one will argue that safe and wholesome food is not important. Everyone wants assurance that the food they purchase is not contaminated with dangerous chemicals or microorganisms. However, there is a growing concern that, as a people, we are being overregulated. Much of this antagonism is aimed at agencies other than those directly concerned with foods. However, the FDA and USDA have received a lot of attention in recent years.

As a Congressman, Miller (1956) stated that unwise and excessive regulations coupled with unwise and excessive administration, can do harm to any industry. Also, free enterprise must be encouraged in its constant research for a better, more attractive, more abundant and safer food supply. He commented that the progress of industry must not be blocked by an unwise delegation of authority to any government agency.

One of the concerns about government regulations is the cost of the regulations compared to the benefits.

Cost

How much are we willing to pay for the protection of our food? In 1960, the FDA had 2,000 employees and a budget of $20 million. By 1970, this agency had grown to 4,500 employees and a budget of $70 million. For 1976, the budget had ballooned to $203 million.

In 1969, the USDA had 4,600 employees in the meat and poultry inspection program. The budget was $31 million. By 1970, this had mushroomed to 7,750 employees and a budget of $110 million. For fiscal year 1977-78, the budget for poultry and meat inspection was almost $242 million.

If these two enforcement agencies continue to grow at these rates, by the year 2000, their combined budgets will approach $15 billion.

The cost of these regulators is a small part of the total cost of regulations. There are many other federal enforcers who are involved in the food industry. The total cost to states, counties and cities for enforcement of regulations imposed upon the food industry also should be considered.

Various surveys have indicated that every American family pays about $2,000 per year for federal regulations.

Industry has increased costs due to the regulators. It has been estimated that the time needed to fill out and file all of the required government forms and other information is equivalent to at least one full-time employee for each establishment. This means about 30,000 people in the food industry are busy just to satisfy the paperwork of the government regulators. Reportedly, a government official has estimated that small business people spend $40 billion a year on paperwork to comply with regulations that are forcing them out of business.

Who pays for all of this regulation? The taxpayer pays the cost of the regulators and the consumer pays higher prices for regulated products. The consumers should ask if they are getting the service for which they are paying.

The costs of government have become so excessive that, as Assistant Secretary of Agriculture, Lyng (1970) stated, "The time has come to begin to look at ways to keep government food inspection costs within some sort of reason." He suggested that regulations should concentrate in areas of extreme risk of safety and health with less government interference when only modest economic loss is the issue. He found that many of our food laws are duplicated, inconsistent and involve unnecessary costs by federal, state and local agencies.

Safety

Both industry and regulators agree that, in the United States, foods are safer than ever before in history.

Since we have regulations and regulators, it is no wonder that consumers expect the foods they eat will not cause illness. The lack of proper enforcement of food laws may be worse than having no laws at all. With no laws, at least the consumer would be wary of the consequences of eating.

Is the FDA doing the job which the consumer expects? Even representatives of the FDA admit that there are insufficiencies in the ability of FDA to properly enforce the laws for which they are responsible. It has been suggested that an inventory should be made of regulations appearing in the Code of Federal Regulations which are not being enforced (Wylie 1977). These, as well as trivial regulations, might be repealed. If a regulation is not enforced, there is a tendency to overlook other regulations in the belief that they will not be enforced. Wylie (1977) stated that an agency's actions should be scrutinized for its efficiency and effectiveness rather than being measured by the regulations that are written, the size of the budget or number of people on the staff.

In order to achieve the desired objectives, the FDA expects the food industry to be self-regulating. This will help reduce the burden of the FDA, but there are surely some people who will question the food industry regulating itself.

Foster (1972) pointed out some areas in which our present food regulation activities are deficient. He suggested that more scientific research is needed. The information acquired could be used to make decisions based on science rather than on the emotional whim of consumer advocates, in order to obtain safe foods.

Product Development

There is a general belief that regulations tend to stifle innovation and the development of new products or processes. In a discussion of the effect of regulations on products, Sayer (1976) used margarine as one example of government interference. By means of laws forbidding the addition of color, and by taxation, the federal, state and local governments attempted to stifle margarine. However, despite all of this adversity, margarine has prevailed. One of the means by which the government continues to stifle product development is by means of standards.

Due to the proliferation of regulations by the various government agencies, research and product development has been slowed. Much of the money and effort formerly put into research is now going into projects to satisfy the requirements of the government for present activities or for future products.

Various problems that are caused by government regulations were discussed by Albrecht (1975) and Briskey (1976). These include such things as multiplicity of regulations, effluent regulations, capricious decisions and regulations, such as banning of DDT, sugar substitutes and possibly diethylstilbestrol and nitrites, noise abatement, drained weight labeling and nutrition labeling. Nutrition labeling has cost the food industry millions of dollars. One company alone spent more than $750,000 for initial analysis and labeling and expects to spend more than $300,000 each year to monitor the products (Albrecht 1975).

Summary

Protecting the safety of our food supply is an important attribute of the regulators. In an interview (Kimm and Angelotti 1977), Angelotti stated, "There will always be regulations because there are always segments of any part of the sociology . . . that are self-centered and interested in self-aggrandizement . . . ." This statement also applies to the regulators, as well as to those being regulated.

Although economic frauds of foods are not desirable, many people pay more for taxes than they do for food. This is not to infer that most tax money is used to regulate food, but the amount is increasing.

Because companies are under the jurisdiction of several different government agencies, they find total compliance difficult with contradictory rulings.

Perhaps one regulatory agency that would base decisions and judgments on scientific evidence would be sufficient for any one industry.

FOOD STANDARDS

Section 401 of the FD&C Act provides that whenever, in the judgment of the Secretary, such action will promote honesty and fair dealing in the interest of consumers, he shall promulgate regulations fixing and establishing for any food, under its common or usual name, a reasonable definition and standard of identity, standard of quality and/or standard fill of container. Fresh and dried fruits and vegetables are exempt, except for avocados, cantaloupes, melons and citrus fruits. Also, butter is exempt since it was defined by an Act of Congress in 1923. Meat and poultry products are in the domain of the USDA.

The Hale Amendment of 1954 simplified the procedures for establishing food standards. A standard constitutes the official specification for that food. Supposedly, these standards are important to the consumers and food manufacturers, as well as the enforcers. Standards prevent certain types of debasement and promote honesty and fairness to the consumer, and serve to prevent unfair competition. Unfortunately, the methods by which standards have been applied have resulted in consumer confusion and have stifled innovation by the food industry. Once a food is standardized, there is little incentive to spend money and time for research to improve the product.

The Codex Alimentarius Commission is concerned with establishing food standards on a worldwide basis rather than having many regional standards. Establishing international standards for foods is complicated. Over the years many countries have developed their own standards and codes of practice. The differences in these standards must be reconciled by participating countries with those being developed by the Commission.

The FDA has stated that it will be their policy to accept the standards recommended by the Codex Alimentarius Commission insofar as the requirements of those standards can be shown to be reasonable and calculated to promote honesty and fair dealing in the interest of the American consumers.

Although these standards should promote better international trade, they, like any other standards, must be somewhat flexible and subject to change when new technology requires a change.

The regulatory agencies attempt to improve sanitary standards through the development of ordinances, codes, manuals and guides to safe practices. This is done at the local, state and/or federal level. Although the FD&C Act does not specifically mention sanitary standards, these are written on the premise of increasing food safety. Reportedly, the sanitary standards are used to prevent foodborne illness in the U.S.

The GMP's of the FDA and the Regulations of the USDA for operating meat, poultry and egg products plants can be considered sanitary standards for food processors. Some food processors have in-house standards which exceed the requirements of the federal agencies.

The sanitary standards usually include observation of the overall plant and facilities, raw material, plant operations, clean-up operations and finished products. Many also include the handling and storage of the finished products, especially if the product is perishable.

MICROBIOLOGICAL CRITERIA

A criterion is a standard that can be used to evaluate the correctness of a judgment. When the microbial load of a food is determined, we then need to make a judgment as to whether the food is satisfactory or not. Hence, microbial criteria are established in order to help make a valid judgment concerning the safety and keeping quality of a food.

Another reason for adoption of microbial standards for foods is to compel people in the food industry to adopt quality control procedures to protect the microbial quality of food from producer to consumer.

Most microbiological criteria are developed to protect the safety of the food supply and reduce the risk to the consumer. Since the FD&C Act states that harmful or deleterious substances are prohibited in food, the position of the FDA is that standards are needed to evaluate quality. In previous chapters, several potential pathogens and toxins were discussed. Of all of these, only salmonellae are determined in foods on a regular basis.

It is well known that salmonellae are present on much of our raw red meat and poultry. Although these products are not under the jurisdiction of the FDA, they are an example of foods which are known to contain pathogens and, at present, they are tolerated.

Since the local health authorities and the FDA cannot adequately inspect all of the food processing establishments in the U.S., the establishment of microbiological criteria for the finished product is the only method to estimate the microbial quality of food.

The functions of microbiological criteria should be to:

1. Control the risk to the consumer from pathogenic organisms or their toxins,
2. Reveal if gross contamination has occurred,
3. Assure that a reasonable shelf life of the food is attainable.

There are various criteria that can be established for foods. A microbiological criterion of a food should include the microorganisms and/or their toxins which are of concern, the methods for analyzing the food for detection and quantification, a sampling plan which includes the number of samples to be obtained and the size of the sample unit, the limits for the microorganisms and/or toxins which are appropriate for the specific food and the proportion of sample units that should conform to these limits.

The microbiological load of any food is influenced by various production and processing procedures. These variations should be considered in establishing microbial limits. The limits should be attainable by using good manufacturing practices. There should be tolerances due to the inherent errors in sampling and microbial analyses.

Types of Criteria

In order to evaluate the microbiological condition of foods, the criteria which have been used are guidelines, recommended limits, specifications and standards.

Guidelines.—Microbiological guidelines have been developed for many foods. A guideline is used when no official standard exists for a particular food. Guidelines might be considered as administrative standards.

A guideline usually is an estimate of the microbial load that can be attained readily in a satisfactory plant using satisfactory methods of processing and materials.

Although not an official standard, a guideline can be useful for quality control personnel to make judgments about the acceptability of raw materials, finished products or operations during processing.

Both processors and regulators can establish microbiological guidelines. The FDA has used guidelines in conjunction with processing plant inspec-

TABLE 13.2

MICROBIAL DEFECT ACTION LEVELS IN SELECTED HUMAN FOODS

Product	Defect Action Level
Potato chips	6% of the chips by weight contain rot.
Dried whole eggs	Decomposed as determined by direct microscopic count of 100,000,000 bacteria per gram.
Frozen eggs	Two cans contain decomposed eggs; and subsamples examined from cans classed as decomposed have counts of 5,000,000 bacteria per gram.
Citrus fruit juices (canned)	Microscopic mold count average of 10%.
Pineapple (canned, crushed)	Microscopic mold count average of 30%.
Plums (canned)	Average of 5% by count of plums with rot spots larger than the area of a circle 12 mm in diameter.
Prunes (dried and dehydrated low moisture)	Average of 5% by count insect infested and/or showing mold and/or dirty fruits or pieces of fruit.
Raisins	Average of 5% by count of natural raisins showing mold.
Strawberries (frozen) (whole or sliced)	Microscopic mold count average of 45% and mold count of 55% in one-half of the subsamples.
Apple butter	Microscopic mold count average of 12%.
Cherry jam	Microscopic mold count average of 30%.
Black currant jam	Microscopic mold count average of 75%.
Allspice, bay leaves, capsicum, cassia or cinnamon	Average of 5% moldy berries by weight.
Beets (canned)	Average of 5% by weight of pieces with dry rot.
Greens (canned)	Average of 10% of leaves by count or weight showing mildew ½″ in diameter.
Spinach	Average of 10% leaves by count or weight show mildew or other type of decomposition ½″ in diameter.

Source: Food and Drug Administration (1973)

tions. When the microbial load exceeds the guidelines, it supports the observations of poor sanitation and failure to follow good manufacturing practices (Read and Baer 1974). These guidelines have been useful at the processing level but less useful at the retail level. The analysis of food at the retail level has merit if the consumer is considered. Abuses that might occur between processing and retailing could influence the consumer risk.

The defect action levels for molds or mold fragments established by the FDA for certain fruits and various other foods can be considered as microbiological guidelines. Some defect action levels for microorganisms are listed in Table 13.2.

Recommended Limits.—These are the suggested maximum acceptable number of microorganisms or of specific types of microorganisms, as determined by prescribed methods in a food (NAS 1964). They are similar to guidelines.

There are many recommended microbiological limits. Most of these recommendations are based on too few samples or other considerations to be of much value. The ranges of recommended limits for various foods as found in the literature are listed in Table 13.3. Due to the wide variation, it is evident that most of these are not valid.

TABLE 13.3

RANGES OF SUGGESTED LIMITS FOR TOTAL VIABLE AEROBES IN VARIOUS FOODS

Food Product	Microbial Range per g
Frozen precooked foods	2,000 to 100,000
Precooked meats	10 to 10,000
Raw ground meat	100,000 to 50,000,000
Frozen whole eggs	200,000 to 10,000,000
Fish	100,000 to 10,000,000
Shellfish	50,000 to 1,000,000
Vegetables	50,000 to 500,000
Frozen desserts	5,000 to 1,000,000

Source: NAS (1964)

Specifications.—A specification is a document that provides a detailed description of a material. Included in a specification of a food may be acceptable microbiological quality. A microbiological specification has been defined as the maximum acceptable number of microorganisms or of specific types of microorganisms, as determined by prescribed methods, in a food being purchased by a firm or agency for its own use (NAS 1964).

The acceptable microbiological limit that is written into a specification is not always easy to derive. A microbiological specification should consider both the buyer and the capabilities of the supplier. If the specification is too strict there may not be enough acceptable material to meet the need of the buyer. If too lenient, the buyer may obtain rather poor quality material. By analyzing many samples, or using information obtained through guidelines, a compromise may be made between the microbial quality available and what is considered to be acceptable. To obtain a high quality product, the purchaser may expect to pay a premium price for the material.

TABLE 13.4

MICROBIOLOGICAL SPECIFICATIONS FOR MILITARY AND FEDERAL PURCHASES—BEEF AND PORK PRODUCTS

Product	Aerobic Plate Count	Coliform	*E. coli*	Salmonellae (25 g sample)
Beef, cooked, dried	150,000	40	–	–
Beef stew, cooked, dried	75,000	–	neg.	–
Beef with rice, cooked, dried	75,000	–	neg.	–
Swiss steak w/gravy, frozen	100,000	100	neg.	neg.
Pork slices, cooked, dried	110,000	20	–	–
Pork sausage, cooked, dried	200,000	40	–	–
Pork loin, sliced, w/gravy, frozen	100,000	100	neg.	neg.
Pork and beef chop suey, frozen	100,000	100	neg.	neg.
Meat balls and meat ball products, cooked, dried	150,000	40	–	–
Spaghetti w/meat sauce, dried	75,000	–	neg.	–
Escalloped potatoes w/pork, cooked, dried	75,000	–	neg.	–
Beef hash, cooked, dried	75,000	–	neg.	–

A practical specification should have tolerance limits set wide enough to allow permissible deviations. However, the limits should be narrow enough to ensure that the material will be acceptable consistently.

The specification should include sampling and analytical procedures so that there is no question of the system that is to be used.

The microbiological limits that appear in various military and federal specifications for food products are listed in Tables 13.4-13.6.

TABLE 13.5

MICROBIOLOGICAL SPECIFICATIONS FOR MILITARY AND FEDERAL PURCHASES—MILK AND DAIRY PRODUCTS

Product	Aerobic Plate Count	Coliform	Yeast & Mold	Salmonellae	Direct Microscopic Count
Milk and milk products, fresh, fluid, concentrated and frozen	20,000	10	–	–	–
Milk, whole, dry (premium)	–	10	–	–	40,000,000
(extra grade)	–	10	–	–	75,000,000
Milk, nonfat, dry	50,000	–	–	neg.	–
Processed American cheese, dehydrated	50,000	90	–	–	–
Milk, fat	5,000	10	20	–	–
Frozen fudge bar	50,000	20	–	–	–
Flavored dairy drink, dry chocolate or coffee flavored	20,000	10	–	neg.	–
Cream substitute	20,000	10	–	–	–
Cottage cheese	–	10	10	–	–
Cream, sour, cultured, or acidified	–	10	10	–	–
Ice cream, ice milk:					
vanilla	50,000	10	–	–	–
chocolate, nuts, fruits	50,000	20	–	–	–
Malted milk	30,000	10	–	–	–
Buttermilk	–	10	–	–	–

TABLE 13.6

MICROBIOLOGICAL SPECIFICATIONS FOR MILITARY AND FEDERAL PURCHASES—POULTRY PRODUCTS

Product	Maximum Count per Gram				
	Aerobic Plate Count	Coliforms	*E. coli*	Salmonellae (25 g sample)	Yeasts & Mold
Frozen chicken a la king	100,000	100	neg.	neg.	–
Frozen chicken cacciatore	100,000	100	neg.	neg.	–
Frozen turkey with gravy	100,000	100	neg.	neg.	–
Cooked turkey, dried	200,000	40	–	–	–
Cooked chicken with rice, dried	75,000	–	neg.	–	–
Cooked chicken and cooked chicken products, dried	75,000	–	neg.	–	–
Frozen eggs and egg products, whole egg, egg white, yolk	50,000	–	–	neg.	50
table grade	20,000	–	–	neg.	50
Egg mix, dried	25,000	10	–	neg.	–

Standards.—Of these criteria, only microbiological standards have a regulatory function. The National Academy of Sciences (1964) defined a microbiological standard as that part of a law or administrative regulation designating the maximum acceptable number of microorganisms or of specific types of microorganisms, as determined by prescribed methods, in a food produced, packed, stored or imported into the area of jurisdiction of an enforcement agency.

A microbiological standard should not be established haphazardly. A standard must be meaningful and attainable by what is considered to be good manufacturing practice. It should be subject to reevaluation as new technologies are developed. Above all, to be of value, a microbiological standard must be enforceable. When enforced, a standard should reduce the public health hazard of a food.

When the need for microbial standards for food is discussed, an example often used is how the establishment of standards eliminated or reduced the involvement of milk in foodborne diseases. Microbiological standards, per se, do nothing to increase the safety or quality of the food. It is the acceptance of the standards by all segments of industry and the enforcement by the regulators that is necessary before a standard has value.

Microbial Groups and Species

Microbiological criteria could include the determination of many groups or species of microorganisms. Keeping the number of required tests to a minimum has advantages. There is less expense and time involved in the analysis and the interpretation of the results is less complex when fewer tests are needed.

The microbial limits might include the aerobic plate count for mesophiles, as well as coliforms, fecal coliforms, *E. coli*, salmonellae, *Staphylococcus aureus, Clostridium botulinum, C. perfringens*, group D streptococci, Enterobacteriaceae, yeasts and molds.

The standard aerobic plate count of mesophilic bacteria is a rough measure of the bacterial content and the sanitary conditions during processing and/or temperature abuses of a food. The aerobic plate count cannot be related unequivocally either to a health hazard or to spoilage.

Some foods normally have high microbial loads. Depending on the raw material and the processing conditions, a particular number of organisms may represent a good or bad product. Precooked or other heat treated foods would be expected to have lower counts than raw foods. High microbial numbers in a heated food may be due to poor processing control, contamination after heating, poor holding temperature allowing muliplication or a combination of these factors. The viable microorganisms tend to decrease in frozen or dried foods and to increase in refrigerated foods. Hence, the microbial number that is obtained will depend on the time of analysis after processing.

The coliform group is determined in food mainly because this group is important in water microbiology and somehow, there was a belief that this importance carried over to food products. The presence of coliforms cannot be related unequivocally to fecal contamination. In a pasteurized or precooked food, the presence of coliforms indicates poor processing controls, post-process contamination, or both.

The presence of *E. coli* reportedly indicates either direct or indirect fecal contamination. However, this organism is part of the normal flora of many raw animal products, even when prepared with acceptable processing techniques, as well as plant products that contact the soil. *E. coli* is heat sensitive. When found in heat processed food, it indicates contamination after heating by fecal material that also could conceivably include enteric pathogens. Certain strains of *E. coli* are enteropathogenic or enterotoxigenic. Distinguishing these types from any other *E. coli* would be too time consuming and costly for practical purposes regarding microbial standards.

The determination of *E. coli* reportedly can be used to estimate the plant sanitation, post heating contamination, poor handling practices, cross contamination from raw products or a combination of these factors. Being pathogenic, salmonellae should not be associated with food.

Staphylococcus aureus has been included in some microbiological criteria. In certain situations, *S. aureus* may be a pathogen, a source of enterotoxin, an indicator of unsanitary practice or of little significance and concern (Bartram 1967). Foods that are handled by humans are likely to contain coagulase positive staphylococci. Standards for *S. aureus* at the processing level would have little value in preventing foodborne illness.

Clostridial spores are prevalent in the environment, especially in soil and dust. Hence, organisms, such as *C. botulinum* and *C. perfringens* may be present in foods. Unless the food is an acceptable environment and is allowed to remain at temperatures that allow growth, a few clostridia or *S. aureus* are not considered to be a health hazard. However, processing and handling should be designed to minimize contamination and prevent the growth of these organisms.

Organisms in the group D streptococci have been considered as indicators related to human contamination of processed meats. They have been proposed as indicators of fecal pollution of various processed foods due to their ability to withstand certain processing and environmental conditions.

The presence of yeasts and molds is used in the microbiological criteria for some foods. A high count (above 50/g) in cocoa powder indicates poor sanitary conditions after roasting. The guidelines for product defects (Table 13.2) consider certain levels of molds unsatisfactory in foods. Tomato products contain high levels of mold due to moldy tomatoes not being properly sorted or trimmed or to contaminated processing equipment. In either case, a high level of mold indicates poor processing technique. In some products, the presence of molds may indicate the possible presence of mycotoxins which may or may not be a health hazard.

The tests that are most useful or used in microbiological criteria are the standard plate count of aerobic mesophiles, coliforms, *E. coli* and salmonellae.

Types of Food

There is more justification for microbiological standards for some foods than for other foods. When there is a risk or hazard, a standard for a food might be needed. There are some foods for which there is no substantive need for microbiological standards.

Those foods which have demonstrated a potential health hazard should be the first ones considered for establishment of any microbial standards. Foods that contain animal products are the primary vehicles for organisms that cause foodborne illness.

Frozen and chilled precooked foods which support the multiplication of pathogenic bacteria when held at growth temperatures are considered to be the main foods that need microbiological standards. However, even though there is a potential hazard, these foods have been involved in relatively few known outbreaks of foodborne illness.

Foods under consideration for international standards by the Codex Commission include egg products, dried milk, molluscan shellfish, frozen froglegs, precooked frozen shrimps and prawns, food for infants and children, and frozen desserts. Those considered for prompt attention include chilled frozen poultry and meat, precooked frozen crabmeat and lobster, cheese and coconut.

Microbiological standards have been applied to milk for a number of years. Standards for bottled drinking water and local standards for a few foods have been established. Milk and water were involved in many disease outbreaks before microbial standards were established. These liquids are homogeneous and can be subjected readily to heating, filtration or chemicals to control the microbial load. Hence, a microbial limit can be attained relatively easily and the standard can be enforced fairly easily. These problems are increased with heterogeneous solid foods.

In 1973, the FDA proposed microbial standards for food grade gelatin and frozen ready-to-eat banana, coconut, chocolate or lemon cream-type pies. These were rescinded in 1978. In 1977, the FDA planned microbiological standards for fish cakes, fish sticks and crab cakes. Information has been obtained for several other foods with the intention of establishing more microbial standards.

Sampling Plan

Microbiological criteria should include a sampling plan, as well as methods for analysis. Since the microbial contamination in most foods is heterogeneous, the testing of as many samples as possible is necessary to satisfy statistical requirements. Statistical sampling plans can ensure that the tests which are done are sufficient to give good assessment of the microbial condition of the food. It is important to minimize the chances of rejecting an acceptable lot of food (producer risk) or accepting a substandard lot (consumer risk).

Sampling plans for food have been amply described by Thatcher and Clark (1968) and ICMSF (1974). Sampling of food is discussed briefly in Chapter 2. The sample units should be representative of the lot and be obtained in an unbiased manner.

The number and size of samples to be analyzed should be included in the microbiological standard or other criterion.

Methods for Analysis

Standard methods for analysis must be selected to determine if foods meet the requirements of microbiological standards. Methods published by AOAC (1975), APHA (1976), FDA (1976A) or Thatcher and Clark (1968) could be acceptable. The AOAC methods are standardized and are accepted in courts of law in the United States.

On an international basis, the standardization of methods is more difficult. There are different media manufacturers in different countries. The organisms that are prevalent vary from continent to continent. Food from a tropical country would undoubtedly contain more thermophiles than would the food from colder climates. However, standardized methods would help facilitate international trade in foods.

Whatever methods are designated, they should not require the use of unusual and expensive equipment not found in most laboratories. The procedure should be easy to follow by a technician and not be so complex that only a research scientist can master the techniques. However, the method must give reproducible results so that the findings from different laboratories will agree within reasonable limits.

Methods which are applicable to many commodities or groups of commodities are preferred to methods which apply to only one or a few foods.

Microbial Limits and Variables

There are various ways in which the limits of microbiological criteria can be developed. One method is to make an extensive survey of the microbiological condition of the finished product. The samples should be obtained from several processors that operate with what are considered acceptable sanitation controls. With these data, the expected level of microorganisms in a particular product can be estimated.

A system can be used in which an inspection team visits the processing plant and evaluates the apparent sanitary conditions on various days. Samples of the finished product are obtained during these visits and analyzed for the microbial condition. In this way, it may be determined if there is a correlation between sanitation and microbial condition.

The microbial limits must be achievable when good manufacturing practices are used in processing the food. In establishing limits, there should be a consideration of the risk associated with the presence of the organism(s) and the number of organisms which would affect the acceptability of the food.

Any microbiological limit that is established by a regulatory agency may be challenged in court. It may be difficult to prove that a certain number of bacteria above a certain limit is a health hazard or an indication of filth and decomposition.

It is well recognized that there are variations in the microbial content between sample units of food. These variations must be considered when microbial limits are applied to a food product.

The good manufacturing practices that are applied today might be considered poor or unacceptable in a few years. Also, the factors that are considered as health hazards are continually changing. To be effective, microbiological criteria must be reevaluated and altered as experience is gained and as technology is changed.

Decision Criteria.—These are part of the overall sampling plan. However, they are used in the application of the data obtained from analyzing sample units to the microbial limits that have been established.

We can assume that the limit for the aerobic plate count (APC) is 100,000/g. This limit is designated as m. In sampling by variables plans, the lot is accepted if the mean or average of the APC of the sample units is m or less. If the mean is greater than m, the lot is substandard. Variable sampling requires some knowledge of the distribution of microorganisms within the product (Thatcher and Clark 1968).

The attribute sampling plan can be used with any food that is tested. With this system, either the APC of a sample unit exceeds m or is equal to or less than m, the microbial limit. If ten sample units are analyzed and the APC of eight are equal to or less than m and two are greater than m, what decision can one make regarding the acceptability of the lot?

The Two-class Plan.—In a two-class plan, several sample units are analyzed. The number of sample units which must be examined from a lot of food to satisfy the requirements of a particular sampling plan is designated as n. If 10 sample units are needed, then n = 10. With a two-class sample plan, a certain number of nonacceptable or defective sample units are allowed. The maximum allowable number of defective sample units is designated as c. If two defective sample units are allowed of the ten that are examined, then the entire lot of a food product that has eight acceptable and two defective units would be considered as acceptable. If there are three defective units, the lot would be rejected.

The Three-class Plan.—In the two-class system, the sample unit is either acceptable or nonacceptable. With a three-class system, a marginal area is included. If we keep m as 100,000, then any sample units of m or less are still considered as acceptable. An upper limit is established which we can assume to be 5,000,000, which is designated as M. In this example, sample units with APC above m (100,000) and equal to or less than M (5,000,000) are considered marginal. A product with any sample unit with an APC over M is unacceptable. With a three-class system, of 10 sample units, 2 could be between m and M, with no sample units greater than M. Our plan can be formulated as: $n = 10$, $c = 2$, $m = 10^5$, $M = 5 \times 10^6$. For salmonellae, which are not tolerated in food, the plan would be $n = 10$, $c = 0$, $m = 0$. There would be other values for m, M or c if *E. coli*, coliforms or other types of microorganisms were in the standard.

Different foods might have different types of organisms listed in the standards, as well as different values for n, c, m and M; these values depend on the type of food and the microorganism or group of microorganisms. The value of M should be so high that it constitutes a definite hazard or evidence of overt spoilage. Hence, no food could be tolerated with sample units above this level. With high potential health hazards, the value of n would be high and c would be low. For cases in which there is essentially no hazard and leniency is designated, n may be low and c relatively high, or n might be 5 with a c of 3. In other words, 3 of 5 samples could exceed the limit m without rejection of the lot.

With any sampling plan, there is a probability that acceptable products will be rejected and products not in compliance will be accepted. The probability of these occurrences with various sampling plans is discussed by ICMSF (1974), Reed and Baer (1974) and Thatcher and Clark (1968).

Disposition of Substandard Food

Foods which exceed the acceptable microbiological limits in a standard are not necessarily wasted. The standards as proposed by the FDA were for quality determination. Foods that did not qualify for the standards were to be labeled "Below Standard in Quality—Contains Excessive Bacteria." It is evident that this statement on a food will decrease sales. If there are excessive bacteria in a food, it may be subject to seizure on the basis that it is adulterated by having been "prepared, packed or held under unsanitary conditions."

According to Elliot (1970), when poultry and red meat products exceed the microbiological criteria, production of further products is halted to determine the cause of the contamination. The disposition of the product depends upon the extent of contamination, the existence of a hazard, presence of decompositon or adulteration and the nature of the microorganisms. After considering these factors, the product may be released, sorted, reprocessed or destroyed. If the food is not acceptable for human consumption, depending upon the contamination, it might be fed to animals. Food should not be destroyed unnecessarily. However, if the hazard is such that it cannot be remedied, destruction may be the only solution.

Feasibility of Microbiological Standards

Greenberg *et al.* (1974), as well as others, have suggested that the imposition of microbiological quality standards can result only in less food, higher costs and a dilution of effort to protect the public against foodborne illness. They stated that industry would not object to microbiological standards if actual consumer benefits could be demonstrated.

An understanding of the factors responsible for contamination is needed to establish realistic microbiological standards for foods. Rigid standards based on the aerobic plate count or on indicator organisms are not realistic even when large surveys are used to establish the limits (Anema 1975). Plating or MPN systems yield only a representation of microbial levels, not actual facts.

A microbial standard should be established only when there is a definite need for such a regulation. The establishment of arbitrary standards by local authorities can result in barriers to the free flow of foods, especially if the standards are different in the various local areas. National or international standards would be better than thousands of local standards.

Safety and Quality.—Although microbial standards are supposed to assure that the food is safe, wholesome and of good quality, the microbiological limits for standards are established on the basis of practical attainability which may or may not relate to safety or wholesomeness of the food.

The absolute absence of pathogens in food is difficult both to attain and to determine. The only way in which complete safety of a food can be assured is to analyze all of the food. This would be technically impossible and there would be no food left to eat.

The occurrence of staphylococcal enterotoxin in dried milk which contained no viable cells of *S. aureus* indicates that even the absence of toxigenic organisms in the finished product is no assurance that foods are without hazard. Due to the complexity of the tests, organisms such as viruses and rickettsiae or microbial toxins are not determined routinely. Usually they are not included in microbiological criteria.

Evaluating the Food.—In order to establish the microbiological limits for standards, the FDA obtained and analyzed food samples from retail stores. It is the general opinion that, when possible, microbial limits should be established in conjunction with plant inspection for sanitary conditions. Basing the standards only on retail product may cause certain problems.

Dried and Frozen Foods.—During the storage of dried or frozen foods, the number of associated viable microorganisms tends to decline. Hence, the age of the food could influence the number of organisms detected in the FDA survey. If these counts were low due to microbial death in the stored products, the microbial limits might be unrealistic. Food processors might decide to hold the food off the market until the death of cells is sufficient to make the product acceptable according to the standards. This type of treatment would not offer any benefit to the consumer, since the quality of food tends to decline during storage.

If the FDA decides to sample freshly processed product at the processing plant, it would be unfair to expect this product to compare with the older product which was used to establish the microbiological limits.

Highly Perishable Foods.—The FDA is considering sampling food at the retail level. An aerobic plate count requires incubation of the prepared plate for 48 hr. Other tests may require five days or more to complete. The present microbiological methods do not give us results as rapidly as needed to make any standards for perishable foods effective.

Fermented Foods.—These foods are developed by the metabolism of microorganisms. This process often includes the addition of a starter culture. The establishment of microbial limits for certain types of these foods might be questioned. Reportedly, the establishment of standards for foods in which fermented foods are ingredients would not be practical. An example that has been given is pizza, due to cheese being an ingredient. However, the Codex Alimentarius Commission is considering microbiological standards for cheese.

Technical Aspects.—Stephens (1970) reported that the enormous volume, the range of products and the rapid flow of foods from processor to stores make it impossible to rely on a system of inspection at the retail level.

If a minimum of five samples is to be analyzed for each lot of each food at the retail level, the enormous task should be apparent. The alternative is to use the standards as a threat. This would be a poor system for enforcement of standards.

Even if the FDA does not analyze all of the food that could be subject to standards, the food industry will be obligated. This will increase the need for technicians, laboratory equipment and supplies. This will be a satisfactory situation for food microbiologists. Will there be enough technical people available to satisfy the demand?

Costs.—The costs that will accrue from microbiological standards are merely estimates. Greenberg *et al.* (1974) conservatively estimated the cost for establishing microbiological limits at $50,000 for each food. Originally, it was planned to set standards on six foods per year, but apparently this has been increased. Greenberg *et al.* (1974) suggested that through 1975, some 43 foods would be tested to obtain information for microbial standards. There are several hundred different foods in retail stores. The cost for establishing standards could be several million dollars. Other costs that were suggested by Greenberg *et al.* (1974) were due to increased sampling, lower action levels, storage of foods, disposing of product not in compliance and increased processing cost. According to them, small companies that cannot afford the cost of all of the required analyses will be forced out of business. This will lead to less competition. The FTC then will have to investigate to find out why small companies went out of business.

Corlett (1976) estimated the cost of implementing FDA microbiological standards for the frozen food industry for one year would be over $1 billion. The cost to the entire food industry would be much greater and even an estimate would be impossible to make. The cost of the microbiological standards will be borne by the consumers and the taxpayers. Is the suggestion that the quality of food will be increased worth the expense?

Miscellaneous Standards

Although specifications or guidelines have been established for most foods and food ingredients, there are very few official standards. Many of the so-called standards in the literature are either specifications, guidelines or suggested standards. Most of the microbiologial standards adopted in the U.S. have been of the single "fixed" type. That is, if the maximum aerobic plate count is 100,000/g, any lot that exceeds this value is unacceptable or substandard.

Water.—In 1973, the United States Public Health Service revised the standards for drinking water in water systems (tap water) and the USFDA proposed microbiological standards for bottled water used for drinking.

Tap Water.—Standards have been established for the standard plate count (SPC) and coliforms. The SPC limit is 500 organisms per ml, based on the arithmetic average of samples collected in a calendar month.

Either the membrane filter method or the fermentation tube method can be used to determine coliforms. For the membrane filter method, a 100 ml sample is used. The arithmetic mean coliform density of all samples examined per month shall not exceed 1/100 ml. Per sample, the coliform colonies shall not exceed 4/ml in more than one sample if fewer than 20 samples are examined per month, or more than 5% of the samples if more than 20 samples are examined.

For the fermentation tube method, five portions of either 10 ml or 100 ml are tested. When 10 ml portions are tested, not more than 10% in any one month shall show the presence of the coliform group. The presence of coliforms in three or more 10 ml portions of any one sample are not allowed if this occurs in either more than one sample per month, if fewer than 20 samples are examined, or more than 5% of the samples, if over 20 samples are examined during the month.

When 100 ml portions are tested by the fermentation tube method, not more than 60% in any one month shall show the presence of coliforms. The presence of coliforms in all five of the 100 ml portions in a sample shall not be allowed if this occurs in either more than one sample per month, when fewer than five samples are examined per month, or more than 20% of the samples, when five or more are examined per month.

There are provisions for analyzing more samples if the level of organisms is above the standards. Besides microorganisms, limits are set for various chemicals. These standards are used as guidelines by many states and cities in setting up limits for potable water.

The World Health Organization standard for drinking water includes coliform limits. In 90% of the samples, there should be fewer than 1 count/100 ml and for all samples, there should be fewer than 10 counts/100 ml.

Bottled Water.—The standard for bottled water includes the coliform index. The sample used is 10 containers (sample units) of water selected from the lot.

The coliforms can be determined by either the membrane or fermentation method. For the membrane filter system, not more than one of the sample units shall have more than four or more coliforms/100 ml and the arithmetic mean of the coliform density shall not exceed one coliform/100 ml.

For the fermentation tube method, not more than one of the sample units shall have an MPN of 2.2 or more coliforms/100 ml, and no analytical unit shall have an MPN of 9.2 coliforms/100 ml.

Milk.—Due to its widespread consumption by infants and children, and its potential for containing health hazards, milk was the first food for which microbiological standards were established.

The microbiological standards for milk and other dairy products are listed in Table 13.7. The SPC of pasteurized grade A milk must not exceed 20,000 organisms/ml and more than 10 coliforms/ml. Restrictions imposed by individual dairy plants are frequently much more rigorous than those imposed by public health requirements. No limit is established for the SPC of fermented products, due to the addition of a starter culture. However, there is a limit established for coliforms.

For milk used in manufacturing, such as in making cheese, the USDA recommended standards are:

Grade	Standard Plate Count Organisms/ml
1	not over 500,000
2	not over 3,000,000
3	over 3,000,000

The bacterial level in manufacturing milk is much higher than that allowed in grade A milk.

TABLE 13.7

MICROBIOLOGICAL STANDARDS FOR SOME GRADE A DAIRY PRODUCTS

	Maximum Number per Gram	
Product	Standard Plate Count	Coliform Count
Raw milk, at pick-up	100,000	–
Raw milk, commingled	300,000	–
Pasteurized milk	20,000	10
Pasteurized, condensed milk	30,000	10
Pasteurized, condensed whey	30,000	10
Nonfat dry milk	30,000	10
Dry whey	30,000	10

Other countries have other standards. Quite often, a reductase test using resazurin or methylene blue is used in place of the SPC to estimate the bacterial quality of the milk or milk product (Cox 1970).

In Canada, the standard for raw market milk is an SPC of less than 75,000/ml and a coliform count of less than 50/100 ml. Sampling plans and microbiological limits for dairy products and other foods in Canada were discussed by Pivnick (1978).

Oysters.—The National Shellfish Sanitation Program established a microbiological criterion at the wholesale level. A satisfactory sample should not have a fecal coliform density of more than 78/100 g, with occasional values to 230/100 g (MPN) and a standard plate count of no more than 100,000/g, with occasional samples at 500,000/g.

Ayres (1975) stated that, for bivalve shellfish, there is no justification for a standard based on total plate counts which often exceed 1,000,000/g. He suggested that the relation of non-specific bacteria growing at 20° and 37°C might be useful in assessing the potential health risk of raw shellfish.

Meat.—There are no official federal microbiological standards for meats.

In 1973, Oregon established standards for meat as shown in Table 13.8 (Oregon State Department of Agriculture Regulation-CH603.28-400). The single aerobic plate count standard of 5,000,000/g of meat caused a great deal of controversy. A single all or nothing standard is always a problem to meet as well as to enforce. The 5,000,000 total count standard applied to ground meat, roasts and steaks, whether fresh or frozen. It is evident that ground meat will tend to have higher counts than meat cuts and that freezing will lower the bacterial count. The Oregon experiment is an example of unacceptable microbiological standards that should never have been established.

Westhoff and Feldman (1976) and Wehr (1978) listed guidelines, proposed standards, and standards for several states for meat, as well as for other foods.

TABLE 13.8

MICROBIOLOGICAL STANDARDS FOR MEATS AT RETAIL LEVEL (OREGON)

	Maximum Number per Gram	
Type of Meat	Aerobic Plate Count	*E. coli* (MPN)
Fresh or frozen	5,000,000	50
Processed (bologna types)	1,000,000	10

TABLE 13.9

CANADIAN MICROBIOLOGICAL STANDARDS FOR FRESH AND FROZEN GROUND BEEF

				Limit per Gram	
	Test	n[a]	c	m	M
Fresh	Aerobic plate count	5	3	1 (10^7)	5 (10^7)
Frozen	Aerobic plate count	5	2	1 (10^6)	1 (10^7)
Fresh or frozen	*E. coli*	5	3	100	500
	S. aureus (coagulase positive)	5	2	100	1000
	Salmonellae (25 g sample units)	5	0	0	0

[a]n = number of sample units
c = number of defectives that are allowed greater than m and less than M
m = number of organisms separating acceptable from marginal product
M = number of organisms separating marginal product from unacceptable product

Canada has established microbiological standards for ground beef based on the ICMSF sampling plan. This type of standard is based on the fact that there are statistical variations due to sampling, analysis and the heterogeneous distribution of cells in solid food samples.

Sugar.—This ingredient may contain spores of thermophilic bacteria which are important in the spoilage of low-acid canned foods. In 1931, the National Canners Association (NCA) formulated and published standards for sugar for use in canning of low-acid canned foods. These standards have been applied by industry, as well as federal and state laboratories. By definition, these microbiological criteria for sugar are specifications, since there are no official federal standards for sugar. However, due to the importance of these limits, they are included in this section.

Limits have been established for total thermophilic spores, flat sour spores, thermophilic anaerobic spores and sulfide spoilage spores. The sample from a lot or shipment consists of 5 sample units of about 225 g each taken from each of 5 bags or containers.

For the 5 sample units, there shall be a maximum of 150 total thermophilic spores and an average of not more than 125 spores/10 g of sugar. There shall be a maximum of 75 flat sour spores and an average of not more than 50 spores/10 g of sugar. Thermophilic anaerobic spores shall be present in not more than 3 of the 5 sample units and in any 1 sample unit to the extent of not more than 4 of 6 inoculated tubes. For this test, an amount of solution equivalent to 4 g of sugar is distributed equally among 6 tubes containing liver infusion broth.

The sulfide spoilage spores shall be present in not more than 2 of the 5 sample units and not more than 5 spores in 10 g of any 1 sample unit.

Starch.—This ingredient can cause spoilage when added to low-acid canned foods due to the presence of thermophilic spores. The microbial limits for starch are the same as for sugar.

Peanuts.—There are no standards for the number of microorganisms that can be present in peanuts or peanut products. However, there is a tolerance of aflatoxin of less than 20 μg/g of peanuts.

BIBLIOGRAPHY

ALBRECHT, J.J. 1975. The cost of government regulations to the food industry. Food Technol. *29*, No. 10, 61, 64-65.

ANEMA, P.J. 1975. The microbiological monitoring of food in the food processing industry. Antonie van Leeuwenhoek *41*, 375-376.

APHA. 1976. Compendium of Methods for the Microbiological Examination of Foods. American Public Health Association, Washington, D.C.

A.O.A.C. 1975. Official Methods of Analysis of the Association of Official Analytical Chemists, 12th Edition. Association of Official Analytical Chemists, Washington, D.C.

AYRES, P.A. 1975. The quantitative bacteriology of some commercial bivalve shellfish entering British markets. J. Hyg. Camb. *74,* 431-440.

BARTRAM, M.T. 1967. International microbiological standards for foods. J. Milk Food Technol. *30*, 349-351.

BRISKEY, E.J. 1976. Impact of government regulations on the food industry. Food Technol. *30*, No. 5, 38, 40, 42, 44.

CORLETT, D.A., JR. 1976. Evaluating the need for microbial standards for refrigerated and frozen foods. Food Prod. Develop. *10*, No. 7, 95-98, 100.

COX, W.A. 1970. Microbiological standards for dairy products. Chem. Ind. *1970*, 223-229.

ELLIOT, R.P. 1970. Microbiological criteria in USDA regulatory programs for meat and poultry. J. Milk Food Technol. *33*, 173-177.

FDA. 1973. Current Levels for Natural or Unavoidable Defects in Food for Human Use that Present No Health Hazard. Food and Drug Administration, Department of Health, Education and Welfare, Rockville, Md.

FDA. 1976A. Bacteriological Analytical Manual. U.S. Food and Drug Administration, Washington, D.C.

FDA. 1976B. Recall procedure. *In* Regulatory Procedures Manual. U.S. Food and Drug Administration, Washington, D.C.

FDA. 1977. Code of Federal Regulations. Title 21. Food and Drugs. General Services Administration, Washington, D.C.

FOSTER, E.M. 1972. The need for science in food safety. Food Technol. *26*, No. 8, 81-87.

GREENBERG, R.A., TOMPKIN, R.B., and GEISTER, R.S. 1974. Who will pay for microbiological quality standards? Food Technol. *28*, No. 10, 48, 50.

ICMSF. 1974. Microorganisms in Foods. 2. Sampling for Microbiological Analysis: Principles and Specific Applications. International Commission on Microbiological Specifications for Foods. University of Toronto Press, Toronto.

KIMM, V.J., and ANGELOTTI, R. 1977. Who safeguards your drinking water? Prof. Nutr. *9*, No. 1, 1-5.

LYNG, R.E. 1970. The government's role in quality assurance. Food Technol. *24*, 1094, 1096.

MILLER, A.L. 1956. Proposed legislation pertaining to chemicals in food. Food Technol. *10*, 337-339.

MORRISON, A.B. 1976. How risk/benefit decisions are made in Canada. Food Prod. Develop. *10*, No. 10, 36-37.

NAS. 1964. An Evaluation of Public Health Hazards from Microbiological Contamination of Foods. National Academy of Sciences, National Research Council Publication *1195.* Washington, D.C.

PIVNICK, H. 1978. Canadian microbiological standards for foods. Food Technol. *32,* No. 1, 58-62.

PIVNICK, H. *et al.* 1976. Proposed microbiological standards for ground beef based on a Canadian study. J. Milk Food Technol. *39,* 408-412.

READ, R.B., JR., and BAER, E.F. 1974. Role of the regulatory in setting microbiological quality standards. Food Technol. *28,* No. 10, 42, 44, 46.

SAYER, M.D. 1976. Food development—innovation or stagnation. Food Technol. *30,* No. 8, 64, 66-69.

SLOAN, A.E. 1976. Educators, regulators, industry discuss product development—food safety problems. Food Prod. Develop. *10,* No. 6, 66, 68, 72.

STEPHENS, R.L. 1970. Bacteriological standards for foods: a retailer's point of view. Chem. Ind. *1970,* 220-222.

THATCHER, F.S., and CLARK, D.S. 1968. Microorganisms in Foods: Their Significance and Methods of Enumeration. University of Toronto Press, Toronto.

WEHR, H.M. 1978. Attitudes and policies of state governments. Food Technol. *32,* No. 1, 63-67.

WESTHOFF, D., and FELDSTEIN, F. 1976. Bacteriological analysis of ground beef. J. Milk Food Technol. *39,* 401-404.

WYLIE, R. 1977. Paving the way for faster and better agency decisions and possibly fewer regulations. Food Prod. Develop. *11,* No. 4, 85, 87.

Index